Basic Mathematics
FOR College Students
CONCEPTS AND APPLICATIONS

GREGORY FIORE
Dundalk Community College

ADDISON-WESLEY

An imprint of Addison Wesley Longman, Inc.

Reading, Massachusetts • Menlo Park, California • New York • Harlow, England
Don Mills, Ontario • Sydney • Mexico City • Madrid • Amsterdam

This book is dedicated to my parents,
Frances and Nick, who understood the
value of an education.

Sponsoring Editors: Anne Kelly, Karin E. Wagner
Development Editor: Elaine Silverstein
Project Editor: Ann-Marie Buesing
Art Director: Julie Anderson
Text Design: Lesiak/Crampton Design Inc: Lucy Lesiak
Cover Design: Lesiak/Crampton Design Inc: Cynthia Crampton
Cover Photo: Hans Neleman
Photo Researcher: Rosemary Hunter
Production Administrator: Brian Branstetter
Compositor: TSI Graphics
Printer and Binder: Courier Corporation
Cover Printer : Courier Corporation

Basic Mathematics for College Students: Concepts and Applications

Library of Congress Cataloging-in-Publication Data

Fiore, Gregory.
    Basic math for college students : concepts and applications /
Gregory Fiore.
        p.   cm.
    Includes index.
    ISBN 0-321-01305-0
    1. Mathematics.   1. Title.
QA39.2.F54   1992
510--dc20

                                                                    92-27047
                                                                    CIP

03 04 05 CK 9 8 7 6 5 4 3 2

CONTENTS

PREFACE

Basic Mathematics for College Students: Concepts and Applications is an invitation to think, to try, and to learn. This is not a rote text. It is conceptual. Mathematical principles are developed and explained to help students grasp the reasons behind procedures. The diversity of applications, both worked and practice, appeal to student experience and to their intellects. Arousing student interest and exciting curiosity has been my source of inspiration and motivation for writing this text.

This book is designed to teach students to be problem solvers by providing comprehensive coverage of basic computation skills and their application. My goal has been to challenge students with problems that are not solved just by applying algorithms. Many problems require thinking and reasoning, and they encourage students to be inventive and to evaluate the information. The book also helps students learn from mistakes by presenting problems containing common errors that must be corrected. Throughout the book, the clear link between arithmetic and algebra is illustrated and reinforced. Word problems and procedures for solving them are introduced early in Chapter 1 and reinforced throughout the text.

Students entering an introductory college-level mathematics course bring with them considerable variety in their level of preparation, their experience, and their needs. This book is written to accommodate both those students who need extensive preparation and those students who need only a review and more of a challenge. The book's clear explanations and extensive pedagogy make it appropriate for both traditional lecture classes or individualized instruction.

Organization and Special Features

Basic Mathematics for College Students: Concepts and Applications contains twelve chapters, each of which is divided into sections that cover one or more related objectives. Each objective teaches a single concept, thus allowing students to master one concept before moving on to another.

Each chapter is consistent in its format, and the following two pages help illustrate many of the features found throughout this text.

The text incorporates many of the arithmetic and basic algebra and geometry requirements for the Florida College Level Academic Skills Test (CLAST), as well as many of the basic geometry and reasoning requirements for the Texas Academic Skills Program (TASP). In addition, many of NCTM's Curriculum and Evaluation Standards For School Mathematics (1989) are addressed.

Supplements

The *Annotated Instructor's Edition* includes the basic student text plus all the answers to every drill and Skillsfocus exercise and Extend Your Thinking exercise. The answers are printed in color next to or below the corresponding exercise. Calculator exercises are specially marked in both the student's and instructor's editions.

The *Instructor's Manual: Tests and Solutions* includes worked-out solutions to all even-numbered problems in the text, including the Extend Your Thinking and Writing to Learn problem sets. Five versions of each chapter test also appear: two open response, two multiple choice, and one combination of open response and multiple choice. Two versions of the final examination in an open-response format are also included. All answers to chapter tests and cumulative tests are included at the end of the *Instructor's Manual*.

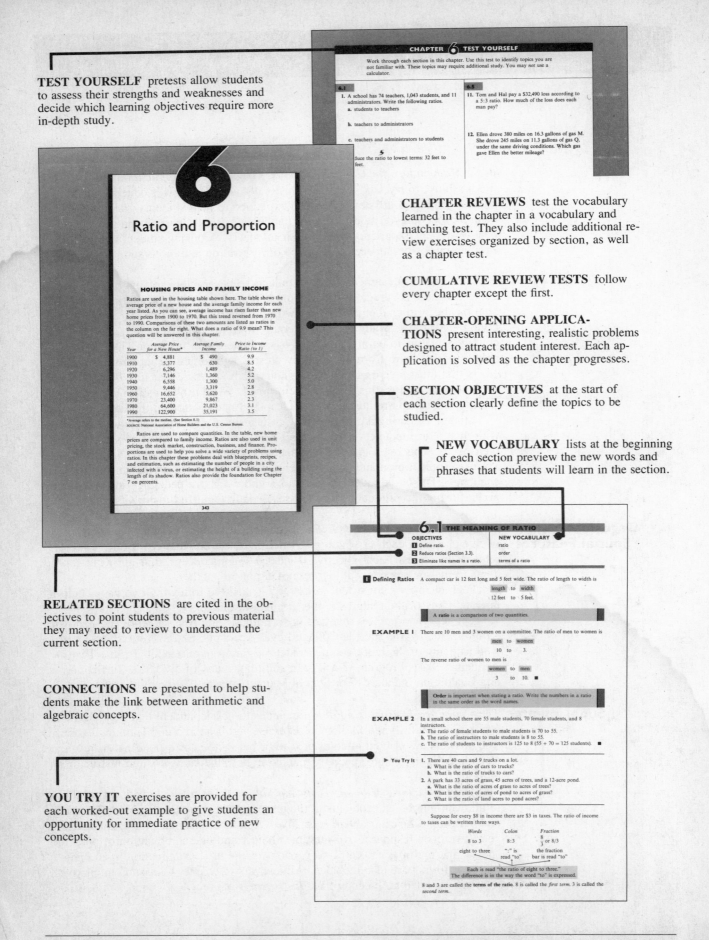

TEST YOURSELF pretests allow students to assess their strengths and weaknesses and decide which learning objectives require more in-depth study.

CHAPTER REVIEWS test the vocabulary learned in the chapter in a vocabulary and matching test. They also include additional review exercises organized by section, as well as a chapter test.

CUMULATIVE REVIEW TESTS follow every chapter except the first.

CHAPTER-OPENING APPLICATIONS present interesting, realistic problems designed to attract student interest. Each application is solved as the chapter progresses.

SECTION OBJECTIVES at the start of each section clearly define the topics to be studied.

NEW VOCABULARY lists at the beginning of each section preview the new words and phrases that students will learn in the section.

RELATED SECTIONS are cited in the objectives to point students to previous material they may need to review to understand the current section.

CONNECTIONS are presented to help students make the link between arithmetic and algebraic concepts.

YOU TRY IT exercises are provided for each worked-out example to give students an opportunity for immediate practice of new concepts.

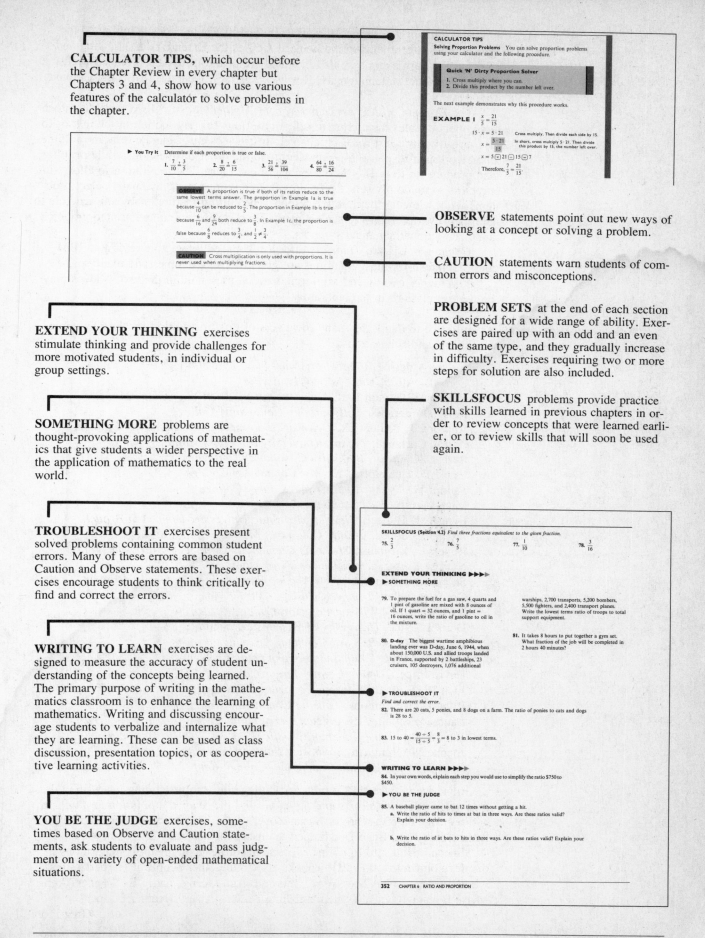

CALCULATOR TIPS, which occur before the Chapter Review in every chapter but Chapters 3 and 4, show how to use various features of the calculator to solve problems in the chapter.

OBSERVE statements point out new ways of looking at a concept or solving a problem.

CAUTION statements warn students of common errors and misconceptions.

PROBLEM SETS at the end of each section are designed for a wide range of ability. Exercises are paired up with an odd and an even of the same type, and they gradually increase in difficulty. Exercises requiring two or more steps for solution are also included.

EXTEND YOUR THINKING exercises stimulate thinking and provide challenges for more motivated students, in individual or group settings.

SKILLSFOCUS problems provide practice with skills learned in previous chapters in order to review concepts that were learned earlier, or to review skills that will soon be used again.

SOMETHING MORE problems are thought-provoking applications of mathematics that give students a wider perspective in the application of mathematics to the real world.

TROUBLESHOOT IT exercises present solved problems containing common student errors. Many of these errors are based on Caution and Observe statements. These exercises encourage students to think critically to find and correct the errors.

WRITING TO LEARN exercises are designed to measure the accuracy of student understanding of the concepts being learned. The primary purpose of writing in the mathematics classroom is to enhance the learning of mathematics. Writing and discussing encourage students to verbalize and internalize what they are learning. These can be used as class discussion, presentation topics, or as cooperative learning activities.

YOU BE THE JUDGE exercises, sometimes based on Observe and Caution statements, ask students to evaluate and pass judgment on a variety of open-ended mathematical situations.

The *Student's Solution Manual* (0-06-501908-3) includes solutions to the odd-numbered problems in the book as well as the Extend Your Thinking and Writing to Learn exercises; also included are solutions to Test Yourself pretests, Vocabulary and Matching tests, Review Exercises, Chapter Tests, and Cumulative Tests.

The *HarperCollins Test Generator for Mathematics* for IBM and Macintosh computers enables instructors to select questions for any section in the text or to use a ready-made test for each chapter. Instructors can generate tests in multiple-choice or open-response formats, scramble the order of questions while printing, and produce up to twenty-five versions of each test. The program also allows instructors to randomly select problems from a section or to manually select problems while viewing them on the screen. The editing feature enables instructors to add their own problems to the chapters. They software features printed graphics and accurate mathematical symbols.

Computer-Assisted Tutorials offer a self-paced interactive review in IBM and Macintosh formats. Solutions are given for all examples and exercises.

Videotapes review and reinforce material presented in the text. Videos may be used in class or in the laboratory.

Acknowledgments

I would like to thank the many reviewers of this first edition for their valuable comments and suggestions.

Josette Ahlering, *Central Missouri State University*
Carol L. Atnip, *University of Louisville*
LaVerne Blagmon-Earl, *University of the District of Columbia*
Betsy Borderieux, *Hillsborough Community College*
Anthony J. Brunswick, *Delaware Technical and Community College*
Carmy Carranza, *Indiana University of Pennsylvania*
Barbara Conway, *Berkshire Community College*
Grace D. DeVelbiss, *Sinclair Community College*
Sandra W. Evans, *Cumberland County College*
Mike Farrell, *Carl Sandburg College*
Lynne M. Hartsell, *New Mexico State University-Doña Ana Branch*
Jeffrey C. Jones, *County College of Morris*
Robert C. Limburg, *St. Louis Community College at Florissant Valley*
Nenette Loftsgaarden, *University of Montana*
Sharon MacKendrick, *New Mexico State University-Grants Branch*
Grace Malaney, *Donnelly College*
Carl Mancuso, *William Paterson College*
Doris L. Nice, *University of Wisconsin-Parkside*
Peggy Phillips, *Brevard Community College*
Frank W. Post, *South Seattle Community College*
David Price, *Tarrant County Junior College*
Jack W. Rotman, *Lansing Community College*
Roberta L. Simmons, *Lansing Community College*
Laurence Small, *Los Angeles Pierce College*
John Steele, *Lane Community College*
Joanne W. Vaughan, *Southwest Baptist University*
Cora S. West, *Florida Community College-Kent*

Irene Doo of Austin Community College did an outstanding job writing the *Instructor's Manual: Tests and Solutions* and the *Student's Solution Manual*. In addition, the careful work of Susan Boyer of the University of Maryland-Baltimore County and Irene Doo helped to ensure the accuracy of the answers.

Special thanks go to Elaine Silverstein for her many helpful suggestions in the development of this text. Thanks also to the other people at HarperCollins whose assistance has been very important: Anne Kelly, Karin E. Wagner, Ann-Marie Buesing, Linda Youngman, Julie Anderson, and Heather Cooper.

Greg Fiore

1

Introduction to Whole Number Operations

COMPUTER MEMORY

When you buy a computer, you are concerned with how much memory it has. One unit of computer memory is called a *byte*. It is the memory a computer needs to store a single letter, punctuation mark, or other keyboard character. Storing the word *computer* requires 8 bytes of computer memory because the word is 8 characters long.

One kilobyte, or 1 K, of computer memory is

$$2^{10} \text{ bytes.}$$

The more memory your computer has, the more sophisticated the applications it can run. If you purchase a computer with 256 K of memory, then your computer has

$$256 \cdot 2^{10} \text{ bytes of memory.}$$

One megabyte of memory (1 Meg of RAM) is 1 K of K, or a kilobyte of kilobytes. You will learn how to evaluate these expressions in this chapter.

You are about to begin your study of arithmetic. This is a crucial body of knowledge in your education. The rules you learn here will help you solve the most practical problems you meet at home and in the workplace. The many different applications you will see in this text will reinforce this point. In addition, the rules in arithmetic carry over into algebra. Learn them well here, and you will also be preparing yourself for later study of that subject.

Work through each section in this chapter. Use this test to identify topics you are not familiar with. These topics may require additional study. You may *not* use a calculator.

1.1

1. Write all the digits from 1 to 5 exclusive.

2. Insert the correct symbol $<$, $=$, or $>$ between the numbers.
   a. 6   4   **b.** 10   15   **c.** 2   2

1.2

3. Write 6,072 in expanded notation.

4. Write in standard notation: four thousand, twenty-eight.

1.3

5. Solve by trial $5 + w = 7$.

6. Solve by trial $y + 2 = 10$.

7. Add $478 + 609 + 3,217$.

8. Add $92,417 + 69 + 7,815 + 6 + 285$.

1.4

9. Write the related addition for $9 - 2 = 7$.

10. Subtract $4,000 - 2,105$.

11. Subtract $49,250 - 19,372$.

1.5

12. Round 4,916 to tens.

13. Round 62,981 to hundreds.

14. Seven people were on an elevator. Their weights were 109, 88, 185, 42, 131, 210, and 78 pounds, respectively. Estimate the total weight in the elevator.

1.6

15. The balance in Juanita's checkbook is $649. She writes checks for $17, $9, $253, and $88. What is her new balance?

16. Eric started his cross-country trip with an odometer reading of 47,306 miles. The reading at the end of the trip was 56,874 miles. How many miles did Eric travel on his trip?

1.7

17. Solve by trial $8y = 56$.

18. Solve by trial $5v = 45$.

1.8

19. Multiply $58 \cdot 639$.

20. Multiply $5,807 \cdot 306$.

21. Multiply $1,000 \cdot 251$.

22. Multiply $3,400 \cdot 200$.

23. Simplify $4 + 2 \cdot 5$.

24. Estimate $712 \cdot 48$.

1.9

25. Write $7 \cdot 7 \cdot 7 \cdot 7$ in exponential notation.

26. Write 4^2 in standard notation.

27. Write 5^3 in standard notation.

28. Write 2^6 in standard notation.

29. Find $\sqrt{81}$.

VOCABULARY *Explain the meaning of each term. Use your own examples.*

30. odd digit   **31.** product

1.1

OBJECTIVES

1 Define terminology.
2 Order whole numbers.
3 Use letters to represent numbers.

NEW VOCABULARY

digit	odd number	whole numbers	inequality
even digit	inclusive	number line	variable
even number	exclusive	graph	
odd digit	counting numbers	order	

---

1 Defining Terms   You build numbers using the ten **digits**

$$0, 1, 2, 3, 4, 5, 6, 7, 8, 9.$$

EXAMPLE 1

5 is a one digit number.
26 is a two digit number.
648 is a three digit number. ∎

▶ **You Try It**   How many digits are in each number?

1. 437   **2.** 8   **3.** 92   **4.** 6,051

The **even digits** are 0, 2, 4, 6, and 8. An **even number** has an even digit in the rightmost position. The numbers 4, 38, and 246 are even. The **odd digits** are 1, 3, 5, 7, and 9. An **odd number** has an odd digit in the rightmost position. The numbers 7, 53, and 845 are odd.

The digits 3 through 7 inclusive are 3, 4, 5, 6, and 7. **Inclusive** means include the end numbers 3 and 7. The digits 3 through 7 exclusive are 4, 5, and 6. **Exclusive** means exclude the end numbers 3 and 7.

The **counting numbers** or **natural numbers** are

$$1, 2, 3, 4, 5, 6, 7, 8, 9, 10, 11, 12, 13, \ldots.$$

The smallest counting number is 1. There is no largest counting number. The three dots "..." mean the set of counting numbers goes on without end.

The **whole numbers** are the counting numbers with the number zero, 0. The whole numbers are

$$0, 1, 2, 3, 4, 5, 6, 7, 8, 9, 10, 11, 12, 13, \ldots.$$

A **number line** is a line with equally spaced marks. Each mark is labeled with a number. A ruler is an example of a number line. The whole numbers are represented on the number line below.

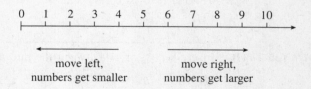

Whole numbers can be pictured, or **graphed**, as points on the number line.

EXAMPLE 2   Graph the number 4.

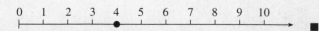

EXAMPLE 3 Graph the odd digits.

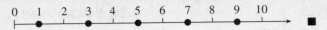

▶ **You Try It** Graph the following numbers using a number line.

5. the number 3   **6.** the even digits   **7.** 1, 2, 5, and 6

2 Ordering Whole Numbers The number line can be used to compare, or **order**, whole numbers. Just graph the numbers on the line. The number farther to the right is larger.

EXAMPLE 4 Use the number line to compare 5 and 8.

Because 8 lies to the right of 5, 8 is greater. The expression *8 is greater than 5* is written **8 > 5**. You can also write *5 is less than 8* as **5 < 8**. ■

Symbols commonly used to order numbers are shown below.

> = is read 'equals' or 'is equal to.'
> > is read 'is greater than.'
> < is read 'is less than.'

The symbols > and < are called *inequality symbols*.

| Wider side of symbol faces larger number | > | Pointed side faces smaller number |

OBSERVE > and < resemble open mouths about to eat the larger of two desserts.

EXAMPLE 5 6 > 2 is read "6 is greater than 2."
2 < 6 is read "2 is less than 6."
0 < 5 is read "0 is less than 5." ■

▶ **You Try It** Insert >, <, or = between the numbers. Then write the expression in words.

8. 4 __ 3   **9.** 5 __ 6   **10.** 2 __ 2
11. 7 __ 1   **12.** 8 __ 9

CONNECTIONS

Ordering Negative Numbers   Is $-4 < -2$? To decide, use familiar units with the numbers. For example, you have seen negative temperatures. On a thermometer, $-4°$ is a lower temperature than $-2°$. So, $-4 < -2$.

Is $-5 > -10$? Think in terms of money. A checkbook balance of $-\$5$ is higher than a balance of $-\$10$. So, $-5 > -10$.

The slash, /, is read 'not' when drawn through =, >, and <.

$\neq$ **is read 'not equal to.'**

$\not<$ **is read 'not less than.'**

$\not>$ **is read 'not greater than.'**

OBSERVE   In computer programming, the symbol $< >$ is used to mean not equal to.

EXAMPLE 6 $6 \neq 8$ is read "6 is not equal to 8."

$9 \not< 5$ is read "9 is not less than 5."

$2 \not> 3$ is read "2 is not greater than 3." ∎

▶ **You Try It**   Write in words.

13. $5 \not> 9$ **14.** $3 \neq 0$ **15.** $6 \not< 1$

EXAMPLE 7 $6 \_\_ 8$ Write three different symbols that make this statement true.

$6 < 8$, $6 \neq 8$, and $6 \not> 8$ are all true. ∎

▶ **You Try It**   Write three different symbols that make each statement true.

16. $4 \_\_ 2$ **17.** $7 \_\_ 7$

---

3 Using Letters to Represent Numbers

Let $r =$ number of runs.
Let $h =$ number of hits.
Let $e =$ number of errors.
There are n people in a room.
Tom weighs w pounds.

Letters are often used to represent numbers. Suppose r stands for number of runs. If your team scores 4 runs, write $r = 4$.

If Tom's weight, w, is greater than 150 pounds, write $w > 150$. The expression $w > 150$ is called an **inequality**. Can Tom weigh $w = 160$ pounds? Yes, because $160 > 150$ is true. But $w \neq 130$ pounds because $130 \not> 150$.

EXAMPLE 8 Is the inequality $x < 5$ true or false for each value of x below?

a. $x = 1$. True because $1 < 5$.
b. $x = 7$. False because $7 \not< 5$.
c. $x = 5$. False because $5 \not< 5$. ∎

Letters such as r, w, and x are called **variables** because they can represent many numbers. (Their value can vary.)

▶ **You Try It** Is the inequality $y < 8$ true or false for each value of y below?

18. $y = 2$ **19.** $y = 10$ **20.** $y = 8$ **21.** $y = 0$

▶ **Answers to You Try It** 1. 3 2. 1 3. 2 4. 4

5.
```
0  1  2  3  4  5  6  7  8  9  10
```

6.
```
0  1  2  3  4  5  6  7  8  9  10
```

7.
```
0  1  2  3  4  5  6  7  8  9  10
```

8. $>$; 4 is greater than 3   9. $<$; 5 is less than 6   10. $=$; 2 is equal to 2
11. $>$; 7 is greater than 1   12. $<$; 8 is less than 9   13. 5 is not greater than 9
14. 3 is not equal to 0   15. 6 is not less than 1   16. $>$, $\neq$, $\not<$ 17. $=$, $\not>$, $\not<$
18. true   19. false   20. false   21. true

1. What is the smallest even whole number?

2. What is the smallest odd counting number?

3. What is the smallest two digit whole number?

4. What is the smallest even counting number?

5. What is the smallest odd whole number?

6. Write all the digits greater than 7.

7. Write all the digits less than 4.

8. Write all the odd digits greater than 3.

9. Write the digits 0 through 6 inclusive.

10. Write the digits 5 through 8 exclusive.

11. Write the even numbers 8 through 16 exclusive.

12. Write the odd numbers 7 through 14 inclusive.

2 *Write each true statement in words. If the statement is false, put a / through the symbol, then write the corrected statement in words.*

13. $7 > 2$

14. $6 < 3$

15. $4 = 4$

16. $3 < 8$

17. $10 = 11$

18. $9 < 1$

19. $9 > 12$

20. $0 < 7$

21. $6 > 4$

22. $4 < 8$

23. $8 > 7$

24. $3 < 0$

25. $8 = 10$

26. $6 > 6$

Find at least three out of the six symbols ($=$, $<$, $>$, $\neq$, $\not<$, $\not>$) that make each statement true.

27. $6 \_\_ 4$

28. $5 \_\_ 5$

29. $8 \_\_ 0$

30. $3 \_\_ 6$

31. $8 \_\_ 2$

32. $5 \_\_ 9$

33. $0 \_\_ 0$

34. $1 \_\_ 0$

3 *When the letter in the inequality is replaced by each number, state whether the statement is true or false.*

35. $t > 5$
    a. $t = 6$
    b. $t = 10$
    c. $t = 3$

36. $f < 7$
    a. $f = 2$
    b. $f = 7$
    c. $f = 10$

37. $y < 2$
    a. $y = 0$
    b. $y = 1$
    c. $y = 8$

38. $n > 6$
    a. $n = 7$
    b. $n = 0$
    c. $n = 9$

EXTEND YOUR THINKING ▶▶▶▶
▶ SOMETHING MORE

39. The Unlimited Dating Service Write the counting numbers above the even counting numbers. A numerical dating service matched each counting number with its double directly below. 1 was matched with 2, 2 with 4, 3 with 6, and so forth.

Counting numbers: 1, 2, 3, 4, 5, 6, 7, 8, 9, 10, 11, 12, ...
Even counting ↑ ↑ ↑ ↑ ↑ ↑ ↑ ↑ ↑ ↑ ↑ ↑
numbers (doubles): 2, 4, 6, 8, 10, 12, 14, 16, 18, 20, 22, 24, ...

a. Was any number left without a date? Explain.

b. You Be the Judge Does this mean there are just as many even counting numbers as counting numbers? Explain.

▶ TROUBLESHOOT IT
Find and correct the error.

40. $6 > 10$

41. $7 \neq 7$

42. $0 \not< 8$

▶ CONNECTIONS
Think of each number as a temperature or dollar amount. Then insert the correct symbol, $<$ or $>$, between the numbers.

43. -1 ___ -6

44. -5 ___ -2

45. -45 ___ -10

46. -200 ___ -500

WRITING TO LEARN ▶▶▶▶

47. Explain what " ... " means with 1, 2, 3, ...

48. Explain what $>$, $=$, and $<$ mean. How does "/" change their meaning?

49. Some number lines are circular. An example is the old fashioned nondigital produce scale in a grocery store.
   a. Give two more examples of circular number lines.

   b. Explain how a circular number line is like a straight number line.

   c. Explain how it is unlike a straight number line.

   d. Could a straight number line be used as a clock? Explain.

▶ YOU BE THE JUDGE

50. Do you ever have a need for leading zeros in a number? For example, do you ever write 9 as 009? Explain your decision with your own examples.

1.2 STANDARD AND EXPANDED NOTATIONS

OBJECTIVES

1 Define standard notation and place value.

2 Read whole numbers.

3 Write whole numbers.

4 Write whole numbers in expanded notation.

NEW VOCABULARY

standard notation   expanded notation

place value   period

place value key

---

1 Standard Notation and Place Value

A whole number written using the digits 0, 1, 2, 3, 4, 5, 6, 7, 8, 9 is written in **standard notation**.

EXAMPLE 1

24, 351, and 1,686 are written in standard notation. ∎

Our number system is a **place value** system. This means the position of a digit in a number determines its place value.

The **place value key** in Figure 1.1 gives the first fifteen place value names for whole numbers.

FIGURE 1.1
Place Value Key

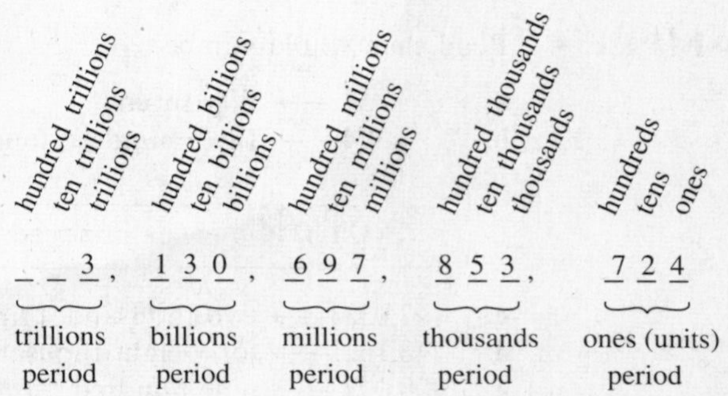

EXAMPLE 2

Use the place value key in Figure 1.1 to find the place value name for the digit 2 in each number.

4,17**2** ⟶ 2 is in the units or ones place. It is read 2 ones.

5,6**2**7 ⟶ 2 is in the tens place. It is read 2 tens.

219 ⟶ 2 is in the hundreds place. It is read 2 hundreds.

2,803 ⟶ 2 is in the thousands place. It is read 2 thousands. ∎

▶ **You Try It**   Use Figure 1.1 to find the place value name for the digit 7 in each number.

1. 375   **2.** 7,240   **3.** 687   **4.** 14,708

---

2 Reading Whole Numbers

Commas are used to separate groups of three digits (see Figure 1.1). Each group of three digits is called a **period**. In each period, the numbers range from 000 to 999. The name of each period is the place name of the rightmost digit in that period.

EXAMPLE 3   The whole number in Figure 1.1 is read:

three trillion

one hundred thirty billion

six hundred ninety-seven million

eight hundred fifty-three thousand

seven hundred twenty-four   ∎

OBSERVE   The U.S. national debt surpassed this number in 1991.

> To read a whole number
> 1. Read the number in each period, followed by the period name.
> 2. Do not read the name of the ones period.
> 3. The word *and* is never used when reading a whole number.

EXAMPLE 4   Read each whole number.

a.   57 ⟶ fifty-seven
b.   304 ⟶ three hundred four

CAUTION   Three hundred and four is wrong. Why?

---

c.   2,038 ⟶ two thousand, thirty-eight
d.   48,162 ⟶ forty-eight thousand, one hundred sixty-two
e.   900,007 ⟶ nine hundred thousand, seven
f. 2,080,500 ⟶ two million, eighty thousand, five hundred   ∎

OBSERVE   Use the hypen (-) only when writing the rightmost two digits in any period in words: 21 to 29, 31 to 39, up to 91 to 99.

▶ **You Try It**   Read each whole number and write it in words.

5. 38

6. 697

7. 5,804

8. 800,016

9. 97,346

10. 7,010,600

---

3 Writing Whole Numbers

Write one thousand, eight hundred thirty-four in standard notation.

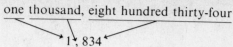

one thousand, eight hundred thirty-four

1 , 834

To write a whole number in standard notation when it is given in words

1. Write the whole number for each period.
2. Replace each period name with a comma.

EXAMPLE 5 Write each number in standard notation.

a. seven hundred sixteen thousand, four hundred sixty-one

716 , 461

b. twenty-seven thousand, ninety-two

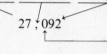

27 , 092

The 0 is a place holder. It says there are no hundreds in the units period.

c. two million, seven hundred

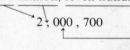

2 , 000 , 700

000 is a place holder. It says there are no thousands in the thousands place. ■

▶ **You Try It** Write the following expressions in standard notation.

11. seven hundred nine
12. sixty-one thousand, two hundred fifty-eight
13. eight thousand, four hundred ninety
14. ten million, fifty thousand

---

4 Writing Whole Numbers in Expanded Notation

The number 368 is in standard notation. Write it in **expanded notation** using the place value key in Figure 1.1.

$$3 \quad 6 \quad 8 = 3 \text{ hundreds} + 6 \text{ tens} + 8 \text{ ones}$$

$$368 \quad = 300 + 60 + 8$$

standard notation

expanded notation

EXAMPLE 6   Write 40,206 in expanded notation.

$$\begin{array}{c} \text{ten thousands} \\ \text{thousands} \\ \text{hundreds} \\ \text{tens} \\ \text{ones} \end{array}$$

4 0 , 2 0 6 = 4 ten thousands + 0 thousands
            + 2 hundreds + 0 tens + 6 ones
          = 40,000 + 0 + 200 + 0 + 6
          = 40,000 + 200 + 6   ■

▶ **You Try It**   Write each number in expanded notation.

15. 397

16. 12

17. 6,735

18. 94,060

▶ **Answers to You Try It**   1. tens   2. thousands   3. ones
4. hundreds   5. thirty-eight   6. six hundred ninety-seven
7. five thousand, eight hundred four   8. eight hundred thousand, sixteen
9. ninety-seven thousand, three hundred forty-six
10. seven million, ten thousand, six hundred   11. 709   12. 61,258   13. 8,490
14. 10,050,000   15. 300 + 90 + 7   16. 10 + 2   17. 6,000 + 700 + 30 + 5
18. 90,000 + 4,000 + 60

SECTION 1.2 EXERCISES

1 2 *Read then write the whole number in words.*

1. 42   **2.** 68   **3.** 904   **4.** 602

5. 511   **6.** 713   **7.** 2,345   **8.** 5,378

9. 45,824   **10.** 93,154   **11.** 60,306   **12.** 90,003

13. 790,213   **14.** 205,070   **15.** 27,000,000   **16.** 4,000,300

3 *Write the whole number in standard notation.*

17. The distance from the center of the Earth to the equator is three thousand, nine hundred sixty-one miles.

18. Mount Everest, the highest point on Earth, is twenty-nine thousand, twenty-eight feet above sea level.

19. The lowest land surface on Earth is the shore of the Dead Sea, at one thousand, three hundred ten feet below sea level.

20. The deepest part of the ocean is the Marianas Trench in the Pacific Ocean. It is thirty-six thousand, one hundred ninety-eight feet deep.

21. The sun is ninety-three million miles away.

22. The Earth is four billion, five hundred million years old.

23. Light travels about six trillion miles in one year.

24. The new stadium will cost ninety-eight million, four hundred twenty thousand, two hundred dollars.

4 *Write each number in expanded notation.*

25. 63   **26.** 79   **27.** 719   **28.** 214

29. 901   **30.** 803   **31.** 1,739   **32.** 7,268

33. 8,002   **34.** 4,030   **35.** 71,348   **36.** 17,306

37. 100,001   **38.** 709,300   **39.** 3,406,556   **40.** 8,080,800

SKILLSFOCUS (Section 1.1) *Insert* $>$, $=$, *or* $<$.

41. 5 __ 2   **42.** 8 __ 8   **43.** 10 __ 25   **44.** 1 __ 0

EXTEND YOUR THINKING ▶▶▶▶

▶TROUBLESHOOT IT

Find and correct the error.

45. $453 = 400 + 30 + 5$ **46.** $6,052 = 6,000 + 500 + 2$

WRITING TO LEARN ▶▶▶▶

47. Explain the difference between standard notation and expanded notation. Use your own examples in your explanation.

▶YOU BE THE JUDGE

48. a. Explain why $2,500 is sometimes read twenty-five hundred dollars, instead of two thousand, five hundred dollars.

b. Recently a comedian, when asked her age, said she was in her late twenty-tens. Explain what she means.

1.3 ADDITION OF WHOLE NUMBERS

OBJECTIVES

1 Define addition and its properties (Section 1.1, Objectives 1, 2).

2 Solve equations by trial.

3 Add whole numbers without carrying.

4 Add whole numbers with carrying.

NEW VOCABULARY

addition	additive identity	associative
addends	commutative	equation
sum	parentheses	carry

1 Addition and Its Properties

Addition is a way of combining two or more numbers to get a total. The numbers being combined are called **addends**. The total is called the **sum**.

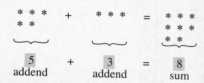

$$\underbrace{5}_{\text{addend}} \quad + \quad \underbrace{3}_{\text{addend}} \quad = \quad \underbrace{8}_{\text{sum}}$$

Addition can be performed on a number line using arrows (see Section 1.1, Objectives 1, 2). Each addend is an arrow pointing to the right. To add the arrows, attach the tail of one arrow to 0. Attach the tail of the next arrow to the head of the first. The total length of the two arrows is the sum.

EXAMPLE 1 Add 5 + 3 using the number line.

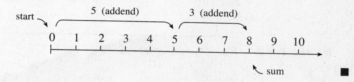

▶ **You Try It** Add using the number line.

1. 2 + 6   **2.** 4 + 5

Figure 1.2 summarizes the basic addition facts for the digits 0, 1, 2, 3, 4, 5, 6, 7, 8, and 9. Memorize this table.

FIGURE 1.2
Addition Facts

+	0	1	2	3	4	5	6	7	8	9
0	0	1	2	3	4	5	6	7	8	9
1	1	2	3	4	5	6	7	8	9	10
2	2	3	4	5	6	7	8	9	10	11
3	3	4	5	6	7	8	9	10	11	12
4	4	5	6	7	8	9	10	11	12	13
5	5	6	7	8	9	10	11	12	13	14
6	6	7	8	9	10	11	12	13	14	15
7	7	8	9	10	11	12	13	14	15	16
8	8	9	10	11	12	13	14	15	16	17
9	9	10	11	12	13	14	15	16	17	18

EXAMPLE 2   Use Figure 1.2 to add 5 + 3.

Find 5 in the left-hand column. Move to the right. Find 3 in the top row. Move downward. You intersect at the answer, 8. (See the shaded area in Figure 1.2.) ■

▶ **You Try It**   Add using Figure 1.2.

3. 4 + 7   **4.** 9 + 6

Addition Property of Zero
In Figure 1.2, notice that 0 added to any digit gives the same digit back again.

EXAMPLE 3 $0 + 4 = 4$ $7 + 0 = 7$

$4 + 0 = 4$ $0 + 7 = 7$ ■

> **The Addition Property of Zero**
>
> Adding 0 to any number does not change the number. If a represents any number, then
>
> $$a + 0 = a \quad \text{and} \quad 0 + a = a.$$
>
> 0 is called the **additive identity**. (Adding 0 to any number does not change the identity of that number.)

▶ **You Try It**   Add using Figure 1.2.

5. 6 + 0   **6.** 0 + 2

Commutative Property of Addition
What pattern do you see with the following sums?

EXAMPLE 4 $2 + 7 = 9$ and $7 + 2 = 9$

$5 + 3 = 8$ and $3 + 5 = 8$

$0 + 4 = 4$ and $4 + 0 = 4$ ■

Adding two numbers in any order gives the same answer.

> **The Commutative Property of Addition**
>
> Two numbers added in any order give the same sum. If a and b represent any two numbers, then
>
> $$a + b = b + a.$$

▶ **You Try It**   Use Figure 1.2 to verify that these sums are the same.

7. 5 + 8 and 8 + 5   **8.** 3 + 7 and 7 + 3

Associative Property of Addition

To add three numbers together, does it matter which two you add first? For example, add $5 + 2 + 4$. There are two ways to write this addition. **Parentheses** are used to *group* numbers to tell you which two numbers to add first.

EXAMPLE 5

$$
\begin{array}{c|c}
(5 + 2) + 4 & 5 + (2 + 4) \\
7 \quad + 4 & 5 + \quad 6 \\
11 & 11
\end{array}
$$

Because the answers are the same, it does not matter which two numbers are added first. The two expressions are equal.

$$(5 + 2) + 4 = 5 + (2 + 4) \quad \blacksquare$$

> **The Associative Property of Addition**
>
> When adding three numbers together, it does not matter which two you add first. The answer will be the same.
>
> $$(a + b) + c = a + (b + c)$$

▶ **You Try It** Use Figure 1.2 to verify that these sums are the same.

9. $3 + (1 + 6)$ and $(3 + 1) + 6$ **10.** $(7 + 2) + 4$ and $7 + (2 + 4)$

The commutative and associative properties of addition say that you can add numbers together in any order or grouping, and you will get the same sum.

2 Solving Equations by Trial

An **equation** is a statement that two amounts are equal.

EXAMPLE 6 $x + 3 = 7$ is an example of an equation. $\blacksquare$

To solve an equation, replace the letter with a number that makes the equation true. The following examples are solved by trial and error.

EXAMPLE 7 Solve $x + 3 = 7$.

Since $4 + 3 = 7$, the solution to $x + 3 = 7$ is $x = 4$. $\blacksquare$

EXAMPLE 8 Solve $y + 5 = 8$.

Since $3 + 5 = 8$, the solution to $y + 5 = 8$ is $y = 3$. $\blacksquare$

EXAMPLE 9 Solve $9 + t = 9$.

Since $9 + 0 = 9$, the solution to $9 + t = 9$ is $t = 0$. This problem illustrates the addition property of zero. $\blacksquare$

Solve each equation.

11. $x + 4 = 9$ **12.** $6 + y = 7$

13. $4 + z = 4$ **14.** $16 = 7 + w$

---

3 **Adding Whole Numbers without Carrying**

To add whole numbers, add like place values to like place values. To do this, write the addends so that digits with the same place value names are lined up vertically in the same column.

EXAMPLE 10   Add $14 + 23$.

$$\begin{array}{r} \overset{\text{tens}}{1} \; \overset{\text{ones}}{4} \\ +2 \quad 3 \\ \hline \end{array}$$

Write the addends so the ones digits are in the same column, and the tens digits are in the same column.

$$\begin{array}{r} 1 \quad \boxed{4} \\ +2 \quad \boxed{3} \\ \hline \boxed{7} \end{array}$$

Add the ones digits.
4 ones + 3 ones = 7 ones.
Write 7 in the ones column in the sum.

$$\begin{array}{r} \boxed{1} \quad 4 \\ +\boxed{2} \quad 3 \\ \hline \boxed{3} \quad 7 \end{array}$$

Add the tens digits.
1 ten + 2 tens = 3 tens.
Write 3 in the tens column in the sum.

The sum of $14 + 23$ is 37. ■

EXAMPLE 11   Find the sum of $1,320 + 214 + 5,212 + 32$.

The commutative and associative laws say you can add numbers in any order or grouping, and the sum will be the same.

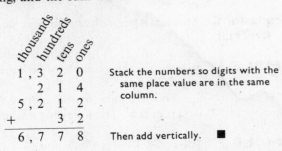

$$\begin{array}{r} \overset{\text{thousands}}{1} , \overset{\text{hundreds}}{3} \; \overset{\text{tens}}{2} \; \overset{\text{ones}}{0} \\ 2 \quad 1 \quad 4 \\ 5 , 2 \quad 1 \quad 2 \\ + \qquad 3 \quad 2 \\ \hline 6 , 7 \quad 7 \quad 8 \end{array}$$

Stack the numbers so digits with the same place value are in the same column.

Then add vertically. ■

▶ **You Try It**   Add.

15. $\begin{array}{r} 42 \\ +37 \\ \hline \end{array}$ **16.** $\begin{array}{r} 613 \\ +144 \\ \hline \end{array}$

17. $41 + 106$ **18.** $1,100 + 232 + 6,010 + 14$

---

4 **Adding Whole Numbers with Carrying**

In Examples 10 and 11, all like named digits added to 9 or less. When like digits add to 10 or more, you must **carry**. Carrying is explained in Example 12 using expanded notation.

EXAMPLE 12   Add 26 + 17.

$$2\ 6 = \quad 2 \text{ tens} + 6 \text{ ones}$$
$$+1\ 7 = +1 \text{ ten} + 7 \text{ ones}$$

To see why you must carry, write each addend in expanded notation.

Then add the ones digits:
6 ones + 7 ones = 13 ones.

$$\begin{array}{l}1\\2\ 6 = \quad 2 \text{ tens} + 6 \text{ ones}\\+1\ 7 = +1 \text{ ten} + 7 \text{ ones}\\ \hline 3 \qquad\qquad\qquad 3 \text{ ones}\end{array}$$

1 ten

Now, 13 ones = 10 ones + 3 ones
= 1 ten + 3 ones.
Write the 3 ones in the ones column in the sum.
Carry the 1 ten to the top of the tens column.

Next add the tens digits, including the carry.
1 ten + 2 tens + 1 ten = 4 tens.

$$\begin{array}{l}1\\2\ 6 = \quad 2 \text{ tens} + 6 \text{ ones}\\+1\ 7 = +1 \text{ ten} + 7 \text{ ones}\\ \hline 4\ 3 \quad 4 \text{ tens} + 3 \text{ ones}\end{array}$$

1 ten

Write 4 tens in the tens place in the sum.
When you solve the problem, just write this. ∎

▶ **You Try It**   **19.** Add 47 + 39.   **20.** Add 35 + 17.

EXAMPLE 13   Add 365 + 281.

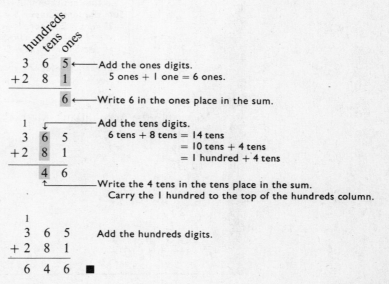

$$\begin{array}{ccc} \text{hundreds} & \text{tens} & \text{ones} \\ 3 & 6 & 5 \\ +2 & 8 & 1 \\ \hline & & 6 \end{array}$$

Add the ones digits.
5 ones + 1 one = 6 ones.

Write 6 in the ones place in the sum.

$$\begin{array}{ccc} 1 & & \\ 3 & 6 & 5 \\ +2 & 8 & 1 \\ \hline & 4 & 6 \end{array}$$

Add the tens digits.
6 tens + 8 tens = 14 tens
= 10 tens + 4 tens
= 1 hundred + 4 tens

Write the 4 tens in the tens place in the sum.
Carry the 1 hundred to the top of the hundreds column.

$$\begin{array}{ccc} 1 & & \\ 3 & 6 & 5 \\ +2 & 8 & 1 \\ \hline 6 & 4 & 6 \end{array}$$

Add the hundreds digits. ∎

▶ **You Try It**   **21.** Add 437 + 281.   **22.** Add 586 + 354.

EXAMPLE 14   Add 6,725 + 5,837.

Write the addends so like digits are in the same vertical column.

$$\begin{array}{l}\quad 1 \quad 1 \quad \longleftarrow \text{ carries}\\ \ 6{,}7\ 2\ 5\\ +5{,}8\ 3\ 7\\ \hline 1\ 2{,}5\ 6\ 2\end{array}$$ ∎

▶ **You Try It**   **23.** Add 4,863 + 2,518.   **24.** Add 8,465 + 3,159.

EXAMPLE 15   Add 5,702 + 697 + 8,847 + 73.

Write the addends so like digits are in the same vertical column.

```
    2 2 1    ← carries
    5,7 0 2
      6 9 7
    8,8 4 7
  +     7 3
  1 5,3 1 9  ∎
```

▶ **You Try It**   **25.** Add 3,846 + 792 + 9,085 + 68.   **26.** Add 895 + 5,273 + 44 + 6,497.

SOMETHING MORE

The Accountant's Method   The **Accountant's Method** is used to add columns of numbers. The carry is written below each column.

```
    5 7
    4 2
    6 5
    3 9
  +7 3
  ────
    2 6
  2 5 0
  ────
  2 7 6
```

⌐Sum the ones column: 7 + 2 + 5 + 9 + 3 = 26.
  Since 26 ones = 2 tens + 6 ones, write 6 below the
  ones column. Write the carry, 2 tens, below the
  tens column (not above it). Note that the 2 tens is still
  in the tens column!
⌐Sum the tens column: 5 + 4 + 6 + 3 + 7 = 25.
  25 tens = 2 hundreds + 5 tens. Write 5 in the tens column.
  Write 2 in the hundreds column. If desired, write a 0
  place holder.
← Add vertically for the answer.

Add using the Accountant's Method.

a.   47
         38
         95
         64
         19
         83
          7
       +63

b.   341
          95
         929
         168
          47
         567
         386
       +274

▶ **Answers to You Try It**

1.

2.   4   5

     0 1 2 3 4 5 6 7 8 9 10

     4 + 5 = 9

3. 11   **4.** 15   **5.** 6   **6.** 2   **7.** 13 and 13   **8.** 10 and 10   **9.** 10 and 10
10. 13 and 13   **11.** $x = 5$ **12.** $y = 1$ **13.** $z = 0$ **14.** $w = 9$ **15.** 79
16. 757   **17.** 147   **18.** 7,356   **19.** 86   **20.** 52   **21.** 718   **22.** 940
23. 7,381   **24.** 11,624   **25.** 13,791   **26.** 12,709.

▶ **Answers to Something More**   **a.** 416   **b.** 2,807

SECTION 1.3 EXERCISES

1 *Add.*

1. $8 + 0$

2. $0 + 7$

3. $(2 + 4) + 3$

4. $6 + (1 + 1)$

5. $(4 + 2) + 3$

6. $4 + (2 + 3)$

7. $(0 + 4) + 1$

8. $0 + (4 + 1)$

Replace the letter with the number that makes the equation true. Then identify each as a statement of a) the addition property of zero, b) the commutative property of addition, or c) the associative property of addition.

9. $0 + x = 4$

10. $6 + 0 = x$

11. $7 + 3 = 3 + n$

12. $n + 4 = 4 + 2$

13. $5 + c = 5$

14. $q + 7 = 7$

15. $(p + 4) + 5 = 1 + (4 + 5)$

16. $2 + (p + 6) = (2 + 3) + 6$

17. $(0 + 4) + 8 = 0 + (4 + s)$

18. $7 + y = 7$

19. $0 + 1 = 1 + x$

20. $2 + (2 + 6) = (2 + 2) + b$

2 *Solve each equation by trial.*

21. $6 + x = 13$

22. $3 + t = 11$

23. $y + 7 = 14$

24. $z + 9 = 16$

25. $k + 5 = 10$

26. $q + 8 = 8$

27. $5 + p = 5$

28. $8 + y = 12$

3 *Add.*

29. $23 + 45$

30. $71 + 17$

31. $123 + 456$

32. $712 + 235$

33. $132 + 21 + 5,132 + 4$

34. $123 + 321 + 31 + 1,111$

| 35. | 427
+ 501 | 36. | 823
+ 166 | 37. | 8,415
+ 1,403 | 38. | 7,153
+ 2,524 |

| 39. | 3,524
124
1,221
+ 10 | 40. | 5,123
2,512
321
+ 32 | 🖩 41. | 12,342
23,412
+ 1,134 | 🖩 42. | 34,132
12,523
+ 1,213 |

43. Jackie wrote checks for $32, $20, and $45. What total did she spend?

44. Tom drove 140 miles on Friday, 516 miles on Saturday, and 231 miles on Sunday. What total did he drive for the three days?

4

45. 56 + 24

46. 68 + 83

47. 527 + 269

48. 382 + 765

49. 69 + 957

50. 916 + 98

| 51. | 73
+ 27 | 52. | 42
+ 68 | 53. | 547
+ 68 |

| 54. | 925
+ 78 | 55. | 647
+ 176 | 56. | 273
+ 497 |

| 57. | 379
+ 843 | 58. | 482
+ 948 | 59. | 1,435 + 49 |

60. 9,367 + 84

61. 5,829 + 488

62. 3,978 + 756

63. 5,006 + 997

64. 7,017 + 988

65. 67 + 3,745 + 9 + 416

66. 98 + 3,508 + 718 + 4

67. 60 + 905 + 12,054 + 8 + 9,073 + 855 + 45,607

68. 30 + 303 + 3,300 + 30,033 + 3,003 + 3,030

69. 3,576
        + 217

70. 9,365
        + 528

71. 5,737
        + 896

72. 9,046
        + 967

73. 6,835
        +3,128

74. 2,634
        +6,192

75. 43,908
        + 8,159

76. 83,706
        + 6,298

77. 6,803
         978
       2,704
       + 803

78. 67,904
       32,087
       89,431
      +21,315

79. 6,897,034
       1,840,225
       7,454,662
      +6,334,817

80. 8,900,725
         99,046
          8,903
       +   88

SKILLSFOCUS (Section 1.2) *Write in expanded notation.*

81. 426

82. 65

83. 30,408

84. 4,508

EXTEND YOUR THINKING ▶▶▶▶

▶SOMETHING MORE

85. Add using the Accountant's Method.
   a. $56 + 87 + 23 + 84 + 77 + 65 + 9 + 34$

   b. $370 + 285 + 199 + 78 + 715 + 88 + 632 + 444$

86. Susan knows $x = 4$ in the equation $x + y = 13$. Then what number does she replace for y to make the equation true?

87. Jack is told $h = 6$ in the equation $1 + t = h$. What number does he put in place of t to make the equation true?

▶TROUBLESHOOT IT

Find and correct the error.

88.
```
  1 1
  9,407
 +6,094
 ------
 16,401
```

89.
```
      2
     59
    278
  +  14
  -----
    331
```

90.
```
     21
  45,956
      95
 + 6,071
 -------
  51,122
```

91. Explain how to use Figure 1.2 to demonstrate the commutative property of addition.

92. Are washing your hands and drying your hands commutative activities? Explain.

93. How are putting on a ring, a watch, and a pendant associative activities?

▶ YOU BE THE JUDGE

94. Todd claimed putting on socks, shoes, and a tie are associative activities. Is he right? Explain your decision.

95. Amy claimed she found 9 digits to fill the grid on the right. Furthermore, she claimed there were no carries when she added. Exactly three of the digits were 9's. Is Amy correct? Explain your decision.

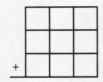

1.4 SUBTRACTION OF WHOLE NUMBERS

OBJECTIVES

1 Define subtraction and its properties.

2 Check subtraction with addition.

3 Subtracting without borrowing.

4 Subtracting with borrowing.

NEW VOCABULARY

subtraction

related addition

borrow

---

1 Subtraction and Its Properties

Subtraction is the opposite of addition. It undoes addition.

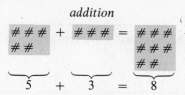
addition

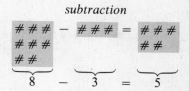
subtraction

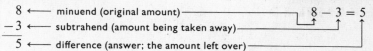

Subtraction means taking one amount away from a second amount. There are two ways to write subtraction.

Vertical Subtraction

8 ← minuend (original amount)

−3 ← subtrahend (amount being taken away)

5 ← difference (answer; the amount left over)

Horizontal Subtraction

$8 - 3 = 5$

> **OBSERVE** In vertical subtraction, the subtrahend is written below the minuend. The prefix *sub* means "under," or "beneath."

The number line can be used to perform subtraction.

EXAMPLE 1 Use the number line to subtract $8 - 3$.

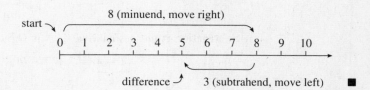

▶ **You Try It** Subtract using the number line.

1. $7 - 3$

2. $9 - 6$

Distance

In Example 1, $8 - 3$ gives the distance from 3 to 8 on the number line.

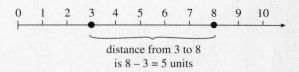

distance from 3 to 8
is $8 - 3 = 5$ units

> To find the distance between two numbers on the number line, subtract the numbers.

EXAMPLE 2   Find the distance between 4 and 7 on the number line.

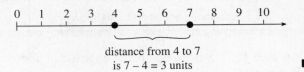

distance from 4 to 7
is $7 - 4 = 3$ units ∎

▶ **You Try It**   Use the number line and subtraction to find the distance between the two numbers.

3. 6 and 1   **4.** 9 and 7

The Zero Property of Subtraction
What pattern do you see in the following subtractions?

EXAMPLE 3 $6 - 0 = 6$
$1 - 0 = 1$
$4 - 0 = 4$ ∎

> Subtracting 0 from a number does not change the number.
> $$a - 0 = a$$

Subtraction is not commutative. This is shown below.

$$\underbrace{7 - 2}_{5} \quad \neq \quad \underbrace{2 - 7}_{\text{not a whole number}}$$

Subtraction is not associative. This is shown below.

$$\underbrace{(7 - 4)}_{3} - 2 \neq 7 - \underbrace{(4 - 2)}_{2} \left. \right\} \begin{array}{l}\text{Simplify within} \\ \text{parentheses first.}\end{array}$$
$$3 - 2 \neq 7 - 2$$
$$1 \quad \neq \quad 5$$

2 Check Subtraction with Addition

Use Figure 1.2 on page 17 to do subtraction. To show $8 - 3 = 5$, locate 3 in the left-hand column. Move to the right to 8. Move upward to the answer, 5, in the top row. You use the same shaded region to add $3 + 5 = 8$. Therefore, addition can be used to check subtraction.

Every subtraction problem has a **related addition** problem. You can use the related addition to check the subtraction.

subtraction problem	related addition (check)
8	3
−3	+5
5	8

You can combine both operations in one diagram.

$$\text{subtraction problem} \left\{\begin{array}{r} 8 \\ -3 \\ +5 \\ \hline 8 \end{array}\right\} \text{related addition (check)}$$

To check subtraction

1. Add the subtrahend to the difference.
2. The answer should be the minuend. If not, resolve the problem.

EXAMPLE 4

subtraction	related addition (check)
$9 - 4 = 5$	$\longrightarrow$ $4 + 5 = 9$
$7 - 6 = 1$	$\longrightarrow$ $6 + 1 = 7$
$4 - 0 = 4$	$\longrightarrow$ $0 + 4 = 4$ ∎

▶ **You Try It** Write the related addition for each subtraction.

5. $8 - 2 = 6$ **6.** $6 - 1 = 5$ **7.** $2 - 0 = 2$

3 Subtraction without Borrowing

To subtract whole numbers, subtract like place values from like place values. To do this, write the subtraction so that digits with the same place value names are in the same vertical column.

EXAMPLE 5 Subtract $68 - 52$.

$$\begin{array}{cc} \text{tens} & \text{ones} \\ 6 & 8 \\ -5 & 2 \\ \hline & 6 \end{array}$$

Write the subtraction with like named digits in the same vertical column.
Subtract the ones.
8 ones − 2 ones = 6 ones.
← Write 6 in the ones column in the difference.

$$\begin{array}{cc} 6 & 8 \\ -5 & 2 \\ \hline 1 & 6 \end{array}$$

Subtract the tens.
6 tens − 5 tens = 1 ten.
Write 1 in the tens column in the difference.

Check:
$$\begin{array}{r} 6\ 8 \\ -5\ 2 \\ +1\ 6 \\ \hline 6\ 8 \end{array}$$

These match, so the answer is correct. ∎

▶ **You Try It** Subtract.

8. $47 - 31$ **9.** $89 - 56$

EXAMPLE 6 Subtract 6,385 − 182.

First, write the numbers so digits with like place value names are in the same vertical column.

$$
\begin{array}{r}
6,385 \\
-182 \\
\hline
6,203
\end{array}
$$

└─ There are 0 thousands in the subtrahend.
Since 6 − 0 = 6, just bring down the 6.

Check:
$$
\begin{array}{r}
6,385 \leftarrow \\
-182 \\
+6,203 \\
\hline
6,385 \leftarrow
\end{array}
$$

These match, so the answer is correct. ∎

▶ **You Try It**   Subtract.

10. 459 − 253   **11.** 8,564 − 3,132

EXAMPLE 7 What number makes the equation $x - 5 = 4$ true?

Since $9 - 5 = 4$, the number is $x = 9$. ∎

EXAMPLE 8 What number makes the equation $6 - n = 5$ true?

Since $6 - 1 = 5$, the number is $n = 1$. ∎

▶ **You Try It**   Find the number that makes the equation true.

12. $x - 3 = 6$ **13.** $y - 1 = 4$ **14.** $9 - z = 2$

4 Subtraction with Borrowing

When the digit in the subtrahend is larger than the corresponding digit in the minuend, you must **borrow** to subtract.

EXAMPLE 9 Subtract 34 − 19.

$$
\begin{array}{rl}
\overset{\text{tens \; ones}}{} & \\
3\;\;4 &= 3 \text{ tens} + 4 \text{ ones} \\
-1\;\;9 &= 1 \text{ ten} + 9 \text{ ones}
\end{array}
$$

Write minuend and subtrahend in expanded notation. You cannot subtract 9 from 4. So you must borrow.

$$
\begin{array}{rl}
3\;\;4 &= (2 \text{ tens} + 1 \text{ ten}) + 4 \text{ ones} \\
-1\;\;9 &= 1 \text{ ten} \phantom{+ 1 \text{ ten}) +} 9 \text{ ones}
\end{array}
$$

Borrow 1 ten from 3 tens.

$$
\begin{array}{rl}
3\;\;4 &= (2 \text{ tens} + 10 \text{ ones}) + 4 \text{ ones} \\
-1\;\;9 &= 1 \text{ ten} \phantom{+ 10 \text{ ones}) +} 9 \text{ ones}
\end{array}
$$

$$
\begin{array}{rl}
\overset{2\;\;10}{\cancel{3}}\;\;4 &= 2 \text{ tens} + (10 \text{ ones} + 4 \text{ ones}) \\
-1\;\;9 &= 1 \text{ ten} \phantom{+ (10 \text{ ones} +} 9 \text{ ones}
\end{array}
$$

Write the 1 ten as 10 ones. Use the associative property for addition to group the 10 ones with the 4 ones.

$$
\begin{array}{rl}
\overset{2\;\;14}{\cancel{3}\;\;\cancel{4}} &= 2 \text{ tens} + 14 \text{ ones} \\
-1\;\;9 &= 1 \text{ ten} + 9 \text{ ones} \\
\hline
1\;\;5 &= 1 \text{ ten} + 5 \text{ ones}
\end{array}
$$

You now have 14 ones, enough to subtract the ones. Subtract like place values for the answer.

└─ You write only this last step. ∎

The completed Example 9 and the check are shown below. Notice that carrying and borrowing act like opposite operations.

```
    2  14 ←————————— Here you borrowed I ten to subtract the ones.
  → 3   4          1 ←——— Here you carried I ten to add the ones.
  − 1   9  ——→   19
    1   5  ——→  + 15
                  ————
                   34
```

Check

▶ **You Try It**   Subtract.

15. $53 - 27$ **16.** $82 - 34$

EXAMPLE 10   Subtract $605 - 279$.

```
    6   0   5
  − 2   7   9
```
Subtract the ones. Since $9 > 5$, you must borrow. There are no tens in 605. So you must borrow from 6 in the hundreds column.

```
    5  10
    6   0   5
  − 2   7   9
```
Borrow I hundred from 6 hundreds. This leaves 5 in the hundreds column. Since I hundred = 10 tens, write 10 in the tens column.

```
        9
    5  10  15
    6   0   5
  − 2   7   9
    ———————
    3   2   6  ← Subtract.
```
Borrow I ten from 10 tens. This leaves 9 in the tens column. Since I ten = 10 ones, add 10 ones + 5 ones = 15 ones. Write 15 in the ones column.

Check:
```
     605 ←
    −279
    +326
    —————
     605
```
■

EXAMPLE II   Subtract $5,367 - 3,548$.

```
    4  13   5  17
    5,  3   6   7
  − 3,  5   4   8
    ——————————————
    1,  8   1   9
```
To subtract ones, borrow I ten.
I ten + 7 ones = 17 ones.

To subtract hundreds, borrow I thousand.
I thousand + 3 hundreds = 13 hundreds.   ■

EXAMPLE 12   Subtract $40,000 - 8,637$.

You cannot subtract 7 from 0 in the ones column. So borrow 1 ten-thousand from 4 ten-thousands, and write it

1 ten-thousand = 9 thousands + 9 hundreds + 9 tens + 10 ones.
10,000   =   9,000   +   900   +   90   +   10

```
    3   9   9   9  10 ←—— borrows
    4   0 , 0   0   0
  −     8 , 6   3   7
    ———————————————————
    3   1 , 3   6   3    ■
```

▶ **You Try It**   Subtract.

17. $807 - 429$

18. $723 - 248$

19. $7,295 - 2,638$

20. $4,183 - 2,895$

21. $700 - 239$

22. $30,000 - 6,417$

▶ **Answers to You Try It**

1.
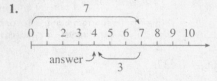

2.
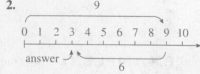

3.
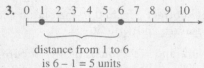

distance from 1 to 6
is $6 - 1 = 5$ units

4.
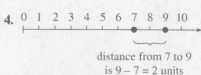

distance from 7 to 9
is $9 - 7 = 2$ units

5. $2 + 6 = 8$ **6.** $5 + 1 = 6$ **7.** $2 + 0 = 2$ **8.** 16   **9.** 33   **10.** 206
11. 5,432   **12.** $x = 9$ **13.** $y = 5$ **14.** $z = 7$ **15.** 26   **16.** 48   **17.** 378
18. 475   **19.** 4,657   **20.** 1,288   **21.** 461   **22.** 23,583

SECTION 1.4 EXERCISES

1 *Answer true or false. If true, state the property used.*

1. $4 + 1 = 1 + 4$

2. $5 - 2 = 2 - 5$

3. $6 - 0 = 0$

4. $6 - (3 - 2) = (6 - 3) - 2$

5. $(3 - 3) - 3 = 3 - (3 - 3)$

6. $8 - 0 = 8$

7. $5 - 7 = 7 - 5$

8. $0 + 9 = 0$

9. $5 + (2 + 6) = (5 + 2) + 6$

10. $9 + 3 = 3 + 9$

11. $4 + 0 = 4$

12. $0 + 2 = 2 + 0$

13. $9 - 6 = 6 - 3$

14. $(5 + 5) + 5 = 5 + (5 + 5)$

15. $3 + (6 + 0) = (0 + 4) + 5$

16. $7 - 2 = 6 - 1$

17. $2 + (5 + 3) = 8$

18. $7 + 2 = 3 - 1 - 2 + 9$

2 *Write one related addition for each subtraction, or one related subtraction for each addition.*

19. $4 + 5 = 9$

20. $6 + 2 = 8$

21. $7 - 3 = 4$

22. $8 - 2 = 6$

23. $1 - 0 = 1$

24. $7 - 7 = 0$

25. $0 + 4 = 4$

26. $7 + 0 = 7$

3 *Subtract and check.*

27. $254 - 42$

28. $739 - 27$

29. $438 - 216$

30. $134 - 122$

31. $5,673 - 5,241$

32. $2,182 - 1,172$

33.
$$\begin{array}{r} 59 \\ -24 \\ \hline \end{array}$$

34.
$$\begin{array}{r} 17 \\ -12 \\ \hline \end{array}$$

35.
$$\begin{array}{r} 938 \\ -417 \\ \hline \end{array}$$

36.
$$\begin{array}{r} 805 \\ -603 \\ \hline \end{array}$$

37.
$$\begin{array}{r} 3,067 \\ -2,040 \\ \hline \end{array}$$

38.
$$\begin{array}{r} 5,000 \\ -2,000 \\ \hline \end{array}$$

39.
$$\begin{array}{r} 56,872 \\ -26,040 \\ \hline \end{array}$$

40.
$$\begin{array}{r} 54,047 \\ -51,026 \\ \hline \end{array}$$

Solve each equation by trial.

41. $x - 7 = 2$ **42.** $v - 5 = 6$ **43.** $m + 3 = 4$ **44.** $k + 1 = 1$

45. $6 - f = 6$ **46.** $7 - h = 4$ **47.** $z - 9 = 0$ **48.** $a - 0 = 6$

4 *Subtract and check.*

49. $34 - 17$ **50.** $37 - 19$ **51.** $40 - 21$ **52.** $60 - 37$

53. $78 - 39$ **54.** $82 - 47$ **55.** $101 - 49$ **56.** $520 - 18$

57. $503 - 291$ **58.** $948 - 653$ **59.** $700 - 3$ **60.** $800 - 361$

61. Subtract 278 from 455.

62. Subtract 289 from 678.

63. There were 3,265 people at a basketball game. If 78 had free tickets, how many people paid for their tickets?

64. John has $8,034 in his checking account. He wrote a check for $57 to pay for car repairs. What is his new balance?

65.   37
     -18

66.   56
     -47

67.   465
     $-\ 78$

68.   932
     $-\ 84$

69.   503
     -168

70.   804
     -276

71.   5,670
     $-\ \ 179$

72.   8,451
     $-\ \ 273$

73.   3,603
     $-1,394$

74.   9,406
     $-5,188$

75. 4,006
         − 792

76. 1,037
         − 951

77. 8,050
         −5,683

78. 5,815
         −2,991

79. 4,000
         −1,572

80. 2,000
         −1,785

81. 70,000
         − 8,963

82. 58,049
         − 9,761

83. 70,501
         −69,298

84. 80,000
         −17,199

SKILLSFOCUS (Section 1.3) *Add.*

85. 4,388
         +9,375

86. 499
         +599

87. 2,452
         + 548

88. 9,381
         +7,688

EXTEND YOUR THINKING ▶▶▶▶
▶SOMETHING MORE

89. If $x − y = 6$ and $y = 2$, what is x?

90. If $b − a = 1$ and $b = 4$, what is a?

91. If $f − g = 0$ and $g = 7$, what is f?

92. If $s − t = 3$ and $t = 0$, what is s?

Find and correct the error.

$$\begin{array}{r} \overset{9\ \ 16}{4\ \ 0\ \ 6} \\ -1\ \ 3\ \ 8 \\ \hline 3\ \ 6\ \ 8 \end{array}$$

93.

$$\begin{array}{r} \overset{0\ \ 7\ \ 13}{8\ ,\ 1\ \ 8\ \ 3} \\ -2\ ,\ 0\ \ 3\ \ 5 \\ \hline 6\ ,\ 0\ \ 4\ \ 8 \end{array}$$

94.

$$\begin{array}{r} \overset{5\ \ 10\ \ 10\ \ 10}{6\ ,\ 0\ \ 0\ \ 0} \\ -2\ ,\ 1\ \ 9\ \ 4 \\ \hline 3\ ,\ 9\ \ 1\ \ 6 \end{array}$$

95.

WRITING TO LEARN ▶▶▶▶

96. Use your own example to show subtraction is not commutative.

► YOU BE THE JUDGE

97. Trina claims the box of numbers can be made into a magic square. This means adding the three numbers in any row, column, or diagonal gives the same sum. Can you find the five missing numbers to prove Trina's case?

		50
45	35	25

1.5 ROUNDING AND ESTIMATION

OBJECTIVES

1 Round off whole numbers (Section 1.2, Objective 1).

2 Estimate answers.

NEW VOCABULARY

rounding off
estimation

---

1 Rounding Off Whole Numbers

You buy a coat for $82. In conversation, you round the price to the nearest ten dollars, and say the coat cost you $80. The Census Bureau says the population of the United States is 253,783,496. Since this number is closer to 254,000,000 than 253,000,000, a newspaper reports the population as 254 million.

It is sometimes more convenient to express numbers as approximate values. **Rounding off** is finding approximate values for exact values. You round off to a given place value.

EXAMPLE 1   Round 28 to the nearest ten. On the number line, 28 is closer to 30 (3 tens) than to 20 (2 tens).

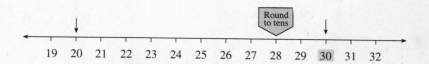

Therefore, 28 rounded to the nearest ten is 30. ∎

EXAMPLE 2   Round 23 to the nearest ten. On the number line, 23 is closer to 20 (2 tens) than to 30 (3 tens).

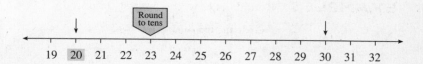

Therefore, 23 rounded to the nearest ten is 20. ∎

▶ **You Try It**   Use the number line to round to the nearest ten.

1. 77   **2.** 72   **3.** 49   **4.** 44

Round 25 to the nearest ten. Since 25 is midway between 20 and 30, you could round up or down. The accepted convention is to round up. Therefore, 25 rounded to the nearest ten is 30.

> To round off whole numbers
>
> **1.** Circle the digit in the place you are rounding to.
> **2.** If the digit on its right is
>   **a.** 5 or larger (5, 6, 7, 8, or 9), add 1 to the circled digit.
>   **b.** 4 or smaller (4, 3, 2, 1, or 0), do not change the circled digit.
> **3.** Replace all digits to the right of the circled digit with zeros.

The symbol ≐ means "is approximately equal to." It is used to say the rounded number is approximately equal to the original number.

EXAMPLE 3   Round 573 to the nearest ten.

Circle the tens digit. ———————┐   ┌——————— 7 does not change because the digit
                                    on its right, 3, is less than 5.

$$5⑦3 ≐ 570$$

Replace the 3 with a 0.

573 ≐ 570 to the nearest ten. This is pictured on the number line. You can see that 573 is closer to 570 (57 tens) than 580 (58 tens).

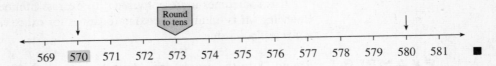

EXAMPLE 4   Round 8,352 to the nearest hundred.

Circle the hundreds digit. ———————┐   ┌——————— Add 1 to the 3 since the digit on
                                      its right, 5, is 5 or larger.

$$8,③52 ≐ 8,400$$

Replace 5 and 2 with 0's.

8,352 ≐ 8,400 to the nearest hundred. On the number line, 8,352 is closer to 8,400 (84 hundreds) than 8,300 (83 hundreds).

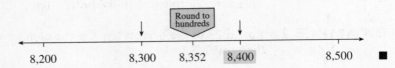

EXAMPLE 5   Round 7,499 to the nearest thousand.

Circle the thousands place. ———————┐   ┌——————— 7 remains the same because the digit
                                       on its right, 4, is less than 5.

$$⑦,499 ≐ 7,000$$

Replace 4, 9, and 9 with 0's.

7,499 ≐ 7,000 to the nearest thousand. Observe, 499 is less than half a thousand. So 7,499 is closer to 7,000 than 8,000. ■

▶ **You Try It**   Round off to the indicated place.

5. 362 to tens   **6.** 879 to tens   **7.** 5,813 to hundreds
8. 3,054 to hundreds   **9.** 3,299 to thousands   **10.** 46,922 to thousands

---

EXAMPLE 6   Round 4,972 to the nearest hundred.

Circle the hundreds digit. ———————┐   ┌——————— Add 1 to the 9 since the digit on its right,
                                      7, is 5 or larger. Since 1 + 9 = 10,
                                      write 0 in the hundreds place and
                                      carry 1 to the thousands place.

$$4,⑨72 ≐ 5,000$$

Replace 7 and 2 with 0's.

4,972 ≐ 5,000 to the nearest hundred. Observe, 4,972 is closer to 5,000 (50 hundreds) than 4,900 (49 hundreds). ■

► **You Try It**   Round to the indicated place.

11. 8,972 to hundreds   **12.** 16,499 to hundreds

2 Estimation   A young couple decides to build a home. The land will cost them $28,560. The house will cost $93,825. They make a quick mental estimate of the total cost as follows.

                         land: $28,560 ⟶ $30,000 estimate for land
                         house: $93,825 ⟶ + $90,000 estimate for house
                    actual cost: $122,385 $120,000 estimate of cost

$28,560 is close to $30,000. $93,825 is close to $90,000. The $120,000 estimate is close to the actual cost, $122,385. Note that the couple rounded both costs to the ten-thousands place.

Estimation is not meant to be exact. You estimate because you want a quick, approximate answer. An estimate is a "ballpark figure."

> To estimate an answer to an addition or subtraction
>
> **1.** Round all numbers to the same place value.
> **2.** Add or subtract the rounded values.

Choose to round to a place value that gives a fast, though practical, answer. In this text, you will round to the digit farthest left, or second from the left, in the largest number to be added or subtracted. There are many different ways to estimate, and many different answers depending on how you estimate.

EXAMPLE 7   Five passengers are on an elevator. Their weights are 247, 115, 189, 36, and 78 pounds, respectively. Estimate the total weight in the elevator.

Since most of the weights are in the hundreds, round all weights to the hundreds place. Then add.

        Actual weights: 247 + 115 + 189 + 36 + 78
                            ↓   ↓   ↓   ↓   ↓
        Estimated weights: 200 + 100 + 200 + 0 + 100 = 600 lbs is the
        (rounded to the   estimated weight
        hundreds place)   in the elevator

Verify that the actual total weight is 665 pounds.   ■

> **OBSERVE**   In Example 7, 36 is less than 50, or half a hundred. Therefore, 36 rounded to the nearest hundred is 0 hundreds, or 0.

► **You Try It**   **13.** Six people are in a van. They weigh 180, 214, 48, 136, 75, and 155 pounds, respectively. Estimate the total weight.

EXAMPLE 8   The population of California in 1940 was 6,907,387. In 1980, it was 23,667,826. Estimate the growth in population from 1940 to 1980.

Both numbers are in the millions. Round to the millions place. Then subtract.

$$
\begin{array}{lll}
1980: & 23,667,826 & \xrightarrow[\text{millions}]{\text{round to}} & 24,000,000 \\
1940: & 6,907,387 & & -\ 7,000,000 \\
\hline
& & & 17,000,000 \text{ estimated growth}
\end{array}
$$

Verify that the actual growth in population was 16,760,439. ■

> **OBSERVE** If you round both numbers in Example 8 to ten-millions, the estimate becomes $20,000,000 - 10,000,000 = 10,000,000$, which is much less accurate. You must use your judgment when selecting the place value to round to.

▶ **You Try It**   **14.** The population in Texas in 1970 was 11,198,655. In 1980 it was 14,227,574. Estimate the growth in population from 1970 to 1980.

Estimation can help you catch errors.

EXAMPLE 9   Use estimation to decide if the answer below is reasonable.

$$647 + 3,219 + 1,159 + 986 + 860 + 2,180 \stackrel{?}{=} 6,751$$

$$\downarrow \qquad \downarrow \qquad \downarrow \qquad \downarrow \qquad \downarrow \qquad \downarrow$$

Estimate: $600 + 3,200 + 1,200 + 1,000 + 900 + 2,200 = 9,100$ estimate of
(round all   the sum
addends to
hundreds)

The estimate, 9,100, is not reasonably close to 6,751. Rechecking your work shows the actual sum is 9,051. ■

> **OBSERVE** In Example 9, the fact that the answer and the estimate do not agree does not tell you which one is wrong. It only tells you to recheck your work.

▶ **You Try It**   **15.** Use estimation to decide if the answer is reasonable. $4,695 + 815 + 2,751 + 367 + 7,216 \stackrel{?}{=} 15,844$

▶ **Answers to You Try It**   1. 80   2. 70   3. 50   4. 40   5. 360   6. 880   7. 5,800   8. 3,100   9. 3,000   10. 47,000   11. 9,000   12. 16,500   13. about 800 lb (round to hundreds)   14. about 3,000,000 (round to millions)   15. estimate = 16,000 (round to thousands); answer = 15,844, so the answer is reasonable

SECTION 1.5 EXERCISES

I *Round off each number to the indicated place.*

1. 56   tens

2. 72   tens

3. 175   hundreds

4. 416   hundreds

5. 1,782   tens

6. 2,428   tens

7. 549   hundreds

8. 751   hundreds

9. 4,366   hundreds

10. 7,836   hundreds

11. 3,499   tens

12. 8,994   tens

13. 34   hundreds

14. 63   hundreds

15. 3,723   thousands

16. 9,398   thousands

17. 4,996   tens

18. 7,993   tens

19. 555   thousands

20. 499   thousands

21. 56,834   ten-thousands

22. 7,488   ten-thousands

23. 3,835,921   millions

24. 59,489,999   millions

25. 99,986   tens

26. 9,997   tens

27. 2,499,845   thousands

28. 3,858,705   thousands

Complete the following table by rounding off each number to the three indicated places.

	tens	hundreds	thousands
29. 69			
30. 736			
31. 1,658			
32. 9,989			
33. 7,293			
34. 263			
35. 34,608			
36. 9,999,745			
37. 262,911			
38. 79,567			

39. Round a car's odometer reading of 112,378 miles to the nearest thousand miles.

40. Round the $98,612,450 cost of the new stadium to the nearest million dollars.

41. A sporting event grossed $12,488,122. What is this rounded to the nearest one hundred thousand dollars?

42. Light travels 186,282 miles per second. Round this to the nearest thousand miles per second.

2 *Estimate the answer by first rounding each number to the indicated place.*

43. 36 + 47   tens

44. 68 + 73   tens

45. 268 + 431   hundreds

46. 691 + 380   hundreds

47. 6,192 + 8,398   hundreds

48. 2,820 + 3,746   hundreds

49. 72 − 48   tens

50. 97 − 51   tens

51. 6,924 − 2,318   thousands

52. 3,094 − 950   thousands

53. 1,230 − 467   hundreds

54. 2,880 − 1,343   hundreds

55. $459 + 120 + 785 + 98$ hundreds

56. $77 + 830 + 117 + 457 + 99$ hundreds

57. $75 + 36 + 40 + 98 + 3 + 25$ tens

58. $108 + 56 + 84 + 9 + 33 + 2$ tens

59. $1,416 + 780 + 128 + 45 + 849$ hundreds

60. $6,280 + 12,098 + 2,754 + 860 + 9,812$ thousands

61. Today the printing center spent $712 on paper, $340 on ink, $180 on envelopes, $572 on copying, and $301 on minor repairs. Estimate the amount spent by rounding to hundreds.

62. A real estate agent sold homes this month for $96,423, $87,280, and $156,254. Estimate total sales by rounding to ten-thousands.

63. A family started a cross-country trip with an odometer reading of 36,278 miles. The reading at the end of the trip was 47,819 miles. Estimate the total miles traveled by rounding to thousands.

64. The distance from Baltimore to Atlanta is 675 miles. The distance from Baltimore to Boston is 399 miles. Estimate how much closer Boston is to Baltimore by rounding to hundreds.

65. This month the Lopez family spent $918 on the mortgage, $514 on food, $78 on gas, and $267 on utilities. Estimate the total spent by rounding to tens.

66. Estimate the total area of the six New England states if the actual areas in square miles are 5,009 for Connecticut, 8,257 for Massachusetts, 33,215 for Maine, 9,304 for New Hampshire, 1,214 for Rhode Island, and 9,609 for Vermont. Round to thousands.

Estimate each answer two ways, as indicated.

67. $55 + 78 + 31 + 8 + 93$
   a. tens   **b.** hundreds

68. $89 + 72 + 97 + 8 + 43 + 76$
   a. tens   **b.** hundreds

69. $461 + 790 + 834 + 805$

    a. hundreds   **b.** thousands

70. $943 + 392 + 45 + 958 + 457 + 98$

    a. hundreds   **b.** thousands

71. $8{,}285 + 854 - 6{,}190 + 4{,}120$

    a. thousands   **b.** ten-thousands

72. $4{,}921 + 9{,}605 - 2{,}890 + 812 - 499$

    a. hundreds   **b.** thousands

SKILLSFOCUS (Section 1.3) *Solve by trial.*

73. $x + 9 = 9$ **74.** $x + 4 = 12$ **75.** $y - 2 = 16$ **76.** $w - 12 = 3$

EXTEND YOUR THINKING ▶▶▶▶
▶ TROUBLESHOOT IT

Find and correct the error.

77. 654 rounded to the nearest ten is 660.

78. 29,951 rounded to the nearest hundred is 29,900.

WRITING TO LEARN ▶▶▶▶

79. Explain the difference between $=$ and $\doteq$.

80. Explain why 36 rounded to hundreds is 0 (see Example 7).

81. The attendance for game #1 was 67,306 people. Game #2 was attended by 74,691 people. Estimate how many more people attended game #2 by

    a. Rounding to ten-thousands.

    b. Rounding to thousands.

    c. Explain which answer is more practical and why.

▶ YOU BE THE JUDGE

82. Jack claimed 4,449 rounded to the nearest thousand is 5,000. He reasoned that 4,449 rounded to the nearest ten is 4,450. 4,450 rounded to the nearest hundred is 4,500. And 4,500 rounded to the nearest thousand is 5,000. Is he correct? Explain your decision.

1.6 APPLICATIONS: ADDITION AND SUBTRACTION

OBJECTIVE
Solve word problems (Sections 1.3, 1.4).

Solving Word Problems

To solve word problems, look for key words that tell you what operation to use. Some common key words and phrases meaning add and subtract are listed in Figures 1.3 and 1.4.

FIGURE 1.3
Key Words and Phrases
Meaning Addition

sum $\longrightarrow$	the sum of 3 and 5 is $3 + 5 = 8$
total $\longrightarrow$	the total of 6 and 4 is $6 + 4 = 10$
plus $\longrightarrow$	4 plus 3 is $4 + 3 = 7$
increased by $\longrightarrow$	5 increased by 2 is $5 + 2 = 7$
more than $\longrightarrow$	3 more than 2 is $2 + 3 = 5$
gain $\longrightarrow$	7 with a gain of 2 is $7 + 2 = 9$

FIGURE 1.4
Key Words and Phrases
Meaning Subtraction

difference $\longrightarrow$	the difference between 7 and 5 is $7 - 5 = 2$
minus $\longrightarrow$	8 minus 5 is $8 - 5 = 3$
decreased by $\longrightarrow$	9 decreased by 2 is $9 - 2 = 7$
take away $\longrightarrow$	5 take away 4 is $5 - 4 = 1$
loss $\longrightarrow$	8 with a loss of 2 is $8 - 2 = 6$
less than $\longrightarrow$	4 less than 7 is $7 - 4 = 3$
subtract x from y $\longrightarrow$	subtract 5 from 7 to get $7 - 5 = 2$

OBSERVE In Figure 1.4, the numbers used with *less than* and *subtract x from y* are reversed when you subtract.

To solve word problems

1. *Read the problem carefully.* Understand it. Use a dictionary if needed. *Determine what you are to find.* This is often contained in the question at the end of the problem.
2. *Summarize the givens.* Write each number given in the problem. Next to it, write a brief statement saying what it stands for. If possible, draw and label a diagram.
3. *Decide which operation or operations to use* in the problem. Refer to the key words in Figures 1.3 and 1.4.
4. *Perform the operation(s).* Write your answer, with units, along with a short conclusion. (As a check, estimate your answer and compare.)

EXAMPLE 1

Rita works two jobs. She makes $355 a week as a carpenter, and $164 a week as a school bus driver. What is Rita's total weekly salary?

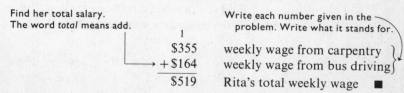

Find her total salary.
The word *total* means add.

Write each number given in the problem. Write what it stands for.

$$
\begin{array}{r}
1 \\
\$355 \\
+\ \$164 \\
\hline
\$519
\end{array}
$$

weekly wage from carpentry
weekly wage from bus driving
Rita's total weekly wage ■

---

1. Carla makes $1,827 per month teaching, and $370 per month giving piano lessons. What is her total monthly salary?

EXAMPLE 2   The Empire State Building in New York City is 1,472 feet high. The Sears Tower in Chicago is 332 feet taller. How high is the Sears Tower?

Draw and label a diagram.

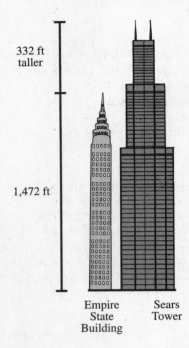

332 ft
taller

1,472 ft

Empire
State
Building

Sears
Tower

The height of the Sears Tower is the height of the Empire State Building plus 332 feet.

$$\begin{array}{r} 1{,}472 \text{ ft} \\ +\quad 332 \text{ ft} \\ \hline 1{,}804 \text{ ft} \end{array}$$

height of Empire State Building
additional height to top of Sears Tower
height of Sears Tower   ■

► **You Try It**

2. The length of a retaining wall is 2,864 feet. An extra 1,084 feet will be added on. What will its length then be?

EXAMPLE 3   **Cholesterol**   Blood contains two kinds of cholesterol, HDL and LDL. Blood cholesterol level is the sum of the HDL and LDL levels. Gene just had his blood cholesterol tested. His HDL level is 68. His LDL level is 113.

a. What is Gene's blood cholesterol level?

The problem says blood cholesterol level is the sum of the HDL and LDL levels. The word *sum* means add.

$$\begin{array}{r} 1 \\ 68 \\ +113 \\ \hline 181 \end{array}$$

HDL level
LDL level
Gene's blood cholesterol level

b. Suppose a blood cholesterol level under 200 is considered low-risk. How far under this level is Gene?

Find the difference between the low-risk level, 200, and Gene's level, 181. *Difference* means subtract.

$$\begin{array}{r} 1\ 9\ 10 \\ \cancel{2}\ \cancel{0}\ \cancel{0} \\ -1\ 8\ 1 \\ \hline 1\ 9 \end{array}$$

low-risk level
Gene's level
points under the low-risk level   ■

▶ **You Try It**　**3.** Lee had his cholesterol level tested. His HDL level was 98 and his LDL level was 163.

　　a. What is Lee's blood cholesterol level?

　　b. How many points is he over the low-risk level?

EXAMPLE 4　A family began their touring vacation in Baltimore. They drove 1,145 miles to New Orleans, then 503 miles to Dallas, 1,405 miles to Los Angeles, 405 miles to San Francisco, 1,264 miles to Denver, and 1,631 miles back home to Baltimore.

a. What was their total mileage for the trip?

Draw and label a diagram.

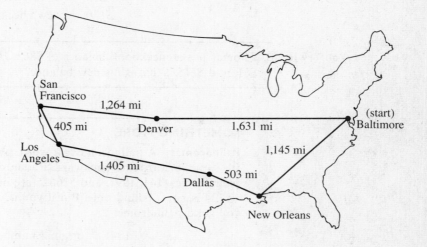

Find the total mileage by adding the individual miles on each leg of the round trip.

$$
\begin{array}{rl}
\overset{2\ 12}{} & \\
1,145 & \text{Baltimore to New Orleans} \\
503 & \text{New Orleans to Dallas} \\
1,405 & \text{Dallas to L.A.} \\
405 & \text{L.A. to San Francisco} \\
1,264 & \text{San Francisco to Denver} \\
+\,1,631 & \text{Denver to Baltimore} \\
\hline
6,353\ \text{mi} & \text{total round trip mileage}
\end{array}
$$

b. Is this answer reasonable? Estimate the mileage by rounding each number to hundreds. Then add.

Actual:　$1,145 + 503 + 1,405 + 405 + 1,264 + 1,631$

Estimated:　$1,100 + 500 + 1,400 + 400 + 1,300 + 1,600 = 6,300$ mi

Since the estimate of 6,300 miles is close to 6,353 miles, the answer is reasonable.　■

▶ **You Try It**　**4.** Victoria made sales of $584, $1,134, $817, $361, and $1,475 for Monday through Friday, respectively.

　　a. What were her total sales for the 5 days?

　　b. Estimate total sales to determine if this answer is reasonable.

Some problems require more than one operation.

EXAMPLE 5   Yesterday Joan had a balance of $362 in her checkbook. Today she wrote checks for $120, $25, and $87 to pay her bills. What is her new balance?

Step #1: Add $120, $25, and $87 to find the total for the checks written.

$$
\begin{array}{r}
11 \\
\$120 \\
25 \\
+\ \ 87 \\
\hline
\$232
\end{array}
\quad \text{total for the checks written}
$$

Step #2: Subtract the check total, $232, from the balance, $362.

$$
\begin{array}{r}
\$362 \\
-\ 232 \\
\hline
\$130
\end{array}
\quad
\begin{array}{l}
\text{original checkbook balance} \\
\text{total of checks written} \\
\text{new checkbook balance}\ \blacksquare
\end{array}
$$

▶ **You Try It**   **5.** Thomas has a checkbook balance of $863. He wrote checks for $38, $146, $61, and $215. What is his new balance?

SOMETHING MORE

Palindromes   A palindrome is a word, phrase, or number that reads the same forward and backward. Radar, dad, and Hannah are word palindromes. 141, 1991, and 526625 are number palindromes.

134 is not a palindrome. But if you add it to its reversal, 431, you get a palindrome.

$$
\begin{array}{r}
134 \\
+431 \\
\hline
565
\end{array}
\quad
\begin{array}{l}
\text{original number} \\
\text{the reversal of 134} \\
\text{This sum is a palindrome.}
\end{array}
$$

134 gave a palindrome after 1 reversal. Many numbers require 2 or more reversals. For example, 68 needs 3 reversals.

$$
\begin{array}{r}
68 \\
+86 \\
\hline
154 \\
+451 \\
\hline
605 \\
+506 \\
\hline
1111
\end{array}
\quad
\begin{array}{l}
\text{original number} \\
\text{reversal of 68} \\
\text{outcome of first reversal} \\
\text{reversal of 154} \\
\text{outcome of second reversal} \\
\text{reversal of 605} \\
\text{outcome of third reversal is a palindrome}
\end{array}
$$

How many reversals are needed to get a palindrome?
a. 65   **b.** 148   **c.** 39   **d.** 79   **e.** 78   **f.** 198

▶ **Answers to You Try It   1.** $2,197   **2.** 3,948 ft   **3. a.** 261   **b.** 61
4. a. answer = $4,371   **b.** estimated (to hundreds) = $4,400, so $4,371 is a reasonable answer   **5.** $403

▶ **Answers to Something More   a.** 121 (1 reversal)   **b.** 989 (1 reversal)
c. 363 (2 reversals)   **d.** 44044 (6 reversals)   **e.** 4884 (4 reversals)
f. 79497 (5 reversals)

1. Nick spent $14 on a tie and $27 on a pair of slacks. What total did Nick spend?

2. Maria spent $36 on a dress and $17 on an umbrella. How much did Maria spend in all?

3. Jack has a $50 bill. He spends $17 on a shirt. What is his change?

4. Amanda has $1,000. She bought a ring for $372. What does she have left?

5. Yellowstone National Park covers 3,468 square miles in area. The area of the state of Rhode Island is 1,214 square miles. How much larger in area is Yellowstone Park than Rhode Island?

6. The area of the state of Delaware is 2,057 square miles. The area of the Grand Canyon National Park is 1,904 square miles. How much larger in area is Delaware than Grand Canyon Park?

7. For breakfast Damon had a doughnut and juice. The doughnut had 238 calories. The juice had 66 calories. How many calories did Damon have for breakfast?

8. Hank had a piece of cake and a glass of milk for an evening snack. The cake had 328 calories. The milk had 165 calories. How many calories were in Hank's snack?

9. Noreen makes $42,000 per year. Brenda makes $17,000 a year more than Noreen. What does Brenda make yearly?

10. Edward weighs 241 pounds. His brother weighs 88 pounds more. What is his brother's weight?

11. Mark's odometer reading at the start of a cross-country trip was 34,217 miles. By the end of the trip the reading was 41,688 miles. How many miles did Mark drive on his trip?

12. The Orioles played 162 games this season, and lost 77 of them. How many games did the Orioles win this season?

13. Lee Ann bought a car for $18,645. She put $3,990 down on the purchase of the car. How much does Lee Ann still owe?

14. Jerry bought a house for $88,600. If the bank requires him to put $17,720 down, how much will Jerry still owe?

15. Cathy's odometer read 56,782 miles this morning. She drove to Chicago, a distance of 472 miles. What did her odometer then read?

16. Arnold's truck weighs 6,245 pounds. Today he is hauling 3,560 pounds of stone. What is the total weight of the truck with the stone?

17. The area of the United States is 3,618,770 square miles. The area of Canada is 3,831,033 square miles.

 a. What is the sum of the areas of the two countries?

 b. What is the difference in the areas of the two countries?

18. In one recent year, the population of Canada was 27,483,000 and the U.S. population was 248,729,000.

 a. What is the difference in the two populations?

 b. What is the combined total of the two populations?

19. A truck has the capacity to carry 4,200 pounds of weight. If 2,880 pounds are now in the truck, how many more pounds can the truck carry?

20. A storage tank can hold 7,500 gallons of oil. If 2,398 gallons are currently in the tank, how many gallons of oil should be ordered to fill the tank?

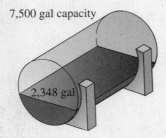

7,500 gal capacity

2,348 gal

21. A rocket has three stages. The first stage is 188 feet high, the second is 127 feet high, and the third is 48 feet high. How tall is the rocket?

22. A student made school purchases of $54, $7, $23, $84, and $37. How much was spent?

23. An empty truck has the capacity to haul 8,500 pounds. At one stop it picked up loads of 2,275 pounds, 920 pounds, and 3,650 pounds.

 a. How many pounds were loaded onto the truck?

 b. How many more pounds can the truck hold without exceeding its capacity?

 c. Is there enough capacity left on the truck to carry two extra loads of 950 pounds and 1,340 pounds respectively?

24. A freight elevator has a capacity of 3,200 pounds. An attendant weighing 195 pounds is on the elevator along with a 435-pound refrigerator and a 1,460-pound generator.

 a. What total weight is on the elevator?

 b. How much more weight can the elevator carry without exceeding its capacity?

 c. Is there enough capacity left on the elevator to also carry a 760-pound condenser along with a 210-pound repairman?

25. David has a balance of $560 in his checking account. He wrote checks for $162, $87, $28, $63, and $104.

    a. What is the total of the checks he wrote?

    b. What is David's new balance?

26. Sarina has an equipment budget of $2,640. She furnished her office with a $760 sofa, two end tables for $180 each, a bookcase for $280, and a chair for $490.

    a. How much has Sarina spent so far to furnish her office?

    b. How much is still left in her equipment budget?

27. Subtract 782 from 2,000.

28. Subtract 1,873 from 100,000.

29. Audry has a balance of $67 in her checkbook. She wrote checks for $32, $45, $17, and $53. She made deposits of $50 and $35. What is Audry's new balance?

30. Quentin weighed 236 pounds. During the year he lost 7 pounds, gained 12 pounds, lost 16 pounds, and gained 13 pounds. What is Quentin's current weight?

31. A photographer is 60 feet from a lion. He moves 23 feet closer to the animal, takes a picture, then moves 41 feet farther away. How far is the photographer from the lion?

32. A hot air balloon is 450 feet above the ground. It drops 178 feet, then rises 310 feet. How high is the balloon?

310 ft

178 ft

450 ft

SKILLSFOCUS (Section 1.2) *Write in standard notation.*

33. Four hundred seven.

34. Fifty thousand, ninety-three.

35. Two hundred eighty-four thousand, six hundred eleven.

EXTEND YOUR THINKING ▶▶▶▶

▶SOMETHING MORE

36. How many reversals are needed to get a palindrome?

  a. 19   **b.** 409   **c.** 97   **d.** 9,077

37. The Acrobat's Ladder   A ladder connected the roof of one building to the roof of another. An acrobat was standing on the middle rung of the ladder. She moved forward 4 rungs. She then backed up 6 rungs. Then she moved forward 9 rungs, tipped her hat, and moved forward the remaining 5 rungs of the ladder to the roof. How many rungs were on the ladder?

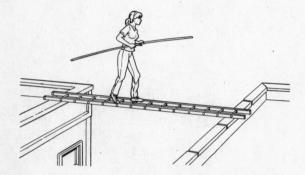

WRITING TO LEARN ▶▶▶▶

38. Write a word problem about balancing a checkbook. You write two checks and make one deposit in the problem. Your new balance is $395. You must specify all other amounts. Then write your solution to the problem.

39. Write down the following four numbers:

   1. The year in which you were born = _____

   2. The happiest year of your life = _____

   3. Your age, by the end of this current year = _____

   4. The number of years since your happiest year = _____

   Add the four numbers. Explain why the answer is always twice the current year.

1.7 MULTIPLICATION AND ITS PROPERTIES

OBJECTIVES
1. Define multiplication.
2. Define product, factor, and multiple.
3. Define the properties of multiplication.
4. Solve equations.

NEW VOCABULARY
implied multiplication   multiplier   multiple

product   factor   multiplicative identity

multiplicand

---

1 **The Meaning of Multiplication**

Multiplication is a shortcut for repeated addition. For example, you can repeatedly add $5 + 5 + 5 + 5 = 20$. Or, since you have four 5's, take the shortcut and multiply $4 \times 5 = 20$.

4×5 is read "four times five"

four 5's

$5 + 5 + 5 + 5 = 4 \times 5$

Multiplication can be used to count items arranged in rows. The same number of items must be in each row.

EXAMPLE 1

5 items in each row

4 rows
$\begin{matrix} * & * & * & * & * \\ * & * & * & * & * \\ * & * & * & * & * \\ * & * & * & * & * \end{matrix}$

4 rows × 5 items per row
$= 4 \times 5 = 20$ items

■

▶ **You Try It**   Write a multiplication to count the items below.

1. * * * * * * *
   * * * * * * *

2. * * * *
   * * * *
   * * * *

Ways to Write Multiplication
Each statement below is read "4 times 5."

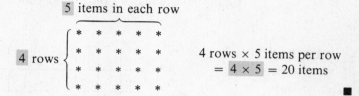

4×5 $\begin{matrix} 5 \\ \times 4 \end{matrix}$ $4 \cdot 5$ $(4)(5)$ $4(5)$ $(4)5$

raised period means multiply

3 ways to write multiplication *without* using a multiplication symbol. This is called **implied multiplication**.

EXAMPLE 2

Different ways to write "7 times 4."

7×4 $\begin{matrix} 4 \\ \times 7 \end{matrix}$ $7 \cdot 4$ $(7)(4)$ $7(4)$ $(7)4$ ■

▶ **You Try It**   Write each multiplication in six ways.

3. 5 times 8

4. 9 times 2

Multiplication is often implied when letters are used to represent numbers.

EXAMPLE 3
$4y$ means $4 \cdot y$
ab means $a \cdot b$
$n7$ means $n \cdot 7$ ∎

▶ **You Try It** Rewrite each multiplication using the multiplication dot.

5. $6y$ **6.** $2m$ **7.** $w5$ **8.** st

2 Products, Factors, and Multiples

When you multiply two numbers, the answer is called the **product**.

$\quad\quad\quad 5 \longleftarrow$ **multiplicand** (number on top)
$\quad\underline{\times 4} \longleftarrow$ **multiplier** (number on bottom)
$\quad\quad 20 \longleftarrow$ **product**

When two numbers are multiplied, each is called a **factor** of the product. The numbers 4 and 5 are called factors of 20.

$$4 \quad\times\quad 5 \quad=\quad 20$$
$$\text{factor} \quad \text{factor} \quad \text{product}$$

Figure 1.5 summarizes the multiplication facts. Memorize this table.

FIGURE 1.5
Multiplication Facts

×	0	1	2	3	4	5	6	7	8	9	10
0	0	0	0	0	0	0	0	0	0	0	0
1	0	1	2	3	4	5	6	7	8	9	10
2	0	2	4	6	8	10	12	14	16	18	20
3	0	3	6	9	12	15	18	21	24	27	30
4	0	4	8	12	16	20	24	28	32	36	40
5	0	5	10	15	20	25	30	35	40	45	50
6	0	6	12	18	24	30	36	42	48	54	60
7	0	7	14	21	28	35	42	49	56	63	70
8	0	8	16	24	32	40	48	56	64	72	80
9	0	9	18	27	36	45	54	63	72	81	90
10	0	10	20	30	40	50	60	70	80	90	100

EXAMPLE 4 Use Figure 1.5 to multiply 6×4.

Find 6 in the left-hand column. Move to the right.
Find 4 in the top row. Move downward. You intersect at the answer, 24.
(See the shading in Figure 1.5.) ∎

▶ **You Try It** Use Figure 1.5 to find the product.

9. $7 \cdot 9$ **10.** $6 \cdot 1$ **11.** $8 \cdot 0$

Since $4 \times 5 = 20$, 20 is called a **multiple** of 4 and a multiple of 5. Each row in Figure 1.5 is a set of multiples for the number on the far left side. For example, the shaded row is a set of multiples of 6. Each column is a set of multiples for the number at the top. The shaded column is a set of multiples of 4.

EXAMPLE 5   State five facts about $3 \times 7 = 21$.

$$3 \times 7 = 21 \begin{cases} 21 \text{ is the product of 3 and 7} \\ 3 \text{ is a factor of 21} \\ 7 \text{ is a factor of 21} \\ 21 \text{ is a multiple of 3} \\ 21 \text{ is a multiple of 7} \quad \blacksquare \end{cases}$$

▶ **You Try It**   **12.** State five facts about the multiplication $6 \cdot 8 = 48$.

3 Properties of Multiplication   In Figure 1.5, multiplying any number in the left-hand column times 0 in the top row gives 0 for an answer. To see why, write multiplication by 0 as repeated addition of 0 to itself.

$$3 \times 0 = \text{three zeros} = 0 + 0 + 0 = 0$$

> **Multiplication Property of Zero**
>
> Multiplying any number by 0 gives 0 for an answer. For any number a,
>
> $$a \times 0 = 0 \quad \text{and} \quad 0 \times a = 0.$$

EXAMPLE 6
$5 \times 0 = 0$
$0 \times 8 = 0$
$0 \cdot n = 0 \quad \blacksquare$

▶ **You Try It**   Find each product.

13. $(3)(0)$ **14.** $0 \cdot w$ **15.** $9(0)$

In Figure 1.5, multiplying any number in the left-hand column times 1 in the top row gives the same number for an answer. To see why, write multiplication by 1 as repeated addition of 1 to itself.

$$4 \times 1 = \text{four ones} = 1 + 1 + 1 + 1 = 4$$

> **Multiplication Property of One**
>
> Multiplying any number by 1 gives that number for an answer. If a is any number,
>
> $$a \times 1 = a \quad \text{and} \quad 1 \times a = a.$$
>
> 1 is called the **multiplicative identity**. Multiplying a number by 1 does not change the identity of that number.

EXAMPLE 7 $7 \times 1 = 7$
$1 \cdot 3 = 3$
$1 \cdot y = y \quad \blacksquare$

Find each product.

16. (4)(1)   **17.** (8)1   **18.** 1 · *w*

What pattern do you see with the following products?

EXAMPLE 8
a. 3 × 5 = 15   and   5 × 3 = 15
b. 2 × 7 = 14   and   7 × 2 = 14
c. 9 × 4 = 36   and   4 × 9 = 36 ■

> ### Commutative Property of Multiplication
> The order in which two numbers are multiplied does not change the product. For any two numbers *a* and *b*,
> $$a \times b = b \times a.$$

EXAMPLE 9
Applications of the commutative property.

a. 3(7) is the same as (7)3

b.
$$\begin{array}{c} 9 \\ \times\,54 \end{array}$$ gives the same answer as $$\begin{array}{c} 54 \\ \times\,9 \end{array}$$

c. *n* · 4 is the same as 4 · *n* (Write *n*4 = 4*n*.)
d. Since multiplication is commutative, you can count the number of stars in the box below in two ways.

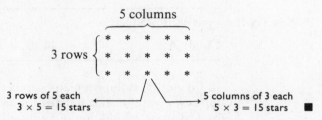

3 rows of 5 each
3 × 5 = 15 stars ⟶   ⟵ 5 columns of 3 each
5 × 3 = 15 stars ■

▶ **You Try It**
Use the commutative property to write each product in a second way.

19. 4(8)   **20.** 12 × 4   **21.** *w* · 9   **22.** $$\begin{array}{c} 45 \\ \times\,326 \end{array}$$

Find the product of 3, 2, and 4. This product can be written in two ways.

$$\underbrace{(3 \times 2)} \times 4 \quad\text{or}\quad 3 \times \underbrace{(2 \times 4)}$$
$$\begin{array}{ccc} 6 & \times 4 & \\ 24 & & \end{array} \qquad \begin{array}{ccc} 3 \times & 8 \\ & 24 \end{array}$$

Use parentheses to group the numbers to be multiplied first.

The product is the same in each case. Therefore,
$$(3 \times 2) \times 4 = 3 \times (2 \times 4).$$

The Associative Property of Multiplication

Changing the grouping when multiplying three numbers together does not change the product. If a, b, and c are any three numbers,

$$(a \times b) \times c = a \times (b \times c).$$

EXAMPLE 10 Use the associative property to regroup each product.

a. $(7 \times 5) \times 2$ may be rewritten $7 \times (5 \times 2)$
b. $2 \times (3 \times 8)$ may be rewritten $(2 \times 3) \times 8$
c. $(9 \times 4) \times n$ may be rewritten $9 \times (4 \times n)$ ■

▶ **You Try It** Use the associative property to regroup each product.

23. $(4 \times 6) \times 5$ **24.** $7 \times (3 \times 8)$ **25.** $(k \cdot 5) \cdot h$

The commutative and associative properties of multiplication say you can multiply numbers together in any order or grouping without changing the final product.

EXAMPLE 11 Replace the letter with the number that makes the statement true.

a. $7 \cdot a = 7$ Since $7 \cdot 1 = 7$, the letter $a = 1$.
This illustrates the multiplication property of one.
b. $4 \cdot 6 = w \cdot 4$ Since $4 \cdot 6 = 6 \cdot 4$, the letter $w = 6$.
This illustrates the commutative property.
c. $8 \cdot v = 0$ Since $8 \cdot 0 = 0$, the letter $v = 0$.
This illustrates the multiplication property of zero.
d. $(y \cdot 2) \cdot 4 = 3 \cdot (2 \cdot 4)$ Since $(3 \cdot 2) \cdot 4 = 3 \cdot (2 \cdot 4)$, the letter $y = 3$.
This illustrates the associative property. ■

▶ **You Try It** Replace the letter with the number that makes the statement true.

26. $3 = a \cdot 3$ **27.** $2 \cdot 9 = 9 \cdot m$
28. $0 = 5 \cdot p$ **29.** $(4 \cdot n) \cdot 6 = 4 \cdot (2 \cdot 6)$

4 Solving Equations Solve the equation by trial.

$$4a = 12$$

The expression $4a$ means "4 times the letter a." To solve by trial, replace the letter a with 0, then with 1, then with 2, and so on, until you find the whole number that makes the equation true.

Replace a with 0: $4 \cdot 0 \neq 12$, so $a \neq 0$.
Replace a with 1: $4 \cdot 1 \neq 12$, so $a \neq 1$.
Replace a with 2: $4 \cdot 2 \neq 12$, so $a \neq 2$.
Replace a with 3: $4 \cdot 3 = 12$ This is true, so $a = 3$
is the solution to $4a = 12$.

EXAMPLE 12   Solve for w by trial: $6w = 42$.

Since $6 \cdot 7 = 42$, the solution is $w = 7$.   ■

EXAMPLE 13   Solve for t: $32 = 4t$.

Since $32 = 4 \cdot 8$, the solution is $t = 8$.   ■

▶ **You Try It**   Solve for x by trial and error.

30. $6x = 18$

31. $3x = 27$

32. $48 = 8x$

33. $63 = 9x$

▶ **Answers to You Try It**   **1.** 7×2 or 2×7 **2.** 4×3 or 3×4
3. 5×8, $8 \ 5 \cdot 8$, $(5)(8)$, $5(8)$, $(5)8$ **4.** 9×2, $2 \ 9 \cdot 2$, $(9)(2)$, $9(2)$, $(9)2$
    $\underline{\times 5}$, $\underline{\times 9}$,
5. $6 \cdot y$ **6.** $2 \cdot m$ **7.** $w \cdot 5$ **8.** $s \cdot t$ **9.** 63 **10.** 6 **11.** 0
12. 48 is the product of 6 and 8; 6 is a factor of 48; 8 is a factor of 48;
48 is a multiple of 6; 48 is a multiple of 8   **13.** 0 **14.** 0 **15.** 0 **16.** 4
17. 8 **18.** w **19.** $(8)4$ **20.** 4×12 **21.** $9 \cdot w$ **22.** 326 **23.** $4 \times (6 \times 5)$
                                                                          $\underline{\times \ 45}$
24. $(7 \times 3) \times 8$ **25.** $k \cdot (5 \cdot h)$ **26.** $a = 1$ **27.** $m = 2$ **28.** $p = 0$
29. $n = 2$ **30.** $x = 3$ **31.** $x = 9$ **32.** $x = 6$ **33.** $x = 7$

1 **2** *Fill in the blank.*

1. With $4 \times 9 = 36$,
   a. 36 is called the _____ .

   b. 4 is called a _____ of 36.

   c. 9 is called a _____ of 36.

   d. 36 is called a _____ of 9.

   e. 36 is called a _____ of 4.

2. With $8 \times 5 = 40$,
   a. 40 is called the _____ .

   b. 8 is called a _____ of 40.

   c. 5 is called a _____ of 40.

   d. 40 is called a _____ of 8.

   e. 40 is called a _____ of 5.

3. With $72 = 8 \times 9$,
   a. 8 is called a _____ of 72.

   b. 72 is called a _____ of 9.

   c. 72 is called the _____ .

   d. 9 is called a _____ of 72.

   e. 72 is called a _____ of 8.

4. With $12 = 2 \times 6$,
   a. 12 is called a _____ of 6.

   b. 6 is called a _____ of 12.

   c. 2 is called a _____ of 12.

   d. 12 is called the _____ .

   e. 12 is called a _____ of 2.

5. With $6 \times 6 = 36$,
   a. 6 is called a _____ of 36.

   b. 36 is called the _____ .

   c. 36 is a _____ of 6.

6. With $8 = 1 \times 8$,
   a. 8 is called either a _____ or a _____ of 8.

   b. 1 is called a _____ of 8.

Using Figure 1.5, find the product.

7. 9×6 **8.** 8×7 **9.** $6(8)$ **10.** $3(9)$

11. 0×5 **12.** 8×0 **13.** $(4)3$ **14.** $(8)2$

15. $(6)(3)$ **16.** $(5)(9)$ **17.** $(4)(0)$ **18.** $(2)(1)$

19. 1×9 **20.** 3×1 **21.** $6(1)$ **22.** $1(7)$

23. $(8)0$ **24.** $0(4)$ **25.** $5 \cdot 2$ **26.** $1 \cdot 1$

27. $(2 \times 4) \times 1$ **28.** $2 \times (3 \times 2)$ **29.** $3(4 \cdot 2)$ **30.** $(3 \cdot 2)7$

31. $(0 \cdot 4)5$ **32.** $(8 \cdot 9)0$ **33.** $8(3 \times 3)$ **34.** $(4 \cdot 1) \times 7$

3 *Replace the letter with the number that makes the statement true. Then identify each statement as an example of a) the multiplicative property of zero, b) the multiplicative property of one, c) the commutative property, or d) the associative property.*

35. $7 \cdot 4 = a \cdot 7$ **36.** $8 \cdot b = 2 \cdot 8$ **37.** $0 \cdot 3 = a$

38. $4 \cdot b = 4$ **39.** $(3 \cdot c)5 = 3(2 \cdot 5)$ **40.** $(5 \cdot 1)7 = 5(1 \cdot t)$

41. $6 \cdot f = 0$ **42.** $x \cdot 5 = 5 \cdot 8$ **43.** $c \cdot 9 = 9$

44. $(r \cdot 2)3 = 4(2 \cdot 3)$ **45.** $0 = 8 \cdot s$ **46.** $1 \cdot d = 6$

47. $(3 \cdot 3)w = 3(3 \cdot 2)$ **48.** $x = 8 \cdot 0$ **49.** $z \cdot 8 = 8$

50. $y(5) = (5)(6)$ **51.** $h(5) = 5$ **52.** $6(t) = 0$

53. $(5)(8) = (x)(5)$ **54.** $(6)3 = x(6)$

Rewrite each statement using the commutative property of multiplication.

55. (9)7   **56.** $7 \cdot 8$ **57.** 6×2 **58.** (5)(1)

59. 4(8)   **60.** (8)(7)   **61.** $8y$ **62.** $n3$

Rewrite each statement using the associative property of multiplication.

63. $(3 \times 2) \times 5$ **64.** $4(6 \cdot 3)$ **65.** $8 \times (7 \times 2)$ **66.** $(6 \cdot 4)9$

67. $1(2 \cdot 3)$ **68.** $(7 \times 6) \times 2$ **69.** $(5 \cdot n)3$ **70.** $y(4 \cdot 9)$

4 *Solve each equation by trial.*

71. $3x = 15$ **72.** $5y = 35$

73. $2t = 6$ **74.** $7f = 14$

75. $4z = 20$ **76.** $8p = 8$

77. $10 = 2w$ **78.** $40 = 8k$

79. $7t = 0$ **80.** $7z = 56$

81. $0 = 5v$ **82.** $4 = 4s$

SKILLSFOCUS (Section 1.4) *Subtract.*

83. $546 - 259$ **84.** $7{,}893 - 5{,}087$

85. $800 - 482$ **86.** $300 - 155$

87. **a.** Write a repeated addition for the number of stars at the right.

   b. Rewrite this repeated addition as a multiplication.

   c. Use the commutative property of multiplication to write this product in a second way.

88. **a.** Write a repeated addition for the number of dollar signs at the right.

   b. Rewrite this repeated addition as a multiplication.

   c. Use the commutative property of multiplication to write this product in a second way.

WRITING TO LEARN ▶▶▶▶

89. Explain the difference between the commutative and associative properties of multiplication. Use your own examples.

90. Explain how to use Figure 1.5 to demonstrate the commutative property of multiplication.

OBJECTIVES

1 Multiply whole numbers.

2 Multiply by powers of 10.

3 Estimate multiplication.

4 Order of operations.

5 Substitute numbers for letters in expressions and equations.

NEW VOCABULARY

partial product

power of 10

order of operations

substitution

---

1 Multiplying Whole Numbers

When multiplying whole numbers, write like named digits in the same vertical column. In the first three examples, the multiplier is a single digit.

EXAMPLE 1 Multiply 4×6.

$$
\begin{array}{cc}
\text{tens} & \text{ones} \\
& 6 \\
\times & 4 \\
\hline
2 & 4 \\
\end{array}
$$

4×6 ones = 24 ones = 20 ones + 4 ones
= 2 tens + 4 ones

← Write 4 in the ones column. Write 2 in the tens column. ∎

EXAMPLE 2 Multiply 7×36.

$$
\begin{array}{ccc}
\text{hundreds} & \text{tens} & \text{ones} \\
& 4 & \\
& 3 & 6 \\
\times & & 7 \\
\hline
& & 2 \\
\end{array}
$$

← 7 × 6 ones = 42 ones
= 4 tens + 2 ones

— Write 2 ones in the ones column. Carry 4 tens to the tens column.

$$
\begin{array}{ccc}
& 4 & \\
& 3 & 6 \\
\times & & 7 \\
\hline
2 & 5 & 2 \\
\end{array}
$$

7 × 3 tens = 21 tens
21 tens + 4 tens (the carry) = 25 tens = 20 tens + 5 tens
= 2 hundreds + 5 tens

— Write 5 in the tens column. Write 2 in the hundreds column. ∎

EXAMPLE 3 What is $6 \cdot 2{,}705$?

$$
\begin{array}{r}
^4 \quad ^3 \\
2{,}7\,0\,5 \\
\times \qquad 6 \\
\hline
1\,6{,}2\,3\,0 \\
\end{array}
$$
∎

▶ **You Try It** Multiply.

1.	2.	3.	4.
9	38	216	4,072
×6	× 4	× 5	× 8

In the next four examples, the multiplier has more than one digit.

EXAMPLE 4   What is 28 × 64?

$$
\begin{array}{r}
6\ 4 \\
\times\ 2\ 8 \\
\hline
5\ 1\ 2 \leftarrow
\end{array}
$$

— Multiply 8 ones times 64.
8 ones × 64 = 512 ones. 512 is called the **ones partial product**.
Write 512 so the digit 2 is in the ones column.

$$
\begin{array}{r}
6\ 4 \\
\times\ 2\ 8 \\
\hline
5\ 1\ 2 \\
1\ 2\ 8\ 0 \leftarrow
\end{array}
$$

— Next, multiply 2 tens times 64.
2 tens × 64 = 128 tens. 128 is called the **tens partial product**.
Write 128 so the digit 8 is in the tens column.

You can write 0 in the ones column to help
you keep the digits in straight vertical columns.

$$
\begin{array}{r}
6\ 4 \\
\times\ 2\ 8 \\
\hline
5\ 1\ 2 \longleftarrow \text{ ones partial product} \\
+1\ 2\ 8\ 0 \longleftarrow \text{ tens partial product} \\
\hline
1{,}7\ 9\ 2 \longleftarrow \text{ add the partial products } \blacksquare
\end{array}
$$

EXAMPLE 5   Multiply 539 × 867.

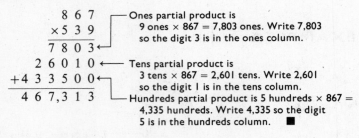

$$
\begin{array}{r}
8\ 6\ 7 \\
\times\ 5\ 3\ 9 \\
\hline
7\ 8\ 0\ 3 \leftarrow \\
2\ 6\ 0\ 1\ 0 \leftarrow \\
+4\ 3\ 3\ 5\ 0\ 0 \leftarrow \\
\hline
4\ 6\ 7{,}3\ 1\ 3
\end{array}
$$

— Ones partial product is
9 ones × 867 = 7,803 ones. Write 7,803
so the digit 3 is in the ones column.

— Tens partial product is
3 tens × 867 = 2,601 tens. Write 2,601
so the digit 1 is in the tens column.

— Hundreds partial product is 5 hundreds × 867 =
4,335 hundreds. Write 4,335 so the digit
5 is in the hundreds column.   ■

▶ **You Try It**

5.
$$
\begin{array}{r}
5\ 7 \\
\times\ 2\ 4 \\
\hline
\end{array}
$$

6.
$$
\begin{array}{r}
2\ 8\ 6 \\
\times\ 6\ 1 \\
\hline
\end{array}
$$

7.
$$
\begin{array}{r}
5\ 4\ 8 \\
\times\ 2\ 7\ 3 \\
\hline
\end{array}
$$

8.
$$
\begin{array}{r}
6{,}7\ 0\ 2 \\
\times\ 8\ 4 \\
\hline
\end{array}
$$

The next example illustrates finding a product when a 0 is in the multiplier.

EXAMPLE 6   What is 308 × 5,947?

$$
\begin{array}{r}
5\ 9\ 4\ 7 \\
\times\ \ \ 3\ 0\ 8 \\
\hline
4\ 7\ 5\ 7\ 6 \\
0\ 0\ 0\ 0\ 0 \leftarrow \\
+1\ 7\ 8\ 4\ 1\ 0\ 0 \\
\hline
1{,}8\ 3\ 1{,}6\ 7\ 6
\end{array}
$$

— The tens partial product is
0 tens × 5,947 = 0 tens. Write a row
of 0's as the tens partial product.

$$
\begin{array}{r}
5\ 9\ 4\ 7 \\
\times\ \ \ 3\ 0\ 8 \\
\hline
4\ 7\ 5\ 7\ 6 \\
+1\ 7\ 8\ 4\ 1\ 0\ 0 \\
\hline
1{,}8\ 3\ 1{,}6\ 7\ 6
\end{array}
$$

Eliminate the row of 0's, and write the solution
as shown. In short, if a 0 is in the multiplier, drop it
straight down into the partial product. Resume
multiplying with the next digit in the multiplier.   ■

EXAMPLE 7 Find 425×300.

$$
\begin{array}{r}
4\ 2\ 5 \\
\times 3\ 0\ 0 \\
\hline
0\ 0\ 0 \\
0\ 0\ 0\ 0 \\
+1\ 2\ 7\ 5\ 0\ 0 \\
\hline
1\ 2\ 7,5\ 0\ 0
\end{array}
$$

As a shortcut, drop both 0's straight down into the product →

$$
\begin{array}{r}
4\ 2\ 5 \\
\times 3\ 0\ 0 \\
\hline
1\ 2\ 7,5\ 0\ 0
\end{array}
$$

▶ **You Try It**   Multiply.

9. $\begin{array}{r} 3{,}247 \\ \times\ \ \ 603 \\ \hline \end{array}$

10. $\begin{array}{r} 871 \\ \times 500 \\ \hline \end{array}$

---

2 Multiplying by Powers of 10

What pattern do you see with the following products?

$$
\begin{array}{r}
3{,}6\ 4\ 2 \\
\times\ \ \ \ \ 1\ 0 \\
\hline
3\ 6{,}4\ 2\ 0
\end{array}
\qquad
\begin{array}{r}
3{,}6\ 4\ 2 \\
\times\ \ \ 1\ 0\ 0 \\
\hline
3\ 6\ 4{,}2\ 0\ 0
\end{array}
\qquad
\begin{array}{r}
3{,}6\ 4\ 2 \\
\times 1{,}0\ 0\ 0 \\
\hline
3{,}6\ 4\ 2{,}0\ 0\ 0
\end{array}
$$

10, 100, 1,000, etc., are called **powers of 10**. Each answer is the multiplicand, 3,642, followed by the number of zeros in the power of 10.

> The product from multiplying a whole number by a power of 10 is the whole number followed by the number of zeros in the power of 10.

EXAMPLE 8   **a.** $4 \times 1{,}000 = 4{,}000$

b. $72 \times 100 = 7{,}200$

c. $365 \times 10{,}000 = 3{,}650{,}000$

d. $67 \times 10 = 670$ ■

▶ **You Try It**   Multiply.

11. 417×100

12. $2{,}491 \times 10$

13. $38 \times 1{,}000$

14. $10{,}000 \times 147$

15. $1{,}000 \times 3{,}216$

---

3 **Estimation** You purchase 4 toys at $19.95 each. Estimate the cost. Since $19.95 is close to $20, you estimate the cost at 4 × $20 = $80. Note that $19.95 rounded to the digit farthest left (tens) is $20.

> To estimate a multiplication
>
> **1.** Round each factor to the place value of its farthest left digit.
> **2.** Multiply the rounded factors.

There are many ways to estimate a multiplication. When estimating answers to multiplication (or addition) problems, the estimate will be high if all numbers are rounded up. The estimate will be low if all numbers are rounded down.

EXAMPLE 9   Estimate the product 43 · 61.

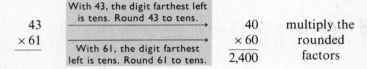

$$
\begin{array}{r} 43 \\ \times 61 \\ \hline \end{array}
\qquad
\begin{array}{r} 40 \\ \times 60 \\ \hline 2,400 \end{array}
\quad
\begin{array}{l} \text{multiply the} \\ \text{rounded} \\ \text{factors} \end{array}
$$

The estimate for 43 · 61 is 2,400. This estimate is low since both factors were rounded down. (The actual product is 2,623.) ■

▶ **You Try It**   Estimate each product.

16. 72 × 91   **17.** 38 × 23

EXAMPLE 10   A men's store owner plans to purchase 58 suits for $183 each. Estimate the total cost of the suits.

The total cost is 58 · $183. Estimate it as follows.

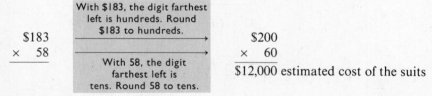

$$
\begin{array}{r} \$183 \\ \times\ \ 58 \\ \hline \end{array}
\qquad
\begin{array}{r} \$200 \\ \times\ \ \ 60 \\ \hline \$12,000 \end{array}
\ \text{estimated cost of the suits}
$$

The $12,000 estimate is high since both factors were rounded up. (The actual cost is $10,614.) ■

In Example 10, to multiply 60 · $200 quickly, save the three zeros. Multiply the nonzero digits 6 · $2 = $12. Attach the three zeros to the right of $12 to get $12,000.

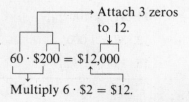

▶ **You Try It** 18. Estimate 483×52.

19. A store owner purchases 82 mowers at $214 each. Estimate the total cost. Is the estimate high or low?

EXAMPLE 11 A car dealer orders 370 cars at a price of $18,426 each. The distributor receives a bill for $10,817,620. Determine if this bill seems correct by estimating the total cost.

The total cost is $370 \cdot \$18,426$. To estimate it, round each factor to its digit farthest left.

$18,426 rounded to the nearest ten-thousand = $20,000
$\times\,370$ rounded to the nearest hundred $\quad=\quad \times\,400$

Attach 6 zeros to the 8.

$400 \cdot \$20,000 = \$8,000,000$ total estimated cost

$\rightarrow 4 \cdot 2 = \quad 8$

The bill the distributor received was for $10,817,620. The estimated cost is $8,000,000. This estimate is high because both factors were rounded up. Therefore, the bill is too high. (The actual cost is $370 \cdot \$18,426 = \$6,817,620$.) ■

> **OBSERVE** There are many ways to estimate. Another way to estimate a multiplication is to round one factor down and the other up. In Example 11, round 370 down to 350. Round $18,426 up to $20,000. Then estimate $370 \cdot \$18,426$ by $350 \cdot \$20,000 = \$7,000,000$.

▶ **You Try It** 20. A truck dealer orders 268 trucks at $32,491 each. The dealer was billed $8,707,588. Use estimation to determine if this bill seems correct.

EXAMPLE 12 One year has 365 days, one day has 24 hours, and one hour has 3,600 seconds. Estimate the number of seconds in one year.

The number of seconds in one year is $3,600 \cdot 24 \cdot 365$.

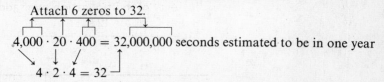

3,600	⟶ rounded to thousands gives ⟶	4,000
24	⟶ rounded to the tens gives ⟶	20
365	⟶ rounded to the hundreds gives ⟶	400

Attach 6 zeros to 32.

$4,000 \cdot 20 \cdot 400 = 32,000,000$ seconds estimated to be in one year

$4 \cdot 2 \cdot 4 = 32$

(The actual answer is $3,600 \cdot 24 \cdot 365 = 31,536,000$ seconds.) ■

▶ **You Try It** 21. One mile has 5,280 feet. One foot has 12 inches. Estimate the number of inches in 8 miles.

$$5 + 2 \times 3.$$

When a problem has two or more operations, these operations must be performed in a definite order. This order is called the **order of operations**.

> When the operations of $+$, $-$, $\times$, and parentheses are mixed in a problem, perform the operations in the following order.
>
> **1.** Simplify operations within parentheses.
> **2.** Do the multiplications.
> **3.** Do the additions and subtractions in order from left to right.

EXAMPLE 13   What is $5 + 2 \times 3$?

$$5 + \underbrace{2 \times 3}$$ First perform the multiplication.

$$5 + \quad 6$$ Then do the addition.

$$11$$ So, $5 + 2 \times 3 = 11$.

> **CAUTION**   If your answer was 21, you added first ($5 + 2 = 7$), and then multiplied ($7 \times 3 = 21$), which is the wrong order for performing the operations.

EXAMPLE 14   Simplify $3 \times 5 - 6 \times 2$.

$$\underbrace{3 \times 5} - \underbrace{6 \times 2}$$ First, do both of the multiplications.

$$15 \quad - \quad 12$$ Then do the subtraction.

$$3$$ So, $3 \times 5 - 6 \times 2 = 3$.   ■

EXAMPLE 15   **a.** $4\underbrace{(2 + 3)}$ Simplify within parentheses first.

$$= 4 \quad (5)$$ Perform the implied multiplication.

$$= \quad 20$$

b. $\underbrace{4 \times 2} + \underbrace{4 \times 3}$ Perform both multiplications.

$$= \quad 8 \quad + \quad 12$$ Add.

$$= \quad\quad 20$$

> **OBSERVE**   The answers in a and b are the same. This means $4(2 + 3) = 4 \times 2 + 4 \times 3$. This property, called the distributive property of multiplication over addition, will be discussed in more detail in Chapter 12.   ■

▶ **You Try It**   Simplify using the order of operations.

22. $8 + 2 \times 5$ **23.** $3 \times 7 - 6$

24. $2 \times 8 + 2 \times 5$ **25.** $2(8 + 5)$

5 Using Substitution to Evaluate Expressions Containing Letters

Replacing a letter with a number is called **substitution**. When you solved equations by trial earlier, you substituted numbers for letters to find the solution.

EXAMPLE 16   What is $5y$ if $y = 3$?

$5y$ means $5 \cdot y$. Substitute 3 for y.

$$5y = 5 \cdot y = 5 \cdot 3 = 15 \quad \blacksquare$$

EXAMPLE 17   What is $2a + 4$ if $a = 3$?

$$
\begin{aligned}
2a + 4 &= 2 \cdot a + 4 &&\text{Substitute } a = 3. \\
&= \underline{2 \cdot 3} + 4 &&\text{Obey the order of operations and multiply first.} \\
&= \quad 6 \; + 4 &&\text{Then add.} \\
&= \quad\quad 10 \quad \blacksquare
\end{aligned}
$$

▶ **You Try It**

26. Evaluate $7t$ when $t = 6$.

27. Evaluate $8w$ when $w = 0$.

28. What is $3m + 7$ if $m = 8$?

29. What is $2 + 5k$ if $k = 2$?

Use Substitution to Check Solutions to Equations

EXAMPLE 18   Which of the three given numbers makes the equation true?

$$3x + 5 = 17 \quad \textbf{a. } x = 3 \quad \textbf{b. } x = 6 \quad \textbf{c. } x = 4$$

Substitute each number for the letter x until you find the one that makes the equation true.

a. False.   If $x = 3$, then
$$
\begin{aligned}
3 \cdot 3 + 5 &\overset{?}{=} 17 \\
9 + 5 &\overset{?}{=} 17 \\
14 &\neq 17
\end{aligned}
$$

b. False.   If $x = 6$, then
$$
\begin{aligned}
3 \cdot 6 + 5 &\overset{?}{=} 17 \\
18 + 5 &\overset{?}{=} 17 \\
23 &\neq 17
\end{aligned}
$$

c. True.   If $x = 4$, then
$$
\begin{aligned}
3 \cdot 4 + 5 &\overset{?}{=} 17 \\
12 + 5 &\overset{?}{=} 17 \\
17 &= 17 \quad \blacksquare
\end{aligned}
$$

▶ **You Try It**

Find the number that makes the equation true.

30. $4y + 9 = 37$ **a.** $y = 3$ **b.** $y = 7$ **c.** $y = 5$

31. $4 + 3N = 28$ **a.** $N = 8$ **b.** $N = 0$ **c.** $N = 10$

SOMETHING MORE

Carl Friedrich Gauss (1777–1855) was one of the greatest mathematicians in history. Before the age of 3, he watched his father tally the payroll for a group of workers. Carl pointed out an addition error. Upon rechecking, his father found that Carl was correct.

When he was 10, his teacher asked his class to add the numbers 1 through 100. Carl immediately wrote the answer, 5050, on his slate. Much later, the rest of the students turned in their work. Gauss had the only correct answer in the class.

Evidently, Gauss noticed that the pairs of numbers 1 and 100, 2 and 99, 3 and 98, up to 50 and 51, each adds up to 101. Since there are 50 such pairs, perform a repeated addition of 101 to itself 50 times for the sum. Since multiplication is a shortcut for repeated addition, multiply $50 \times 101 = 5050$ for the answer.

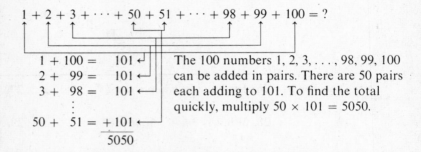

$$1 + 2 + 3 + \cdots + 50 + 51 + \cdots + 98 + 99 + 100 = ?$$

$$
\begin{aligned}
1 + 100 &= 101 \\
2 + 99 &= 101 \\
3 + 98 &= 101 \\
&\vdots \\
50 + 51 &= +101 \\
\hline
&5050
\end{aligned}
$$

The 100 numbers 1, 2, 3, ..., 98, 99, 100 can be added in pairs. There are 50 pairs each adding to 101. To find the total quickly, multiply $50 \times 101 = 5050$.

Using the method Gauss used as a schoolboy, add the following.

a. 1 to 6   **b.** 1 to 10   **c.** 1 to 50   **d.** 1 to 1,000

e. How would you use Gauss's method to add the numbers 51 through 100?

▶ **Answers to You Try It** 1. 54   2. 152   3. 1,080   4. 32,576
5. 1,368   6. 17,446   7. 149,604   8. 562,968   9. 1,957,941   10. 435,500
11. 41,700   12. 24,910   13. 38,000   14. 1,470,000   15. 3,216,000
16. 6,300   17. 800   18. 25,000   19. 16,000; low
20. estimate = 9,000,000, seems correct   21. 500,000   22. 18   23. 15
24. 26   25. 26   26. 42   27. 0   28. 31   29. 12   30. b. $y = 7$
31. a. $N = 8$

▶ **Answers to Something More**   a. 21   b. 55   c. 1,275   d. 500,500
e. add $51 + 100 = 151$, $52 + 99 = 151$, ..., $75 + 76 = 151$; the sum 151 occurs 25 times; 151 added to itself 25 times is $25 \times 151 = 3,775$.

I *Find the product.*

1. 34
    × 2

2. 60
    × 3

3. 357
    × 9

4. 809
    × 5

5. 4,653
    × 3

6. 3,006
    × 8

7. 72,354
    × 4

8. 70,900
    × 6

9. 45
    × 25

10. 58
     × 97

11. 27
     × 30

12. 84
     × 50

13. 372
     × 42

14. 683
     × 65

15. 703
     × 31

16. 220
     × 17

17. 800
     × 67

18. 200
     × 90

19. 2,704
     × 45

20. 7,053
     × 72

21. 327
     × 682

22. 277
     × 134

23. 609
     × 508

24. 721
     × 375

25. 412
     × 900

26. 800
     × 572

27. 7,352
     × 413

28. 3,125
     × 647

29. 2,006
     × 623

30. 9,203
     × 786

31. 4,006
     × 503

32. 30,005
     × 208

🖩 33. 14,957
       × 569

🖩 34. 43,465
       × 763

2

35. 243
     × 10

36. 687
     × 100

37. 58
     × 10,000

38. 26
     × 1,000

39. 3,605
     × 100

40. 7,217
     × 1,000

41. 47,357
     × 10

42. 89,021
     × 10,000

3 *Estimate the answer.*

43. $7 \cdot 82$ **44.** $3 \cdot 62$ **45.** $53 \cdot 91$ **46.** $28 \cdot 53$

47. $631 \cdot 49$ **48.** $85 \cdot 331$ **49.** $5 \cdot 43 \cdot 68$ **50.** $21 \cdot 4 \cdot 57$

51. $6{,}723 \cdot 715$ **52.** $132 \cdot 2{,}889$

---

53. Lena purchased 28 uniforms for the baseball team at a cost of \$63 each. Estimate the cost of the purchase.

54. A department store purchased 42 reclining chairs at a cost of \$378 each. Estimate the cost of the purchase.

55. A car dealership paid \$19,416 each for 22 vans. A salesman said the total cost was \$4,271,520.
 a. Estimate the total cost.

 b. Does the salesman's figure seem correct?

56. Angela purchased 63 dolls at a cost of \$31 each. The invoice said the cost before taxes was \$1,953.
 a. Estimate the total cost.

 b. Do you believe the invoice is approximately correct?

57. Light travels 186,300 miles in one second. There are 31,536,000 seconds in one year. Estimate the distance light travels in one year.

58. Estimate the cost of 8 trimmers at \$129 each.

---

4 *Simplify using the order of operations.*

59. $3 + 2 \times 5$ **60.** $7 + 4 \times 3$ **61.** $8 \times 3 + 1$

62. $6 \times 5 + 5$ **63.** $9 \times 2 - 2$ **64.** $4 \times 9 - 5$

65. $2 \times 3 + 1 \times 6$ **66.** $8 \times 5 + 9 \times 0$ **67.** $4 + 2 \times 7 - 3$

68. $9 + 1 \times 8 - 1$

5 *Substitute the given numbers for the letters and simplify using the order of operations.*

69. What is $2w$ if $w = 5$?   **70.** What is $6b$ if $b = 3$?

71. What is $4 + c$ if $c = 3$?   **72.** What is $g + 8$ if $g = 7$?

73. Find $f - 2$ if $f = 3$.

74. Find $a - 6$ if $a = 10$.

75. Find $3y + 2$ if $y = 4$.

76. Find $5t + 6$ if $t = 0$.

77. Find $6n - 5$ if $n = 2$.

78. Find $3p - 7$ if $p = 9$.

Use substitution to decide which answer is correct.

79. $2x + 1 = 7$ **a.** $x = 0$ **b.** $x = 3$ **c.** $x = 7$

80. $8h + 5 = 21$ **a.** $h = 7$ **b.** $h = 2$ **c.** $h = 5$

81. $3 + 2w = 13$ **a.** $w = 6$ **b.** $w = 3$ **c.** $w = 5$

82. $7 + 4s = 11$ **a.** $s = 1$ **b.** $s = 0$ **c.** $s = 6$

83. $4v - 4 = 0$ **a.** $v = 4$ **b.** $v = 1$ **c.** $v = 2$

84. $5r - 6 = 44$ **a.** $r = 10$ **b.** $r = 20$ **c.** $r = 5$

SKILLSFOCUS (Section 1.6) *Solve.*

85. Alice has $900 in her checkbook. She writes a check for $374. What is her new balance?

86. What is the sum of 1,492 and 688, decreased by 976?

EXTEND YOUR THINKING ▶▶▶▶

87. Sum the numbers 1 through 60 using Gauss's method.

88. Sum the numbers 1 through 200 using Gauss's method.

89. The Double-Half Rule: An Application of the Associative Property of Multiplication   To multiply two numbers using the *double-half rule*, first double one number, halve the other, then multiply. For example, use the double-half rule to multiply 5×16.

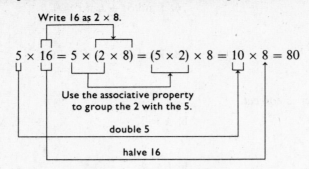

The double-half rule helps you do mental math by writing a multiplication (5×16) as a simpler multiplication (10×8).

EXAMPLE: **a.** $12 \times 15 = 6 \times 30 = 180$
b. $8 \times 16 = 4 \times 32 = 2 \times 64 = 128$
c. $25 \times 40 = 50 \times 20 = 100 \times 10 = 1{,}000$ ∎

Solve the following using the double-half rule.
a. 18×5 **b.** 25×8 **c.** 6×15 **d.** 14×20

90. Your heart beats 63 times per minute. Estimate the number of times your heart will beat during four years of college.

▶ TROUBLESHOOT IT

Find and correct the error.

91. $42 \times 2{,}000 = 8{,}400$ **92.** $5 + 3 \cdot 6 = 8 \cdot 6 = 48$ **93.** $5 \cdot 2 + 2 = 5 \cdot 4 = 20$

WRITING TO LEARN ▶▶▶▶

94. Explain how to multiply $1{,}000 \cdot 472$ using the powers of 10 shortcut.

95. Explain each step you use to estimate $47 \cdot 612$.

▶ YOU BE THE JUDGE

96. Amy claims the commutative property allows you to rewrite a multiplication to make it easier. She cites Example 9b in Section 1.7. Do you agree? Explain your decision.

97. Phil is charged with violating the order of operations. The evidence is a piece of paper with the problem $5 + 5 \cdot 4 + 6$ and the answer 46. Is Phil guilty? If you decide guilty, say which rule he violated and demonstrate how to get the correct answer.

1.9 EXPONENTIAL NOTATION

OBJECTIVES

1. Write repeated multiplication in exponential notation.
2. Write exponential notation in standard notation.
3. Use exponents and bases of 1 and 0; use powers of 10.
4. Find the square root of a number.

NEW VOCABULARY

repeated multiplication   square root

power   radical

exponent   radicand

base   related power sentence

exponential notation

---

1 Writing Repeated Multiplication in Exponential Notation

The expression $2 \times 2 \times 2$ is **repeated multiplication** of the same factor, 2. A short way to write $2 \times 2 \times 2$ is 2^3.

$$2^3 = \overbrace{2 \times 2 \times 2} = 8$$

2^3 **power or exponent** (tells you how many times to write the base 2 in the multiplication)

2^3 **base** (this is the number being repeatedly multiplied)

The expression 2^3 is read "2 raised to the third power," and is written in **exponential notation**. Exponential notation is a short way to write repeated multiplication.

CAUTION 2^3 *never means* 2×3.

EXAMPLE 1   Write each repeated multiplication in exponential notation.

a. $7 \times 7 \times 7 \times 7$

$$7 \times 7 \times 7 \times 7 = 7^4$$

b. $2 \times 2 \times 2 \times 2 \times 2 \times 2$

$$2 \times 2 \times 2 \times 2 \times 2 \times 2 = 2^6$$

c. $3 \times 3 \times 3 \times 3 \times 3 \times 4 \times 4$

$$3 \times 3 \times 3 \times 3 \times 3 \times 4 \times 4 = 3^5 \times 4^2 \quad \blacksquare$$

▶ **You Try It**   Write the following expressions in exponential notation.

1. 5×5 **2.** $8 \times 8 \times 8 \times 8$

3. $10 \times 10 \times 10$ **4.** $3 \cdot 3 \cdot 3 \cdot 3 \cdot 3$

5. $4 \cdot 4 \cdot 4 \cdot 6 \cdot 6$

In algebra, repeated multiplication of letters can be written in exponential notation.

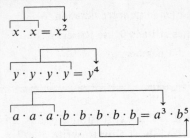

$$x \cdot x = x^2$$

$$y \cdot y \cdot y \cdot y = y^4$$

$$a \cdot a \cdot a \cdot b \cdot b \cdot b \cdot b \cdot b = a^3 \cdot b^5$$

2 Changing Exponential Notation to Standard Notation

Write 3^2 in standard notation.

$$3^2 = 3 \times 3 = 9$$

To change from exponential notation to standard notation

1. Write the base the number of times shown by the exponent.
2. Multiply.

EXAMPLE 2

a. $4^3 = 4 \times 4 \times 4 = 64$

b. $2^5 = 2 \times 2 \times 2 \times 2 \times 2 = 32$

c. $20^2 = 20 \times 20 = 400$

d. $1^7 = 1 \times 1 \times 1 \times 1 \times 1 \times 1 \times 1 = 1$

e. $0^4 = 0 \times 0 \times 0 \times 0 = 0$ ■

CAUTION 4^3 ($= 4 \times 4 \times 4$) means write three 4's, not three multiplication symbols, x. The number of multiplications is one less than the power.

OBSERVE Exponents of 2 and 3 are common, and have special names. 3^2 is read "3 squared." 4^3 is read "4 cubed."

▶ **You Try It** Change the following expressions to standard notation.

6. 7^2 7. 2^4 8. 3^5 9. 8^3 10. 1^{12}

EXAMPLE 3   **Chapter Problem**   Read the problem opening this chapter.

$$1 \text{ kilobyte, or } 1 \text{ K} = 2^{10} \text{ bytes}$$
$$= 2 \times 2 \times 2 \times 2 \times 2 \times 2 \times 2 \times 2 \times 2 \times 2$$
$$= 1{,}024 \text{ bytes of memory}$$
$$256 \text{ K} = 256 \times 1 \text{ K}$$
$$= 256 \times 1{,}024 \text{ bytes}$$
$$= 262{,}144 \text{ bytes}$$
$$1 \text{ megabyte, or } 1 \text{ Meg} = 1 \text{ K of K}$$
$$= 1{,}024 \times 1{,}024 \text{ bytes}$$
$$= 1{,}048{,}576 \text{ bytes} \quad \blacksquare$$

▶ **You Try It**   Write in standard notation.

11. 2 K   **12.** 64 K

CONNECTIONS

In algebra, a letter raised to a power can be written as a repeated multiplication.

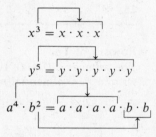

$$x^3 = x \cdot x \cdot x$$

$$y^5 = y \cdot y \cdot y \cdot y \cdot y$$

$$a^4 \cdot b^2 = a \cdot a \cdot a \cdot a \cdot b \cdot b$$

---

3 Exponents of 1 and 0   What pattern do you see as the powers decrease from 3 to 1?

$$4^3 = \text{three 4's multiplied together} = 4 \times 4 \times 4$$
$$4^2 = \text{two 4's multiplied together} = 4 \times 4$$
$$4^1 = \text{one four} = 4$$

> Any number raised to the power of 1 equals the same number. If a is any number, then
> $$a^1 = a.$$

What pattern do you see as the powers decrease from 3 to 0?

$$4^3 = 4 \times 4 \times 4 = 64$$
$$4^2 = 4 \times 4 \quad = 16 \ (= 64 \div 4)$$
$$4^1 = 4 \quad\quad = 4 \ (= 16 \div 4)$$
$$4^0 = \quad\quad\quad = 1 \ (= 4 \div 4)$$

The pattern is: divide the previous answer by 4 and you get the next answer.

> Any number (except 0) raised to a power of 0, equals 1. If a is any number, then
>
> $$a^0 = 1 \qquad (a \neq 0).$$

EXAMPLE 4 **a.** $2^1 = 2$ **b.** $8^1 = 8$ **c.** $64^1 = 64$

d. $4^0 = 1$ **e.** $7^0 = 1$ **f.** $75^0 = 1$ ∎

Bases of 1 and 0

$1^0 = 1$
$1^1 = 1$
$1^2 = 1 \times 1 = 1$
$1^3 = 1 \times 1 \times 1 = 1$, etc.

1 raised to any power equals 1. Why?
No matter how many times 1 is multiplied by 1, the answer is 1.

$0^1 = 0$
$0^2 = 0 \times 0 = 0$
$0^3 = 0 \times 0 \times 0 = 0$, etc.

0 raised to the power of 1, 2, 3, etc., equals 0. Why?
Because 0 times any number is 0.

▶ **You Try It**   Write each expression in standard notation.

13. 6^0 **14.** 5^1 **15.** 12^1

16. 46^0 **17.** 0^5 **18.** 1^{25}

Powers of Ten

The numbers 10, 100, 1000, etc., are called powers of 10 because each can be written in exponential notation with base 10.

$$10^4 = 10 \times 10 \times 10 \times 10 = 10,000 \text{ (ten-thousand)}$$
$$10^3 = 10 \times 10 \times 10 = 1,000 \text{ (thousand)}$$
$$10^2 = 10 \times 10 = 100 \text{ (hundred)}$$
$$10^1 = 10 = 10 \text{ (ten)}$$
$$10^0 = 1 = 1 \text{ (one)}$$

Notice the pattern as you move top to bottom. Each power of 10 is 1 followed by the number of 0's in the exponent.

10^4 is 1 followed by 4 zeros $= 10,000$
10^3 is 1 followed by 3 zeros $= 1,000$
10^2 is 1 followed by 2 zeros $= 100$
10^1 is 1 followed by 1 zero $= 10$
10^0 is 1 followed by no zeros $= 1$

4 Square Root   **Square roots** are written using the **radical** symbol, $\sqrt{}$. The square root of 9, written $\sqrt{9}$, is equal to the number you square to get 9.

$$\sqrt{9} = 3 \qquad \text{because} \qquad 3^2 = 9.$$

The square root of 9 is 3. 3 squared gives 9. Therefore, square root and squaring are opposite operations.

CAUTION   Write $\sqrt{9} = 3$, not $\sqrt{9} = 3^2$.

> To find the square root of the number a, use the rule
>
> $$\sqrt{a} = b \qquad \text{if} \qquad b^2 = a.$$
>
> The number a is called the **radicand**.

To find the square root of a number by trial, square 0, then square 1, then square 2, and so on, until you get the number.

EXAMPLE 5   Find $\sqrt{25}$.

Square whole numbers beginning with 0, until you get 25. $0^2 = 0$. $1^2 = 1$. $2^2 = 4$. $3^2 = 9$. $4^2 = 16$. $5^2 = 25$. Therefore,

$$\sqrt{25} = 5 \qquad \text{because} \qquad 5^2 = 25. \quad \blacksquare$$

Another way to find square roots by trial is to square a number you think is close to the answer. If the square of the number is too small, try a larger number. If too large, try a smaller one.

EXAMPLE 6   Find $\sqrt{169}$.

Try 10.	$10^2 = 100$.	Since $100 < 169$, 10 is too small.
Try 20.	$20^2 = 400$.	Since $400 > 169$, 20 is too large. Now, since 169 is closer to 100 than 400, try a number closer to 10 than to 20.
Try 12.	$12^2 = 144$.	Since $144 < 169$, 12 is close but still too small.
Try 13.	$13^2 = 169$.	13 is the answer.

$$\sqrt{169} = 13 \text{ because } 13^2 = 169. \quad \blacksquare$$

EXAMPLE 7
a. $\sqrt{36} = 6$ because $6^2 = 36$.
b. $\sqrt{81} = 9$ because $9^2 = 81$.
c. $\sqrt{1} = 1$ because $1^2 = 1$.
d. $\sqrt{16} = 4$ because $4^2 = 16$.
e. $\sqrt{0} = 0$ because $0^2 = 0$. $\quad \blacksquare$

▶ **You Try It**   Evaluate.

19. $\sqrt{4}$ **20.** $\sqrt{49}$ **21.** $\sqrt{361}$

Addition and subtraction are opposite operations. Multiplication and division are opposite operations. Similarly, square root and square are opposite operations. They undo each other.

Square undoes square root. For example, start with 9. Take the square root of 9 to get $\sqrt{9} = 3$. Then square 3 to get $3^2 = 9$. You end with 9, the same number you started with.

Square root undoes square. For example, start with 3. Square 3 to get $3^2 = 9$. Then take the square root of 9 to get $\sqrt{9} = 3$. You end with 3, the same number you started with.

In Example 7, each root sentence on the left has its **related power sentence** to its right. For example, the root sentence $\sqrt{36} = 6$ has the related power sentence $6^2 = 36$.

Use Your Calculator to Find Square Root

The square root button, $\sqrt{}$, on your calculator can be used to find square roots of large numbers.

EXAMPLE 8 **a.** Find $\sqrt{256}$.   256 $\sqrt{}$ [**16**]   Check: $16^2 = 256$

answer

b. Find $\sqrt{1,849}$.   1,849 $\sqrt{}$ [**43**]   Check: $43^2 = 1,849$ ∎

▶ **You Try It**   Evaluate each expression using a calculator, and check.

22. $\sqrt{1,089}$ **23.** $\sqrt{529}$ **24.** $\sqrt{7,396}$ **25.** $\sqrt{24,336}$

Some square roots are not whole numbers. Use your calculator to verify the following.

EXAMPLE 9   Find $\sqrt{20}$.   20 $\sqrt{}$ [**4.4721359**]

Check: $(4.4721359)^2 = $ [**19.999999**] ← close to 20

OBSERVE Because the calculator only displays part of the answer, the check is approximately correct. 4.4721359 and 19.999999 are decimal numbers. Decimal numbers will be studied in Chapter 5. ∎

▶ **Answers to You Try It** 1. 5^2 2. 8^4 3. 10^3 4. 3^5 5. $4^3 \cdot 6^2$ 6. 49
7. 16 8. 243 9. 512 10. 1 11. 2,048 12. 65,536 13. 1 14. 5
15. 12 16. 1 17. 0 18. 1 19. 2 20. 7 21. 19 22. 33 23. 23
24. 86 25. 156

1 *Write each expression in exponential notation.*

1. $4 \times 4 \times 4$

2. $3 \times 3 \times 3 \times 3 \times 3$

3. 7×7

4. $1 \times 1 \times 1 \times 1$

5. $12 \times 12 \times 12$

6. $0 \times 0 \times 0 \times 0$

7. 6

8. $8 \times 8 \times 8 \times 8 \times 8 \times 8$

9. $121 \times 121 \times 121$

10. $2 \times 2 \times 2 \times 2 \times 2 \times 2 \times 2 \times 2 \times 2 \times 2$

2 **3** *Write each expression in standard notation.*

11. 3^1

12. 7^2

13. 0^5

14. 1^4

15. 6^0

16. 11^2

17. 5^2

18. 2^6

19. 12^0

20. 4^5

21. 100^2

22. 9^1

23. 15^1

24. 0^8

25. 30^3

26. 10^7

27. 50^2

28. 11^4

29. 1^{34}

30. 20^4

31. 3^5

32. 5^4

33. 9^3

34. 2^{12}

35. 212^0

36. 6^1

37. 7^3

38. 4^0

39. 1^5

40. 0^4

41. 4^8

42. 2^{16}

43. 8^4

44. 3^{12}

45. 25^5

46. 2^{25}

4 *Find each square root by trial. Write the related power sentence.*

47. $\sqrt{25}$

48. $\sqrt{64}$

49. $\sqrt{100}$

50. $\sqrt{121}$

51. $\sqrt{225}$

52. $\sqrt{144}$

53. $\sqrt{400}$

54. $\sqrt{289}$

55. $\sqrt{256}$

56. $\sqrt{196}$

57. $\sqrt{324}$

58. $\sqrt{900}$

Use your calculator to find the square root. If the answer is not a whole number, write the entire answer shown on your display. Then use your calculator to check each answer.

59. $\sqrt{625}$

60. $\sqrt{4,624}$

61. $\sqrt{121,801}$

62. $\sqrt{4,900}$

63. $\sqrt{8}$

64. $\sqrt{2}$

65. $\sqrt{10}$

66. $\sqrt{50}$

SKILLSFOCUS (Section 1.8) *Simplify.*

67. $6 + 3 \cdot 4$

68. $9(8 + 2)$

69. $4 + 3 \cdot 2 - 1$

EXTEND YOUR THINKING ▶▶▶▶

▶SOMETHING MORE

70. Use Example 3 to help you answer each question.

    a. A computer has 32 K of memory. How many bytes is this?

    b. How many bytes are in 640 K of computer memory?

    c. How many K of memory are in 4 Meg?

    d. One edition of the novel *Crime and Punishment* is 574 pages long with an average of 1,780 characters per page. Will this novel fit into 1 Meg of computer memory?

71. A google is 1 followed by 100 zeros. Write a google in exponential notation.

▶TROUBLESHOOT IT

Find and correct the error.

72. $5^3 = 5 \times 3 = 15$ **73.** $7^0 = 7 \times 0 = 0$ **74.** $1^{20} = 20$

▶CONNECTIONS

Write each expression in exponential notation.

75. $t \cdot t \cdot t$ **76.** $k \cdot k$ **77.** $y \cdot y \cdot y \cdot y \cdot y \cdot y$

Write each expression in standard notation.

78. p^4 **79.** x^7 **80.** w^3

WRITING TO LEARN ▶▶▶▶

81. Explain in writing how to evaluate 3^4. Use the words *base* and *exponent* in your explanation.

82. Explain why $\sqrt{9} = 3$ is correct, but $\sqrt{9} = 3^2$ is incorrect.

▶YOU BE THE JUDGE

83. Rick claims that $\sqrt{100} = 50$ because $2 \times 50 = 100$. Is he correct? Explain your decision.

CALCULATOR TIPS

Clearing Your Calculator   Before 1975, the slide rule and pencil and paper were the tools students used to calculate. Since then the electronic calculator has dominated the computation market.

There are many different brands of calculator. Before attempting to use yours, read your manual thoroughly.

[c] or [CE/C] [CE/C] or [ON/C] [ON/C]   Press before starting a new calculation. Clears any calculation in progress, and clears the display to 0.

[CE] or [CE/C] or [ON/C]   Only clears the last number entered.

Try these examples on your calculator.

EXAMPLE 1   Use [c] to clear an entire problem and start over again. Find 40 + 25.

└─ Clears all previous entries.   ■

EXAMPLE 2   Use [CE] (clear entry) to clear the last number entered. Find 5×8.

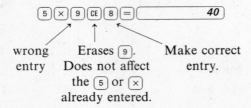

wrong entry   Erases [9]. Does not affect the [5] or [×] already entered.   Make correct entry.   ■

Note: In the remainder of this text, only operations will be enclosed in boxes. The numbers being operated on will not be boxed.

CHAPTER 1 REVIEW

VOCABULARY AND MATCHING: Sections 1.1 to 1.4

New words and phrases introduced in Sections 1.1 to 1.4 are shown in the left-hand column.
Match each term on the left with the phrase or sentence on the right that best describes it.

A. digits   _____ rightmost digit is odd

B. even digits   _____ natural numbers 1, 2, 3, . . .

C. even number   _____ used to compare numbers

D. odd digits   _____ 7 + 0 = 7

E. odd number   _____ take 1 from the digit on the left in the minuend

F. inclusive   _____ a group of three digits with a name

G. exclusive   _____ 3 − 0 = 3

H. counting numbers   _____ a related subtraction is 9 − 5 = 4

I. whole numbers   _____ a statement that two quantities are equal

J. number line   _____ 0, 1, 2, 3, 4, 5, 6, . . .

K. <, =, >   _____ 4, 0, 6, 2, and 8

L. standard notation   _____ 728 = 700 + 20 + 8

M. place value names   _____ done when place value sums are greater than 9

N. period   _____ 6 + 4 = 4 + 6

O. expanded notation   _____ 5 + (3 + 2) = (5 + 3) + 2

P. addition property of zero   _____ ten characters used to build numbers

Q. commutative property of addition   _____ numbers written in the form 243, 4,957, or 68

R. parentheses   _____ 7, 1, 5, 3, and 9

S. associative property of addition   _____ when 3 through 7 means 4, 5, and 6

T. equation   _____ rightmost digit is even

U. carrying   _____ examples are ruler, scale, and thermometer

V. zero property of subtraction   _____ used to indicate which operation to do first

W. 4 + 5 = 9   _____ ones, tens, hundreds, thousands, and so on

X. borrow   _____ when 3 through 7 means 3, 4, 5, 6, and 7

VOCABULARY AND MATCHING: Sections 1.5 to 1.9

New words and phrases introduced in Sections 1.5 to 1.9 are shown in the left-hand column.
Match each term on the left with the phrase or sentence on the right that best describes it.

A. rounding off

B. estimate addition and subtraction

C. indicate addition

D. indicate subtraction

E. repeated addition

F. implied multiplication

G. product

H. factors

I. multiples

J. multiplication property of zero

K. multiplication property of one

L. multiplicative identity

M. commutative property of multiplication

N. associative property of multiplication

O. substitution

P. indicate multiplication

Q. estimate multiplication

R. repeated multiplication

S. base

T. power or exponent

U. exponential notation

V. square root

W. radicand

X. exponent of 1

Y. exponent of 0

Z. base of 1 or 0

_____ what 7 and 4 are called in $7 \times 4 = 28$

_____ sum, total, increase, addend

_____ answer to a multiplication problem

_____ finding approximate values for exact values

_____ what 25 is in $\sqrt{25}$

_____ first round all numbers to the digit farthest left

_____ small number written on the upper right-hand corner of the base

_____ answer is always 1, as long as the base is not 0

_____ 1

_____ multiplication is its shortcut

_____ written in the form 2^3, 4^1, 7^2, not in the form 8, 4, 49

_____ in exponential notation, this number is repeatedly multiplied

_____ what 6, 12, 18, 24, 30, 36, and 42 are to 6

_____ $7 \cdot 1 = 7$ or $1 \cdot 7 = 7$

_____ $(2 \times 5) \times 9 = 2 \times (5 \times 9)$

_____ first round all numbers to the same place value

_____ difference, decrease, loss, minuend, subtrahend

_____ $5 \cdot 0 = 0$ or $0 \cdot 5 = 0$

_____ opposite of squaring

_____ answer is always the base

_____ the answer is always 1 or 0, with any counting number as an exponent

_____ replace a number with a letter

_____ examples are (8)(3), (8)3, and 8(3)

_____ $8 \times 5 = 5 \times 8$

_____ product, repeated addition

_____ written in shortcut form as a power

REVIEW EXERCISES

1.1 Definitions and Order

1. Find three symbols that make each statement true.
   a. 7 _?_ 4
   b. 3 _?_ 3
   c. 4 _?_ 5

2. Write all the digits greater than 7.

3. Write the whole numbers 8 through 12 inclusive.

4. Write the digits 0 through 5 exclusive.

5. Which numbers make the inequality $x > 4$ true?
    a. $x = 7$ **b.** $x = 0$ **c.** $x = 4$

6. Which numbers make the inequality $y < 3$ true?
    a. $y = 0$ **b.** $y = 3$ **c.** $y = 8$

1.2 Standard and Expanded Notations

7. Write each whole number in words.
    a. 47   **b.** 3,502   **c.** 4,628,531

8. Write each number in expanded notation.
    a. 72   **b.** 639   **c.** 28,503

9. Write in standard form: four thousand, six hundred twenty-nine.

10. Write in standard form: eight hundred three.

11. Write in standard form: forty-six thousand, six.

12. Write in standard form: two million, three hundred thirty-one thousand, five hundred nineteen.

1.3 Addition of Whole Numbers

13. Solve each equation by trial.
    a. $4 + t = 7$ **b.** $y + 5 = 9$ **c.** $9 + f = 9$

14. $38 + 41$ **15.** $1,645 + 3,230$

16. $674 + 76$ **17.** $56,847 + 74,348$

18. $34,509 + 681 + 3,416 + 9 + 47$

19. $4{,}659 + 9{,}034 + 5{,}045$

20. $34 + 56 + 345 + 78 + 6 + 27 + 84 + 69 + 560 + 3$

1.4 Subtraction of Whole Numbers

21. Write a related addition for each subtraction.

    a. $7 - 4 = 3$ **b.** $9 - 2 = 7$ **c.** $5 - 0 = 5$

22. Solve each equation by trial.

    a. $6 - x = 2$ **b.** $t - 5 = 4$ **c.** $n - 1 = 7$

23. $84 - 52$ **24.** $4{,}745 - 1{,}621$

25. $587 - 293$ **26.** $6{,}254 - 658$

27. $82{,}503 - 28{,}261$ **28.** $6{,}002 - 3{,}527$

1.5 Rounding and Estimation

29. Round 657 to tens.   **30.** Round 1,322 to hundreds.

31. Round 7,185,299 to millions.   **32.** Round 37,995 to tens.

33. Round 568,199 to thousands.   **34.** Round 7,060 to hundreds.

Estimate each answer.

35. $243 + 860 + 408$ **36.** $5{,}623 + 8{,}291 + 715$

37. $695 - 316$ **38.** $24{,}812 - 8{,}154$

39. Sally sold three sofas for $675, $419, and $1,267, respectively. Estimate her total sales.

40. Jean purchased a boat for $84,916. She received $31,251 for a trade-in. Estimate what Jean still owes for the boat.

1.6 Applications: Addition and Subtraction

41. Sam weighed 235 pounds. He went on a diet and lost 57 pounds in 6 months. What does Sam now weigh?

44. The balance in Jay's checkbook is $412. He wrote checks for $12, $47, and $172. What is Jay's new balance?

42. Alida bought a blouse for $24, a skirt for $43, a pair of shoes for $38, and a tennis racket for $61. How much did she spend before taxes?

45. Four men weigh 180 pounds, 234 pounds, 191 pounds, and 205 pounds, respectively. What is their total weight?

43. Jan bought her home for $56,000 ten years ago. Today it is worth $76,500 more than she originally paid for it. What is her home now worth?

46. An oil tank has a 650-gallon capacity. It currently holds 478 gallons. How many gallons of oil are needed to fill the tank?

1.7 Multiplication and Its Properties

47. Find the product.

    a. $8 \cdot 0$ **b.** $6(1)$ **c.** $(4 \cdot 2)7$

48. Replace the letter with the number that makes the statement true. Identify the property.

    a. $7 \cdot y = 7$ **b.** $6 \cdot 3 = 3 \cdot w$ **c.** $0 = 9 \cdot r$

49. Solve the equation.

    a. $7v = 35$ **b.** $64 = t \cdot 8$ **c.** $10z = 90$

1.8 Multiplication of Whole Numbers and Estimation

50. 6×47

51. $9 \times 5,308$

52. 38×95

53. 72×483

54. $697(3,680)$

55. $(1,607)(5,086)$

56. $1,000 \times 581$

57. $4,035 \times 600$

58. Estimate $94 \cdot 26$.

59. Estimate $564 \cdot 387$.

60. Estimate 582×78.

61. Estimate $391 \cdot 2{,}109$.

62. Estimate the total cost for 237 round-trip tickets if each ticket sells for $512.

63. There are 43 people each paying $225 for a bus trip. Estimate the total paid. Is this estimate high or low?

64. $9 + 1 \times 6$

65. $3 \times 7 - 4$

66. $6 \times 3 + 8 \times 7$

67. $8 + 4 \times (2 - 2)$

68. Substitute the number for the letter and simplify.
  a. $4t$ if $t = 5$
  b. $3k + 2$ if $k = 8$

  c. $4 - 4h$ if $h = 0$
  d. $8p - 7$ if $p = 5$

69. Which number makes the equation $5t - 6 = 9$ true?
  a. $t = 4$
  b. $t = 3$
  c. $t = 8$

70. Which number makes the equation $8 + 5n = 43$ true?
  a. $n = 7$
  b. $n = 4$
  c. $n = 6$

1.9 Exponential Notation

Write each expression in exponential notation.

71. $9 \cdot 9 \cdot 9$

72. $6 \cdot 6$

73. $5 \cdot 5 \cdot 5 \cdot 5$

74. $40 \cdot 40 \cdot 40$

75. $11 \cdot 11$

76. $8 \cdot 8 \cdot 8 \cdot 8 \cdot 8 \cdot 8$

Write each expression in standard notation.

77. 5^3

78. 4^2

79. 8^0

80. 12^1

81. 3^6

82. 10^9

Evaluate.

83. $\sqrt{64}$

84. $\sqrt{36}$

85. $\sqrt{81}$

86. $\sqrt{225}$

87. $\sqrt{1}$

88. $\sqrt{484}$

Allow yourself 50 minutes to complete this test. Write the work for each problem. When done, check your answers. Rework each problem solved incorrectly.

1. Write all the digits from 3 through 8 exclusive.

2. Which numbers make the inequality $5 < y$ true?
 a. $y = 5$ **b.** $y = 3$ **c.** $y = 9$

3. Write 805 in expanded notation.

4. Write 47,692 in expanded notation.

5. Write in standard notation: seventy thousand, four hundred twenty-three.

Solve each equation by trial.
 6. $6 + t = 6$ **7.** $t - 7 = 0$

 8. $8n = 56$ **9.** $y \div 5 = 8$

10. $8,674 + 1,694$ **11.** $58,210 + 894 + 9,607 + 9 + 71$

12. $8,000 - 2,169$ **13.** $78,036 - 9,725$

14. Write the related addition for the subtraction $8 - 2 = 6$.

15. The balance in Jane's checkbook is $832. She writes checks for $65, $328, $19, and $126. What is her new balance?

16. Round 9,407 to the nearest ten.

17. Round 68,492 to the nearest thousand.

18. Round 5,649 to the nearest hundred.

19. $47 \cdot 609$ **20.** $6,792 \times 385$ **21.** $1,000(5,812)$

22. Ellen has 500 candies. She gave 15 candies to each child in a class of 27 children. How many candies did she have left?

23. Five items were purchased to furnish an office. The items cost $92, $427, $38, $1,056, and $795. Estimate the total cost for the five items by rounding to hundreds.

24. Al breathes about 16 times each minute. There are 60 minutes in an hour, 24 hours in a day, and 31 days in May. Estimate the number of breaths Al will take in the month of May.

---

25. Write $4 \cdot 4 \cdot 4 \cdot 4 \cdot 4$ in exponential notation.

26. Write $5 \cdot 5 \cdot 8 \cdot 8 \cdot 8 \cdot 8 \cdot 8$ in exponential notation.

27. Write 4^4 in standard notation.

28. Write 7^3 in standard notation.

29. Evaluate $\sqrt{25}$.

30. Evaluate $\sqrt{144}$.

Explain the meaning of each term. Use your own examples.
31. even number

32. expanded notation

33. factor

34. power or exponent

2

Additional Whole Number Operations

HIGHWAYS

A road engineer will paint median strips, each 12 feet long, in the center of a road. The distance between strips is 20 feet. How many strips are painted per mile? Remember that 1 mile = 5,280 feet.

The distance from the beginning of one strip to the beginning of the next is

$$12 \text{ feet} + 20 \text{ feet}.$$

Divide this sum into 5,280 feet to find the number of strips per mile. This problem will be solved in this section.

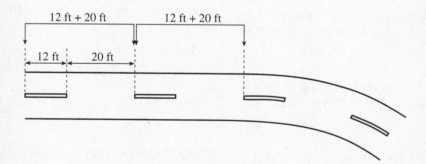

Work through each section in this chapter. Use this test to identify topics you are not familiar with. These topics may require additional study. You may *not* use a calculator.

2.1

1. Write the related multiplication for $36 \div 9 = 4$.

2. Write the related multiplication for $\frac{42}{7} = 6$.

3. Solve by trial $\frac{t}{3} = 8$.

4. Find each quotient, or write not legal.
   a. $6 \div 6$
   b. $72 \div 9$
   c. $0 \div 4$
   d. $7 \div 0$

2.2

5. Divide $912 \div 24$.

6. Divide $12,417 \div 68$.

7. Divide $735 \div 15$.

8. Divide $9,361 \div 23$.

2.3

9. Carl has 450 tickets. He gives 12 tickets to each of 37 employees. How many tickets does he have left?

10. There are 36 tour buses each carrying 54 people. Estimate the total number of people on the buses.

11. Elaine has $473 to purchase textbooks for a library. Each book costs $26.
   a. How many books can she purchase?

   b. How much money is left over?

12. Find the average of 69, 88, 92, 75, and 76.

2.4 *Simplify.*

13. $10 - 4 \cdot 2$

14. $16 \div 4 \cdot 4$

15. $7^0 + 7^2$

16. $2 \cdot 3^2$

17. $5\sqrt{9} + 2(4 + 1)$

2.5 *Change each word sentence into an arithmetic expression. Then simplify.*

18. Four plus six times five.

19. Seven times the sum of nine and two.

Evaluate each expression using the given values.

20. $2y$ when $y = 9$

21. $3 + 5x$ when $x = 6$

22. Find P in $P = 2L + 2W$ when $L = 6$ and $W = 13$.

2.6 *Solve each equation using opposite operations.*

23. $x + 62 = 403$

24. $k - 36 = 91$

25. $18y = 324$

26. $\frac{x}{12} = 15$

VOCABULARY *Explain the meaning of each term. Use your own examples.*

27. quotient

28. average

OBJECTIVES

1 Define division.
2 Check division with multiplication.
3 Define the properties of division.
4 Use 0 in a division.
5 Solve equations.

NEW VOCABULARY

calculator form	dividend
fraction form	related multiplication
long division form	undefined
divisor	indeterminate
quotient	

1 The Meaning of Division

Suppose 2 children have 5 model cars each. All together they have $2 \times 5 = 10$ cars. You multiply to find the total.

Division is the opposite of multiplication. Suppose 10 cars are split equally between 2 children. Each child gets $10 \div 2 = 5$ cars. Division is used to separate the total into smaller equal groups.

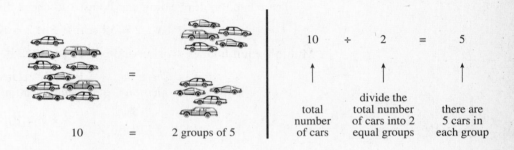

$10 \div 2 = 5$		
total number of cars	divide the total number of cars into 2 equal groups	there are 5 cars in each group

$10 = 2$ groups of 5

A division can be written in four common forms. Each is read "10 divided by 2."

$$10 \div 2 \qquad 10/2 \quad \frac{10}{2} \qquad 2\overline{)10}$$

calculator form   **fraction form**   **long division form**

Each term in a division has a name.

$$5 \longleftarrow \textbf{quotient} \text{ (answer to a division)}$$
$$\textbf{divisor} \longrightarrow 2\overline{)10} \longleftarrow \textbf{dividend}$$

EXAMPLE 1 Write 21 divided by 7 in four forms. Identify the divisor and dividend in each form.

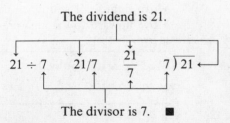

The dividend is 21.

$$21 \div 7 \qquad 21/7 \qquad \frac{21}{7} \qquad 7\overline{)21}$$

The divisor is 7. ■

▶ **You Try It**   Write each division four ways. Identify the divisor and dividend.

1. 18 divided by 3   **2.** 50 divided by 10

2 **Checking Division with Multiplication**   Since division and multiplication are opposite operations, every division problem has a **related multiplication**.

EXAMPLE 2

Division	Related Multiplication

$10 \div 2 = 5 \longrightarrow 5 \times 2 = 10$

$12 \div 4 = 3 \longrightarrow 3 \times 4 = 12$

$18 \div 3 = 6 \longrightarrow 6 \times 3 = 18$ ■

▶ **You Try It**   Write the related multiplication for each division.

3. $6 \div 2 = 3$ **4.** $45 \div 5 = 9$ **5.** $28 \div 4 = 7$

The related multiplication can be used to check division.

$$10 \div 2 = 5 \text{ because } 5 \times 2 = 10$$

Multiplication undoes division and is used to check it.

$$\begin{array}{cc} \textit{Check the} & \textit{With the related} \\ \textit{division} \longrightarrow & \textit{multiplication} \end{array}$$

$$2\overline{)10}^{\,5} \qquad 5 \times 2 = 10$$

Check division:	Using multiplication:
$\text{divisor}\overline{)\text{dividend}}^{\text{quotient}}$	quotient × divisor = dividend

EXAMPLE 3   Check each division using multiplication.

a. $30 \div 6 = 5$ because $5 \times 6 = 30$
b. $24 \div 8 = 3$ because $3 \times 8 = 24$
c. $54 \div 9 = 6$ because $6 \times 9 = 54$ ■

▶ **You Try It**   Check each division using multiplication.

6. $72 \div 8 = 9$ **7.** $10 \div 10 = 1$ **8.** $42 \div 7 = 6$

The multiplication table in Figure 1.5 on page 52 can be used to divide.

EXAMPLE 4   Use Figure 1.5 to divide $24 \div 6$.
Locate the divisor 6 in the left-hand column.
Move right until you reach the dividend 24.
Move upward to the answer, 4, in the top row.   ■

▶ **You Try It**   Use Figure 1.5 to divide.

9. $63 \div 7$ **10.** $54 \div 9$

3 **Properties of Division**

What pattern do you see with the following divisions? In each case, the divisor is 1.

EXAMPLE 5   **a.** $8 \div 1 = 8$ **b.** $5/1 = 5$

c. $\dfrac{3}{1} = 3$ **d.** $1 \overline{)7}^{\,7}$ ■

> **Division Property of One**
>
> Any number divided by 1 gives the same number for an answer.

Division is not commutative. In the following example, changing the order in the division changes the answer.

EXAMPLE 6

$$8 \div 4 \neq 4 \div 8$$

$$8 \div 4 = 2 \qquad 4 \div 8 \text{ is not a whole number}$$

The answers are different. Therefore, division is not commutative. This means you must divide two numbers in the correct order to get the right answer.   ■

EXAMPLE 7   The *correct* way to write "18 divided by 6" is

$$18 \div 6 \quad \text{or} \quad 6 \overline{)18}$$

The *wrong* way to write "18 divided by 6" is

$$\cancel{6 \div 18} \quad \text{or} \quad \cancel{18 \overline{)6}}$$

_____
OBSERVE   18 ÷ 6 is also read "6 divided into 18."   ■
_____

▶ **You Try It**   Write each expression in long division form.

11. 24 divided by 4   **12.** $40 \div 5$ **13.** 30 divided by 6

Division is not associative. When three numbers are divided, how you group the numbers can change the answer.

EXAMPLE 8

$$(8 \div 4) \div 2 \neq 8 \div (4 \div 2)$$

$$
\begin{array}{c|c}
(8 \div 4) \div 2 & 8 \div (4 \div 2) \\
= \quad (2) \ \div 2 & = 8 \div \quad (2) \\
= 1 & = 4
\end{array}
$$

Simplify within parentheses first.

Since the answers are different, division is not associative.   ■

▶ **You Try It**   Show that division is not associative by showing that the two expressions are unequal.

14. $(12 \div 6) \div 2$ and $12 \div (6 \div 2)$ **15.** $(16 \div 4) \div 4$ and $16 \div (4 \div 4)$

4 Division Involving Zero: The Divisor May Never Be 0

$0 \div 4$ **is a legal division. The dividend may be 0.** To see why, write $0 \div 4$ in long division form.

EXAMPLE 9

$$0 \div 4 \longrightarrow 4 \overline{)\,0}^{\,0} \xrightarrow{\text{check}} 0 \times 4 = 0$$

$0 \div 4 = 0$ because $0 \times 4 = 0$. ∎

$4 \div 0$ **is an illegal division. The divisor may never be 0.** To see why, write $4 \div 0$ in long division form.

EXAMPLE 10

$$4 \div 0 \longrightarrow 0 \overline{)\,4}^{\,n} \xrightarrow{\text{check fails}} 0 \times n = 4$$

0 times any number
equals 0, not 4.

Suppose n represents the answer to this division. No matter what number you substitute for n, the check fails, because 0 times any number equals 0, not 4. $4 \div 0$ is called **undefined.** ∎

$0 \div 0$ **is an illegal division. The divisor may never be 0.** There are many different answers to $0 \div 0$. And each one checks out as correct!

EXAMPLE 11 Show $0 \div 0$ has many answers.

$$0 \overline{)\,0}^{\,0} \text{ because } 0 \times 0 = 0 \qquad 0 \overline{)\,0}^{\,1} \text{ because } 1 \times 0 = 0$$

$$0 \overline{)\,0}^{\,2} \text{ because } 2 \times 0 = 0 \qquad 0 \overline{)\,0}^{\,8} \text{ because } 8 \times 0 = 0$$

And so on. 0, 1, 2, 8, or any whole number is an answer to $0 \div 0$, because it checks out. Since you cannot determine what number to write as the correct answer, $0 \div 0$ is called **indeterminate.** It is not a legal division. ∎

Division by 0 is illegal

The divisor may never be 0.

illegal	*legal, for $a \neq 0$*
$a \div 0, \ a/0, \ \dfrac{a}{0}, \ 0 \overline{)\,a}$	$0 \div a = 0, \ 0/a = 0$
$0 \div 0, \ 0/0, \ \dfrac{0}{0}, \ 0 \overline{)\,0}$	$\dfrac{0}{a} = 0, \ a \overline{)\,0}^{\,0}$

▶ **You Try It** Give the answer, or write not legal.

16. $0 \div 6$ **17.** $8 \div 0$ **18.** $0 \div 0$

19. $12 \div 0$ **20.** $0 \div 1$ **21.** $\dfrac{0}{0}$

22. $\dfrac{9}{0}$ **23.** $3 \overline{)\,0}$ **24.** $0 \overline{)\,2}$

Using the Calculator to Divide

The calculator is a tool to help you do mathematical calculations fast. It is not a substitute for knowledge. To effectively use a calculator, you must first learn the rules of mathematics.

For example, division is not commutative. Therefore, to divide using a calculator you must enter the numbers in the correct order. The dividend is always entered into the calculator first.

EXAMPLE 12   Solve $12 \div 3$ using your calculator.

Enter this sequence of key presses.

$$12 \boxed{\div} 3 \boxed{=} 4$$

> **OBSERVE** $12 \div 3$ is the **calculator form for a division**. Enter this division into your calculator exactly as you read it, left to right.   ∎

▶ **You Try It**   Use your calculator to divide.

25. $30 \div 5$ **26.** $48 \div 6$ **27.** $7\overline{)84}$ **28.** $12\overline{)168}$

Use your calculator to show you cannot divide by 0.

EXAMPLE 13   Divide using your calculator.

a. $0 \div 4$: $0 \boxed{\div} 4 \boxed{=} 0$

b. $4 \div 0$: $4 \boxed{\div} 0 \boxed{=}$ Error

c. $0 \div 0$: $0 \boxed{\div} 0 \boxed{=}$ Error

Why does your calculator give an error message for $4 \div 0$ and $0 \div 0$? Because your calculator knows that division using a 0 divisor is illegal.   ∎

▶ **You Try It**   Use your calculator to divide. Write your answer, or write not legal.

29. $6 \div 0$ **30.** $0 \div 9$ **31.** $0\overline{)7}$ **32.** $7\overline{)0}$

5 Solving Equations   Solve by trial.

$$\frac{x}{4} = 3$$

The expression $\frac{x}{4}$ is read "the letter x divided by 4." To solve by trial, replace the letter x by 0, 1, 2, and so on, until you get a true equation. The answer is $x = 12$ because $12/4 = 3$.

A shortcut to find x is to write the related multiplication for the division. The related multiplication for the whole number division

$$\frac{12}{4} = 3 \quad \text{is} \quad 12 = 4 \cdot 3.$$

The related multiplication for the equation

$$\frac{x}{4} = 3 \quad \text{is} \quad \begin{aligned} x &= 4 \cdot 3 \\ &= 12 \end{aligned}$$

$x = 12$ is the solution to the equation.

EXAMPLE 14 Solve each equation by writing the related multiplication.

a. Find y if $\frac{y}{3} = 5$. $y = 3 \cdot 5 = 15$ Check: $\frac{15}{3} = 5$

b. Find n if $\frac{n}{8} = 6$. $n = 8 \cdot 6 = 48$ Check: $\frac{48}{8} = 6$

c. Find t if $\frac{t}{10} = 15$. $t = 10 \cdot 15 = 150$ Check: $\frac{150}{10} = 15$

d. Find p if $p \div 4 = 8$. $p = 4 \cdot 8 = 32$ Check: $32 \div 4 = 8$ ■

▶ **You Try It** Solve each equation by writing the related multiplication. Check each answer.

33. $\frac{x}{4} = 5$ **34.** $\frac{y}{6} = 7$ **35.** $\frac{w}{12} = 14$ **36.** $s \div 10 = 5$

▶ **Answers to You Try It** 1. $18 \div 3$, $18/3$, $\frac{18}{3}$, $3\overline{)18}$; divisor $= 3$, dividend $= 18$

2. $50 \div 10$, $50/10$, $\frac{50}{10}$, $10\overline{)50}$; divisor $= 10$, dividend $= 50$ 3. $2 \times 3 = 6$

4. $5 \times 9 = 45$ 5. $4 \times 7 = 28$ 6. $8 \times 9 = 72$ 7. $10 \times 1 = 10$ 8. $7 \times 6 = 42$
9. 9 10. 6 11. $4\overline{)24}$ 12. $5\overline{)40}$ 13. $6\overline{)30}$ 14. $1 \neq 4$ 15. $1 \neq 16$
16. 0 17. not legal 18. not legal 19. not legal 20. 0 21. not legal
22. not legal 23. 0 24. not legal 25. 6 26. 8 27. 12 28. 14
29. not legal 30. 0 31. not legal 32. 0 33. $x = 20$ 34. $y = 42$
35. $w = 168$ 36. $s = 50$

1 *Write each division in four different ways.*

1. eight divided by two

2. seven divided by one

3. 6 divided by 3

4. 0 divided by 3

5. 42 divided by 7

6. 72 divided by 9

2 *Identify the divisor, dividend, and quotient. Write a related multiplication for each problem.*

7. $10 \div 2 = 5$

8. $18/9 = 2$

9. $1 \overline{)6}$ with 6 above

10. $\dfrac{63}{9} = 7$

11. $\dfrac{4}{4} = 1$

12. $8 \overline{)24}$ with 3 above

13. $25/5 = 5$

14. $15 \div 5 = 3$

3 **4** *Find the quotient. Check each answer using related multiplication. Write NL (Not Legal) if you cannot perform the division.*

15. $12 \div 6$

16. $27/9$

17. $7 \overline{)35}$

18. $\dfrac{32}{4}$

19. $54/6$

20. $2 \overline{)14}$

21. $\dfrac{28}{4}$

22. $40 \div 5$

23. $30 \div 5$

24. $56 \div 8$

25. $8/8$

26. $5/1$

27. $\dfrac{64}{8}$

28. $\dfrac{16}{4}$

29. $45 \div 5$

30. $18 \div 6$

31. $0/8$

32. $9 \div 1$

33. $35 \div 5$

34. $7 \div 0$

35. $\dfrac{48}{6}$

36. $0 \div 0$

37. $0 \overline{)4}$

38. $\dfrac{0}{6}$

39. $5/0$

40. $35 \div 7$

41. $42 \div 6$

42. $6 \div 3$

Determine if a, b, or c has the same answer as the problem above.

43. $12 \div 2$

  a. $6/2$

  b. $4\overline{)24}$

  c. $\dfrac{30}{5}$

44. $0 \div 9$

  a. $4 \div 4$

  b. $2 \times 3 - 5$

  c. $0/3$

45. $16 \div (8 \div 2)$

  a. $(16 \div 8) \div 2$

  b. $\dfrac{8}{2}$

  c. $18 \div 3$

46. $(18 \div 3) \div 3$

  a. $5\overline{)10}$

  b. $18/6$

  c. $\dfrac{18}{9}$

47. $40/8$

  a. $5 \div (5 \div 1)$

  b. $20 \div 5$

  c. $15/5$

48. $45/5$

  a. $1 \times 2 \times 3 + 4$

  b. $9 - (4 - 4)$

  c. $54/9$

49. $\dfrac{32}{8}$

  a. $4\overline{)20}$

  b. $12/4$

  c. $12 \div 3$

50. $\dfrac{63}{9}$

  a. $49 \div (7 \div 1)$

  b. $14/7$

  c. $27 \div 9$

51. $6\overline{)18}$

  a. $81/9$

  b. $9\overline{)27}$

  c. $28 \div 4$

52. $5\overline{)0}$

  a. $0 \times 5 + 1$

  b. $5/5$

  c. $0 \div 5$

5 *Solve each equation by writing the related multiplication, and check.*

53. $\dfrac{s}{3} = 18$

54. $\dfrac{p}{2} = 8$

55. $\dfrac{b}{10} = 8$

56. $\dfrac{c}{5} = 12$

57. $\dfrac{f}{6} = 0$

58. $\dfrac{z}{3} = 0$

59. $x \div 3 = 21$

60. $x \div 1 = 9$

SKILLSFOCUS (Section 1.7) *Solve by trial.*

61. $5y = 40$

62. $9r = 81$

63. $6p = 42$

64. $15z = 15$

EXTEND YOUR THINKING ▶▶▶▶
▶TROUBLESHOOT IT
Find and correct the error.

65. $0 \div 6 = 6$

66. $3 \div 0 = 0$

67. $0 \div 0 = 1$

WRITING TO LEARN ▶▶▶▶

68. Explain why division is not commutative. Use your own example in your explanation.

69. Explain why division is not associative. Use your own example in your explanation.

70. Explain what the answers are to the divisions $9 \div 0$ and $0 \div 9$, and why.

OBJECTIVES

1 Divide using a one digit divisor.

2 Divide using a divisor with two or more digits.

3 Cross out zeros and divide by powers of 10.

4 Estimate solutions to problems (Section 1.8, Objective 3).

NEW VOCABULARY

remainder

trial divisor

---

1 Divide Using a One Digit Divisor

Multiplication is a shortcut for repeated addition. Division is a shortcut for repeated subtraction. For example, show $15 \div 5 = 3$ using repeated subtraction.

$$
\begin{array}{r}
15 \\
- 5 \\
\hline
10 \\
- 5 \\
\hline
5 \\
- 5 \\
\hline
0
\end{array}
$$

There are 3 fives in 15. So, $15 \div 5 = 3$

$$
\begin{array}{r}
3 \\
5\overline{)\ 15} \\
-15 \longleftarrow 3 \times 5 = 15 \\
\hline
0 \longleftarrow \text{Remainder} = 0.
\end{array}
$$

In the next two examples, place-value names are used to stress the importance of keeping like named digits in the same vertical column.

EXAMPLE 1 Find $68 \div 4$.

$$
\begin{array}{r}
\overset{tens\ ones}{1} \\
4\overline{)\ 6\ \ 8} \\
-4 \\
\hline
2
\end{array}
$$

Step 1: Divide 4 into 6 tens. 6 tens ÷ 4 gives 1 ten. Write 1 ten directly above 6 tens in the quotient.

Step 2: 1 ten × 4 = 4 tens. Write 4 tens directly below 6 tens.

Step 3: Subtract the tens. 6 − 4 = 2. Write 2 directly below the 4 in the tens column.

$$
\begin{array}{r}
1\ \ 7 \\
4\overline{)\ 6\ \ 8} \\
-4 \\
\hline
2\ \ 8 \\
-2\ \ 8 \\
\hline
0
\end{array}
$$

Step 4: Bring down the next digit, 8 ones.

Step 5: Divide 4 into 28 ones. 28 ones ÷ 4 = 7 ones. Write 7 ones in the quotient directly above 8 ones.

Step 6: 7 ones × 4 = 28 ones. Write 28 below 28.

Step 7: Subtract 28 − 28 = 0. There are no more digits left to bring down in the dividend. The division is done.

Therefore, $68 \div 4 = 17$. The check is: $17 \times 4 = 68$. ■

In long division, every digit you bring down from the dividend must have a digit written above it in the quotient.

EXAMPLE 2   Find 2,695 ÷ 7.

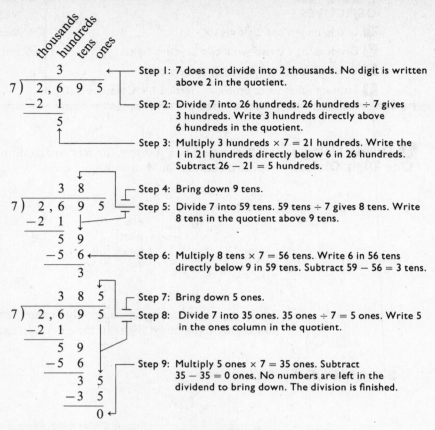

Step 1: 7 does not divide into 2 thousands. No digit is written above 2 in the quotient.

Step 2: Divide 7 into 26 hundreds. 26 hundreds ÷ 7 gives 3 hundreds. Write 3 hundreds directly above 6 hundreds in the quotient.

Step 3: Multiply 3 hundreds × 7 = 21 hundreds. Write the 1 in 21 hundreds directly below 6 in 26 hundreds. Subtract 26 − 21 = 5 hundreds.

Step 4: Bring down 9 tens.

Step 5: Divide 7 into 59 tens. 59 tens ÷ 7 gives 8 tens. Write 8 tens in the quotient above 9 tens.

Step 6: Multiply 8 tens × 7 = 56 tens. Write 6 in 56 tens directly below 9 in 59 tens. Subtract 59 − 56 = 3 tens.

Step 7: Bring down 5 ones.

Step 8: Divide 7 into 35 ones. 35 ones ÷ 7 = 5 ones. Write 5 in the ones column in the quotient.

Step 9: Multiply 5 ones × 7 = 35 ones. Subtract 35 − 35 = 0 ones. No numbers are left in the dividend to bring down. The division is finished.

Therefore, 2,695 ÷ 7 = 385.   Check: 385 × 7 = 2,695.   ■

▶ **You Try It**   Divide and check.

1. 91 ÷ 7   **2.** 135 ÷ 5   **3.** 836 ÷ 4   **4.** 6,201 ÷ 9

Use repeated subtraction to find 29 ÷ 6.

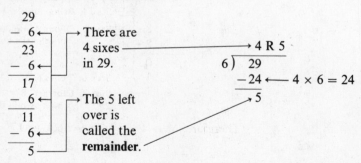

There are 4 sixes in 29.

The 5 left over is called the **remainder**.

The answer to 29 ÷ 6 is 4 with a remainder of 5, written 4 R 5. To check the answer, use the following rule.

quotient × divisor + remainder = dividend

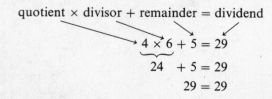

$$4 \times 6 + 5 = 29$$
$$24 + 5 = 29$$
$$29 = 29$$

EXAMPLE 3   Divide 523 ÷ 8.

$$
\begin{array}{r}
6\ 5\ \text{R}\ 3 \\
8\overline{)\ 5\ 2\ 3} \\
-4\ 8 \quad \longleftarrow\ 6 \times 8 = 48 \\
\hline
4\ 3 \\
-4\ 0 \longleftarrow\ 5 \times 8 = 40 \\
\hline
3
\end{array}
$$

Check: $65 \times 8 + 3$
$= \underbrace{\quad}\ 520 \ + 3$
$= 523$

Therefore, 523 ÷ 8 = 65 R 3.   ■

> **OBSERVE**   In Example 3, you can write the answer 65 R 3 as $65\frac{3}{8}$ by writing the remainder 3 over the divisor 8. 3/8 is called the fraction form for a remainder.

▶ **You Try It**   Divide and check.

5. 488 ÷ 7

6. 941 ÷ 6

Each time you bring down a digit when dividing, you must write a digit above it in the quotient. As shown in the next two examples, the digit you write in the quotient may be 0.

EXAMPLE 4   What is 43 ÷ 4?

$$
\begin{array}{r}
1\ 0\ \text{R}\ 3 \\
4\overline{)\ 4\ 3} \\
-4 \\
\hline
0\ 3 \\
-\ 0 \\
\hline
3
\end{array}
$$

← Bring down the 3 ones. Divide 4 into 3. Since 3 ones ÷ 4 gives 0 ones, *you must write 0 ones in the quotient above 3 ones.* 0 is a placeholder. It says there are no ones in the quotient. The remainder is 3.

Therefore, 43 ÷ 4 = 10 R 3.   Check: $10 \times 4 + 3$
$= \quad 40 \ + 3$
$= 43$ ■

EXAMPLE 5   Find $\dfrac{23,027}{5}$.

$$
\begin{array}{r}
4,6\ 0\ 5\ \text{R}\ 2 \\
5\overline{)\ 2\ 3,0\ 2\ 7} \\
-2\ 0 \\
\hline
3\ 0 \\
-3\ 0 \\
\hline
0\ 2 \\
-\ 0 \\
\hline
2\ 7 \\
-2\ 5 \\
\hline
2
\end{array}
$$

← Bring down the 2 tens. 2 tens ÷ 5 gives a quotient of 0. *You must write 0 in the quotient above the 2.* 0 is a placeholder, meaning there are no tens in the quotient.

Multiply 0 tens × 5 = 0 tens. Write 0 tens below 2 tens and subtract.

Therefore, $\dfrac{23{,}027}{5} = 4{,}605$ R 2.   Check: $4{,}605 \times 5 + 2$

$$= \; 23{,}025 \; + 2$$
$$= 23{,}027 \quad \blacksquare$$

▶ **You Try It**   Divide and check.

7. $65 \div 6$ **8.** $21{,}549 \div 7$

2 Divide Using a Divisor with Two or More Digits

When the divisor has two or more digits, use a **trial divisor** to help you find the quotient. Use the trial divisor found by rounding the divisor to the digit farthest left.

EXAMPLE 6

Divisor		Trial Divisor
54	$\longrightarrow$	50
783	$\longrightarrow$	800
17	$\longrightarrow$	20 $\blacksquare$

EXAMPLE 7   Divide $379 \div 54$.

$54\overline{\smash{)}379}$ 54 does not divide into 3. Do not write a digit above 3. 54 does not divide into 37. Do not write a digit above 7. 54 does divide into 379.

$54\overline{\smash{)}379} \longrightarrow$ Think: $5\,0\overline{\smash{)}\overset{7}{3\,7}\,9}$ The trial divisor is 50. 5 into 37 goes 7 times. Try 7 in the quotient.

$$\begin{array}{r} 7\ \text{R}\ 1 \\ 54\overline{\smash{)}\ 379} \\ -378 \\ \hline 1 \end{array}$$

Therefore, $379 \div 54 = 7$ R 1.   Check: $7 \cdot 54 + 1$
$$= \; 378 \; + 1$$
$$= 379 \quad \blacksquare$$

EXAMPLE 8   Find $37\overline{\smash{)}1{,}526}$.

$37\overline{\smash{)}1{,}526} \longrightarrow$ Think: $4\,0\overline{\smash{)}\overset{3}{15}\,2}$ The trial divisor is 40. 4 into 15 goes 3 times. Try 3 in the quotient.

$$\begin{array}{r} 3 \\ 37\overline{\smash{)}\ 1{,}526} \\ -1\,11 \\ \hline 41 \end{array}$$

$\longleftarrow 3 \times 37 = 111$

$\longrightarrow$ Since $41 > 37$, the trial divisor 3 is too small. Try 4.

$$\begin{array}{r} 41\ \text{R}\ 9 \\ 37\overline{\smash{)}\ 1{,}526} \\ -1\,48 \\ \hline 46 \\ -37 \\ \hline 9 \end{array}$$

$\longleftarrow 4 \times 37 = 148$

$\longleftarrow$ 37 divides into 46, 1 time

Therefore, $1{,}526 \div 37 = 41$ R 9.   Check: $37 \times 41 + 9$
$$= \; 1{,}517 \; + 9$$
$$= 1{,}526 \quad \blacksquare$$

▶ **You Try It**   Divide and check.

9. $379 \div 76$ **10.** $3,279 \div 43$

---

3 **Crossing Out Zeros and Dividing By Powers of 10**

Suppose both divisor and dividend are whole numbers with zeros on their right sides. Simplify long division by first crossing out an equal number of zeros from the right sides of both divisor and dividend.

$$50\overline{)\,300} \quad \begin{array}{c}6\\ \\ -300 \\ \hline 0\end{array}$$

Crossing out one zero on the right side of both divisor and dividend gives the same answer.

$$5\cancel{0}\,\overline{)\,30\cancel{0}} \quad \text{or} \quad 5\,\overline{)\,30}^{\,6}$$

Zeros may be crossed out regardless of how the division is written.

$$300 \div 50 = 30\cancel{0} \div 5\cancel{0} = 6$$

$$300/50 = 30\cancel{0}/5\cancel{0} = 6$$

$$\frac{300}{50} = \frac{30\cancel{0}}{5\cancel{0}} = 6$$

EXAMPLE 9   **a.** $8,000 \div 200 = 8,0\cancel{00} \div 2\cancel{00} = 80 \div 2 = 40$

b. $\dfrac{72,000}{9,000} = \dfrac{72,0\cancel{00}}{9,0\cancel{00}} = \dfrac{72}{9} = 8$

c. $60,000\,\overline{)\,4,800,000} = 6\cancel{0,000}\,\overline{)\,480\cancel{0,000}} = 6\,\overline{)\,480}^{\,80}$ ■

▶ **You Try It**   Divide by first crossing out zeros.

11. $9,000 \div 30$ **12.** $\dfrac{56,000}{7,000}$ **13.** $3,000\,\overline{)\,120,000}$

What pattern do you see when you divide by powers of 10 (10, 100, 1,000, and so on)?

EXAMPLE 10   **a.** $\dfrac{56,000}{10} = \dfrac{56,00\cancel{0}}{1\cancel{0}} = \dfrac{5,600}{1} = 5,600$

b. $\dfrac{56,000}{100} = \dfrac{56,0\cancel{00}}{1\cancel{00}} = \dfrac{560}{1} = 560$

c. $\dfrac{56,000}{1,000} = \dfrac{56,\cancel{000}}{1,\cancel{000}} = \dfrac{56}{1} = 56$ ■

> To divide by a power of 10, cross out the same number of zeros on the right side of the dividend as are in the power of 10.

▶ **You Try It**   Divide by first crossing out zeros.

14. $\dfrac{82,500}{100}$ **15.** $\dfrac{400,000}{10,000}$ **16.** $\dfrac{680}{10}$

---

4 Estimation   The rules for estimating a multiplication are used to estimate division.

> To estimate an answer to a multiplication or division problem
> 1. Round each number to the place value of its digit farthest left.
> 2. Multiply or divide the rounded numbers.

EXAMPLE 11   Estimate $836 \div 38$.

$$836 \div 38$$

round to hundreds   round to tens   Estimate $836 \div 38$ by $800 \div 40$.

$$800 \div 40$$

$$\rightarrow 40\overline{)800} = 4\cancel{0}\,\overline{)80\cancel{0}} = 4\overline{)80}^{\,20}$$

Estimate $836 \div 38$ to be about 20. (Actual answer = 22.) ■

EXAMPLE 12   The cost to take 47 people on a weekend bus trip to Niagara Falls is $2,838. Estimate the cost per person.

The cost per person is $2,838 \div 47$.

round to thousands   round to tens   Estimate $2,838 \div 47$ by $3,000 \div 50$.

$$\$3,000 \div 50$$

$$\rightarrow \$3,000 \div 50 = \frac{\$3,00\cancel{0}}{5\cancel{0}} = \frac{\$300}{5} = \$60 \text{ estimated cost per person}$$ ■

EXAMPLE 13   The defense department plans to spend $729,837,406 to build 184 high-tech tanks. Estimate the cost per tank.

The cost per tank is $729,837,406 \div 184$.

round to hundred millions   round to hundreds

$$\$700,000,000 \div 200$$

$$= \frac{\$700,000,0\cancel{0}\cancel{0}}{2\cancel{0}\cancel{0}} = \frac{\$7,000,000}{2} = \$3,500,000$$

The estimated cost is about $3,500,000 per tank. ■

▶ **You Try It**   **17.** Estimate $783 \div 18$.

18. A plane was chartered by 27 people for a total of $3,217. Estimate the cost per person.

19. The county budgeted $5,875,418 to hire 191 new teachers. Estimate the amount budgeted per teacher.

▶ **Answers to You Try It**   1. 13   2. 27   3. 209   4. 689   5. 69 R 5
6. 156 R 5   7. 10 R 5   8. 3,078 R 3   9. 4 R 75   10. 76 R 11
11. 300   12. 8   13. 40   14. 825   15. 40   16. 68   17. 40   18. $100
19. $30,000

SECTION 2.2 EXERCISES

1 *Find the quotient and remainder for each division, and check.*

1. $14 \div 3$ **2.** $17 \div 2$ **3.** $21 \div 4$ **4.** $11 \div 5$

5. $\dfrac{26}{8}$ **6.** $\dfrac{19}{6}$ **7.** $\dfrac{32}{8}$ **8.** $\dfrac{27}{3}$

9. $38 \div 9$ **10.** $45 \div 6$ **11.** $50 \div 7$ **12.** $61 \div 8$

13. $\dfrac{3,052}{8}$ **14.** $\dfrac{8,407}{2}$ **15.** $513 \div 6$ **16.** $944 \div 9$

17. $6,304 \div 7$ **18.** $7,003/8$ **19.** $4\overline{)20,107}$ **20.** $9\overline{)60,003}$

2 *Find the quotient and remainder. Check each answer.*

21. $96 \div 12$ **22.** $90 \div 15$ **23.** $127 \div 31$ **24.** $217 \div 43$

25. $678 \div 52$ **26.** $690 \div 31$ **27.** $1,598 \div 47$ **28.** $4,066 \div 19$

29. $5,007 \div 76$ **30.** $7,009 \div 84$ **31.** $17,385 \div 65$ **32.** $56,803 \div 67$

33. $67,021 \div 52$ **34.** $28,209 \div 28$ **35.** $89,264 \div 603$

36. $48,167 \div 380$ **37.** $340 \div 10$ **38.** $430 \div 10$

3

39. $600 \div 40$ **40.** $800 \div 50$ **41.** $700 \div 100$

42. $400 \div 100$ **43.** $4,500 \div 150$ **44.** $7,940 \div 20$

45. $34,600 \div 200$ **46.** $77,000 \div 1,000$ **47.** $30,000 \div 1,000$

48. $390,000 \div 100$ **49.** $1,600,000 \div 80,000$ **50.** $3,600,000 \div 150,000$

4 *Estimate the answer.*

51. $942 \div 87$ **52.** $384 \div 21$ **53.** $870 \div 33$ **54.** $610 \div 28$

55. $3,280 \div 615$ **56.** $8,617 \div 276$ **57.** $7,724 \div 43$ **58.** $5,290 \div 98$

59. A lottery jackpot of $2,238,162 was shared equally by 37 people. About how much will each person receive?

60. A 472-page book contains 184,692 words. Estimate the number of words per page.

61. A company paid a total of $6,728,544 for 72 tractors. Someone claimed this amounts to about $68,000 per tractor.
   a. Estimate the cost of each tractor.

b. Is the claim approximately correct?

62. A total of $12,569 is spent to train 215 students. A paper claims this amounts to about $500 per student.
   a. Estimate the answer.

   b. Does the paper's claim seem correct?

SKILLSFOCUS (Section 1.8) *Multiply.*

63. $68 \cdot 406$

64. $314 \cdot 200$

65. $5,319 \cdot 1,000$

66. Estimate $894 \cdot 37$

67. Estimate $139 \cdot 422$

EXTEND YOUR THINKING ▶▶▶▶

▶ SOMETHING MORE

Figure the answer mentally.

68. $\dfrac{(61 + 61 + 61)}{3}$

69. $\dfrac{(18 + 18 + 18 + 18 + 18)}{18}$

▶ TROUBLESHOOT IT

Find and correct the error.

70.
$$
\begin{array}{r}
3\,2\,4 \\
3\,)\overline{\,9,0\,7\,2} \\
-9 \\
\hline
0\,0\,7 \\
-6 \\
\hline
1\,2 \\
-1\,2 \\
\hline
0
\end{array}
$$

71. $\dfrac{60,200}{100} = \dfrac{60,20\cancel{0}}{1\cancel{0}\cancel{0}} = 620$

72. Estimate $261 \div 46 \doteq 200 \div 50$
$= 4$

WRITING TO LEARN ▶▶▶▶

73. Explain the procedure for crossing out zeros in a division. Then explain how you would apply it to the following problem.

$$\frac{450,000}{30,000} =$$

2.3 APPLICATIONS: MULTIPLICATION, DIVISION, AND AVERAGE

OBJECTIVES
1 Solve word problems (Section 1.6).
2 Solve average problems.

NEW VOCABULARY
average

1 Solving Word Problems

Solving a word problem is at first a reading and comprehension problem, not a math problem. First read the problem carefully. Understand it. Then use the key words to help you decide which operations to use to solve it. Review the procedure for solving word problems in Section 1.6 beginning on page 43.

FIGURE 2.1
Key Words and Phrases
Meaning Multiplication

product ⟶ the product of 4 and 7 is $4 \times 7 = 28$
times ⟶ 5 times 6 is $5 \times 6 = 30$
@, read "at" ⟶ 5 pies @ \$4 is $5 \times \$4 = \$20$
an indication of ⟶ 6 rows of seats with 10 seats per row
repeated addition   is 10 added to itself 6 times, or
                         $6 \times 10 = 60$ seats

FIGURE 2.2
Key Words and Phrases
Meaning Division

quotient ⟶ the quotient of 12 by 2 is $12 \div 2 = 6$
divided by ⟶ 24 divided by 6 is $24 \div 6 = 4$
divided into ⟶ 7 divided into 35 is written $35 \div 7 = 5$
how many of one ⟶ how many 5's are in 40
number is in another   is written $40 \div 5 = 8$

OBSERVE In Figure 2.2, the numbers used with *divided into* and *how many of one number is in another* are reversed when you divide.

EXAMPLE 1 Tom makes \$250 each week. How much will be make over the next 6 weeks?

Add the weekly salary of \$250 to itself 6 times. Since this is repeated addition, multiply for the answer.

$$\begin{array}{rl} \$250 & \text{weekly salary} \\ \times\ \ \ 6 & \text{weeks} \\ \hline \$1,500 & \text{Tom's total salary for 6 weeks} \ \blacksquare \end{array}$$

EXAMPLE 2 Each week, Susan makes \$8 per hour for the first 40 hours, and \$12 per hour for each hour over 40. This week she worked 46 hours. What was Susan's salary for the week?

Susan worked 46 hours. She made

$$\begin{array}{rl} \$8 \text{ per hour for the first 40 hours} = \$8 \times 40 = & \$320 \\ \$12 \text{ per hour for the next 6 hours} = \$12 \times 6 = + & 72 \\ \hline \text{Susan's total salary for the week} = & \$392 \end{array}$$

> **OBSERVE** You can write Susan's total salary as $\underbrace{\$8 \times 40}$ + $\underbrace{\$12 \times 6}$
>
> $\qquad\qquad\qquad\qquad\qquad\qquad\;\; = \quad \$320 \quad + \quad \$72$
>
> $\qquad\qquad\qquad\qquad\qquad\qquad\;\; = \qquad\qquad \$392.$ ■

▶ **You Try It**
1. Earl makes $425 per week. How much will he make in 8 weeks?

2. Each week Juanita makes $14 per hour for the first 40 hours, and $21 per hour for each hour over 40. She worked 53 hours this week. What is Juanita's salary for the week?

EXAMPLE 3 Ellen's home mortgage payment is $915 per month. She will make this payment for 30 years. What total will she pay for the home over the next 30 years?

Ellen will pay $915 per month, for 12 months per year, for 30 years. In one year she will pay $915 · 12.

$$
\begin{array}{rl}
\$915 & \text{per month} \\
\times\;\;\;12 & \text{months in one year} \\
\hline
1,830 & \\
+\;915 & \\
\hline
\$10,980 & \text{paid in one year}
\end{array}
$$

Over 30 years she will pay $10,980 · 30.

$$
\begin{array}{rl}
\$10,980 & \\
\times\;\;\;\;\;\;30 & \text{years} \\
\hline
\$329,400 & \text{total of Ellen's mortgage payments over 30 years}
\end{array}
$$
■

EXAMPLE 4 Ron plans to pay off a $5,460 debt by putting $1,200 down, and paying the rest in 15 equal payments. How much is each payment?

Step 1: Subtract the $1,200 down payment from $5,460.

$$
\begin{array}{rl}
\$5,460 & \text{total owed} \\
-\;1,200 & \text{down payment} \\
\hline
\$4,260 & \text{remaining to be paid}
\end{array}
$$

Step 2: Divide $4,260 into 15 equal parts.

$$
\begin{array}{l}
\$284 \longrightarrow \text{Each payment will be } \$284. \\
15\,\overline{)\,\$4,260}
\end{array}
$$

Check: Total paid = $\underbrace{\$284 \times 15}$ + $1,200

$\qquad\qquad\quad = \quad \$4,260 \quad + \$1,200$

$\qquad\qquad\quad = \$5,460$ ■

▶ **You Try It**   **3.** Kelly's mortgage payment is $640 per month. She will make this payment for 20 years. What total will she pay over 20 years?

4. Carla plans to pay off an $8,016 debt by putting $2,400 down and paying the rest in 36 equal payments. How much is each payment?

EXAMPLE 5   Jayne's car gets 21 miles to a gallon on the highway. She plans to drive from New York to San Francisco, a distance of 3,024 miles. How many gallons of gas will she need?

Jayne gets 21 miles on one gallon. She will drive 3,024 miles. To find how many gallons of gas she needs, find how many times 21 divides into 3,024.

$$\overset{144}{21\,)\overline{3,024}} \quad \text{gallons of gas needed for the trip}$$

Check: $144 \times 21 = 3,024$ ■

EXAMPLE 6   Yana earns $28,762 per year. Estimate how much this is per week.

There are 52 weeks in one year. Estimate $28,762 ÷ 52. To estimate a division, round each number to the digit farthest left.

$$52\,)\overline{\$28,762} \xrightarrow[\text{estimate}]{\text{round to}} \overset{6\,00}{5\cancel{0}\,)\overline{\$30,00\cancel{0}}}$$

Yana will earn about $600 per week. ■

▶ **You Try It**   **5.** Loretta gets 12 miles to the gallon with her mobile home. How many gallons of gas will she need for a 5,652-mile touring vacation?

6. Eva earns $39,072 per year. Estimate how much this is per week.

EXAMPLE 7   **Chapter Problem**   Read the problem opening this chapter.

The distance from the beginning of one strip to the beginning of the next is 12 feet + 20 feet = 32 feet. (32 feet contains 1 strip plus 1 gap between strips.) To find the number of strips per mile, find how many 32's are in 5,280.

$$
\begin{array}{r}
165 \text{ strips per mile} \\
32\,)\overline{5,280} \\
-3\,2 \\
\hline
2\,08 \\
-1\,92 \\
\hline
160 \\
-160 \\
\hline
0 \quad ■
\end{array}
$$

▶ **You Try It**   **7.** Solve Example 7 again, given that the length of one strip is 18 feet, and the distance between strips is 30 feet.

2 **Average**   In their last 4 games, a baseball team scored 2, 5, 1, and 4 runs, respectively. They scored a total of

$$2 + 5 + 1 + 4 = 12 \text{ runs.}$$

Divide the total of 12 runs by the number of games played. This gives the **average** number of runs scored per game.

$$\text{average} = \frac{\text{total runs}}{\text{games played}} = \frac{12}{4} = 3 \text{ runs per game}$$

3 is the average of 2, 5, 1, and 4 runs. Replace each of these four numbers by the average, 3, and you get the same total number of runs.

	number of runs scored in each game	*average number of runs per game*
game #1	2	3
game #2	5	3
game #3	1	3
game #4	+4	+3
—total—→	$12 \div 4 = 3$	$12 \div 4 = 3$

└─Same total number─┘
of runs and same average, 3.

To find the average for a set of numbers

1. Add the numbers together.
2. Divide this total by how many numbers you have.

EXAMPLE 8   What is the average of 12, 40, 32, 28, 50, 35, 16, and 27?

Add the 8 numbers together. Then divide this total by 8.

$$\text{average} = \frac{12 + 40 + 32 + 28 + 50 + 35 + 16 + 27}{8} = \frac{240}{8} = 30 \quad \blacksquare$$

EXAMPLE 9   Find the average of the numbers 0, 0, 1, 0, 3, 1, and 2.

0 counts as a number. There are 7 numbers. Add the numbers together. Divide this sum by 7.

$$\text{average} = \frac{0 + 0 + 1 + 0 + 3 + 1 + 2}{7} = \frac{7}{7} = 1$$

The average of 0, 0, 1, 0, 3, 1, and 2 is 1.   ∎

▶ **You Try It**   **8.** Find the average of 68, 75, 90, 82, 94, 71, 85, and 91.

9. Find the average of 0, 17, 11, 0, 6, 20, and 2.

EXAMPLE 10   Terri will take a total of 5 tests in her history class. Her scores on the first four tests are 76, 83, 70, and 89.

a. What must she score on the fifth test to have an 80 average? 100 is the highest score.

If she scored 80 on each of her 5 tests, her average would be 80. To see why, add 80 five times.

$$80 + 80 + 80 + 80 + 80 = 5 \times 80 = 400$$

Then her average is $400 \div 5 = 80$. Therefore, to get an 80 average, her test scores must add up to 400 points. The sum of her first four tests is

$$76 + 83 + 70 + 89 = 318.$$

Terri needs to score

$$\begin{array}{r} 400 \\ -318 \\ \hline 82 \end{array}$$ on her fifth test to average 80.

Check: $\text{average} = \dfrac{76 + 83 + 70 + 89 + \boxed{82}}{5} = \dfrac{400}{5} = 80$

b. What must Terri score on her fifth test to have a 90 average?

An average of 90 on 5 tests means Terri needs a total of $5 \times 90 = 450$ points. She has 318 points from her first four tests. She needs to score $450 - 318 = 132$ on her last test. Since 100 is the highest score, it is not possible for Terri to average 90. ■

▶ **You Try It**   **10.** Janet will take 4 tests in her art class. Her scores on the first 3 tests were 75, 91, and 82. What must she score on her fourth test to average 80?

CONNECTIONS

There is another way to solve Example 10a. Determine how much each score differs from the desired average, 80.

scores: 76   83   70   89
          ↓   ↓   ↓   ↓
    4 below 80   3 above 80,   10 below 80,   9 above 80,
                   gives a   gives a   gives a
                  balance of   balance of   balance of
                   1 below   11 below   2 below
                                                  ↓
                                    Overall, you are 2 points
                                    short of an 80 average. So
                                    you need 80 + 2 = 82 on
                                    the last test to average 80.

If −4 means 4 points below 80, and +3 means 3 points above 80, then you can also write this problem as follows.

scores: 76   83   70   89
           ↓   ↓   ↓   ↓
    4 below + 3 above + 10 below + 9 above = 2 below
      (−4) + (+3) + (−10) + (+9) = (−2)

▶ **Answers to You Try It** 1. $3,400 2. $833 3. $153,600 4. $156
5. 471 gal 6. about $800 per week 7. 110 stripes per mile 8. 82
9. 8 10. 72

SECTION 2.3 EXERCISES

I

1. Lauren makes $375 per week. How much will she make in 8 weeks?

2. George sold 12 air conditioners today. If each sold for $289, what were his total sales for the day?

3. Ed makes $23,556 a year. What does Ed make each week?

4. If 25 lamps cost a store owner $675, what is the cost of one lamp?

5. Irma's car gets 28 miles on a gallon of gas. How many miles will she get on a full tank of 18 gallons?

6. Tom's mortgage payment is $752 per month. How much does he pay each year for his mortgage?

7. A 30-foot camper gets 6 miles to a gallon of gas. How much gas is needed to drive 402 miles from Los Angeles to San Francisco?

8. A 26-foot motorboat gets 4 miles to a gallon of gas. How much gas is needed to travel 128 miles?

9. Diane can seed 990 square feet of lawn in one minute. Her favorite TV show is on in 45 minutes. Does she have time to seed her 1-acre lawn? 1 acre = 43,560 square feet.

10. A swimming pool requires 6,000 gallons of water. A hose pumps 8 gallons of water per minute. In how many minutes will the pool be filled?

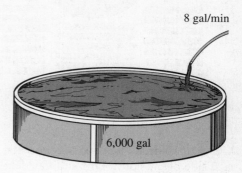

8 gal/min

6,000 gal

11. You count 521 words on one page of a novel. Estimate how many words the 462-page novel contains.

12. A case of soda has 24 bottles. Each bottle has 16 ounces. Estimate how many ounces of soda are in the case.

13. Sandy makes $13 an hour. What will she make if she works 27 hrs?

14. Jim makes $9 an hour. What will he make if he works 40 hrs?

15. Each week, Joyce is paid $12 an hour for the first 40 hours, and $18 for each hour she works over 40. What will Joyce make if she works 62 hours this week?

16. Each week, Jayne makes $14 an hour for the first 40 hours, and $21 an hour for each hour over 40. What will her salary be if she works 43 hours this week?

17. An auditorium has 3 groups of chairs. The middle group is 24 chairs wide and 42 chairs deep. Each side group is 13 chairs wide and 35 chairs deep. How many people can be seated in the auditorium?

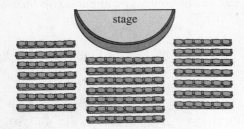

18. A basketball arena has two sets of seats, one on each side of the court. One side can seat 54 people across and 31 people deep. The other side can seat 50 people across and 28 deep. How many seats are in the arena?

19. In one year, the Earth makes one complete revolution around the sun, traveling a total of 584,000,000 miles. There are about 31,560,000 seconds in one year. Estimate the number of miles the Earth travels each second as it moves around the sun.

20. A development corporation plans to construct 475 homes. The total cost for this project is $82,937,000. Estimate what each home must be sold for to recover this cost.

21. A wedding caterer purchased 36 cases of regular soda @ $7 per case and 28 cases of diet soda @ $6 per case. What is the total cost?

22. The last performance of a play is sold out. If 40 box seats sold @ $28, 54 mezzanine seats sold @ $21, and 36 balcony seats sold @ $19, how much money did the theater collect from the last performance?

23. Susan, Diane, Nancy, and Joyce purchase a sailboat for $32,648. If they decide to evenly split the cost of the boat, how much does each woman pay?

24. Al, Bob, and Ted own a business. In the first year of operation, the business lost $4,659. If each man pays an equal share of this loss, how much will each man pay?

25. The county recreation budget is $45,626. It is split evenly among 7 recreation councils. How much does each council get?

26. A department store has 78 bikes to display in 6 equal rows. How many bikes will be in each row?

27. Bob bought a TV for $684. He put $120 down, and paid the rest in 12 equal monthly payments. What was the monthly payment?

28. Jack bought a stereo for $2,597. He put $500 down and paid the rest in 9 equal monthly payments. Find the monthly payment.

29. Angela bought a car for $15,400. She put $3,400 down, and paid the rest in 48 equal monthly payments. Find her monthly payment.

30. Hal bought a jade ring for $260. He put $40 down, and paid the rest in 5 equal monthly payments. Find Hal's monthly payment.

2

31. Find the average of 4, 6, 7, 7, 9, and 15.

32. Find the average of 88, 70, 78, 90, 67, 76, 100, 90, 88, and 73.

33. Ron sold 12 vacuum cleaners in March, 9 in April, 13 in May, and 6 in June. What average number did he sell per month?

34. A baseball team scored 3, 6, 0, 1, 2, 3, 0, 4, and 8 runs over the last 9 games. Find the average number of runs scored per game.

35. Recently homes sold in Al's neighborhood for $124,000, $130,500, $121,250, $128,000, and $129,750. Find the average selling price.

36. A basketball team scored 77, 98, 102, 89, 85, 91, 112, 90, 96, 77, 79, and 84 points on a 12-game road trip. What is the average number of points scored per game?

37. Don took a five part exam. He needs an average score of 45 to pass. His scores on the first four parts were 41, 46, 40, and 47. What must he score on the fifth part to pass?

38. Phil must take 6 tests in his biology class. His scores on the first 5 were 87, 67, 94, 83, and 75. What must Phil score on his last test to have an 80 average?

SKILLSFOCUS (Section 1.10) *Simplify.*

39. $1 + 3 \cdot 5$

40. $30 \cdot 2 + 15$

41. $4(2 + 7)$

42. $7 \cdot 10 - 5 \cdot 4$

EXTEND YOUR THINKING ▶▶▶▶
▶**SOMETHING MORE**

43. Voyager II The *Voyager II* spacecraft, launched on August 20, 1977, reached the planet Neptune 12 years later. It sent a variety of radio messages back to Earth containing video images of Neptune and its moons. These radio signals travel at the speed of light, 186,000 miles each second. If Neptune is about 2,790,000,000 miles from Earth,

a. how many seconds would it take a video image of Neptune's polar cap to reach Earth?

b. how many minutes is this?

c. how many hours and minutes is this? (1 hour = 60 minutes)

44. Cancer According to a recent report, someone in the United States dies of cancer every 62 seconds. Estimate how many people this is per year.

45. A pumpkin farmer planted 16 acres with pumpkins. He plans to harvest 420 pumpkins per acre. He can pack 12 pumpkins per crate for shipment to market. If he sells his pumpkins for $13 per crate, how much can the farmer expect to make?

46. A company employs 280 workers. Each employee works 8 hours each day, and is paid the exact same hourly wage. If the total payroll for one day is $20,160, what hourly wage are these workers paid?

47. A United Nations report projects the world's population will increase by 1 billion people from 1987 to 1999. Estimate how many people per day this amounts to.

48. The Earth travels 584,000,000 miles in one revolution around the sun. If the Earth is 4,500,000,000 years old, how far has it traveled in its lifetime?

49. What time is it exactly 89 hours after 5:04 P.M.?

▶ **CONNECTIONS**

50. Solve Exercises 37 and 38 using the method described in CONNECTIONS in this section.

WRITING TO LEARN ▶▶▶▶

51. You take five quizzes. One quiz score is 0. Explain why the score of 0 must be figured into your average.

52. Explain why the number 30 in Exercise 7 does not enter into the solution of the problem.

▶ **YOU BE THE JUDGE**

53. Is it possible to completely submerge yourself in a lake whose average depth is 1 inch? Explain your decision.

54. Zack claims that every palindrome (see Section 1.6, page 46) with an even number of digits is exactly divisible by 11, with no remainder. Do the following palindromes support this claim?

   a. 2442   **b.** 8118   **c.** 744447   **d.** 1238998321

   e. Do 252 or 62226 contradict this claim? (Careful!)

2.4 ORDER OF OPERATIONS

OBJECTIVES

Evaluate expressions using the order of operations (Section 1.8, Objective 4).

NEW VOCABULARY

order of operations

Order of Operations

The expression $6 + 4 \cdot 5$ is an arithmetic expression. It has two operations. When two or more operations are in the same expression, performing them in the right order is essential to getting the correct answer. Mathematicians have agreed that the following order be used. This order is called the **order of operations**.

The order of operations was discussed briefly in Section 1.8. Here this order is updated to include the operations of division and exponents.

Order of Operations

Begin by reading the problem. Identify the operations being used. You may find it helpful to write the multiplication dot, $\cdot$, in any position where multiplication is implied. Evaluate the operations in the following order.

1. Simplify operations within parentheses.
2. Simplify exponents (powers and roots).
3. Do multiplications and divisions in the order you see them from left to right.
4. Do additions and subtractions in the order you see them from left to right.

OBSERVE This order of operations is the same order used by computers, programmable calculators, and in algebra.

EXAMPLE 1   Simplify $4 + 7 - 5 + 1$.

Addition and subtraction are performed in order from left to right.

$$
\begin{aligned}
& \underbrace{4 + 7} - 5 + 1 && \text{Add } 4 + 7. \\
= & \underbrace{11 - 5} + 1 && \text{Subtract } 11 - 5. \\
= & \quad \underbrace{6 + 1} && \text{Add } 6 + 1. \\
= & \qquad 7 && \text{The answer is 7.} \quad \blacksquare
\end{aligned}
$$

CAUTION It is wrong to say $4 + 7 - 5 + 1 = 11 - 6 = 5$. Why? Because the addition on the right was performed before the subtraction in the middle. This violates the left to right order for doing addition and subtraction.

EXAMPLE 2   Simplify $6 + 4 \cdot 5$.

$$6 + \underbrace{4 \cdot 5}\qquad \text{Multiply first.}$$
$$= 6 + 20 \qquad \text{Add.}$$
$$= \quad 26 \qquad \text{The answer is 26.} \quad \blacksquare$$

CAUTION   It is wrong to say $6 + 4 \cdot 5 = 10 \cdot 5 = 50$. Why?

▶ **You Try It**   Simplify the following expressions.

1. $6 + 4 - 2 + 3$

2. $10 - 5 + 3 - 1$

3. $1 + 3 \cdot 4$

4. $4 + 7 \cdot 2$

EXAMPLE 3   Evaluate $20 \div 4 \cdot 5$.

Perform multiplication and division in order from left to right.

$$20 \div 4 \cdot 5 = \underbrace{(20 \div 4)} \cdot 5 \qquad \text{Division is on the left. Do division first. (You may write your own parentheses around } 20 \div 4 \text{ to indicate that you divide first.)}$$
$$= \quad 5 \quad \cdot 5 \qquad \text{Multiply.}$$
$$= \quad 25 \quad \blacksquare$$

CAUTION   To say $20 \div 4 \cdot 5 = 20 \div 20 = 1$ is wrong because the operations were not performed in order from left to right.

EXAMPLE 4   Evaluate $3 \cdot 4 + 2 \cdot 5$.

$$\underbrace{3 \cdot 4} + \underbrace{2 \cdot 5} \qquad \text{Perform both multiplications.}$$
$$= 12 + 10 \qquad \text{Add.}$$
$$= \quad 22 \quad \blacksquare$$

EXAMPLE 5   Simplify $80 \div 20 \div 2$.

$$80 \div 20 \div 2 = \underbrace{(80 \div 20)} \div 2 \qquad \text{Perform the divisions in order from left to right.}$$
$$= \quad 4 \quad \div 2$$
$$= \quad 2 \quad \blacksquare$$

▶ **You Try It**   Simplify the following expressions.

5. $15 \div 3 \cdot 5$

6. $28 \div 4 \cdot 7$

7. $5 \cdot 6 + 2 \cdot 9$

8. $45 \div 15 \div 3$

EXAMPLE 6 Simplify $5\sqrt{49} + 15 \div 3$.

means multiply

$$5\sqrt{49} + 15 \div 3 = 5 \cdot \sqrt{49} + 15 \div 3 \qquad \text{Perform square root before multiplication and division.}$$

$$= \underbrace{5 \cdot 7}_{} + \underbrace{15 \div 3}_{} \qquad \text{Multiply and divide.}$$

$$= \quad 35 \quad + \quad 5 \qquad \text{Add.}$$

$$= \qquad 40 \quad \blacksquare$$

EXAMPLE 7 Evaluate $6(3 + 4)$.

means multiply

$$6\ (3 + 4) = 6 \cdot (3 + 4) \qquad \text{First simplify the addition within parentheses.}$$

$$= 6 \cdot \quad (7) \qquad \text{Then multiply.}$$

$$= \quad 42 \ \blacksquare$$

EXAMPLE 8 Simplify $1 + 5(4^2)$.

means multiply

$$1 + 5\ (4^2) = 1 + 5 \cdot (4^2) \qquad \text{First, evaluate the power.}$$

$$= 1 + 5 \cdot (16) \qquad \text{Next, multiply.}$$

$$= 1 + \quad 80 \qquad \text{Finally, add.}$$

$$= \quad 81 \ \blacksquare$$

▶ **You Try It** Simplify the following expressions.

9. $8\sqrt{4} + 32 \div 8$

10. $9(3 + 7)$

11. $3 + 2(5^2)$

EXAMPLE 9 Evaluate $4 \cdot 3^2 + 30 \div \sqrt{100} + 2^3 \div (6 - 2)$.

$$4 \cdot 3^2 + 30 \div \sqrt{100} + 2^3 \div (6 - 2) \qquad \text{First, simplify within parentheses.}$$

$$= 4 \cdot 3^2 + 30 \div \sqrt{100} + 2^3 \div (4) \qquad \text{Second, do powers and roots.}$$

$$= 4 \cdot 9 + 30 \div 10 + 8 \div (4) \qquad \text{Third, multiply and divide from left to right.}$$

$$= 36 + 3 + 2 \qquad \text{Finally, add.}$$

$$= \qquad 41 \ \blacksquare$$

Evaluate the expression.

12. $2 \cdot 3^2 + 40 \div \sqrt{64} + 4^3 \div (12 - 4)$

SOMETHING MORE

Use the order of operations to determine if each equation is true or false.

a. _____ $\sqrt{(16 + 9)} \overset{?}{=} \sqrt{16} + \sqrt{9}$

b. _____ $\sqrt{(25 - 16)} \overset{?}{=} \sqrt{25} - \sqrt{16}$

c. _____ $\sqrt{(9 \cdot 4)} \overset{?}{=} \sqrt{9} \cdot \sqrt{4}$

d. _____ $\sqrt{\left(\dfrac{100}{25}\right)} \overset{?}{=} \dfrac{\sqrt{100}}{\sqrt{25}}$

▶ **Answers to You Try It**   1. 11   2. 7   3. 13   4. 18   5. 25   6. 49   7. 48   8. 1   9. 20   10. 90   11. 53   12. 31

▶ **Answers to Something More**   a. false, $5 \neq 4 + 3$ b. false, $3 \neq 5 - 4$ c. true, $6 = 3 \cdot 2$ d. true, $2 = \dfrac{10}{5}$

Use the order of operations to simplify each expression.

1. $4 - 3 + 2$

2. $7 - 5 + 1$

3. $9 + 5 - 5 + 7$

4. $4 + 8 - 3 + 2$

5. $1 + 4 \cdot 5$

6. $4 + 3 \cdot 6$

7. $4 + 8 \div 4$

8. $9 + 3 \div 3$

9. $3 \cdot 4^2$

10. $6 \cdot 3^2$

11. $2 + 3^2$

12. $5 + 2^3$

13. $25 \div 5^2$

14. $36 \div 3^2$

15. $100 \div 50 \div 2$

16. $90 \div 30 \div 3$

17. $100 \div (50 \div 2)$

18. $90 \div (30 \div 3)$

19. $5 \cdot 6 + 8 \cdot 3$

20. $5 \cdot 5 + 2 \cdot 2$

21. $40 \div 10 + 10 \div 2$

22. $64 \div 8 + 8 \div 4$

23. $30 \div 2 \cdot 5$

24. $8 \div 4 \cdot 2$

25. $3^0 + 6$

26. $5 + 5^0$

27. $4^2 + 4^1 + 4^0$

28. $3^1 + 3^2 + 3^0$

29. $20 \div 5 + 5$

30. $12 \div 4 - 1$

31. $9 + 6 \div 3 + 7$

32. $4 + 8 \div 2 + 1$

33. $40^6 \cdot 0$

34. $17(127)^5 \cdot 0$

35. $60 \div 10(2) \cdot 3$

36. $20 \div 4(3) \cdot 5$

37. $4\sqrt{49} + 2 \cdot 6^2$

38. $8\sqrt{100} - 9 \cdot 3^1$

39. $4(5 + 6)$

40. $3(9 + 2)$

41. $(2 + 6)(4 + 1)$

42. $(17 - 13)(50 - 47)$

43. $3 + 2(4^2)$

44. $100 + 6(2^3)$

45. $7 + 2\sqrt{81} + 9$

46. $8 + 7\sqrt{4} - 20$

47. $10(2 + 3)^2$

48. $7(4 + 6)^2$

49. $4 \cdot 2^3 + 28 \div \sqrt{49}$

50. $5 \cdot 3^2 + 72 \div \sqrt{36}$

Three similar expressions are given. The only difference is in the placement of the parentheses. Evaluate each expression, and determine which two give the same answer.

51. a. $2 + 3 \cdot 5$ **b.** $(2 + 3)5$ **c.** $2 + (3 \cdot 5)$

52. a. $4 \cdot (6 - 3)$ **b.** $(4 \cdot 6) - 3$ **c.** $4 \cdot 6 - 3$

53. a. $(20/4)5$ **b.** $20/4 \cdot 5$ **c.** $20/(4 \cdot 5)$

54. a. $60 \div 30 \cdot 2$ **b.** $60 \div (30 \cdot 2)$ **c.** $(60/30) \cdot 2$

55. a. $4 \cdot 5/2 \cdot 10$ **b.** $(4 \cdot 5)/(2 \cdot 10)$ **c.** $(4 \cdot 5)/2 \cdot 10$

56. a. $(2 \cdot 5)^2$ **b.** $2 \cdot 5^2$ **c.** $2(5)^2$

SKILLSFOCUS (Section 1.8) *Multiply.*

57. $56 \cdot 472$

58. $30 \cdot 46$

59. $1,000 \cdot 78$

60. $600 \cdot 80$

EXTEND YOUR THINKING ▶▶▶▶

▶SOMETHING MORE

61. Use the order of operations to determine if each equation is true or false.

a. $\sqrt{(25 + 144)} \stackrel{?}{=} \sqrt{25} + \sqrt{144}$ **b.** $\sqrt{(169 - 25)} \stackrel{?}{=} \sqrt{169} - \sqrt{25}$

c. $\sqrt{(16 \cdot 25)} \stackrel{?}{=} \sqrt{16} \cdot \sqrt{25}$ **d.** $\sqrt{\left(\dfrac{36}{9}\right)} \stackrel{?}{=} \dfrac{\sqrt{36}}{\sqrt{9}}$

62. Make the expression true by inserting the correct operational symbols on the left side of each equal sign.

a. 8   8   8   8 = 3   **b.** 2   2   2   2 = 2

c. 6   6   6   6 = 13   **d.** 4   4   4   4 = 4

e. 1   1   1 = 0   **f.** 9   9   9   9   9 = 0

g. 2   1   7 = 7   **h.** 3   2   4   1 = 25

i. 1   0   0   1 = 1   **j.** 4   3   2   1 = 5

63. Do You Remember? Simplify each expression on the left. Then match the answer to the movie, book, or show on the right.

A. $2 + 5 \cdot 2^2$ _____: _____: *A Space Odyssey*

B. $8 \cdot 50 + 3 \cdot 17$ _____: "Rescue _____"

C. $2 \cdot 10^3 + 10^0$ _____: *Fahrenheit* _____

D. $200 \div 2 + 3(1 + 2)$ _____: *Catch* _____

E. $7\sqrt{36} - 2^3$ _____: "Car _____, Where Are You"

F. $120 \div 4 \cdot 30 + 11$ _____: *Miracle on* _____*th Street*

G. $10(8 - 2) - 6^1$ _____: *PT* _____

64. Solve each problem, or write not possible.

    a. $(4-4) \div 0$ **b.** $5/(6-6)$

    c. $(8-8) \div (6-1)$ **d.** $(3-3)/(3+3)$

▶TROUBLESHOOT IT

Find and correct the error.

65. $7 + 2 \cdot 3 = 9 \cdot 3 = 27$

66. $3 \cdot 6 - 3 = 3 \cdot 3 = 9$

67. $10 \div 5 \cdot 2 = 10 \div 10 = 1$

68. $20 \div 10 \div 2 = 20 \div 5 = 4$

69. $10 \cdot 10 \div 10 \cdot 10 = 100 \div 100 = 1$

70. $8 + 3 - 4 + 2 = 11 - 6 = 5$

71. $2 \cdot 3^2 = 6^2 = 36$

72. $3^0 + 3^1 = 0 + 3 = 3$

73. $3 \cdot 2 + 4 \cdot 7 = 6 + 4 \cdot 7 = 10 \cdot 7 = 70$

74. $2 + 2 \cdot 2 - 2 = 4 \cdot 2 - 2 = 4 \cdot 0 = 0$

75. $5(6 + 2) = 5 \cdot (6 + 2) = 30 + 2 = 32$

76. $5 + 2\sqrt{16} - 7 = 5 + 2\sqrt{9} = 5 + 2 \cdot 3 = 5 + 6 = 11$

77. $4 + 6 \div 2 = 10 \div 2 = 5$

WRITING TO LEARN ▶▶▶▶

78. Write an expression for your age. It must include an addition, a power, a root, and one other operation. Then evaluate it.

79. The sentence "Please Excuse My Dear Aunt Sally" has been used for years by students to help them remember the order of operations.

        P in Please means do parentheses first
        E in Excuse means do exponents (powers and roots) next
        M in My means do multiplication next
        D in Dear means do division next
        A in Aunt means do addition next
        S in Sally means do subtraction last

Explain why the strict use of this memory aid violates the order of operations.

OBJECTIVES

1 Change word sentences into arithmetic expressions (Sections 1.6 and 1.7, Figures 1.3, 1.4, 1.6, and 1.7).

2 Evaluate expressions containing numbers and letters.

3 Evaluate formulas containing numbers and letters.

NEW VOCABULARY

variable

formula

1 Changing Word Sentences into Arithmetic Expressions

To change a word sentence into an arithmetic expression, change each word or phrase into a number or operation. Refer to the key word lists in Sections 1.6 and 2.3 (pages 43 and 109) before going on.

EXAMPLE 1 Change each word sentence into an arithmetic expression, then simplify.

a. Four plus nine.

$$4 + 9 = 13.$$

b. Twenty less 2 with a gain of 5.

$$20 - 2 + 5 = \underbrace{(20 - 2)}_{} + 5$$
$$= 18 + 5 = 23$$

OBSERVE The parentheses are used to indicate the operation to be performed first.

---

c. Two plus three times four.

$$2 + 3 \cdot 4 = 2 + (3 \cdot 4)$$
$$= 2 + 12 = 14$$

CAUTION Do not add first. By the order of operations, multiply first, then add.

---

d. Thirty divided by five times six.

$$30 \div 5 \cdot 6 = \underbrace{(30 \div 5)}_{} \cdot 6$$
$$= 6 \cdot 6 = 36$$

e. Five squared decreased by two times 8.

$$5^2 - 2 \cdot 8 = 25 - 2 \cdot 8$$
$$= 25 - \underbrace{16}_{} = 9 \quad \blacksquare$$

▶ **You Try It** Change each word sentence into an arithmetic expression and simplify.

1. Seven minus two.

2. Ten increased by eight, with a loss of three.

3. Seven plus five times six.

4. Ninety divided by thirty times three.

5. Six squared less four times seven.

The words *sum, difference, product,* and *quotient,* when combined with other operations, may require the use of parentheses to be properly written.

EXAMPLE 2 **a.** What is four times three plus five?

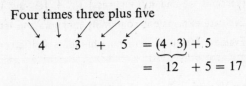

$$4 \cdot 3 + 5 = \underbrace{(4 \cdot 3)} + 5$$
$$= 12 + 5 = 17$$

b. What is 4 times the sum of 3 and 5?

4 times the sum of 3 and 5

$$4 \cdot (3 + 5) = 4 \cdot \underbrace{(3 + 5)}$$
$$= 4 \cdot (8) = 32$$

In 2b, 4 is being multiplied times a sum. Therefore, add first to get the sum. Then multiply. To indicate that you add before multiplying, write parentheses around the sum. ■

EXAMPLE 3 What is twenty divided by the product of 4 and 5?

Twenty divided by the product of 4 and 5

$$20 \div (4 \cdot 5) = 20 \div \underbrace{(4 \cdot 5)}$$
$$= 20 \div (20) = 1$$

In this expression, 20 is being divided by a product. Therefore, multiply first to get the product. Then divide. To indicate that you multiply first, write parentheses around the product. ■

▶ **You Try It**   Write the expression and simplify.

6. Six times two plus three.
7. Six times the sum of two and three.
8. Eighty divided by the product of eight and ten.

2 Evaluating Expressions   Expressions containing numbers, letters, and operations are commonly seen in mathematics. To evaluate such an expression, replace the letters by the given numbers, then simplify.

EXAMPLE 4   Evaluate xy if $x = 5$ and $y = 8$.

xy means $x \cdot y$.

$$xy = x \cdot y \qquad \text{Replace } x \text{ with 5 and } y \text{ with 8.}$$
$$= 5 \cdot 8$$
$$= 40 \quad ■$$

> To evaluate expressions containing numbers and letters
>
> 1. Rewrite the expression with any implied multiplication replaced by the multiplication dot, ·.
> 2. Replace letters with the given numbers.
> 3. Simplify according to the order of operations.

EXAMPLE 5 Evaluate $2L + 2W$ if $L = 7$ and $W = 10$.

$$2L + 2W = 2 \cdot L + 2 \cdot W \qquad \text{Replace } L \text{ with 7 and } W \text{ with 10.}$$
$$= 2 \cdot 7 + 2 \cdot 10 \qquad \text{Multiply first.}$$
$$= 14 + 20 \qquad \text{Then add.}$$
$$= 34 \quad \blacksquare$$

▶ **You Try It**   9. Evaluate $3mk$ if $m = 6$ and $k = 5$.

10. Evaluate $2K + 3R$ if $K = 4$ and $R = 7$.

EXAMPLE 6 Evaluate $P(6 + ST)$ if $P = 100$, $S = 2$, and $T = 7$.

$$P(6 + ST) = P \cdot (6 + S \cdot T) \qquad \text{Replace } P \text{ with 100, } S \text{ with 2,}$$
$$\text{and } T \text{ with 7.}$$
$$= 100 \cdot (6 + 2 \cdot 7) \qquad \text{First simplify within parentheses.}$$
$$\text{Since there are two operations}$$
$$\text{in parentheses, apply the order}$$
$$\text{of operations. Multiply first.}$$
$$= 100 \cdot (6 + 14) \qquad \text{Then add.}$$
$$= 100 \cdot (20)$$
$$= 2{,}000 \quad \blacksquare$$

EXAMPLE 7 Evaluate $3 + 7w$ for each value of w.

a. $w = 2$: $3 + 7w = 3 + 7 \cdot w = 3 + 7 \cdot 2 = 3 + 14 = 17$.
b. $w = 5$: $3 + 7w = 3 + 7 \cdot w = 3 + 7 \cdot 5 = 3 + 35 = 38$.
c. $w = 0$: $3 + 7w = 3 + 7 \cdot w = 3 + 7 \cdot 0 = 3 + 0 = 3$. $\blacksquare$

> **OBSERVE** In Example 7, each time the letter w is replaced by a different number, you get a different answer. Since the number you substitute for the letter w can vary, w is called a **variable**.

▶ **You Try It**   11. Evaluate $S(4 + WQ)$ for $S = 200$, $W = 3$, and $Q = 5$.

12. Evaluate $5 + 2y$ for **a.** $y = 2$. **b.** $y = 6$. **c.** $y = 10$.

3 Evaluating Formulas

A **formula** is an equation. It usually consists of a letter on one side of the equal sign, set equal to an expression consisting of letters, numbers, and operations on the other side. For example,

$$F = 3 + 4H$$

is a formula. It is read "the letter F equals 3 plus 4 times H." Evaluate a formula the same way you evaluate an expression.

EXAMPLE 8   Find F in the formula $F = 3 + 4H$ when $H = 5$.

$$F = 3 + 4H \qquad \text{Write the dot for the implied multiplication.}$$
$$F = 3 + 4 \cdot H \qquad \text{Replace } H \text{ with 5.}$$
$$\downarrow$$
$$F = 3 + 4 \cdot 5 \qquad \text{Multiply first.}$$
$$F = 3 + 20$$
$$F = 23 \quad \blacksquare$$

> **OBSERVE**   The difference between a formula and an expression is that the answer in a formula has a name. In Example 8, F is the name for the answer, 23.

▶ **You Try It**   **13.** Find L if $L = 10 + 4P$ and $P = 6$.

EXAMPLE 9   Evaluate the formula $A = (X + Y + Z)/3$ when $X = 6$, $Y = 8$, and $Z = 13$.

$$A = (X + Y + Z)/3$$
$$A = (6 + 8 + 13)/3 \qquad \text{First, simplify within parentheses.}$$
$$A = (27)/3$$
$$A = 9 \quad \blacksquare$$

EXAMPLE 10   Find S in $S = 6E^2$ when $E = 4$.

$$S = 6E^2 = 6 \cdot E^2 \qquad \text{Replace } E \text{ with 4.}$$
$$= 6 \cdot 4^2 \qquad \text{Evaluate the power first.}$$
$$= 6 \cdot 16$$
$$= 96 \quad \blacksquare$$

▶ **You Try It**   **14.** Evaluate $A = (X + Y)/(Z + 4)$ for $X = 6$. $Y = 8$, and $Z = 3$.
15. Find S in $S = 5B^3$ when $B = 2$.

▶ **Answers to You Try It**   1. $7 - 2 = 5$ 2. $10 + 8 - 3 = 15$
3. $7 + 5 \cdot 6 = 37$ 4. $90 \div 30 \cdot 3 = 9$ 5. $6^2 - 4 \cdot 7 = 8$ 6. $6 \cdot 2 + 3 = 15$
7. $6(2 + 3) = 30$ 8. $80 \div (8 \cdot 10) = 1$ 9. 90 10. 29 11. $3,800$ 12. a. 9
b. 17 c. 25 13. $L = 34$ 14. $A = 2$ 15. $S = 40$

SECTION 2.5 EXERCISES

1 *Change each word sentence into an arithmetic expression, then simplify.*

1. Seven plus ten.

2. Five more than sixteen.

3. 7 plus 10 decreased by five.

4. 23 more than 14 less nine.

5. The product of twelve and ten.

6. The product of 30 and six.

7. Four times six less eleven.

8. Seven times ten minus 13.

9. The quotient of forty by 8.

10. The quotient of 100 and 20.

11. 3 plus 4 times 7.

12. The sum of 3 and 4, times 7.

13. 25 more than 6 divided by 2.

14. One plus 50 divided by 5.

15. Three squared plus seven.

16. Two cubed minus eight.

17. 5 times 8 added to 6 times 2.

18. 11 times two take away 5 times three.

19. The sum of 12 and 8, times 3.

20. 12 plus 8 times 3.

21. 56 divided by the product of four and 2.

22. Two hundred divided by the product of twenty-five and 8.

23. The product of 12 and 15 less the quotient of 200 and 4.

24. 70 times three decreased by 800 divided by 8.

25. Five times the square root of 49.

26. The product of ten and the square root of 81.

2 *Evaluate each expression using the values $x = 2$, $y = 5$, $z = 100$, $a = 8$, and $b = 10$.*

27. $3y$

28. $7b$

29. $x + 3y$

30. $6x + 5$

31. $4a - 3$

32. $2z - 50$

33. $3x + by$

34. $ab + 8z$

35. $a(2 + b)$

36. $(x + 4)y$

37. $4a - 3b$

38. $2z - 10b$

39. axy

40. $3az$

41. y^2

42. z^2

43. $4x^3$

44. $7b^4$

45. $b(a - 4x)$

46. $z(2y - a)$

47. $3z/b$

48. $\dfrac{4b}{y}$

49. $(x + y)(b - a)$

50. $(z - b)(y - x)$

3 *Evaluate each formula.*

51. Find A if $A = LW$ and $L = 8$ and $W = 6$.

52. Find D if $D = ST$ and $S = 50$ and $T = 3$.

53. Find P if $P = A + B + C$ if $A = 7$, $B = 9$, and $C = 4$.

54. Find E if $E = F + V - 2$ if $F = 6$ and $V = 8$.

55. Find P if $P = 2L + 2W$ and $L = 7$ and $W = 9$.

56. Find M if $M = 4Q + 5T$ and $Q = 5$ and $T = 8$.

57. Find F if $F = 9C/5 + 32$ and $C = 10$.

58. Find F if $F = 9C/5 + 32$ and $C = 45$.

59. Evaluate $D = GT^2$ if $G = 3$ and $T = 2$.

60. Evaluate $D = GT^2$ if $G = 10$ and $T = 1$.

61. Evaluate $L = 4X + R$ if $X = 7$ and $R = 5$.

62. Evaluate $L = 4X + R$ if $X = 50$ and $R = 100$.

SKILLSFOCUS (Section 1.3, 1.4) *Solve each equation by trial.*

63. $x + 4 = 11$ **64.** $y + 9 = 17$ **65.** $t - 6 = 2$ **66.** $10 - y = 6$

EXTEND YOUR THINKING ▶▶▶▶
▶TROUBLESHOOT IT

Find and correct the error.

67. 4 plus 3 times $5 = (4 + 3) \cdot 5 = (7) \cdot 5 = 35$.

68. 20 divided by the product 2 times $5 = 20 \div 2 \cdot 5 = 10 \cdot 5 = 50$.

69. Evaluate $N(P + 3)$ when $N = 5$ and $P = 8$.

$$N(P + 3) = 5 \cdot (8 + 3) = 40 + 3 = 43$$

70. Find W in $W = 3R + 5T$ if $R = 3$ and $T = 5$.

$$W = 3R + 5T = 3 \cdot 5 + 5 \cdot 3 = 15 + 15 = 30$$

WRITING TO LEARN ▶▶▶▶

71. Explain in your own words how to change the following into an arithmetic expression. Then simplify it.

Ten increased by five times the sum of two and six.

72. Explain how to evaluate $A + 5B$ when $A = 3$ and $B = 8$. Show each step you use. Write the reason for each step.

2.6 SOLVING EQUATIONS USING OPPOSITE OPERATIONS

OBJECTIVES	NEW VOCABULARY
1 Solve $x + a = b$ for x (Section 1.3, Objective 2).	equation
2 Solve $x - a = b$ for x (Section 1.4, Objective 3).	solution
3 Solve $ax = b$ for x (Section 1.7, Objective 4).	
4 Solve $x/a = b$ for x (Section 2.1, Objective 5).	

In the last chapter you learned to solve equations by trial. But many equations, such as $x + 267 = 703$, are not easily solved by trial. In this section you take the next step and learn to solve equations using opposite operations. A more formal discussion of equation solving will be given in Chapter 12.

From Chapter 1, an **equation** is a statement that two quantities are equal.

$$x + 3 = 8$$

left side of the equation ⟵ ⟶ right side of the equation

The equal sign makes a statement of balance.
What number do you replace the letter x with
so the left side equals the right side?
This number is called the **solution**
to the equation.

1 Solving $x + a = b$ Solve $x + 3 = 8$ by trial. Since $5 + 3 = 8$, the solution is $x = 5$.

You get the same solution using opposite operations. In the equation $x + 3 = 8$, 3 is being added to x. Perform the opposite operation. The opposite of addition is subtraction.

Solve the addition equation

$$x + 3 = 8$$
$$x + 3 - 3 = 8 - 3$$
$$x + 0 = 8 - 3$$

In algebra, you include this step. Subtract 3 from each side to keep the equation balanced.

by writing the related subtraction equation.

$$x = 8 - 3$$
$$x = 5$$

This is the same solution as by trial. To check, replace x with 5 in the original equation and see if you get a true statement.

Check:
$$x + 3 = 8$$
$$5 + 3 = 8 \quad \text{True}$$

Therefore, $x = 5$ is the correct solution.

> The solution to the addition equation $x + a = b$ is the related subtraction $x = b - a$.

EXAMPLE 1   Solve each addition equation by writing the related subtraction.

a. Solve $x + 7 = 9$.
$$x = 9 - 7$$
$$x = 2$$
Check: $x + 7 = 9$
$$2 + 7 = 9$$

b. Solve $x + 267 = 703$.
$$x = 703 - 267$$
$$x = 436$$
Check: $x + 267 = 703$
$$436 + 267 = 703 \quad \blacksquare$$

Solve the addition equation by writing the related subtraction.

1. $x + 16 = 41$ **2.** $y + 415 = 692$

2 Solving $x - a = b$

Solve $x - 2 = 5$ by trial. Since $7 - 2 = 5$, the solution is $x = 7$.

You get the same solution by using opposite operations. In the equation $x - 2 = 5$, 2 is being subtracted from x. Perform the opposite operation. The opposite of subtraction is addition.

Solve the subtraction equation

$x - 2 = 5$

$x - 2 + 2 = 5 + 2$ In algebra, you include this step. Add 2 to each side to keep
$x + 0 = 5 + 2$ the equation balanced.

by writing the related addition equation.

$x = 5 + 2$
$x = 7$

This is the same solution as by trial. To check, replace x with 7 in the original equation and see if you get a true statement.

Check: $x - 2 = 5$
                $7 - 2 = 5$ True

Therefore, $x = 7$ is the correct solution.

> The solution to the subtraction equation $x - a = b$ is the related addition $x = b + a$.

EXAMPLE 2

Solve each subtraction equation by writing the related addition.

a. Solve $x - 9 = 5$
$x = 5 + 9$
$x = 14$
Check: $x - 9 = 5$
$14 - 9 = 5$

b. Solve $x - 83 = 97$
$x = 97 + 83$
$x = 180$
Check: $x - 83 = 97$
$180 - 83 = 97$ ∎

► **You Try It** Solve the subtraction equation by writing the related addition.

3. $x - 13 = 51$ **4.** $w - 146 = 75$

3 Solving $ax = b$

Solve $5x = 30$ by trial. Recall, $5x$ means $5 \cdot x$. Since $5 \cdot 6 = 30$, the solution is $x = 6$.

You get the same solution by using opposite operations. In the equation $5x = 30$, 5 is being multiplied times x. Perform the opposite operation. The opposite of multiplication is division.

Solve the multiplication equation

$$5x = 30$$

$$\frac{5 \cdot x}{5} = \frac{30}{5}$$

In algebra, you include this step. Divide both sides by 5 to keep the equation balanced. Note, 5/5 = 1.

$$1 \cdot x = \frac{30}{5}$$

by writing the related division equation.

$$x = \frac{30}{5}$$

$$x = 6$$

This is the same solution as by trial. To check, replace x with 6 in the original equation and see if you get a true statement.

Check:
$$5x = 30$$
$$5 \cdot 6 = 30 \qquad \text{True}$$

Therefore, $x = 6$ is the correct solution.

> The solution to the multiplication equation $ax = b$ is the related division $x = \dfrac{b}{a}$.

EXAMPLE 3   Solve each multiplication equation by writing the related division.

a. Solve $7x = 28$

$$x = \frac{28}{7}$$

$$x = 4$$

Check: $7x = 28$

$7 \cdot 4 = 28$

b. Solve $29y = 1,827$

$$y = \frac{1,827}{29}$$

$$y = 63$$

Check: $29y = 1,827$

$29 \cdot 63 = 1,827$ ∎

▶ **You Try It**   Solve each multiplication equation by writing the related division.

5. $9x = 45$ **6.** $31z = 1,767$

4 Solving $\dfrac{x}{a} = b$ Solve the equation $\dfrac{x}{4} = 6$ by trial. Since $\dfrac{24}{4} = 6$, the answer is $x = 24$.

You get the same solution by using opposite operations. In the equation $\dfrac{x}{4} = 6$, x is being divided by 4. Perform the opposite operation. The opposite of division is multiplication.

Solve the division equation

$$\frac{x}{4} = 6$$

$$\frac{4}{1} \cdot \frac{x}{4} = 4 \cdot 6$$

In algebra, you include this step. Multiply both sides by $4 = 4/1$ to keep the equation balanced.

$$1 \cdot x = 4 \cdot 6$$

Note, $\frac{4}{1} \cdot \frac{x}{4} = \frac{4 \cdot x}{4} = 1 \cdot x$.

by writing the related multiplication equation.

$$x = 4 \cdot 6$$

$$x = 24$$

This is the same solution as by trial. To check, replace x with 24 in the original equation and see if you get a true statement.

Check:

$$\frac{x}{4} = 6$$

$$\frac{24}{4} = 6 \qquad \text{True}$$

Therefore, $x = 24$ is the correct solution.

> The solution to the division equation $\frac{x}{a} = b$ is the related multiplication $x = a \cdot b$.

EXAMPLE 4   Solve each division equation by writing the related multiplication.

a. Solve $\dfrac{x}{8} = 5$

$$x = 8 \cdot 5$$

$$x = 40$$

Check: $\dfrac{x}{8} = 5$

$$\frac{40}{8} = 5$$

b. Solve $\dfrac{x}{12} = 16$

$$x = 12 \cdot 16$$

$$x = 192$$

Check: $\dfrac{x}{12} = 16$

$$\frac{192}{12} = 16 \quad \blacksquare$$

▶ **You Try It**   Solve each division equation by writing the related multiplication.

7. $\dfrac{x}{5} = 10$

8. $\dfrac{t}{34} = 19$

▶ **Answers to You Try It**   **1.** $x = 25$ **2.** $y = 277$ **3.** $x = 64$ **4.** $w = 221$
5. $x = 5$ **6.** $z = 57$ **7.** $x = 50$ **8.** $t = 646$

SECTION 2.6 EXERCISES

1 **2** *Solve each equation by writing the related operation, and check.*

1. $x + 7 = 8$ **2.** $x + 3 = 7$ **3.** $x + 2 = 4$

4. $x + 6 = 6$ **5.** $x - 4 = 3$ **6.** $x - 2 = 0$

7. $x - 6 = 3$ **8.** $x - 1 = 8$ **9.** $x + 14 = 22$

10. $y + 33 = 40$ **11.** $w - 12 = 10$ **12.** $t - 35 = 6$

13. $k + 67 = 112$ **14.** $h + 30 = 60$ **15.** $71 + s = 100$

16. $154 + v = 243$ **17.** $p + 412 = 700$ **18.** $c + 1,324 = 4,239$

19. $x - 580 = 732$ **20.** $v - 3,127 = 491$ **21.** $w - 900 = 700$

22. $j - 401 = 399$ **23.** $x + 407 = 600$ **24.** $c + 4,205 = 7,852$

25. $v + 37 = 803$ **26.** $y + 347 = 1,100$ **27.** $z - 655 = 1,345$

28. $n - 990 = 990$ **29.** $s + 515 = 515$ **30.** $a + 64 = 64$

31. $t - 918 = 0$ **32.** $x - 41 = 0$

3 **4** *Solve each equation by writing the related operation, and check.*

33. $7x = 14$ **34.** $3x = 24$ **35.** $6x = 42$

36. $8x = 40$ **37.** $9x = 63$ **38.** $4x = 40$

39. $\frac{x}{3} = 15$

40. $\frac{x}{8} = 24$

41. $\frac{x}{5} = 45$

42. $\frac{x}{9} = 30$

43. $\frac{x}{6} = 25$

44. $\frac{x}{10} = 9$

45. $\frac{x}{15} = 24$

46. $\frac{y}{20} = 12$

47. $8x = 96$

48. $11k = 176$

49. $23v = 644$

50. $120x = 600$

51. $347x = 75{,}993$

52. $1{,}450x = 43{,}500$

53. $\frac{x}{5} = 36$

54. $\frac{v}{8} = 40$

55. $\frac{n}{16} = 45$

56. $\frac{x}{225} = 36$

57. $\frac{y}{450} = 2{,}560$

58. $\frac{z}{491} = 578$

59. $63x = 0$

60. $725v = 725$

61. $\frac{h}{58} = 0$

62. $\frac{x}{84} = 1$

Solve each equation by writing the related operation, and check.

63. $x + 47 = 351$

64. $z + 400 = 620$

65. $h - 72 = 346$

66. $t - 80 = 145$

67. $7x = 574$

68. $15y = 960$

69. $\frac{p}{35} = 28$

70. $\frac{s}{9} = 685$

71. $36 + x = 58$

72. $86 + v = 423$ **73.** $72x = 2{,}232$ **74.** $8x = 448$

75. $t - 57 = 271$ **76.** $h - 672 = 672$ **77.** $45x = 900$

78. $600y = 1{,}800$ **79.** $\dfrac{t}{85} = 60$ **80.** $\dfrac{v}{145} = 22$

SKILLSFOCUS (Section 1.9) *Write in exponential notation.*

81. $3 \times 3 \times 3$ **82.** 6×6 **83.** $5 \times 5 \times 5 \times 5 \times 5 \times 5$

EXTEND YOUR THINKING ▶▶▶▶

▶ TROUBLESHOOT IT

Find and correct the error.

84. Solve: $x - 6 = 9$
$$x = 9 - 6$$
$$x = 3$$

85. Solve: $\dfrac{x}{8} = 8$
$$x = 8 \div 8$$
$$x = 1$$

WRITING TO LEARN ▶▶▶▶

86. Explain each step you would use to solve $z - 29 = 47$ using related operations.

87. Explain each step you would use to solve $6x = 54$ using related operations.

CALCULATOR TIPS

The Order of Operations   Evaluate $3 + 2 \times 5$ using your calculator. This expression has two operations. Apply the order of operations. Multiplication must be performed before addition.

$$2 \;\boxed{\times}\; 5 \;\boxed{=}\; \boxed{+}\; 3 \;\boxed{=}\; \boxed{\qquad\quad 13}$$

$\quad\quad\quad$└Calculator displays 10.

Some calculators have the order of operations built in. Your calculator does if you can enter $3 + 2 \times 5$ directly, press $\boxed{=}$, and get 13 on the display. Any other answer is incorrect, and indicates that your calculator does not have a built-in order of operations.

In this example, calculators without the order of operations built in allow you to eliminate the first $\boxed{=}$ keypress and get the same answer.

$$2 \;\boxed{\times}\; 5 \;\boxed{+}\; 3 \;\boxed{=}\; \boxed{\qquad\quad 13}$$

Display shows 10.┘

The calculator can be used to evaluate expressions and formulas containing letters, numbers, and operations. For example, evaluate $6E^2$ when $E = 3$.

$$6E^2 = 6 \cdot E \cdot E \qquad \text{Replace } E \text{ with 3.}$$
$$= 6 \;\boxed{\times}\; 3 \;\boxed{\times}\; 3 \;\boxed{=}\; 54$$

Give the answer without using a calculator.

1. $5 \;\boxed{\times}\; 8 \;\boxed{=}\; \boxed{+}\; 6 \;\boxed{=}$

2. $8 \;\boxed{-}\; 5 \;\boxed{=}\; \boxed{\times}\; 4 \;\boxed{=}$

3. $6 \;\boxed{\times}\; 2 \;\boxed{=}\; \boxed{\times}\; 3 \;\boxed{=}\; \boxed{\div}\; 9 \;\boxed{=}$

4. $60 \;\boxed{\div}\; 30 \;\boxed{=}\; \boxed{\times}\; 2 \;\boxed{=}$

Evaluate each expression or formula using your calculator.

5. $3 + 9 \times 8$ **6.** $8(15 - 7)$

7. $6T^3$ when $T = 5$.   **8.** $18 + 30/M$ when $M = 6$.

The only working keys on your calculator are 5, 8, $\boxed{+}$, $\boxed{\div}$, and $\boxed{=}$. Using only these keys, find a sequence of keypresses that gives the following answers.

9. 18   **10.** 1   **11.** 2

12. 6   **13.** 33   **14.** 3

VOCABULARY AND MATCHING

New words and phrases introduced in this chapter are shown in the left-hand column. Match each term on the left with the phrase or sentence on the right that best describes it.

A. fraction form for a division _____ add the numbers, divide this sum by how many numbers were added

B. long division form _____ $x + 2 = 5$ and $w/6 = 42$ are examples

C. divisor _____ how many of one number is in another; quotient; repeated subtraction

D. quotient _____ what $t = 5$ is to $4t = 20$

E. dividend _____ can never be a divisor

F. checks long division _____ when this is the divisor, the answer is always the dividend

G. related multiplication _____ what 3 is in $19 \div 4$

H. 1 _____ $12\overline{)96}$

I. remainder _____ what 3 is to 15, or 8 is to 40

J. trial divisor _____ what 36 is in $36 \div 9$ or $6\overline{)36}$

K. average _____ quotient × divisor + remainder = dividend

L. indicate division _____ for $582\overline{)33,271}$, it is 600

M. 0 _____ what $8 \times 5 = 40$ is to $40 \div 5 = 8$

N. order of operations _____ the answer to $28 \div 7$ or $6 \div 2$

O. equations _____ $\dfrac{40}{8}$

P. solution to the equation _____ parentheses first, then exponents, then multiplication and division left to right, then addition and subtraction left to right

REVIEW EXERCISES

2.1 Division and Its Properties; Zero

Write a related multiplication for each division.

1. $6 \div 2 = 3$ **2.** $18/9 = 2$ **3.** $450 \div 15 = 30$

Find each quotient. If not possible, write NP.

4. $16 \div 4$ **5.** $15 \div 15$ **6.** $24 \div 8$ **7.** $60 \div 10$

8. $0/6$ **9.** $48 \div 6$ **10.** $5 \div 0$ **11.** $72 \div 9$

Solve each equation and check your answer.

12. $\dfrac{y}{5} = 5$ **13.** $\dfrac{k}{4} = 8$ **14.** $\dfrac{x}{9} = 3$ **15.** $\dfrac{h}{7} = 10$

Substitute to find the number making the equation true.

16. $\dfrac{s}{3} + 7 = 14$ **a.** $s = 14$ **b.** $s = 21$ **c.** $s = 7$

17. $\dfrac{y}{8} - 5 = 3$ **a.** $y = 64$ **b.** $y = 40$ **c.** $y = 80$

2.2 Division of Whole Numbers and Estimation

18. $56 \div 8$

19. $38 \div 9$

20. $317 \div 9$

21. $562 \div 7$

22. 824/24

23. 5,207/83

24. $45,062 \div 803$

25. $417,498 \div 1,704$

26. $67,000,000 \div 5,000$

27. 56,700/900

28. Estimate $868 \div 290$.

29. Estimate $4,521 \div 173$.

30. Estimate $8,216 \div 38$.

31. Estimate 923/32.

32. There were 1,288 people who paid a total of $23,672 to attend a homecoming. Estimate what each person paid.

33. A homeowner's association said that 181 homeowners must share the cost of a $55,940 sewer line improvement. Estimate the cost per household.

2.3 Applications: Multiplication, Division, and Average

34. Tom makes $1,350 a month. How much does he make in 12 months?

35. A school plans to buy folding chairs. They have $3,528 to spend. If each chair costs $42, how many can they buy?

36. Each week Kate makes $16 per hour for the first 35 hours, and $24 per hour for each hour over 35. She worked 52 hours this week. How much did she make?

37. Harry gets 35 miles to a gallon of gas. He plans to go on a 10,500-mile cross-country vacation. How many gallons of gas will he need?

38. Arnold has $1,870. He plans to purchase 12 disk drives at a price of $132 each. How much money will he have left after the purchase?

39. Hal purchased a riding mower for $1,650. He put $450 down, and will pay the rest in 15 equal monthly payments. Find the monthly payment.

40. What is the average of the following 8 test scores: 65, 89, 77, 84, 90, 100, 78, and 73?

41. Shirley must take 4 tests in her biology class. She scored 96, 100, and 87 on the first three tests. What does she need on the fourth test to have an average of 90?

2.4 Order of Operations

42. $6 + 3 - 4$

43. $9 - 5 + 4 - 1$

44. $6 + 1 \cdot 7$

45. $2 + 3 \cdot 4$

46. $12 \div 2 \cdot 3$

47. $81 \div 9 \cdot 9$

48. $4 \cdot 2 + 7 \div 1$

49. $3 + 5 \cdot 2 + 6$

50. $5^2 + 5^1 + 5^0$

51. $6 \cdot 5^2$

52. $7(4 + 5)$

53. $8 + 2(8 - 3)$

54. $5 \cdot 2^3 \div 4 + 4$

55. $6(5 + 2) + 3(9 - 4) - 5$

56. $3 + 4 \div \sqrt{16} - 2^2$

57. $8 \cdot 5^2 + 70 \div \sqrt{100} + 2(6 - 6)$

2.5 Evaluating Expressions Containing Numbers and Letters

Change each word sentence into an arithmetic expression, then simplify it.

58. What is seven plus ten less eight?

59. Evaluate five plus four times ten.

60. What is nine times one plus forty divided by 8?

61. Evaluate 3 increased by 12 divided by 3 take away two.

62. What is seven times the sum of twenty and four?

63. What is 300 divided by the product of ten and 6?

64. Evaluate the product of 4 and 10 minus the quotient of 20 and 2.

65. Evaluate 70 decreased by 20 times 2.

66. What is six times two squared?

67. What is the product of six times two, squared?

Evaluate the expression using the given values.

68. ab if $a = 3$ and $b = 8$

69. $7x + 2$ if $x = 5$

70. $5hk$ if $h = 10$ and $k = 8$

71. $c^2 + 3$ if $c = 4$

72. $x(2 + 4m)$ if $x = 3$ and $m = 4$

73. $3v + 5z$ if $v = 5$ and $z = 2$

74. Find X if $X = 2RW$ and $W = 10$ and $R = 12$.

75. Find S if $S = P(1 + RT)$ and $P = 100$, $R = 1$, and $T = 2$.

76. Find P if $P = 2L + 2W$ and $L = 22$ and $W = 14$.

77. Find F if $F = 9C/5 + 32$ and $C = 30$.

2.6 Solving Equations Using Opposite Operations

Solve each equation.

78. $x + 5 = 12$

79. $x + 48 = 82$

80. $x - 6 = 2$

81. $x - 74 = 91$

82. $t + 551 = 672$

83. $v + 294 = 600$

84. $k - 591 = 823$

85. $h - 5,704 = 28$

86. $260 + w = 544$

87. $724 + x = 902$

88. $y - 90 = 0$

89. $s + 62 = 62$

Solve each equation.

90. $7x = 63$

91. $15x = 120$

92. $\dfrac{x}{5} = 9$

93. $\dfrac{x}{12} = 17$

94. $142x = 3,976$

95. $500v = 0$

96. $\dfrac{n}{120} = 5$

97. $\dfrac{h}{692} = 20$

98. $900 = 30p$

99. $18,055 = 785x$

100. $600 = \dfrac{m}{250}$

101. $28 = \dfrac{x}{12}$

Allow yourself 50 minutes to complete this test. Write the work for each problem. When done, check your answers. Rework each problem solved incorrectly.

1. $486 \div 18$

2. $15,619/568$

3. $480,000 \div 1,600$

4. Write the related multiplication for $24 \div 6 = 4$.

Simplify each expression.

5. $30 \div 6 \cdot 5$

6. $30 \cdot 2 + 10 \div 5$

7. $6 \cdot 7 + 4 \cdot 2 - 1$

8. There are 364 senior citizens going on a bus trip.
   a. How many buses are needed if 42 seniors can be seated on each bus?

   b. How many seniors will be on the partially filled bus?

9. Find the average of the seven numbers 5, 2, 0, 7, 9, 2, and 3.

10. $4 \div 2 + 60 \cdot 2$

11. $4 \cdot 3^2 + 6(9 - 5)$

12. Evaluate $3ST$ if $S = 7$ and $T = 15$.

13. Evaluate $2x + 6(y + 2)$ if $x = 4$ and $y = 13$.

14. Find L if $L = 6S^2$ and $S = 8$.

15. Find P if $P = 2L + 2W$ and $L = 120$ and $W = 45$.

16. Solve $24x = 408$.

17. Solve $x + 87 = 416$.

18. Solve $w - 603 = 499$.

19. Solve $\dfrac{x}{28} = 72$.

Explain the meaning of each term. Use your own examples.

20. divisor

21. check for long division

22. average

1. Write 4,705 in expanded notation.

2. Write in standard notation: seven million, four hundred eighty-two thousand, twenty-six.

3. Add.   4,895
          7,068
       +   571

4. Subtract.   4,076
            −   694

5. Multiply.   483
            ×   67

6. Multiply.   1,000
            ×   452

7. An elevator contains 5 people weighing 175, 56, 241, 115, and 82 pounds, respectively.
   a. Estimate the weight in the elevator.

   b. What is the actual weight in the elevator?

8. A business had $42,569 to pay salaries. Thirty-two employees were each paid $1,240. How much money was left?

9. Write $3 \cdot 3 \cdot 3 \cdot 3 \cdot 3$ in exponential notation.

10. Write 2^4 in standard notation.

11. Divide. $24\overline{)7,392}$

12. Divide. $41\overline{)973}$

13. Simplify $40 \div 10 \cdot 2$.

14. Simplify $4 + 6 \div 2$.

15. Find the average of 14, 19, 16, 16, 20, and 17.

16. Next Saturday, 578 people will be transported to a ball game by bus. One bus holds 38 people.
    a. How many buses are needed?

    b. How many people will be on the last bus?

17. Evaluate $3a + 2b$ when $a = 6$ and $b = 10$.

18. Find d in $d = 16g^2$ when $g = 5$.

Solve each equation.

19. $12 + x = 29$

20. $n - 56 = 118$

21. $6p = 108$

22. $\dfrac{w}{15} = 42$

Explain the meaning of each term. Use your own examples.

23. multiple

24. base

Multiplication and Division of Fractions

CONSTRUCTION

A truck can deliver $6\frac{1}{4}$ tons of stone in one trip. How many trips are needed to deliver 85 tons of stone? To solve this problem, find how many $6\frac{1}{4}$-ton loads are contained in 85 tons. This is written

$$85 \div 6\frac{1}{4}.$$

In this chapter you will see how to solve this problem. You will also learn how to correctly interpret the answer.

In this chapter and the next you will study a new kind of number, the fraction. Fractions are used in recipe books, in carpentry, with tools to mark wrench and drill bit sizes, in stock quotations on the financial page of the newspaper, and in other areas where a part of a whole is used. You will learn how the operations applied to whole numbers in the last two chapters can also be applied to fractions. You will solve a variety of real world applications so you can see for yourself the role fractions play in everyday life.

Work through each section in this chapter. Use this test to identify topics you are not familiar with. These topics may require additional study. You may *not* use a calculator.

3.1

1. How many ninths are in one whole?

2. Draw and shade a figure to represent $\frac{5}{8}$.

3.2

3. Change $3\frac{4}{7}$ to an improper fraction.

4. Change $10\frac{1}{5}$ to an improper fraction.

5. Change $\frac{31}{6}$ to a mixed number.

6. Change $\frac{83}{15}$ to a mixed number.

3.3

7. Find all of the divisors of 48.

8. Prime factor 30.

9. Prime factor 42.

3.4

10. Reduce $\frac{25}{40}$.

11. Reduce $3\frac{12}{10}$.

12. Reduce $\frac{21}{105}$.

3.5

13. Multiply $\frac{5}{8} \times \frac{4}{9}$.

14. Multiply $6 \times \frac{5}{3}$.

15. Multiply $3\frac{3}{4} \times 2\frac{14}{15}$.

16. Multiply $2\frac{2}{9} \times 12$.

3.6

17. Divide $\frac{4}{9} \div \frac{1}{3}$.

18. Divide $\frac{7}{10} \div \frac{21}{30}$.

19. Divide $4\frac{1}{2} \div 1\frac{2}{5}$.

20. Find $5\frac{3}{7} \div 2$.

21. Solve $\frac{4}{5}w = \frac{1}{4}$.

3.7

22. A mower uses $3\frac{1}{4}$ gallons of gas each week. How much gas is needed to operate the mower for the 13 weeks of summer?

23. How many $\frac{2}{3}$ are in $5\frac{3}{5}$?

24. What is $\frac{7}{10}$ of $280?

25. How many $4\frac{3}{8}$-inch wires can be cut from a 51-inch length of wire?

VOCABULARY *Explain the meaning of each term. Use your own examples.*

26. denominator

27. improper fraction

28. composite number

3.1 THE MEANING OF A FRACTION

OBJECTIVE

Define a fraction (Section 2.1, Objectives 1,4).

NEW VOCABULARY

fraction   denominator

numerator   terms

fraction bar

Defining a Fraction

In Chapter 2 you saw that the division $8 \div 4$ can be written in *fraction form* as $\frac{8}{4}$. $8 \div 4 = \frac{8}{4} = 2$, a whole number.

In like manner, the division $3 \div 4$ can be written in fraction form as $\frac{3}{4}$. But the answer to this division is not a whole number. The number $\frac{3}{4}$ is a fraction. It is a part of a whole. To understand what $\frac{3}{4}$ means, cut a pie into 4 equal pieces.

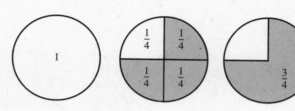

whole pie   Cut into 4 equal pieces. Shade 3 of them.
The shaded area is $\frac{3}{4}$ of the pie.

Each piece represents 1 part out of 4, or $\frac{1}{4}$ of the pie. If you take 3 pieces, you are taking 3 parts out of 4, or $\frac{3}{4}$ of the pie.

$$\frac{3}{4} \begin{cases} \text{means divide the whole into} \\ \text{4 equal pieces and take 3 of them} \end{cases}$$

The entire pie is 4 parts out of 4 equal parts, or 1 whole.

$$\begin{matrix} \text{4 out of 4} \\ \text{equal pieces} \end{matrix} = \frac{4}{4} \text{ of a pie} = 1 \text{ whole pie}$$

EXAMPLE 1   Read each fraction and write its meaning.

a. $\frac{2}{3}$ is read "two thirds." It means divide the whole into 3 equal pieces and take 2 of them.

b. $\frac{1}{8}$ is read "one eighth." It means divide the whole into 8 equal parts and take 1 of them. ■

▶ **You Try It**   Read each fraction and write its meaning.

1. $\frac{4}{5}$ **2.** $\frac{1}{6}$

A **fraction** is a division. The fraction $\frac{a}{b} = a \div b$. It means divide the whole into b equal parts, and take a of them.

Numerator, or top number. Tells how many parts to take from the whole.

$\frac{a}{b}$ ← **Fraction bar.** Indicates division.

Denominator, or bottom number. Tells how many equal parts the whole is to be divided into. Since division by 0 is not possible, $b \neq 0$.

a and b are also called the **terms** of the fraction.

EXAMPLE 2   Represent each fraction with a shaded geometric figure.

a. $\frac{3}{5}$ means 3 out of 5 equal parts: $= \frac{3}{5}$

b. $\frac{5}{9}$ means 5 out of 9 equal parts: $= \frac{5}{9}$

c. $\frac{2}{3}$ means 2 out of 3 equal parts: $= \frac{2}{3}$

d. $\frac{0}{4}$ means no shaded parts out of 4 equal parts: $= \frac{0}{4} = 0$

e. $\frac{6}{6}$ means 6 shaded parts out of 6: $= \frac{6}{6} = 1$ whole   ■

CAUTION In Section 2.1 you learned you cannot divide by 0. Since $\frac{4}{0} = 4 \div 0$, this means 0 may not be in the denominator.

 versus $\frac{0}{4}$

This means divide a pie into 0 equal pieces and take 4 of them. This is meaningless.

From Example 2d, $\frac{0}{4}$ means divide a pie into 4 equal pieces, and take 0 of them. So you get $\frac{0}{4} = 0$, or no pieces of the pie. This statement makes sense.

▶ **You Try It**   What fraction of each figure is shaded?

3.

4.

5.

6.

Draw and shade a geometric figure to represent each fraction.

7. $\frac{1}{5}$ 8. $\frac{3}{8}$ 9. $\frac{4}{4}$ 10. $\frac{0}{3}$

EXAMPLE 3   On a team of 18 sprinters, 5 are ranked as Olympic hopefuls.

a. What fraction of the team is this?

5 sprinters are ranked as Olympic hopefuls out of 18 sprinters $= \frac{5}{18}$ of the team.

b. What fraction of the team is not ranked?

Of 18 team members, 5 are ranked. The rest, $18 - 5 = 13$, are not ranked. $\frac{13}{18}$ of the team is not ranked.   ■

▶ **You Try It**   **11.** Of 12 tennis balls, 5 are yellow.
   a. What fraction are yellow?
   b. What fraction are not yellow?

SECTION 3.1 EXERCISES

1. How many thirds are in one whole?

2. How many eighths are in one whole?

3. How many hundredths are in one whole?

4. How many tenths are in one whole?

5. Divide a pie into 7 equal pieces.

   a. Each piece represents what fraction of the total pie?

   b. Seven pieces represent what fraction of the pie?

6. A treasure is divided into 12 equal piles.

   a. Four piles represent what fraction of the treasure?

   b. One pile represents what fraction of the treasure?

Write the fraction that describes the portion of the figure that is shaded.

7.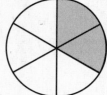

8.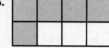

9.

10.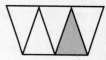

11.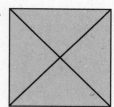

12.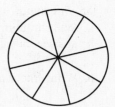

13.

14.

Write the fraction in words. Draw and shade a figure that represents the fraction.

15. $\dfrac{4}{5}$

16. $\dfrac{1}{3}$

17. $\dfrac{12}{12}$

18. $\dfrac{3}{7}$

19. $\dfrac{0}{6}$

20. $\dfrac{11}{15}$

21. $\dfrac{1}{10}$

22. $\dfrac{9}{16}$

Solve each problem by writing the appropriate fraction.

23. In a tennis tournament, the first and second place finishers will split all of the prize money. First place will get five equal parts of the money. Second place will get two equal parts.

    a. How many parts will the prize money be divided into?

    b. What fraction will the first place finisher get?

    c. What fraction will second place get?

    d. What fraction will third place get?

24. Ed picked 38 bushels of apples. He sold 26 bushels to the local market. He split the remaining bushels between his two aunts Alice and Brie.

    a. What fraction of the apples were sold to the market?

    b. What fraction did Alice receive?

    c. What fraction did Brie get?

    d. What fraction of the apples were not sold to the market?

25. The Earth is seven parts water and three parts land.

    a. What fraction of the Earth is water?

    b. What fraction of the Earth is land?

26. Of 24 children in a class, eight are boys.

    a. What fraction of the class are boys?

    b. What fraction of the class are girls?

27. Toni's monthly take-home pay is $2,145. Her monthly mortgage payment is $861.

    a. What fraction of her take-home pay is her mortgage payment?

    b. What fraction of her take-home pay is left over?

28. You answered 16 out of 20 questions correctly on an exam.

    a. What fraction of the questions did you answer incorrectly?

    b. What fraction did you answer correctly?

SKILLSFOCUS (Section 2.4) *Simplify each expression.*

29. $4 \times 6 + 5$ **30.** $7 + 3 \times 4$ **31.** $(5 \times 3 + 1) \div 2$ **32.** $(6 + 2 \times 6) \div 9$

WRITING TO LEARN ▶▶▶▶

33. Explain the difference between the numerator and the denominator.

34. Explain the meaning of $\frac{3}{5}$. Use a cake in your explanation.

35. Explain why $\frac{0}{2}$ makes sense but $\frac{2}{0}$ does not. Use a jar of 100 pennies in your explanation.

3.2 PROPER AND IMPROPER FRACTIONS; MIXED NUMBERS

OBJECTIVES

1. Define proper and improper fractions.
2. Define mixed numbers.
3. Change a mixed number to an improper fraction.
4. Change an improper fraction to a mixed number.

NEW VOCABULARY

proper fraction

improper fraction

mixed number

---

1 Proper and Improper Fractions

A **proper fraction** is any fraction whose numerator is smaller than its denominator.

EXAMPLE 1 Each fraction below is proper.

$$\frac{4}{5}, \frac{1}{9}, \frac{0}{4}, \frac{2}{6}, \frac{5}{8}, \frac{22}{23} \quad \blacksquare$$

Every proper fraction is less than 1 whole.

EXAMPLE 2 $\frac{4}{5}$ is less than 1 whole. As you can see, less than the whole figure is shaded.

$$\frac{4}{5} = $$ 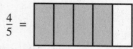 $\blacksquare$

An **improper fraction** is any fraction whose numerator is greater than or equal to its denominator.

EXAMPLE 3 Each fraction below is improper.

$$\frac{5}{3}, \frac{8}{5}, \frac{7}{1}, \frac{4}{4}, \frac{63}{62} \quad \blacksquare$$

Every improper fraction is greater than or equal to 1 whole.

EXAMPLE 4 $\frac{5}{3}$ is greater than 1 whole. As you can see in graphic form, more than one whole is needed to represent it.

$$\frac{5}{3} = \qquad + \qquad$$

shade 3 parts out of   shade 2 parts out of
3 in one whole   3 in a second whole $\blacksquare$

▶ **You Try It** Determine if the fraction is proper or improper.

1. $\frac{5}{7}$ 2. $\frac{10}{9}$ 3. $\frac{9}{2}$ 4. $\frac{6}{11}$ 5. $\frac{5}{5}$ 6. $\frac{85}{86}$

A whole number can be written as an improper fraction by writing it with a denominator of 1.

EXAMPLE 5 Write each whole number as an improper fraction.

a. $3 = \frac{3}{1}$ **b.** $8 = \frac{8}{1}$ **c.** $64 = \frac{64}{1}$ $\blacksquare$

▶ **You Try It**   Write each whole number as an improper fraction.

7. 5   **8.** 24   **9.** 250

2 Mixed Numbers   A **mixed number** is a whole number plus a proper fraction.

EXAMPLE 6   Examples of mixed numbers.

$$3\frac{1}{2}, \quad 1\frac{5}{6}, \quad 37\frac{7}{8}, \quad 2\frac{0}{5} \quad \blacksquare$$

A mixed number is an addition. The plus sign, $+$, is usually omitted when writing a mixed number.

EXAMPLE 7 $3\frac{1}{2}$ means $3 + \frac{1}{2}$.

$$3\frac{1}{2} = 3 + \frac{1}{2} = \boxed{} + \boxed{} + \boxed{} + \boxed{|}$$

$$3 \text{ wholes } + \frac{1}{2} \text{ of a whole} \quad \blacksquare$$

CAUTION	$3\frac{1}{2}$ never means $3 \cdot \frac{1}{2}$.

EXAMPLE 8   **a.** $6\frac{2}{3} = 6 + \frac{2}{3}$ **b.** $4\frac{7}{8} = 4 + \frac{7}{8}$ **c.** $72\frac{1}{2} = 72 + \frac{1}{2}$ $\quad \blacksquare$

▶ **You Try It**   Write each mixed number as an addition.

10. $1\frac{3}{4}$ **11.** $3\frac{1}{5}$ **12.** $39\frac{7}{8}$ **13.** $5\frac{1}{10}$ **14.** $65\frac{12}{17}$

3 Changing a Mixed Number to an Improper Fraction   In order to multiply or divide with mixed numbers, you must first change them to improper fractions.

EXAMPLE 9   Change $2\frac{3}{4}$ to an improper fraction.

$$2\frac{3}{4} = 2 + \frac{3}{4} = 1 + 1 + \frac{3}{4} = \frac{4}{4} + \frac{4}{4} + \frac{3}{4} = \frac{11}{4}$$

$$= \boxed{} + \boxed{} + \boxed{} = \boxplus + \boxplus + \boxplus = \quad \blacksquare$$

Note $1 = \frac{4}{4}$. 2 wholes of 4 quarters each ($2 \times 4 = 8$ quarters) plus 3 quarters of a third whole gives $2 \times 4 + 3 = 11$ shaded quarters. A shortcut way to write this is

$$2\frac{3}{4} = 2 \overset{+}{\underset{\times}{}} \frac{3}{4} = \frac{2 \times 4 + 3}{4} = \frac{11}{4}.$$

> To change a mixed number to an improper fraction
>
> **1.** Multiply the denominator by the whole number.
> **2.** To this add the numerator.
> **3.** Write this answer over the denominator.

EXAMPLE 10   Change $5\frac{3}{8}$ to an improper fraction.

Step 1: Multiply the denominator 8 by the whole number 5: $5 \times 8 = 40$.

Step 2: To this add the numerator 3: $40 + 3 = 43$.

Step 3: Write 43 over the denominator 8: $\frac{43}{8}$.

In summary, $5\frac{3}{8} = \frac{5 \times 8 + 3}{8} = \frac{43}{8}$. ■

EXAMPLE 11   Change each mixed number to an improper fraction.

a. $4\frac{2}{5} = \frac{4 \times 5 + 2}{5} = \frac{22}{5}$

b. $1\frac{7}{12} = \frac{1 \times 12 + 7}{12} = \frac{19}{12}$

c. $9\frac{13}{16} = \frac{9 \times 16 + 13}{16} = \frac{157}{16}$ ■

▶ **You Try It**   Change each mixed number to an improper fraction.

15. $2\frac{5}{6}$ **16.** $5\frac{3}{7}$ **17.** $10\frac{8}{9}$ **18.** $1\frac{1}{2}$

4 Changing an Improper Fraction to a Mixed Number

If an answer to a problem is an improper fraction, you are often asked to write the answer as a mixed number.

EXAMPLE 12   Change $\frac{8}{3}$ to a mixed number.

$\frac{8}{3}$ means divide the whole into 3 equal pieces, and take 8 of them. This means you need more than 1 whole.

$$\frac{8}{3} = \frac{3}{3} + \frac{3}{3} + \frac{2}{3} = 1 + 1 + \frac{2}{3} = 2\frac{2}{3}$$

2 wholes $\frac{2}{3}$ of a whole

$\frac{8}{3}$ means $3\overline{)8}$.
Divide to get the same answer.
Write the remainder 2 over the divisor 3.

$$3\overline{)\begin{array}{r} 2 \\ 8 \\ -6 \\ \hline 2 \end{array}} = 2\frac{2}{3}$$

■

EXAMPLE 13   Change $\frac{19}{5}$ to a mixed number.

Step 1: Divide the numerator 19
   by the denominator 5. $\longrightarrow$

$$5 \overline{)\ 19} \quad 3 \text{ R } 4 = 3\frac{4}{5}$$

Step 2: Write the remainder 4
   over the denominator 5.

$$\begin{array}{r} 3 \text{ R } 4 = 3\frac{4}{5} \\ 5\overline{)\ 19\ } \\ -15 \\ \hline 4 \end{array}$$

Check: $\quad 3\frac{4}{5} = \frac{3 \times 5 + 4}{5} = \frac{19}{5}$ ∎

EXAMPLE 14   Change each improper fraction to a mixed number.

a. $\frac{13}{3} = \quad 3\overline{)13}^{\,4 \text{ R } 1} = 4\frac{1}{3}$

b. $\frac{58}{9} = \quad 9\overline{)58}^{\,6 \text{ R } 4} = 6\frac{4}{9}$

c. $\frac{126}{7} = \quad 7\overline{)126}^{\,18 \text{ R } 0} = 18\frac{0}{7}$ or 18 ∎

▶ **You Try It**   Change each improper fraction to a mixed number.

19. $\frac{15}{8}$ **20.** $\frac{7}{6}$ **21.** $\frac{29}{4}$ **22.** $\frac{137}{10}$

EXAMPLE 15   A farmer split 107 acres of land equally among his four children. How much land did each child receive?

To split 107 acres into 4 equal pieces, divide 107 by 4.

$$4\overline{)107}^{\,26 \text{ R } 3} = 26\frac{3}{4} \text{ acres per child} \quad ∎$$

▶ **You Try It**   **23.** A mother splits 16 cookies equally among 3 children. How many does each child get? Express your answer as a mixed number.

▶ **Answers to You Try It**   1. proper   2. improper   3. improper

4. proper   5. improper   6. proper   7. $\frac{5}{1}$ 8. $\frac{24}{1}$ 9. $\frac{250}{1}$

10. $1 + \frac{3}{4}$ 11. $3 + \frac{1}{5}$ 12. $39 + \frac{7}{8}$ 13. $5 + \frac{1}{10}$ 14. $65 + \frac{12}{17}$ 15. $\frac{17}{6}$

16. $\frac{38}{7}$ 17. $\frac{98}{9}$ 18. $\frac{3}{2}$ 19. $1\frac{7}{8}$ 20. $1\frac{1}{6}$ 21. $7\frac{1}{4}$ 22. $13\frac{7}{10}$ 23. $5\frac{1}{3}$ cookies

1 **2** *Identify each fraction as proper, improper, or mixed. Then draw and shade a geometric figure representing each fraction.*

1. $\dfrac{10}{9}$
2. $\dfrac{1}{5}$
3. $4\dfrac{1}{2}$
4. $\dfrac{17}{12}$
5. $1\dfrac{5}{6}$

6. $10\dfrac{5}{8}$
7. $\dfrac{25}{6}$
8. $\dfrac{3}{1}$
9. $7\dfrac{3}{4}$
10. $\dfrac{7}{8}$

3 *Change the mixed number to an improper fraction.*

11. $2\dfrac{1}{2}$
12. $1\dfrac{2}{3}$
13. $2\dfrac{5}{7}$
14. $1\dfrac{7}{8}$
15. $3\dfrac{2}{3}$
16. $4\dfrac{5}{6}$

17. $7\dfrac{1}{3}$
18. $5\dfrac{3}{4}$
19. $8\dfrac{1}{2}$
20. $6\dfrac{4}{5}$
21. $12\dfrac{5}{9}$
22. $10\dfrac{2}{5}$

23. $3\dfrac{4}{21}$
24. $28\dfrac{2}{11}$
25. $14\dfrac{7}{10}$
26. $5\dfrac{0}{6}$
27. $3\dfrac{8}{8}$
28. $11\dfrac{1}{4}$

29. $30\dfrac{1}{6}$
30. $15\dfrac{5}{9}$
31. $7\dfrac{1}{4}$
32. $2\dfrac{3}{5}$
33. $3\dfrac{2}{7}$
34. $4\dfrac{3}{8}$

4 *Change each improper fraction to a mixed number.*

35. $\dfrac{5}{4}$
36. $\dfrac{7}{3}$
37. $\dfrac{5}{2}$
38. $\dfrac{9}{5}$
39. $\dfrac{11}{4}$
40. $\dfrac{4}{3}$

41. $\dfrac{13}{2}$
42. $\dfrac{8}{5}$
43. $\dfrac{23}{6}$
44. $\dfrac{17}{8}$
45. $\dfrac{39}{4}$
46. $\dfrac{47}{6}$

47. $\dfrac{31}{3}$
48. $\dfrac{24}{7}$
49. $\dfrac{73}{10}$
50. $\dfrac{57}{5}$
51. $\dfrac{96}{8}$
52. $\dfrac{91}{12}$

53. $\dfrac{60}{11}$
54. $\dfrac{171}{20}$
55. $\dfrac{91}{14}$
56. $\dfrac{101}{16}$
57. $\dfrac{41}{8}$
58. $\dfrac{31}{9}$

59. Ellen paid $325 for 8 shares of stock. How much did one share cost? Express your answer as a mixed number.

60. Express $62 \div 5$ as a mixed number.

61. Find the average of the numbers 6, 8, 3, 12, 5, and 13.

62. What is the average of the test scores 81, 77, and 90?

63. An 18-foot board is cut into 7 equal pieces. How long is each piece?

64. Write $\dfrac{215}{19}$ as a mixed number.

65. A small car traveled 317 miles on 6 gallons of gas. How many miles did the car travel on each gallon?

66. Amanda cycled 83 miles in 5 hours. How many miles did she cycle in one hour?

SKILLSFOCUS (Section 1.5) *Round to the indicated place.*

67. 473 (tens)

68. 6,492 (hundreds)

69. 4,997 (tens)

70. 560,700 (thousands)

EXTEND YOUR THINKING ▶▶▶▶
▶SOMETHING MORE

71. **The Cats** An ancient Egyptian problem dating from 1700 B.C. can be phrased as follows: In each of seven houses there are seven cats. Each cat kills seven mice. Each mouse would have eaten seven ears of corn. Each ear of corn would have produced seven ounces of grain. How many pounds of grain are saved by the cats?

▶TROUBLESHOOT IT
Find and correct the error.

72. $5\dfrac{3}{3} = 5 \cdot 1 = 5$

73. $\dfrac{17}{5} = 5\overline{)17}\;\;\begin{array}{r} 3 \\ \hline 17 \\ -15 \\ \hline 2 \end{array} = 5\dfrac{2}{3}$

WRITING TO LEARN ▶▶▶▶

74. Explain the difference between proper fractions, improper fractions, and mixed numbers. Use your own examples.

75. Explain how to change $\dfrac{19}{6}$ to a mixed number.

76. Explain how to change $2\dfrac{4}{9}$ to an improper fraction.

77. If one dollar is the whole, is a proper fraction less than, equal to, or greater than a dollar? How about an improper fraction? Explain.

3.3 PRIME NUMBERS AND PRIME FACTORING

OBJECTIVES

1 Find all the divisors of a counting number (Sections 1.7, 2.1).

2 Determine if a number is prime or composite.

3 Use the divisibility rules.

4 Prime factor a counting number.

NEW VOCABULARY

divisor	2-3-5-7 rule
divisible	prime factor
prime number	prime factorization
composite number	successive division by primes

---

1 Finding the Divisors of a Counting Number

The numbers 3 and 4 are called factors of 12 because $3 \cdot 4 = 12$. The word *factor* is used when referring to a product.

Since $12 \div 3$ has a zero remainder, 3 is called a **divisor** of 12. 4 is also a divisor of 12. 5 is not a divisor of 12 because $12 \div 5$ has a remainder of 2. The word *divisor* is used when referring to a quotient.

Divisors are found in pairs. If a number is a divisor for a counting number, its quotient is also a divisor.

EXAMPLE 1 The number 3 is a divisor of 12 because $12 \div 3 = 4$. Then the quotient 4 is also a divisor of 12 ($12 \div 4 = 3$). ■

Since 3 and 4 are divisors of 12, 12 is said to be **divisible** by 3 and divisible by 4.

The words *factor* and *divisor* are closely related. A factor of a number is also a divisor of the number. A divisor of a number is also a factor of the number.

EXAMPLE 2 The numbers 3 and 5 are factors of 15 because $3 \cdot 5 = 15$. 3 and 5 are also divisors of 15 because

$$15 \div 3 = 5 \quad \text{and} \quad 15 \div 5 = 3. \quad ■$$

> **CAUTION** The word *divisor* can be used in two contexts. In the division $5\overline{)12}$, 5 is called *the divisor*. However, 5 is not *a divisor of* 12 because $12 \div 5$ gives a remainder of 2.

> To find all the divisors of a counting number *n*
>
> 1. Divide *n* by 1, 2, 3, 4, and so on.
> 2. When you get a zero remainder, both that number and its quotient are divisors of *n*.
>   When you get a nonzero remainder, that number is not a divisor of *n*. Skip it. Test the next number.
> 3. Stop when the list of divisors begins to repeat (or when the square of the number being tested exceeds *n*).

EXAMPLE 3 Find all of the divisors of 12.

Since $12 \div 1 = 12$,   1 and 12 are divisors.

Since $12 \div 2 = 6$,   2 and 6 are divisors.

Since $12 \div 3 = 4$,   3 and 4 are divisors.

Since $12 \div 4 = 3$,   stop because the set of divisors is beginning to repeat. The divisors 4 and 3 were found in the previous step.

The divisors of 12 are 1, 2, 3, 4, 6, and 12. ■

EXAMPLE 4 Find all the divisors of 15.

$15 \div 1 = 15,$ 1 and 15 are divisors.

$15 \div 2 = 7 \text{ R } 1,$ 2 is not a divisor. The remainder is not zero.

$15 \div 3 = 5,$ 3 and 5 are divisors.

$15 \div 4 = 3 \text{ R } 3,$ 4 is not a divisor.

$15 \div 5 = 3,$ stop. The divisors are repeating.

The divisors of 15 are 1, 3, 5, and 15. ■

Every counting number (except 1) has at least two divisors, 1 and the number itself. The number 1 is the smallest divisor of every counting number. The largest divisor is the number itself.

EXAMPLE 5 Find all the divisors of 7.

$7 \div 1 = 7,$ 1 and 7 are divisors.

$7 \div 2 = 3 \text{ R } 1,$ 2 is not a divisor.

$7 \div 3,$ stop. Since $3^2 > 7$, 7 has no more divisors.

The divisors of 7 are 1 and 7. ■

▶ **You Try It**   **2.** Find all the divisors of 30.   **3.** Find all the divisors of 13.

2 Prime and Composite Numbers

The number 7 has exactly two divisors, 1 and 7. Therefore, 7 is called a prime number. A **prime number** is any counting number greater than 1 having exactly two divisors, 1 and the number itself. The smallest prime number is 2. There is no largest prime. The first fifteen prime numbers are shown below.

2, 3, 5, 7, 11, 13, 17, 19, 23, 29, 31, 37, 41, 43, 47, . . .

EXAMPLE 6 **a.** 3 is a prime number. Its only divisors are 1 and 3.
b. 17 is a prime number. Its only divisors are 1 and 17.
c. 53 is a prime number. 1 and 53 are its only divisors. ■

If a counting number larger than 1 is not prime, it is called a **composite number**. Every composite number has three or more divisors: 1, itself, and at least one other number in between. The smallest composite number is 4. It has three divisors: 1 and 4, as well as 2. The first fifteen composite numbers are shown below.

4, 6, 8, 9, 10, 12, 14, 15, 16, 18, 20, 21, 22, 24, 25, . . .

EXAMPLE 7 **a.** 15 is a composite number. It has four divisors: 1, 3, 5, and 15.
b. 9 is composite. It has three divisors: 1, 3, and 9.
c. 32 is composite. Its divisors are 1, 2, 4, 8, 16, and 32. ■

The number 1 is neither prime nor composite. It has exactly one divisor, itself.

3 Divisibility Rules

Is 111 a prime number? No, because 3 is a divisor of 111. In fact, $111 \div 3 = 37$. To help you determine if larger numbers are prime or composite, use the following divisibility rules. These rules will be helpful in Chapters 3 and 4 when reducing fractions.

2 is a divisor of a number if its ones digit is even: 0, 2, 4, 6, or 8.

EXAMPLE 8   4, 26, 50, 288, and 5,732 are all divisible by 2 because each number has an even ones digit.   ■

3 is a divisor of a number if 3 is a divisor of the sum of its digits.

EXAMPLE 9
a. 3 is a divisor of 15 because 3 is a divisor of the sum of its digits $1 + 5 = 6$.
b. 3 is a divisor of 174 because 3 is a divisor of the sum of its digits $1 + 7 + 4 = 12$.
c. 3 is not a divisor of 53 because 3 is not a divisor of $5 + 3 = 8$.   ■

5 is a divisor of any number if its ones digit is 0 or 5.

EXAMPLE 10   10, 25, 40, 700, and 8,245 are all divisible by 5 because each number has 0 or 5 in the ones place.   ■

7 is a divisor of a number if 7 is a divisor of the difference between *the number left when the ones digit is crossed out* and *2 times the ones digit.*

EXAMPLE 11
a. 7 is a divisor of 35 because 7 is a divisor of the difference between 3$\cancel{5}$ and $2 \cdot 5 = 2 \cdot 5 - 3 = 10 - 3 = 7$.
b. 7 is a divisor of 476 because 7 is a divisor of the difference between 47$\cancel{6}$ and $2 \cdot 6 = 47 - 2 \cdot 6 = 35$.
c. 7 is not a divisor of 185 because 7 is not a divisor of $18 - 2 \cdot 5 = 8$.
d. 7 is a divisor of 63 because 7 is a divisor of $6 - 2 \cdot 3 = 0$.   ■

The **2-3-5-7 rule** is useful for finding divisors of counting numbers. Simply test the primes 2, 3, 5, and 7 first for divisibility, before trying larger primes.

▶ **You Try It**   Determine if the number is divisible by 2, 3, 5, 7, or none of these.

4. 28   **5.** 39   **6.** 60   **7.** 157

4 Prime Factoring

To **prime factor** a counting number, write it as a product of prime numbers only. Prime factoring will be used in Chapter 4 to help you add and subtract fractions.

EXAMPLE 12   6 is prime factored as $2 \cdot 3$ because $6 = 2 \cdot 3$, and 2 and 3 are both prime numbers.   ■

The expression $2 \cdot 3$ is called the **prime factorization** of 6. Every counting number greater than 1 can be prime factored in only one way, though the order of the factors may differ.

EXAMPLE 13   30 is prime factored as $2 \cdot 3 \cdot 5$ because $30 = 2 \cdot 3 \cdot 5$, and 2, 3, and 5 are prime numbers.

CAUTION   30 is not prime factored as $5 \cdot 6$ because even though $30 = 5 \cdot 6$, 6 is not prime.   ■

> Prime factor a counting number using the method of **successive division by primes**.
>
> 1. Divide the number by the primes 2, 3, 5, 7, 11, and so on, starting with the smallest prime, 2. Only primes giving a zero remainder are used as factors.
> 2. Stop when the quotient is a prime number.
> 3. The prime factorization is the product of all the prime divisors and the final prime quotient.

EXAMPLE 14   Prime factor 20.

First, divide 20 by the smallest prime, 2.
$20 \div 2 = 10$. Write the quotient 10 above 20.

$$\begin{array}{r} 10 \\ 2 \overline{\smash{)}20} \end{array}$$

Since 10 is even, divide by 2 again. $10 \div 2 = 5$.
Write the quotient 5 above 10.
Since the quotient 5 is prime, you are done.

$$\begin{array}{r} 5 \\ 2 \overline{\smash{)}10} \\ 2 \overline{\smash{)}20} \end{array}$$

Multiply the prime divisors (2 and 2), and the final quotient (5), to get the prime factorization of 20.

$$\begin{array}{r} 5 \\ 2 \overline{\smash{)}10} \\ 2 \overline{\smash{)}20} \end{array}$$

20 is prime factored as:
$20 = 2 \cdot 2 \cdot 5$
$= 2^2 \cdot 5$. ■

EXAMPLE 15   Prime factor 105.

2 is not a divisor of 105, because 105 ends in an odd digit. Try the next prime, 3.
3 is a divisor of 105 because 3 divides $1 + 0 + 5 = 6$. $105 \div 3 = 35$.

$$\begin{array}{r} 35 \\ 3 \overline{\smash{)}105} \end{array}$$

3 does not divide the quotient 35.
Try the next prime, 5.
5 divides 35. $35 \div 5 = 7$. Since the quotient 7 is prime, you are done.

$$\begin{array}{r} 7 \\ 5 \overline{\smash{)}35} \\ 3 \overline{\smash{)}105} \end{array}$$

$105 = 3 \cdot 5 \cdot 7$

The prime factorization of $105 = 3 \cdot 5 \cdot 7$. ■

▶ **You Try It**   Prime factor.   **8.** 45   **9.** 72   **10.** 140

EXAMPLE 16   Prime factor 54.

$$\begin{array}{r} 3 \\ 3 \overline{\smash{)}9} \\ 3 \overline{\smash{)}27} \\ 2 \overline{\smash{)}54} \end{array}$$

The prime factorization of 54 is
$54 = 2 \cdot 3 \cdot 3 \cdot 3$
$= 2 \cdot 3^3$. ■

▶ **You Try It**   Prime factor.   **11.** 84   **12.** 270   **13.** 127

▶ **Answers to You Try It**   **1.** 1, 21, 3, 7   **2.** 1, 30, 2, 15, 3, 10, 5, 6
3. 1, 13   **4.** 2, 7   **5.** 3   **6.** 2, 3, 5   **7.** none   **8.** $3^2 \cdot 5$ **9.** $2^3 \cdot 3^2$
10. $2^2 \cdot 5 \cdot 7$ **11.** $2^2 \cdot 3 \cdot 7$ **12.** $2 \cdot 3^3 \cdot 5$ **13.** prime

1 2 3 *Find all of the divisors for the given number. Use the 2-3-5-7 rule as a guide.*
Identify each number as prime or composite.

1. 6	**2.** 9	**3.** 16	**4.** 10
5. 14	**6.** 18	**7.** 13	**8.** 29
9. 26	**10.** 30	**11.** 33	**12.** 42
13. 80	**14.** 100	**15.** 24	**16.** 125
17. 79	**18.** 47	**19.** 142	**20.** 181
21. 51	**22.** 39	**23.** 59	**24.** 75

25. What is the smallest odd prime number?

26. What is the only even prime number?

27. What is the smallest two-digit prime number?

28. What is the smallest prime number?

29. Find the largest two-digit prime number.

30. Find the largest one-digit prime number.

31. What is the smallest composite number?

32. What is the smallest odd composite number?

33. Find the smallest two-digit composite number.

34. Find the largest two-digit composite number.

35. What is the largest one-digit composite number?

36. (True/False) Every composite number is divisible by 2.

37. (True/False) All prime numbers are odd.

38. (True/False) A number cannot be both prime and composite.

4 *Prime factor each number below. Use the 2-3-5-7 rule as a guide.*
If a number is prime, write prime.

39. 4	**40.** 9	**41.** 13	**42.** 29	**43.** 14
44. 18	**45.** 16	**46.** 40	**47.** 33	**48.** 10
49. 80	**50.** 100	**51.** 81	**52.** 42	**53.** 144
54. 225	**55.** 51	**56.** 26	**57.** 66	**58.** 71
59. 77	**60.** 666	**61.** 210	**62.** 39	

63. $34 \div 3$ **64.** $54 \div 9$ **65.** $8 \div 16$ **66.** $109 \div 18$

EXTEND YOUR THINKING ▶▶▶▶

▶ SOMETHING MORE

67. Perfect Numbers   A perfect number is a counting number equal to the sum of all of its divisors less than itself. For example, 6 is a perfect number. The divisors of 6 are 1, 2, 3, and 6, and $1 + 2 + 3 = 6$. Which of the following are perfect numbers?

a. 20   **b.** 28   **c.** 100   **d.** 12   **e.** 496

Consider the following two sets of numbers:

$$1 \quad 2 \quad 4 \quad 8 \quad 16 \quad 32 \quad \cdots$$
$$1 \quad 3 \quad 7 \quad 15 \quad 31 \quad 63 \quad \cdots$$

The top row of numbers corresponds to powers of 2: $2^0 = 1$, $2^1 = 2$, $2^2 = 4$, $2^3 = 8$, and so on. Each number in the bottom row is the sum of the number above it with all the numbers above and to the left. For example, $15 = 8 + 4 + 2 + 1$. The ancient mathematician Euclid (c.300 B.C.) claimed that whenever the lower number is a prime number, the product of it and the number above is a perfect number. For example, since 3 in the lower row is prime, then $3 \cdot 2 = 6$ is perfect. Was Euclid right for the other prime numbers shown in the bottom row?

▶ TROUBLESHOOT IT

Find and correct the error.

68. Prime factor 48.

$$
\begin{array}{r}
6 \\
\hline
2 \,|\, 12 \\
2 \,|\, 24 \\
2 \,|\, 48 \\
\end{array}
$$

Therefore, $48 = 2^3 \cdot 6$.

WRITING TO LEARN ▶▶▶▶

69. Explain the difference between a prime and a composite number.

70. Explain in words how to check a number for divisibility by 3.

▶ YOU BE THE JUDGE

71. Dan wanted to impress his younger brother Kile. He claimed that in the ten expressions that follow,

$$2^2 - 1 \qquad 2^3 - 1 \qquad 2^4 - 1 \qquad 2^5 - 1 \qquad 2^6 - 1$$
$$2^7 - 1 \qquad 2^8 - 1 \qquad 2^9 - 1 \qquad 2^{10} - 1 \qquad 2^{11} - 1$$

if the power on 2 is a prime number, the expression simplifies to a prime. For example, 3 is prime, and $2^3 - 1 = 8 - 1 = 7$ is prime. If the power on 2 is a composite number, it simplifies to a composite number. For example, 4 is composite, and $2^4 - 1 = 16 - 1 = 15$ is composite. Did Kile have reason to be impressed?

72. Jan claims that the divisibility rules for 2, 3, 5, and 7, can be used to check divisibility by the numbers 4, 6, 8, 9, or 10. Do you agree with Jan? Explain your decision.

73. Rob claims if a number is divisible by 3, then any number you get by arranging its digits in a different order is also divisible by 3. Is he correct? Explain your decision.

3.4 REDUCING FRACTIONS

OBJECTIVES

1. Define equivalent fractions and lowest terms form.
2. Reduce fractions by inspection.
3. Reduce fractions by prime factoring.
4. Reduce mixed numbers and improper fractions.

NEW VOCABULARY

equivalent fractions

lowest terms

reducing

3reducing by inspection

reducing by prime factoring

greatest common factor

1 Equivalent Fractions and Lowest Terms Form

A fraction can be written in many ways. For example, the fraction one half can be written in each of the following ways.

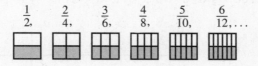

$$\frac{1}{2}, \quad \frac{2}{4}, \quad \frac{3}{6}, \quad \frac{4}{8}, \quad \frac{5}{10}, \quad \frac{6}{12}, \ldots$$

All of these fractions are equal. In each case, half of the figure is shaded. Fractions that are equal are called **equivalent fractions**. In this example, the fraction $\frac{1}{2}$ is in **lowest terms** form. This means both terms in the fraction are the smallest numbers that can be used to make one half.

> A fraction is in lowest terms form if the only counting number that divides exactly into *both* numerator and denominator is the number 1.

The process of writing a fraction in lowest terms form is called **reducing**. In this text all fractional answers will be reduced to lowest terms.

2 Reducing Fractions by Inspection

> To reduce a fraction to lowest terms by **inspection**
>
> 1. Divide both numerator and denominator by any counting number that you can see is a divisor of both. Use the divisibility rules for primes to help you find divisors.
> 2. Keep performing these divisions until the only number that divides both numerator and denominator exactly is 1.

EXAMPLE 1 Reduce $\frac{6}{12}$ to lowest terms by inspection.

A number that divides 6 and 12 exactly is the number 6.

$$\frac{6}{12} = \frac{6 \div 6}{12 \div 6} = \frac{1}{2} \quad \text{(lowest terms)}$$

The fraction $\frac{1}{2}$ is in lowest terms form because the only number that divides both the numerator 1 and the denominator 2 exactly is the number 1. ∎

Fractions that can be reduced to the same lowest terms answer are equivalent. The fractions $\frac{1}{2}$ and $\frac{6}{12}$ are equivalent fractions. Reducing a fraction to lowest terms means finding an equivalent fraction with the smallest possible terms.

EXAMPLE 2   Reduce $\dfrac{15}{24}$ to lowest terms by inspection.

3 is a divisor of both 15 and 24. Reduce using 3.

$$\dfrac{15}{24} = \dfrac{15 \div 3}{24 \div 3} = \dfrac{5}{8}$$

$\dfrac{5}{8}$ is in lowest terms form because the only number that divides both 5 and 8 exactly is 1. ∎

▶ **You Try It**   Reduce to lowest terms by inspection.

1. $\dfrac{8}{24}$ **2.** $\dfrac{12}{30}$ **3.** $\dfrac{32}{42}$

EXAMPLE 3   Reduce $\dfrac{36}{60}$ to lowest terms using inspection.

$$\dfrac{36}{60} = \dfrac{36 \div 2}{60 \div 2} = \dfrac{18}{30} = \dfrac{18 \div 6}{30 \div 6} = \dfrac{3}{5} \quad \text{(lowest terms)}$$

reduce using 2   reduce using 6   ∎

The reduction in Example 3 may also be written vertically.

$$\dfrac{\overset{3}{\cancel{\overset{18}{\cancel{36}}}}}{\underset{5}{\cancel{\underset{30}{\cancel{60}}}}} = \dfrac{3}{5} \begin{cases} \text{First reduce 36 and 60 using 2.} \\ \text{Then reduce 18 and 30 using 6.} \end{cases}$$

EXAMPLE 4   Reduce $\dfrac{63}{147}$ to lowest terms by inspection.

Use the divisibility rules to find a divisor for 63 and 147.
3 is a divisor of 63 because 3 divides $6 + 3 = 9$.
3 is a divisor of 147 because 3 divides $1 + 4 + 7 = 12$.

$$\dfrac{63}{147} = \dfrac{63 \div 3}{147 \div 3} = \dfrac{21}{49} = \dfrac{21 \div 7}{49 \div 7} = \dfrac{3}{7} \quad \text{(lowest terms)}$$

reduce using 3   reduce using 7   ∎

▶ **You Try It**   Reduce to lowest terms by inspection.

4. $\dfrac{42}{63}$ **5.** $\dfrac{231}{264}$

3 Reducing Fractions by Prime Factoring

To reduce fractions to lowest terms using **prime factoring**

1. Prime factor both numerator and denominator (Section 3.3).
2. Reduce numerator and denominator by all factors common to both prime factorizations.
3. Simplify for the lowest terms answer.

EXAMPLE 5   Reduce $\dfrac{6}{10}$ to lowest terms using prime factoring.

6 is prime factored as $6 = 2 \times 3$.
10 is prime factored as $10 = 2 \times 5$.

$$\frac{6}{10} = \frac{2 \times 3}{2 \times 5}$$

2 is the only factor common to both numerator and denominator.

$$\frac{6}{10} = \frac{\overset{1}{\cancel{2}} \times 3}{\underset{1}{\cancel{2}} \times 5} = \frac{1 \times 3}{1 \times 5} = \frac{3}{5}$$

Reduce both numerator and denominator by the common factor 2. Since $2 \div 2 = 1$, write 1 in place of each 2.

EXAMPLE 6   Reduce $\dfrac{30}{75}$ to lowest terms using prime factoring.

30 is prime factored as $30 = 2 \times 3 \times 5$.
75 is prime factored as $75 = 3 \times 5 \times 5$.

$$\frac{30}{75} = \frac{2 \times 3 \times 5}{3 \times 5 \times 5}$$

3 and 5 are common factors of both numerator and denominator.

$$\frac{30}{75} = \frac{2 \times \overset{1}{\cancel{3}} \times \overset{1}{\cancel{5}}}{\underset{1}{\cancel{3}} \times \underset{1}{\cancel{5}} \times 5} = \frac{2}{5}$$

Reduce both numerator and denominator by 3, then by 5. Replace all reduced factors by 1.

EXAMPLE 7   Reduce $\dfrac{60}{96}$ to lowest terms using prime factoring.

60 is prime factored as $60 = 2 \times 2 \times 3 \times 5$.
96 is prime factored as $96 = 2 \times 2 \times 2 \times 2 \times 2 \times 3$.

$$\frac{60}{96} = \frac{\overset{1}{\cancel{2}} \times \overset{1}{\cancel{2}} \times \overset{1}{\cancel{3}} \times 5}{2 \times 2 \times 2 \times \underset{1}{\cancel{2}} \times \underset{1}{\cancel{2}} \times \underset{1}{\cancel{3}}}$$

Reduce numerator and denominator by the common factors 2, 2, and 3.

$$= \frac{5}{2 \times 2 \times 2} = \frac{5}{8}$$

OBSERVE   In Example 7, the factors common to both 60 and 96 are 2, 2, and 3. The product of these common factors is called the **greatest common factor,** or **GCF.**

$$\text{GCF of 60 and 96} = 2 \times 2 \times 3 = 12$$

The GCF is the number that will reduce the fraction to lowest terms in one step.

$$\frac{60}{96} = \frac{60 \div 12}{96 \div 12} = \frac{5}{8}$$ lowest terms in one step

▶ **You Try It** Reduce to lowest terms using prime factoring.

6. $\dfrac{15}{18}$ 7. $\dfrac{40}{64}$ 8. $\dfrac{72}{120}$

4 Reducing Mixed Numbers and Improper Fractions

Reduce a mixed number by reducing its fractional part.

EXAMPLE 8 Reduce the mixed number $5\dfrac{8}{12}$.

$$5\frac{8}{12} = 5 + \frac{8}{12} = 5 + \frac{\overset{2}{\cancel{8}}}{\underset{3}{\cancel{12}}} = 5 + \frac{2}{3} = 5\frac{2}{3} \qquad \left(\text{In short, } 5\frac{\overset{2}{\cancel{8}}}{\underset{3}{\cancel{12}}} = 5\frac{2}{3}\right) \quad \blacksquare$$

As seen in Example 8, to reduce a mixed number, just reduce the proper fraction. Example 9 shows how to reduce an improper fraction.

EXAMPLE 9 Reduce $\dfrac{35}{30}$ to lowest terms.

$$\frac{35}{30} = 1\frac{5}{30} = 1\frac{\overset{1}{\cancel{5}}}{\underset{6}{\cancel{30}}} = 1\frac{1}{6}$$

Change the improper fraction to a mixed number.   Reduce the proper fraction using 5.   ■

▶ **You Try It** Reduce to lowest terms.   9. $3\dfrac{21}{24}$ 10. $\dfrac{40}{25}$

EXAMPLE 10 Simplify $5\dfrac{24}{9}$.

$$5\frac{24}{9} = 5 + \frac{24}{9} = 5 + 2\frac{6}{9} = (5+2) + \frac{6}{9} = 7 + \frac{\overset{2}{\cancel{6}}}{\underset{3}{\cancel{9}}} = 7\frac{2}{3}$$

This mixed number contains an improper fraction.   Change the improper fraction to a mixed number.   Add the whole numbers.   Reduce.   ■

▶ **You Try It** Simplify.   11. $4\dfrac{15}{6}$

▶ **Answers to You Try It**   1. $\dfrac{1}{3}$ 2. $\dfrac{2}{5}$ 3. $\dfrac{16}{21}$ 4. $\dfrac{2}{3}$ 5. $\dfrac{7}{8}$ 6. $\dfrac{5}{6}$ 7. $\dfrac{5}{8}$

8. $\dfrac{3}{5}$ 9. $3\dfrac{7}{8}$ 10. $1\dfrac{3}{5}$ 11. $6\dfrac{1}{2}$

SECTION 3.4 EXERCISES

1 2 *Reduce to lowest terms using inspection.*

1. $\dfrac{6}{8}$ 2. $\dfrac{4}{8}$ 3. $\dfrac{6}{9}$ 4. $\dfrac{5}{10}$ 5. $\dfrac{6}{14}$

6. $\dfrac{14}{21}$ 7. $\dfrac{15}{20}$ 8. $\dfrac{8}{24}$ 9. $\dfrac{12}{20}$ 10. $\dfrac{7}{28}$

11. $\dfrac{12}{16}$ 12. $\dfrac{21}{24}$ 13. $\dfrac{9}{36}$ 14. $\dfrac{10}{30}$

3 *Reduce to lowest terms using prime factoring.*

15. $\dfrac{21}{35}$ 16. $\dfrac{25}{30}$ 17. $\dfrac{32}{40}$ 18. $\dfrac{24}{42}$ 19. $\dfrac{33}{77}$

20. $\dfrac{42}{56}$ 21. $\dfrac{24}{32}$ 22. $\dfrac{9}{33}$ 23. $\dfrac{45}{75}$ 24. $\dfrac{49}{84}$

25. $\dfrac{52}{65}$ 26. $\dfrac{60}{72}$ 27. $\dfrac{48}{120}$ 28. $\dfrac{25}{35}$

Reduce to lowest terms using any method.

29. $\dfrac{24}{64}$ 30. $\dfrac{18}{90}$ 31. $\dfrac{34}{44}$ 32. $\dfrac{28}{49}$ 33. $\dfrac{57}{78}$

34. $\dfrac{54}{81}$ 35. $\dfrac{15}{35}$ 36. $\dfrac{12}{80}$ 37. $\dfrac{60}{100}$ 38. $\dfrac{120}{150}$

39. $\dfrac{90}{270}$ 40. $\dfrac{300}{480}$ 41. $\dfrac{75}{225}$ 42. $\dfrac{135}{270}$

43. What fractional part of an hour is 15 minutes?

44. What fractional part of a day is 18 hours?

45. A person making $30,000 a year pays $5,000 in taxes. What fractional part of the salary pays for taxes?

46. A person making $1,680 a month in wages pays $630 a month for rent. What fraction of the monthly salary is rent?

47. Out of 12,000 students attending a university, 8,400 are women. What fraction are women?

48. 64,000 out of 80,000 inhabitants live in villages. What fraction live in villages?

49. In a high school of 371 students, 318 plan to attend college.
   a. What fraction plan to attend college?

   b. What fraction do not plan to attend college?

50. Out of a sample of 710 people, 639 are holding full-time jobs.
   a. What fraction of the sample is this?

   b. What fraction are not holding full-time jobs?

51. Which fraction does not reduce to $\frac{1}{3}$?

$$\frac{3}{9}, \frac{14}{42}, \frac{12}{36}, \frac{18}{48}, \text{ or } \frac{13}{39}$$

52. Which fraction does reduce to $\frac{3}{8}$?

$$\frac{15}{24}, \frac{16}{32}, \frac{15}{40}, \frac{24}{72}, \text{ or } \frac{36}{48}$$

4 *Express each number as a mixed number reduced to lowest terms.*

53. $4\frac{32}{40}$ **54.** $2\frac{18}{24}$ **55.** $5\frac{15}{21}$ **56.** $7\frac{12}{28}$ **57.** $\frac{33}{27}$

58. $\frac{25}{20}$ **59.** $\frac{57}{33}$ **60.** $\frac{108}{48}$ **61.** $\frac{75}{40}$ **62.** $\frac{720}{300}$

63. $6\frac{15}{10}$ **64.** $10\frac{24}{18}$ **65.** $7\frac{38}{9}$ **66.** $12\frac{27}{13}$

SKILLSFOCUS (Section 2.6) *Solve each equation using opposite operations.*

67. $8x = 96$ **68.** $24x = 168$ **69.** $\frac{y}{12} = 15$ **70.** $\frac{z}{9} = 324$

EXTEND YOUR THINKING ▶▶▶▶

▶ TROUBLESHOOT IT

Find and correct the error.

71. $\frac{36}{54} = \frac{36 \div 6}{54 \div 6} = \frac{6}{8} = \frac{3}{4}$

72. $\frac{48}{72} = \dfrac{\overset{1}{\cancel{2}} \times \overset{1}{\cancel{2}} \times 2 \times 2 \times \overset{1}{\cancel{3}}}{\underset{1}{\cancel{2}} \times \underset{1}{\cancel{2}} \times 3 \times 3 \times \underset{1}{\cancel{3}}} = \frac{4}{9}$

WRITING TO LEARN ▶▶▶▶

73. Explain why $\frac{3}{8}$ is reduced to lowest terms, but $\frac{6}{16}$ is not.

▶ YOU BE THE JUDGE

74. Jim claims that $\frac{10}{16}, \frac{25}{40}$, and $\frac{15}{24}$, are three different ways to write $\frac{5}{8}$. Is he correct? Explain how you arrived at your decision.

OBJECTIVES

1 Multiply proper fractions without reducing.

2 Multiply proper fractions with reducing.

3 Multiply mixed numbers.

4 Evaluate fractions raised to powers (Section 1.9).

1 Multiplying Proper Fractions

EXAMPLE 1

$$\frac{1}{3} \times \frac{2}{5} = \frac{1 \times 2}{3 \times 5} = \frac{2}{15} \quad \blacksquare$$

> To multiply two proper fractions
> 1. Multiply the two numerators for the new numerator.
> 2. Multiply the two denominators for the new denominator.
> 3. Reduce your answer, if possible.

EXAMPLE 2

$$\frac{3}{4} \times \frac{5}{7} = \frac{3 \times 5}{4 \times 7} = \frac{15}{28} \quad \blacksquare$$

EXAMPLE 3

$$\frac{5}{9} \times \frac{2}{3} \times \frac{4}{7} = \frac{5 \times 2 \times 4}{9 \times 3 \times 7} = \frac{40}{189} \quad \blacksquare$$

▶ **You Try It**   **1.** $\frac{2}{3} \times \frac{4}{9}$ **2.** $\frac{7}{8} \times \frac{3}{5}$ **3.** $\frac{5}{6} \times \frac{1}{2} \times \frac{7}{9}$

2 Multiplying Proper Fractions with Reducing

In the next example, the answer must be reduced to lowest terms.

EXAMPLE 4

$$\frac{3}{4} \times \frac{2}{5} = \frac{3 \times 2}{4 \times 5} = \frac{6}{20} = \frac{\overset{3}{\cancel{6}}}{\underset{10}{\cancel{20}}} = \frac{3}{10}$$

Reduce 6 and 20 using 2.   ■

In Example 4, you reduced after multiplying. You can reduce fractions before multiplying. If a numerator and a denominator have a common factor, reduce both using the factor. In Example 4, the numerator 2 and the denominator 4 have a factor of 2 in common. Reduce 2 and 4 using 2 before you multiply.

$$\frac{3}{\underset{2}{\cancel{4}}} \times \frac{\overset{1}{\cancel{2}}}{5} = \frac{3 \times 1}{2 \times 5} = \frac{3}{10} \quad \text{(same lowest terms answer as in Example 4)}$$

The advantage to reducing before you multiply is that you avoid reducing a fraction with a very large numerator or denominator.

EXAMPLE 5 Multiply $\dfrac{7}{8} \times \dfrac{36}{45}$.

Solution #1: Multiply first, then reduce.

$$\frac{7}{8} \times \frac{36}{45} = \frac{7 \times 36}{8 \times 45} = \frac{\overset{\displaystyle 7}{\overset{\displaystyle \cancel{63}}{\overset{\displaystyle \cancel{126}}{\cancel{252}}}}}{\underset{\displaystyle 10}{\underset{\displaystyle \cancel{90}}{\underset{\displaystyle \cancel{180}}{\cancel{360}}}}} = \frac{7}{10}$$

Reduce 252 and 360 using 2 (/).
Reduce 126 and 180 using 2 (−).
Reduce 63 and 90 using 9 (\).

Solution #2: Now solve the same problem by reducing first, then multiplying. Reduce any numerator with any denominator that share a common factor.

Reduce 36 and 45 using 9 (/).
Reduce 8 and 4 using 4 (−).

$$\frac{7}{\underset{2}{\cancel{8}}} \times \frac{\overset{\overset{1}{\cancel{4}}}{\cancel{36}}}{\underset{5}{\cancel{45}}} = \frac{7 \times 1}{2 \times 5} = \frac{7}{10} \quad \blacksquare$$

The answer is the same in each case. By reducing before multiplying (see Solution #2), the numbers you reduced and multiplied were smaller. In Solution #2, you may reduce *vertically*, as with 36 and 45, or *diagonally* as with 4 and 8. The order in which the terms are reduced will not change the answer.

> To reduce before multiplying fractions
>
> 1. If a numerator has a factor in common with a denominator, reduce both by the common factor.
> 2. Repeat this process until the only factor any nuumerator has in common with any denominator is 1.
> 3. Multiply across. Your answer will be in lowest terms.

EXAMPLE 6

$$\frac{10}{15} \times \frac{7}{24} = \frac{\overset{\overset{1}{\cancel{2}}}{\cancel{10}}}{\underset{3}{\cancel{15}}} \times \frac{7}{\underset{12}{\cancel{24}}} = \frac{1 \times 7}{3 \times 12} = \frac{7}{36} \quad \blacksquare$$

Reduce 10 and 15 using 5 (/).
Reduce 2 and 24 using 2 (−).

CAUTION You may not reduce 15 and 24 using 3, because 15 and 24 are both denominators.

▶ **You Try It** Reduce, then multiply.

4. $\dfrac{5}{8} \times \dfrac{4}{7}$

5. $\dfrac{10}{12} \times \dfrac{2}{5}$

EXAMPLE 7

$$\frac{15}{21} \times \frac{28}{35} = \overset{\underset{\displaystyle 7}{\cancel{15}}}{\cancel{21}} \times \overset{\overset{\displaystyle 4}{\cancel{28}}}{\underset{1}{\cancel{35}}} = \frac{1 \times 4}{7 \times 1} = \frac{4}{7}$$

Reduce 28 and 35 using 7 (/).
Reduce 15 and 21 using 3 (—).
Reduce 5 and 5 using 5 (\\). ■

CAUTION Always check to see if your answer can still be reduced. If so, you did not reduce completely before multiplying across.

EXAMPLE 8

$$\frac{5}{9} \times \frac{16}{35} \times \frac{3}{40} = \frac{\cancel{5}}{\cancel{9}} \times \frac{\cancel{16}}{35} \times \frac{\cancel{3}}{\cancel{40}} = \frac{1 \times 2 \times 1}{3 \times 35 \times 1} = \frac{2}{105}$$

Reduce 3 and 9 using 3 (/).
Reduce 16 and 40 using 8 (—).
Reduce 5 and 5 (\\). ■

▶ **You Try It**   6. $\dfrac{28}{42} \times \dfrac{12}{16}$ 7. $\dfrac{5}{9} \times \dfrac{3}{10}$ 8. $\dfrac{3}{8} \times \dfrac{6}{9} \times \dfrac{8}{15}$

---

3 Multiplying Mixed Numbers

To multiply mixed numbers, first write each mixed number as an improper fraction.

EXAMPLE 9

$$3\frac{1}{2} \times \frac{5}{6} = \frac{7}{2} \times \frac{5}{6} = \frac{7 \times 5}{2 \times 6} = \frac{35}{12} = 2\frac{11}{12}$$

Rewrite $3\frac{1}{2}$ as $\frac{7}{2}$.   Rewrite $\frac{35}{12}$ as $2\frac{11}{12}$. ■

To multiply two or more mixed numbers

1. Change each mixed number to an improper fraction.
2. Reduce, if possible.
3. Multiply across.

EXAMPLE 10

$$2\frac{4}{5} \times 4\frac{2}{7} = \frac{\cancel{14}}{\cancel{5}} \times \frac{\cancel{30}}{\cancel{7}} = \frac{2 \times 6}{1 \times 1} = \frac{12}{1} = 12$$

Reduce 14 and 7 using 7 (/).
Reduce 30 and 5 using 5 (—). ■

▶ **You Try It**   9. $7\frac{1}{3} \times \frac{2}{5}$ 10. $4\frac{3}{5} \times 3\frac{1}{3}$

To multiply a fraction by a whole number, first write the whole number as an improper fraction with a denominator of 1.

EXAMPLE 11
$$6 \times \frac{3}{5} = \frac{6}{1} \times \frac{3}{5} = \frac{6 \times 3}{1 \times 5} = \frac{18}{5} = 3\frac{3}{5}$$

Rewrite 6 as $\frac{6}{1}$.

EXAMPLE 12
$$3\frac{9}{10} \times \frac{5}{12} \times 5\frac{1}{3} = \frac{\overset{13}{\cancel{39}}}{\cancel{10}} \times \frac{\overset{1}{\cancel{5}}}{\cancel{12}} \times \frac{\overset{2}{\cancel{16}}}{\cancel{3}} = \frac{13 \times 1 \times 2}{1 \times 3 \times 1} = \frac{26}{3} = 8\frac{2}{3}$$

Reduce 39 and 3 using 3 (/).
Reduce 12 and 16 using 4 (−).
Reduce 10 and 5 using 5 (\).
Reduce 2 and 4 using 2 (//).

▶ **You Try It** 11. $8 \times \frac{5}{6}$ 12. $2\frac{5}{8} \times \frac{3}{7} \times 1\frac{1}{9}$

4 Evaluating a Fraction Raised to a Power

Recall from Section 1.9 that a power on a number means repeated multiplication of that number. When a fraction is raised to a power, write the fraction the number of times indicated by the power, then multiply.

EXAMPLE 13
$$\left(\frac{3}{5}\right)^2 = \frac{3}{5} \times \frac{3}{5} = \frac{9}{25}$$

EXAMPLE 14
$$\left(2\frac{1}{4}\right)^2 = 2\frac{1}{4} \times 2\frac{1}{4} = \frac{9}{4} \times \frac{9}{4} = \frac{81}{16} \quad \text{or} \quad 5\frac{1}{16}$$

EXAMPLE 15
$$\left(\frac{1}{2}\right)^4 = \frac{1}{2} \times \frac{1}{2} \times \frac{1}{2} \times \frac{1}{2} = \frac{1}{16}$$

▶ **You Try It** Evaluate.

13. $\left(\frac{3}{5}\right)^2$ 14. $\left(3\frac{1}{2}\right)^2$ 15. $\left(\frac{5}{4}\right)^3$

▶ **Answers to You Try It** 1. $\frac{8}{27}$ 2. $\frac{21}{40}$ 3. $\frac{35}{108}$ 4. $\frac{5}{14}$ 5. $\frac{1}{3}$ 6. $\frac{1}{2}$
7. $\frac{1}{6}$ 8. $\frac{2}{15}$ 9. $\frac{14}{15}$ 10. $15\frac{1}{3}$ 11. $6\frac{2}{3}$ 12. $1\frac{1}{4}$ 13. $\frac{9}{25}$ 14. $12\frac{1}{4}$ 15. $1\frac{61}{64}$

1 **2** *Multiply and express all answers in lowest terms.*

1. $\frac{3}{5} \times \frac{2}{3}$

2. $\frac{3}{4} \times \frac{1}{2}$

3. $\frac{4}{7} \times \frac{8}{9}$

4. $\frac{5}{6} \times \frac{7}{8}$

5. $\frac{4}{7} \times \frac{3}{4}$

6. $\frac{5}{6} \times \frac{1}{5}$

7. $\frac{3}{4} \times \frac{4}{7}$

8. $\frac{3}{5} \times \frac{4}{9}$

9. $\frac{7}{10} \times \frac{4}{5}$

10. $\frac{8}{9} \times \frac{5}{12}$

11. $\frac{6}{11} \times \frac{11}{6}$

12. $\frac{7}{10} \times \frac{10}{7}$

13. $\frac{3}{5} \times \frac{25}{36}$

14. $\frac{9}{10} \times \frac{2}{3}$

15. $\frac{15}{28} \times \frac{35}{70}$

16. $\frac{24}{60} \times \frac{30}{36}$

17. $\frac{27}{45} \times \frac{6}{15}$

18. $\frac{24}{32} \times \frac{12}{16}$

19. $\frac{60}{90} \times \frac{20}{30}$

20. $\frac{25}{63} \times \frac{56}{35}$

21. $\frac{5}{6} \times \frac{2}{15} \times \frac{20}{24}$

22. $\frac{4}{5} \times \frac{3}{8} \times \frac{6}{9}$

23. $\frac{50}{75} \times \frac{16}{28} \times \frac{21}{24}$

24. $\frac{18}{42} \times \frac{36}{45} \times \frac{55}{64}$

3 *Multiply and express each answer in lowest terms.*

25. $\frac{5}{6} \times 4\frac{1}{2}$

26. $\frac{5}{8} \times 3\frac{1}{5}$

27. $5\frac{1}{2} \times \frac{4}{5}$

28. $6\frac{3}{4} \times 1\frac{1}{7}$

29. $2\frac{1}{4} \times 3\frac{1}{2}$

30. $4\frac{2}{3} \times 1\frac{3}{5}$

31. $2\frac{1}{3} \times 4\frac{1}{2}$

32. $6\frac{1}{8} \times 3\frac{1}{7}$

33. $5\frac{1}{9} \times 2\frac{2}{3}$

34. $3\frac{3}{4} \times 1\frac{1}{10}$

35. $4 \times 3\frac{1}{5}$

36. $8 \times 2\frac{1}{4}$

37. $5\frac{1}{4} \times \frac{6}{7}$

38. $2\frac{2}{9} \times \frac{4}{5}$

39. $2\frac{2}{3} \times 6$

40. $4\frac{2}{5} \times 3$

41. $\frac{2}{3} \times 10 \times 4\frac{1}{5}$

42. $3 \times 4\frac{1}{5} \times \frac{5}{7}$

43. $4\frac{1}{2} \times 3\frac{2}{3} \times 5\frac{5}{9}$

44. $1\frac{1}{4} \times 4\frac{1}{3} \times 2\frac{1}{2}$

45. $4\frac{1}{5} \times 2\frac{1}{7} \times 3\frac{2}{3}$

46. $4\frac{1}{12} \times \frac{3}{70} \times \frac{8}{14}$

47. What is 2 times $\frac{1}{2}$?

48. What is $5\frac{5}{6}$ times $\frac{12}{35}$?

4 *Evaluate each power.*

49. $\left(\dfrac{3}{5}\right)^2$ 　　　　**50.** $\left(\dfrac{1}{4}\right)^2$ 　　　　**51.** $\left(2\dfrac{5}{6}\right)^2$ 　　　　**52.** $\left(4\dfrac{1}{2}\right)^2$

53. $\left(\dfrac{6}{7}\right)^3$ 　　　　**54.** $\left(\dfrac{2}{5}\right)^3$ 　　　　**55.** $\left(\dfrac{3}{10}\right)^4$ 　　　　**56.** $\left(\dfrac{1}{3}\right)^4$

57. $\left(\dfrac{5}{3}\right)^4$ 　　　　**58.** $\left(\dfrac{5}{2}\right)^3$ 　　　　**59.** $\left(3\dfrac{1}{3}\right)^2$ 　　　　**60.** $\left(6\dfrac{1}{2}\right)^3$

61. What is $\dfrac{7}{8}$ squared? 　　　　　　　　**62.** What is $\dfrac{3}{4}$ cubed?

SKILLSFOCUS (Section 1.9) *Evaluate:*

63. $\sqrt{36}$ 　　　　**64.** $\sqrt{81}$ 　　　　**65.** $\sqrt{1}$ 　　　　**66.** $\sqrt{100}$

EXTEND YOUR THINKING ▶▶▶▶

▶ TROUBLESHOOT IT

Find and correct the error.

67. $\dfrac{3}{8} \times \dfrac{1}{4} = \dfrac{3}{\overset{}{\underset{2}{\cancel{8}}}} \times \dfrac{1}{\overset{}{\underset{1}{\cancel{4}}}} = \dfrac{3 \times 1}{2 \times 1} = \dfrac{3}{2}$

68. $2\dfrac{1}{2} \times 5\dfrac{3}{4} = (2 \times 5) + \left(\dfrac{1}{2} \times \dfrac{3}{4}\right) = 10 + \dfrac{3}{8} = 10\dfrac{3}{8}$

69. $3\dfrac{1}{4} \times \dfrac{2}{5} = 3\dfrac{1}{\underset{2}{\cancel{4}}} \times \dfrac{\overset{1}{\cancel{2}}}{5} = 3\dfrac{1}{2} \times \dfrac{1}{5} = \dfrac{7}{2} \times \dfrac{1}{5} = \dfrac{7}{10}$

WRITING TO LEARN ▶▶▶▶

70. Explain each step you would use to multiply a mixed number times a whole number. Demonstrate using your own example.

71. Explain the advantage to reducing before multiplying fractions. Demonstrate your point using your own example.

▶ YOU BE THE JUDGE

72. Ron claims $\left(\dfrac{24}{40}\right)^3$ can only be evaluated by reducing within parentheses first, then cubing. Barb says no, you must write $\dfrac{24}{40}$ times itself 3 times first. Then you reduce and multiply. Settle the dispute.

3.6 DIVIDING FRACTIONS

OBJECTIVES
1 Find the reciprocal of a fraction.
2 Divide fractions.
3 Divide mixed numbers.
4 Solve equations (Section 2.6).

NEW VOCABULARY
reciprocal

---

1 The Reciprocal of a Fraction

In order to divide fractions, you need to understand reciprocals. The **reciprocal** of a fraction is found by interchanging the numerator with the denominator.

EXAMPLE 1 **a.** The reciprocal of $\frac{2}{3}$ is $\frac{3}{2}$. **b.** The reciprocal of $\frac{7}{12}$ is $\frac{12}{7}$.

c. The reciprocal $\frac{1}{4}$ is $\frac{4}{1}$, or 4. ■

To find the reciprocal of a counting number, first write the counting number with a denominator of 1.

EXAMPLE 2 **a.** Since $6 = \frac{6}{1}$, the reciprocal of 6 is $\frac{1}{6}$. **b.** The reciprocal of $25 = \frac{25}{1}$ is $\frac{1}{25}$. ■

> **CAUTION** Since the reciprocal of $0 = \frac{0}{1}$ is $\frac{1}{0}$, 0 has no reciprocal. $\frac{1}{0}$ is division by 0, which is not possible.

▶ **You Try It** Find the reciprocal.

1. $\frac{4}{7}$ 2. $\frac{9}{10}$ 3. $\frac{1}{8}$ 4. 2   5. 1   6. 18

To find the reciprocal of a mixed number, first write the mixed number as an improper fraction.

EXAMPLE 3 **a.** Since $3\frac{1}{2} = \frac{7}{2}$, the reciprocal of $3\frac{1}{2}$ is $\frac{2}{7}$.

b. Since $2\frac{7}{9} = \frac{25}{9}$, the reciprocal of $2\frac{7}{9}$ is $\frac{9}{25}$. ■

▶ **You Try It** Find the reciprocal.

7. $2\frac{1}{5}$ 8. $5\frac{3}{8}$

> The product of a fraction and its reciprocal is 1.

EXAMPLE 4 **a.** The reciprocal of $\frac{2}{5}$ is $\frac{5}{2}$, and $\frac{2}{5} \times \frac{5}{2} = \frac{10}{10} = 1$.

b. The reciprocal of $4 = \frac{4}{1}$ is $\frac{1}{4}$, and $4 \times \frac{1}{4} = \frac{4}{1} \times \frac{1}{4} = \frac{4}{4} = 1$.

c. The reciprocal of $2\frac{3}{4} = \frac{11}{4}$ is $\frac{4}{11}$, and $2\frac{3}{4} \times \frac{4}{11} = \frac{11}{4} \times \frac{4}{11} = \frac{44}{44} = 1$. ■

▶ **You Try It**   Find the reciprocal. Then find the product of each fraction and its reciprocal.

9. $\frac{4}{9}$ **10.** 6   **11.** $7\frac{2}{3}$

---

2 Dividing Fractions   How do you divide the fractions below?

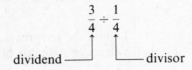

$$\frac{3}{4} \div \frac{1}{4}$$

dividend —⌐   ⌐— divisor

OBSERVE   This is read "$\frac{3}{4}$ divided by $\frac{1}{4}$." Intuitively, the answer is 3. There are three $\frac{1}{4}$ cups in $\frac{3}{4}$ cups of milk.

To see how to divide, first write the division in fraction form.

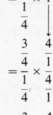

$\frac{3}{4} \div \frac{1}{4} = \dfrac{\frac{3}{4}}{\frac{1}{4}}$ Recall, $a \div b = \frac{a}{b}$. Replace the letter a with $\frac{3}{4}$.
Replace b with $\frac{1}{4}$.

$= \dfrac{\frac{3}{4}}{\frac{1}{4}} \times 1$ Multiplication property of one.

$= \dfrac{\frac{3}{4}}{\frac{1}{4}} \times \dfrac{\frac{4}{1}}{\frac{4}{1}}$ Write 1 as the reciprocal of $\frac{1}{4}$ (the denominator), divided by itself.

$= \dfrac{\frac{3}{4} \times \frac{4}{1}}{\frac{1}{4} \times \frac{4}{1}}$ In the denominator, a fraction times its reciprocal equals 1.

$= \dfrac{\frac{3}{4} \times \frac{4}{1}}{1}$

$= \frac{3}{4} \times \boxed{\frac{4}{1}}$ To divide fractions, multiply the dividend by the reciprocal of the divisor.

$= 3$

> To divide two proper fractions
>
> 1. Multiply the dividend by the reciprocal of the divisor. (You also say: invert the divisor, then multiply. To invert a fraction means write its reciprocal.)
>
> $$\frac{a}{b} \div \frac{c}{d} = \frac{a}{b} \times \frac{d}{c}$$
>
> 2. Reduce *after* the problem has been rewritten as a multiplication.

EXAMPLE 5 a. $\dfrac{4}{5} \div \dfrac{3}{7} = \dfrac{4}{5} \times \boxed{\dfrac{7}{3}} = \dfrac{28}{15}$ or $1\dfrac{13}{15}$

b. $\dfrac{14}{15} \div \dfrac{7}{10} = \dfrac{\overset{}{\cancel{14}}}{\underset{3}{\cancel{15}}} \times \dfrac{\overset{2}{\cancel{10}}}{\underset{1}{\cancel{7}}} = \dfrac{4}{3}$ or $1\dfrac{1}{3}$ ∎

EXAMPLE 6 a. $4 \div \dfrac{2}{5} = \dfrac{4}{1} \div \dfrac{2}{5} = \dfrac{\overset{2}{\cancel{4}}}{1} \times \dfrac{5}{\underset{1}{\cancel{2}}} = \dfrac{10}{1}$ or 10

b. $\dfrac{2}{9} \div 3 = \dfrac{2}{9} \div \dfrac{3}{1} = \dfrac{2}{9} \times \dfrac{1}{3} = \dfrac{2}{27}$ ∎

▶ **You Try It** Divide. **12.** $\dfrac{3}{8} \div \dfrac{2}{5}$ **13.** $\dfrac{3}{4} \div \dfrac{5}{6}$ **14.** $\dfrac{5}{9} \div \dfrac{10}{21}$ **15.** $9 \div \dfrac{3}{7}$ **16.** $\dfrac{4}{5} \div 2$

You can divide whole numbers using fraction notation.

EXAMPLE 7 $12 \div 3 = \dfrac{12}{1} \div \dfrac{3}{1} = \dfrac{\overset{4}{\cancel{12}}}{1} \times \dfrac{1}{\underset{1}{\cancel{3}}} = \dfrac{4}{1} = 4$ ∎

▶ **You Try It** Divide using fraction notation. **17.** $18 \div 6$

CAUTION When dividing fractions, if you make the mistake of multiplying by the reciprocal of the dividend, your answer will be the reciprocal of the correct answer. Using Example 5,

$$\frac{4}{5} \div \frac{3}{7} = \frac{5}{4} \times \frac{3}{7} = \frac{15}{28}$$

is **wrong**. Observe, this answer is the reciprocal of the correct answer.

Division of fractions is not commutative. Changing the order of the fractions in the following division changes the answer.

$$\frac{3}{4} \div \frac{2}{5} \neq \frac{2}{5} \div \frac{3}{4}$$

$$\frac{3}{4} \times \frac{5}{2} = \frac{15}{8} \qquad \neq \qquad \frac{2}{5} \times \frac{4}{3} = \frac{8}{15}$$

In the exercises at the end of this section (Problem 78), you will see for yourself that *division of fractions is not associative.*

3 Dividing Mixed Numbers

To divide with mixed numbers

1. First change each mixed number to an improper fraction.
2. Then follow the same procedure used to divide fractions.

EXAMPLE 8   **a.** $3\frac{2}{5} \div \frac{3}{4} = \frac{17}{5} \div \frac{3}{4} = \frac{17}{5} \times \frac{4}{3} = \frac{68}{15}$ or $4\frac{8}{15}$

b. $5\frac{5}{8} \div 2\frac{1}{4} = \frac{45}{8} \div \frac{9}{4} = \frac{\overset{5}{\cancel{45}}}{\underset{2}{\cancel{8}}} \times \frac{\overset{1}{\cancel{4}}}{\underset{1}{\cancel{9}}} = \frac{5}{2}$ or $2\frac{1}{2}$ ■

EXAMPLE 9   **a.** $2\frac{5}{6} \div 4 = \frac{17}{6} \div \frac{4}{1} = \frac{17}{6} \times \frac{1}{4} = \frac{17}{24}$

b. $12 \div 1\frac{1}{5} = \frac{12}{1} \div \frac{6}{5} = \frac{\overset{2}{\cancel{12}}}{1} \times \frac{5}{\underset{1}{\cancel{6}}} = \frac{10}{1}$ or 10 ■

▶ **You Try It**   Divide.   **18.** $1\frac{2}{7} \div \frac{5}{6}$ **19.** $3\frac{3}{10} \div 2\frac{1}{5}$ **20.** $5\frac{1}{8} \div 8$ **21.** $16 \div 1\frac{1}{3}$

OBSERVE   Every division of fractions problem is solved by first rewriting it as a multiplication problem.

---

4 Solving Equations

Recall that multiplication and division are opposite operations. Solve a multiplication equation by writing the related division.

EXAMPLE 10   Solve $\frac{3}{8}y = 2\frac{1}{4}$.

Since $\frac{3}{8}y$ means $\frac{3}{8} \cdot y$, solve for y by writing the related division.

$$y = 2\frac{1}{4} \div \frac{3}{8} = \frac{9}{4} \div \frac{3}{8} = \frac{\overset{3}{\cancel{9}}}{\underset{1}{\cancel{4}}} \times \frac{\overset{2}{\cancel{8}}}{\underset{1}{\cancel{3}}} = \frac{6}{1} = 6$$

Substitute to check. $\frac{3}{8}y = \frac{3}{8} \cdot 6 = \frac{3}{\underset{4}{\cancel{8}}} \cdot \frac{\overset{3}{\cancel{6}}}{1} = \frac{9}{4} = 2\frac{1}{4}$ ■

▶ **You Try It**   **22.** Solve for w. $\frac{7}{10}w = \frac{2}{5}$

▶ **Answers to You Try It**   1. $\frac{7}{4}$ 2. $\frac{10}{9}$ 3. $\frac{8}{1}$ or 8   4. $\frac{1}{2}$ 5. 1   6. $\frac{1}{18}$

7. $\frac{5}{11}$ 8. $\frac{8}{43}$ 9. $\frac{9}{4}$; 1   10. $\frac{1}{6}$; 1   11. $\frac{3}{23}$; 1   12. $\frac{15}{16}$ 13. $\frac{9}{10}$ 14. $1\frac{1}{6}$

15. 21   16. $\frac{2}{5}$ 17. 3   18. $1\frac{19}{35}$ 19. $1\frac{1}{2}$ 20. $\frac{41}{64}$ 21. 12   22. $\frac{4}{7}$

SECTION 3.6 EXERCISES

1 *Find the reciprocal. Then show the product of the fraction and its reciprocal is 1.*

1. $\dfrac{2}{7}$ **2.** $\dfrac{5}{6}$ **3.** 8   **4.** 3   **5.** $\dfrac{1}{10}$ **6.** $\dfrac{1}{6}$

7. $\dfrac{11}{15}$ **8.** $\dfrac{9}{14}$ **9.** $5\dfrac{1}{8}$ **10.** $2\dfrac{8}{9}$ **11.** $\dfrac{7}{4}$ **12.** $\dfrac{6}{25}$

2 *Divide and reduce all answers to lowest terms.*

13. $\dfrac{3}{4} \div \dfrac{5}{7}$ **14.** $\dfrac{2}{9} \div \dfrac{3}{5}$ **15.** $\dfrac{4}{7} \div \dfrac{1}{3}$ **16.** $\dfrac{1}{2} \div \dfrac{4}{9}$ **17.** $\dfrac{2}{3} \div \dfrac{5}{7}$

18. $\dfrac{1}{8} \div \dfrac{1}{3}$ **19.** $\dfrac{7}{9} \div \dfrac{5}{6}$ **20.** $\dfrac{4}{5} \div \dfrac{1}{5}$ **21.** $\dfrac{7}{10} \div \dfrac{3}{5}$ **22.** $\dfrac{3}{4} \div \dfrac{6}{7}$

23. $\dfrac{8}{11} \div \dfrac{4}{5}$ **24.** $\dfrac{7}{12} \div \dfrac{3}{14}$ **25.** $\dfrac{5}{9} \div \dfrac{5}{12}$ **26.** $\dfrac{3}{7} \div \dfrac{4}{21}$ **27.** $\dfrac{5}{18} \div \dfrac{2}{9}$

28. $\dfrac{9}{10} \div \dfrac{4}{5}$ **29.** $\dfrac{4}{5} \div 3$ **30.** $\dfrac{5}{8} \div 8$ **31.** $6 \div \dfrac{2}{3}$ **32.** $2 \div \dfrac{3}{5}$

33. $\dfrac{15}{20} \div \dfrac{3}{20}$ **34.** $\dfrac{24}{40} \div \dfrac{36}{50}$ **35.** $\dfrac{6}{11} \div \dfrac{30}{55}$ **36.** $\dfrac{35}{49} \div \dfrac{10}{14}$

3 *Divide and reduce all answers to lowest terms.*

37. $3\dfrac{3}{4} \div 2\dfrac{1}{3}$ **38.** $9\dfrac{2}{3} \div 2\dfrac{1}{2}$ **39.** $2\dfrac{1}{10} \div 1\dfrac{1}{6}$ **40.** $10\dfrac{1}{2} \div 6\dfrac{3}{8}$ **41.** $4\dfrac{1}{2} \div \dfrac{1}{4}$

42. $5\dfrac{1}{6} \div \dfrac{3}{4}$ **43.** $4 \div 1\dfrac{1}{3}$ **44.** $9 \div 2\dfrac{3}{4}$ **45.** $4\dfrac{3}{8} \div 2$ **46.** $3\dfrac{4}{5} \div 5$

47. $11 \div 2\dfrac{1}{5}$ **48.** $5 \div 6\dfrac{7}{8}$ **49.** $\dfrac{13}{20} \div 2\dfrac{3}{5}$ **50.** $7\dfrac{1}{8} \div \dfrac{2}{3}$ **51.** $16\dfrac{1}{2} \div 4\dfrac{1}{8}$

52. $5\dfrac{2}{7} \div 1\dfrac{3}{10}$ **53.** $1\dfrac{1}{6} \div 3\dfrac{1}{2}$ **54.** $5\dfrac{1}{3} \div 16$ **55.** $4\dfrac{7}{11} \div 7\dfrac{1}{6}$ **56.** $3\dfrac{5}{9} \div 8\dfrac{4}{9}$

57. $3\dfrac{3}{20} \div 2\dfrac{4}{5}$ **58.** $17\dfrac{1}{2} \div 11\dfrac{2}{3}$ **59.** $20\dfrac{7}{10} \div 2\dfrac{1}{4}$ **60.** $13\dfrac{3}{8} \div 7\dfrac{5}{6}$

4 *Solve each equation by writing the related operation, and check.*

61. $\frac{2}{3}s = \frac{4}{7}$

62. $\frac{3}{4}t = \frac{1}{2}$

63. $\frac{5}{8}k = \frac{7}{9}$

64. $\frac{6}{7}n = \frac{2}{3}$

65. $\frac{3}{10}s = \frac{11}{15}$

66. $\frac{10}{13}y = \frac{7}{26}$

67. $\frac{4}{15}m = 4$

68. $\frac{1}{2}p = 5\frac{1}{2}$

SKILLSFOCUS (Section 2.3) *Solve each word problem.*

69. What is the product of 5 and 8?

70. What is 9 added to itself 25 times?

71. How many $7 tapes can be purchased for $252?

EXTEND YOUR THINKING ▶▶▶▶

▶ TROUBLESHOOT IT

Find and correct the error.

72. $\frac{4}{5} \div \frac{3}{8} = \frac{\cancel{4}^{1}}{5} \times \frac{\cancel{8}^{2}}{3} = \frac{1 \times 2}{5 \times 3} = \frac{2}{15}$

73. $\frac{7}{9} \div \frac{1}{6} = \frac{\cancel{9}^{3}}{7} \times \frac{1}{\cancel{6}_{2}} = \frac{3}{14}$

74. $3\frac{3}{4} \div 1\frac{5}{9} = \frac{\cancel{15}}{4} \div \frac{\cancel{14}}{\cancel{9}} = \frac{5}{2} \div \frac{7}{3} = \frac{5}{2} \times \frac{3}{7} = \frac{15}{14}$ or $1\frac{1}{14}$

75. $\frac{1}{2} \div 2 = \frac{1}{\cancel{2}_{1}} \times \frac{\cancel{2}^{1}}{1} = \frac{1}{1} = 1$

WRITING TO LEARN ▶▶▶▶

76. Explain how to find the reciprocal of a mixed number. Use your own example to demonstrate your point.

77. Explain why division of fractions is not commutative. Use your own example to demonstrate your point.

78. Explain why division of fractions is not associative using the statement below.

$$\left(\frac{3}{4} \div \frac{2}{5}\right) \div \frac{1}{3} \neq \frac{3}{4} \div \left(\frac{2}{5} \div \frac{1}{3}\right)$$

▶ YOU BE THE JUDGE

79. Dave claims that every number has a reciprocal. Is he correct? Explain your decision.

3.7 APPLICATIONS: MULTIPLICATION AND DIVISION OF FRACTIONS

OBJECTIVES

1 Find a fraction of an amount (Sections 1.6, 2.3).

2 Solve repeated addition problems.

3 Find how many times one amount is in another.

4 Split an amount *a* into *b* equal parts.

Refer to the key words in Sections 1.6 and 2.3, and the procedure for solving word problems in Section 1.6.

1 Finding a Fraction of an Amount

Taking a fraction *of* an amount means *multiply* the fraction times the amount. To see why, suppose there are 6 rooms in a house. One half of the rooms are painted. This means 3 rooms are painted because $\frac{1}{2}$ of 6 rooms is 3 rooms.

Multiply $\frac{1}{2}$ times 6 to get the same result.

$$\frac{1}{2} \text{ of 6 rooms} = \frac{1}{2} \times 6 = \frac{1}{\cancel{2}} \times \frac{\cancel{6}^{3}}{1} = 3 \text{ rooms}$$

6 rooms

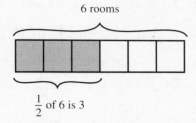

$\frac{1}{2}$ of 6 is 3

Taking a fraction of another fraction also means multiply. For example,

$$\frac{1}{2} \text{ of } \frac{3}{4} = \frac{1}{2} \times \frac{3}{4} = \frac{3}{8}.$$

To see why this is true, represent $\frac{3}{4}$ using the figure.

$$\frac{3}{4} = \boxed{}$$

You can see that $\frac{1}{2}$ of $\frac{3}{4}$ is half of the shaded figure. This is $\frac{3}{8}$ of it, as shown in the next figure.

$$\frac{1}{2} \text{ of } \frac{3}{4} = \boxed{} = \frac{3}{8}$$

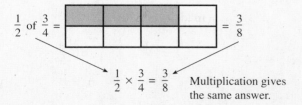

$\frac{1}{2} \times \frac{3}{4} = \frac{3}{8}$ Multiplication gives the same answer.

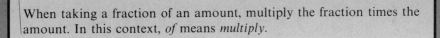

When taking a fraction of an amount, multiply the fraction times the amount. In this context, *of* means *multiply*.

EXAMPLE 1   The gas tank on a riding mower holds $2\frac{1}{4}$ gallons of gas. If the tank is $\frac{2}{3}$ full, how many gallons of gas are in the tank?

A full tank contains $2\frac{1}{4}$ gallons of gas. When $\frac{2}{3}$ full, the tank contains $\frac{2}{3}$ of $2\frac{1}{4}$ gallons. Since this is a *fraction of an amount*, multiply the fraction times the amount.

$$\frac{2}{3} \text{ of } 2\frac{1}{4} = \frac{2}{3} \times 2\frac{1}{4} = \frac{\overset{1}{\cancel{2}}}{\underset{1}{\cancel{3}}} \times \frac{\overset{3}{\cancel{9}}}{\underset{2}{\cancel{4}}}$$

$$= \frac{3}{2} \text{ or } 1\frac{1}{2} \text{ gallons of gas left in the tank} \quad \blacksquare$$

▶ **You Try It**   **1.** A gas tank has a capacity of $16\frac{1}{2}$ gallons. The tank is $\frac{2}{3}$ full. How many gallons of gas are in the tank?

EXAMPLE 2   A real estate agent sold a home for $145,500. The agent makes a commission of $\frac{7}{100}$ of the selling price of the home. What commission is made on this sale?

The agent makes $\frac{7}{100}$ of the $145,500 selling price. Since you are taking a *fraction of an amount*, multiply the fraction times the amount.

$$\frac{7}{100} \text{ of } \$145,500 = \frac{7}{100} \times \$145,500 = \frac{7}{\underset{1}{\cancel{100}}} \times \frac{\overset{\$1,455}{\cancel{\$145,500}}}{1} = \$10,185$$

The agent made a $10,185 commission on the sale of the house.   ■

▶ **You Try It**   **2.** A real estate agent sold 5 acres of land for $79,300. The agent made a commission of $\frac{9}{100}$ of the selling price. What commission did the agent make on this sale?

EXAMPLE 3   Kim makes $104 for working an 8-hour day. She only worked 5 hours today. How much did she make?

Kim worked 5 hours out of 8, or $\frac{5}{8}$ of a day. She is paid $104 a day. So she made $\frac{5}{8}$ of $104.

$$\frac{5}{8} \text{ of } \$104 = \frac{5}{\underset{1}{\cancel{8}}} \times \frac{\overset{\$13}{\cancel{\$104}}}{1} = \$65 \quad \blacksquare$$

▶ **You Try It** **3.** Irma makes $150 for working a 12-hour shift. How much will she make if she only works 7 hours of her shift?

2 Solving Repeated Addition Problems

You buy 5 gifts costing $8 each. Find the total cost by adding $8 to itself 5 times.

$$\$8 + \$8 + \$8 + \$8 + \$8 = \$40$$

This is repeated addition. Recall from Section 2.3 that repeated addition can be solved quickly by multiplying.

$$5 \times \$8 = \$40$$

Now suppose Nancy uses $\frac{2}{3}$ of a tank of gas to drive to work each week. How many tanks will she use in 6 weeks? Add $\frac{2}{3}$ to itself 6 times. Since this is repeated addition, just multiply $\frac{2}{3}$ times 6.

$$\frac{2}{3} \times 6 = \frac{2}{\underset{1}{3}} \times \frac{\overset{2}{6}}{1} = 4 \text{ tanks used in 6 weeks}$$

> A problem involving repeated addition of the same number n to itself p times can be solved by multiplying $n \cdot p$.

EXAMPLE 4 One share of GulfOil stock costs $\$29\frac{3}{8}$. You purchase 20 shares. What do you pay?

One share costs $\$29\frac{3}{8}$. The cost of 20 shares is $\$29\frac{3}{8}$ added to itself 20 times. Since this is *repeated addition*, just multiply 20 times $\$29\frac{3}{8}$.

$$20 \times 29\frac{3}{8} = \frac{\overset{5}{20}}{1} \times \frac{235}{\underset{2}{8}} = \frac{5 \times 235}{1 \times 2} = \$587\frac{1}{2} \text{ for 20 shares} \quad \blacksquare$$

EXAMPLE 5 One inch on a map represents $2\frac{3}{5}$ miles. How many miles long is a road that measures $7\frac{1}{2}$ inches on the map?

One inch on the map represents $2\frac{3}{5}$ miles. $7\frac{1}{2}$ inches equals $2\frac{3}{5}$ miles added to itself $7\frac{1}{2}$ times. This is *repeated addition*. Multiply to find the length of the road.

$$7\frac{1}{2} \times 2\frac{3}{5} = \frac{\overset{3}{15}}{2} \times \frac{13}{\underset{1}{5}} = \frac{3 \times 13}{2 \times 1} = \frac{39}{2} = 19\frac{1}{2} \text{ miles long} \quad \blacksquare$$

4. One share of ABH stock costs $12\dfrac{3}{4}$. What do you pay for 80 shares?

5. One inch on a map represents $7\dfrac{1}{2}$ miles. A road measures 5 inches on the map. How long is the road?

3 Finding How Many Times *b* Is Contained in *a*

To find how many 3's are in 12, write $12 \div 3 = 4$. In the same way, to find how many $\dfrac{1}{4}$ are in $\dfrac{3}{4}$ write:

$$\text{How many } \dfrac{1}{4} \text{ are in } \dfrac{3}{4} = \dfrac{3}{4} \div \dfrac{1}{4} = \dfrac{3}{\cancel{4}} \times \dfrac{\overset{1}{\cancel{4}}}{1} = 3.$$

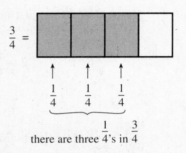

$\dfrac{3}{4} =$

$\dfrac{1}{4}$ $\dfrac{1}{4}$ $\dfrac{1}{4}$

there are three $\dfrac{1}{4}$'s in $\dfrac{3}{4}$

> To find how many times the number *b* is contained in the number *a*, divide
>
> $$a \div b.$$

EXAMPLE 6 How many shelves $1\dfrac{1}{2}$ feet long can be cut from a board 12 feet long?

To find *how many* $1\dfrac{1}{2}$-foot shelves *are contained in* a 12-foot board, divide 12 by $1\dfrac{1}{2}$.

$$\text{How many } 1\dfrac{1}{2} \text{ in } 12 = 12 \div 1\dfrac{1}{2} = \dfrac{12}{1} \div \dfrac{3}{2} = \dfrac{\overset{4}{\cancel{12}}}{1} \times \dfrac{2}{\cancel{3}} = \dfrac{4 \times 2}{1 \times 1} = 8 \text{ shelves.}$$

8 shelves each $1\dfrac{1}{2}$ feet long can be cut from a 12-foot board. ■

▶ **You Try It** **6.** How many pieces of wire $2\dfrac{1}{2}$ inches long can be cut from a 35-inch piece of wire?

The next example illustrates the difference between the answer to a division problem, and the answer to its associated word problem.

EXAMPLE 7 **Chapter Problem**   Read the problem at the beginning of this chapter. A dump truck can deliver $6\frac{1}{4}$ tons of stone in one trip. How many trips are needed to deliver 85 tons of stone?

$6\frac{1}{4}$ tons can be delivered in one trip. To find how many trips are needed to deliver 85 tons, find *how many times* $6\frac{1}{4}$ *is contained in* 85.

$$\text{How many } 6\frac{1}{4} \text{ in } 85 = 85 \div 6\frac{1}{4} = \frac{85}{1} \div \frac{25}{4} = \frac{\overset{17}{\cancel{85}}}{1} \times \frac{4}{\underset{5}{\cancel{25}}} = \frac{17 \times 4}{1 \times 5} = 13\frac{3}{5}.$$

$$13\frac{3}{5} \text{ loads} = \underbrace{13 \text{ full loads}}_{13 \text{ trips}} + \underbrace{\frac{3}{5} \text{ of a load}}_{1 \text{ trip}} = 14 \text{ trips needed}$$

CAUTION $13\frac{3}{5}$ is the answer to the division problem. It is not the answer to the word problem. The answer to the word problem is 14 trips. ∎

▶ **You Try It**   7. A truck can haul $5\frac{1}{3}$ cubic yards of topsoil. How many trips must be made to deliver 42 cubic yards?

4 Splitting an Amount *a* into *b* Equal Parts

To split \$20 into 4 equal parts, write \$20 ÷ 4 = \$5 per part.

> To split an amount *a* into *b* equal parts, write
>
> $$a \div b$$
>
> total amount to ⟋   ⟍ number of equal
> be divided up   parts wanted

EXAMPLE 8   A $4\frac{1}{2}$-foot rope is cut into 3 equal pieces. How long is each piece?

$4\frac{1}{2}$ feet is the *amount to be split into 3 equal parts.*

$$4\frac{1}{2} \div 3 = \frac{9}{2} \div \frac{3}{1} = \frac{\overset{3}{\cancel{9}}}{2} \times \frac{1}{\underset{1}{\cancel{3}}} = \frac{3 \times 1}{2 \times 1} = \frac{3}{2} = 1\frac{1}{2} \text{ ft}$$

Each piece of rope will be $1\frac{1}{2}$ feet long. This is pictured in the diagram.

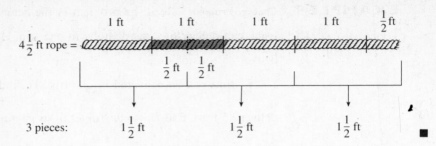

EXAMPLE 9 Tom purchased $7\frac{1}{2}$ pounds of ground beef to make 30 hamburgers. How much ground beef will be in each hamburger?

Tom wants to *split* $7\frac{1}{2}$ pounds of ground beef *into* 30 *equal parts*.

$$7\frac{1}{2} \div 30 = \frac{15}{2} \div \frac{30}{1} = \frac{\overset{1}{\cancel{15}}}{2} \times \frac{1}{\underset{2}{\cancel{30}}} = \frac{1 \times 1}{2 \times 2} = \frac{1}{4} \text{ lbs}$$

Each hamburger will contain $\frac{1}{4}$ pound of hamburger meat. ■

▶ **You Try It** 8. $3\frac{1}{2}$ tons of mulch is split evenly among 5 customers. How much mulch does each customer get?

9. Susan purchased $10\frac{3}{4}$ pounds of ham to make 43 sandwiches. How much ham can she put in each sandwich?

Division Summary	
Fractions stay in same order:	Fractions reverse order:
1. $\frac{2}{3}$ divided by $\frac{4}{5} = \frac{2}{3} \div \frac{4}{5}$.	**4.** How many $\frac{1}{5}$ are in $\frac{7}{8} =$
	$\frac{7}{8} \div \frac{1}{5}$.
2. Divide $\frac{2}{3}$ by $\frac{4}{5} = \frac{2}{3} \div \frac{4}{5}$.	
3. Split $\frac{3}{4}$ into 3 parts $= \frac{3}{4} \div 3$.	**5.** Divide $\frac{1}{5}$ into $\frac{7}{8} = \frac{7}{8} \div \frac{1}{5}$.

▶ **Answers to You Try It**   1. 11 gal   2. $7,137   3. 87\frac{1}{2}$ 4. $1,020

5. $37\frac{1}{2}$ mi   6. 14 pieces   7. 8 trips   8. $\frac{7}{10}$ of a ton   9. $\frac{1}{4}$ lb

---

■ ② ③ ④

1. What is $\frac{1}{3}$ of 600?

2. What is $\frac{3}{4}$ of 800?

3. Find $\frac{4}{7}$ of 350 men.

4. Find $\frac{7}{10}$ of 32,000 people.

5. If $\frac{2}{9}$ of a cow's milk is butterfat, how much butterfat is contained in 12 gallons of cow's milk?

6. Mr. Malacki purchased 32 stamps and used $\frac{3}{8}$ of them to send a package to his daughter in college. How many stamps did he use?

7. How many $\frac{1}{8}$ are in 1?

8. How many $\frac{2}{5}$ are in 4?

9. How many $\frac{7}{10}$ are in $2\frac{4}{5}$?

10. How many $4\frac{3}{8}$ are in $30\frac{5}{8}$?

11. How many $\frac{2}{5}$-quart packages can be made from 8 quarts of berries?

13. Justin weighs 156 pounds. On the moon he weighs $\frac{1}{6}$ of what he weighs on Earth. What would Justin weigh on the moon?

12. A family orders $16\frac{1}{4}$ tons of coal. They use $\frac{5}{8}$ tons each week. For how many weeks will the coal last?

14. A real estate agent makes a commission of $\frac{7}{100}$ of the selling price of a home. Find the commission on a $162,500 sale.

15. How many $\frac{1}{4}$-pound hamburgers can be made from $18\frac{3}{4}$ pounds of ground beef?

16. How many small bags of flour can be packed from the large sack of flour?

$2\frac{2}{3}$ lb each

17. What is 7 divided by $\frac{4}{9}$?

18. What is $2\frac{1}{2}$ divided by $\frac{3}{10}$?

19. What is $\frac{5}{18}$ of $2\frac{4}{7}$?

20. What is $\frac{16}{21}$ of $7\frac{4}{9}$?

21. Divide $3\frac{1}{6}$ by $2\frac{2}{3}$.

22. Divide $8\frac{5}{8}$ by $3\frac{1}{4}$.

23. Lettuce is $\frac{24}{25}$ water. How much water is in 50 pounds of lettuce?

27. A steak weighs $2\frac{1}{2}$ pounds. It costs $3\frac{1}{2}$ dollars per pound. What is the total cost of the steak?

24. Peaches, when dried, lose $\frac{7}{10}$ of their fresh weight. How much weight will 45 pounds of peaches lose when dried?

28. A plane flies 175 miles in one hour. How far will it go in $\frac{5}{8}$ hours?

25. You want to tile a hallway 24 feet long. Each tile is $\frac{3}{4}$ feet long. How many tiles must be placed end to end to extend the length of the hallway?

29. What is $\frac{5}{9}$ divided by $\frac{1}{9}$?

26. A road $17\frac{5}{8}$ miles long has signposts placed after each $\frac{3}{8}$ of a mile. How many signposts are needed?

30. What is $\frac{1}{9}$ divided by $\frac{5}{9}$?

31. Chris has $3\frac{1}{2}$ pies and wants to serve 14 people. What part of a pie will each person be served?

32. How many shelves can be cut from the board shown here?

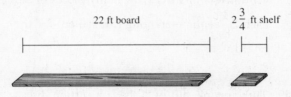

22 ft board $2\frac{3}{4}$ ft shelf

33. Al worked from noon to 8 P.M. on his history project. He spent $\frac{4}{5}$ of this time doing research. How many hours of research did he do?

34. One inch on a map represents $3\frac{3}{8}$ miles. How many miles long is a road that measures $6\frac{2}{3}$ inches on the map?

35. How many trips must the truck make to haul all of the dirt shown in the figure?

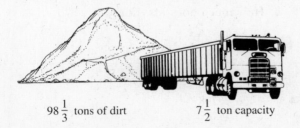

$98\frac{1}{3}$ tons of dirt $7\frac{1}{2}$ ton capacity

36. A total of 138 gallons of oil must be carried 40 feet. If a $4\frac{1}{5}$-gallon container is used to carry the oil, how many trips are needed?

37. Using 93 pints of cola, how many $\frac{3}{4}$-pint cups can be filled?

38. A 10-acre piece of land is subdivided into $\frac{2}{5}$-acre lots. How many lots will you get?

39. A family can save $\frac{1}{8}$ of its $1,240 monthly income.
   a. How long will it take to save $1,085 for a dining room set?

   b. How long will it take to save $2,635 for a used car?

40. A theater that can hold 2,730 people is $\frac{2}{3}$ full. How many people are seated in the theater?

41. What is the thickness of a stack of 24 boards if each board is $1\frac{1}{4}$ inches thick?

42. A stock sells for $16\frac{3}{4}$ per share. What will 45 shares cost?

43. There are 450 people who will be inoculated. If each inoculation contains $\frac{2}{9}$ of an ounce of serum, how many ounces of the serum are needed?

44. A hospital has 72 ounces of a drug. If each patient receives $\frac{3}{8}$ of an ounce, how many patients can be treated?

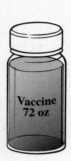

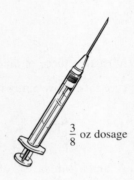

$\frac{3}{8}$ oz dosage

45. A couple applies for a mortgage. Their monthly payment can be no larger than $\frac{1}{4}$ of their monthly income. If their monthly income is $2,860 what maximum mortgage payment are they allowed?

46. A basketball team scored 112 points. Christie scored $\frac{2}{7}$ of the points. How many points did Christie score?

47. One tie requires $\frac{3}{16}$ of a yard of material. How many ties can be made from 24 yards of material?

48. Ellen is building a circuit board. She has 189 centimeters of wire. How many pieces of wire each $2\frac{5}{8}$ centimeters long can she cut for the board?

49. Three people out of 80 have a certain disease. In a town of 72,000 people, how many will have the disease?

50. Five out of every 8 ounces of a solution is acid. How much acid is contained in 32 ounces of the solution?

51. Tom's insurance will pay $\frac{4}{5}$ of a $2,670 medical bill.
a. How much will the insurance pay?

b. How much will Tom pay?

52. A jury awards Ms. Edwards $87,000 for injuries. Her lawyers take $\frac{1}{3}$ of this amount for their legal fees.
a. How much do the lawyers take?

b. How much does Ms. Edwards get?

SKILLSFOCUS (Section 2.6) *Solve each equation.*

53. $x + 56 = 147$

54. $y + 426 = 798$

55. $x - 56 = 300$

56. $w - 79 = 34$

EXTEND YOUR THINKING ▶▶▶▶
▶ **SOMETHING MORE**

57. Snow A snow gauge collected $9\frac{3}{8}$ inches of snow.

   a. A very dry snow gives 1 inch of water to every 30 inches of snow. How many inches of water will be in the gauge if the snow is very dry?

   b. A very wet snow gives 1 inch of water to every 6 inches of snow. How many inches of water will be in the gauge if the snow is very wet?

58. One inoculation uses $\frac{3}{8}$ ounce of antibiotic and costs \$2. You plan to inoculate 160,000 people.
   a. How many ounces of antibiotic are needed?

   b. What is the cost?

59. Two children need dessert in their lunch for 5 school days. Dad purchased a package of 12 cupcakes.

   a. Is there enough to give each child $1\frac{1}{2}$ cupcakes per day for 5 days?

   b. At this rate, exactly how long will the cupcakes last?

60. Ed gets 20 miles to a gallon of gas. His gas tank has a capacity of $14\frac{2}{5}$ gallons. Ed fills his tank, and uses $9\frac{3}{10}$ gallons to drive to the beach. Driving home that evening, he ran out of gas. How far was Ed from home when he ran out of gas?

61. Recipes A recipe that serves 6 people calls for the following.

$1\frac{1}{2}$ pints pistachio ice cream $\frac{3}{4}$ cup whipping cream

6 maraschino cherries $\frac{1}{2}$ cup sugar

$\frac{1}{3}$ cup pistachio nuts   1 package raspberries

   a. How much of each ingredient is needed to serve 2 people?

   b. How much of each ingredient is needed to serve 24 people?

62. The Fractional Pie Chart and Budget The Gaynor family lives on a \$1,920 monthly budget. The money is spent according to the pie chart on the right. For example, $\frac{1}{6}$ of \$1,920 is spent on food each month. How much of the monthly budget is spent on each of the five listed categories?

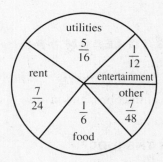

63. Mrs. Smith left \$96,000 to be divided among her 3 sons Tom, Dick, and Harry. She left $\frac{1}{4}$ to Tom, $\frac{5}{12}$ to Dick, and $\frac{1}{3}$ to Harry. Tom kept $\frac{1}{3}$ of his share and gave $\frac{1}{3}$ to each of his daughters Terry and Teresa. Dick kept $\frac{1}{2}$ of his share and split the rest evenly between his 2 sons Dan and David. Harry kept \$2,000 of his share and gave the rest to his daughters Helen, Hazel, and Harriet in an even split. How much did each person receive?

64. There are 16,000 people attending a concert. Of those attending, $\frac{4}{5}$ are from California. Of the Californians, $\frac{3}{8}$ are from L.A. How many people from L.A. are attending the concert?

65. Sue uses $1\frac{7}{10}$ gallons of gas to drive to work one way. Her gas tank has a capacity of $16\frac{1}{2}$ gallons. If Sue fills her tank, will she have enough gas to drive back and forth to work for 5 days?

▶ **TROUBLESHOOT IT**

Find and correct the error.

66. How many $\frac{5}{8}$ are in $\frac{2}{3} = \frac{5}{8} \div \frac{2}{3} = \frac{5}{8} \times \frac{3}{2} = \frac{15}{16}$.

67. Split $\frac{5}{6}$ pounds into 3 equal parts $= \frac{\overset{1}{\cancel{3}}}{1} \times \frac{5}{\underset{2}{\cancel{6}}} = 2\frac{1}{2}$ lb per part.

68. Divide $2\frac{1}{2}$ into $6\frac{1}{2} = 6\frac{1}{2} \div 2\frac{1}{2} = (6 \div 2) + \left(\frac{1}{2} \div \frac{1}{2}\right) = 3 + 1 = 4$.

WRITING TO LEARN ▶▶▶▶

69. Write a word problem about 5 sisters. The answer is $7\frac{1}{2}$. The problem must include division of fractions, and involve a supermarket. Then write the solution.

▶ **YOU BE THE JUDGE**

70. Yana claims if you multiply two proper fractions, the answer will be smaller than both fractions. Is she right? Explain your decision. $\left(\text{Hint: Use } \frac{1}{2} \times \frac{1}{2} = \frac{1}{4} \text{ as an example,}\right.$ and the fact that $\frac{1}{2} \times \frac{1}{2} = \frac{1}{2}$ of $\frac{1}{2}.\Big)$

CHAPTER 3 REVIEW

VOCABULARY AND MATCHING

New words and phrases introduced in this chapter are shown in the left-hand column. Match each term on the left with the phrase or sentence on the right that best describes it.

A. fraction   _____   the numerator and denominator

B. numerator   _____   what 6 is to 30

C. fraction bar   _____   write the reciprocal of the fraction

D. denominator   _____   like 1/2, 5/8, or 3/5; not like 2/4, 10/16, or 6/10

E. terms of a fraction   _____   the process of writing a fraction in lowest terms

F. proper fraction   _____   has 3 or more divisors

G. improper fraction   _____   write each term as a product of prime factors, then reduce common factors

H. mixed number   _____   indicates division

I. divisor   _____   whole number plus a fraction

J. successive division by primes   _____   writing a counting number as a product of primes only

K. prime number   _____   examples are 2/3, 4/6, 6/9, and 8/12

L. composite number   _____   numerator < denominator; it is always less than 1

M. prime factorization   _____   method used to prime factor a composite number

N. equivalent fractions   _____   product of the numerators divided by the product of the denominators

O. lowest terms fraction   _____   a portion of a whole amount; may be proper, improper, or mixed

P. reducing   _____   when used to indicate a fraction of an amount, it means multiply

Q. reducing by inspection   _____   number of parts you divide the whole into; bottom number of a fraction

R. reducing by prime factoring   _____   numerator > denominator or numerator = denominator

S. multiplying fractions   _____   multiply the dividend by the reciprocal of the divisor

T. reciprocal   _____   the number of parts you take; the top number of a fraction

U. invert a fraction   _____   find a common factor by looking, then divide numerator and denominator by it

V. dividing fractions   _____   has exactly two divisors

W. of   _____   for 2/5 is 5/2; for 7 is 1/7

REVIEW EXERCISES

3.1 The Meaning of a Fraction

1. How many sixths are in one whole?

2. How many twentieths are in one whole?

What fraction is represented by the shading in each figure?

3.

4.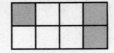

5.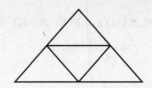

Draw and shade a figure for each fraction.

6. $\dfrac{5}{6}$

7. $\dfrac{7}{10}$

8. $\dfrac{9}{4}$

Simplify each fraction, or write not possible.

9. $\dfrac{14}{7}$

10. $\dfrac{11}{11}$

11. $\dfrac{0}{7}$

12. $\dfrac{2}{0}$

3.2 Proper and Improper Fractions; Mixed Numbers

Change each mixed number to an improper fraction.

13. $4\dfrac{1}{5}$.

14. $2\dfrac{9}{10}$

15. $5\dfrac{3}{7}$

16. $15\dfrac{2}{3}$

Change each improper fraction to a mixed number.

17. $\dfrac{19}{4}$

18. $\dfrac{11}{9}$

19. $\dfrac{45}{8}$

20. $\dfrac{107}{12}$

3.3 Prime Numbers and Prime Factoring

Find all the divisors. Identify each as prime or composite.

21. 49

22. 26

23. 280

24. 73

25. 167

26. 68

27. 94

28. 89

Prime factor.

29. 117　　　　　　　**30.** 94　　　　　　　**31.** 56　　　　　　　**32.** 71

33. 365　　　　　　　**34.** 1,200　　　　　　　**35.** 273　　　　　　　**36.** 38

3.4 Reducing Fractions

Reduce each fraction to lowest terms.

37. $\dfrac{8}{12}$　　　　　**38.** $\dfrac{9}{15}$　　　　　**39.** $\dfrac{21}{24}$　　　　　**40.** $\dfrac{24}{30}$

41. $\dfrac{136}{90}$　　　　　**42.** $\dfrac{200}{125}$　　　　　**43.** $\dfrac{153}{171}$　　　　　**44.** $\dfrac{840}{1,200}$

3.5 Multiplying Fractions

45. $\dfrac{4}{5} \times \dfrac{3}{7}$　　　　**46.** $\dfrac{7}{10} \times \dfrac{15}{24}$　　　　**47.** $\dfrac{25}{36} \times \dfrac{30}{54}$　　　　**48.** $\dfrac{15}{8} \times \dfrac{9}{12}$

49. $3\dfrac{2}{3} \times 2\dfrac{1}{4}$　　　　**50.** $2\dfrac{5}{8} \times 4\dfrac{1}{6}$　　　　**51.** $3\dfrac{5}{12} \times 9$　　　　**52.** $6 \times 1\dfrac{7}{9}$

53. $\left(\dfrac{7}{9}\right)^3$　　　　**54.** $\left(\dfrac{3}{10}\right)^3$　　　　**55.** $\left(\dfrac{13}{20}\right)^2$　　　　**56.** $\left(4\dfrac{1}{3}\right)^2$

3.6 Dividing Fractions

57. $\dfrac{5}{8} \div \dfrac{1}{4}$　　　　**58.** $\dfrac{3}{10} \div \dfrac{7}{12}$　　　　**59.** $\dfrac{7}{9} \div \dfrac{14}{15}$　　　　**60.** $\dfrac{16}{25} \div \dfrac{1}{30}$

61. $3\dfrac{1}{5} \div \dfrac{2}{3}$　　　　**62.** $6\dfrac{3}{4} \div 2\dfrac{5}{6}$　　　　**63.** $6 \div \dfrac{2}{5}$　　　　**64.** $1\dfrac{7}{8} \div 10$

65. Solve for *x.* $\dfrac{5}{6}x = \dfrac{3}{8}$　　　　　　　**66.** Solve for *w.* $\dfrac{9}{10}w = \dfrac{18}{25}$

3.7 Applications: Multiplication and Division of Fractions

67. Twenty-four cars need an oil change. If each car requires $4\frac{2}{3}$ quarts of oil, how much oil is needed altogether?

68. What will $3\frac{1}{2}$ pounds of beef cost if one pound costs $\$2\frac{3}{4}$?

69. How many $\frac{4}{5}$ are in 25?

70. Tom uses $\frac{9}{20}$ of a gallon of gas to mow his lawn. He mows his lawn once a week. If he has $6\frac{3}{4}$ gallons of gas stored in his garage, for how many weeks can he mow his lawn before needing more gas?

71. Fred purchased $5\frac{2}{3}$ pounds of plums. If there are 6 plums per pound, how many plums did he buy?

72. Elvin's basketball team scored 135 points. If Elvin scored $\frac{2}{5}$ of his team's points, how many points did he score?

73. How many $2\frac{3}{4}$-inch wires can be cut from a 77-inch length of wire?

74. A dress can be made from $2\frac{1}{4}$ yards of fabric. How many dresses can be made from 27 yards of fabric?

Allow yourself 50 minutes to complete this test. Write the work for each problem. When done, check your answers. Rework each problem solved incorrectly.

Change each mixed number to an improper fraction.

1. $5\frac{1}{6}$

2. $12\frac{2}{9}$

3. $3\frac{7}{15}$

4. $1\frac{7}{8}$

Change each improper fraction to a mixed number.

5. $\frac{21}{4}$

6. $\frac{45}{16}$

7. $\frac{67}{9}$

8. $\frac{101}{3}$

9. Find all the divisors of 200.

10. Find all the divisors of 88.

11. Prime factor 114.

12. Prime factor 70.

13. $\frac{7}{8} \cdot \frac{3}{10}$

14. $8\frac{1}{6} \cdot 3\frac{3}{7}$

15. $9 \cdot 4\frac{1}{6}$

16. $5\frac{5}{8} \cdot 8 \cdot \frac{3}{10}$

17. $\frac{7}{9} \div \frac{15}{21}$

18. $2\frac{3}{5} \div 3\frac{9}{10}$

19. $12 \div 1\frac{1}{5}$

20. $3\frac{7}{16} \div \frac{21}{32}$

21. $\left(\frac{5}{8}\right)^3$

22. Solve $\frac{7}{8}w = \frac{1}{2}$.

23. A circuit board requires 64 wires each $1\frac{7}{16}$ inches long. What total length of wire is needed?

24. Mario budgeted \$4,000 for furniture. Suppose $\frac{3}{8}$ of it can be spent on den furniture. How much money is budgeted for the den?

25. Ellen uses $\frac{3}{10}$ of a gallon of gas when she mows her lawn. How many times can she mow her lawn on 6 gallons of gas?

26. A new development has $16\frac{3}{4}$ acres of land. This land will be subdivided into half-acre lots.

a. How many lots will there be?

b. How much land is left over?

Explain the meaning of each term. Use your own examples.

27. numerator

28. lowest terms fraction

29. reciprocal

NAME   DATE   HOUR

1. Add.   45,209
               8,352
              71,984
          +   695

2. Subtract.   54,021
                  − 6,263

3. Multiply.   603
                  × 294

4. Divide.   45)‾6,302‾

5. Round 5,816 to the nearest ten.

6. Round 62,948 to the nearest hundred.

7. Simplify $4 + 32 \div 8 - 7$.

8. Find P in $P = 2L + 2W$ when $L = 5$ and $W = 12$.

9. What is nine plus the product of six and four?

10. What is seven times the sum of ten and thirty?

11. Find all the divisors of 96.

12. Prime factor 63.

13. $\dfrac{4}{5} \cdot \dfrac{8}{10}$

14. $\dfrac{21}{30} \cdot 1\dfrac{7}{18}$

15. $4\dfrac{1}{2} \div 3$

16. $5\dfrac{1}{4} \div \dfrac{7}{8}$

17. Solve $\dfrac{2}{3} \cdot s = \dfrac{4}{9}$.

18. $z - 198 = 402$

19. Two-fifths of the monthly budget is used to pay the mortgage. If the monthly budget is $2,650, how large is the mortgage payment?

20. How many lengths of wire each $3\dfrac{5}{8}$ centimeters long can be cut from a $43\dfrac{1}{2}$-centimeter length of wire?

Explain the meaning of each term. Use your own examples.

21. solution to an equation

22. square root

4

Addition and Subtraction of Fractions

THE NEW YORK STOCK EXCHANGE

Fractions are used extensively in finance. See this for yourself in the following segment taken from the financial page of a popular daily newspaper.

Donna purchased 400 shares of GulfOil stock at the Low price. Two hours later she sold them at the High for the day. What was her gain from the transaction?

From the NYSE summary of the 20 most active stocks, the Low price for GulfOil was $27\frac{3}{4}$ per share. The High was $30\frac{3}{8}$ per share.

Her gain per share was $\qquad \$30\frac{3}{8} - \$27\frac{3}{4}$.

Her gain for 400 shares was

$$400\left(\$30\frac{3}{8} - \$27\frac{3}{4}\right).$$

In this chapter you will see how to evaluate this expression.

NYSE SUMMARY 20 MOST ACTIVE

	Sales	High	Low	Last	Chg.
Exxon s	5,338,400	$29\frac{3}{8}$	$27\frac{3}{4}$	29	$+ \frac{7}{8}$
IBM	4,188,100	$59\frac{5}{8}$	$57\frac{1}{8}$	$57\frac{7}{8}$	$- \frac{5}{8}$
RCA	4,012,800	$21\frac{5}{8}$	$19\frac{1}{8}$	$19\frac{1}{4}$	$- \frac{1}{8}$
Tandy s	3,391,500	29	25	27	$-1\frac{1}{2}$
Mobil s	2,820,600	$23\frac{1}{2}$	$20\frac{3}{4}$	$23\frac{1}{8}$	$+2\frac{3}{8}$
ATT	2,712,200	$57\frac{1}{8}$	$53\frac{1}{3}$	57	$+ \frac{3}{8}$
MaOil	2,647,600	$76\frac{1}{4}$	$74\frac{1}{4}$	$75\frac{5}{8}$	$+ 8$
Sears	2,466,600	$18\frac{1}{4}$	$17\frac{3}{4}$	$18\frac{1}{4}$	$- \frac{1}{2}$
TexInt	2,411,500	$13\frac{3}{4}$	$10\frac{3}{4}$	$13\frac{1}{2}$	$+1\frac{1}{4}$
SFeInd s	2,347,700	$14\frac{3}{8}$	13	$13\frac{7}{8}$	$- \frac{1}{2}$
Schlimb s	2,330,200	$44\frac{1}{2}$	40	$42\frac{5}{8}$	$- \frac{1}{4}$
Digital	2,131,300	$78\frac{3}{8}$	$71\frac{3}{8}$	$72\frac{1}{4}$	$-5\frac{5}{8}$
Deltaa s	2,040,000	31	$28\frac{1}{4}$	$28\frac{1}{2}$	$-1\frac{7}{8}$
StOInd	2,026,300	38	$34\frac{1}{8}$	$37\frac{3}{4}$	$+2\frac{1}{2}$
K mart	1,972,400	$18\frac{3}{4}$	$17\frac{3}{4}$	$18\frac{1}{4}$	$- \frac{3}{8}$
WrnCm	1,935,400	$55\frac{1}{4}$	$50\frac{1}{4}$	$51\frac{5}{8}$	$-2\frac{1}{2}$
Pennzol	1,932,800	$42\frac{1}{2}$	$33\frac{5}{8}$	34	$-6\frac{1}{2}$
Halbtn	1,856,900	$35\frac{3}{4}$	32	$35\frac{1}{2}$	$+1\frac{1}{2}$
GulfOil	1,812,500	$30\frac{3}{8}$	$27\frac{3}{4}$	$29\frac{3}{8}$	$+1\frac{5}{8}$
PhilPet	1,799,700	$29\frac{5}{8}$	27	$28\frac{1}{4}$	$- \frac{7}{8}$

Work through each section in this chapter. Use this test to identify topics you are not familiar with. These topics may require additional study. You may *not* use a calculator.

4.1

1. Add. $\dfrac{2}{5} + \dfrac{4}{5}$

2. Add. $3\dfrac{1}{8} + 6$

3. Add. $4\dfrac{5}{9} + 2\dfrac{8}{9}$

4. Subtract. $\dfrac{7}{10} - \dfrac{3}{10}$

5. Subtract. $5\dfrac{4}{7} - 2\dfrac{1}{7}$

4.2

6. Are $\dfrac{6}{14}$ and $\dfrac{15}{35}$ equivalent?

7. Are $\dfrac{21}{24}$ and $\dfrac{49}{64}$ equivalent?

8. Write four fractions equivalent to $\dfrac{6}{7}$.

9. Find the missing numerator. $\dfrac{3}{8} = \dfrac{}{48}$

10. Find the missing numerator. $\dfrac{5}{6} = \dfrac{}{72}$

4.3 *Find the least common denominator for each set of fractions.*

11. $\dfrac{7}{8}$ and $\dfrac{1}{6}$

12. $\dfrac{5}{12}$ and $\dfrac{3}{16}$

13. $\dfrac{5}{9}, \dfrac{11}{18},$ and $\dfrac{7}{45}$

4.4

14. $\dfrac{7}{8} + \dfrac{1}{6}$

15. $1\dfrac{5}{12} + 2\dfrac{3}{16}$

16. $\dfrac{5}{9} + 6\dfrac{1}{3} + 1\dfrac{3}{4}$

4.5

17. $\dfrac{7}{10} - \dfrac{1}{3}$

18. $5\dfrac{3}{8} - 2\dfrac{4}{5}$

19. $7 - 2\dfrac{5}{16}$

4.6

20. A share of stock sold for $\$46\dfrac{3}{4}$ yesterday. Today it sells for $\$1\dfrac{7}{8}$ more. What is today's price for one share?

21. A board is 16 feet long. A shelf $8\dfrac{7}{12}$ feet long is cut from it. What length of board is left?

22. Ed's old car used $12\dfrac{3}{10}$ gallons of gas each week. His new car uses $2\dfrac{4}{5}$ more gallons per week. How many gallons per week does his new car use?

4.7

23. Simplify. $\dfrac{2}{3} + \dfrac{1}{3} \cdot \dfrac{3}{5}$

24. Arrange the fractions in order from smallest to largest. $\dfrac{5}{6}, \dfrac{8}{9}, \dfrac{2}{3}$

VOCABULARY *Explain the meaning of each term. Use your own examples.*

25. unlike fractions

26. expand a fraction

4.1 ADDING AND SUBTRACTING LIKE FRACTIONS

OBJECTIVES

1 Add like fractions and mixed numbers (Section 3.4).

2 Subtract like fractions and mixed numbers.

3 Solve equations (Section 2.6, Objectives 1, 2).

NEW VOCABULARY

like fractions

---

1 Adding Like Fractions

You can add quantities that have the same name.

$$2 \text{ cars} + 3 \text{ cars} = 5 \text{ cars}$$

$$5 \text{ men} + 4 \text{ men} = 9 \text{ men}$$

$$1 \text{ fourth} + 2 \text{ fourths} = 3 \text{ fourths}$$

$$\frac{1}{4} + \frac{2}{4} = \frac{3}{4}$$

Like fractions have the same denominator. $\frac{1}{4}$ and $\frac{2}{4}$ are like fractions. To add like fractions, add the numerators ($1 + 2 = 3$). Place this sum over the common denominator, 4. This addition is pictured in the figure.

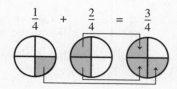

$$\frac{1}{4} + \frac{2}{4} = \frac{3}{4}$$

To add like fractions

1. Add the numerators.
2. Write this sum over the common denominator.
3. Reduce.

EXAMPLE 1

$$\frac{3}{5} + \frac{1}{5} = \frac{3+1}{5} = \frac{4}{5} \quad \blacksquare$$

EXAMPLE 2

$$\frac{3}{20} + \frac{7}{20} + \frac{13}{20} = \frac{3+7+13}{20} = \frac{23}{20} = 1\frac{3}{20} \quad \blacksquare$$

EXAMPLE 3

$$\frac{2}{9} + \frac{4}{9} = \frac{2+4}{9} = \frac{\overset{2}{\cancel{6}}}{\underset{3}{\cancel{9}}} = \frac{2}{3} \quad \blacksquare$$

You may add fractions vertically.

EXAMPLE 4 Add $\frac{11}{12} + \frac{5}{12}$.

$$\begin{array}{r} \frac{11}{12} \\ +\frac{5}{12} \\ \hline \frac{16}{12} \end{array}$$

Add the numerators, $11 + 5 = 16$.
Write 16 over the common denominator 12, and reduce.

$$\frac{16}{12} = \frac{\overset{4}{\cancel{16}}}{\underset{3}{\cancel{12}}} = \frac{4}{3} = 1\frac{1}{3} \quad \blacksquare$$

CAUTION Never add the two denominators together. The addition below shows why this gives the wrong answer.

$$\cancel{\dfrac{1}{2} + \dfrac{1}{2} = \dfrac{1+1}{2+2} = \dfrac{2}{4} = \dfrac{1}{2}}$$

$\dfrac{1}{2} + \dfrac{1}{2}$ equals 1 whole, not $\dfrac{1}{2}$:

▶ **You Try It**   **1.** $\dfrac{1}{6} + \dfrac{3}{6}$ **2.** $\dfrac{5}{8} + \dfrac{7}{8}$ **3.** $\dfrac{4}{11} + \dfrac{2}{11}$ **4.** $\dfrac{7}{10} + \dfrac{9}{10}$ **5.** $\dfrac{2}{5} + \dfrac{1}{5} + \dfrac{7}{5} + \dfrac{4}{5}$

Adding Like Mixed Numbers

Like mixed numbers have proper fractions with the same denominator. To add them, use the fact that addition is commutative and associative.

$$3\dfrac{2}{7} + 1\dfrac{4}{7} = \left(3 + \dfrac{2}{7}\right) + \left(1 + \dfrac{4}{7}\right) = (3+1) + \left(\dfrac{2}{7} + \dfrac{4}{7}\right) = 4 + \dfrac{6}{7} = 4\dfrac{6}{7}$$

Write each mixed number as an addition.   Add the whole numbers.   Add the proper fractions.

> To add like mixed numbers
>
> **1.** Add the whole numbers.   **2.** Add the like proper fractions.
> **3.** Reduce.

EXAMPLE 5 $4\dfrac{1}{5} + 2\dfrac{3}{5} = (4+2) + \left(\dfrac{1}{5} + \dfrac{3}{5}\right) = 6 + \dfrac{4}{5} = 6\dfrac{4}{5}$ ■

EXAMPLE 6 $7 + 2\dfrac{5}{6} = (7+2) + \dfrac{5}{6} = 9 + \dfrac{5}{6} = 9\dfrac{5}{6}$ ■

EXAMPLE 7 $3\dfrac{1}{8} + \dfrac{5}{8} = 3 + \left(\dfrac{1}{8} + \dfrac{5}{8}\right) = 3 + \dfrac{\overset{3}{\cancel{6}}}{\underset{4}{\cancel{8}}} = 3\dfrac{3}{4}$ ■

You may add mixed numbers vertically.

EXAMPLE 8

Write the whole numbers in the same column.
Write the proper fractions in the same column.

$$
\begin{aligned}
8\dfrac{7}{10}& \\
+\,5\dfrac{9}{10}& \\
\hline
13\dfrac{16}{10}& = 13 + 1\dfrac{\overset{3}{\cancel{6}}}{\underset{5}{\cancel{10}}} = 14\dfrac{3}{5} \quad ■
\end{aligned}
$$

▶ **You Try It**   **6.** $1\dfrac{6}{7} + 4\dfrac{5}{7}$ **7.** $6\dfrac{1}{3} + 4\dfrac{1}{3}$ **8.** $2 + 5\dfrac{2}{9}$ **9.** $7\dfrac{1}{4} + \dfrac{3}{4}$

2 Subtracting Like Fractions

You can subtract quantities that have the same name.

$$5 \text{ cabs} - 2 \text{ cabs} = 3 \text{ cabs}$$

$$8 \text{ girls} - 3 \text{ girls} = 5 \text{ girls}$$

$$7 \text{ tenths} - 4 \text{ tenths} = 3 \text{ tenths}$$

$$\frac{7}{10} - \frac{4}{10} = \frac{3}{10}$$

> **To subtract like fractions**
>
> 1. Subtract the two numerators.
> 2. Write this difference over the common denominator.
> 3. Reduce.

EXAMPLE 9

$$\frac{6}{7} - \frac{2}{7} = \frac{6-2}{7} = \frac{4}{7} \ \blacksquare$$

EXAMPLE 10

$$\frac{7}{8} - \frac{3}{8} = \frac{7-3}{8} = \frac{\overset{1}{\cancel{4}}}{\underset{2}{\cancel{8}}} = \frac{1}{2} \ \blacksquare$$

EXAMPLE 11

$$\frac{13}{5} - \frac{6}{5} = \frac{13-6}{5} = \frac{7}{5} = 1\frac{2}{5} \ \blacksquare$$

EXAMPLE 12

$$\begin{array}{r} \dfrac{7}{12} \\[2mm] -\dfrac{5}{12} \\[1mm] \hline \dfrac{2}{12} \end{array} = \frac{\overset{1}{\cancel{2}}}{\underset{6}{\cancel{12}}} = \frac{1}{6} \ \blacksquare$$

> **You Try It**
>
> **10.** $\dfrac{5}{7} - \dfrac{1}{7}$ **11.** $\dfrac{9}{10} - \dfrac{3}{10}$ **12.** $\dfrac{16}{3} - \dfrac{8}{3}$ **13.** $\dfrac{11}{16} - \dfrac{7}{16}$

Subtracting Like Mixed Numbers

EXAMPLE 13

First subtract the proper fractions.
Then subtract the whole numbers.

$$\begin{array}{r} 3\dfrac{5}{8} \\[2mm] -1\dfrac{2}{8} \\[1mm] \hline 2\dfrac{3}{8} \ \blacksquare \end{array}$$

> **To subtract two like mixed numbers**
>
> 1. Subtract the like proper fractions.   2. Subtract the whole numbers.
> 3. Reduce.

EXAMPLE 14 **a.**

Drop down the whole number, 5.
Subtract the like proper fractions.

$$5\frac{6}{7}$$
$$-\ \frac{4}{7}$$
$$\overline{\quad 5\frac{2}{7}}$$

b.

$$8\frac{7}{10}$$
$$-2\frac{3}{10}$$
$$\overline{\quad 6\frac{4}{10}} = 6\frac{\overset{2}{\cancel{4}}}{\underset{5}{\cancel{10}}} = 6\frac{2}{5}\ \blacksquare$$

▶ **You Try It**   **14.** $4\frac{7}{10} - 1\frac{3}{10}$ **15.** $6\frac{8}{9} - \frac{4}{9}$ **16.** $14\frac{7}{12} - 6\frac{1}{12}$

3 Solving Equations   Recall from Section 2.6, to solve an addition equation, write the related subtraction. To solve a subtraction equation, write the related addition.

EXAMPLE 15   Solve. $x + \frac{1}{5} = \frac{4}{5}$

Solve the addition equation $x + \frac{1}{5} = \frac{4}{5}$ by writing the related subtraction equation.

$$x = \frac{4}{5} - \frac{1}{5} = \frac{3}{5}$$

Check: $\frac{3}{5} + \frac{1}{5} = \frac{4}{5}$ $\blacksquare$

EXAMPLE 16   Solve. $s - \frac{2}{7} = \frac{3}{7}$

To solve a subtraction equation, write the related addition.

$$s = \frac{3}{7} + \frac{2}{7} = \frac{5}{7}$$

Check: $\frac{5}{7} - \frac{2}{7} = \frac{3}{7}$ $\blacksquare$

▶ **You Try It**   Solve for x.

17. $x + \frac{4}{11} = \frac{10}{11}$ **18.** $x - \frac{2}{9} = \frac{5}{9}$

▶ **Answers to You Try It**   **1.** $\frac{2}{3}$ **2.** $1\frac{1}{2}$ **3.** $\frac{6}{11}$ **4.** $1\frac{3}{5}$ **5.** $2\frac{4}{5}$ **6.** $6\frac{4}{7}$

7. $10\frac{2}{3}$ **8.** $7\frac{2}{9}$ **9.** 8   **10.** $\frac{4}{7}$ **11.** $\frac{3}{5}$ **12.** $2\frac{2}{3}$ **13.** $\frac{1}{4}$ **14.** $3\frac{2}{5}$ **15.** $6\frac{4}{9}$

16. $8\frac{1}{2}$ **17.** $\frac{6}{11}$ **18.** $\frac{7}{9}$

SECTION 4.1 EXERCISES

1 *Add the fractions, and reduce your answers.*

1. $\dfrac{2}{9} + \dfrac{5}{9}$
2. $\dfrac{3}{7} + \dfrac{2}{7}$
3. $\dfrac{3}{8} + \dfrac{1}{8}$
4. $\dfrac{2}{5} + \dfrac{2}{5}$

5. $\dfrac{7}{10} + \dfrac{6}{10}$
6. $\dfrac{10}{13} + \dfrac{8}{13}$
7. $\dfrac{1}{18} + \dfrac{5}{18}$
8. $\dfrac{9}{16} + \dfrac{23}{16}$

9. $\dfrac{13}{24} + \dfrac{7}{24}$
10. $\dfrac{13}{30} + \dfrac{11}{30}$
11. $\dfrac{7}{25} + \dfrac{10}{25}$
12. $\dfrac{27}{50} + \dfrac{33}{50}$

13. $\dfrac{1}{8} + \dfrac{9}{8} + \dfrac{4}{8}$
14. $\dfrac{5}{6} + \dfrac{7}{6} + \dfrac{3}{6}$
15. $\dfrac{3}{9} + \dfrac{7}{9} + \dfrac{8}{9}$
16. $\dfrac{1}{4} + \dfrac{3}{4} + \dfrac{2}{4}$

17. $\dfrac{16}{25} + \dfrac{21}{25} + \dfrac{18}{25}$
18. $\dfrac{4}{16} + \dfrac{9}{16} + \dfrac{7}{16}$
19. $\dfrac{2}{7} + \dfrac{6}{7} + \dfrac{6}{7}$
20. $\dfrac{26}{45} + \dfrac{14}{45} + \dfrac{32}{45}$

Add the mixed numbers, and reduce your answers.

21. $3\dfrac{2}{5} + 4\dfrac{1}{5}$
22. $2\dfrac{1}{8} + 6\dfrac{5}{8}$
23. $1\dfrac{1}{3} + 3\dfrac{2}{3}$
24. $6\dfrac{1}{6} + 9\dfrac{5}{6}$

25. $\begin{aligned} 5\dfrac{4}{9} \\ +7\dfrac{2}{9} \\ \hline \end{aligned}$
26. $\begin{aligned} 10\dfrac{4}{7} \\ +13\dfrac{1}{7} \\ \hline \end{aligned}$
27. $\begin{aligned} 4\dfrac{3}{10} \\ +2\dfrac{9}{10} \\ \hline \end{aligned}$
28. $\begin{aligned} 7\dfrac{11}{14} \\ +6\dfrac{13}{14} \\ \hline \end{aligned}$
29. $\begin{aligned} 6\dfrac{7}{12} \\ +11\dfrac{11}{12} \\ \hline \end{aligned}$
30. $\begin{aligned} 9\dfrac{3}{10} \\ +5\dfrac{7}{10} \\ \hline \end{aligned}$

31. $3\dfrac{8}{9} + 4\dfrac{3}{9} + 1\dfrac{7}{9}$
32. $10\dfrac{5}{7} + 4 + \dfrac{4}{7} + 8\dfrac{6}{7}$

33. $23\dfrac{5}{12} + 14\dfrac{11}{12} + 10\dfrac{8}{12}$
34. $7\dfrac{5}{6} + 4\dfrac{1}{6} + 3\dfrac{7}{6}$

2 *Subtract the fractions, and reduce all answers.*

35. $\dfrac{5}{7} - \dfrac{2}{7}$
36. $\dfrac{3}{5} - \dfrac{2}{5}$
37. $\dfrac{3}{8} - \dfrac{1}{8}$
38. $\dfrac{4}{5} - \dfrac{2}{5}$

39. $\dfrac{9}{8} - \dfrac{5}{8}$
40. $\dfrac{7}{6} - \dfrac{3}{6}$
41. $\dfrac{7}{10} - \dfrac{6}{10}$
42. $\dfrac{10}{13} - \dfrac{8}{13}$

43. $\dfrac{11}{12} - \dfrac{7}{12}$
44. $\dfrac{9}{11} - \dfrac{4}{11}$
45. $\dfrac{7}{18} - \dfrac{1}{18}$
46. $\dfrac{23}{16} - \dfrac{2}{16}$

Subtract the mixed numbers, and reduce all answers.

47. $4\frac{4}{7} - 3\frac{2}{7}$ **48.** $6\frac{7}{8} - 3\frac{1}{8}$ **49.** $3\frac{2}{3} - 3\frac{1}{3}$ **50.** $9\frac{5}{6} - 7\frac{5}{6}$

51. $9\frac{8}{9}$ **52.** $16\frac{4}{7}$ **53.** $7\frac{11}{14}$ **54.** $12\frac{7}{10}$ **55.** $8\frac{3}{4}$ **56.** $2\frac{21}{25}$

$\quad\ -7\frac{2}{9}$ $\quad\ -13\frac{1}{7}$ $\quad\ -6\frac{4}{14}$ $\quad\ -6\frac{2}{10}$ $\quad\ -\frac{1}{4}$ $\quad\ -\frac{16}{25}$

57. Subtract $\frac{7}{10}$ from $\frac{9}{10}$. **58.** Subtract $\frac{5}{16}$ from $\frac{14}{16}$.

59. Subtract $7\frac{5}{12}$ from $16\frac{10}{12}$. **60.** Subtract $2\frac{1}{6}$ from $4\frac{3}{6}$.

3 *Solve the equation. Check each answer.*

61. $s + \frac{2}{9} = \frac{7}{9}$ **62.** $t + \frac{3}{11} = \frac{10}{11}$ **63.** $k - \frac{1}{7} = \frac{4}{7}$

64. $y - \frac{9}{13} = \frac{3}{13}$ **65.** $x + 3\frac{2}{5} = 7\frac{3}{5}$ **66.** $h - 4\frac{1}{3} = 5\frac{1}{3}$

SKILLSFOCUS (Section 1.9) *Evaluate.*

67. 7^2 **68.** 10^1 **69.** 2^4 **70.** 6^0

EXTEND YOUR THINKING ▶▶▶▶

▶SOMETHING MORE

71. Four-eighteenths of a class had an A average. One-eighteenth had a D average or lower. Eight-eighteenths had a C average. What fraction of the class had a B average?

72. Two families want to split a bushel of steamed crabs in a fair way. One family has 3 adults and 2 children. The other has 6 adults and 2 children. A child will eat half as many crabs as an adult. Crabs cost $77 a bushel. How much should each family pay?

▶TROUBLESHOOT IT

Find and correct the error.

73. $\frac{5}{8} + \frac{7}{8} = \frac{12}{16} = \frac{3}{4}$ **74.** $\frac{7}{9} - \frac{2}{9} = \frac{7-2}{9-9} = \frac{5}{0} = 0$

WRITING TO LEARN ▶▶▶▶

75. Explain why $\frac{1}{2} + \frac{1}{2} \neq \frac{2}{4}$. Relate your explanation to your own experience.

4.2 EQUIVALENT FRACTIONS

OBJECTIVES

1 Determine if two fractions are equivalent (Section 3.4).

2 Expand a fraction to get an equivalent fraction (Section 1.7, Objective 3).

3 Rename a fraction in terms of a larger denominator.

NEW VOCABULARY

equivalent fractions   lower terms

expanding fractions   higher terms

rename a fraction

---

1 Determining if Fractions Are Equivalent

Equivalent fractions are equal fractions. The fractions $\frac{2}{3}$ and $\frac{4}{6}$ are equivalent because they represent the same portion of the whole.

$$\frac{2}{3} = \qquad\qquad \left.\begin{array}{c} \\ \\ \end{array}\right\} \text{Therefore,}$$

$$\frac{4}{6} = \qquad\qquad \frac{2}{3} = \frac{4}{6}$$

$\frac{2}{3}$ is in lowest terms. $\frac{4}{6}$ reduces to $\frac{2}{3}$.

$$\frac{4}{6} = \frac{\overset{2}{\cancel{4}}}{\underset{3}{\cancel{6}}} = \frac{2}{3}$$

If two fractions reduce to the same lowest terms answer, they are equivalent.

> **To determine if two fractions are equivalent**
>
> 1. Reduce both to lowest terms.
> 2. They are equivalent if they reduce to the same lowest terms answer. Otherwise, they are unequal.

EXAMPLE 1 Are the fractions $\frac{6}{8}$ and $\frac{15}{20}$ equivalent?

Reduce both fractions to lowest terms.

$$\frac{6}{8} = \frac{\overset{3}{\cancel{6}}}{\underset{4}{\cancel{8}}} = \frac{3}{4} \qquad\qquad \frac{15}{20} = \frac{\overset{3}{\cancel{15}}}{\underset{4}{\cancel{20}}} = \frac{3}{4}$$

$\frac{6}{8} = \frac{15}{20}$ because both reduce to the same lowest terms answer, $\frac{3}{4}$. ∎

EXAMPLE 2 Are the fractions $\frac{16}{28}$ and $\frac{24}{40}$ equivalent?

Reduce both fractions to lowest terms.

$$\frac{16}{28} = \frac{\overset{4}{\cancel{16}}}{\underset{7}{\cancel{28}}} = \frac{4}{7} \qquad\qquad \frac{24}{40} = \frac{\overset{3}{\cancel{24}}}{\underset{5}{\cancel{40}}} = \frac{3}{5}$$

Since $\frac{4}{7} \neq \frac{3}{5}$, the fractions $\frac{16}{28}$ and $\frac{24}{40}$ are not equal. ■

▶ **You Try It**   Determine if the fractions are equivalent.

1. $\frac{6}{14}$ and $\frac{15}{36}$
2. $\frac{12}{27}$ and $\frac{20}{45}$

---

2 **Expanding Fractions**

Reducing a fraction writes the fraction with a smaller numerator and denominator. **Expanding a fraction** writes it with a larger numerator and denominator. The fraction resulting from either reducing or expanding is equivalent to the original fraction. Expanding fractions will be used later to add and subtract fractions with unlike denominators.

The number 1 can be written as any counting number divided by itself.

$$1 = \frac{1}{1} = \frac{2}{2} = \frac{3}{3} = \frac{4}{4} = \frac{5}{5} = \frac{6}{6} = \cdots$$

You expand a fraction by multiplying it by 1 written in any of these forms. The number 1 is the multiplicative identity (Section 1.7). Multiplying by 1 does not change the value of a fraction.

EXAMPLE 3   Expand $\frac{2}{7}$ using $\frac{5}{5}$.

$$\frac{2}{7} = \frac{2}{7} \times \boxed{\frac{5}{5}} = \frac{10}{35}$$

Check by reducing: $\frac{10}{35} = \frac{10 \div 5}{35 \div 5} = \frac{2}{7}$ ■

EXAMPLE 4   Expand $\frac{7}{9}$ using $\frac{6}{6}$.

$$\frac{7}{9} = \frac{7}{9} \times \boxed{\frac{6}{6}} = \frac{42}{54}$$

Check: $\frac{42}{54} = \frac{42 \div 6}{54 \div 6} = \frac{7}{9}$ ■

▶ **You Try It**   **3.** Expand $\frac{5}{8}$ using $\frac{6}{6}$.
4. Expand $\frac{1}{10}$ using $\frac{3}{3}$.

Any fraction can be written in an unlimited number of equivalent ways. For example, $\frac{1}{2}$ can be written in the following equivalent ways by multiplying it by 1 written as $\frac{2}{2}, \frac{3}{3}, \frac{4}{4}$, and so on.

$$\frac{1}{2} \times \frac{1}{1} = \frac{1}{2}$$

$$\frac{1}{2} \times \frac{2}{2} = \frac{2}{4}$$

$$\frac{1}{2} \times \frac{3}{3} = \frac{3}{6}$$

$$\frac{1}{2} \times \frac{4}{4} = \frac{4}{8}$$

$$\frac{1}{2} \times \frac{5}{5} = \frac{5}{10}$$

Each fraction in this column equals 1. Multiplying by 1 does not change the value of a fraction.

Each fraction in this column is equivalent to $\frac{1}{2}$. As a check, reduce each of these fractions to lowest terms. You get $\frac{1}{2}$.

. . . and so on.

EXAMPLE 5   Write four fractions equivalent to $\frac{3}{5}$.

$$\frac{3}{5}, \frac{3}{5} \times \frac{2}{2} = \frac{6}{10}, \quad \frac{3}{5} \times \frac{3}{3} = \frac{9}{15}, \quad \frac{3}{5} \times \frac{4}{4} = \frac{12}{20}, \quad \frac{3}{5} \times \frac{5}{5} = \frac{15}{25}$$

fractions equivalent to $\frac{3}{5}$: $\frac{3}{5}, \frac{6}{10}, \frac{9}{15}, \frac{12}{20}, \frac{15}{25}$ ∎

EXAMPLE 6   Write five fractions equivalent to $\frac{3}{8}$.

$$\frac{3}{8}, \frac{6}{16}, \frac{9}{24}, \frac{12}{32}, \frac{15}{40}, \frac{18}{48}$$ ∎

▶ **You Try It**   **5.** Write four fractions equivalent to $\frac{1}{6}$.

6. Write five fractions equivalent to $\frac{5}{7}$.

To **rename a fraction** means to write an equivalent fraction. You can rename a fraction in two ways.

1. Reducing: Divide numerator and denominator by the same counting number, n. This gives an equivalent fraction in **lower terms**.

$$\frac{a}{b} = \frac{a \div n}{b \div n}$$

2. Expanding: Multiply numerator and denominator by the same counting number, n. This gives an equivalent fraction in **higher terms** (with a larger numerator and denominator).

$$\frac{a}{b} = \frac{a \cdot n}{b \cdot n}$$

3 Renaming Fractions in Terms of a Larger Denominator

Multiplying a fraction by 1 gives an equivalent fraction. To write a fraction in higher terms, multiply the fraction by 1 written in one of the forms

$$1 = \frac{1}{1} = \frac{2}{2} = \frac{3}{3} = \frac{4}{4} = \frac{5}{5} = \frac{6}{6} = \cdots.$$

To write a fraction in terms of a larger denominator

1. Divide the larger denominator by the smaller.
2. Multiply the original fraction by 1 written as the quotient in Step 1 divided by itself.

EXAMPLE 7 Find the missing numerator. $\frac{2}{3} = \frac{?}{12}$

Divide the denominators. $12 \div 3 = 4$. Multiply $\frac{2}{3}$ by $1 = \frac{4}{4}$.

$$\frac{2}{3} = \frac{2}{3} \times \frac{4}{4} = \frac{8}{12}.$$

The missing numerator is 8. ∎

▶ **You Try It** 7. Find the missing numerator. $\frac{4}{7} = \frac{?}{28}$

EXAMPLE 8 Find the missing numerator. $\frac{5}{6} = \frac{?}{42}$

Divide the denominators. $42 \div 6 = 7$. Multiply $\frac{5}{6}$ by $1 = \frac{7}{7}$.

$$\frac{5}{6} = \frac{5}{6} \times \frac{7}{7} = \frac{35}{42}.$$

The missing numerator is 35. ∎

EXAMPLE 9 Find the missing numerator. $7\frac{3}{8} = 7\frac{?}{40}$

Find the missing numerator so that $\frac{3}{8} = \frac{?}{40}$. Divide the denominators.

$40 \div 8 = 5$. Multiply $\frac{3}{8}$ by $1 = \frac{5}{5}$.

$$7\frac{3}{8} = 7 + \frac{3}{8} \times \frac{5}{5} = 7 + \frac{15}{40} = 7\frac{15}{40}.$$

The missing numerator is 15. ∎

▶ **You Try It** Find the missing numerator. 8. $\frac{3}{5} = \frac{?}{45}$ 9. $5\frac{3}{10} = 5\frac{?}{60}$

▶ **Answers to You Try It** 1. no: $\frac{3}{7} \neq \frac{5}{12}$ 2. yes: $\frac{4}{9} = \frac{4}{9}$ 3. $\frac{30}{48}$
4. $\frac{3}{30}$ 5. $\frac{2}{12}, \frac{3}{18}, \frac{4}{24}, \frac{5}{30}$ 6. $\frac{10}{14}, \frac{15}{21}, \frac{20}{28}, \frac{25}{35}, \frac{30}{42}$ 7. 16   8. 27   9. 18

1 *Determine if the fractions are equivalent or not by reducing each fraction to lowest terms.*

1. $\dfrac{1}{2}, \dfrac{7}{14}$

2. $\dfrac{2}{3}, \dfrac{8}{12}$

3. $\dfrac{3}{4}, \dfrac{16}{24}$

4. $\dfrac{1}{3}, \dfrac{6}{18}$

5. $\dfrac{2}{8}, \dfrac{12}{40}$

6. $\dfrac{12}{36}, \dfrac{18}{27}$

7. $\dfrac{6}{10}, \dfrac{21}{35}$

8. $\dfrac{12}{15}, \dfrac{21}{24}$

9. $\dfrac{4}{12}, \dfrac{8}{18}$

10. $\dfrac{20}{28}, \dfrac{24}{42}$

11. $\dfrac{12}{14}, \dfrac{18}{21}$

12. $\dfrac{20}{32}, \dfrac{24}{40}$

13. $\dfrac{36}{60}, \dfrac{32}{48}$

14. $\dfrac{44}{55}, \dfrac{12}{15}$

15. $\dfrac{35}{50}, \dfrac{60}{100}$

16. $\dfrac{64}{80}, \dfrac{15}{20}$

17. $\dfrac{150}{200}, \dfrac{80}{120}$

18. $\dfrac{14}{24}, \dfrac{21}{36}$

19. $\dfrac{8}{20}, \dfrac{10}{25}$

20. $\dfrac{6}{16}, \dfrac{16}{40}$

2 *Expand each fraction using the form given for 1.*

21. $\dfrac{3}{4}$ using $\dfrac{6}{6}$

22. $\dfrac{1}{3}$ using $\dfrac{7}{7}$

23. $\dfrac{5}{6}$ using $\dfrac{4}{4}$

24. $\dfrac{2}{5}$ using $\dfrac{12}{12}$

25. $\dfrac{7}{8}$ using $\dfrac{8}{8}$

26. $\dfrac{5}{9}$ using $\dfrac{3}{3}$

27. $\dfrac{3}{8}$ using $\dfrac{6}{6}$

28. $\dfrac{7}{10}$ using $\dfrac{7}{7}$

29. $\dfrac{8}{15}$ using $\dfrac{2}{2}$

30. $\dfrac{7}{12}$ using $\dfrac{5}{5}$

31. $\dfrac{6}{7}$ using $\dfrac{3}{3}$

32. $\dfrac{16}{25}$ using $\dfrac{4}{4}$

Write five fractions equivalent to the given fraction.

33. $\dfrac{1}{3}$

34. $\dfrac{4}{5}$

35. $\dfrac{7}{10}$

36. $\dfrac{5}{8}$

37. $\dfrac{3}{4}$

38. $\dfrac{5}{6}$

3 *Find the missing numerator. Check each answer by reducing.*

39. $\dfrac{2}{3} = \dfrac{?}{6}$

40. $\dfrac{3}{4} = \dfrac{?}{8}$

41. $\dfrac{4}{5} = \dfrac{?}{20}$

42. $\dfrac{1}{6} = \dfrac{?}{18}$

43. $\dfrac{4}{9} = \dfrac{?}{36}$

44. $\dfrac{7}{8} = \dfrac{?}{24}$

45. $\dfrac{2}{7} = \dfrac{?}{35}$

46. $\dfrac{3}{10} = \dfrac{?}{50}$

47. $\dfrac{1}{3} = \dfrac{?}{18}$

48. $\dfrac{5}{7} = \dfrac{?}{28}$

49. $\dfrac{2}{9} = \dfrac{?}{45}$

50. $\dfrac{5}{8} = \dfrac{?}{56}$

51. $\dfrac{12}{11} = \dfrac{?}{33}$

52. $\dfrac{17}{10} = \dfrac{?}{50}$

53. $\dfrac{8}{5} = \dfrac{?}{35}$

54. $\dfrac{25}{12} = \dfrac{?}{60}$

55. $\dfrac{5}{2} = \dfrac{?}{70}$

56. $\dfrac{7}{4} = \dfrac{?}{28}$

57. $\dfrac{9}{6} = \dfrac{?}{54}$

58. $\dfrac{5}{4} = \dfrac{?}{52}$

59. $\dfrac{11}{8} = \dfrac{?}{48}$

60. $\dfrac{7}{5} = \dfrac{?}{30}$

61. $\dfrac{17}{6} = \dfrac{?}{54}$

62. $\dfrac{41}{10} = \dfrac{?}{90}$

63. $3\dfrac{7}{9} = 3\dfrac{?}{27}$

64. $2\dfrac{5}{8} = 2\dfrac{?}{64}$

65. $6\dfrac{2}{3} = 6\dfrac{?}{24}$

66. $5\dfrac{7}{12} = 5\dfrac{?}{84}$

SKILLSFOCUS (Section 3.3) *Prime factor.*

67. 40

68. 72

69. 18

70. 21

71. 33

EXTEND YOUR THINKING ▶▶▶▶
▶ TROUBLESHOOT IT

Find and correct the error.

72. The following fractions are equivalent to $\dfrac{3}{7}$: $\dfrac{6}{14}, \dfrac{18}{35}, \dfrac{9}{21}, \dfrac{36}{63}, \dfrac{21}{49}$.

WRITING TO LEARN ▶▶▶▶

73. Explain the difference between expanding a fraction and reducing a fraction.

OBJECTIVES

1 Define common denominator and why you need it.
2 Find the LCD by inspection.
3 Find the LCD by prime factoring (Section 3.3).
4 Find the LCD by successive division by primes.

NEW VOCABULARY

common denominator

least common denominator (LCD)

common multiple

least common multiple (LCM)

inspection

1 The Common Denominator and Why You Need It

You cannot add 3 dimes to 2 quarters and get 5 dimes, or 5 quarters, because the quantities are unlike. To add dimes to quarters you must first write both in terms of some common name, such as cents.

$$
\begin{aligned}
3 \text{ dimes} &= \boxed{30 \text{ cents}} \\
+\ 2 \text{ quarters} &= \boxed{50 \text{ cents}} \\
\hline
3 \text{ dimes} + 2 \text{ quarters} &= \boxed{80 \text{ cents}}
\end{aligned}
$$

In the same way, you cannot add unlike fractions. For example, add $\frac{1}{2} + \frac{1}{3}$. The problem is pictured in the figure.

How do you add unlike shapes?

$\frac{1}{2}$ + $\frac{1}{3}$ = ?

Halves and thirds are as different as dimes and quarters. You cannot add them until you first rewrite each in terms of a common name, or common denominator. Begin by writing fractions equivalent to $\frac{1}{2}$ and $\frac{1}{3}$.

$$\text{one half} = \frac{1}{2} = \frac{2}{4} = \frac{3}{6} = \frac{4}{8} = \frac{5}{10} = \frac{6}{12} = \frac{7}{14} = \frac{8}{16} = \frac{9}{18} = \cdots$$

$$\text{one third} = \frac{1}{3} = \frac{2}{6} = \frac{3}{9} = \frac{4}{12} = \frac{5}{15} = \frac{6}{18} = \frac{7}{21} = \frac{8}{24} = \cdots$$

To add $\frac{1}{2}$ to $\frac{1}{3}$, add a fraction equivalent to $\frac{1}{2}$ to a fraction equivalent to $\frac{1}{3}$ that have the same denominator. Three possible choices are shown using the denominators 6, 12, and 18.

$$\frac{1}{2} = \frac{3}{6} = $$

$$+\ \frac{1}{3} = \frac{2}{6} = $$

$$\frac{5}{6} = $$

$$\frac{1}{2} = \frac{6}{12}$$

$$+\ \frac{1}{3} = \frac{4}{12}$$

$$\frac{10}{12} = \frac{\overset{5}{\cancel{10}}}{\underset{6}{\cancel{12}}} = \frac{5}{6}$$

$$\frac{1}{2} = \frac{9}{18}$$

$$+\ \frac{1}{3} = \frac{6}{18}$$

$$\frac{15}{18} = \frac{\overset{5}{\cancel{15}}}{\underset{6}{\cancel{18}}} = \frac{5}{6}$$

To add irregularly shaped figures, divide them into pieces of the same size and shape. Divide each circle into sixths and add the sixths.

Same answer as on the left. But the terms are larger, and each answer had to be reduced.

In each addition, the answer is the same. The numbers 6, 12, and 18 are called common denominators. A **common denominator** for two fractions is any number divisible by both denominators.

The number 6 is called the least common denominator because it is the smallest. The **least common denominator (LCD)** for two fractions is the smallest number that both denominators divide into exactly.

Use the LCD to add and subtract fractions. It is easier to use because it is the smallest possible common denominator. Smaller numbers mean there is less chance of an addition or reducing error. Furthermore, when you use the LCD, your answer will sometimes be in lowest terms. If you use a denominator larger than the LCD, you will *always* have to reduce your answer.

A common denominator is also called a common multiple. 6, 12, and 18 are multiples of 2. They are also multiples of 3. As such, 6, 12, and 18 are called **common multiples** for 2 and 3. Because 6 is the smallest, it is called the **least common multiple (LCM)**. Finding the LCD for two fractions is the same as finding the LCM for the two denominators. Three methods are discussed in this section.

2 Finding the LCD by Inspection

If by reading a problem you can determine the LCD for two fractions, you are using the method of **inspection**.

EXAMPLE I   The LCD for $\frac{1}{2}$ and $\frac{3}{5}$ is 10 because 10 is the smallest number that both 2 and 5 divide into exactly.   ∎

EXAMPLE 2   The LCD for $\frac{3}{4}$ and $\frac{2}{7}$ is 28 because 28 is the smallest number that both 4 and 7 divide into exactly.   ∎

▶ **You Try It**   Find the LCD by inspection.

1. $\frac{2}{3}$ and $\frac{5}{6}$ ⬛⬛⬛⬛⬛⬛⬛⬛⬛⬛⬛⬛⬛ **2.** $\frac{1}{4}$ and $\frac{2}{5}$

3 Finding the LCD by Prime Factoring

Find the LCD for $\frac{7}{12}$ and $\frac{11}{15}$. It is the smallest number that both 12 and 15 divide into exactly. To find it, prime factor both denominators.

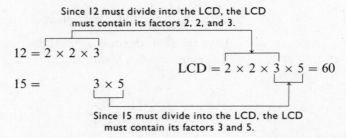

Since 12 must divide into the LCD, the LCD must contain its factors 2, 2, and 3.

$12 = 2 \times 2 \times 3$

$15 = \quad\quad 3 \times 5$

$LCD = 2 \times 2 \times 3 \times 5 = 60$

Since 15 must divide into the LCD, the LCD must contain its factors 3 and 5.

Notice that 12 and 15 each contain one factor of 3. But you did not need to include the factor 3 twice in the LCD. The LCD contains the least number of factors so that both prime factorizations are contained in it.

To find the LCD for two fractions using prime factoring

1. Prime factor both denominators.
2. Write like prime factors in the same column.
3. With each like prime, circle the greatest number of times it occurs in its column.
4. The LCD is the product of the circled factors.

EXAMPLE 3 Find the LCD for $\dfrac{7}{12}$ and $\dfrac{11}{15}$.

$$
\begin{aligned}
12 &= \boxed{2 \times 2} \times \boxed{3} \\
15 &= 3 \times \boxed{5} \\
\hline
\text{LCD} &= 2 \times 2 \times 3 \times 5 = 60
\end{aligned}
$$

Prime factor both denominators. Since 3 is the only factor common to both, only circle one factor of 3. Drop the circled factors down into the LCD, and multiply. ∎

▶ **You Try It**   3. Use prime factoring to find the LCD for $\dfrac{7}{9}$ and $\dfrac{5}{12}$.

EXAMPLE 4 Find the LCD for $\dfrac{17}{24}$ and $\dfrac{9}{40}$.

$$
\begin{aligned}
24 &= \boxed{2 \times 2 \times 2} \times \boxed{3} \\
40 &= 2 \times 2 \times 2 \times \boxed{5} \\
\hline
\text{LCD} &= 2 \times 2 \times 2 \times 3 \times 5 = 120 \quad ∎
\end{aligned}
$$

Prime factor both denominators. 2 is a factor 3 times in both prime factorizations. Circle one occurrence.

EXAMPLE 5 Find the LCD for $\dfrac{11}{30}$ and $\dfrac{4}{45}$.

$$
\begin{aligned}
30 &= \boxed{2} \times 3 \times \boxed{5} \\
45 &= \boxed{3 \times 3} \times 5 \\
\hline
\text{LCD} &= 2 \times 3 \times 3 \times 5 = 90 \quad ∎
\end{aligned}
$$

Since 3 occurs twice in 45, circle the 3×3 in 45. Do not circle the single occurence of 3 in 30.

▶ **You Try It**   Find the LCD by prime factoring.

4. $\dfrac{1}{36}$ and $\dfrac{7}{45}$ 5. $\dfrac{3}{20}$ and $\dfrac{9}{28}$

EXAMPLE 6 Find the LCD for $\dfrac{2}{45}$, $\dfrac{5}{36}$, and $\dfrac{13}{75}$.

$$
\begin{aligned}
36 &= \boxed{2 \times 2} \times \boxed{3 \times 3} \\
45 &= 3 \times 3 \times 5 \\
75 &= 3 \times \boxed{5 \times 5} \\
\hline
\text{LCD} &= 2 \times 2 \times 3 \times 3 \times 5 \times 5 = 900 \quad ∎
\end{aligned}
$$

Prime factor the three denominators. Since 3 occurs twice in two denominators, only circle one 3×3.

▶ **You Try It**   6. Find the LCD by prime factoring.

$$\frac{7}{21}, \frac{4}{35}, \text{ and } \frac{9}{30}$$

---

4 Finding the LCD by Successive Division by Primes

This procedure is a graphic approach to prime factorization.

> To find the LCD by using successive division by primes
>
> 1. Write the denominators on the same horizontal line.
> 2. Find a prime number that exactly divides two (or more) of the denominators. Write the quotients on the line below. (If a denominator is not divisible by the prime number, move it down unchanged.)
> 3. Keep repeating division by primes until no prime can be found that divides two (or more) of the quotients.
> 4. Multiply "around" for the LCD (that is, multiply all the divisors times the final quotients for the LCD).

EXAMPLE 7   Solve Example 3 using successive division by primes. Find the LCD for $\dfrac{7}{12}$ and $\dfrac{11}{15}$.

$$3 \,\big|\, \underline{12 \quad 15}$$
$$ 4 \quad 5$$

Write the denominators on the same line.
The prime 3 divides into both 12 and 15.

Write the quotients 4 and 5 directly below. Since no prime divides 4 and 5, you are done. Multiply "around" for the LCD.

$$\text{LCD} = 3 \times 4 \times 5 = 60 \quad \blacksquare$$

OBSERVE   The factors 3, 4 (as 2 × 2), and 5 used in Example 7 to get the LCD are the same factors found in Example 3 using prime factoring.

▶ **You Try It**   7. Use successive division by primes to find the LCD.

$$\frac{7}{15} \text{ and } \frac{4}{27}$$

EXAMPLE 8   Find the LCD for $\dfrac{5}{18}$ and $\dfrac{11}{24}$.

$$2 \,\big|\, \underline{18 \quad 24}$$
$$3 \,\big|\, \underline{9 \quad 12}$$
$$ 3 \quad 4$$

The prime 2 divides 18 and 24 exactly. Write the quotients 9 and 12 below.
The prime 3 divides 9 and 12 exactly. Write the quotients 3 and 4 below.
Since no prime divides both 3 and 4, you are done. Multiply around for the LCD.

$$\text{LCD} = 2 \times 3 \times 3 \times 4 = 72 \quad \blacksquare$$

EXAMPLE 9   Find the LCD for $\dfrac{2}{7}$ and $\dfrac{8}{9}$.

$$\big|\, \underline{7 \quad 9}$$

Since no prime divides both 7 and 9, the LCD is just the product of the two denominators.

$$\text{LCD} = 7 \times 9 = 63 \quad \blacksquare$$

▶ **You Try It** Use successive division by primes to find the LCD.

8. $\dfrac{2}{9}$ and $\dfrac{5}{18}$ **9.** $\dfrac{9}{32}$ and $\dfrac{11}{24}$ **10.** $\dfrac{13}{36}$ and $\dfrac{1}{60}$

The next two examples demonstrate how to find the LCD for 3 or more fractions.

EXAMPLE 10 Find the LCD for $\dfrac{5}{48}, \dfrac{3}{20},$ and $\dfrac{19}{30}.$

2	48	20	30	The prime 2 divides all three exactly.
2	24	10	15	2 divides 24 and 10 exactly. 2 does not divide 15. Drop down 15 as is.
3	12	5	15	3 divides 12 and 15. 3 does not divide the 5. Drop 5 down as is.
5	4	5	5	5 divides the two fives. 5 does not divide the 4. Drop down the 4.
	4	1	1	Since 4, 1, and 1 have no common factors, multiply around for the LCD.

LCD = $2 \times 2 \times 3 \times 5 \times 4 \times 1 \times 1 = 240$ ∎

> **OBSERVE** In Example 10, you clearly see the value of finding the *least* common denominator. Multiplying the original denominators always gives a common denominator. But $48 \times 20 \times 30 = 28{,}800$! Compared to 240, the denominator 28,800 is much too large to be practical.

EXAMPLE 11 Find the LCD for $\dfrac{3}{14}, \dfrac{10}{21},$ and $\dfrac{4}{5}.$

| 7 | 14 | 21 | 5 | No prime divides all three denominators. The prime 7 divides 14 and 21. |
| | 2 | 3 | 5 | Write the quotients 2 and 3 below. Drop down the 5. |

LCD = $7 \times 2 \times 3 \times 5 = 210$ ∎

▶ **You Try It** Use successive division by primes to find the LCD.

11. $\dfrac{3}{40}, \dfrac{7}{25},$ and $\dfrac{11}{90}$ **12.** $\dfrac{5}{22}, \dfrac{7}{33},$ and $\dfrac{1}{66}$

▶ **Answers to You Try It** 1. 6 2. 20 3. 36 4. 180 5. 140 6. 210 7. 135 8. 18 9. 96 10. 180 11. 1,800 12. 66

■ **1** **2** **3** **4** *Find the LCD for the given denominators using any method.*

1. 8 and 16   **2.** 21 and 15   **3.** 12 and 15   **4.** 10 and 18

5. 6 and 10   **6.** 9 and 16   **7.** 8 and 24   **8.** 7 and 14

9. 10 and 12   **10.** 14 and 22   **11.** 16 and 20   **12.** 15 and 25

13. 12 and 36   **14.** 20 and 40   **15.** 10 and 21   **16.** 6 and 25

17. 15 and 20   **18.** 10 and 36   **19.** 6 and 9   **20.** 14 and 16

21. 8 and 40   **22.** 4 and 24   **23.** 4 and 7   **24.** 5 and 11

25. 25 and 50   **26.** 33 and 11   **27.** 18 and 30   **28.** 24 and 36

29. 36 and 60   **30.** 30 and 45   **31.** 42 and 63   **32.** 39 and 26

33. 80 and 64   **34.** 90 and 120   **35.** 240 and 400   **36.** 77 and 121

37. 3, 5, and 10   **38.** 4, 8, and 12   **39.** 4, 5, and 7   **40.** 2, 3, and 5

41. 3, 4, and 6   **42.** 6, 8, and 18   **43.** 10, 15, and 20   **44.** 8, 16, and 40

45. 36, 24, and 30   **46.** 72, 40, and 54   **47.** 10, 17, and 12

48. 160, 144, and 96   **49.** 4, 8, 12, and 16   **50.** 12, 15, 90, and 45

SKILLSFOCUS (Section 3.2) *Change to an improper fraction.*

51. $4\frac{1}{6}$ **52.** $2\frac{7}{8}$ **53.** $6\frac{3}{4}$ **54.** $12\frac{2}{5}$

▶TROUBLESHOOT IT

Find and correct the error.

55. By inspection, the LCD for $\dfrac{3}{4}$ and $\dfrac{5}{8}$ is 32.

56. Find the LCD for 6, 9, and 12.

$$\begin{array}{c|ccc} 3 & 6 & 9 & 12 \\ \hline & 2 & 3 & 4 \end{array} \quad \text{LCD} = 3 \times 2 \times 3 \times 4 = 72$$

WRITING TO LEARN ▶▶▶▶

57. Explain how successive division by primes is similar to the method of prime factoring when finding the LCD.

58. Explain the difference between a common denominator and a least common denominator.

59. Make up your own problem to find the LCD for two fractions using successive division by primes. Then solve it. Explain each step.

OBJECTIVES

1 Add unlike proper fractions.

2 Add unlike mixed numbers.

---

1 Adding Unlike Proper Fractions

To add unlike fractions, rewrite them as like fractions. To do this, find the LCD, then rewrite each fraction as an equivalent fraction with the LCD as its denominator.

EXAMPLE 1 Add $\frac{1}{3} + \frac{2}{5}$.

By inspection, the LCD for 3 and 5 is 15. Rewrite each fraction as an equivalent fraction with a denominator of 15.

$$\frac{1}{3} = \frac{1}{3} \times \frac{5}{5} = \frac{5}{15} \qquad \text{Since } 15 \div 3 = 5, \text{ multiply } \frac{1}{3} \text{ by } 1 = \frac{5}{5}.$$

$$+\frac{2}{5} = \frac{2}{5} \times \frac{3}{3} = \frac{6}{15} \qquad \text{Since } 15 \div 5 = 3, \text{ multiply } \frac{2}{5} \text{ by } 1 = \frac{3}{3}.$$

$$\frac{11}{15} \qquad \text{Add the like fractions.} \quad \blacksquare$$

To add unlike fractions

1. Find the LCD for the denominators.
2. Rewrite each fraction as an equivalent fraction with the LCD as its denominator.
3. Add the like fractions. Reduce if needed.

EXAMPLE 2 What is $\frac{3}{4} + \frac{5}{7}$?

By inspection, the LCD for 4 and 7 is 28.

$$\frac{3}{4} = \frac{3}{4} \times \frac{7}{7} = \frac{21}{28}$$

$$+\frac{5}{7} = \frac{5}{7} \times \frac{4}{4} = \frac{20}{28}$$

$$\frac{41}{28} = 1\frac{13}{28} \quad \blacksquare$$

▶ **You Try It**   **1.** Add $\frac{5}{6} + \frac{3}{8}$.   **2.** Add $\frac{2}{9} + \frac{3}{5}$.

EXAMPLE 3 Add $\frac{7}{30} + \frac{5}{18}$.

$$30 = ② \times 3 \qquad \times ⑤ \qquad \text{Find the LCD for 30 and 18}$$
$$18 = 2 \times ③ \times 3 \qquad \text{by prime factoring.}$$
$$\text{LCD} = 2 \times 3 \times 3 \times 5 = 90$$

Rewrite each fraction as an equivalent fraction with a denominator of 90.

$$\frac{7}{30} = \frac{7}{30} \times \frac{3}{3} = \frac{21}{90}$$

$$+\frac{5}{18} = \frac{5}{18} \times \frac{5}{5} = \frac{25}{90}$$

$$\frac{46}{90} = \frac{\overset{23}{\cancel{46}}}{\underset{45}{\cancel{90}}} = \frac{23}{45} \quad \blacksquare$$

▶ **You Try It**   **3.** Add $\dfrac{7}{12} + \dfrac{11}{30}$.

EXAMPLE 4   Add $\dfrac{2}{3} + \dfrac{1}{4} + \dfrac{5}{6}$.

By inspection, the LCD for 3, 4, and 6 is 12.

$$\frac{2}{3} = \frac{2}{3} \times \frac{4}{4} = \frac{8}{12}$$

$$\frac{1}{4} = \frac{1}{4} \times \frac{3}{3} = \frac{3}{12}$$

$$+\frac{5}{6} = \frac{5}{6} \times \frac{2}{2} = \frac{10}{12}$$

$$\frac{21}{12} = \frac{\overset{7}{\cancel{21}}}{\underset{4}{\cancel{12}}} = \frac{7}{4} \quad \text{or} \quad 1\frac{3}{4} \quad \blacksquare$$

EXAMPLE 5   Add $\dfrac{7}{24} + \dfrac{1}{18} + \dfrac{4}{15}$.

Find the LCD using successive division by primes.

$$
\begin{array}{r|ccc}
2 & 24 & 18 & 15 \\
\hline
3 & 12 & 9 & 15 \\
\hline
 & 4 & 3 & 5
\end{array}
$$

$$\text{LCD} = 2 \times 3 \times 4 \times 3 \times 5 = 360$$

$$\frac{7}{24} = \frac{7}{24} \times \frac{15}{15} = \frac{105}{360}$$

$$\frac{1}{18} = \frac{1}{18} \times \frac{20}{20} = \frac{20}{360}$$

$$+\frac{4}{15} = \frac{4}{15} \times \frac{24}{24} = \frac{96}{360}$$

$$\frac{221}{360} \quad \blacksquare$$

▶ **You Try It**   **4.** Add $\dfrac{1}{6} + \dfrac{4}{5} + \dfrac{7}{15}$.   **5.** Add $\dfrac{4}{45} + \dfrac{1}{30} + \dfrac{3}{50}$.

You may reduce a fraction to lowest terms before adding.

EXAMPLE 6   Add $\dfrac{9}{16} + \dfrac{21}{24}$.

$$\frac{9}{16} \qquad = \frac{9}{16} \qquad = \frac{9}{16}$$

$$+\frac{21}{24} = \frac{\overset{7}{\cancel{21}}}{\underset{8}{\cancel{24}}} = \frac{7}{8} \times \frac{2}{2} = \frac{14}{16}$$

$$\frac{23}{16} = 1\frac{7}{16} \quad \blacksquare$$

6. Add $\dfrac{20}{36} + \dfrac{4}{27}$.

---

2 Adding Unlike Mixed Numbers

To add unlike mixed numbers, add the proper fractions. Then add the whole numbers.

EXAMPLE 7 Add $2\dfrac{1}{4} + 6\dfrac{3}{5}$.

Rewrite each proper fraction with LCD = 20.

$$2\dfrac{1}{4} = 2 + \dfrac{1}{4} = 2 + \dfrac{1}{4} \times \dfrac{5}{5} = 2 + \dfrac{5}{20}$$

$$+6\dfrac{3}{5} = 6 + \dfrac{3}{5} = 6 + \dfrac{3}{5} \times \dfrac{4}{4} = 6 + \dfrac{12}{20}$$

$$8 + \dfrac{17}{20} = 8\dfrac{17}{20} \quad \blacksquare$$

To add mixed numbers

1. Find the LCD for the proper fractions.
2. Rewrite each proper fraction as an equivalent fraction with the LCD as denominator.
3. Add the equivalent fractions.
4. Add the whole numbers.
5. Reduce if possible.

EXAMPLE 8 Add $4\dfrac{2}{3} + 1\dfrac{5}{6}$.

The LCD for the proper fractions 3 and 6 is 6.

$$4\dfrac{2}{3} = 4 + \dfrac{2}{3} \times \dfrac{2}{2} = 4 + \dfrac{4}{6}$$

$$+1\dfrac{5}{6} = 1 + \dfrac{5}{6} \qquad = 1 + \dfrac{5}{6}$$

$$5 + \dfrac{9}{6} = 5 + 1\dfrac{\overset{1}{\cancel{3}}}{\underset{2}{\cancel{6}}} = 5 + 1\dfrac{1}{2} = 6\dfrac{1}{2} \quad \blacksquare$$

▶ **You Try It** **7.** Add $4\dfrac{5}{7} + 6\dfrac{2}{3}$. **8.** Add $5\dfrac{4}{9} + 3\dfrac{7}{15}$.

EXAMPLE 9 Add $5\dfrac{3}{16} + 3\dfrac{7}{12}$.

$$
\begin{array}{r|cc}
2 & 16 & 12 \\
2 & 8 & 6 \\
\hline
 & 4 & 3
\end{array}
$$

$$5\dfrac{3}{16} = 5 + \dfrac{3}{16} \times \dfrac{3}{3} = 5 + \dfrac{9}{48}$$

$$+3\dfrac{7}{12} = 3 + \dfrac{7}{12} \times \dfrac{4}{4} = 3 + \dfrac{28}{48}$$

LCD = 2 × 2 × 4 × 3 = 48

$$8 + \dfrac{37}{48} = 8\dfrac{37}{48} \quad \blacksquare$$

EXAMPLE 10 Add $2\frac{1}{6} + 4\frac{5}{9} + 5\frac{11}{24}$.

$$
\begin{array}{c|ccc}
3 & 6 & 9 & 24 \\
2 & 2 & 3 & 8 \\
\hline
 & 1 & 3 & 4
\end{array}
$$

$\text{LCD} = 3 \times 2 \times 1 \times 3 \times 4$
$= 72$

$2\frac{1}{6} = 2 + \frac{1}{6} \times \frac{12}{12} = 2 + \frac{12}{72}$

$4\frac{5}{9} = 4 + \frac{5}{9} \times \frac{8}{8} = 4 + \frac{40}{72}$

$+5\frac{11}{24} = 5 + \frac{11}{24} \times \frac{3}{3} = 5 + \frac{33}{72}$

$11 + \frac{85}{72} = 12\frac{13}{72}$ ∎

▶ **You Try It**   **9.** Add $14\frac{5}{18} + 16\frac{1}{12}$.   **10.** Add $1\frac{3}{4} + 2\frac{5}{8} + 5\frac{3}{10}$.

Adding Unlike Mixed Numbers by First Writing Them as Improper Fractions

Another way to add unlike mixed numbers is to first change them to improper fractions. Then add the improper fractions just as you added proper fractions. Solve Example 7 again.

EXAMPLE 11   Add $2\frac{1}{4} + 6\frac{3}{5}$ by first changing each mixed number to an improper fraction.

The LCD for 4 and 5 is 20.

$$2\frac{1}{4} = \frac{9}{4} = \frac{9}{4} \times \frac{5}{5} = \frac{45}{20}$$

$$+6\frac{3}{5} = \frac{33}{5} = \frac{33}{5} \times \frac{4}{4} = \frac{132}{20}$$

$$\frac{177}{20} = 8\frac{17}{20}$$

This is the same answer as in Example 7. ∎

Compare Example 11 to Example 7. Adding by first changing mixed numbers to improper fractions will always give the correct answer. The disadvantage is having to work with larger numerators.

▶ **You Try It**   **11.** Add by first changing each mixed number to an improper fraction.

$$3\frac{1}{2} + 2\frac{4}{5}$$

▶ **Answers to You Try It**   **1.** $1\frac{5}{24}$ **2.** $\frac{37}{45}$ **3.** $\frac{19}{20}$ **4.** $1\frac{13}{30}$ **5.** $\frac{41}{225}$

6. $\frac{19}{27}$ **7.** $11\frac{8}{21}$ **8.** $8\frac{41}{45}$ **9.** $30\frac{13}{36}$ **10.** $9\frac{27}{40}$ **11.** $6\frac{3}{10}$

1 *Add the fractions. Reduce all answers to lowest terms.*

1. $\dfrac{2}{3} + \dfrac{1}{2}$

2. $\dfrac{3}{4} + \dfrac{1}{2}$

3. $\dfrac{3}{5} + \dfrac{1}{4}$

4. $\dfrac{2}{3} + \dfrac{5}{6}$

5. $\dfrac{1}{3} + \dfrac{2}{9}$

6. $\dfrac{1}{6} + \dfrac{5}{12}$

7. $\dfrac{3}{4} + \dfrac{1}{3}$

8. $\dfrac{3}{8} + \dfrac{5}{16}$

9. $\dfrac{4}{5} + \dfrac{7}{10}$

10. $\dfrac{11}{12} + \dfrac{5}{6}$

11. $\dfrac{3}{7} + \dfrac{5}{14}$

12. $\dfrac{5}{9} + \dfrac{5}{12}$

13. $\dfrac{2}{3} + \dfrac{7}{12}$

14. $\dfrac{1}{6} + \dfrac{4}{15}$

15. $\dfrac{3}{10} + \dfrac{8}{15}$

16. $\dfrac{3}{16} + \dfrac{5}{24}$

17. $\dfrac{1}{12} + \dfrac{1}{18}$

18. $\dfrac{14}{15} + \dfrac{17}{30}$

19. $\dfrac{13}{20} + \dfrac{3}{32}$

20. $\dfrac{27}{40} + \dfrac{13}{24}$

21. $\dfrac{1}{36} + \dfrac{1}{48}$

22. $\dfrac{8}{21} + \dfrac{4}{35}$

23. $\dfrac{5}{36} + \dfrac{10}{27}$

24. $\dfrac{17}{20} + \dfrac{3}{50}$

25. $\dfrac{7}{10} + \dfrac{13}{30}$

26. $\dfrac{4}{25} + \dfrac{8}{75}$

27. $\dfrac{9}{16} + \dfrac{3}{20}$

28. $\dfrac{1}{100} + \dfrac{3}{250}$

29. $\dfrac{5}{16} + \dfrac{3}{32}$

30. $\dfrac{31}{48} + \dfrac{17}{80}$

31. $\dfrac{1}{4} + \dfrac{1}{6} + \dfrac{1}{2}$

32. $\dfrac{2}{3} + \dfrac{5}{6} + \dfrac{1}{9}$

33. $\dfrac{1}{4} + \dfrac{5}{8} + \dfrac{11}{16}$

34. $\dfrac{1}{6} + \dfrac{7}{9} + \dfrac{5}{18}$

35. $\dfrac{3}{25} + \dfrac{13}{50} + \dfrac{7}{15}$

36. $\dfrac{4}{9} + \dfrac{1}{18} + \dfrac{25}{36}$

37. $\dfrac{11}{12} + \dfrac{17}{18} + \dfrac{1}{24}$

38. $\dfrac{9}{10} + \dfrac{7}{20} + \dfrac{1}{30}$

39. $\dfrac{5}{8} + \dfrac{7}{12} + \dfrac{9}{20}$

40. $\dfrac{13}{24} + \dfrac{11}{36} + \dfrac{31}{60}$

2 *Add the mixed numbers. Reduce all answers to lowest terms.*

41. $2\dfrac{1}{3} + 3\dfrac{1}{2}$

42. $4\dfrac{3}{4} + 1\dfrac{2}{5}$

43. $3\dfrac{4}{7} + 5\dfrac{2}{3}$

44. $1\dfrac{4}{5} + 2\dfrac{1}{4}$

45. $5\dfrac{7}{8} + 3\dfrac{3}{4}$

46. $4\dfrac{2}{9} + 1\dfrac{1}{3}$

47. $2\dfrac{3}{8} + 4\dfrac{7}{12}$

48. $5\dfrac{1}{6} + 2\dfrac{7}{10}$

49. $5\dfrac{1}{2} + \dfrac{3}{4}$

50. $4\dfrac{5}{12} + \dfrac{7}{8}$

51. $16\dfrac{3}{8} + 1\dfrac{3}{4}$

52. $2\dfrac{4}{9} + 5\dfrac{7}{27}$

53. $4 + 3\frac{5}{7}$

54. $3\frac{5}{31} + 7$

55. $7\frac{2}{5} + 4\frac{3}{8}$

56. $10\frac{5}{8} + 12\frac{11}{18}$

57. $25\frac{1}{2} + 13\frac{3}{5}$

58. $50\frac{5}{7} + 40\frac{1}{8}$

59. $4\frac{7}{9} + 8$

60. $12 + 16\frac{13}{20}$

61. $6\frac{16}{25} + 8\frac{3}{10}$

62. $5\frac{4}{15} + 7\frac{7}{9}$

63. $3\frac{5}{6} + 9\frac{8}{21}$

64. $60\frac{2}{5} + 4\frac{1}{8}$

65. $3\frac{1}{2} + 5\frac{3}{4} + 2\frac{2}{3}$

66. $5\frac{3}{4} + 6\frac{1}{3} + 4\frac{4}{5}$

67. $4\frac{5}{25} + 1\frac{7}{60} + 3\frac{9}{40}$

68. $7\frac{7}{20} + 4\frac{9}{10} + 3\frac{1}{5}$

SKILLSFOCUS (Section 2.6) *Solve the equation.*

69. $x + 29 = 61$

70. $w + 16 = 71$

71. $y - 45 = 80$

72. $z - 7 = 78$

EXTEND YOUR THINKING ▶▶▶▶
▶ TROUBLESHOOT IT

Find and correct the error.

73.
$$\frac{2}{5} = \frac{2}{5} \times \frac{3}{3} = \frac{6}{15}$$
$$+\frac{2}{3} = \frac{2}{3} \times \frac{5}{5} = \frac{10}{15}$$
$$\overline{}$$
$$\frac{16}{30} = \frac{8}{15}$$

74.
$$5\frac{1}{2} = 5 + \frac{1}{2} \times \frac{5}{5} = 5 + \frac{6}{10}$$
$$+2\frac{4}{5} = 2 + \frac{4}{5} \times \frac{2}{2} = 2 + \frac{6}{10}$$
$$\overline{}$$
$$7 + \frac{12}{10} = 8\frac{1}{5}$$

WRITING TO LEARN ▶▶▶▶

75. Create a word problem requiring you to add two mixed numbers. It must involve gallons of gasoline and driving to work. Then solve it, explaining each step you use in the solution.

76. You carry when you add whole numbers. Explain how you carry when you add fractions. Use your own example to show your point.

▶ YOU BE THE JUDGE

77. A perfect number (see Exercise 67, Section 3.3 on page 166) equals the sum of all its divisors less than itself. Six is perfect because the divisors of 6 are 1, 2, 3, and 6, and 6 $= 1 + 2 + 3$.

    a. Len claims that the sum of the reciprocals of all the divisors of the perfect number 6 equals 2. That is, $\frac{1}{1} + \frac{1}{2} + \frac{1}{3} + \frac{1}{6} = 2$. Is he correct? Support your decision with calculations.

    b. Another perfect number is 28. Can the same assertion be made for 28? Support your decision with calculations.

4.5 SUBTRACTING FRACTIONS WITH UNLIKE DENOMINATORS

OBJECTIVES

1. Subtract unlike proper fractions.
2. Subtract unlike mixed numbers.
3. Subtract mixed numbers with borrowing.
4. Subtract by changing mixed numbers to improper fractions (Section 3.2, Objective 3).

NEW VOCABULARY

borrowing

1 Subtracting Unlike Proper Fractions

EXAMPLE 1 Subtract $\dfrac{5}{8} - \dfrac{1}{3}$.

The LCD for 8 and 3 is 24.

Rewrite each fraction with an LCD of 24.
Then subtract the like fractions.

$$\dfrac{5}{8} = \dfrac{5}{8} \times \dfrac{3}{3} = \dfrac{15}{24}$$

$$-\dfrac{1}{3} = \dfrac{1}{3} \times \dfrac{8}{8} = \dfrac{8}{24}$$

$$\dfrac{7}{24} \quad \blacksquare$$

To subtract fractions with unlike denominators

1. Find the LCD for the denominators.
2. Rewrite each fraction as an equivalent fraction with the LCD as denominator.
3. Subtract the like fractions. Reduce if needed.

EXAMPLE 2 Subtract $\dfrac{5}{6} - \dfrac{7}{10}$.

$$6 = \textcircled{2} \times \textcircled{3}$$
$$10 = 2 \quad\quad \times \textcircled{5}$$
$$\overline{\text{LCD} = 2 \times 3 \times 5}$$
$$= 30$$

$$\dfrac{5}{6} = \dfrac{5}{6} \times \dfrac{5}{5} = \dfrac{25}{30}$$

$$-\dfrac{7}{10} = \dfrac{7}{10} \times \dfrac{3}{3} = \dfrac{21}{30}$$

$$\dfrac{4}{30} = \dfrac{\overset{2}{\cancel{4}}}{\underset{15}{\cancel{30}}} = \dfrac{2}{15} \quad \blacksquare$$

▶ **You Try It**   **1.** Subtract $\dfrac{3}{7} - \dfrac{1}{4}$.   **2.** Subtract $\dfrac{7}{8} - \dfrac{5}{6}$.

EXAMPLE 3 Subtract $\dfrac{7}{16} - \dfrac{3}{8}$.

The LCD for 16 and 8 is 16.

$$
\begin{array}{r}
\dfrac{7}{16} = \phantom{\dfrac{3}{8}} = \dfrac{7}{16} \\[2mm]
-\dfrac{3}{8} = \dfrac{3}{8} \times \dfrac{2}{2} = \dfrac{6}{16} \\[2mm]
\hline
\dfrac{1}{16}
\end{array}
$$ ■

EXAMPLE 4 Subtract $\dfrac{11}{24} - \dfrac{7}{30}$.

$$
\begin{array}{c|cc}
2 & 24 & 30 \\
3 & 12 & 15 \\
\hline
  & 4 & 5
\end{array}
$$

$$\text{LCD} = 2 \times 3 \times 4 \times 5 = 120$$

$$
\begin{array}{r}
\dfrac{11}{24} = \dfrac{11}{24} \times \dfrac{5}{5} = \dfrac{55}{120} \\[2mm]
-\dfrac{7}{30} = \dfrac{7}{30} \times \dfrac{4}{4} = \dfrac{28}{120} \\[2mm]
\hline
\dfrac{27}{120} = \dfrac{\overset{9}{\cancel{27}}}{\underset{40}{\cancel{120}}} = \dfrac{9}{40}
\end{array}
$$ ■

▶ **You Try It**   **3.** Subtract $\dfrac{11}{15} - \dfrac{4}{9}$.   **4.** Subtract $\dfrac{13}{27} - \dfrac{17}{45}$.

---

2 Subtracting Unlike Mixed Numbers

EXAMPLE 5 Subtract $3\dfrac{2}{5} - 2\dfrac{1}{4}$.

The LCD for 5 and 4 is 20.

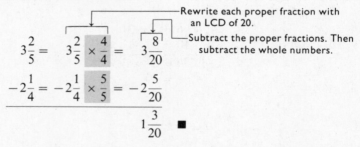

— Rewrite each proper fraction with an LCD of 20.

— Subtract the proper fractions. Then subtract the whole numbers.

$$
\begin{array}{r}
3\dfrac{2}{5} = \quad 3\dfrac{2}{5} \times \dfrac{4}{4} = \quad 3\dfrac{8}{20} \\[2mm]
-2\dfrac{1}{4} = -2\dfrac{1}{4} \times \dfrac{5}{5} = -2\dfrac{5}{20} \\[2mm]
\hline
1\dfrac{3}{20}
\end{array}
$$ ■

> **To subtract mixed numbers**
>
> 1. Find the LCD for the proper fractions.
> 2. Rewrite each proper fraction as an equivalent fraction with the LCD as the denominator.
> 3. Subtract the equivalent fractions.
> 4. Subtract the whole numbers.
> 5. Reduce if possible.

EXAMPLE 6 Subtract $24\frac{5}{6} - 17\frac{3}{8}$.

The LCD for 6 and 8 is 24.

$$24\frac{5}{6} = \quad 24\frac{5}{6} \times \frac{4}{4} = \quad 24\frac{20}{24}$$

$$-17\frac{3}{8} = -17\frac{3}{8} \times \frac{3}{3} = -17\frac{9}{24}$$

$$\overline{\qquad\qquad\qquad 7\frac{11}{24}} \quad \blacksquare$$

▶ **You Try It**   **5.** Subtract $4\frac{2}{3} - 2\frac{1}{8}$.   **6.** $7\frac{9}{10} - 2\frac{4}{15}$.

3 Borrowing   The next three examples demonstrate **borrowing** a 1 from the whole number in order to subtract.

EXAMPLE 7 Subtract $5\frac{1}{8} - 2\frac{5}{8}$.

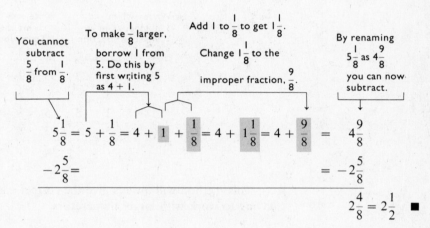

You cannot subtract $\frac{5}{8}$ from $\frac{1}{8}$.

To make $\frac{1}{8}$ larger, borrow 1 from 5. Do this by first writing 5 as $4 + 1$.

Add 1 to $\frac{1}{8}$ to get $1\frac{1}{8}$. Change $1\frac{1}{8}$ to the improper fraction, $\frac{9}{8}$.

By renaming $5\frac{1}{8}$ as $4\frac{9}{8}$ you can now subtract.

$$5\frac{1}{8} = 5 + \frac{1}{8} = 4 + 1 + \frac{1}{8} = 4 + 1\frac{1}{8} = 4 + \frac{9}{8} = \quad 4\frac{9}{8}$$

$$-2\frac{5}{8} = \qquad\qquad\qquad\qquad\qquad\qquad\qquad = -2\frac{5}{8}$$

$$\overline{\qquad\qquad\qquad\qquad 2\frac{4}{8} = 2\frac{1}{2}} \quad \blacksquare$$

EXAMPLE 8 Subtract $8\frac{1}{6} - 5\frac{7}{9}$.

The LCD is 18.

$$8\frac{1}{6} = 8\frac{1}{6} \times \frac{3}{3} = 8\frac{3}{18} = 7\frac{21}{18}$$

$$-5\frac{7}{9} = 5\frac{7}{9} \times \frac{2}{2} = 5\frac{14}{18} = 5\frac{14}{18}$$

$$\overline{\qquad\qquad\qquad\qquad 2\frac{7}{18}} \quad \blacksquare$$

You can't subtract $\frac{14}{18}$ from $\frac{3}{18}$. To make $\frac{3}{18}$ larger, borrow 1 from the 8, leaving 7. Add the 1 to $\frac{3}{18}$ to get $1\frac{3}{18} = \frac{21}{18}$. Now you can subtract.

▶ **You Try It**   **7.** Subtract $8\frac{4}{9} - 2\frac{7}{9}$.   **8.** Subtract $2\frac{3}{10} - 1\frac{3}{4}$.

EXAMPLE 9 Find $7 - 2\frac{3}{5}$.

$$7 = 6\frac{5}{5}$$

$$-2\frac{3}{5} = -2\frac{3}{5}$$

$$\overline{4\frac{2}{5}} \quad \blacksquare$$

To subtract, first write a fraction above $\frac{3}{5}$. To do this, borrow 1 from 7. This leaves 6. Write the 1 as $\frac{5}{5}$, using the same denominator as $\frac{3}{5}$. Now you can subtract the like fractions.

▶ **You Try It**   **9.** Subtract $15 - 9\frac{5}{12}$.

4 Subtracting Mixed Numbers by First Changing Them to Improper Fractions

You can subtract two mixed numbers by first writing both as improper fractions. You do not have to borrow 1 from the whole number if you use this method.

EXAMPLE 10 Subtract by changing to improper fractions. $4\frac{1}{8} - 1\frac{2}{3}$

Write each mixed number as an improper fraction.

Write each fraction in terms of the LCD = 24. Then subtract.

$$4\frac{1}{8} = \frac{33}{8} = \frac{33}{8} \times \frac{3}{3} = \frac{99}{24}$$

$$-1\frac{2}{3} = \frac{5}{3} = \frac{5}{3} \times \frac{8}{8} = \frac{40}{24}$$

$$\overline{\frac{59}{24} = 2\frac{11}{24}} \quad \blacksquare$$

This method will always give the correct answer. The disadvantage is having to work with larger numerators.

EXAMPLE 11 Subtract $8\frac{11}{12} - 5\frac{7}{9}$.

$$8\frac{11}{12} = \frac{107}{12} = \frac{107}{12} \times \frac{3}{3} = \frac{321}{36}$$

$$-5\frac{7}{9} = \frac{52}{9} = \frac{52}{9} \times \frac{4}{4} = \frac{208}{36}$$

$$\overline{\frac{113}{36} = 3\frac{5}{36}} \quad \blacksquare$$

▶ **You Try It**   Subtract by changing to improper fractions.

10. $3\frac{1}{2} - 1\frac{3}{5}$ **11.** $20\frac{1}{6} - 12\frac{5}{8}$

▶ **Answers to You Try It**   1. $\frac{5}{28}$ 2. $\frac{1}{24}$ 3. $\frac{13}{45}$ 4. $\frac{14}{135}$ 5. $2\frac{13}{24}$ 6. $5\frac{19}{20}$

7. $5\frac{2}{3}$ 8. $\frac{11}{20}$ 9. $5\frac{7}{12}$ 10. $1\frac{9}{10}$ 11. $7\frac{13}{24}$

SECTION 4.5 EXERCISES

1 *Subtract the fractions. Reduce all answers to lowest terms.*

1. $\dfrac{3}{4} - \dfrac{1}{2}$

2. $\dfrac{5}{6} - \dfrac{2}{3}$

3. $\dfrac{7}{8} - \dfrac{1}{4}$

4. $\dfrac{4}{5} - \dfrac{3}{10}$

5. $\dfrac{3}{4} - \dfrac{3}{8}$

6. $\dfrac{9}{10} - \dfrac{2}{5}$

7. $\dfrac{1}{3} - \dfrac{1}{6}$

8. $\dfrac{1}{2} - \dfrac{1}{4}$

9. $\dfrac{1}{6} - \dfrac{1}{8}$

10. $\dfrac{8}{9} - \dfrac{5}{6}$

11. $\dfrac{5}{8} - \dfrac{1}{10}$

12. $\dfrac{7}{8} - \dfrac{1}{2}$

13. $\dfrac{11}{12} - \dfrac{3}{8}$

14. $\dfrac{9}{14} - \dfrac{2}{7}$

15. $\dfrac{5}{9} - \dfrac{5}{12}$

16. $\dfrac{7}{12} - \dfrac{3}{20}$

17. $\dfrac{23}{36} - \dfrac{5}{9}$

18. $\dfrac{5}{16} - \dfrac{9}{32}$

19. $\dfrac{17}{18} - \dfrac{5}{12}$

20. $\dfrac{1}{2} - \dfrac{7}{30}$

21. $\dfrac{5}{7} - \dfrac{2}{5}$

22. $\dfrac{8}{13} - \dfrac{2}{9}$

23. $\dfrac{20}{27} - \dfrac{5}{18}$

24. $\dfrac{49}{60} - \dfrac{24}{36}$

25. $\dfrac{24}{35} - \dfrac{8}{21}$

26. $\dfrac{7}{24} - \dfrac{2}{15}$

27. $\dfrac{18}{24} - \dfrac{9}{18}$

28. $\dfrac{45}{80} - \dfrac{16}{40}$

29. $\dfrac{28}{35} - \dfrac{15}{25}$

30. $\dfrac{16}{21} - \dfrac{15}{28}$

31. $\dfrac{11}{30} - \dfrac{5}{36}$

32. $\dfrac{27}{50} - \dfrac{9}{40}$

2 **3** **4** *Subtract the mixed numbers. Reduce answers to lowest terms.*

33. $3\dfrac{2}{3} - 1\dfrac{1}{6}$

34. $4\dfrac{3}{4} - 1\dfrac{5}{8}$

35. $8\dfrac{4}{7} - 5\dfrac{1}{3}$

36. $11\dfrac{4}{5} - 2\dfrac{1}{4}$

37. $6\dfrac{1}{7} - 2\dfrac{5}{7}$

38. $8\dfrac{1}{9} - 3\dfrac{7}{9}$

39. $5\dfrac{7}{8} - 3\dfrac{3}{4}$

40. $4\dfrac{5}{9} - 1\dfrac{1}{3}$

41. $12\dfrac{3}{4} - 4\dfrac{2}{3}$

42. $5\dfrac{2}{3} - 4\dfrac{3}{5}$

43. $6\dfrac{3}{8} - 4\dfrac{7}{12}$

44. $5\dfrac{1}{6} - 2\dfrac{7}{10}$

45. $5\dfrac{1}{2} - \dfrac{3}{4}$

46. $4\dfrac{5}{12} - \dfrac{7}{8}$

47. $6\dfrac{5}{9} - \dfrac{1}{3}$

48. $2\dfrac{4}{5} - \dfrac{4}{9}$

49. $5\dfrac{1}{6} - 3\dfrac{5}{6}$

50. $8\dfrac{15}{21} - 5\dfrac{20}{21}$

51. $16\dfrac{3}{8} - 1\dfrac{3}{4}$

52. $20\dfrac{2}{9} - 5\dfrac{7}{27}$

53. $4 - 3\dfrac{5}{7}$

54. $13 - 8\dfrac{5}{31}$

55. $7\dfrac{2}{5} - 4\dfrac{3}{8}$

56. $16\dfrac{5}{8} - 12\dfrac{5}{18}$

57. $25\dfrac{1}{2} - 13\dfrac{3}{5}$

58. $50\dfrac{5}{7} - 40\dfrac{13}{14}$

59. $15\dfrac{7}{9} - 8$

60. $16\dfrac{13}{20} - 11$

61. $8\dfrac{16}{25} - 8\dfrac{3}{10}$

62. $7\dfrac{4}{15} - 7\dfrac{2}{9}$

63. $9\dfrac{5}{6} - 2\dfrac{8}{21}$

64. $60\dfrac{2}{5} - 4\dfrac{1}{8}$

65. $24 - 5\dfrac{7}{12}$

66. $50 - 37\dfrac{1}{2}$

67. $7\dfrac{5}{12} - 4\dfrac{4}{7}$

68. $15\dfrac{6}{13} - 3\dfrac{3}{4}$

69. $8\dfrac{3}{7} - 4\dfrac{6}{14}$

70. $4\dfrac{15}{18} - 4\dfrac{10}{12}$

SKILLSFOCUS (Section 1.6) *Solve each problem.*

71. What is the sum of 35 and 26?

72. Find 10 increased by 3.

73. Subtract 37 from 52.

74. What is 9 less than 9?

EXTEND YOUR THINKING ▶▶▶▶
▶ TROUBLESHOOT IT

Find and correct the error.

75.
$$3\dfrac{1}{5} = 3\dfrac{1}{5} \times \dfrac{8}{8} = 3\dfrac{8}{40} = 2\dfrac{18}{40}$$
$$-1\dfrac{3}{8} = 1\dfrac{3}{8} \times \dfrac{5}{5} = 1\dfrac{15}{40} = 1\dfrac{15}{40}$$
$$\rule{4cm}{0.4pt}$$
$$1\dfrac{3}{40}$$

76. $\dfrac{5}{8} - \dfrac{2}{5} = \dfrac{5-2}{8-5} = \dfrac{3}{3} = 1$

WRITING TO LEARN ▶▶▶▶

77. Create a subtraction problem where you must borrow. Solve the problem. Explain why you borrow, and show how it is done.

78. How is borrowing in the whole number subtraction $25 - 19$ like borrowing in the fraction subtraction $3\dfrac{1}{4} - 1\dfrac{2}{3}$? How is it different?

4.6 APPLICATIONS: ADDITION AND SUBTRACTION OF FRACTIONS

OBJECTIVE

Solve word problems (Section 1.6).

Solving Word Problems

EXAMPLE 1 Yesterday one share of XYZ stock sold for $27\frac{1}{2}$. Today the price for one share increased by $2\frac{5}{8}$. What is the new price for one share?

The phrase *increased by* indicates addition.

$$\text{yesterday's price:} \quad 27\frac{1}{2} = 27 + \frac{1}{2} \times \frac{4}{4} = 27\frac{4}{8}$$

$$\text{increase in price:} \quad +2\frac{5}{8} = \qquad\qquad = +2\frac{5}{8}$$

$$\text{today's price:} \qquad\qquad\qquad\qquad \$29\frac{9}{8} = \$30\frac{1}{8} \text{ per share} \quad \blacksquare$$

▶ **You Try It** 1. Yesterday 1 share of TRW stock sold for $64\frac{3}{4}$. Today the price for one share increased by $1\frac{7}{8}$. What is the current price per share?

EXAMPLE 2 Three pieces of clothing require 3, $1\frac{3}{4}$, and $\frac{5}{6}$ yards of material, respectively.

a. How many yards of material are needed to make the clothing?

To find the total amount of material needed, add the three amounts together. The LCD for 4 and 6 is 12.

$$\begin{aligned}
\text{Write 3 yards} \quad & 3 = 3 && = 3 \\
\text{in the whole} \\
\text{number column.} \quad & 1\frac{3}{4} = 1 + \frac{3}{4} \times \frac{3}{3} = 1\frac{9}{12} \\
& +\frac{5}{6} = \quad \frac{5}{6} \times \frac{2}{2} = +\frac{10}{12} \\
\hline
& \qquad\qquad\qquad 4\frac{19}{12} = 5\frac{7}{12} \text{ yards of material needed}
\end{aligned}$$

b. What do you pay for the material if it costs $15 per yard?

You want to buy $5\frac{7}{12}$ yards of material *at* $15 per yard. The word *at* indicates multiplication.

$$5\frac{7}{12} \text{ yds } at \text{ } \$15 \text{ per yard} = 5\frac{7}{12} \times 15 = \frac{67}{12} \times \frac{\cancel{15}^{5}}{1} = \$83\frac{3}{4} \quad \blacksquare$$

▶ **You Try It**   **2.** Three pieces of clothing require 8, $3\frac{5}{8}$, and $\frac{7}{10}$ yards of material, respectively.

    a. How many total yards of material are needed?

    b. What is the cost of the material at $40 per yard?

EXAMPLE 3   A newly designed cylinder has a volume of $63\frac{4}{9}$ cubic inches. The old model had a volume of $55\frac{7}{10}$ cubic inches. How much more volume does the new cylinder have?

    To find the difference in the two volumes, subtract them. The LCD for 9 and 10 is 90.

$$\text{new model} \quad 63\frac{4}{9} = 63 + \frac{4}{9} \times \frac{10}{10} = 63\frac{40}{90} = \quad 62\frac{130}{90} \quad \leftarrow \text{Borrow 1 from 63.}$$

$$\quad\quad\quad\quad\quad\quad\quad\quad\quad\quad\quad\quad\quad\quad\quad\quad\quad\quad\quad 1\frac{40}{90} = \frac{130}{90}$$

$$\text{old model} -55\frac{7}{10} = 55 + \frac{7}{10} \times \frac{9}{9} = 55\frac{63}{90} = -55\frac{63}{90}$$

$$\rule{6cm}{0.4pt}$$

$$7\frac{67}{90} \text{ cubic inches difference}$$

The new model cylinder has $7\frac{67}{90}$ cubic inches more volume. ∎

▶ **You Try It**   **3.** A newly designed freezer has a volume of $10\frac{1}{6}$ cubic feet. The old model had a volume of $9\frac{2}{5}$ cubic feet. How much more volume does the new model have?

EXAMPLE 4   **Chapter Problem**   Read the problem at the beginning of this chapter.

    Donna purchased 400 shares at $27\frac{3}{4}$ each. She then sold them for $30\frac{3}{8}$ each. Finding her gain is a two-part problem.

Step #1: Find the gain for 1 share. This is $30\frac{3}{8} - \$27\frac{3}{4}$.

$$30\frac{3}{8} = \quad\quad\quad\quad 30\frac{3}{8} = \quad 29\frac{11}{8} \quad \leftarrow \text{Borrow 1 from 30.}$$

$$\quad\quad\quad\quad\quad\quad\quad\quad\quad\quad\quad\quad\quad\quad\quad\quad 1\frac{3}{8} = \frac{11}{8}$$

$$-27\frac{3}{4} = 27 + \frac{3}{4} \times \frac{2}{2} = 27\frac{6}{8} = -27\frac{6}{8}$$

$$\rule{6cm}{0.4pt}$$

$$\$2\frac{5}{8} \text{ gain per share}$$

Step #2: The gain on one share is $\$2\frac{5}{8}$. The gain on 400 shares is 400 times this amount.

$$400 \times \$2\frac{5}{8} = \frac{\overset{50}{\cancel{400}}}{1} \times \frac{21}{\cancel{8}} = \frac{50 \times 21}{1 \times 1} = \$1{,}050$$

$$\underset{1}{}$$

Donna gained $1,050 on this transaction. ∎

▶ **You Try It**   **4.** Read the problem at the beginning of this chapter. Danielle bought 400 shares of Sears stock at the High. She sold them at the Low. How much did she lose on this transaction?

EXAMPLE 5   It takes Susan 4 hours to paint a room. It takes Beth 6 hours to paint the same room. If they work together, how fast can they paint the room?

Step #1: If Susan can paint the room in 4 hours, in one hour she can paint $\frac{1}{4}$ of it.

If Beth can paint the room in 6 hours, in one hour she can paint $\frac{1}{6}$ of it.

If they work together, in one hour they can paint $\frac{1}{4} + \frac{1}{6}$ of the room.

$$\frac{1}{4} = \frac{1}{4} \times \frac{3}{3} = \frac{3}{12}$$
$$+\frac{1}{6} = \frac{1}{6} \times \frac{2}{2} = \frac{2}{12}$$
$$\frac{5}{12}$$

In one hour they can paint $\frac{5}{12}$ of the room.

Step #2: To find how many hours it takes to paint 1 whole room, find how many $\frac{5}{12}$ are in 1 whole.

How many $\frac{5}{12}$ are in $1 = 1 \div \frac{5}{12} = \frac{1}{1} \times \frac{12}{5} = \frac{12}{5} = 2\frac{2}{5}$ hrs

Working together, Susan and Beth can paint the room in $2\frac{2}{5}$ hours.

OBSERVE   The time it takes them to paint the room $\left(\frac{12}{5}\right)$ is the reciprocal of the fraction of the room they can paint in one hour $\left(\frac{5}{12}\right)$. ∎

▶ **You Try It**   **5.** It takes Tom 5 hours to paint a room. It takes Steve 8 hours to paint the same room. If they work together, how fast can they paint the room?

▶ Answers to You Try It   1. $66\frac{5}{8}$ per share   2. a. $12\frac{13}{40}$ yards   b. $493

3. $\frac{23}{30}$ cubic feet   4. $350   5. $3\frac{1}{13}$ hours

4.6 APPLICATIONS: ADDITION AND SUBTRACTION OF FRACTIONS   **241**

SECTION 4.6 EXERCISES

Solve each word problem. Reduce answers to lowest terms.

1. Ron purchased $16\frac{9}{10}$ gallons of gas this morning and another $13\frac{1}{5}$ gallons this afternoon. How much gasoline did Ron buy today?

2. In one week a family drank $3\frac{4}{5}$ gallons of milk. They drank $4\frac{2}{3}$ gallons the following week. How much milk did they drink for the two weeks?

3. A nail $3\frac{1}{6}$ inches long is driven into a board $2\frac{5}{8}$ inches thick. How much of the nail will protrude from the other side of the board?

4. A share of PG stock sold for $\$62\frac{1}{8}$ one month ago. Today it sells for $\$15\frac{3}{4}$ less. What is today's price for one share of PG?

5. A carpenter needs $22\frac{5}{8}$ feet of wood to build a table and $16\frac{7}{12}$ feet to build a bench. How much wood is needed altogether?

6. A sofa requires $16\frac{5}{6}$ yards of material to be reupholstered. A love seat requires $10\frac{3}{4}$ yards. How much material is needed for the two pieces?

7. A tank contains $7\frac{1}{2}$ gallons of gas.

$24\frac{3}{8}$ gal capacity

$7\frac{1}{2}$ gal

a. How much more gas is needed to fill the tank?

b. What will this additional gas cost if one gallon costs $\$1\frac{1}{5}$?

8. A tractor requires $8\frac{2}{3}$ quarts of oil. It now holds $5\frac{1}{2}$ quarts.

a. How many more quarts of oil does the tractor need?

b. What will this additional oil cost if one quart runs for $\$1\frac{1}{2}$?

9. One share of BMI sold for $\$46\frac{3}{4}$ at 10 A.M. One share sold for $\$1\frac{3}{8}$ more at 3 P.M. the same day. Sandra owns 160 shares.

a. How much did one share sell for at 3 P.M.?

b. How much more were Sandra's shares worth at 3 P.M. compared to 10 A.M.?

10. A share of XRT stock sold for 24\frac{5}{8}$ yesterday. Today one share is selling for 5\frac{1}{4}$ less. James owns 1,200 shares.
   a. How much is one share selling for today?

   b. How much less are James's shares worth today compared to yesterday?

11. A recipe calls for $\frac{1}{3}$ cup of walnuts, $\frac{3}{4}$ cups of almonds, and $\frac{1}{2}$ cup of peanuts. How many cups of nuts are used in the recipe?

12. A new business purchased $\frac{5}{8}$ of a page of advertising in Monday's paper, $\frac{5}{12}$ of a page in Tuesday's paper, and $\frac{1}{4}$ of a page in Wednesday's paper. How many pages of advertising did the business buy?

13. What must be added to $3\frac{2}{5}$ to get $5\frac{1}{6}$?

14. What must $\frac{2}{3}$ be increased by to get $\frac{9}{10}$?

15. Subtract $\frac{5}{12}$ from $\frac{1}{2}$.

16. Subtract $\frac{3}{8}$ from $\frac{11}{15}$.

17. Hector purchased 12 gallons of punch for a party. By midnight, $3\frac{2}{5}$ gallons were left. How much punch was consumed by midnight?

18. A $5\frac{7}{12}$-foot piece of wood is cut from a 16-foot board. What length of wood is left?

19. Two boards each $\frac{7}{8}$ inches thick are stacked on a third board $2\frac{1}{4}$ inches thick. How thick is the stack?

20. A bedroom required $2\frac{1}{3}$ gallons of paint. An adjoining sitting room needed $1\frac{3}{4}$ gallons. How much paint did the two rooms need?

21. Al owns $\frac{1}{3}$ of a business. Bob owns $\frac{2}{5}$. Cy owns the rest. What fraction of the business does Cy own?

22. In a student election, $\frac{1}{2}$ of the people voted for Cheryl. $\frac{3}{7}$ voted for Jayne. The rest abstained. What fraction of the voters abstained?

23. The City Fair record for eating ice cream is $9\frac{1}{8}$ quarts in one hour. In a practice session George ate $7\frac{2}{3}$ quarts. How many quarts was George short of the record?

24. A customer purchased $42\frac{7}{10}$ yards of wire from a roll that was 210 yards long. How much wire was left on the roll?

25. Tom can paint a room in 8 hours. Ed can paint it in 10 hours. How long will it take them to paint the room if they work together?

26. Pat can erect a jungle gym in 12 hours. Nancy can do it in 9 hours. It takes Rowan 10 hours. How long will it take them to erect the jungle gym if they work together?

27. Find the missing dimensions for the figure below.

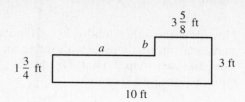

28. What is the inside diameter of the pipe in the figure?

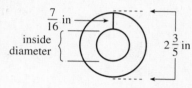

29. Anna inflated 60 balloons for a party. Of these, $\frac{2}{5}$ were popped, $\frac{4}{15}$ were used as party favors,

and $\frac{1}{6}$ were given to the children next door. The rest are hanging from the chandelier.

a. What fraction of the balloons are hanging from the chandelier?

b. How many balloons are hanging?

30. Jayne spent $\frac{1}{3}$ of her day working, $\frac{1}{4}$ sleeping, and $\frac{1}{6}$ with her family. The rest was devoted to attending classes and doing homework.

a. What fraction of Jayne's day was devoted to her education?

b. How many hours is this out of a 24-hour day?

SKILLSFOCUS (Section 2.4) *Simplify each expression.*

31. $4 + 3 \times 5$ **32.** $25 \div 5 \times 5$ **33.** $4(9 - 2)$ **34.** $5 \cdot 2^2$

EXTEND YOUR THINKING ▶▶▶▶

35. A cook has four cup measures of 1, $\frac{1}{2}$, $\frac{1}{3}$, and $\frac{1}{4}$ cups, respectively. Describe the steps the cook uses to measure

a. exactly $\frac{1}{6}$ of a cup of oil.

b. exactly $\frac{1}{12}$ of a cup of lime juice.

36. Foresight in the Stock Market Sheila owns 480 shares of TLQ stock. Anticipating a downturn

in the market soon, she sold them for $\$45\frac{3}{8}$ per share. The next week the stock market did turn down, and one share of TLQ dropped to $\$30\frac{1}{4}$. Sheila then spent all the money from the previous week's sale (assume no commissions or taxes were deducted from her sale) and bought all TLQ stock at the lower price.

a. How many shares of TLQ does Sheila now own?

b. In six months, the market recovered to the point that TLQ was again selling for $\$45\frac{3}{8}$ per share. By how much did the value of Sheila's investment in TLQ stock increase as a result of her anticipating the market downturn?

37. A father owns 5 acres of land. The land is in the shape of a square. He keeps $\frac{1}{4}$ of it for himself. He splits the rest among his four sons.

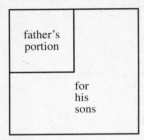

a. How many acres does each son get?

b. How does the father subdivide the land so that each son gets a piece with the exact same size and shape? (Hint: Divide the entire square into sixteenths.)

WRITING TO LEARN ▶▶▶▶

38. Write a word problem whose answer is $12\frac{3}{4}$. The problem must be about food, and must include a subtraction of fractions.

▶ YOU BE THE JUDGE

39. George flew his single-engine plane from Baltimore to Richmond and used $\frac{3}{8}$ of a tank of gas. He used another $\frac{1}{6}$ of a tank to get from Richmond to Norfolk. At the halfway point in his flight from Norfolk directly back to Baltimore, he used another $\frac{11}{24}$ of a tank of gas. George started with a full tank, and has not refueled since he left Baltimore. Does he have reason to be alarmed? Explain your decision.

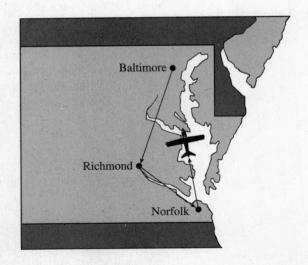

4.7 COMBINED OPERATIONS; ORDER

OBJECTIVES

1 Apply the order of operations to fractions (Section 1.9, Objectives 1, 2, 3; Section 2.4; Section 2.5, Objective 2).

2 Order fractions.

NEW VOCABULARY

simple fraction

complex fraction

---

1 **Applying the Order of Operations to Fractions**

The order of operations applied to whole numbers in Section 2.4 also applies to fractions. You may insert parentheses to highlight the operation to be performed first.

EXAMPLE 1

$$\frac{2}{3} + \frac{1}{8} \times \frac{4}{5}$$

$$= \frac{2}{3} + \left(\frac{1}{\cancel{8}_2} \times \frac{\cancel{4}^1}{5}\right)$$

Multiply before you add. You may insert parentheses around the multiplication to indicate it is done first.

$$= \frac{2}{3} + \frac{1}{10}$$

Add. The LCD for 3 and 10 is 30.

$$= \frac{20}{30} + \frac{3}{30} = \frac{23}{30} \quad \blacksquare$$

EXAMPLE 2

$$\frac{4}{27} \div \frac{5}{9} \times \frac{3}{5}$$

$$= \left(\frac{4}{\cancel{27}_3} \times \frac{\cancel{9}^1}{5}\right) \times \frac{3}{5}$$

Multiply and divide in order from left to right. Insert parentheses to indicate you divide first.

$$= \frac{4}{\cancel{15}_5} \times \frac{\cancel{3}^1}{5} = \frac{4}{25} \quad \blacksquare$$

EXAMPLE 3

$$3\frac{1}{2}\left(\frac{3}{4} + \frac{7}{8}\right)$$

$$= \frac{7}{2} \cdot \left(\frac{6}{8} + \frac{7}{8}\right)$$

Write the dot for implied multiplication. Simplify within parentheses first. The LCD = 8.

$$= \frac{7}{2} \cdot \frac{13}{8} = \frac{91}{16} \quad \text{or} \quad 5\frac{11}{16} \quad \blacksquare$$

▶ **You Try It**

1. $\dfrac{5}{6} + \dfrac{3}{4} \cdot \dfrac{2}{9}$

2. $\dfrac{9}{10} \div \dfrac{6}{5} \cdot \dfrac{5}{6}$

3. $6\dfrac{3}{5}\left(\dfrac{4}{5} + \dfrac{7}{10}\right)$

EXAMPLE 4

$$\left(2\frac{1}{3}\right)^2 - 3 \cdot 1\frac{3}{5}$$

$$= \left(\frac{7}{3}\right)^2 - \frac{3}{1} \cdot \frac{8}{5} \qquad \text{Write the mixed numbers as improper fractions. Evaluate the power.}$$

$$= \frac{49}{9} - \frac{3}{1} \cdot \frac{8}{5} \qquad \text{Multiply.}$$

$$= \frac{49}{9} - \frac{24}{5} \qquad \text{Subtract. The LCD for 9 and 5 is 45.}$$

$$= \frac{49}{9} \cdot \boxed{\frac{5}{5}} - \frac{24}{5} \cdot \boxed{\frac{9}{9}}$$

$$= \frac{245}{45} - \frac{216}{45} = \frac{29}{45} \quad \blacksquare$$

EXAMPLE 5

$$2 \cdot 3 + 5 \div 10$$

$$= 2 \cdot 3 + \frac{5}{10} = 6 + \frac{\overset{1}{\cancel{5}}}{\underset{2}{\cancel{10}}} = 6\frac{1}{2} \quad \blacksquare$$

▶ **You Try It**
 4. $\left(2\frac{1}{2}\right)^2 - 3 \cdot 1\frac{2}{9}$

 5. $4 \cdot 5 + 10 \div 6$

EXAMPLE 6 Evaluate $16t + 3$ when $t = \frac{3}{4}$.

$$16t + 3 = 16 \cdot t + 3 \qquad \text{Write the dot for the implied multiplication.}$$

$$= 16 \cdot \frac{3}{4} + 3 \qquad \text{Substitute } \frac{3}{4} \text{ for } t.$$

$$= \left(\frac{\overset{4}{\cancel{16}}}{1} \cdot \frac{3}{\underset{1}{\cancel{4}}}\right) + 3$$

$$= 12 + 3 = 15 \quad \blacksquare$$

▶ **You Try It**
 6. Evaluate $5m - 2$ when $m = 1\frac{3}{8}$.

A **simple fraction** has whole numbers for its numerator and denominator. A **complex fraction** has a fraction in its numerator, or denominator, or both. Therefore, a complex fraction has more than one fraction bar.

Recall that $\frac{12}{6}$ means $12 \div 6$. If you replace the numerator 12 with $\frac{3}{4}$ and the denominator 6 with $\frac{2}{5}$, you can write the following complex fraction.

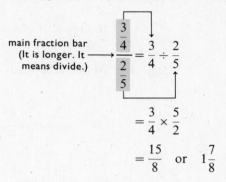

main fraction bar
(It is longer. It
means divide.)

$$\frac{\frac{3}{4}}{\frac{2}{5}} = \frac{3}{4} \div \frac{2}{5}$$

$$= \frac{3}{4} \times \frac{5}{2}$$

$$= \frac{15}{8} \quad \text{or} \quad 1\frac{7}{8}$$

OBSERVE In short, when a fraction is in the denominator, multiply the numerator by the reciprocal of the denominator.

$$\frac{\frac{3}{4}}{\frac{2}{5}} = \frac{3}{4} \times \frac{5}{2}$$

Multiply by reciprocal.

EXAMPLE 7
$$\frac{8}{\frac{4}{5}} = 8 \div \frac{4}{5} = \frac{\overset{2}{\cancel{8}}}{1} \cdot \frac{5}{\underset{1}{\cancel{4}}} = 10 \quad \blacksquare$$

EXAMPLE 8
$$\frac{6\frac{1}{4}}{\frac{3}{10}} = 6\frac{1}{4} \div \frac{3}{10} = \frac{25}{\underset{2}{\cancel{4}}} \cdot \frac{\overset{5}{\cancel{10}}}{3} = \frac{125}{6} \quad \text{or} \quad 20\frac{5}{6} \quad \blacksquare$$

▶ **You Try It**   Simplify.

7. $\dfrac{\frac{7}{10}}{\frac{3}{5}}$

8. $\dfrac{6}{\frac{7}{9}}$

9. $\dfrac{2\frac{5}{12}}{\frac{1}{3}}$

EXAMPLE 9 Three bottles contain $\frac{5}{8}$ ounce, $1\frac{1}{2}$ ounces, and $\frac{3}{4}$ ounce of the same drug. If a pharmacist combines the contents of the three bottles, how many $\frac{1}{4}$-ounce doses of the drug can she make?

Step #1: Find the total number of ounces of the drug.

$$\text{total ounces of drug} = \frac{5}{8} + 1\frac{1}{2} + \frac{3}{4} = \frac{5}{8} + \frac{3}{2} + \frac{3}{4} = \frac{5}{8} + \frac{12}{8} + \frac{6}{8} = \frac{23}{8} \text{ oz}$$

Step #2: To find how many $\frac{1}{4}$-ounce doses can be made, divide this sum by $\frac{1}{4}$.

$$\text{number of } \frac{1}{4}\text{-ounce doses} = \frac{23}{8} \div \frac{1}{4} = \frac{23}{\overset{}{\underset{2}{8}}} \cdot \frac{\overset{1}{4}}{1} = \frac{23}{2} \quad \text{or} \quad 11\frac{1}{2}$$

Note, $11\frac{1}{2}$ doses = 11 doses + $\frac{1}{2}$ of a dose. 11 doses of $\frac{1}{4}$ ounces each can be made. There is only enough drug left over to make $\frac{1}{2}$ of a twelfth dose. The answer is 11 doses. ■

▶ **You Try It** 10. Three bottles contain $\frac{9}{10}$ ounce, $4\frac{3}{4}$ ounces, and $1\frac{7}{8}$ ounces of the same drug. How many $\frac{1}{3}$-ounce doses of the drug can be made if the contents of the three bottles are combined?

---

2 Ordering Fractions To order fractions, first write each fraction with the same common denominator. Then order the numerators.

EXAMPLE 10 Which fraction is larger, $\frac{4}{7}$ or $\frac{3}{5}$?

Rewrite each fraction in terms of the LCD 35.

$$\frac{4}{7} = \frac{4}{7} \cdot \frac{5}{5} = \frac{20}{35}$$

$$\frac{3}{5} = \frac{3}{5} \cdot \frac{7}{7} = \frac{21}{35}$$

Compare the numerators. Since $\frac{20}{35} < \frac{21}{35}$, then $\frac{4}{7} < \frac{3}{5}$. $\frac{3}{5}$ is larger. ■

Given two or more fractions, to put them in order

1. Rewrite each fraction as an equivalent fraction with the same common denominator.
2. Compare the numerators. The largest fraction has the largest numerator. The smallest fraction has the smallest numerator.

EXAMPLE 11   Arrange the fractions $\frac{11}{14}, \frac{5}{7}$, and $\frac{3}{4}$ in order from smallest to largest.

Find the LCD.

$$
\begin{array}{c|ccc}
7 & 14 & 7 & 4 \\
2 & 2 & 1 & 4 \\
\hline
 & 1 & 1 & 2
\end{array}
$$

$$
\begin{aligned}
\text{LCD} &= 7 \times 2 \times 1 \times 1 \times 2 \\
&= 28
\end{aligned}
$$

Rewrite each fraction in terms of the LCD 28.

$$
\frac{11}{14} = \frac{11}{14} \cdot \frac{2}{2} = \frac{22}{28}
$$

$$
\frac{5}{7} = \frac{5}{7} \cdot \frac{4}{4} = \frac{20}{28}
$$

$$
\frac{3}{4} = \frac{3}{4} \cdot \frac{7}{7} = \frac{21}{28}
$$

Compare the numerators. Since $\frac{20}{28} < \frac{21}{28} < \frac{22}{28}$, then $\frac{5}{7} < \frac{3}{4} < \frac{11}{14}$. ■

▶ **You Try It**   Order the fractions from smallest to largest.

11. $\frac{3}{4}$ and $\frac{7}{10}$

12. $\frac{2}{5}, \frac{3}{10}$, and $\frac{7}{25}$

SOMETHING MORE

The Number Line and Midpoint   The number exactly in the middle of two given numbers on the number line is called the *midpoint*. The midpoint for two numbers is the average of the numbers. For example, the midpoint for 1 and 7 is the average of 1 and 7.

$$
\text{midpoint for 1 and 7} = \frac{1 + 7}{2} = \frac{8}{2} = 4
$$

4 is the midpoint for 1 and 7. 4 lies exactly in the middle of 1 and 7. This is shown on the graph.

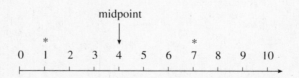

Find the midpoint for each pair of numbers. Graph the two numbers and their midpoint on a number line.

a. 4 and 12   **b.** 3 and 8

c. 4 and $7\frac{1}{2}$ **d.** $\frac{1}{8}$ and $\frac{3}{4}$

▶ **Answers to You Try It** 1. 1 2. $\frac{5}{8}$ 3. $9\frac{9}{10}$ 4. $2\frac{7}{12}$ 5. $21\frac{2}{3}$ 6. $4\frac{7}{8}$
7. $1\frac{1}{6}$ 8. $7\frac{5}{7}$ 9. $7\frac{1}{4}$ 10. 22 doses 11. $\frac{7}{10} < \frac{3}{4}$ 12. $\frac{7}{25} < \frac{3}{10} < \frac{2}{5}$

▶ **Answers to Something More** a. 8 b. $5\frac{1}{2}$ c. $5\frac{3}{4}$ d. $\frac{7}{16}$

SECTION 4.7 EXERCISES

I *Simplify each expression using the order of operations.*

1. $\dfrac{3}{5} + \dfrac{2}{3} - \dfrac{1}{6}$

2. $\dfrac{3}{8} + \dfrac{1}{2} - \dfrac{1}{4}$

3. $\dfrac{5}{8} + \dfrac{1}{2} \cdot \dfrac{3}{4}$

4. $\dfrac{5}{9} + \dfrac{2}{3} \cdot \dfrac{5}{6}$

5. $\dfrac{1}{6} \cdot 3\dfrac{1}{3} + 1\dfrac{1}{5}$

6. $\dfrac{5}{8} \cdot \dfrac{4}{15} + 2\dfrac{2}{9}$

7. $\dfrac{7}{10} \cdot \dfrac{2}{5} \div \dfrac{3}{8}$

8. $\dfrac{5}{6} \cdot 2\dfrac{1}{4} \div 5\dfrac{1}{3}$

9. $8 \div 1\dfrac{3}{4} \cdot 2$

10. $10 \div \dfrac{5}{8} \cdot 2\dfrac{1}{4}$

11. $7\dfrac{1}{2} \div 2 \div \dfrac{3}{4}$

12. $\dfrac{7}{9} \div \dfrac{2}{3} \div \dfrac{1}{6}$

13. $6 \cdot \dfrac{3}{4} + 1\dfrac{5}{6} \cdot 2$

14. $\dfrac{5}{4} \cdot 3\dfrac{3}{7} + \dfrac{3}{8} \cdot 4$

15. $4\dfrac{1}{2} \cdot 4^2$

16. $\dfrac{5}{8} \cdot 5^2$

17. $\dfrac{1}{2} + \dfrac{1}{2} \cdot \dfrac{1}{2} + \dfrac{1}{2}$

18. $\dfrac{2}{5} + \dfrac{3}{5} \cdot \dfrac{1}{5} + \dfrac{4}{5}$

19. $\left(2\dfrac{3}{5}\right)^2 + \dfrac{4}{5} \cdot \dfrac{3}{10}$

20. $\left(7\dfrac{1}{2}\right)^2 + 3\dfrac{1}{4} \cdot \dfrac{5}{2}$

21. $\dfrac{1}{8}\left(\dfrac{3}{8} + \dfrac{5}{8}\right)$

22. $\dfrac{3}{10}\left(\dfrac{7}{10} + \dfrac{8}{10}\right)$

23. $6\left(\dfrac{4}{5} - \dfrac{2}{15}\right)$

24. $14\left(\dfrac{3}{7} - \dfrac{5}{21}\right)$

25. Evaluate $3n + 4$ when $n = 2\dfrac{1}{3}$.

26. Evaluate $7p - 4$ when $p = \dfrac{5}{6}$.

27. Find E if $E = 4h + 5$ and $h = \dfrac{7}{8}$.

28. Find F if $F = 4A$ and $A = 5\dfrac{1}{3}$.

29. What is the product of $\dfrac{3}{7}$ and 280, increased by 50?

30. The income tax is $\dfrac{2}{7}$ of \$28,000 plus \$1,250. Find the tax.

Simplify.

31. $\dfrac{\frac{2}{3}}{\frac{1}{2}}$

32. $\dfrac{\frac{5}{8}}{\frac{15}{16}}$

33. $\dfrac{\frac{12}{4}}{\frac{5}{}}$

34. $\dfrac{\frac{7}{3}}{14}$

35. $\dfrac{\frac{3}{8}}{30}$

36. $\dfrac{\frac{7}{9}}{21}$

37. $\dfrac{2\frac{5}{6}}{1\frac{1}{6}}$

38. $\dfrac{5\frac{8}{21}}{3\frac{2}{21}}$

39. $\dfrac{5\frac{5}{8}}{\frac{5}{6}}$

40. $\dfrac{\frac{4}{9}}{2\frac{6}{7}}$

Solve.

41. What is the average of $\frac{3}{4}$, $1\frac{5}{8}$, $\frac{1}{6}$, and $\frac{2}{3}$?

42. What is the average of $3\frac{1}{2}$, $2\frac{3}{4}$, 3, and $\frac{7}{8}$?

---

43. Three bottles contain $3\frac{1}{4}$ ounces, $1\frac{5}{8}$ ounces, and 2 ounces of the same drug. If a pharmacist combines the contents of the three bottles, how many $\frac{2}{3}$-ounce doses of the drug can she make?

45. Tom owns 40 shares of EZ stock. One share sells for $\$16\frac{3}{8}$. Tom tells his broker to sell all the shares. His broker deducts a \$25 fee for making the sale. Tom splits the remainder of the money with his brother, Dick. How much does each man get?

44. Ed has $4\frac{3}{8}$ pounds of hamburger in his refrigerator, and $2\frac{1}{3}$ pounds in his freezer. If he combines the two amounts, how many hamburgers weighing $\frac{1}{4}$ pound each can he make?

46. 250 shares of GF stock are sold for $\$62\frac{1}{4}$ each. A broker deducts a \$300 fee for making the sale. Three sisters split the remainder of the money. How much does each sister receive?

---

2 *Order the fractions from smallest to largest.*

47. $\frac{3}{5}, \frac{5}{8}$

48. $\frac{4}{7}, \frac{1}{2}$

49. $\frac{11}{12}, \frac{9}{10}$

50. $\frac{2}{9}, \frac{1}{6}$

51. $\frac{4}{5}, \frac{5}{6}, \frac{3}{4}$

52. $\frac{1}{4}, \frac{2}{5}, \frac{1}{3}$

53. $\frac{2}{4}, \frac{7}{15}, \frac{1}{5}$

54. $\frac{10}{15}, \frac{15}{25}, \frac{14}{20}$

55. $\frac{5}{16}, \frac{8}{20}, \frac{1}{8}$

56. $\frac{7}{12}, \frac{13}{24}, \frac{1}{3}$

57. $\frac{5}{8}, \frac{1}{4}, \frac{3}{5}, \frac{2}{3}$

58. $\frac{1}{3}, \frac{1}{6}, \frac{3}{10}, \frac{1}{4}$

59. $\frac{3}{4}, \frac{7}{12}, \frac{5}{6}, \frac{5}{8}$

60. $\frac{2}{5}, \frac{1}{4}, \frac{3}{8}, \frac{3}{10}$

---

61. Three drill bits have sizes $\frac{5}{8}, \frac{11}{16}$, and $\frac{3}{4}$ inches, respectively. Which is the smallest?

62. Screws are purchased with sizes $\frac{1}{4}, \frac{5}{16}$, and $\frac{3}{8}$ inches, respectively. Which is the largest?

63. Company A pays a $\frac{7}{60}$ commission on sales. Company B pays a $\frac{9}{80}$ commission. Which company pays the higher commission?

64. Company A gave $\frac{3}{8}$ of its employees a promotion. Company B gave $\frac{4}{9}$ of its employees a promotion. Which company gave a larger portion of its employees a promotion?

SKILLSFOCUS (Section 2.2) *Divide.*

65. $3,400 \div 10$

66. $52,000 \div 100$

67. $640 \div 10$

68. $327,000 \div 1,000$

EXTEND YOUR THINKING ▶▶▶▶

▶ SOMETHING MORE

69. Graph each pair of numbers on the number line. Then find and graph the midpoint for each pair.

a. 2 and 8

b. 6 and 9

c. 1 and $\frac{2}{5}$

d. $1\frac{3}{4}$ and $\frac{1}{2}$

70. Al made a commission of $\frac{1}{7}$ of the sale price, plus $10, on the sale of a $350 bike. Bob made a commission of $\frac{1}{10}$ of the sale price, plus $25, on the sale of a $360 bike. Who made the greater commission?

▶ TROUBLESHOOT IT

Find and correct the error.

71. $\frac{1}{2} + \frac{1}{2} \cdot 6 = 1 \cdot 6 = 6$

72. $\frac{(5+3)}{(11+3)} = \frac{(5+\overset{1}{\cancel{3}})}{(11+\underset{1}{\cancel{3}})} = \frac{(5+1)}{(11+1)} = \frac{6}{12} = \frac{1}{2}$

WRITING TO LEARN ▶▶▶▶

73. Explain the difference between a simple fraction and a complex fraction.

74. Explain what happens to the size of a fraction when the denominator gets smaller, but the numerator stays the same. Use your own examples.

75. Calculate the answer to each problem.

a. $\frac{10}{(2 \cdot 5)}$

b. $10 \div (2 \cdot 5)$

c. $10 \div 2 \cdot 5$

Explain why the answer in c is different from the answer in a and b.

CHAPTER **4** REVIEW

VOCABULARY AND MATCHING

*New words and phrases introduced in this chapter are shown in the left-hand column. Match
each term on the left with the phrase or sentence on the right that best describes it.*

A. like fractions   _____ involves dividing the denominators by prime factors

B. equivalent fractions   _____ rewrite a fraction by either expanding or reducing

C. expanding a fraction   _____ any number that 2 or more denominators divide into exactly

D. rename a fraction   _____ numerator and denominator are whole numbers

E. lower terms   _____ done with a mixed number, to make its proper fraction improper

F. higher terms   _____ write each fraction in terms of the LCD, then find the difference between the numerators

G. common denominator   _____ the smallest number that 2 or more denominators divide into exactly

H. least common denominator   _____ look different, but have the same value

I. find LCD by inspection   _____ has more than one fraction bar

J. find LCD by prime factoring   _____ when a fraction is reduced, its numerator and denominator are in this

K. find LCD by successive division by primes   _____ involves multiplying the most frequently occurring prime factors in each prime factorization

L. unlike fractions   _____ have the same name, or denominator

M. adding fractions   _____ requires looking at the denominators, and *seeing* what it is

N. subtracting fractions   _____ writing an equivalent fraction with a larger numerator and denominator

O. borrowing   _____ arrange them by first writing each in terms of the same denominator

P. simple fraction   _____ fractions with different denominators

Q. complex fraction   _____ write each fraction in terms of the LCD, then sum the numerators

R. ordering fractions   _____ when a fraction is expanded, its numerator and denominator are in this

REVIEW EXERCISES

4.1 Adding and Subtracting Like Fractions

1. $\frac{3}{8} + \frac{7}{8}$

2. $\frac{7}{10} + \frac{5}{10}$

3. $\frac{2}{7} + \frac{5}{7}$

4. $\frac{3}{20} + \frac{7}{20}$

5. $4\frac{2}{5} + 1\frac{1}{5}$

6. $6\frac{4}{9} + 5\frac{8}{9}$

7. $4 + 1\frac{7}{12}$

8. $3\frac{5}{6} + 4$

9. $\frac{7}{9} - \frac{5}{9}$

10. $\frac{13}{15} - \frac{4}{15}$

11. $\frac{8}{25} - \frac{3}{25}$

12. $\frac{5}{16} - \frac{1}{16}$

13. $5\frac{7}{8} - 2\frac{3}{8}$

14. $4\frac{2}{3} - 1\frac{1}{3}$

15. $3\frac{11}{18} - \frac{5}{18}$

16. $1\frac{37}{60} - \frac{23}{60}$

4.2 Equivalent Fractions

Determine which fractions are equivalent by reducing each to lowest terms.

17. $\dfrac{12}{18}$ and $\dfrac{15}{24}$

18. $\dfrac{12}{15}$ and $\dfrac{20}{25}$

19. $\dfrac{15}{20}$ and $\dfrac{27}{36}$

20. $\dfrac{25}{60}$ and $\dfrac{20}{45}$

Expand each fraction using the form for 1 given.

21. $\dfrac{2}{5}$ using $\dfrac{6}{6}$

22. $\dfrac{2}{3}$ using $\dfrac{8}{8}$

23. $\dfrac{7}{10}$ using $\dfrac{3}{3}$

24. $\dfrac{11}{12}$ using $\dfrac{15}{15}$

Write four fractions equivalent to the given fraction.

25. $\dfrac{1}{8}$

26. $\dfrac{6}{7}$

27. $\dfrac{3}{5}$

28. $\dfrac{7}{9}$

Find the new numerator for the given denominator.

29. $\dfrac{5}{6} = \dfrac{}{30}$

30. $\dfrac{5}{9} = \dfrac{}{27}$

31. $\dfrac{7}{8} = \dfrac{}{72}$

32. $\dfrac{7}{10} = \dfrac{}{70}$

4.3 Finding the Least Common Denominator

The numbers shown are denominators for fractions. Find the LCD for each group of denominators using any method.

33. 12 and 18

34. 15 and 25

35. 20 and 32

36. 60 and 84

37. 12, 18, and 24

38. 15, 30, and 75

39. 7, 8, and 21

4.4 Adding Fractions with Unlike Denominators

40. $\dfrac{5}{6} + \dfrac{7}{8}$

41. $\dfrac{3}{10} + \dfrac{11}{12}$

42. $\dfrac{5}{9} + \dfrac{5}{6}$

43. $\dfrac{11}{18} + \dfrac{4}{45}$

44. $3\dfrac{1}{5} + 1\dfrac{3}{4}$

45. $5\dfrac{7}{10} + 6\dfrac{1}{6}$

46. $4\dfrac{8}{9} + 6\dfrac{7}{8}$

47. $2\dfrac{5}{18} + 7\dfrac{11}{24}$

4.5 Subtracting Fractions with Unlike Denominators

48. $\dfrac{5}{6} - \dfrac{3}{8}$

49. $\dfrac{4}{9} - \dfrac{1}{4}$

50. $\dfrac{7}{12} - \dfrac{3}{8}$

51. $\dfrac{15}{28} - \dfrac{4}{21}$

52. $3\dfrac{4}{7} - 1\dfrac{1}{4}$

53. $5\dfrac{2}{9} - 4\dfrac{8}{15}$

54. $6\dfrac{3}{8} - 4\dfrac{5}{6}$

55. $8 - 5\dfrac{3}{10}$

4.6 Applications: Addition and Subtraction of Fractions

56. A share of stock sold for 23\frac{7}{8}$ a month ago. Today it sells for 26\frac{3}{4}$. How much has the price for one share increased?

57. One board is 12 feet long. The other is $8\frac{7}{12}$ feet long. What is the difference in length between the two boards?

58. On vacation, Beth filled her gas tank twice, buying $10\frac{4}{5}$ and $12\frac{7}{10}$ gallons of gas respectively. How much gas did Beth buy on her vacation?

59. On four successive days, a jogger ran $6\frac{3}{4}$, $2\frac{7}{10}$, $5\frac{3}{5}$, and 8 miles. What total mileage did the jogger run?

60. One-quarter of a fortune was invested in real estate, and $\frac{4}{9}$ was invested in the stock market. The rest was given to charity. What fraction of the fortune was given to charity?

61. Marty purchased 400 shares of a stock at 56\frac{1}{2}$ a share, and sold them at 48\frac{7}{8}$ a share. How much money did Marty lose?

4.7 Combined Operations; Order

62. $\dfrac{5}{6} + \dfrac{2}{3} \cdot \dfrac{4}{5}$

63. $\dfrac{6}{8} \div \dfrac{3}{4} \div \dfrac{1}{8}$

64. $\left(\dfrac{2}{5}\right)^2 + \dfrac{1}{6} \cdot \dfrac{3}{4}$

65. $4 \cdot 3\dfrac{1}{2} - 5$

66. $\dfrac{\frac{3}{4}}{\frac{5}{9}}$

67. $\dfrac{\frac{6}{2}}{7}$

Arrange each set of fractions in order from smallest to largest.

68. $\dfrac{1}{4}, \dfrac{3}{8}, \dfrac{1}{6}$

69. $\dfrac{2}{3}, \dfrac{7}{9}, \dfrac{11}{15}$

70. $\dfrac{3}{10}, \dfrac{8}{25}, \dfrac{7}{20}$

71. $\dfrac{5}{18}, \dfrac{9}{32}, \dfrac{7}{24}$

Allow yourself 50 minutes to complete this test. Write the work for each problem. When done, check your answers. Rework each problem solved incorrectly.

1. $3\frac{2}{7} + 4\frac{1}{8}$

2. $\frac{4}{21} + \frac{8}{49}$

3. $6 + 3\frac{5}{6} + \frac{7}{9}$

4. $\frac{11}{25} + \frac{9}{10} + \frac{2}{5}$

5. $\frac{9}{10} - \frac{5}{16}$

6. $7\frac{5}{8} - 3\frac{11}{12}$

7. $10\frac{13}{16} - 2\frac{5}{12}$

8. $\frac{15}{40} - \frac{8}{24}$

9. $27\frac{1}{12}$ yards of carpet are cut from a 40-yard roll. How many yards of carpet are left on the roll?

10. On three successive days a jogger runs $12\frac{7}{8}$, $16\frac{2}{3}$, and $18\frac{1}{2}$ miles. What is the jogger's total mileage for the three days?

Determine which fractions are equivalent by reducing.

11. $\frac{24}{40}$ and $\frac{15}{25}$

12. $\frac{96}{144}$ and $\frac{68}{102}$

13. A car's gas tank was $\frac{7}{8}$ full at the start of a trip and $\frac{1}{4}$ full at the end.

a. What fraction of a full tank was used on the trip?

b. If the tank holds 24 gallons when full, how many gallons were used on the trip?

14. A developer has 18 acres of land. He will divide the land into $\frac{1}{4}$-acre and $\frac{1}{2}$-acre lots, using all the available land. There are to be $34\frac{1}{4}$-acre lots. How many $\frac{1}{2}$-acre lots will there be?

15. $3\frac{1}{2} + \frac{4}{9} \cdot \frac{6}{10}$

16. $4 \div \left(1\frac{2}{9} + \frac{14}{18}\right)$

17. $\dfrac{\frac{25}{30}}{\frac{75}{16}}$

18. Solve $x + \frac{2}{5} = \frac{1}{2}$.

19. Solve $\frac{5}{9}w = \frac{5}{6}$.

20. Order the fractions from smallest to largest. $\frac{1}{12}, \frac{3}{40}, \frac{5}{48}$

Explain the meaning of each term. Use your own examples.

21. equivalent fractions

22. least common denominator

23. Does every whole number have an even number of divisors? Explain by using examples. Start by finding the divisors of 10. Then of 9. When will a whole number have an odd number of divisors?

1. Write 5,209 in expanded notation.

2. Write in standard notation: seventy-eight thousand, six hundred six.

3. Solve each equation.
   a. $y + 7 = 10$ **b.** $h - 3 = 3$ **c.** $6n = 42$ **d.** $h - 4 = 8$

4. $3,406 + 562 + 3 + 94 + 807$

5. $5,183 - 2,804$

6. Yana's checkbook balance is $659. She writes checks for $56, $37, and $162. She deposits $85. What is her new balance?

7. Round 538 to the nearest ten.

8. Round 45,263 to the nearest thousand.

9. 591×56 10. $10,000 \times 304$

11. $910 - 7$ 12. $45,000 \div 750$

13. $5 \times 6 + 4$ 14. $7 + 5 \times 3 + 2 \times 8 - 1$

15. Julia has $300 in her budget. She purchases 15 textbooks at $17 each. How much money does she have left?

16. What is the average of 89, 76, 91, 84, 90, 77, 86, 95, and 68?

17. Write $5 \times 5 \times 5 \times 5 \times 7 \times 7$ in exponential notation.

18. Write 3^4 in standard notation.

19. 6^0 20. $20(7 - 5) + 12 \times 3$

21. Evaluate $K^2 + 3$ if $K = 7$.

22. Find Y if $Y = 3T + 4M$ and $T = 5$ and $M = 8$.

23. Find x. $45x = 765$

24. Find y. $y + 143 = 679$

25. Find h. $h - 68 = 12$

26. Find t. $\dfrac{t}{24} = 15$

27. Find all the divisors for 80.

28. Prime factor 98.

29. Change $5\dfrac{4}{7}$ to an improper fraction.

30. Change $\dfrac{45}{14}$ to a mixed number.

31. $\dfrac{10}{21} \times \dfrac{27}{35}$

32. $4\dfrac{2}{9} - 2\dfrac{3}{8}$

33. Veronica plans to invest $\dfrac{1}{6}$ of her annual income of \$42,150 in stocks. How much does she plan to invest?

34. $5\dfrac{2}{9} + 3\dfrac{5}{6}$

35. $4\dfrac{5}{8} - \dfrac{9}{10}$

36. A stock selling for \$60 a share one month ago is selling for $\$52\dfrac{3}{8}$ per share today. How much value has the stock lost during the last month?

37. $\dfrac{\frac{5}{9}}{\frac{3}{10}}$

38. $4 + 5 \cdot 3$

39. $\left(\dfrac{4}{5}\right)^2$

Explain the meaning of each term. Use your own examples.

40. numerator

41. prime number

42. factor

43. square root

44. variable

45. equation

Decimals

MORTGAGE PAYMENTS

Nancy and Dave are purchasing a home. They plan to borrow $100,000 at 10.875% interest. Their monthly payments for a 20-year and a 30-year mortgage are shown here.

Term	Monthly Payment
20 years	$1,023.69
30 years	$ 942.89

A mortgage payment consists of principal plus interest. Each time the bank receives your payment, they first deduct the interest you owe on the loan. They then apply the rest toward reducing your principal balance. The couple wants to know how much interest they will save with the 20-year mortgage compared to the 30-year mortgage.

With the 30-year loan, you pay $942.89 per month for 12 months a year for 30 years, or $942.89 × 12 × 30 over 30 years. The total interest paid with the 30-year mortgage is this product less the $100,000 borrowed.

$$\text{total interest for 30-year loan} = \$942.89 \times 12 \times 30 - \$100,000$$

Next, calculate the interest for the 20-year loan, and compare. This problem will be solved completely in this chapter.

In this chapter you will study decimal numbers. You will learn the rules to add, subtract, divide, and multiply decimals, which will allow you to solve practical problems like the one here. You will also learn to estimate answers to decimal problems. You will see the operations on decimal numbers applied to a wide variety of everyday problems including balancing a checkbook, making a grocery store purchase, reading sports statistics, and estimating the cost of a job.

Work through each section in this chapter. Use this test to identify topics you are not familiar with. These topics may require additional study. You may *not* use a calculator.

5.1

1. How many decimal places does 42.7009 have?

2. Write 29.074 in expanded notation.

3. Write 285.33 in expanded notation.

4. Write 4.062 in words.

5. Write in decimal notation: two hundred ninety and four hundred sixty-seven ten-thousandths.

5.2

6. Round 1,675.0964 to hundredths.

7. Round 582.736 to tens.

5.3

8. Add $0.47 + 5.602 + 907.3 + 6 + 0.008$.

9. Subtract $5.62 - 2.074$.

5.4

10. Sally had a bank balance of $649.73. She wrote checks for $9.78, $140.60, $312.66, and $42.50. She made a $280.17 deposit. What is her new balance?

5.5

11. Multiply $5.064 \cdot 28.93$.

5.6

12. Divide $4.69 \div 12.7$ (round to thousandths).

5.7

13. Phil traveled 490 miles on 17.8 gallons of gas. How many miles per gallon did he get? Round to tenths.

14. What will 1.83 pounds of cheese cost at $3.19 per pound? Round to the nearest cent.

5.8

15. Change $\dfrac{3}{20}$ to a decimal.

16. Change 0.025 to a fraction. Express in lowest terms.

5.9

17. Simplify $3.7 + 4 \cdot 1.52$.

18. Arrange the numbers in order from largest to smallest. 0.06, 0.0660, 0.106, 0.0096

VOCABULARY Explain the meaning of each term. Use your own examples.

19. repeating decimal

20. thousandths place

5.1 DECIMAL NOTATION

OBJECTIVES

1 Define decimal notation and place value (Section 1.2).

2 Write a decimal in expanded notation (Section 1.2, Objective 4).

3 Read a decimal number.

4 Write a decimal number.

NEW VOCABULARY

decimal notation

place holding zero

---

1 Decimal Notation and Place Value

Decimal numbers are an extension of the way you write whole numbers. A decimal point and additional place values are added to the right of the ones place (see Figure 5.1). Numbers written using a decimal point are written in **decimal notation**. The numbers 4.6, 3.708, and 456.51 are written in decimal notation.

FIGURE 5.1
Place Value Names

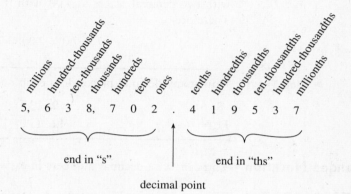

EXAMPLE 1 What is the place value name for the digit 4 in each number?

a. 5.4 ┌─tenths place

b. 25.0437 ┌─hundredths place

c. 3,249.06 ┌─tens place

d. 2.001549 ┌─hundred-thousandths place ■

▶ **You Try It** What is the place value name for the digit 6 in each number?

1. 2.06   **2.** 6.913   **3.** 293.612   **4.** 5.846

The decimal point for a whole number is written to the right of the ones place.

EXAMPLE 2

$$1 = 1.$$
$$7 = 7.$$
$$46 = 46. \quad ■$$

▶ **You Try It** Write each whole number with its decimal point.

5. 3   **6.** 240   **7.** 62

The *number of decimal places* in a number is the number of digits to the right of the decimal point.

EXAMPLE 3   How many decimal places does each number have?

a. 4.98 ⌐ two decimal places

b. 8,467.3 ⌐ one decimal place

c. 0.004801 ⌐ six decimal places

d. 16.000 ⌐ three decimal places

e. 7 has no decimal places. Since 7 = 7., there are no digits to the right of the decimal point. ■

Though 7 has no decimal places, write it with one decimal place as 7.0, with two decimal places as 7.00, with three decimal places as 7.000, and so on. Therefore,

$$7 = 7.0 = 7.00 = 7.000 = 7.0000 \ldots.$$

▶ **You Try It**   How many decimal places does each number have?

8. 6.07   **9.** 13.5   **10.** 0.0073

11. 4   **12.** 15.0

2 Expanded Notation   You can write decimal numbers in expanded notation.

EXAMPLE 4   Write each decimal in expanded notation.

a. 6.8 = 6 ones + 8 tenths

$$= 6 + \frac{8}{10}$$

b. 52.317 = 5 tens + 2 ones + 3 tenths + 1 hundredth + 7 thousandths

$$= 50 + 2 + \frac{3}{10} + \frac{1}{100} + \frac{7}{1,000}$$

c. 0.0409 = 0 tenths + 4 hundredths + 0 thousandths + 9 ten-thousandths

$$= \frac{0}{10} + \frac{4}{100} + \frac{0}{1,000} + \frac{9}{10,000}$$

$$= \frac{4}{100} + \frac{9}{10,000} \quad ■$$

▶ **You Try It**   Write each decimal in expanded notation.

13. 5.73   **14.** 82.1   **15.** 0.406

3 Reading Decimal Numbers   How do you read 0.26? First, write 0.26 in expanded notation.

$$0.2\underline{6} = \frac{2}{10} + \frac{6}{100} = \frac{2}{10} \times \frac{10}{10} + \frac{6}{100} = \frac{20}{100} + \frac{6}{100} = \frac{26}{100} = 26 \ \text{hundredths}$$

6, the digit farthest right, is in the hundredths place. The part of a decimal number to the right of the decimal point is named by the place value of the digit farthest right.

> **To read a decimal number**
>
> 1. Read the number to the left of the decimal point as you read a whole number.
> 2. Read the decimal point as "and."
> 3. Read the number to the right of the decimal point as you read a whole number, followed by the place value name of the digit farthest right.

EXAMPLE 5

a. 0.3 is read "three tenths."

    3 is in the tenths place

b. 0.107 is read "one hundred seven thousandths."

    7 is in the thousandths place

c. 0.0045 is read "forty-five ten-thousandths."

    5 is in the ten-thousandths place

d. 0.20 is read "twenty hundredths."

    This 0 is the digit farthest right. It is in the hundredths place. ∎

> **OBSERVE** In Examples 5a–d, a zero is in the ones place. It tells you the whole number part of the decimal is 0. Writing this zero is optional. Both 0.3 and .3 are read "three tenths."

▶ **You Try It** Read each decimal, then write it in words.

16. 0.8   **17.** 0.095   **18.** 0.41

19. 0.2077   **20.** 0.300

EXAMPLE 6

a. 2.4 is read "two and four tenths."
b. 56.018 is read "fifty-six and eighteen thousandths."
c. 300.000004 is read "three hundred and four millionths."
d. 682.4257 is read "six hundred eighty-two and four thousand two hundred fifty-seven ten-thousandths."

> **CAUTION** The decimal point is read "and." The word "and" is used in no other place when you read or write a decimal. It is wrong to read 682.4257 as "six hundred *and* eighty-two and four thousand two hundred *and* fifty-seven ten-thousandths." ∎

▶ **You Try It** Read each decimal and write it in words.

21. 5.7   **22.** 38.04   **23.** 50.2008   **24.** 4.156

4 Writing Decimal Numbers

Write nine and five tenths in decimal notation.

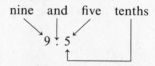

> Given a decimal number written in words, to write it in decimal notation
>
> 1. If the word *and* is used, read the whole number to the left of the word *and*. Write it as you read it. Then write the decimal point for *and*.
>   If the word *and* is not used, and the decimal number ends in "ths," there is no whole number part for this decimal. Write 0 in the ones place, followed by the decimal point.
> 2. Read the rest of the number as a whole number. Write it as you read it, placing the rightmost digit in the place named by the decimal. This may require the use of place holding zeros.

EXAMPLE 7   Write in decimal notation.

a. seventy-eight and four hundredths

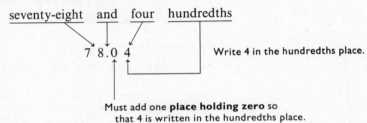

b. six thousand four hundred ten and two hundred ninety-five thousandths

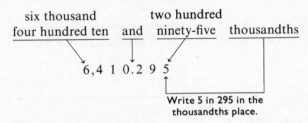

c. thirty-two ten-thousandths

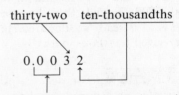

To write 2 in 32 in the ten-thousandths place, write two place holding zeros between the decimal point and 32. ■

▶ **You Try It**   Write in decimal notation.

25. thirty-two and seventeen thousandths
26. four hundred ninety and one hundred sixty-eight thousandths
27. three hundred fifty ten-thousandths

The next example demonstrates the importance of the hyphen and the word *and* when writing numbers in decimal notation.

EXAMPLE 8   Write in decimal notation.

a. four hundred ten thousandths

The word *and* does not appear, so the whole number part of this decimal is 0.

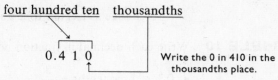

b. four hundred ten-thousandths

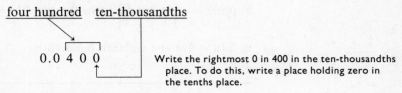

c. four hundred and ten thousandths

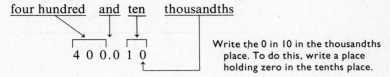

d. four hundred ten ten-thousandths

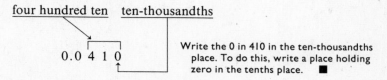

▶ **You Try It**   Write in decimal notation.

28. nine hundred ten thousandths
29. nine hundred ten-thousandths
30. nine hundred and ten thousandths
31. nine hundred ten ten-thousandths

If a fraction has a power of 10 for a denominator, change it to a decimal by reading it, then writing it the way you read it.

EXAMPLE 9   Write each fraction in decimal notation.

a. $\dfrac{7}{10}$ = seven tenths = 0.7

b. $\dfrac{649}{1,000}$ = six hundred forty-nine thousandths = 0.649

c. $\dfrac{275}{100} = 2\dfrac{75}{100}$ = two and seventy-five hundredths = 2.75

d. $\dfrac{416}{10,000}$ = four hundred sixteen ten-thousandths = 0.0416   ∎

Write each fraction in decimal notation.

32. $\dfrac{17}{100}$ **33.** $\dfrac{4,253}{10,000}$ **34.** $\dfrac{9}{10}$ **35.** $\dfrac{5}{1,000}$

To change a decimal to a fraction, read it, then write it as you read it.

EXAMPLE 10   Write each decimal in fraction notation. Do not reduce.

a. $0.3 = $ three tenths $= \dfrac{3}{10}$

b. $0.802 = $ eight hundred two thousandths $= \dfrac{802}{1,000}$

c. $5.18 = $ five and eighteen hundredths $= 5\dfrac{18}{100}$ or $\dfrac{518}{100}$ ∎

▶ **You Try It**   Write each decimal in fraction notation. Do not reduce.

36. 0.72   **37.** 3.9   **38.** 14.080

▶ **Answers to You Try It**   1. hundredths   2. ones   3. tenths
4. thousandths   5. 3.   6. 240.   7. 62.   8. two   9. one   10. four   11. none
12. one   13. $5 + \dfrac{7}{10} + \dfrac{3}{100}$ 14. $80 + 2 + \dfrac{1}{10}$ 15. $\dfrac{4}{10} + \dfrac{6}{1,000}$
16. eight tenths   17. ninety-five thousandths   18. forty-one hundredths
19. two thousand seventy-seven ten-thousandths   20. three hundred thousandths
21. five and seven tenths   22. thirty-eight and four hundredths
23. fifty and two thousand eight ten-thousandths
24. four and one hundred fifty-six thousandths   25. 32.017   26. 490.168
27. 0.0350   28. 0.910   29. 0.0900   30. 900.010   31. 0.0910   32. 0.17
33. 0.4253   34. 0.9   35. 0.005   36. $\dfrac{72}{100}$ 37. $3\dfrac{9}{10}$ or $\dfrac{39}{10}$
38. $14\dfrac{80}{1,000}$ or $\dfrac{14,080}{1,000}$

SECTION 5.1 EXERCISES

1 *Find the place value of the digit 6.*

1. 32.6   2. 60.5   3. 3.016   4. 2,603.87   5. 0.0006   6. 25.9671

Find the number of decimal places in each number.

7. 0.5   8. 5.0124   9. 45.92   10. 345.902   11. 8   12. 365

2 *Write each decimal in expanded notation.*

13. 1.7   14. 23.6   15. 0.604   16. 50.444

17. 2.06   18. 90.62   19. 0.2008   20. 0.046

21. 1.60   22. 0.08   23. 6,000.06   24. 0.9

3 *Read the decimal, then write it in words.*

25. 5.2   26. 9.1   27. 0.04   28. 0.12

29. 65.4   30. 70.08   31. 0.6   32. 12.0054

33. 265.2   34. 1,250.0506   35. 6.05   36. 300.002

4 *Write in decimal notation.*

37. nine tenths

38. five tenths

39. six and seven hundredths

40. one and one hundredth

41. twenty-eight and four tenths

42. seventy and seven tenths

43. three hundred twelve thousandths

44. seventy-four thousandths

45. two hundred ten thousandths

46. two hundred ten-thousandths

47. two hundred and ten thousandths

48. two hundred ten ten-thousandths

49. seven hundred twenty-eight thousandths

50. seven hundred and twenty-eight thousandths

51. seventy-five and one hundred thousandths

52. seventy-five and one hundred-thousandth

53. six hundred fifty-nine and thirty-five thousandths

54. one hundred eighty and three hundred

55. eight hundred and four tenths

56. two hundred ninety-eight and sixteen thousandths

57. four million, forty-four thousand, nine and eleven thousandths

58. forty-six and two thousand one ten-thousandths

59. nine hundred sixty-five ten-thousandths   **60.** six and four hundred seventy-five thousandths

61. fifty and fifty hundredths   **62.** eight and four thousandths

Write each fraction in decimal notation.

63. $\dfrac{8}{10}$ **64.** $\dfrac{1}{10}$ **65.** $\dfrac{173}{1,000}$ **66.** $\dfrac{97}{100}$ **67.** $\dfrac{872}{100}$ **68.** $\dfrac{38}{10}$

Write each decimal in fraction notation. Do not reduce.

69. 0.6   **70.** 0.956   **71.** 0.02   **72.** 0.24   **73.** 4.58   **74.** 7.2

SKILLSFOCUS (Section 1.5) *Round to the indicated place.*

75. 928 to tens   **76.** 67,088 to hundreds   **77.** 601 to thousands   **78.** 4,899 to tens

EXTEND YOUR THINKING ▶▶▶▶

▶TROUBLESHOOT IT

Find and correct the error.

79. six hundred nine thousandths = 600.009

80. 7.508 = seven and five hundred and eight thousandths

WRITING TO LEARN ▶▶▶▶

81. A decimal number can only have one decimal point. Explain what happens when you enter 4.2.5, or 0.68.4, into your calculator.

▶YOU BE THE JUDGE

82. Which decimal is larger, 0.08 or 0.4, and by how much? Explain how you arrive at your answer. (Hint: First express each decimal as money.)

5.2 ROUNDING DECIMALS

OBJECTIVE
Round off decimals (Section 1.5).

NEW VOCABULARY
truncation

Rounding Decimals

You round off to estimate answers. You buy seven party favors for $4.95 each. You round $4.95 to $5 and estimate you paid about $7 \cdot \$5 = \$35$. You also round off for convenience. You pay $1,992.50 for a piano, but in conversation you say you paid $2,000. You also round repeating calculator answers such as $4.3333333 to the nearest cent, giving $4.33.

> To round off decimal numbers
>
> 1. Circle the digit in the place you are rounding to.
> 2. If the digit on its right is
>   a. 5 or larger (5, 6, 7, 8, or 9), add 1 to the circled digit.
>   b. 4 or smaller (4, 3, 2, 1, or 0), do not change the circled digit.
> 3. Drop all digits to the right of the circled digit. Replace each dropped digit to the left of the decimal point, if any, with 0.

EXAMPLE 1 Round off 5.68 to the nearest tenth.

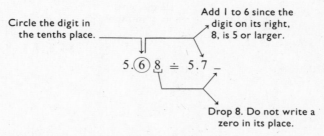

$5.68 \doteq 5.7$ rounded to the nearest tenth. As seen on the number line below, 5.68 is closer to 5.7 than to 5.6.

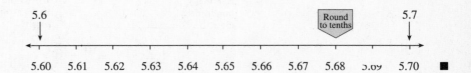

EXAMPLE 2 Round 672.98 to the nearest ten.

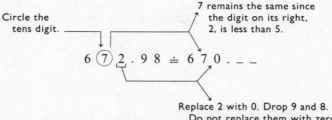

$672.98 \doteq 670$ rounded to the nearest ten. In terms of dollars, $672.98 is closer to $670 than $680. ■

EXAMPLE 3   Round 183.65 to the nearest one.

Circle the
ones digit.

Add 1 to 3 since the digit
on its right, 6, is 5
or larger.

$$1\ 8\ ③.6\ 5 \doteq 1\ 8\ 4\ .\ \_\ \_$$

Drop 6 and 5. Do not
replace them with zeros.

183.65 $\doteq$ 184 rounded to the nearest one. In terms of money, $183.65 is closer to $184 than $183.   ■

▶ **You Try It**   1. Round 4.829 to tenths.
2. Round 869.738 to tens.
3. Round 59.3986 to ones.

To round dollar amounts to the nearest cent, round to the hundredths place.

EXAMPLE 4   Round $467.1854 to the nearest cent.

Circle the cents
(hundredths) place.

Add 1 to 8 since the
digit to its right,
5, is 5 or larger.

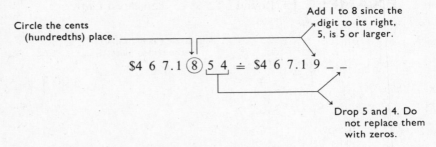

$$\$4\ 6\ 7.1\ ⑧\ 5\ 4 \doteq \$4\ 6\ 7.1\ 9\ \_\ \_$$

Drop 5 and 4. Do
not replace them
with zeros.

$467.1854 $\doteq$ $467.19 rounded to the nearest cent.   ■

EXAMPLE 5   Round 1.399 to the nearest hundredth.

Circle the
hundredths
digit.

Add 1 to the circled 9 since the digit
on its right, also a 9, is 5 or
larger.

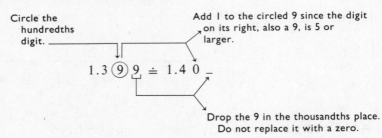

$$1.3\ ⑨\ 9 \doteq 1.4\ 0\ \_$$

Drop the 9 in the thousandths place.
Do not replace it with a zero.

OBSERVE   1.399 $\doteq$ 1.40 rounded to the nearest hundredth. Keep the 0 in the hundredths place. It tells you to the nearest hundredth, there are no hundredths in the answer. In terms of money, suppose the price of one gallon of gas is $1.399. To the nearest penny (hundredth), this is closer to $1.40 than to $1.39.   ■

▶ **You Try It**   Round to the nearest cent.

4. $2.8473

5. $147.995

Rounding to one decimal place means round to tenths. Rounding to two decimal places means round to hundredths. Rounding to three decimal places means round to thousandths. And so on.

EXAMPLE 6   Round 0.007614 to three decimal places.

Rounding to three decimal places means round to thousandths.

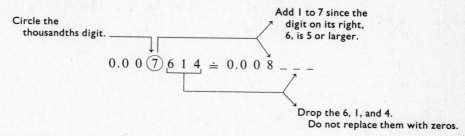

0.007614 ≐ 0.008 rounded to three decimal places.   ■

EXAMPLE 7   Round 0.0489 to one decimal place.

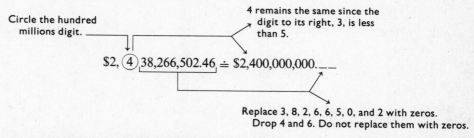

0.0489 ≐ 0.0 rounded to one decimal place. Keep the 0 in the tenths place. It tells you to the nearest tenth, there are 0 tenths.   ■

▶ **You Try It**   **6.** Round 15.9847 to two decimal places.
7. Round 0.80091 to three decimal places.

EXAMPLE 8   The budget for the space program is $2,438,266,502.46. Express this budget rounded to the nearest hundred million dollars.

Circle the hundred millions digit. ———

4 remains the same since the digit to its right, 3, is less than 5.

$2,④38,266,502.46 ≐ $2,400,000,000.__

Replace 3, 8, 2, 6, 6, 5, 0, and 2 with zeros.
Drop 4 and 6. Do not replace them with zeros.

$2,438,266,502.46 ≐ $2,400,000,000 rounded to the nearest hundred million dollars.

OBSERVE   In newspapers and magazines, a large dollar amount such as $2,438,266,502.46 is simply expressed as $2.4 billion.   ■

8. The distance to the sun is 92,746,800 miles. Round this distance to the nearest million miles.

Another way to round off is called **truncation**. To truncate a decimal means to drop all the digits to the right of a given decimal place. Truncating 4.236 to ones gives 4. Truncating 3.99 to ones gives 3. Some calculators truncate answers. For example, a calculator answer such as 66.6666666666 . . . is truncated to six places as 66.666666. It is rounded to six places as 66.666667.

► **Answers to You Try It**   1. 4.8   2. 870   3. 59   4. $2.85   5. $148.00
6. 15.98   7. 0.801   8. 93,000,000, or 93 million, miles

SECTION 5.2 EXERCISES

Round off as indicated.

1. 4.72 to tenths

2. 2.091 to tenths

3. 264.99 to tens

4. 568.09 to tens

5. 4.4821 to hundredths

6. 4,612.517 to hundreds

7. 76.416 to ones

8. 2.7021 to ones

9. 90.0267 to thousandths

10. 9.8261 to thousandths

11. 450.95 to hundreds

12. 2,789.419 to hundredths

13. 6.488 to tenths

14. 50.05 to tenths

15. 27.089 to ones

16. 101.50 to ones

17. 509.5 to hundreds

18. 900.0561 to thousandths

19. 3.0089 to thousandths

20. 56.0293 to hundredths

21. 0.254 to tenths

22. 0.062 to tenths

23. 683.811 to ones

24. 0.49 to ones

25. 38.99 to hundreds

26. 55.015 to hundredths

27. 0.0664 to thousandths

28. 0.07623 to hundredths

29. 427.03 to tenths

30. 58.394 to tenths

31. 2.00475 to ten-thousandths

32. 31.7746 to thousandths

33. 0.00057123 to millionths

34. 0.00000588 to millionths

35. 4.97 to tenths

36. 397.67 to tens

37. 0.0068 to hundredths

38. 69.996 to hundredths

39. 599.9995 to thousandths

40. 54.994 to tenths

41. 35.056 to one decimal place

42. 19.5818 to two decimal places

43. 0.7125 to three decimal places

44. 2.819 to one decimal place

45. 35.01452 to four decimal places

46. 505.5244 to two decimal places

47. 0.000673 to four decimal places

48. 0.50026 to two decimal places

49. 24.9603 to one decimal place

50. 3.0899 to three decimal places

51. The speed of light is 186,282.398 miles per second. Round to hundreds.

52. The length of the year is 365.2422 days. Round to hundreds.

53. The defense budget is $252,476,079,247.69. Round to hundred millions.

54. Sales last year were $44,589,105.21. Round to hundred thousands.

55. The number of children per family is 1.8577. Round to tenths.

56. The average life expectancy is 58.623 years. Round to the nearest year.

57. Round $4,572.0472 to the nearest cent.

58. Round $32.59245 to the nearest cent.

59. Round $0.495 to the nearest cent.

60. Round $0.0749 to the nearest cent.

SKILLSFOCUS (Section 1.4) *Subtract.*

61. $540 - 279$ **62.** $6,000 - 2,508$ **63.** $71 - 19$ **64.** $5,264 - 281$

EXTEND YOUR THINKING ▶▶▶▶
▶ TROUBLESHOOT IT

Find and correct the error.

65. 78.992 (round to tenths) = 78.⑨92 ≐ 79

66. 153.68 (round to tens) = 153.⑥8 ≐ 153.7

WRITING TO LEARN ▶▶▶▶

67. Describe all of the numbers that round off to 10.

68. Describe all of the numbers that round off to 4.8.

69. Explain each step you use to round 42.6985 to hundredths, as if you were explaining it to a friend.

▶ YOU BE THE JUDGE

70. Are 1 inch and 1.0 inch the same amount? For example, would you rather your doctor made her surgical incision accurate to 1 inch or to 1.0 inch? Or doesn't it matter? Explain your decision.

5.3 ADDITION AND SUBTRACTION OF DECIMALS

OBJECTIVES

1 Add decimals and estimate a sum (Sections 1.2, 1.3, 1.5).

2 Subtract decimals and estimate a difference (Section 1.4).

3 Solve equations (Section 2.6).

1 **Adding Decimals**   Use expanded notation to add 3.26 + 4.12.

$$3.26 \longrightarrow 3 \text{ ones} + 2 \text{ tenths} + 6 \text{ hundredths}$$
$$+4.12 \longrightarrow 4 \text{ ones} + 1 \text{ tenth } + 2 \text{ hundredths}$$
$$7.38 \longleftarrow 7 \text{ ones} + 3 \text{ tenths} + 8 \text{ hundredths}$$

You only add like named quantities. Lining the decimal points up vertically keeps digits with the same place value name in the same column.

> **To add decimal numbers**
>
> 1. Write the decimal numbers so the decimal points line up in the same vertical column.
> 2. Add vertically as you add whole numbers.
> 3. Drop the decimal point straight down into your answer, keeping it in the same vertical column with the others.

Addition is commutative and associative. Changing the order or grouping when adding decimal numbers does not change the answer.

EXAMPLE 1   Add 47.3 + 9 + 5.68.

Write 9 as the decimal '9.'.

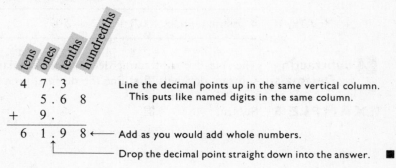

$$
\begin{array}{r}
4\ 7\,.\,3 \\
5\,.\,6\ 8 \\
+\quad 9\,. \\
\hline
6\ 1\,.\,9\ 8
\end{array}
$$

Line the decimal points up in the same vertical column. This puts like named digits in the same column.

Add as you would add whole numbers.

Drop the decimal point straight down into the answer. ■

EXAMPLE 2   Add 7.45 + 1.3.

Correct: The decimal points are lined up in the same vertical column.	**Wrong:** You cannot add 5 hundredths to 3 tenths because they are unlike amounts.
$$\begin{array}{r} \downarrow \\ 7.4\ 5 \\ +1.3 \\ \hline 8.7\ 5 \end{array}$$	$$\begin{array}{r} \downarrow \\ 7.4\ 5 \\ +\quad 1.3 \\ \hline 7.5\ 8 \end{array}$$
Lining up decimal points in the same column means you will add like named digits together.	Not lining up decimal points in the same column means you make the mistake of adding unlike named digits together. ■

Use zeros to fill in for missing digits to the right of the decimal point. This helps reduce addition errors by making it easier to see which column a digit is in.

EXAMPLE 3 Add $36.2 + 415 + 0.006 + 1.73 + 5.8461$.

$$
\begin{array}{r}
3\,6.2 \\
4\,1\,5. \\
0.0\,0\,6 \\
1.7\,3 \\
+\quad 5.8\,4\,6\,1 \\
\hline
4\,5\,8.7\,8\,2\,1
\end{array}
$$

→ Fill in zeros to help you line up like named digits. →

$$
\begin{array}{r}
3\,6.2\,0\,0\,0 \\
4\,1\,5.0\,0\,0\,0 \\
0.0\,0\,6\,0 \\
1.7\,3\,0\,0 \\
+\quad 5.8\,4\,6\,1 \\
\hline
4\,5\,8.7\,8\,2\,1 \quad\blacksquare
\end{array}
$$

▶ **You Try It**   Add.

1. $3.19 + 14 + 0.945$
2. $8.3 + 41.07 + 0.9 + 1 + 2.018$
3. $0.07 + 0.4 + 0.9009 + 0.008$

Recall that you estimate the answer to an addition or subtraction by rounding all numbers to the *same place value*.

EXAMPLE 4 Estimate $5.83 + 2.496 + 0.43 + 7 + 0.8$.

Round all addends to ones, then add.

$$
\begin{array}{lcccccccc}
\text{original addends} \longrightarrow & 5.83 & + & 2.496 & + & 0.43 & + & 7 & + & 0.8 \\
& \downarrow & & \downarrow & & \downarrow & & \downarrow & & \downarrow \\
\text{rounded to ones} \longrightarrow & 6 & + & 2 & + & 0 & + & 7 & + & 1 & = 16
\end{array}
$$

The estimate for the sum is 16. (The actual sum is 16.556.) ∎

▶ **You Try It**   4. Estimate $6.83 + 0.7 + 11.3 + 2.099$.

2 Subtracting Decimals

The rules for subtracting decimals are similar to those for adding decimals. To subtract decimals, first line the decimal points up in the same vertical column.

EXAMPLE 5 Subtract $7.63 - 4.81$.

$$
\begin{array}{r}
\overset{6}{}\overset{16}{} \\
\cancel{7}.\,\cancel{6}\ \ 3 \\
-\,4.\,8\ \ 1 \\
\hline
2.\,8\ \ 2 \quad\blacksquare
\end{array}
$$

> **To subtract two decimal numbers**
>
> 1. Write the two numbers so the decimal points are in the same vertical column.
> 2. Subtract as you would subtract whole numbers.
> 3. Drop the decimal point straight down into your answer so it is in the same vertical column with the others.

EXAMPLE 6   Subtract 7.094 − 5.4.

$$
\begin{array}{r}
7.0\ 9\ 4 \\
-\ 5.4 \\
\hline
1.6\ 9\ 4
\end{array}
$$

Add zeros to the subtrahend. →

$$
\begin{array}{r}
\overset{6\ \ 10}{\cancel{7}.\cancel{0}\ 9\ 4} \\
-\ 5.4\ 0\ 0 \\
\hline
1.6\ 9\ 4 \quad \blacksquare
\end{array}
$$

▶ **You Try It**   Subtract.

5. 14.37 − 9.52

6. 6.194 − 2.5

EXAMPLE 7   Subtract 25.3 − 8.176.

$$
\begin{array}{r}
2\ 5.3 \\
-\ \ \ 8.1\ 7\ 6 \\
\hline
\end{array}
$$

6 and 7 have no digits above them from which to subtract. Attach two zeros to the minuend. →

$$
\begin{array}{r}
\overset{\quad\quad\quad\ 9}{\overset{1\ \ 15\ \ 2\ \ 10\ \ 10}{\cancel{2}\ \cancel{5}.\cancel{3}\ \cancel{0}\ \cancel{0}}} \\
-\ \ \ 8.1\ 7\ 6 \\
\hline
1\ 7.1\ 2\ 4 \quad \blacksquare
\end{array}
$$

EXAMPLE 8   Subtract 0.08 from 1.

$$
\begin{array}{r}
1. \\
-0.0\ 8 \\
\hline
\end{array}
$$

Attach two zeros to the right side of the minuend. →

$$
\begin{array}{r}
\overset{\quad\ 9}{\overset{0\ \ 10\ \ 10}{\cancel{1}.\cancel{0}\ \cancel{0}}} \\
-0.0\ 8 \\
\hline
0.9\ 2 \quad \blacksquare
\end{array}
$$

OBSERVE   It is sometimes helpful to think in terms of money to understand decimal operations. Think of 1 − 0.08 as $1 − $0.08, or 1 dollar minus 8 cents. The answer is 92 cents, or $0.92.

▶ **You Try It**   **7.** 9.4 − 3.187

8. Subtract 0.29 from 4.

EXAMPLE 9   Estimate 14.07 − 5.73.

Round both numbers to the ones place, then subtract.

original numbers: 14.07 − 5.73

rounded to ones: 14 − 6 = 8

The estimate for 14.07 − 5.73 is 8. (The actual answer is 8.34.)   ■

▶ **You Try It**   **9.** Estimate 8.904 − 3.38.

EXAMPLE 10   Solve the equation $x + 3.5 = 6$.

Solve the addition equation $x + 3.5 = 6$ by subtracting 3.5 from each side.

$$
\begin{array}{rll}
x + 3.5 &= 6 \\
-\,3.5 & \;\;-3.5 \\
\hline
x &= \;\;2.5
\end{array}
\qquad \xrightarrow{\text{subtract}} \qquad
\begin{array}{r}
\overset{5\;\;10}{\cancel{6}\,.\,\cancel{0}} \\
-\,3\,.\,5 \\
\hline
2\,.\,5
\end{array}
$$

Check: $x + 3.5 = 6$
  $2.5 + 3.5 = 6$
      $6 = 6$ ∎

EXAMPLE 11   Solve the equation $x - 4.7 = 3.53$.

Solve the subtraction equation $x - 4.7 = 3.53$ by adding 4.7 to each side.

$$
\begin{array}{rl}
x - 4.7 &= 3.53 \\
+\,4.7 & \;\;+4.7 \\
\hline
x &= 8.23
\end{array}
$$

Check: $x - 4.7 = 3.53$
    $8.23 - 4.7 = 3.53$
      $3.53 = 3.53$

$$
\xrightarrow{\text{subtract}}
\begin{array}{r}
\overset{7\;\;12}{8\,.\,\cancel{2}\;\;3} \\
-\,4\,.\,7\;\;0 \\
\hline
3\,.\,5\;\;3
\end{array}
\quad ∎
$$

▶ **You Try It**   **10.** Solve $y + 4.71 = 8$.   **11.** Solve $s - 6.4 = 11.35$.

CONNECTIONS

You add and subtract decimals by lining up decimal points in the same vertical column. You add and subtract like letters in algebra by writing like letters in the same vertical column.

$$
\begin{array}{r}
2x + 2y \\
+\,4x + 7y \\
\hline
6x + 9y
\end{array}
\qquad
\begin{array}{r}
5x + 8y + 2z \\
-\,3x - 5y - 2z \\
\hline
2x + 3y + 0z = 2x + 3y
\end{array}
$$

▶ **Answers to You Try It**   **1.** 18.135   **2.** 53.288   **3.** 1.3789   **4.** 21   **5.** 4.85
6. 3.694   **7.** 6.213   **8.** 3.71   **9.** 6   **10.** $y = 3.29$ **11.** $s = 17.75$

SECTION 5.3 EXERCISES

1

1. $4.62 + 9.7$

2. $54.02 + 17.304$

3. $89.3 + 607.14$

4. $7.051 + 24.7926$

5. $0.014 + 0.5 + 0.58$

6. $0.0034 + 0.6 + 0.882$

7. $12 + 0.068 + 2$

8. $8 + 0.09 + 1.637$

9. $60 + 0.06 + 7.7$

10. $1 + 0.02 + 0.003$

11. $582 + 6.85 + 24.377 + 0.0088$

12. $15.905 + 40 + 6.085 + 2.0006$

13. $300.56 + 0.8 + 34.0056 + 0.0008 + 70 + 200.003$

14. $0.345 + 0.0390 + 0.4 + 3 + 0.007 + 0.00306$

15. $45.9 + 200.5 + 45.2 + 60.0 + 108.2 + 27.1$

16. $3.009 + 6.937 + 4 + 5.004 + 5.65 + 2.905 + 4.6$

17. $9 + 0.09 + 99 + 0.9 + 0.999 + 999 + 0.001$

18. $4 + 0.4 + 404.40 + 44 + 0.404 + 40.04$

19. Estimate $49.95 + 7.50 + 18 + 2.4$ by rounding to tens.

20. Estimate $\$121.89 + \$88.16 + \$275.56 + \$65.70$ by rounding to hundreds.

21. Estimate $1,465.7 + 531.87 + 1,013.7 + 87.03$ by rounding to hundreds.

22. Estimate $\$15.26 + \$7.32 + \$11.68 + \$0.45$ by rounding to ones.

23. 45 feet + 7.25 feet + 0.5 foot + 9.375 feet

24. $\$35 + \$4.65 + \$0.36 + \$9.40 + \$3 + \$0.04$

25. 0.5 inch + 3.25 inches + 1.875 inches + 4 inches + 3.75 inches

26. 100 meters + 45.03 meters + 0.478 meter + 30.3 meters + 16 meters + 0.73 meter

2

27. $4.6 - 3.2$

28. $12.8 - 7.9$

29. $25.07 - 15.42$

30. $5.7 - 2.61$

31. $23.9 - 14.87$

32. $46.239 - 30.4$

33. $6.937 - 3.05$

34. $0.104 - 0.02$

35. $25.66 - 8.912$

36. $7.34 - 2.0056$

37. $0.01 - 0.009$

38. $0.035 - 0.008$

39. $7 - 3.8$

40. $29 - 13.4$

41. $7 - 0.472$

42. $10 - 0.59$

43. $4.97 - 3$

44. $45.904 - 20$

45. $380.6 - 45.91$

46. $2,600 - 56.89$

47. $100.00 - 39.95$

48. $10{,}000 - 139.60$ **49.** $6 - 0.6$ **50.** $1 - 0.004$

51. $10 - 0.1$ **52.** $3 - 0.9$ **53.** Subtract 0.3 from 5.   **54.** Subtract 0.06 from 1.

Estimate each subtraction.

55. $45.7 - 28.09$ **56.** $6.409 - 3.18$ **57.** $\$560.38 - \$256.90$ **58.** $\$0.80 - \$0.39$

3 *Solve the equation.*

59. $x + 2.5 = 4$ **60.** $n - 7.5 = 2$ **61.** $p - 3.6 = 1.2$

62. $x + 6.3 = 8.07$ **63.** $k + 0.064 = 0.3$ **64.** $h - 0.054 = 0.08$

SKILLSFOCUS (Section 3.4) *Multiply.*

65. $\dfrac{4}{5} \times \dfrac{15}{16}$ **66.** $\dfrac{2}{7} \times \dfrac{3}{5}$ **67.** $\dfrac{9}{10} \times \dfrac{23}{100}$ **68.** $\dfrac{51}{100} \times \dfrac{7}{100}$

EXTEND YOUR THINKING ▶▶▶▶

▶ TROUBLESHOOT IT

Find and correct the error.

69. $5.5 + 5.5 = 10.10$ **70.** $5.1 - 2.6 = 3.5$

▶ CONNECTIONS

Add or subtract as indicated.

71.
$$\begin{array}{r} 4x + 2y \\ + 3x + 9y \\ \hline \end{array}$$

72.
$$\begin{array}{r} 5a + 7b + 4c \\ + 9a + 2b + 1c \\ \hline \end{array}$$

73.
$$\begin{array}{r} 7p + 5m \\ - 2p - 3m \\ \hline \end{array}$$

74.
$$\begin{array}{r} 6x + 8y + 7z \\ + 5x - 1y - 6z \\ \hline \end{array}$$

WRITING TO LEARN ▶▶▶▶

75. Write and solve one example demonstrating each of the following.

    a. Addition of decimals is commutative.

    b. Subtraction of decimals is not commutative.

    c. Addition of decimals is associative.

    d. Subtraction of decimals is not associative.

76. Explain why you line the decimal points up in the same vertical column before adding or subtracting decimal numbers.

5.4 APPLICATIONS

OBJECTIVE

Solve applications using addition and subtraction of decimals (Section 1.6).

Applications

Review the key words and phrases for addition and subtraction, and the procedure for solving word problems in Section 1.6, beginning on page 43.

EXAMPLE 1

While Christmas shopping, Sondra purchased four gifts for $25.00, $19.95, $34.50, and $7.95. What total did she spend?

To find the total
spent, add the four
purchase prices.

$$\begin{array}{r} \$25.00 \\ 19.95 \\ 34.50 \\ + 7.95 \\ \hline \$87.40 \end{array}$$
total spent ■

EXAMPLE 2

Sal purchased a tie for $14.69. He paid for it with a $20 bill. What was his change?

To find the change,
subtract the $14.69
purchase price from $20.

$$\begin{array}{r} \$20.00 \\ -14.69 \\ \hline \$5.31 \end{array}$$
change ■

▶ **You Try It**

1. While shopping, Tom wrote checks for $38.42, $16.40, $5.89, and $26.03. What total did he spend?

2. Susan purchased a jacket for $69.95. She paid for it with a $100 bill. What was her change?

EXAMPLE 3

Four women ran the four legs of a relay race in 9.43 seconds, 9.67 seconds, 9.52 seconds, and 9.42 seconds. The world record for this race is 38.41 seconds.

a. What was their time for the race?

Add the four times
together to get
the team's time.

$$\begin{array}{r} 9.43 \\ 9.67 \\ 9.52 \\ +9.42 \\ \hline 38.04 \end{array}$$
seconds,   team time

It took the women's team 38.04 seconds to run the race. The world record is 38.41 seconds. Since 38.04 < 38.41, they broke the record.

b. By how many seconds did the women's team break the world record?

$$\begin{array}{r} 38.41 \text{ seconds,} \\ -38.04 \text{ seconds,} \\ \hline 0.37 \text{ second} \end{array}$$
existing record
team time

The women's team broke the world record by 0.37, or 37 hundredths, of a second. ■

▶ **You Try It**

3. Four swimmers in the four laps of a relay race had times of 24.17 seconds, 28.39 seconds, 25.68 seconds, and 22.90 seconds.
   a. What was the team's time for the race?
   b. The record is 97.84 seconds. By how many seconds were they short of beating the record?

EXAMPLE 4 Jon wants to check the accuracy of the arithmetic in his paycheck. His pay stub states his gross pay is $1,946.51. His net pay is $1,274.09. His deductions are $92.15 for state tax, $176.84 for federal tax, $102.19 for FICA, $126.58 for retirement, $38.60 for medical, and a $136.06 credit union deduction for a car payment. Is Jon's paycheck correct?

Step #1: Add together all the deductions made from his paycheck.

$$
\begin{array}{rl}
\$\ 92.15 & \text{state tax} \\
176.84 & \text{federal tax} \\
102.19 & \text{FICA} \\
126.58 & \text{retirement} \\
38.60 & \text{medical} \\
+\ 136.06 & \text{car payment} \\
\hline
\$672.42 & \text{total deductions}
\end{array}
$$

Step #2: Add net pay to total deductions.

$$
\begin{array}{rl}
\$1,274.09 & \text{net pay} \\
+\ \ \ 672.42 & \text{total deductions} \\
\hline
\$1,946.51 & \text{gross pay}
\end{array}
$$

Since the answer, $1,946.51, is Jon's actual gross pay, the arithmetic in his paycheck is correct. ■

▶ **You Try It**   4. Francis wants to check the accuracy of his paycheck. His gross pay is $1,294.29. His net pay is $742.63. His deductions are $64.82 state tax, $88.18 federal tax, $77.41 for FICA, $13.50 for medical, and a $307.75 deduction for a car payment. Is the arithmetic in Francis's paycheck correct?

EXAMPLE 5 Alan bought a suit for $195, a ring for $549.95, a winter coat for $139.99, and rented a tuxedo for three nights at $85.67 per night. Estimate the total spent.

Since all numbers are close to hundreds, estimate the cost by first rounding all numbers to the nearest hundred. Then add.

actual cost: $195 + $549.95 + $139.99 + $85.67 + $85.67 + $85.67

estimated cost: $200 + $500 + $100 + $100 + $100 + $100 = $1,100

Alan spent approximately $1,100. ■

▶ **You Try It**   5. Phil purchased a sofa for $968, a chair for $410, carpet for $680, bookshelves for $219, and end tables for $1,219. Estimate the total spent.

Equations can help you solve word problems. Let a letter, such as x, represent the amount to be found. Then write an equation for the problem using x.

▶ **Answers to You Try It**   1. $86.74   2. $30.05   3. a. 101.14 sec   b. 3.3 sec
4. yes   5. $3,500 (rounding to hundreds)

1. Sharon purchased a coffee table for $175.42, a sofa for $856.92, and a love seat for $587.03. What total did she spend?

2. When shopping for his son's birthday party, Bill spent $6.75 on a cake, $0.79 on balloons, $12.84 on party favors, and hired a clown for $40. How much did Bill spend?

3. A $7.36 lunch check is paid for with a $10 bill. What is the change?

4. A $29.67 sweater is paid for with a $50 bill. What is the change?

5. Phil has a temperature of 102.3°F. If a normal temperature is 98.6°F, how many degrees above normal is Phil's temperature?

6. Casey's batting average was 0.273 last year. This year it is 0.305. How much higher is his batting average this year?

7. Pieces of wire 5.3 centimeters, 3.75 centimeters, and 4.0 centimeters are cut from a wire 17.5 centimeters long. How much wire is left?

8. This morning the checking account balance was $341.73. During the day checks were written for $30, $17.54, $104.32, and $15.68. What was the balance at the end of the day?

9. Four athletes run a 1 mile relay. Their times for each leg of the relay are 50.23 seconds, 49.8 seconds, 51.68 seconds, and 47.37 seconds.
   a. What was the team's time for the relay?

   b. By how much did they beat the old record of 200.73 seconds?

10. Anna runs 140 miles per week when training for a marathon. This week she has run 18.2, 17, 17.4, 16.9, 20.2, and 19.0 miles.
    a. How many miles has Anna run this week?

    b. How many more miles must she run to make 140 miles?

11. What must be subtracted from 20.013 to get 7.92?

12. What do you subtract from $347.08 to get $280.45?

13. What do you add to 128.07 to get 157.391?

14. What do you add to $5,206.17 to get $8,000?

15. Zack has $45.16. He needs $162.50 to purchase a TV.
    a. Estimate how much more money he needs by rounding to tens.

    b. What is the actual answer?

16. Evan needs $12,451.04 to buy a car. He has $2,578.57.
    a. Estimate how much money he must borrow to purchase the car by rounding to thousands.

    b. What is the actual answer?

17. Helen's gross pay is $2,153.18. Her deductions are $88.34 for state tax, $210.13 for federal tax, $124.05 for FICA, $46.87 for medical insurance, and a $328.22 deduction for a car loan. What is Helen's net pay?

18. Alan's gross pay is $1,658.33. His deductions are $66.54 for state tax, $150.67 for federal tax, $97.04 for FICA, $23.11 for health insurance, and a $50 deduction for a savings plan. What is Alan's net pay?

19. Dean's checking account balance was $670.45 this morning. Today he wrote checks for $26.12, $248.90, $170, and $15.78. He made deposits of $35.56 and $100.
    a. Estimate his new balance.

    b. What is his actual new balance?

20. Jayne has $400.00 in her checking account. She writes checks for $62.34, $45.16, $65.40, $12, and $6.17. She makes deposits of $120.45, $14.25, and $20.

a. Estimate her new balance.

b. What is her actual new balance?

21. On vacation, Vince purchased 12.6 gallons of gas for $14.60, 13.7 gallons for $15.34, 11.9 gallons for $13.80, and 14.1 gallons for $16.

a. How many gallons of gas did Vince purchase on the trip?

b. What was the total cost of this gas?

22. Alicia bought 1.47 pounds of roast beef for $8.34, 0.78 pound of salami for $2.88, and 2.23 pounds of ham for $11.23.

a. How much lunch meat did Alicia buy?

b. What was the total cost?

23. Sal purchased a radio for $69.95, a TV for $329.95, a stereo for $479.95, and four tapes for $10.29 each.

a. Estimate the total spent by rounding to hundreds.

b. Estimate the total spent by rounding to tens.

c. Which estimate is more accurate? Why?

d. How much did Sal actually spend?

24. Erica spent $275 on skis, $89.95 on boots, $32.50 on poles, and $28 for a lift ticket.

a. Estimate the total spent by rounding to hundreds.

b. Estimate the total spent by rounding to tens.

c. Which estimate is more accurate? Why?

d. How much did Erica actually spend?

SKILLSFOCUS (Section 1.9) *Evaluate each power.*

25. 6^2 **26.** 4^3 **27.** 7^1 **28.** 2^5 **29.** 10^3 **30.** 5^0

EXTEND YOUR THINKING ▶▶▶▶

▶SOMETHING MORE

31. **Bacteria**   Bacteria are among the smallest living things. They are one-celled organisms that measure from 0.0000117 to 0.000078 inches across. What is the range in size for the bacteria?

32. Four brothers split a whole pizza. One brother takes 0.125 of the pizza. The second takes 0.5 of it. The third takes 0.375 of it. How much pizza is left for the fourth brother?

33. The large hole is drilled in the center of a board. What is the diameter of the hole? (The diameter is the distance across the center of a circle, as shown in the figure.)

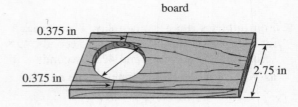

board

0.375 in

0.375 in

2.75 in

WRITING TO LEARN ▶▶▶▶

34. Write a word problem whose answer is 2.71. The word problem must be about a grocery store purchase. It must contain three numbers, one addition, and one subtraction.

35. When you estimate the difference 14.07 − 5.73, explain why you round to ones, and not to tens.

5.5 MULTIPLICATION OF DECIMALS

OBJECTIVES

1 Multiply decimal numbers and estimate a product (Sections 1.8, 3.5).

2 Multiply a decimal by a power of ten (Section 1.8).

1 Multiplying Decimals

Multiply 0.8×0.24. To discover how to multiply decimals, change each decimal to a fraction. Then multiply the fractions.

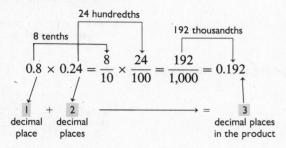

Note that you multiply 8×24 as you would multiply any two whole numbers. The number of decimal places in the product is found by adding the number of decimal places in each factor.

> To multiply two decimal numbers
>
> 1. Multiply as you multiply two whole numbers.
> 2. Count the number of decimal places in each factor being multiplied. Add these numbers together.
> 3. Place the decimal point in the product so it has the same number of decimal places as the sum in Step 2.

Multiplication is commutative and associative. Changing the order or grouping of decimal numbers does not change the product.

EXAMPLE 1 Multiply 0.7×0.3.

$$0.3 \longrightarrow \boxed{1} \text{ decimal place}$$
$$\underline{\times 0.7} \longrightarrow +\boxed{1} \text{ decimal place}$$
$$0.2\,1 \qquad \boxed{2} \text{ decimal places in the product}$$

Count $\boxed{2}$ places from right to left.
Place the decimal point. ∎

EXAMPLE 2 Multiply 4.7×5.32.

$$5.3\,2 \longrightarrow \boxed{2} \text{ decimal places}$$
$$\underline{\times 4.7} \longrightarrow +\boxed{1} \text{ decimal place}$$
$$3\,7\,2\,4 \qquad \boxed{3} \text{ decimal places in the product}$$
$$\underline{2\,1\,2\,8}$$
$$2\,5.0\,0\,4$$

Count $\boxed{3}$ places from right to left.
Place the decimal point. ∎

OBSERVE As seen in Example 2, you do not have to line up decimal points in the same vertical column to *multiply* decimal numbers.

Multiply.

1. 0.9×0.4 **2.** 3.84×6.7

EXAMPLE 3   What is 0.06×0.004?

$$0.0\ 6 \longrightarrow \boxed{2} \text{ decimal places}$$
$$\times 0.0\ 0\ 4 \longrightarrow + \boxed{3} \text{ decimal places}$$
$$\overline{0.0\ 0\ 0\ 2\ 4} \qquad \boxed{5} \text{ decimal places in the product}$$

Count $\boxed{5}$ places right to left. Attach three zeros
to create additional decimal places. ■

EXAMPLE 4   Find 83.9×0.427.

$$8\ 3.9 \longrightarrow \boxed{1} \text{ decimal place}$$
$$\times 0.4\ 2\ 7 \longrightarrow + \boxed{3} \text{ decimal places}$$
$$\overline{5\ 8\ .7\ 3} \qquad \boxed{4} \text{ decimal places in the product}$$
$$1\ 6\ 7\ 8$$
$$3\ 3\ 5\ 6$$
$$\overline{3\ 5.8\ 2\ 5\ 3}$$

Count $\boxed{4}$ places right to left. ■

EXAMPLE 5   Find the product of 0.25×480.

$$4\ 8\ 0 \longrightarrow \boxed{0} \text{ decimal places}$$
$$\times 0.2\ 5 \longrightarrow + \boxed{2} \text{ decimal places}$$
$$\overline{2\ 4\ 0\ 0} \qquad \boxed{2} \text{ decimal places in the product}$$
$$9\ 6\ 0$$
$$\overline{1\ 2\ 0.0\ 0}$$

Count $\boxed{2}$ places right to left. Drop the
two zeros to the right of the decimal
point. The answer is 120. ■

▶ **You Try It**   **3.** 0.07×0.008 **4.** 64.8×4.582 **5.** 0.625×960

Estimate multiplication by first rounding each factor to the place value farthest left. Then multiply the rounded factors.

EXAMPLE 6   Estimate 52.9×4.36.

Round each factor to the place value farthest left.

Rounding 52.9 to tens $\doteq \boxed{50}$.

Rounding 4.36 to ones $\doteq \boxed{4}$.

original problem: 52.9×4.36

estimate: $\boxed{50} \times \boxed{4} = 200$

52.9×4.36 is estimated to be 200. This estimate is low because both factors were rounded down. (The actual answer is 230.644.) ■

▶ **You Try It**   **6.** Estimate 88.57×6.904.

2 Multiplying Decimals by Powers of Ten

Multiply 8.425 by 10, 100, 1,000, and 10,000, respectively. What pattern do you see?

$$\begin{array}{cccc} 8.425 & 8.425 & 8.425 & 8.425 \\ \times\,10 & \times\,100 & \times\,1{,}000 & \times\,10{,}000 \\ \hline 84.250 & 842.500 & 8{,}425.000 & 84{,}250.000 \end{array}$$

In each case, the decimal point in 8.425 moved to the right the number of places equal to the number of zeros in the power of ten.

$$10 \quad \times\ 8.4\ 25 = 84.25$$

| 1 zero | 1 place right |

$$100 \quad \times\ 8.42\ 5 = 842.5$$

| 2 zeros | 2 places right |

$$1{,}000 \quad \times\ 8.425 = 8{,}425$$

| 3 zeros | 3 places right |

$$10{,}000 \quad \times\ 8.4250 = 84{,}250$$

| 4 zeros | 4 places right (add one place holding zero) |

> To multiply a decimal number by a power of ten, move the decimal point to the right the number of places equal to the number of zeros in the power of ten.

EXAMPLE 7 Multiply by the power of ten.

a. $12.4035 \times 100 = 12\ 40.35 = 1{,}240.35$

| 2 zeros | 2 places right |

b. $0.265 \times 10 = 0\ 2.65 = 2.65$

| 1 zero | 1 place right |

c. $0.000058 \times 1{,}000 = 0\ 000.058 = 0.058$

| 3 zeros | 3 places right |

d. $3.65 \times 10^4 = 3.65 \times 10{,}000 = 3\ 6500. = 36{,}500$

| power of 4 | 4 places right | ■

▶ **You Try It** Multiply by moving the decimal point.

7. 4.731×100 **8.** 0.094×10

9. $4.28103 \times 1{,}000$ **10.** 2.03×10^5

The population of the Earth is 5 billion people. A new building will cost $17.4 million. To express these amounts in standard notation, multiply the decimal times the power of ten represented by the decimal's name.

EXAMPLE 8   Write each amount in standard notation.

a. The population of the Earth is 5 billion people.

$$5 \text{ billion} = 5 \times \boxed{1 \text{ billion}}$$
$$= 5 \times 1{,}000{,}000{,}000$$
$$= 5{,}000{,}000{,}000 \text{ people}$$

b. A new building will cost $17.4 million.

$$\$17.4 \text{ million} = \$17.4 \times \boxed{1 \text{ million}}$$
$$= \$17.4 \times 1{,}000{,}000$$
$$= \$17{,}400{,}000$$

c. Light travels 5.88 trillion miles in one year.

$$5.88 \text{ trillion} = 5.88 \times \boxed{1 \text{ trillion}}$$
$$= 5.88 \times 1{,}000{,}000{,}000{,}000$$
$$= 5{,}880{,}000{,}000{,}000 \text{ miles} \quad \blacksquare$$

▶ **You Try It**   Write each amount in standard notation.

11. The population of the United States is 255.6 million people.
12. The country is $3.24 trillion in debt.
13. The average human brain has 60.5 billion cells.

SOMETHING MORE

Are there exactly 52 weeks in one year?
1 week = 7 days. 52 weeks = 52 × 7 days = 364 days. Since 1 year is 365 days, there are 52 weeks + 1 day in a year.

You be the Judge   Suppose you make $26,000 per year. This is $26,000 ÷ 52 = $500 per week. If you make $500 per week for 52 weeks, what are you paid for the extra 1 day? Do you work for free on that day? Explain.

▶ **Answers to You Try It**   **1.** 0.36   **2.** 25.728   **3.** 0.00056   **4.** 296.9136
5. 600   **6.** 630   **7.** 473.1   **8.** 0.94   **9.** 4,281.03   **10.** 203,000
11. 255,600,000   **12.** 3,240,000,000,000   **13.** 60,500,000,000

SECTION 5.5 EXERCISES

1

1. 0.4×0.7 **2.** 0.9×0.2 **3.** 4.5×0.5 **4.** 2.7×0.6

5. 2.83×6.5 **6.** 7.9×1.42 **7.** 8.05×180.3 **8.** 45.82×36.14

9. 0.04×0.6 **10.** 0.8×0.009 **11.** 0.007×0.0002 **12.** 0.01×0.001

13. 58×0.5 **14.** 31×0.62 **15.** 0.056×210 **16.** 0.0054×720

17. $2,000 \times 0.009$ **18.** 800×0.0125 **19.** 312.5×0.01 **20.** 39.3×0.0001

21. 0.08×125 **22.** 0.004×250 🖩 **23.** 129.95×1.06 🖩 **24.** $5,000 \times 0.0002$

25. Estimate 46.2×531.67 **26.** Estimate $5,821.3 \times 805.98$.

27. Estimate 6.83×8.762. Is the estimate high or low?

28. Estimate 432.84×640.1. Is the estimate high or low?

29. What is the cost of 2,400 photocopies at $0.025 per copy?

30. If 22,500 tickets were sold at $11.50 each, how much was made?

2

31. 10×2.34 **32.** 4.8×10 **33.** 100×4.036

34. 55.0786×100 **35.** $10,000 \times 5.23$ **36.** $0.00375 \times 10,000$

37. $1,000 \times 7.2$ **38.** $0.005 \times 1,000$ **39.** $10^2 \times 4.603$

40. 25.9×10^4 **41.** $1,000 \times 40.602$ **42.** 0.06×10

43. 0.0385×10^3 **44.** 0.00054×10^6

45. A store purchased 100 TV sets at $349.95 each. What did they pay?

46. What is the cost of 10,000 photocopies at $0.018 per copy?

47. What is the cost of 1,000 combs at $1.29 each?

48. What will be paid for 10,000,000 tax forms at $0.069 per form?

Write each amount in standard notation.

49. The population surpassed 253.61 million people.

50. The cost of the executive mansion is $630 thousand.

51. The human adult brain contains about 60 billion cells.

52. The national debt exceeds $3.14 trillion.

SKILLSFOCUS (Section 2.2) *Find the quotient and remainder.*

53. $561 \div 7$

54. $98 \div 3$

55. $802 \div 13$

56. $1,259 \div 47$

EXTEND YOUR THINKING ▶▶▶▶

▶SOMETHING MORE

57. To change cents to dollars, multiply cents by 0.01. For example, $69¢ = 69 \times 0.01 = \$0.69$. Use your calculator to change the following to dollars.
   a. 429¢
   b. 89¢
   c. 2,162¢
   d. 50,725¢
   e. 5¢
   f. 200¢

▶TROUBLESHOOT IT

Find and correct the error.

58.

$$
\begin{array}{r}
0.056 \longrightarrow 3 \\
\times\ 0.08 \longrightarrow \times 2 \\
\hline
0.000448 \qquad 6 \\
\end{array}
$$

⌐‾‾‾‾‾⌐
6 places

59.

$$
\begin{array}{r}
0.02 \longrightarrow 3 \\
\times\ 40 \longrightarrow 0 \\
\hline
0.080 \qquad 3 \\
\end{array}
$$

⌐‾‾‾⌐
3 places

WRITING TO LEARN ▶▶▶▶

60. Explain how to multiply two decimal numbers. Illustrate using your own example.

61. Explain why the estimate for 43.6×21.493 is low.

62. Explain why multiplying 0.1×0.1 gives a smaller number, 0.01, for an answer. You may not use multiplication of decimals in your explanation.

OBJECTIVES

1 Divide a decimal by a whole number divisor (Section 3.5).

2 Divide a decimal by a decimal divisor, and estimate a quotient (Sections 2.2, 4.2).

3 Divide by a power of ten (Section 2.2).

4 Solve equations (Section 2.2).

1 Dividing a Decimal by a Whole Number Divisor

What is $1.8 \div 6$? Use division of fractions to get the answer.

$$1.8 \div 6 = 1\frac{8}{10} \div 6 = \frac{18}{10} \div \frac{6}{1} = \frac{\overset{3}{\cancel{18}}}{10} \cdot \frac{1}{\cancel{6}} = \frac{3}{10} = 0.3$$

The answer to $1.8 \div 6$ is 0.3. You can solve this problem using long division, as shown.

$$6)\overline{1.8} \longrightarrow 6)\overline{\begin{array}{l} 0.3 \\ 1.8 \\ -1\,8 \\ \hline 0 \end{array}}$$

Write the decimal point in the quotient directly above the one in the dividend. Then divide as you divide whole numbers. You get the same answer.

OBSERVE The 3 tenths is directly above 8 tenths. Lining up the decimal points keeps digits with like place values in the same column. This is important because you subtract in division problems. And to subtract, like named digits must be in the same column.

To divide a decimal by a whole number divisor

1. Write the decimal point in the quotient directly above the decimal point in the dividend.
2. Divide as you divide whole numbers.

EXAMPLE 1 Divide $2.45 \div 7$.

$$7)\overline{\begin{array}{l} .3\,5 \\ 2.4\,5 \\ -2\,1 \\ \hline 3\,5 \\ -3\,5 \\ \hline 0 \end{array}}$$

Therefore, $2.45 \div 7 = 0.35$. ∎

When dividing decimals, each time you bring down a digit from the dividend, you must write a digit above it in the quotient.

EXAMPLE 2   Divide 26.032 ÷ 4.

```
        6.5 0 8
   4 ) 2 6.0 3 2
      −2 4
        2 0
       −2 0
         0 3 2
          −3 2
             0
```

Bring down the 3. 4 divides into 3, 0 times.
Write 0 above 3 in the quotient.
Bring down the 2.

Therefore, 26.032 ÷ 4 = 6.508.   ■

▶ **You Try It**   Divide.

1. 41.5 ÷ 5   **2.** 42.318 ÷ 9

EXAMPLE 3   Divide 0.1968 ÷ 24.

24 divides into 1, 0 times. Write 0
above 1 in the quotient.

24 divides into 19, 0 times. Write 0
above 9 in the quotient.

```
        .0 0 8 2
   24 ) 0.1 9 6 8
       −1 9 2
           4 8
          −4 8
             0
```

Therefore, 0.1968 ÷ 24 = 0.0082.   ■

▶ **You Try It**   **3.** 1.692 ÷ 36

When dividing decimals, you may add one or more zeros to the right side of the dividend as long as you are also to the right of the decimal point.

EXAMPLE 4   Divide 3 ÷ 8.

```
   8 ) 3  ⟶  8 ) 3.0 0 0
```
Write the decimal point to the right of 3
in the dividend. Attach three zeros.

```
        .3 7 5
    8 ) 3.0 0 0
       −2 4
         6 0
        −5 6
           4 0
          −4 0
             0
```

Therefore, 3 ÷ 8 = 0.375.   ■

▶ **You Try It**   **4.** 9 ÷ 20

How many zeros do you add to the right side of the dividend when dividing? You can clearly answer this question only when the problem tells you what place value to round the answer to.

> Add zeros to the dividend to the place value you are rounding to, *plus* one more zero. This last 0 will give you the digit in the quotient used to do the rounding.

EXAMPLE 5   Divide $13.9 \div 27$. Round to the nearest hundredth.

To round to hundredths, divide until you get a digit in the thousandths place.

$$
27 \overline{)13.9} \rightarrow 27 \overline{)13.900}
$$

```
        .514
27 )1 3.9 0 0
   -1 3 5
       4 0
      -2 7
       1 3 0
      -1 0 8
          2 2
```

Attach zeros to the dividend out to the thousandths place.

← Stop. No more dividing is needed. You already have a digit in the quotient in the thousandths place.

The answer is $.514 \doteq 0.51$ rounded to the nearest hundredth. ∎

▶ **You Try It**   **5.** $4.7 \div 13$ (round to hundredths)   **6.** $56.2 \div 71$ (round to thousandths)

2 **Dividing a Decimal by a Decimal Divisor**

What is $0.6 \overline{)0.24}$? You just learned how to divide a decimal by a whole number divisor. To divide by a decimal divisor, first change the divisor 0.6 into a whole number. Then divide as before. To see how this is done, first write $0.6 \overline{)0.24}$ as a fraction. Then multiply both terms of the fraction by the power of ten making the divisor 0.6 into a whole number.

$$
0.6 \overline{)0.24} = \frac{0.24}{0.6} = \frac{0.24}{0.6} \times \frac{10}{10} = \frac{2.4}{6.} = 6. \overline{)2.4}
$$

In short, move the decimal point in both dividend and divisor one place to the right.

$$
0.6 \overline{)0.2\,4} = 6 \overline{)2.4}
$$

The answer to a division is not changed when the decimal points in both divisor and dividend are moved the same number of places to the right.

> To divide a number by a decimal divisor
>
> 1. Count the number of decimal places to the right of the decimal point in the divisor.
> 2. Move the decimal point in the divisor this number of places to the right. This makes the divisor into a whole number.
> 3. Move the decimal point in the dividend the same number of places to the right. Attach zeros if needed.
> 4. Place the decimal point in the quotient directly above the one in the dividend. Divide by the whole number divisor.

EXAMPLE 6   Divide 0.24 ÷ 0.6.

$$0.6 \overline{)0.2\,4} \longrightarrow 6 \overline{)\begin{array}{r} .4 \\ 2.4 \\ -2\,4 \\ \hline 0 \end{array}}$$

Therefore, 0.24 ÷ 0.6 = 0.4.   ■

EXAMPLE 7   Divide 1.504 ÷ 0.47.

$$0.4\,7 \overline{)1.5\,0\,4} \longrightarrow 47 \overline{)\begin{array}{r} 3.2 \\ 1\,5\,0.4 \\ -1\,4\,1 \\ \hline 9\,4 \\ -9\,4 \\ \hline 0 \end{array}}$$

Therefore, 1.504 ÷ 0.47 = 3.2.   ■

▶ **You Try It**   **7.** 0.56 ÷ 0.8   **8.** 3.654 ÷ 0.63

EXAMPLE 8   Divide 4.2 by 0.015.

$$0.0\,1\,5 \overline{)4.2\,0\,0} \longrightarrow 15 \overline{)\begin{array}{r} 2\,8\,0. \\ 4\,2\,0\,0. \\ -3\,0 \\ \hline 1\,2\,0 \\ -1\,2\,0 \\ \hline 0\,0 \end{array}}$$

Attach two zeros.

Bring down this 0. Each time you bring a digit down, write one above it in the quotient. Since 0 ÷ 15 = 0 write 0 in the quotient.

Therefore, 4.2 ÷ 0.015 = 280.   ■

EXAMPLE 9   What is 0.9254 ÷ 2.63? Round to thousandths.

$$2.6\,3 \overline{)0.9\,2\,5\,4} \longrightarrow 263 \overline{)\begin{array}{r} .3\,5\,1\,8 \\ 9\,2.5\,4\,0\,0 \\ -7\,8\,9 \\ \hline 1\,3\,6\,4 \\ -1\,3\,1\,5 \\ \hline 4\,9\,0 \\ -2\,6\,3 \\ \hline 2\,2\,7\,0 \\ -2\,1\,0\,4 \\ \hline 1\,6\,6 \end{array}}$$

To round to thousandths, attach zeros out to the ten-thousandths place.

←Stop.

The answer is .3518 ≐ 0.352 rounded to the nearest thousandth.   ■

▶ **You Try It**   **9.** 31.39 ÷ 0.073
10. 28 ÷ 8.75
11. 0.704 ÷ 1.86 (round to hundredths)

EXAMPLE 10   Estimate 578.634 ÷ 11.4.

Estimate the answer to a division by first rounding the divisor and dividend to the digit farthest left.

original problem: $578.634 \div 11.4$

round to hundreds ↓   ↓ round to tens

estimate: $600 \div 10 = 60$

$578.634 \div 11.4$ is estimated to be 60. ∎

▶ **You Try It**

12. Estimate $419.375 \div 38.70$.

3 Dividing Decimals by Powers of Ten

Divide 623.8 by 10, 100, 1,000, and 10,000, respectively. What pattern do you see?

$$10\overline{)623.8}^{\,62.38} \qquad 100\overline{)623.8}^{\,6.238} \qquad 1{,}000\overline{)623.8}^{\,0.6238} \qquad 10{,}000\overline{)623.8}^{\,0.06238}$$

In each answer the decimal point in 623.8 moves to the left the number of places equal to the number of zeros in the power of ten.

$$623.8 \div 10 = 62.3\ 8 = 62.38$$

I zero   I place left

$$623.8 \div 100 = 6.23\ 8 = 6.238$$

2 zeros   2 places left

$$623.8 \div 1{,}000 = .623\ 8 = 0.6238$$

3 zeros   3 places left

$$623.8 \div 10{,}000 = .0623\ 8 = 0.06238$$

4 zeros   4 places left

> To divide a decimal by a power of ten, move the decimal point to the left the number of places equal to the number of zeros in the power of ten.

EXAMPLE 11

Divide by the power of ten.

a. $1{,}643.59 \div 100 = 16.43\ 59 = 16.4359$

2 zeros   2 places left

b. $6.02 \div 10 = 0.6\ 02 = 0.602$

I zero   I place left

c. $34{,}709.5 \div 10^3 = 34{,}709.5 \div 1{,}000 = 34.709\ 5 = 34.7095$

power of 3   3 places left ∎

▶ **You Try It**

Divide by moving the decimal point.

13. $58.2 \div 10$ **14.** $6.04 \div 1{,}000$

15. $382.17 \div 100$ **16.** $421{,}900 \div 10^4$

EXAMPLE 12 A $56,492 lottery prize was split equally among 100 players. How much will each player get?

Since $56,492 = $56,492., each player will receive

$$\$56,492. \div 100 = \$564.92 = \$564.92$$

2 zeros   2 places left ■

▶ You Try It **17.** A lottery jackpot of $6,287,490 was split equally among 1,000 ticket holders. How much does each ticket holder get?

---

4 Solving Equations

EXAMPLE 13 Solve the equation $1.2k = 7.2$.

$$1.2 \cdot k = 7.2$$

Solve this multiplication equation by dividing both sides by 1.2, the number times k.

$$\frac{\overset{1}{\cancel{1.2}} \cdot k}{\underset{1}{\cancel{1.2}}} = \frac{7.2}{1.2}$$

$$\frac{1 \cdot k}{1} = 6$$

Note: $1.2\overline{)7.2} \longrightarrow 12\overline{)72.}$ $\begin{array}{r} 6. \\ \underline{-72} \\ 0 \end{array}$

$$k = 6$$

Check: $1.2 \cdot k = 1.2 \cdot 6$
$= 7.2$ ■

EXAMPLE 14 Solve the equation $\dfrac{n}{1.8} = 0.25$.

$$\frac{n}{1.8} = 0.25$$

Solve this division equation by multiplying both sides by $1.8 = \dfrac{1.8}{1}$, the number divided into n.

$$\frac{\overset{1}{\cancel{1.8}}}{1} \cdot \frac{n}{\underset{1}{\cancel{1.8}}} = 1.8 \cdot 0.25$$

$$\frac{1 \cdot n}{1} = 0.45$$

$$n = 0.45$$

Check: $\dfrac{n}{1.8} = \dfrac{0.45}{1.8} = 1.8\overline{)0.4\ 5} = 18\overline{\smash{)}4.50}^{\,0.25}$ ■

▶ You Try It **18.** Solve $2.6y = 37.7$.   **19.** Solve $\dfrac{s}{7.06} = 8.5$.

▶ **Answers to You Try It** 1. 8.3 2. 4.702 3. 0.047 4. 0.45 5. 0.36
6. 0.792 7. 0.7 8. 5.8 9. 430 10. 3.2 11. 0.38 12. 10 13. 5.82
14. 0.00604 15. 3.8217 16. 42.19 17. $6,287.49 18. $y = 14.5$
19. $s = 60.01$

SECTION 5.6 EXERCISES

1 *Divide.*

1. 4.8 ÷ 6   **2.** 6.4 ÷ 16   **3.** 12.6 ÷ 9   **4.** 41.3 ÷ 7

5. 32.4 ÷ 15   **6.** 214.24 ÷ 26   **7.** 1020.5 ÷ 13   **8.** 45.144 ÷ 18

9. 0.112 ÷ 8   **10.** 0.4347 ÷ 9   **11.** 7 ÷ 8   **12.** 3 ÷ 5

13. 36 ÷ 50   **14.** 12 ÷ 75   **15.** 1.3 ÷ 13   **16.** 0.13 ÷ 13

Divide. Round the quotient to the indicated place.

17. 1.3 ÷ 7   thousandths   **18.** 4.7 ÷ 6   hundredths

19. 24.6 ÷ 5   tenths   **20.** 45.08 ÷ 12   ten-thousandths

21. 128.56 ÷ 32   hundredths   **22.** 1,835.6 ÷ 120   thousandths

23. 0.08 ÷ 31   ten-thousandths   **24.** 0.64 ÷ 24   tenths

25. 2 ÷ 9   hundredths   **26.** 4 ÷ 7   thousandths

27. 56.4 ÷ 42   two decimal places   **28.** 0.08 ÷ 24   four decimal places

2 *Divide.*

29. 3.6 ÷ 0.9   **30.** 2.8 ÷ 0.4   **31.** 7.2 ÷ 0.6   **32.** 8.5 ÷ 0.5

33. 0.56 ÷ 0.8   **34.** 0.35 ÷ 0.7   **35.** 44 ÷ 1.1   **36.** 2.25 ÷ 1.5

37. 2.56 ÷ 0.4   **38.** 72.4 ÷ 0.2   **39.** 0.6 ÷ 12.5   **40.** 33.659 ÷ 6.94

41. 0.711 ÷ 0.45   **42.** 0.04293 ÷ 0.81   **43.** 24.288 ÷ 4.8   **44.** 395.415 ÷ 5.05

45. 39 ÷ 0.13   **46.** 400 ÷ 0.016   **47.** 17 ÷ 0.34   **48.** 29 ÷ 1.16

Divide. Round to the indicated place.

49. 4.8 ÷ 0.7   hundredths   **50.** 8.3 ÷ 0.6   thousandths

51. 12.7 ÷ 0.41   tenths   **52.** 16.4 ÷ 1.18   hundredths

53. 78.3 ÷ 0.14   tens   **54.** 1,250 ÷ 44.6   ones

55. 3.094 ÷ 10.75   three decimal places   **56.** 0.95 ÷ 12.7   four decimal places

3 *Divide by the power of ten.*

57. $12.6 \div 10$ **58.** $351.8 \div 10$ **59.** $417.5 \div 1,000$

60. $4,503.34 \div 1,000$ **61.** $69.301 \div 100$ **62.** $318.7 \div 100$

63. $5,930.16 \div 10,000$ **64.** $2.7 \div 100$ **65.** $6.08 \div 1,000$

66. $6 \div 10$ **67.** $0.0028 \div 1,000$ **68.** $0.039 \div 10$

69. Divide \$256 by 100.   **70.** Divide 2,480 pounds by 1,000.

71. Ten identical toys cost a total of \$176. How much does one toy cost?

72. Suppose 12,560 acres of land is evenly divided among 100 farmers. How much land does each farmer get?

73. A lottery jackpot of \$12,503,410 is evenly divided among 1,000 ticket holders. How much does each get?

74. Suppose 10,000 pencils cost \$900. How much does one pencil cost?

4 *Solve each equation.*

75. $3.2x = 8$ **76.** $7x = 5.6$ **77.** $\dfrac{h}{0.6} = 5.08$ **78.** $\dfrac{s}{0.63} = 12.1$

79. $0.05y = 0.043$ **80.** $40t = 321.2$ **81.** $\dfrac{m}{57.2} = 35$ **82.** $\dfrac{z}{100} = 1.2$

83. $0.68v = 4$ (round to tenths)   **84.** $\dfrac{k}{8.7} = 5.906$ (round to hundredths)

SKILLSFOCUS (Section 2.4) *Simplify.*

85. $4 + 3 \cdot 8$ **86.** $50 \div 10 \cdot 5$ **87.** $6(9 - 7)$ **88.** $4 \cdot 3^2$

EXTEND YOUR THINKING ▶▶▶▶

▶ TROUBLESHOOT IT

Find and correct the error.

89.
$$
\begin{array}{r}
.24 \\
4\,\overline{)\,0.816} \\
-8 \\
\hline
16 \\
-16 \\
\hline
0
\end{array}
$$

90.
$$
\begin{array}{r}
40. \\
0.008\,\overline{)\,3.200} \\
-3\ 2 \\
\hline
00
\end{array}
$$

WRITING TO LEARN ▶▶▶▶

91. Explain why rounding the quotient of $5.6 \div 13$ to the hundredths place requires adding two zeros to the dividend.

▶ YOU BE THE JUDGE

92. Theresa says the following divisions are related. Is she correct? You may use a calculator to prove your point. Explain why you get the answers you get.

$$0.075\,\overline{)\,0.4086} \qquad 75\,\overline{)\,408.6} \qquad 0.75\,\overline{)\,4.086} \qquad 7.5\,\overline{)\,40.86}$$

5.7 APPLICATIONS

OBJECTIVES

1. Solve applications problems using multiplication and division of decimals (Sections 1.6, 2.3).

2. Solve sales tax problems.

NEW VOCABULARY

a per *b*

sales tax

sales tax factor

---

1 Solving Application Problems

Review the key words for multiplication and division (Section 2.3, on page 109), and the procedure for solving word problems (Section 1.6, on page 43).

EXAMPLE 1   Erica purchased 1.53 pounds of lunch meat at $4.69 per pound. What did she pay rounded to the nearest cent?

To find the cost of 1.53 pounds *at* $4.69 per pound, multiply. The word *at* indicates multiplication.

$$
\begin{array}{r}
\$4.6\ 9 \quad \text{per lb} \\
\times 1.5\ 3 \quad \text{lb} \\
\hline
1\ 4\ 0\ 7 \\
2\ 3\ 4\ 5 \\
4\ 6\ 9 \\
\hline
7.1\ 7\ 5\ 7 \doteq \$7.18
\end{array}
$$

Erica paid $7.18 rounded to the nearest cent. ■

▶ **You Try It**   1. Jim purchased 14.827 gallons of gas at $1.239 per gallon. What did he pay, to the nearest cent?

If you paid $100 for 4 tickets, you paid $100 ÷ 4 = $25 per ticket. The word *per* indicates division. Dollars per ticket means divide dollars paid by number of tickets purchased. Miles per gallon means divide total miles by number of gallons, and so on.

> The phrase *a* **per** *b* means $a \div b$.
>
> Examples: miles per gallon = miles ÷ gallons
> dollars per hour = dollars ÷ hours
> price per pound = price ÷ pounds

EXAMPLE 2   Heidi drove 349 miles on 16.3 gallons of gas. How many miles per gallon did Heidi's car get? Round to tenths.

The word *per* indicates division. To find miles per gallon, divide 349 miles ÷ 16.3 gallons.

$$
16.3\,\overline{)\,349.0\,} \longrightarrow 163\,\overline{)\,3490.00\,}
$$

$$
\begin{array}{r}
21.41 \doteq 21.4 \quad \text{miles per gallon} \\
\text{rounded to the} \\
\text{nearest tenth}
\end{array}
$$

$$
\begin{array}{r}
3490.00 \\
-326 \\
\hline
230 \\
-163 \\
\hline
67\,0 \\
-65\,2 \\
\hline
1\,80 \\
-1\,63 \\
\hline
17
\end{array}
$$

Stop. There is a digit in the hundredths place. You can now round to tenths. ■

EXAMPLE 3   Recently, the U.S. national debt was $3.16 trillion. If there are 255 million people in the United States, how much money would each person have to pay to pay off the debt? Round to the nearest hundred dollars.

Find dollars per person. Divide $3.16 trillion by 255 million people.

$$\frac{\$3.16 \text{ trillion}}{255 \text{ million people}} = \frac{\$3.16 \times 1 \text{ trillion}}{255 \times 1 \text{ million}} = \frac{\$3.16 \times 1,000,000,000,000}{255 \times 1,000,000}$$

$$= \frac{\$3,160,000}{255}$$

$$\doteq \$12,400 \text{ per person rounded to hundreds} \quad \blacksquare$$

▶ **You Try It**   **2.** Mark drove his mobile home 234.8 miles on 42.7 gallons of gas. How many miles per gallon did he get? Round to tenths.

3. New York City has a population of 7,071,639 people and an area of 369 square miles. How many people per square mile is this? Round to units.

EXAMPLE 4   A long-distance phone call costs $0.58 for the first minute and $0.41 for each additional minute. What is the cost for an 18-minute call?

The cost for an 18-minute call is the cost for the first minute plus the cost for the next 17 minutes.

Cost for the first minute is $0.58
Cost for the next 17 minutes at $0.41 per minute is
$17 \times \$0.41 = \$6.97$ . . . .   + 6.97
Total cost . . . . $7.55   ■

▶ **You Try It**   **4.** A phone call costs $0.24 for the first minute, and $0.19 for each additional minute. What is the cost of a 36-minute call?

EXAMPLE 5   A new job pays $7.42 an hour plus time and a half for every hour worked over 40 in one week. How much is made for working a 47.5-hour week?

The job pays $7.42 an hour. Time and a half for overtime means you earn one and a half, or 1.5, times the regular hourly pay for each hour worked over 40 hours.

Overtime pay $= 1.5 \cdot \$7.42 = \$11.13$ per overtime hour.

You work 47.5 hours in one week. You are paid $7.42 per hour for the first 40 hours. You are paid $11.13 per hour for the remaining $47.5 - 40 = 7.5$ hours of overtime.

40 hours at $7.42 per hour is
$40 \cdot \$7.42 = \$296.80$ . . . . $296.80 regular pay
7.5 hours at $11.13 per hour
is $7.5 \cdot \$11.13 = \$83.475$ or
$83.48 to the nearest cent . . . . . +$ 83.48 overtime pay
$380.28 total pay   ■

▶ **You Try It**   **5.** You earn $12.28 per hour and make time and a half for each hour worked over 40 in one week. How much do you make for a 58.5-hour week?

EXAMPLE 6 **Chapter Problem** Read the problem at the beginning of this chapter.

30-year loan:

$$\$942.89 \times 12 \times 30 = \begin{array}{r} \$339,440.40 \\ -\,100,000.00 \\ \hline 239,440.40 \end{array} \begin{array}{l} \text{total paid over 30 years} \\ \text{borrowed} \\ \text{interest paid to borrow} \\ \$100,000 \text{ for 30 years} \end{array}$$

20-year loan:

$$\$1,023.69 \times 12 \times 20 = \begin{array}{r} \$245,685.60 \\ -\,100,000.00 \\ \hline 145,685.60 \end{array} \begin{array}{l} \text{total paid over 20 years} \\ \text{borrowed} \\ \text{interest paid to borrow} \\ \$100,000 \text{ for 20 years} \end{array}$$

Savings with a 20-year loan:

You pay:		You make:	You save:	
$1,023.69	(20-yr)	10 years	$239,440.40	(30-yr)
−942.89	(30-yr)	fewer	−145,685.60	(20-yr)
$80.80	more per month	payments	$93,754.80	in interest ■

> **OBSERVE** The $80.80 more per month goes directly towards reducing your principal balance. None of it is used to pay interest on your mortgage loan.

▶ **You Try It** **6.** A couple plan to borrow $60,000 at 12.5% interest. Their monthly payments for a 20-year and a 30-year mortgage are $681.68 and $640.35, respectively. How much interest is saved with the 20-year mortgage when compared to the 30-year mortgage?

EXAMPLE 7 **Running and Miles per Minute** A jogger ran 6.2 miles in 42 minutes and 9 seconds. What was the average time per mile?

Step #1: Change 42 minutes 9 seconds into seconds. Since 1 minute = 60 seconds,

$$\begin{aligned} 42 \text{ minutes } 9 \text{ seconds} &= 42 \cdot 60 \text{ seconds} + 9 \text{ seconds} \\ &= 2,520 \text{ seconds} + 9 \text{ seconds} \\ &= 2,529 \text{ seconds} \end{aligned}$$

Step #2: You want time per mile. Divide 2,529 seconds by 6.2 miles.

$$6.2 \overline{)2,529.0} \longrightarrow 62 \overline{)25,290.0} \qquad \begin{array}{c} 407.9 \doteq 408 \text{ seconds per mile rounded} \\ \text{to the nearest second} \end{array}$$

Step #3: Change 408 seconds back to minutes and seconds. Since there are 60 seconds in one minute, divide 408 by 60.

$$\begin{array}{r} 6 \\ 60 \overline{)408} \\ -360 \\ \hline 48 \end{array}$$

6 ⟵ the quotient is the number of minutes per mile

48 ⟵ the remainder is the number of seconds per mile

The jogger averaged 6 minutes 48 seconds per mile. ■

▶ **You Try It** 7. Beth ran a 10-mile race in 64 minutes 43 seconds. What was her average time per mile?

EXAMPLE 8 **Pricing a Job** 3,000 bricks are needed to brick the front of a home. A bricklayer is paid $18 an hour, and lays 50 bricks per hour. The price of one brick (including mortar) is $0.46. What will it cost to brick the front of the home?

Figure the cost for materials and labor separately.

Cost for materials: The cost of 3,000
    bricks at $0.46 a brick is 3,000 · $0.46 . . . . = $1,380   materials

Cost for labor: To lay 3,000 bricks at 50
    bricks per hour will take 3,000 ÷ 50 =
    60 hours. The cost to hire a bricklayer
    for 60 hours at $18 per hour is 60 · $18 . . . . . . = +$1,080   labor
                                                          $2,460   total

It will cost $2,460 to brick the front of the home. ■

▶ **You Try It** 8. 48,000 bricks are needed for an office building. A bricklayer is paid $21.50 per hour, and lays 60 bricks per hour. The price per brick (including mortar) is $0.39. What will it cost to brick the building?

EXAMPLE 9 A department store chain purchased 78 TV sets for $382.95 each. Estimate the total cost.

The total cost is 78 · $382.95. Estimate this product by first rounding each factor to the digit farthest left.

total cost: 78 · $382.95
             ↓   ↓
estimate: 80 · $400 = $32,000

The total cost is estimated to be $32,000. This estimate is high because both factors were rounded up. ■

OBSERVE There are many ways to estimate answers. Suppose 19 party favors cost $4.52 each. The total cost is 19 · $4.52. A different way to estimate this product is to round the factors to numbers they are close to. By rounding $4.52 to $4.50, and 19 to 20, you get

19 · $4.52 = 20 · $4.50 = 10 · $9 = $90   (Actual = $85.88)

double-half rule   (Section 1.8, Exercise 89, page 72)

▶ **You Try It** 9. A car dealership purchased 44 compact cars at $8,749.52 each. Estimate the total cost.

2 Sales Tax   **Sales tax** is the tax you pay when you buy an item. Sales tax is paid per dollar of selling price, and varies from state to state in the United States.

    For example, the selling price of a coat is $129.95. Suppose the sales tax for this purchase is 5¢ on the dollar. Since 5¢ = $0.05, multiply the selling price by $0.05 to get the sales tax.

$$\begin{array}{rl} \$129.95 & \text{selling price} \\ \times \quad 0.05 & \text{sales tax per dollar} \\ \hline \$6.4975 & \doteq \$6.50 \text{ rounded to the nearest cent} \end{array}$$

Add selling price to sales tax to get the price including tax.

$$\begin{array}{rl} \$129.95 & \text{selling price} \\ + \quad 6.50 & \text{sales tax} \\ \hline \$136.45 & \text{price including tax} \end{array}$$

    The two steps above can be combined into one. If the tax rate is 5¢, multiply selling price by 1.05.

$$\begin{array}{rl} \$129.95 & \text{selling price} \\ \times \quad 1.05 & \text{sales tax factor} \\ \hline \$136.4475 & \doteq \$136.45 \text{ rounded to the nearest cent} \end{array}$$

This gives the same price including tax as before. The number 1.05 is called the **sales tax factor**. If the sales tax is 5¢ per dollar, then for every dollar of selling price you actually pay $1 plus 5¢, or $1.05.

> sales tax factor = $1 + sales tax per dollar
>
> selling price · sales tax factor = price including tax

EXAMPLE 10   Use your calculator to find price including tax for each item.

a. A tie sells for $16.50. The sales tax is 4¢ per dollar.

    The sales tax factor is 1.04.

$$\text{price including tax} = \$16.50 \cdot 1.04 = \$17.16$$

b. A sofa sells for $829.99. The sales tax is 2¢ per dollar.

    The sales tax factor is 1.02.

$$\text{price including tax} = \$829.99 \cdot 1.02 = \$846.5898 \doteq \$846.59$$

c. A candy bar sells for $0.79. The sales tax is 8¢ per dollar.

    The sales tax factor is 1.08.

$$\text{price including tax} = \$0.79 \cdot 1.08 = \$0.8532 \text{ or } \$0.85 \quad \blacksquare$$

▶ **You Try It**   Find price including tax for each item.

10. A jacket sells for $49.95. Sales tax is 5¢.

11. A pair of shoes sells for $79.50. Sales tax is 3¢.

12. A stereo sells for $1,875. Sales tax is 7¢.

13. A pair of socks sells for $2.79. Sales tax is 4¢.

Ernst Mach and the Mach Number   Ernst Mach was an Austrian physicist. He developed a method for measuring the speed of objects in terms of the speed of sound. This measure is called a Mach number.

Mach 1 is the speed of sound. Mach 2 is twice the speed of sound. (A speed faster than sound is called supersonic.) Mach 0.5 is half the speed of sound (called subsonic).

A Mach number is computed as follows.

$$\text{Mach number} = \frac{\text{speed of object}}{\text{speed of sound (at the same altitude)}}$$

For example, a jet is traveling at 1,320 mph. The speed of sound is 740 mph.

$$\text{Mach number} = \frac{1{,}320 \text{ mph}}{740 \text{ mph}} \doteq 1.8 \text{ (rounded to tenths)}$$

The jet is traveling at Mach 1.8, or 1.8 times the speed of sound. Find the Mach number for each speed. Round to tenths.

a. 980 mph   **b.** 550 mph   **c.** 3,200 mph

▶ **Answers to You Try It**   1. $18.37   2. 5.5 mpg   3. 19,164 people per square mile   4. $6.89   5. $831.97   6. save $66,920.80 with the 20-year mortgage   7. 6 min 28 sec per mi   8. material costs $18,720; labor costs $17,200; total cost is $35,920   9. $360,000   10. $52.45   11. $81.89   12. $2,006.25   13. $2.90

▶ **Answers to Something More**   a. 1.3   b. 0.7   c. 4.3

1 2 *Solve each word problem. Round dollar amounts to the nearest cent.*

1. Gas costs $1.179 per gallon. What is the cost to fill a 14.6-gallon tank?

2. Diesel fuel costs $1.349 per gallon. What is the cost to a trucker who purchases 120 gallons?

3. Tom makes $42.08 for working 8 hours. What does he make per hour?

4. Amy makes $16 an hour. How many hours must she work to make $500?

5. You purchase 11 dress shirts selling for $21.69 each.

$21.69

   a. Estimate the total cost.

   b. Is your estimate high or low?

   c. What is the exact cost?

6. Jayne worked 28.5 hours and made $9.48 per hour.
   a. Estimate her total wage.

   b. Is your estimate high or low?

   c. What are her exact earnings?

7. If you earn $28,462.75 per year, how much do you make per week?

8. How many gallons of gas can you buy for $8.30 if gas costs $1.269 per gallon? Round to the nearest tenth of a gallon.

9. If 450 bricks cost $180, what is the cost of one brick?

10. A case of 144 pens costs $122.40. What is the cost of one pen?

11. A doctor needs 200 injections of an antibiotic. One injection uses 3.8 milligrams of antibiotic. The cost of 1 milligram is $0.06. Find the total cost.

12. Alice makes $8.09 per hour. She works 40 hours per week. How much will she make during the 13 weeks of summer?

13. A farm worker gets $3.15 an hour plus $0.92 for every bushel of peaches he picks in a day. If he works 10 hours and picks 52 bushels, how much does he make for the day?

14. A seamstress makes $5.42 an hour plus $0.14 for each piece of clothing she stitches. How much will she make if she works 7 hours and stitches 136 pieces of clothing?

15. Jeff travels 364 miles on 15.8 gallons of gas. How many miles per gallon did he get, rounded to the nearest tenth?

16. Amelia drove from Baltimore to San Francisco, a distance of 2,875 miles. She used a total of 165.2 gallons of gas. How many miles per gallon did she get, rounded to the nearest tenth?

17. Kate purchased 0.78 pound of ham at $3.29 per pound. What did she pay?

18. Ed bought 1.26 pounds of cheese at $3.79/lb. What did he pay?

19. What price including tax do you pay for a wool sweater selling for $48, if the sales tax rate is 6¢ on the dollar?

20. What is the price including tax for a $129 leather jacket, if the sales tax is 4¢ on the dollar?

21. The charge for a long-distance phone call is $0.67 for the first minute and $0.48 for each additional minute.
    a. What is the cost of a 17-minute call?

    b. A federal tax of $0.03 on the dollar is charged for the call. What is the cost of the call including tax?

22. Al called his sister long-distance. He paid $0.45 for the first minute and $0.38 for each additional minute.
    a. What was the charge for a half hour phone call?

    b. What is the cost of the call including a federal tax of $0.07 on the dollar?

23. Arnold makes $10.68 an hour plus time and a half for each hour worked over 40 hours each week. What does he earn for working
    a. a 34.5-hour week?

    b. a 40-hour week?

    c. a 51-hour week?

24. Suzanne makes $14.72 an hour plus time and a half for each hour worked over 37.5 each week. What does she earn for working
    a. a 16-hour week?

    b. a 37.5-hour week?

    c. a 52.5-hour week?

25. Cheryl ran 6.2 miles in 38 minutes and 23 seconds. What was her average time per mile?

26. Phil ran 3.1 miles in 15 minutes and 12 seconds. What was his average time per mile?

27. You pay $3,284.16 in 24 equal monthly payments. Find the payment.

28. You owe $12,683.56, which you plan to pay in 48 equal monthly payments. How large is each payment?

29. Dale paid for a TV on the credit plan. He paid $42.16 per month for 16 months. The selling price for the TV was $499.95.
    a. What total did he pay for the TV?

    b. What was the total interest charge?

30. Lucy paid $237.45 per month for 36 months to pay off a $6,200 car loan.
    a. What total did she pay on the loan?

b. What was the total interest charge?

31. Three bars of soap sell for $0.99. A package of twelve bars costs $3.60.

 a. What is the price per bar in the three-pack?

 b. What is the price per bar in the twelve-pack?

 c. What is the savings per bar with the twelve-pack?

32. Six quarts of oil cost $8.04. A box of 24 quarts costs $30.96.

 a. What is the price per quart in the six-pack?

 b. What is the price per quart in the 24-pack?

 c. What is the savings per quart in the 24-pack?

33. Mailing a package costs $0.28 for the first ounce and $0.17 for each additional ounce.

 a. What will it cost to mail a 32-ounce package?

 b. How heavy a package can you mail for $6.91?

34. Pencils cost $0.12 each for the first 150, $0.10 each for the next 150, and $0.07 each for every pencil thereafter.

 a. What will it cost to purchase 500 pencils?

 b. How many pencils can you purchase for $52.25?

35. A 3¢ sales tax is added to a coat selling for $399.95. What is the price of the coat including tax?

36. A 9¢ sales tax is charged on a $90 jogging suit. What is the price of the suit including tax?

37. Gasoline costs $1.269 per gallon using a credit card. It costs $1.209 per gallon if you pay cash. How much money do you save if you pay cash for 16.3 gallons of gas?

38. A trucker purchases 150 gallons of gasoline. The credit card price is $1.189 per gallon. The cash price is $1.147 per gallon. How much is saved by paying cash?

39. You need 9,000 bricks to build a new home. A bricklayer lays 45 bricks per hour at a cost of $21 per hour. The price of a brick is $0.36. Mortar costs $10 per 500 bricks. What will it cost to brick the house?

40. A contractor needs 75,000 bricks to construct an apartment building. A bricklayer lays 60 bricks per hour at a cost of $16.50 per hour. The price of one brick (including mortar) is $0.33. What will it cost to brick the apartment building?

41. The U.S. national debt was $907,700,000,000 in 1980. The population of the United States was 223,000,000 in 1980. What was the debt per person in 1980? Round to tens.

42. In 1900 the U.S. national debt was $1,263,400,000. If this amounted to $17 for each citizen, what was the U.S. population in 1900? Round to the nearest million.

43. Renting a truck costs $23 a day plus $0.12 per mile. What do you owe if you use the truck for 3 days and drive 1,673 miles?

44. You rent a car for 7 days and drive 512 miles. What do you owe if you pay $16 per day, $0.15 per mile, and $1.50 per day for insurance?

45. Your club has $126.19 to purchase toys for needy youngsters.
   a. If each toy costs $8.62, how many toys can be purchased?

   b. What change is left over?

46. A double-decker hamburger costs $1.69.
   a. How many can you buy for $14?

   b. What change is left over?

47. What is the average of the 8 test scores 78, 90, 65, 80, 98, 91, 100, and 85? Round to the nearest tenth.

48. Five spiders of the same species weighed 0.020, 0.023, 0.0185, 0.019, and 0.027 ounces. Find the average weight, rounded to thousandths of an ounce.

49. Estimate the cost of 62 radios at $39.95 each.

50. Estimate the cost of 207 ounces of gold at $317.62 per ounce.

51. A 6¢ sales tax is added to a coat costing $149.95. What is the price of the coat including tax?

52. A 4¢ sales tax is added to a compact disc costing $12.50. What is the cost of the disc including tax?

53. What is the cost including tax for a pen selling for $2.89, if the sales tax is 2¢ on the dollar?

54. What is the cost including tax for a bike selling for $388, if the sales tax is 9¢ on the dollar?

SKILLSFOCUS (Section 2.5) *Evaluate each expression using $a = 3$ and $b = 5$.*

55. ab **56.** $2a + b$ **57.** $b + 4a$ **58.** $a^2 + b^2$

EXTEND YOUR THINKING ▶▶▶▶
▶SOMETHING MORE

59. Find the Mach number for a jet traveling 1,770 mph.

60. Find the Mach number for a projectile traveling 220 mph.

61. Voyager II   On August 20, 1977, the *Voyager II* spacecraft was launched to explore the solar system. On August 20, 1989, *Voyager II* had traveled 4,400,000,000 miles and had reached Neptune. What was the average speed of *Voyager II* over the 12 years, rounded to the nearest thousand miles per hour?

62. Kim paid $4,560 for a new car. She drove it 173,246 miles over a 12-year period. Over this time, she spent $5,128 for car insurance, $7,461 for gas, and $2,670 for repairs and maintenance. What did the car cost to operate per mile for the 12-year period?

63. Susan makes $14.58 per hour if she works 40 hours or less per week. She makes time and a half for each hour worked over 40, up to 20 hours. She makes double time for each hour worked over 60. If Susan works 64.75 hours in one week, what will she make?

64. On a cross-country vacation, a family traveled a total of 9,370 miles. They used a total of 485.7 gallons of gas costing $585.63.
   a. How many miles per gallon did they get on the trip? Round to tenths.

   b. What was the average cost per gallon of gas on the trip? Round to tenths of a cent.

   c. What was the gasoline cost per mile for the vacation?

65. Diane rented a car for $16 a day plus $0.13 a mile. She paid the rental agency $341.84 for the use of the car on a 1,768-mile trip. For how many days did Diane rent the car? Assume no tax.

66. Paint costs $18.95 per gallon. One gallon covers 450 square feet. You estimate you have 3,475 square feet to paint.
   a. How many gallons of paint do you need?

   b. What will it cost you to purchase the paint? Include a 6¢ sales tax.

67. The Leaning Tower of Pisa leans another 0.008 inches each year.
   a. How much more will it lean over the next 25 years?

   b. In how many years will it lean one more inch?

WRITING TO LEARN ▶▶▶▶

68. Write a word problem whose answer is $24.17 to the nearest cent. The problem must include one multiplication, and the words *tree* and *tax*. The problem may not exceed 40 words, excluding numbers.

69. A turkey is priced at $13.67 and weighs 8.92 pounds. The price per pound is listed at $1.39.

    a. Are these numbers correct? Explain your decision.

    b. If not, can you determine which number is wrong? Explain.

70. Football   Jim Brown once held the record for rushing (running) 1,836 yards in one 12-game football season. Several years later the number of games played in one season increased to 14. O. J. Simpson broke Brown's record by rushing 2,003 yards in one 14-game season.

    a. How many yards per game did each player rush in their respective seasons? Round to tenths.

    b. Who averaged more yards per game in their respective record seasons?

    c. In your opinion, who holds the record? Explain your decision.

5.8 CONVERSIONS BETWEEN FRACTIONS AND DECIMALS

OBJECTIVES

1. Change a fraction to a decimal (Section 3.1).
2. Use your calculator to change a fraction to a decimal.
3. Change a decimal to a fraction.

NEW VOCABULARY

terminating decimal

repeating decimal

---

1 Changing a Fraction to a Decimal

EXAMPLE 1 Change $\frac{4}{5}$ to a decimal.

Recall, a fraction is a division.

$$\frac{4}{5} = 4 \div 5 = 5\overline{)\begin{array}{l}.8\\ 4.0 \\ \underline{-40} \\ 0\end{array}}$$

Therefore, $\frac{4}{5} = 0.8$. ■

> To change a fraction to a decimal, divide the numerator by the denominator.
>
> Memory aid: $\frac{4}{5} = 5\overline{)4.}$

EXAMPLE 2 Change $\frac{3}{8}$ to a decimal.

$$\frac{3}{8} = 8\overline{)\begin{array}{l}.3\ 7\ 5\\ 3.0\ 0\ 0 \\ \underline{-2\ 4} \\ 6\ 0 \\ \underline{-5\ 6} \\ 4\ 0 \\ \underline{-4\ 0} \\ 0\end{array}}$$

Therefore, $\frac{3}{8} = 0.375$. ■

▶ **You Try It** Change each fraction to a decimal.

1. $\frac{3}{10}$ 2. $\frac{7}{16}$

In Examples 1 and 2, the division eventually terminates by giving a remainder of 0. The answers 0.8 and 0.375 are called **terminating decimals**.

Examples 3, 4, and 5, which follow, give **repeating decimals** for answers. This means one or more digits repeats over and over as you carry out the division. Every fraction can be changed into either a terminating or a repeating decimal.

EXAMPLE 3   Change $\frac{2}{3}$ to a decimal. Round to two decimal places.

$$\frac{2}{3} = 3 \overline{)\begin{array}{c} .6\ 6\ 6 \\ 2.0\ 0\ 0 \end{array}}$$
$$\begin{array}{r} -1\ 8 \\ \hline 2\ 0 \\ -1\ 8 \\ \hline 2\ 0 \\ -1\ 8 \\ \hline 2 \end{array}$$

Since 2 keeps repeating as a remainder, 6 will keep repeating in the quotient. You can write the answer in Example 3 as follows.

$$\frac{2}{3} = 0.666666\ldots$$

The three dots indicate that the digit 6 keeps repeating. The answer, rounded to two decimal places, is

$$\frac{2}{3} \doteq 0.67. \quad \blacksquare$$

OBSERVE   A shortcut way to write a repeating decimal as an exact answer is to write a bar over the digit that repeats.

$$\frac{2}{3} = 0.66666666\ldots = 0.\overline{6} \longleftarrow \text{the bar over 6 means the digit 6 keeps repeating}$$

EXAMPLE 4   Change $\frac{7}{11}$ to a decimal. Round to ten-thousandths.

$$\frac{7}{11} = 11 \overline{)\begin{array}{c} .6\ 3\ 6\ 3\ 6 \\ 7.0\ 0\ 0\ 0\ 0 \end{array}}$$
$$\begin{array}{r} -6\ 6 \\ \hline 4\ 0 \\ -3\ 3 \\ \hline 7\ 0 \\ -6\ 6 \\ \hline 4\ 0 \\ -3\ 3 \\ \hline 7\ 0 \\ -6\ 6 \\ \hline 4 \end{array}$$

Therefore, $\frac{7}{11} \doteq 0.6364.$ (Note: $\frac{7}{11} = 0.\overline{63}$) $\blacksquare$

► **You Try It**   Change each fraction to a decimal.

3. $\frac{5}{6}$ (round to thousandths)   4. $\frac{17}{33}$ (round to ten-thousandths)

EXAMPLE 5   Change $6\frac{3}{4}$ to a decimal.

First write the mixed number as an improper fraction.   Then change the improper fraction to a decimal.

$$6\frac{3}{4} = \frac{27}{4} = 4\overline{)27.00}$$

$$\begin{array}{r} 6.75 \\ 4\overline{)27.00} \\ -24 \\ \hline 30 \\ -28 \\ \hline 20 \\ -20 \\ \hline 0 \end{array}$$

Therefore, $6\frac{3}{4} = 6.75$. ∎

► **You Try It**   Change to a decimal.

5. $2\frac{7}{8}$

2 Calculators   The calculator can be used to change a fraction to a decimal. Read the fraction, then enter the numbers into the calculator in the order you read them.

EXAMPLE 6 $\frac{4}{5}$ is read "four fifths." Enter the numerator 4 into the calculator first because it is read first.

$$\frac{4}{5} = 4 \div 5 = 0.8 \quad ∎$$

OBSERVE   Does your calculator round off answers? Change $\frac{2}{3}$ to a decimal using your calculator. From Example 3, the answer is 0.666666666.... If the rightmost digit on your calculator display is 7, then your calculator rounds off answers to the digit farthest right. If the rightmost digit is 6, your calculator does not round.

► **You Try It**   Use your calculator to change each to a decimal.

6. $\frac{13}{20}$ 7. $\frac{8}{9}$ (round to thousandths)

3 Changing a Decimal to a Fraction

EXAMPLE 7   Change 0.6 to a fraction.

$$0.6 = \text{six tenths} = \frac{6}{10} = \frac{3}{5}$$

Read it.   Write it.   Reduce it.   ■

To change a decimal to a fraction

1. Read the decimal.
2. Write the decimal in fraction form as you read it.
3. Reduce to lowest terms.

In short, read it, write it, reduce it.

EXAMPLE 8   Change 0.75 to a fraction.

$$0.75 = \text{seventy-five hundredths} = \frac{75}{100} = \frac{3}{4}$$

Read it.   Write it.   Reduce it.   ■

EXAMPLE 9   Change 0.000219 to a fraction.

$$0.000219 = \text{two hundred nineteen millionths} = \frac{219}{1,000,000}$$ ■

> **OBSERVE**   0.000219 = 219 millionths is also read 219 parts per million, or 219 ppm. Parts per million is used in chemistry. For example, a radon gas concentration of 219 ppm in a residence is considered unhealthy. 219 ppm means 219 molecules of radon gas per 1 million molecules of air.

▶ **You Try It**   Change each decimal to a fraction.

8. 0.5   **9.** 0.64   **10.** 0.00125

EXAMPLE 10   Change 6.3 to a fraction.

$$6.3 = \underline{\text{six}} \text{ and } \underline{\text{three tenths}}$$
$$= 6 + \frac{3}{10} = 6\frac{3}{10}$$ ■

EXAMPLE 11 Change 12.625 to a fraction.

$$12.625 = \underline{\text{twelve}} \text{ and } \underline{\text{six hundred twenty-five thousandths}}$$

$$= \quad 12 \quad + \quad \frac{625}{1,000} = 12\frac{5}{8} \quad \blacksquare$$

▶ **You Try It** Change each decimal to a fraction.

11. 3.08   **12.** 16.325

EXAMPLE 12 Change $0.4\frac{1}{2}$ to a fraction.

The fraction $\frac{1}{2}$ has the same place value name as the digit to its left, 4.

Since 4 is in the tenths place, $0.4\frac{1}{2} = 4\frac{1}{2}$ tenths.

$$0.4\frac{1}{2} = 4\frac{1}{2} \text{ tenths} = \frac{4\frac{1}{2}}{10} = 4\frac{1}{2} \div \frac{10}{1} = \frac{9}{2} \times \frac{1}{10} = \frac{9}{20} \quad \blacksquare$$

▶ **You Try It** Change to a fraction.

13. $0.7\frac{4}{5}$

EXAMPLE 13 Change $0.16\frac{2}{3}$ to a fraction.

The fraction $\frac{2}{3}$ has the same place value name as the digit to its left, 6.

Since 6 is in the hundredths place, $0.16\frac{2}{3} = 16\frac{2}{3}$ hundredths.

$$0.16\frac{2}{3} = 16\frac{2}{3} \text{ hundredths} = \frac{16\frac{2}{3}}{100} = 16\frac{2}{3} \div \frac{100}{1} = \frac{\overset{1}{\cancel{50}}}{3} \times \frac{1}{\underset{2}{\cancel{100}}} = \frac{1}{6} \quad \blacksquare$$

▶ **You Try It** Change to a fraction.

14. $0.28\frac{4}{7}$

▶ **Answers to You Try It**   1. 0.3   2. 0.4375   3. 0.833   4. 0.5152   5. 2.875

6. 0.65   7. 0.889   8. $\frac{1}{2}$ 9. $\frac{16}{25}$ 10. $\frac{1}{800}$ 11. $3\frac{2}{25}$ 12. $16\frac{13}{40}$ 13. $\frac{39}{50}$

14. $\frac{2}{7}$

▶ **Answers to Something More**   a. $\frac{4}{9}$ b. $\frac{25}{99}$ c. 1   d. $\frac{4}{11}$ e. $\frac{100}{333}$ f. $\frac{1}{11}$

g. $5\frac{7}{9}$ h. $1.8\frac{1}{3}$

SECTION 5.8 EXERCISES

1 **2** *Change each fraction to an exact decimal.*

1. $\dfrac{1}{2}$ 2. $\dfrac{5}{16}$ 3. $\dfrac{7}{10}$ 4. $\dfrac{1}{8}$

5. $\dfrac{18}{25}$ 6. $\dfrac{7}{8}$ 7. $\dfrac{1}{4}$ 8. $\dfrac{13}{20}$

9. $\dfrac{7}{4}$ 10. $\dfrac{29}{8}$ 11. $4\dfrac{2}{5}$ 12. $3\dfrac{11}{16}$

13. $9\dfrac{3}{40}$ 14. $6\dfrac{3}{10}$ 15. $15\dfrac{1}{2}$ 16. $12\dfrac{3}{5}$

17. $\dfrac{5}{32}$ 18. $\dfrac{9}{100}$

Change each fraction to a decimal. Round to the indicated place.

19. $\dfrac{5}{6}$ tenths   20. $\dfrac{4}{7}$ thousandths

21. $\dfrac{7}{9}$ ten-thousandths   22. $\dfrac{5}{9}$ hundredths

23. $\dfrac{8}{11}$ thousandths   24. $\dfrac{5}{12}$ tenths

25. $\dfrac{1}{3}$ hundredths   26. $\dfrac{11}{15}$ ten-thousandths

27. $2\dfrac{6}{7}$ thousandths   28. $5\dfrac{2}{9}$ hundredths

29. $4\dfrac{9}{32}$ two decimal places   30. $6\dfrac{1}{8}$ two decimal places

31. $2\dfrac{11}{18}$ one decimal place   32. $3\dfrac{9}{16}$ three decimal places

🔲 **33.** $8\frac{4}{30}$ four decimal places   🔲 **34.** $7\frac{8}{9}$ one decimal place

🔲 **35.** $\frac{4}{75}$ three decimal places   🔲 **36.** $\frac{6}{19}$ five decimal places

3 *Change each decimal to a fraction. Reduce to lowest terms.*

37. 0.2   **38.** 0.7   **39.** 0.18   **40.** 0.65

41. 0.225   **42.** 0.864   **43.** 0.04   **44.** 0.09

45. 0.006   **46.** 0.001   **47.** 0.080   **48.** 0.055

49. 5.7   **50.** 2.8   **51.** 6.08   **52.** 9.02

53. 10.5   **54.** 12.7   **55.** 20.72   **56.** 13.30

57. 0.0004   **58.** 0.00025   **59.** 37.5   **60.** 3.75

61. 0.375   **62.** 0.0375   **63.** 0.00375   **64.** 0.000375

65. 64.17   **66.** 10.064   **67.** 36.36   **68.** 1.01

69. $0.12\frac{1}{2}$ **70.** $0.2\frac{1}{5}$ **71.** $0.3\frac{1}{3}$ **72.** $0.14\frac{2}{7}$

73. $0.9\frac{5}{6}$ **74.** $0.1\frac{2}{3}$

SKILLSFOCUS (Section 1.9) *Evaluate.*

75. $\sqrt{16}$ **76.** $\sqrt{100}$ **77.** $\sqrt{1}$ **78.** $\sqrt{49}$

EXTEND YOUR THINKING ▶▶▶▶

▶SOMETHING MORE

79. Change each repeating decimal to a fraction and reduce. Check each answer using a calculator.

  a. $0.55555\ldots$ **b.** $0.127127\ldots$ **c.** $0.969696\ldots$

80. **Baseball**   A baseball player comes to bat 600 times during a 30-week season. Last year his batting average was 0.250. This year he wants to raise his batting average to 0.300. How many more hits per week does he need to do this? (Note: batting average = number of hits ÷ number of times at bat)

81. **The Calculator and Equivalent Fractions**   To show that two fractions are equivalent, change each fraction to a decimal. The fractions are equal if the decimals are equal.
  a. Use your calculator to decide which fractions are equivalent.

$$\frac{5}{8}, \quad \frac{10}{16}, \quad \frac{15}{24}, \quad \frac{20}{32}, \quad \frac{25}{40}, \quad \frac{30}{48}$$

  b. Use your calculator to decide which fractions are equivalent.

$$\frac{24}{28}, \quad \frac{138}{161}, \quad \frac{16}{21}, \quad \frac{96}{112}, \quad \frac{144}{189}, \quad \frac{54}{63}, \quad \frac{30}{35}$$

82. A marathon is 26 miles 385 yards in length. 1 mile = 1,760 yards. How long is one marathon rounded to the nearest tenth of a mile?

▶TROUBLESHOOT IT

Find and correct the error.

83. $\dfrac{5}{8} = 5\overline{)8.0}^{\;1.6}$ So, $\dfrac{5}{8} = 1.6$.

84. $\dfrac{11}{14} = 11.14$

85. $0.28 = \dfrac{028}{1{,}000} = \dfrac{7}{250}$

86. $1.645 = 1\dfrac{645}{100} = 6\dfrac{45}{100} = 6\dfrac{9}{20}$

87. Explain each step you use to change $\dfrac{7}{25}$ to a decimal.

88. Explain each step you use to change 0.84 to a fraction.

▶ YOU BE THE JUDGE

89. Al, Bob, and Cy equally split a $2 million inheritance. Sly Cy said each man would get $\dfrac{1}{3}$ of $2 million. He said since $\dfrac{1}{3}$ equals 0.33 rounded to the nearest hundredth (see Problem 25), he would multiply $2 million by 0.33 to get each man's share. Impressed by Cy's quick calculations, the two men agreed and left with their share of the money.

a. How much did Al and Bob each get?

b. Cy got the rest. How much did he get?

c. What should have been each man's fair share?

d. How much more than his fair share did Cy get?

e. What mistake did Al and Bob make in agreeing with Cy's calculations?   Explain.

OBJECTIVES

1 Simplify expressions using the order of operations (Sections 1.9, 2.4).

2 Write an arithmetic expression for a word problem, then simplify (Section 2.5).

3 Solve problems involving fractions and decimals.

4 Order decimals.

---

1 Order of Operations

You use the same order of operations with decimals as used with whole numbers and fractions. Review these rules in Section 2.4 (see page 119).

EXAMPLE 1 Simplify $4.2 + 3.1 \cdot 0.6$.

$$4.2 + \underbrace{3.1 \cdot 0.6} \qquad \text{Multiply first.}$$
$$= 4.2 + \quad 1.86 \qquad \text{Then add.}$$
$$= 6.06 \quad \blacksquare$$

EXAMPLE 2 Simplify $0.8 \div 0.5 \cdot 4$.

$$0.8 \div 0.5 \cdot 4 \qquad \text{Divide first. (Recall, multiply and divide in order from}$$
$$= \underbrace{(0.8 \div 0.5)} \cdot 4 \qquad \text{left to right. You may write your own parentheses}$$
$$\text{around the division to indicate it is done first.)}$$
$$= \quad (1.6) \cdot 4$$
$$= 6.4 \quad \blacksquare$$

▶ **You Try It** Simplify.

1. $3.7 + 5.3 \cdot 0.3$

2. $4.8 \div 1.6 \cdot 3$

EXAMPLE 3 Simplify $5 \div 0.8 + 5.4\sqrt{25} - 7.2$.

$$5 \div 0.8 + 5.4\sqrt{25} - 7.2$$
$$= 5 \div 0.8 + 5.4 \cdot \sqrt{25} - 7.2 \qquad \text{Write the dot for the implied multiplication.}$$
$$\text{Evaluate the square root.}$$
$$= \underbrace{5 \div 0.8} + \underbrace{5.4 \cdot 5} \quad - 7.2 \qquad \text{Multiply and divide in order from left to right.}$$
$$= \underbrace{6.25 + 27} \quad - 7.2 \qquad \text{Add and subtract in order from left to right.}$$
$$= \qquad 33.25 \quad - 7.2$$
$$= 26.05 \quad \blacksquare$$

EXAMPLE 4 Simplify $450(1 + 0.04)^2$.

$$450(1 + 0.04)^2$$
$$= 450 \cdot (1 + 0.04)^2 \qquad \text{Simplify within parentheses: } 1 + 0.04 = 1.04.$$
$$= 450 \cdot (1.04)^2 \qquad \text{Evaluate the power:}$$
$$\qquad\qquad\qquad\qquad (1.04)^2 = 1.04 \cdot 1.04 = 1.0816.$$
$$= 450 \cdot (1.0816)$$
$$= 486.72 \quad \blacksquare$$

CAUTION With powers, $1.5^2 \neq 1.25$. Instead, $1.5^2 = 1.5 \times 1.5 = 2.25$.

---

3. $3 \div 0.2 + 3.5\sqrt{16} - 4.1$ **4.** $280(1 - 0.08)^2$

2 Writing an Arithmetic Expression for a Word Problem, Then Simplifying

You use the order of operations when solving many practical problems.

EXAMPLE 5 A cab ride costs $1.80 plus $0.90 a mile. How much do you pay for a 16.5-mile ride across town?

The charge is $1.80 plus $0.90 a mile times 16.5 miles.

$$\text{Charge} = \$1.80 + \underbrace{\$0.90 \cdot 16.5}_{} \qquad \text{Multiply first.}$$

$$= \$1.80 + \quad \$14.85 \qquad \text{Then add.}$$

$$= \$16.65 \quad \blacksquare$$

EXAMPLE 6 Each week Tom makes $6.32 per hour for the first 40 hours, plus $10.16 for each hour he works over 40. How much will he make if he works a 48-hour week?

| Tom makes $6.32 per hour for the first 40 hours. | | Tom makes $10.16 per hour for the remaining (48 − 40) hours. |

$$\text{Salary} = \$6.32 \cdot 40 \quad + \quad \$10.16 \cdot (48 - 40)$$

$$= \$6.32 \cdot 40 \quad + \quad \$10.16 \cdot \quad (8)$$

$$= \$252.80 \quad + \quad \$81.28$$

$$= \$334.08 \quad \blacksquare$$

Write the expression for each problem, then simplify.

5. A cab ride costs $1.40 to get in plus $0.80 a mile. How much do you pay for a 7.5-mile ride?

6. Each week Sandra earns $8.56 for the first 37.5 hours and $12.84 for each hour over 37.5. How much will she make if she works a 52-hour week?

3 Solving Problems Involving Fractions and Decimals

To add or subtract a fraction and a decimal, first change the fraction to a decimal, or change the decimal to a fraction.

EXAMPLE 7 What is $0.4 + \frac{1}{2}$?

Solution #1: Change 0.4 to the fraction $\frac{4}{10}$. Then add.

$$0.4 + \frac{1}{2} = \frac{4}{10} + \frac{1}{2} = \frac{4}{10} + \frac{1}{2} \cdot \frac{5}{5} = \frac{4}{10} + \frac{5}{10} = \frac{9}{10} \quad \text{or} \quad 0.9$$

Solution #2: Change $\frac{1}{2}$ to the decimal 0.5. Then add.

$$0.4 + \frac{1}{2} = 0.4 + 0.5 = 0.9 \quad \text{or} \quad \frac{9}{10} \quad \blacksquare$$

▶ **You Try It**   **7.** $0.6 + \dfrac{3}{4}$ **8.** $3.91 + \dfrac{7}{10}$

To multiply a fraction and a decimal, first write the decimal over 1. Then multiply as you would multiply two fractions.

EXAMPLE 8   A mechanic charges $28.75 per hour for labor. It takes $4\dfrac{1}{3}$ hours to overhaul an engine. What is the labor charge?

Multiply $28.75 per hour times $4\dfrac{1}{3}$ hours of labor.

$$\text{Labor charge} = \$28.75 \cdot 4\dfrac{1}{3} = \dfrac{\$28.75}{1} \cdot \dfrac{13}{3} = \dfrac{\$373.75}{3} \doteq \$124.58 \quad \blacksquare$$

EXAMPLE 9   How many $\dfrac{5}{6}$-ounce mini-bags of chips can be filled from a large 12.5-ounce bag?

Find how many $\dfrac{5}{6}$-ounce bags are in 12.5 ounces. Divide 12.5 by $\dfrac{5}{6}$.

$$\text{Number of } \dfrac{5}{6}\text{-ounce bags in 12.5 ounces} = 12.5 \div \dfrac{5}{6} = \dfrac{12.5}{1} \div \dfrac{5}{6} = \dfrac{\overset{2.5}{\cancel{12.5}}}{1} \cdot \dfrac{6}{\underset{1}{\cancel{5}}} = 15 \text{ bags} \quad \blacksquare$$

▶ **You Try It**   **9.** An electrician charges $34.80 per hour for labor. It takes $12\dfrac{5}{6}$ hours for him to wire a clubroom. Find the labor charge.

10. One share of stock costs $\$9\dfrac{1}{8}$. How many shares can you buy for $2,591.50?

4 Order Decimals   You order decimals to find the largest or smallest one. Which is larger, 0.1 or 0.09?

$$0.1 \ = 1 \text{ tenth}$$
$$0.09 = 9 \text{ hundredths}$$

You cannot compare tenths to hundredths. They are unlike amounts. To make them like, attach zeros until each has the same number of places.

$$\left. \begin{array}{l} 0.1 \ = 0.10 = 10 \ \text{hundredths (larger)} \\ 0.09 = 0.09 = \ 9 \ \text{hundredths} \end{array} \right\} \text{So, } 0.1 > 0.09$$

Now each decimal has two decimal places, and the same name. Since 10 hundredths > 9 hundredths, then 0.1 > 0.09.

> **OBSERVE** It may help to think in terms of money. 0.1 = $0.10 and 0.09 = $0.09. Since $0.10 > $0.09, you have 0.1 > 0.09.

To arrange decimals in order according to size

1. Attach zeros until each decimal has the same number of places.
2. Each decimal now has the same name. Arrange them in order as you would arrange whole numbers in order.

EXAMPLE 10   Arrange the decimals in order from largest to smallest: 0.1, 0.011, 0.101, 0.01.

0.011 and 0.101 have the greatest number of decimal places with three each.

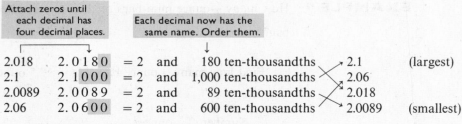

Attach zeros until each decimal has three decimal places.

Each decimal now has the same name, thousandths. Order them.

0.1	0.100	= 100 thousandths	0.101	(largest)
0.011	0.011	= 11 thousandths	0.1	
0.101	0.101	= 101 thousandths	0.011	
0.01	0.010	= 10 thousandths	0.01	(smallest)

Therefore, 0.101 > 0.1 > 0.011 > 0.01.   ■

EXAMPLE 11   Arrange the decimals in order from largest to smallest: 2.018, 2.1, 2.0089, 2.06.

Attach zeros until each decimal has four decimal places.

Each decimal now has the same name. Order them.

2.018	2.0180	= 2 and 180 ten-thousandths	2.1	(largest)
2.1	2.1000	= 2 and 1,000 ten-thousandths	2.06	
2.0089	2.0089	= 2 and 89 ten-thousandths	2.018	
2.06	2.0600	= 2 and 600 ten-thousandths	2.0089	(smallest)

Therefore, 2.1 > 2.06 > 2.018 > 2.0089.   ■

▶ **You Try It**   Arrange in order from largest to smallest.

11. 0.03, 0.303, 0.0033, 0.1   **12.** 7.064, 7.009, 7.11, 7.0599

▶ **Answers to You Try It**   1. 5.29   2. 9   3. 24.9   4. 236.992
5. $1.40 + $0.80 · 7.5 = $7.40   6. $8.56 · 37.5 + $12.84(52 − 37.5) = $507.18
7. 1.35   8. 4.61   9. $446.60   10. 284   11. 0.303 > 0.1 > 0.03 > 0.0033
12. 7.11 > 7.064 > 7.0599 > 7.009

SECTION 5.9 EXERCISES

I *Simplify using the order of operations.*

1. $5.6 + 4.2 \cdot 6.8$

2. $6.3 + 2.4 \cdot 8.5$

3. $7.2 \cdot 4.1 - 12.3$

4. $5.7 \cdot 16.2 - 38.13$

5. $4.8 \div 0.12 \cdot 10$

6. $6.9 \div 23 \cdot 0.1$

7. $12.5 \cdot 8 \div 4$

8. $6.25 \cdot 4 \div 2$

9. $0.9 \div 0.3 \div 0.1$

10. $6.4 \div 3.2 \div 1.6$

11. $0.85(4.2 + 0.53)$

12. $10.5(1 + 4.6)$

13. $3.5(2.6 - 0.6) + 3$

14. $40.8(7.6 - 7.5) + 0.92$

15. $(6.38 + 13.62)2.25$

16. $(0.91 + 0.9)5.7$

17. $3.02 + 45 \cdot 0.45 + 2.21$

18. $60.5 + 15.2 \cdot 7.5 + 12.6$

19. $2.36 \cdot 10^2$

20. $5.6 \cdot 10^0$

21. $2 \cdot 7.5 + 2 \cdot 5.25$

22. $4 \cdot 12.3 + 6 \cdot 7.2$

23. $4.0176 \cdot 10^3$

24. $8.036 \cdot 10^5$

25. $1.06 \cdot 10^3 + 8.63 \cdot 10^2$

26. $4.827 \cdot 10^4 + 2.9 \cdot 10^1$

27. $1,000(1 + 0.03)^2$

28. $580(1 + 0.08)^3$

29. $7 \div 8 + 6.5\sqrt{64} - 52.875$

30. $3.6 \div 0.36 + 0.1\sqrt{100} - 1$

2 *Write an expression for each word problem. Then simplify.*

31. A cab ride costs $1.90 plus $0.80 a mile. What is the charge for an 11-mile ride?

32. A cab ride costs $1.60 plus $0.90 a mile. What is the cost for a 3.5-mile ride?

33. What is the total cost of 3 pairs of slacks at $24 each and 6 blouses at $14.95 each?

34. What is the total bill for 16 dinners at $18.25 each and 18 desserts at $3.50 apiece?

35. A long-distance phone call is $0.83 for the first minute and $0.52 for each additional minute. What is the cost of a 12-minute call?

36. Each week Sally makes $8.52 an hour for the first 40 hours and $12.78 for each hour over 40. What does she make if she works a 52-hour week?

3 *Perform the indicated operation.*

37. $0.5 + \dfrac{3}{4}$

38. $4\dfrac{1}{2} + 3.75$

39. $5\dfrac{1}{4} - 3.85$

40. $1.2 - \dfrac{4}{5}$

41. $\dfrac{1}{2} \cdot 0.62$

42. $3\dfrac{5}{8} \cdot 45.2$

43. $25.8 \div \dfrac{6}{7}$

44. $8.05 \div 3\dfrac{1}{2}$

45. $4.6 \cdot \dfrac{1}{3}$ round to tenths

46. $4\dfrac{3}{7} \cdot 56.1$ round to hundredths

47. $7.2 \div 2\dfrac{1}{3}$ round to hundredths

48. $0.451 \div 2\dfrac{5}{6}$ round to thousandths

49. What will $2\dfrac{1}{3}$ pounds of Swiss cheese cost at $4.59 a pound?

50. Alice worked $14\dfrac{5}{6}$ hours at $8.64 per hour. What did she earn?

51. A refrigerator normally selling for $659 is on sale for $\dfrac{1}{4}$ off.

   a. How much is being taken off the price of the refrigerator?

b. What is the sale price?

52. A dress normally selling for $39.50 is on sale for $\dfrac{1}{3}$ off.

   a. How much is being taken off the price? Round to cents.

   b. What is the sale price

4 *Order the decimals from smallest to largest.*

53. 0.8, 0.089

54. 0.04, 0.015

55. 0.0971, 0.103

56. 0.78, 0.8

57. 0.00091, 0.002

58. 5.016, 5.0201

59. 0.603, 0.6104, 0.6

60. 9.03, 9.105, 9.0084

61. 3.03, 3.0063, 3.026

62. 0.044, 0.0404, 0.042

63. 0.254, 0.2042, 0.26, 0.205

64. 0.082, 0.09, 0.0815, 0.085

65. 4.03, 4.1, 4.0069, 4

66. 1.53, 1.053, 1.503, 1.5

67. 0.02, 0.0061, 0.019, 0.1

68. 5.071, 5.1, 5.0708, 5.07

69. 7.07, 7.018, 7.2, 7.09, 7.089

70. 23.05, 23.108, 23.009, 23.3, 23.0802

SKILLSFOCUS (Section 3.3) *Reduce to lowest terms.*

71. $\dfrac{9}{15}$

72. $\dfrac{25}{40}$

73. $\dfrac{28}{63}$

74. $4\dfrac{72}{90}$

EXTEND YOUR THINKING ▶▶▶▶
▶SOMETHING MORE

75. Insert the appropriate operations to get the answer shown. Use a calculator.

   a. 0.9   0.9   0.9 = 1.9

   b. 2.5   2.5   2.5 = 0

   c. 1.8   0.2   1 = 2

   d. 5   5   5 = 0.5

   e. 0.08   0.05   0.07 = 0.2

   f. 10   0.1   0.1 = 1,000

76. The Taxicab Problem   A cab costs \$1.80 to get in plus \$0.10 for every $\dfrac{1}{8}$ of a mile.

   a. Write an expression for total cab fare if you ride *m* miles.

   b. What is the total charge if the cab takes you 9.5 miles?

c. You have $20 and must take a cab to a destination 23 miles away. Do you have enough money?

77. A family has their TV set on an average of $6\frac{2}{3}$ hours per day. Suppose the cost to operate a TV is $0.024 per hour.

    a. What does the family pay per day to operate their TV?

    b. What do they pay monthly (assume 1 month = 30 days)?

    c. What do they pay per year (1 year = 365 days)?

▶ TROUBLESHOOT IT

Find and correct the error.

78. $2.8^2 = 2.64$

79. $2.4 \div 1.2 \cdot 2 = 2.4 \div 2.4 = 1$

WRITING TO LEARN ▶▶▶▶

80. a. Explain in your own words how to evaluate $5.7(4)^2$.

    b. Evaluate this expression using pencil and paper.

    c. Write the sequence of keypresses that must be made to evaluate this expression using a calculator.

81. Write a word problem about a trip to the bakery. The cost of your purchase is given by the expression $4 \times 2.95 + 3.15$. In your problem, explain what each number represents.

▶ YOU BE THE JUDGE

82. Donald claims the following expression

$$9.78 + 9.78 + 9.78 - 9.78 + 9.78 + 9.78 - 9.78 - 9.78 + 9.78$$

can be represented by two products and a difference. What does he mean? Is he correct? Explain your decision.

CALCULATOR TIPS

Using Memory Your calculator has two areas for doing calculations. The first is shown on your calculator's display. The second is hidden and is called *memory*. Memory is a separate area used to add and subtract numbers.

The memory keys do not affect any calculations in progress, so you can use them any time. On many calculators, a small "M" appears on the display when you press a memory key. This tells you memory is being used.

$\boxed{\text{M+}}$ or $\boxed{\text{M}\pm}$ or $\boxed{\text{SUM}}$	Adds the number currently displayed on your calculator to the number in memory.
$\boxed{\text{M}-}$ or $\boxed{\text{M}\equiv}$ or $\boxed{+/-}$ $\boxed{\text{SUM}}$	Subtracts the number currently displayed on your calculator from the number in memory.
$\boxed{\text{MR}}$ or $\boxed{\text{RM}}$ or $\boxed{\text{MRC}}$ or $\boxed{\text{RCL}}$	Memory Recall. Pressing this key displays the number currently in memory.
$\boxed{\text{CM}}$ or $\boxed{\text{MC}}$	Clear Memory. Clears the value in memory to 0. $\boxed{\text{CM}}$ does not affect the calculator display. Pressing $\boxed{\text{MR}}$ followed by $\boxed{\text{M}-}$ will also clear memory to 0. Why?

Use memory to hold a calculation for later use.

EXAMPLE I Evaluate $4 \times 5 + 6 \times 2$.

$\boxed{\text{CM}}$ 4 $\boxed{\times}$ 5 $\boxed{=}$ $\boxed{\text{M+}}$ 6 $\boxed{\times}$ 2 $\boxed{=}$ $\boxed{\text{M+}}$ $\boxed{\text{MR}}$

Clears memory to 0.	Answer on display is 20.	Adds 20 to memory.	Answer on display is 12.	Adds 12 to the 20 in memory.	Displays the answer in memory, 32. ∎

EXAMPLE 2 Evaluate $\dfrac{4.5 - 1.7}{3.6 - 2.93}$.

$\boxed{\text{CM}}$ 3.6 $\boxed{-}$ 2.93 $\boxed{=}$ $\boxed{\text{M+}}$ 4.5 $\boxed{-}$ 1.7 $\boxed{=}$ $\boxed{\div}$ $\boxed{\text{MR}}$ $\boxed{=}$

Evaluate the denominator first. $3.6 - 2.93 = 0.67$	Stores 0.67 in memory.	Evaluate the numerator. $4.5 - 1.7 = 2.8$	Recalls 0.67 from memory.	Gives the answer, 4.1791045. ∎

Memory can be used to perform operations on fractions.

EXAMPLE 3 Add $\dfrac{3}{4} + \dfrac{5}{6}$.

3 $\boxed{\div}$ 4 $\boxed{=}$ $\boxed{\text{M+}}$ 5 $\boxed{\div}$ 6 $\boxed{=}$ $\boxed{\text{M+}}$ $\boxed{\text{MR}}$ 1.5833333 ∎

EXAMPLE 4 Subtract $\dfrac{4}{5} - \dfrac{3}{8}$.

4 $\boxed{\div}$ 5 $\boxed{=}$ $\boxed{\text{M+}}$ 3 $\boxed{\div}$ 8 $\boxed{=}$ $\boxed{\text{M}-}$ $\boxed{\text{MR}}$ 0.425 ∎

EXAMPLE 5 Multiply $\dfrac{7}{10} \times \dfrac{15}{28}$.

7 $\boxed{\div}$ 10 $\boxed{\times}$ 15 $\boxed{\div}$ 28 $\boxed{=}$ 0.375 ∎

EXAMPLE 6 Divide $\dfrac{3}{8} \div \dfrac{1}{4}$.

1 ÷ 4 = M+ 3 ÷ 8 = ÷ MR = 1.5 ∎

Memory is useful for keeping a running total during a calculation.

EXAMPLE 7 A department store purchased 34 ties @ $14.95, 26 shirts @ $22.00, 18 slacks @ $28.50, 40 socks @ $2.69, and 14 sports jackets @ $79.95. What total was paid?

quantity	unit cost		calculations
34 ties	@ $14.95	⟶	CM 34 × 14.95 = 508.3 M+
26 shirts	@ $22.00	⟶	26 × 22 = 572 M+
18 slacks	@ $28.50	⟶	18 × 28.5 = 513 M+
40 socks	@ $2.69	⟶	40 × 2.69 = 107.6 M+
14 jackets	@ $79.95	⟶	14 × 79.95 = 1119.3 M+ MR

total cost = $2,820.20 ∎

EXAMPLE 8 Use memory to evaluate $4.2X^2 + 0.6X + 2.79$ when $X = 3.8$.

Replace X with 3.8: $4.2 \cdot (3.8)^2 + 0.6 \cdot (3.8) + 2.79$

CM 4.2 × 3.8 × 3.8 = M+ 0.6 × 3.8 = M+ 2.79 M+ MR 65.718 ∎

Use your calculator and memory to solve each problem.

1. $6 \times 9 + 7 \times 4$
2. $12 \times 3.5 + 11 \times 647$
3. $6 \times 3 - 4.7 \times 1.6$
4. $230 \times 48 - 87.3 \times 65.8$
5. $\dfrac{182.45 - 8.6}{36.7 - 28.9}$ round to ones
6. $\dfrac{200 - 143.06}{98 - 7.034}$ round to hundredths
7. $\dfrac{3}{5} + \dfrac{7}{8}$
8. $\dfrac{7}{12} - \dfrac{1}{3}$
9. $5\dfrac{1}{4} \times \dfrac{6}{7}$
10. $\dfrac{9}{15} \div \dfrac{3}{10}$

11. Find the total cost of
    24 scarves @ $18.50,
    35 hats @ 22.88,
    7 down coats @ $69.95,
    14 pairs of boots @ $48.99,
    and 56 pairs of wool socks @ $7.95.

12. Find the total cost of
    40 math books @ $32.95,
    65 history texts @ $26.95,
    28 biology texts @ $42.50,
    and 120 literature books @ $35.60.

13. Evaluate $32P + 41Q$ when $P = 57.913$ and $Q = 106.88$.

14. Find S if $S = 2LW + 2WH + 2LH$ and $L = 2.5$, $W = 6$, and $H = 1.75$.

CHAPTER **5** REVIEW

VOCABULARY AND MATCHING

New words and phrases introduced in this chapter are shown in the left-hand column in the order they appeared. Match each term on the left with the phrase or sentence on the right that best describes it.

A. decimal place value names

B. number of decimal places

C. decimal notation

D. expanded notation

E. and

F. 0.0200

G. 0.210

H. 200.010

I. addition of decimals

J. subtraction of decimals

K. 6.7 − 2.015

L. rounding off decimals

M. multiplication of decimals

N. division of decimals

O. 6.7 miles per 2.015 hours

P. sales tax

Q. sales tax factor

R. change a fraction to a decimal

S. terminating decimals

T. repeating decimals

U. change a decimal to a fraction

V. comparing decimals

_____ examples are 2.373737 . . . and $0.\bar{8}$; $\frac{1}{3}$ can be turned into one

_____ placing the decimal point requires adding the number of decimal places in each factor

_____ line decimal points up in same vertical column; is not commutative

_____ read it, write it, reduce it

_____ two hundred and ten thousandths

_____ must add zeros to minuend before you can do this

_____ requires a whole number divisor

_____ equals 1.04 for 4¢ on the dollar

_____ tenths, thousandths, and millionths are examples

_____ the way decimals are written in your checkbook or on a bill

_____ two hundred ten thousandths

_____ charged per dollar of selling price; varies from state to state

_____ means $6.7 \div 2.015$

_____ when 1.24 is written $1 + \frac{2}{10} + \frac{4}{100}$

_____ 4.6284 becomes 4.6 to the tenths, 5 to the ones, or 4.63 to hundredths

_____ decimal point, "."

_____ two hundred ten-thousandths

_____ number of digits present to the right of the decimal point

_____ first add zeros so each decimal has the same number of decimal places

_____ line decimal points up in same vertical column; is commutative

_____ divide numerator by denominator

_____ examples are 4.6, 0.0863, and 12.05; $\frac{3}{4}$ can be turned into one; $\frac{1}{3}$ cannot

REVIEW EXERCISES

5.1 Decimal Notation

1. Find the place value of the digit 2 in each decimal.
  a. 0.0245
  b. 320.067

  c. 43.206
  d. 7.8032

2. Find the number of decimal places in each number.
  a. 4.302
  b. 70.08

  c. 3
  d. 6.0

3. Write each decimal in expanded notation.

    a. 1.6   **b.** 54.097

    c. 5.0062   **d.** 0.08

4. Write each decimal in words.

    a. 3.5   **b.** 0.02

    c. 500.005   **d.** 0.0406

5. Write each number in decimal notation.

    a. four tenths   **b.** six and twenty-four thousandths   **c.** eighteen hundredths

    d. six hundred twenty-four thousandths   **e.** five thousandths

    f. twenty and one thousand two hundred nine ten-thousandths

5.2 Rounding Decimals

Round each decimal to the indicated place.

    6. 5.082   tenths   **7.** 200.84   ones   **8.** 0.0462   thousandths

    9. 28.285   hundredths   **10.** 435.92   tens   **11.** 0.04   tenths

    12. 0.997   hundredths   **13.** 1.80482   thousandths

5.3 Addition and Subtraction of Decimals

    14. $4.92 + 0.4$ **15.** $26.061 + 7.352$ **16.** $2.03 + 50 + 3.084$

    17. $7 + 0.04 + 31.9402$ **18.** $0.87 + 2.5 + 1 + 0.5$ **19.** $300.3 + 25.98 + 5.092 + 6$

    20. $40.06 + 3.9 + 200.54 + 47.036 + 0.0075 + 80 + 0.7$

    21. $70.5 - 4.06$ **22.** $40 - 3.79$ **23.** $10 - 0.39$

    24. $365.057 - 47.06$ **25.** $5,804.6 - 309.064$ **26.** $0.0052 - 0.00068$

5.4 Applications

27. Samantha had a batting average of 0.279 last season. This season her average is 0.332. By how many points did her average rise?

28. The purchase price for a tie is $17.53. If you pay for the tie with a $20 bill, what is your change?

29. Paul had a bank balance of $110.32. He made deposits of $13.48 and $56.03. He wrote checks for $6.15, $28.04, $40, and $58.92. What is his new balance?

30. Four sprinters in a 400-meter relay race had times of 9.27, 9.43, 9.06, and 9.61 seconds. What was the team time for the race?

31. Helen deposited checks for $300, $69.57, and $175.92 into her checking account this month. What were her total deposits?

32. Hank's gross pay is $1,578.04. His deductions are $185.13 for federal tax, $76.24 for state tax, $88.67 for FICA, $200 for the credit union, and $56.70 for medical. What is Hank's net pay?

5.5 Multiplication of Decimals

33. $3.8 \cdot 0.56$ **34.** $45.9 \cdot 8.92$ **35.** $0.052 \cdot 0.8$ **36.** $783.65 \cdot 24$

37. $1,000 \cdot 4.0893$ **38.** $45.6 \cdot 100$ **39.** $0.0065 \cdot 10$ **40.** $10,000 \cdot 10.68$

41. What is the cost of 2,540 photocopies at $0.018 each?

42. What is the product of 305.76 and 4.82?

5.6 Division of Decimals

43. $45.6 \div 3$ **44.** $299.52 \div 72$ **45.** $68.202 \div 5.4$

46. $0.052 \div 0.65$ **47.** $0.36 \div 0.05$ **48.** $334.138 \div 82.3$

49. $17.04 \div 100$ **50.** $7.72 \div 1,000$

51. $56.93 \div 3.48$ (round to hundredths)

52. $0.054 \div 0.67$ (round to thousandths)

53. $0.07 \div 9.2$ (round to four decimal places)

54. $560.54 \div 6.15$ (round to two decimal places)

5.7 Applications

55. 0.653 of the weight of the human body is water. If the heaviest man on Earth weighs 862 pounds, how many pounds of water does he carry?

56. What is the cost of 6 gallons of ice cream at $2.89 per gallon?

57. If gas costs $1.349 per gallon, what is the cost of 11.3 gallons to the nearest cent?

58. What is the cost of 12.5 yards of material at $8.29 per yard? Round to the nearest cent.

59. If you pay $27.16 a month on a bill, how long will it take you to pay off $651.84?

60. You purchased 2.46 pounds of cheese for $7.85. What is the price per pound to the nearest cent?

61. You travel 314 miles on 38.2 gallons of gas. How many miles per gallon did you get? Round to the nearest tenth.

62. A consultant made $567.34 for one 8-hour day. What did she make per hour rounded to the nearest dollar?

63. What is the cost of 500 balloons at $0.035 each?

64. Felicia purchased 12 albums on sale for $8.67 each. What did she pay to the nearest cent if she was charged a 5¢ sales tax?

65. Nick must drive 96 miles to his mother's house. He has 4.8 gallons of gas in his tank and gets 18.6 miles to the gallon. Will he make it without having to purchase more gas?

66. You put $1,600 down on a car and agree to pay $216.58 per month for 48 months. What total will you pay for the car?

67. Gas costs $1.239 if you pay by credit card, or $1.199 if you pay cash. To the nearest cent, how much do you save if you pay cash for 16.3 gallons of gas?

68. You rent a car for $19.95 a day plus $0.14 per mile. You use the car for 6 days and drive 1,208 miles. What do you pay?

69. You must inoculate 200 people. Each inoculation contains 180 milligrams of antibiotic that costs $0.048 per milligram. Find the total cost.

70. You must pay $13,878.30 over a 30-month period. What is the monthly payment?

5.8 Conversions Between Fractions and Decimals

Change each fraction to an exact decimal.

71. $\dfrac{9}{20}$ **72.** $\dfrac{1}{16}$ **73.** $\dfrac{9}{10}$ **74.** $4\dfrac{7}{8}$ **75.** $12\dfrac{11}{25}$

Change each fraction to a decimal. Round as requested.

76. $\dfrac{7}{13}$ hundredths **77.** $\dfrac{4}{17}$ thousandths **78.** $6\dfrac{1}{9}$ tenths **79.** $\dfrac{19}{24}$ ten-thousandths

Change each decimal to a fraction. Reduce to lowest terms.

80. 0.2 **81.** 0.81 **82.** 6.4 **83.** 2.03

84. 0.808 **85.** 20.5 **86.** 6.0032 **87.** 100.3125

5.9 Combining Operations; Order

88. $2.4 + 3.6 \cdot 1.5$ **89.** $4.8 \div 2.4 \div 0.2$ **90.** $0.5(0.24 + 3.8) - 1$

91. $7.502 \cdot 10^5$ **92.** $0.06 \cdot 12.6 + 6.83 \cdot 2$ **93.** $3.052 + 10 \cdot 6.03 - 6^2$

94. $4.73 + \dfrac{5}{8}$ **95.** $2\dfrac{3}{5} - 1.972$ **96.** $\dfrac{6}{7} \cdot 4.732$ **97.** $5.86 \div \dfrac{2}{9}$

98. What will $2\dfrac{3}{4}$ pounds of roast beef cost at $5.29 a pound? Round to the nearest cent.

Arrange the decimals in order from smallest to largest.

99. 0.062, 0.0619 **100.** 0.51, 0.508, 0.513

101. 0.038, 0.0359, 0.0402, 0.03 **102.** 7.602, 7.58, 7.6101, 7.6

This test will measure your understanding of decimals. Allow yourself 50 minutes to complete it. Write the work for each problem. When done, check your answers. Rework each problem solved incorrectly.

1. Write in words: 0.052.

2. Write in words: 100.0710.

3. Write in decimal notation: three and three hundredths.

4. Write in decimal notation: eight hundred nine ten-thousandths.

5. Write in decimal notation: four thousand sixty and two tenths.

6. Round 52.7 to ones.

7. Round 50.906 to tenths.

8. Round 0.8995 to thousandths.

9. Round 5,847.904 to hundreds.

10. $4.7 + 0.492 + 6$

11. $124.76 + 8.093 + 4.1 + 38 + 0.0036$

12. $20 − 3.14

13. $45.07 − 6.2853$

14. At noon, the temperature of a substance was 12.6°C. At 3:00 P.M. the temperature was 2.59°C higher. What was the temperature at 3:00 P.M.?

15. $61.9 \cdot 470.04$

16. $1,000 \cdot 0.63$

17. $36.176 \div 9.52$

18. $0.0681 \div 0.83$ (round to thousandths)

19. The capacity of a dump truck is 14.5 tons. How many trips will be necessary to transport 203 tons of stone?

21. Cindy paid $16.75 to play 18 holes of golf. She shot a 92. What was the cost per shot to the nearest cent?

20. What will you pay if 1.28 pounds of pastrami costs $4.39 a pound? Round to the nearest cent.

22. A cab costs $1.80 to get in plus $0.80 per mile. What will Gail pay if she takes a cab to work, a distance of 14.5 miles?

23. Change $\dfrac{7}{16}$ to an exact decimal.

24. Change $4\dfrac{3}{7}$ to a decimal. Round to three places.

25. Change 0.45 to a fraction in lowest terms

26. Change 4.082 to a fraction in lowest terms.

27. $3.05 + 4 \cdot 0.63$

28. $8.03 \cdot 10^1$

29. $3\dfrac{7}{9} \cdot 2.697$ (round to hundredths)

30. Solve $2.4w = 17.4$

31. Solve $y + 0.84 = 2.74$.

32. Solve $z \div 4.07 = 6$.

33. Arrange the following numbers in order from smallest to largest. 2.0614, 2.065, 2.06, 2.0609

Explain the meaning of each term. Use your own examples.

34. complement

35. terminating decimal

1. Subtract $6{,}200 - 4{,}800$.

2. Multiply 750×359.

3. Divide $56\overline{)15{,}344}$.

4. Simplify $40 \div 10 \cdot 4$.

5. Solve $x - 7 = 13$.

6. Solve $\dfrac{x}{6} = 8$.

7. Write as an arithmetic expression and simplify: four times the sum of twelve and eight, less six.

8. Add $\dfrac{2}{3} + \dfrac{7}{9}$.

9. Subtract $3\dfrac{1}{8} - 1\dfrac{2}{5}$.

10. Multiply $\dfrac{5}{8} \cdot \dfrac{4}{7} \cdot \dfrac{21}{25}$.

11. Divide $10\dfrac{1}{2} \div 2\dfrac{3}{4}$.

12. Manuel filled his gas tank 6 times driving from Baltimore to Orlando. If he purchased $12\dfrac{3}{10}$ gallons on each fill-up, how many gallons of gas did he buy?

13. Simplify $\left(\dfrac{3}{5}\right)^2 + 4 \cdot \dfrac{5}{6}$.

14. Simplify $\dfrac{7}{8} \cdot 3.28$.

15. Add $4.2 + 0.09 + 53.72 + 6$.

16. Subtract $3.8 - 1.09$.

17. Multiply $0.06 \cdot 5.83$.

18. Divide $11 \div 2.64$. Round to tenths.

19. Simplify $4(0.6) + 7(2.3)$.

20. Simplify $0.08 \div 8 \cdot 0.1$

Explain the meaning of each term. Use your own examples.

21. quotient

22. reducing fractions

Ratio and Proportion

HOUSING PRICES AND FAMILY INCOME

Ratios are used in the housing table shown here. The table shows the average price of a new house and the average family income for each year listed. As you can see, average income has risen faster than new home prices from 1900 to 1970. But this trend reversed from 1970 to 1990. Comparisons of these two amounts are listed as ratios in the column on the far right. What does a ratio of 9.9 mean? This question will be answered in this chapter.

Year	Average Price for a New House*	Average Family Income	Price to Income Ratio (to 1)
1900	$ 4,881	$ 490	9.9
1910	5,377	630	8.5
1920	6,296	1,489	4.2
1930	7,146	1,360	5.2
1940	6,558	1,300	5.0
1950	9,446	3,319	2.8
1960	16,652	5,620	2.9
1970	23,400	9,867	2.3
1980	64,600	21,023	3.1
1990	122,900	35,191	3.5

*Average refers to the median. (See Section 8.1)

SOURCE: National Association of Home Builders and the U.S. Census Bureau.

Ratios are used to compare quantities. In the table, new home prices are compared to family income. Ratios are also used in unit pricing, the stock market, construction, business, and finance. Proportions are used to help you solve a wide variety of problems using ratios. In this chapter these problems deal with blueprints, recipes, and estimation, such as estimating the number of people in a city infected with a virus, or estimating the height of a building using the length of its shadow. Ratios also provide the foundation for Chapter 7 on percents.

Work through each section in this chapter. Use this test to identify topics you are not familiar with. These topics may require additional study. You may *not* use a calculator.

6.1

1. A school has 74 teachers, 1,043 students, and 11 administrators. Write the following ratios.
   a. students to teachers

   b. teachers to administrators

   c. teachers and administrators to students

2. Reduce the ratio to lowest terms: 32 feet to 20 feet.

3. Eliminate the units and reduce: 5 minutes to 40 seconds.

6.2

4. Reduce the rate: 180 miles to 12 hours.

5. Change yards to feet, and reduce: $30 to 7 yards.

6.3

6. Write as a ratio of whole numbers: $7\frac{1}{2}$ to $2\frac{1}{4}$.

7. Write as a ratio of whole numbers: 4.2 hours to 0.35 hour.

6.4

8. Write as a unit ratio: $45 to $18.

9. Write as a unit rate: 375 miles to 2.5 hours.

10. A 42-ounce can of mixed fruit costs $1.98. Find the unit price.

6.5

11. Tom and Hal pay a $32,490 loss according to a 5:3 ratio. How much of the loss does each man pay?

12. Ellen drove 380 miles on 16.3 gallons of gas M. She drove 245 miles on 11.3 gallons of gas Q, under the same driving conditions. Which gas gave Ellen the better mileage?

6.6   *Is the proportion true or false?*

13. $\dfrac{8}{13} \overset{?}{=} \dfrac{6}{11}$

14. $\dfrac{15}{25} \overset{?}{=} \dfrac{27}{45}$

6.7   *Solve each proportion.*

15. $\dfrac{x}{10} = \dfrac{36}{45}$

16. $\dfrac{4.72}{y} = \dfrac{0.6}{1.3}$
    Round to tenths.

6.8

17. Tom used 14.8 gallons of gas to drive 291 miles. How many gallons of gas will he need to drive 3,700 miles? Round to tenths.

VOCABULARY *Explain the meaning of each term. Use your own examples.*

18. ratio

19. proportion

6.1 THE MEANING OF RATIO

OBJECTIVES	NEW VOCABULARY
1 Define ratio.	ratio
2 Reduce ratios (Section 3.3).	order
3 Eliminate like names in a ratio.	terms of a ratio

1 Defining Ratios

A compact car is 12 feet long and 5 feet wide. The ratio of length to width is

length to width

12 feet to 5 feet.

> A **ratio** is a comparison of two quantities.

EXAMPLE 1 There are 10 men and 3 women on a committee. The ratio of men to women is

men to women

10 to 3.

The reverse ratio of women to men is

women to men

3 to 10. ■

> **Order** is important when stating a ratio. Write the numbers in a ratio in the same order as the word names.

EXAMPLE 2 In a small school there are 55 male students, 70 female students, and 8 instructors.
a. The ratio of female students to male students is 70 to 55.
b. The ratio of instructors to male students is 8 to 55.
c. The ratio of students to instructors is 125 to 8 ($55 + 70 = 125$ students). ■

▶ **You Try It**

1. There are 40 cars and 9 trucks on a lot.
   a. What is the ratio of cars to trucks?
   b. What is the ratio of trucks to cars?
2. A park has 33 acres of grass, 45 acres of trees, and a 12-acre pond.
   a. What is the ratio of acres of grass to acres of trees?
   b. What is the ratio of acres of pond to acres of grass?
   c. What is the ratio of land acres to pond acres?

Suppose for every $8 in income there are $3 in taxes. The ratio of income to taxes can be written three ways.

Words	*Colon*	*Fraction*
8 to 3	8:3	$\frac{8}{3}$ or 8/3
eight to three	":" is read "to"	the fraction bar is read "to"

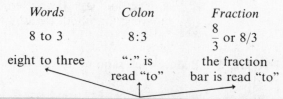

> Each is read "the ratio of eight to three."
> The difference is in the way the word "to" is expressed.

8 and 3 are called the **terms of the ratio**. 8 is called the *first term*. 3 is called the *second term*.

> Any ratio can be written in the following three ways.
>
> **1.** with the word "to"
> **2.** by use of a colon, ":"
> **3.** as a fraction

EXAMPLE 3 Ken made 30 hits in his last 50 times at bat. Write the ratio of *hits* to *at bats* in three ways.

using words: 30 to 50, or thirty to fifty
using a colon: 30:50
as a fraction: 30/50 ■

▶ **You Try It** **3.** There are 4 CD players to every 3 tape players in a music store. Write the ratio of CD players to tape players in three ways.

4. Terri threw the horseshoe 21 times and made 8 ringers. Write the ratio of ringers to throws in three ways.

Any fraction can be written as a ratio. In fact, fractions are called *rational numbers*.

The fraction $\frac{a}{b}$ is read as "the ratio of the numerator a to the denominator b."

$$\frac{a}{b} = a \text{ to } b$$

EXAMPLE 4 **a.** $\frac{5}{8}$ is the ratio 5 to 8 or 5:8.

b. $5\frac{1}{4} = \frac{21}{4}$ is the ratio 21 to 4 or 21:4. ■

▶ **You Try It** Write each fraction as a ratio in two other ways.

5. $\frac{7}{10}$ **6.** $4\frac{3}{7}$

2 Reducing Ratios Any ratio can be written as a fraction. Therefore, it can be reduced to lowest terms in the same way fractions are reduced. For example,

$$6 \text{ to } 4 = \frac{6}{4} = \frac{6 \div 2}{4 \div 2} = \frac{3}{2} = 3 \text{ to } 2.$$

The ratio 6 to 4 is equivalent to the lowest terms ratio 3 to 2.

> To reduce a ratio to lowest terms
>
> **1.** Divide both terms of the ratio by the same factor.
> **2.** The ratio is in lowest terms when the only number that divides evenly into both terms is 1.

EXAMPLE 5 Reduce each ratio to lowest terms.

a. $20 \text{ to } 15 = \overset{4}{2\!\!\!/0} \text{ to } \overset{3}{1\!\!\!/5} = 4 \text{ to } 3$ reduce using 5

b. $40:100 = \cancel{40:100} = 2:5$ $\overset{2\quad 5}{}$ $\boxed{\text{reduce using 20}}$

c. $\dfrac{9}{54} = \dfrac{9 \div 9}{54 \div 9} = \dfrac{1}{6}$ ∎

▶ **You Try It**   Reduce each ratio to lowest terms.

7. 35 to 25   **8.** 24:18   **9.** $\dfrac{12}{30}$

Reducing ratios gives smaller numbers that are easier to understand.

EXAMPLE 6   Reduce the ratio of 6,000 students to 240 instructors to lowest terms. Interpret the answer.

students	to	instructors	
6,000	to	240	reduce using 10
600	to	24	reduce using 6
100	to	4	reduce using 4
25	to	1	

A ratio of 6,000 students to 240 instructors means there are 25 students per instructor. ∎

▶ **You Try It**   **10.** Write the ratio of 435 representatives to 100 senators in lowest terms.

---

3 Eliminating Like Names in a Ratio

If both terms in a ratio have the same name, eliminate the names by reducing. You reduce names in a ratio in the same way you reduce common factors when you reduce a fraction.

$$8 \text{ feet to } 5 \text{ feet} = \frac{8 \cancel{\text{feet}}}{5 \cancel{\text{feet}}} = \frac{8}{5} = 8 \text{ to } 5$$

EXAMPLE 7 $25 \text{ yards} : 10 \text{ yards} = \dfrac{\overset{5}{\cancel{25 \text{ yards}}}}{\underset{2}{\cancel{10 \text{ yards}}}} = 5:2$ ∎

▶ **You Try It**   Express in simplest form.

11. 40 miles to 13 miles   **12.** 32 hours:20 hours

If the names in a ratio are of the same type, write one name in terms of the other. Then eliminate the names.

EXAMPLE 8   Eliminate the names in the ratio 7 inches to 1 foot.

7 inches	to	1 foot	Inches and feet are names of the same type. Both are lengths. Note, 1 foot = 12 inches.
7 ~~inches~~	to	12 ~~inches~~	Replace 1 foot with 12 inches and eliminate like names.
7	to	12	

The ratio 7 inches to 1 foot is a 7 to 12 ratio. ∎

EXAMPLE 9   Eliminate the names in the ratio 2 hours to 45 minutes.

2 hours	to	45 minutes	Since 1 hour = 60 minutes, 2 hours = 2 × 60 minutes = 120 minutes.
120 ~~minutes~~	to	45 ~~minutes~~	Replace 2 hours with 120 minutes and eliminate the names.
120	to	45	Reduce using 5.
24	to	9	Reduce using 3.
8	to	3	

The ratio 2 hours to 45 minutes is an 8 to 3 ratio. ■

▶ **You Try It**   Eliminate the names in each ratio.

13. 2 yards to 4 feet (Note: 1 yard = 3 feet)

14. 4 minutes to 30 seconds (Note: 1 minute = 60 seconds)

EXAMPLE 10   **Blueprint Scaling**   A wall is 9 inches long on a blueprint. The actual wall is 12 feet long. What is the scale for the blueprint?

The blueprint scale is the ratio of blueprint measurement to actual measurement.

blueprint	to	actual	
9 inches	to	12 feet	Reduce using 3.
3 inches	to	4 feet	Since 1 foot = 12 inches, 4 feet = 4 × 12 inches = 48 inches.
3 ~~inches~~	to	48 ~~inches~~	Eliminate names. Reduce using 3.
1	to	16	Scale for blueprint. ■

> **OBSERVE**   1 inch on the blueprint corresponds to 16 inches on the actual wall. The final ratio 1 to 16, or 1 : 16, is called the *scale* of the blueprint. If the blueprint could be blown up to 16 times its size, it would then have the exact same dimensions as the building it represents.
>
> In addition, the ratio of blueprint to actual is 1 to 16. This means 1 inch on the blueprint corresponds to 16 inches on the actual wall. It also means 1 foot on the blueprint corresponds to 16 feet on the actual wall, or 1 blueprint meter corresponds to 16 meters on the wall. When the terms in a ratio have no name, you may insert your own names, as long as you give each term the same name.

▶ **You Try It**   **15.** An entrance 2 inches wide on a blueprint will actually be 6 feet wide. Calculate the scale for the blueprint.

▶ **Answers to You Try It**   **1. a.** 40 to 9   **b.** 9 to 40   **2. a.** 33 to 45
b. 12 to 33   **c.** 78 to 12   **3.** 4 to 3, 4:3, $\frac{4}{3}$ **4.** 8 to 21, 8:21, $\frac{8}{21}$

5. 7 to 10, 7:10   **6.** 31 to 7, 31:7   **7.** 7 to 5   **8.** 4:3   **9.** $\frac{2}{5}$

10. 87 to 20   **11.** 40 to 13   **12.** 8:5   **13.** 3 to 2   **14.** 8 to 1
15. 1 to 36 or 1:36

I

1. There were 23 mothers, 21 fathers, and 51 children at a picnic. Write the ratio of
   a. fathers to mothers.
   b. mothers to fathers.

   c. mothers to children.
   d. children to adults.

2. A debutante's wardrobe consists of 33 gowns, 32 hats, 26 pairs of evening shoes, and 11 pairs of lounging shoes. Write the ratio of
   a. gowns to hats.
   b. hats to gowns.

   c. pairs of lounging shoes to pairs of evening shoes.

   d. gowns to total pairs of shoes.

3. During visiting hours, a maternity ward consisted of 24 mothers, 23 fathers, 12 male babies, and 17 female babies. Write the ratio of
   a. male babies to female babies.
   b. fathers to mothers.

   c. babies to parents.
   d. mothers to female babies.

4. An art exposition consists of 9 artists, 7 critics, and 250 guests. Write the ratio of
   a. critics to artists.
   b. guests to critics.

   c. artists to guests.
   d. guests to artists.

---

5. Forty boys and 27 girls went on a canoe trip. Write the ratio of boys to girls in three ways.

6. A builder ordered 4,000 bricks and 633 blocks. Write the ratio of bricks to blocks in three ways.

7. Tom has 15 suits and 49 shirts. Write the ratio of shirts to suits in three ways.

8. Amanda has 10 dolls and 17 doll dresses. Write the ratio of dresses to dolls in three ways.

9. A grandmother is 80 years old. Her granddaughter is 21. Write the ratio of their ages (older to younger) in three ways.

10. Justin has 3 A's and 5 B's on his report card. Write the ratio of A's to B's in three ways.

11. A baseball team consists of 24 players and 5 coaches.

   a. Write the ratio of players to coaches as a fraction.

   b. Write the ratio of coaches to players using a colon.

   c. Write the ratio of players to players and coaches, in words.

12. A military base has 11 ships, 103 planes, and 240 tanks stationed nearby.
   a. Write the ratio of planes to ships in words.

   b. Write the ratio of ships to tanks as a fraction.

   c. Write the ratio of ships to planes and tanks using a colon.

Write each ratio in two other forms.

13. 5 to 3   **14.** 14 to 11   **15.** 6/5   **16.** 4/21

17. 40:1   **18.** 5:18   **19.** $\dfrac{1}{64}$ **20.** $\dfrac{35}{2}$

21. 1 to 1   **22.** 7 to 3   **23.** 8:9   **24.** 27:10

2 *Reduce each ratio to lowest terms.*

25. 22 to 16   **26.** 90 to 20   **27.** 38 to 20   **28.** 35 to 42

29. 48:12   **30.** 40:5   **31.** 15:45   **32.** 24:28

33. $\dfrac{30}{9}$ **34.** $\dfrac{64}{36}$ **35.** $\dfrac{52}{65}$ **36.** $\dfrac{14}{16}$

37. 7 to 4   **38.** 61 to 30   **39.** 100 to 28   **40.** 150 to 90

41. 900 to 250   **42.** 450 to 180   **43.** 128:640   **44.** 37:111

45. Tom weighs 180 pounds on Earth. NASA says he will weigh 30 pounds on the moon. Write the ratio of Earth weight to moon weight in lowest terms.

46. A 540-page textbook has 80 pages of problems. Write the lowest terms ratio of pages in the book to pages of problems.

47. $6,000 in profit is made on a $2,500 investment. Write the ratio of investment to profit in lowest terms.

48. A building 600 feet tall casts an 840-foot shadow. Write the ratio of building height to shadow length in lowest terms.

---

3 *Eliminate the names in each ratio and reduce to lowest terms.*

49. 9 feet to 10 feet

50. 4 inches to 3 inches

51. 4 feet to 9 inches

52. 15 inches to 2 feet

53. 5 yards:6 feet

54. 1 foot:3 yards

55. 12 seconds to 5 minutes

56. 1 minute to 40 seconds

57. 25 minutes/1 hour

58. 5 hours/20 minutes

59. 7 weeks to 14 days

60. 52 weeks to 364 days

61. 3 years:9 months

62. 2 months:2 years

63. $4 to 60 cents

64. 35 cents to $3.15

65. 4 hours 20 minutes to 30 minutes

66. 1 hour 15 minutes to 2 hours

67. 2 feet 4 inches to 7 inches

68. 5 feet 3 inches to 1 yard

69. 3 yards 2 feet 10 inches to 2 feet

70. 2 hours 15 minutes 30 seconds to 1 hour

---

71. John's waist measures 30 inches. His height is 6 feet. What is the ratio of John's waist measure to his height?

72. Sally can build a model in 2 hours, and paint it in 25 minutes. What is the ratio of build time to paint time?

73. A room 3 inches wide on a blueprint is actually 12 feet long. What is the scale for the blueprint?

74. A building is 160 feet long. On its blueprint, it is 8 inches long. What is the scale of the blueprint?

Find three fractions equivalent to the given fraction.

75. $\dfrac{2}{3}$ **76.** $\dfrac{7}{5}$ **77.** $\dfrac{1}{10}$ **78.** $\dfrac{3}{16}$

EXTEND YOUR THINKING ▶▶▶▶
▶ SOMETHING MORE

79. To prepare the fuel for a gas saw, 4 quarts and 1 pint of gasoline are mixed with 8 ounces of oil. If 1 quart = 32 ounces, and 1 pint = 16 ounces, write the ratio of gasoline to oil in the mixture.

80. D-day The biggest wartime amphibious landing ever was D-day, June 6, 1944, when about 150,000 U.S. and allied troops landed in France, supported by 2 battleships, 23 cruisers, 105 destroyers, 1,076 additional warships, 2,700 transports, 5,200 bombers, 5,500 fighters, and 2,400 transport planes. Write the lowest terms ratio of troops to total support equipment.

81. It takes 8 hours to put together a gym set. What fraction of the job will be completed in 2 hours 40 minutes?

▶ TROUBLESHOOT IT

Find and correct the error.

82. There are 20 cats, 5 ponies, and 8 dogs on a farm. The ratio of ponies to cats and dogs is 28 to 5.

83. 15 to 40 $= \dfrac{40 \div 5}{15 \div 5} = \dfrac{8}{3} = 8$ to 3 in lowest terms.

WRITING TO LEARN ▶▶▶▶

84. In your own words, explain each step you would use to simplify the ratio $750 to $450.

▶ YOU BE THE JUDGE

85. A baseball player came to bat 12 times without getting a hit.
    a. Write the ratio of hits to times at bat in three ways. Are these ratios valid? Explain your decision.

    b. Write the ratio of at bats to hits in three ways. Are these ratios valid? Explain your decision.

OBJECTIVES	NEW VOCABULARY
1 Define a rate (Section 3.3). **2** Change a unit name in a rate.	rate

1 Defining Rates

When the names in a ratio are of different types, you cannot write one name in terms of the other. So you cannot eliminate the names. Such a ratio is called a **rate**.

For example, the ratio

150 miles   to   3 hours

is a rate. Miles cannot be written in terms of hours. Therefore, you cannot eliminate the names. This rate can be written using the word *per*.

150 miles to 3 hours = 150 miles per 3 hours

It is also common to write a rate in fraction form.

$$150 \text{ miles } per \text{ 3 hours} = 150 \text{ miles}/3 \text{ hours} = \frac{150 \text{ miles}}{3 \text{ hours}}$$

fraction bar is read "per"

Other common rates are miles per gallon, dollars per pound, and dollars per hour.

> A ratio is called a rate when the names of the terms cannot be eliminated. You can write any rate using the word *per*.

EXAMPLE 1 Examples of rates.

a. 55 miles per hour = 55 miles per 1 hour = $\dfrac{55 \text{ miles}}{1 \text{ hour}}$

b. \$3.29 per pound = \$3.29 per 1 pound = $\dfrac{\$3.29}{1 \text{ pound}}$

c. Meg drove 210 miles on 13 gallons of gas. This is the rate

210 miles per 13 gallons = $\dfrac{210 \text{ miles}}{13 \text{ gallons}}$. ■

► You Try It Write each rate using the fraction bar.   **1.** 25 students per course
2. \$1.19 per gallon   **3.** 43 seconds per commercial

When the second term of a rate is 1, the rate can be written without the 1 by using *per* or a slash, /.

EXAMPLE 2 Examples of rates with a second term of 1.

a. \$5 to 1 hr = \$5 per 1 hr
              = \$5 per hr or \$5/hr
b. 30 mi to 1 hr = 30 mi per 1 hr
                  = 30 mi per hr or 30 mi/hr or 30 mph
c. \$7.35 to 1 lb = \$7.35 per 1 lb
                = \$7.35 per lb or \$7.35/lb ■

► You Try It Write each rate using *per*, and using a slash.

4. 24.2 miles to 1 gallon   **5.** \$32 to 1 ounce   **6.** 50 people to 1 room

A rate is reduced in the same way a ratio is reduced.

EXAMPLE 3   Write $450 for 20 radios as a rate in lowest terms.

$450 for 20 radios = $450 per 20 radios   `Reduce using 10.`
$45 per 2 radios   `Lowest terms.` ■

EXAMPLE 4   Write $72 for 8 hours work as dollars per hour.

$72 for 8 hours = $72 per 8 hours   `Reduce using 8.`
$9 per 1 hour   `Lowest terms.`
$9 per hour ■

EXAMPLE 5   Jayne traveled 240 miles on 16 gallons. Write this as a rate in fraction form.

$$240 \text{ mi on } 16 \text{ gal} = \frac{240 \text{ mi}}{16 \text{ gal}} = \frac{15 \text{ mi}}{1 \text{ gal}} = 15 \text{ mi/gal} \quad \text{or} \quad 15 \text{ mpg} \quad ■$$

▶ **You Try It**   Write the following rates in lowest terms using "per."

7. Write the rate in lowest terms using *per*: $60 to 18 pounds.

8. $120 for 4 hours   **9.** The satellite travels 24,000 miles in 90 minutes.

2 **Changing the Names in a Rate**   Though you cannot reduce the names in a rate, you can change the name of one or both terms in a rate.

EXAMPLE 6   Change feet to inches in the rate 5 feet to 6 seconds.

$$5 \text{ ft to } 6 \text{ sec} = \frac{5 \text{ ft}}{6 \text{ sec}} = \frac{60 \text{ in}}{6 \text{ sec}} = \frac{10 \text{ in}}{1 \text{ sec}} = 10 \text{ in per sec}$$

Since 1 ft = 12 in,
5 ft = 5 × 12 in = 60 in. ■

EXAMPLE 7   Liz has 6 hours to pack 20 boxes. Write the rate of hours per boxes, and change hours to minutes.

6 hrs per 20 boxes   `Reduce using 2.`
3 hrs per 10 boxes   `Change hours to minutes. Since 1 hr = 60 min, 3 hr = 3 × 60 min = 180 min.`

180 min per 10 boxes   `Reduce using 10.`
18 min per 1 box
18 min /box

If Liz has 6 hours to pack 20 boxes, she must pack at a rate of 18 minutes per box. ■

▶ **You Try It**   **10.** Change yards to feet, and reduce: 60 yards to 9 seconds.

11. A supermarket paid $360 for 240 pounds of butter. Write the rate of dollars to pounds, change dollars to cents, and reduce.

▶ **Answers to You Try It**   **1.** $\dfrac{25 \text{ students}}{1 \text{ course}}$ **2.** $\dfrac{\$1.19}{1 \text{ gal}}$ **3.** $\dfrac{43 \text{ sec}}{1 \text{ commercial}}$
4. 24.2 mi per gal, 24.2 mi/gal   **5.** $32 per oz, $32/oz   **6.** 50 people per room, 50 people/room   **7.** $10 per 3 lb   **8.** $30 per hr, $30/hr   **9.** 800 mi per 3 min   **10.** 20 ft per 1 sec, 20 ft/sec   **11.** 150¢ per 1 lb, 150¢/lb

SECTION 6.2 EXERCISES

1 *Reduce each rate to lowest terms.*

1. 40 engines to 12 cars

2. 15 cars to 6 trucks

3. $200 to 15 hours

4. $280 to 40 hours

5. 400 miles to 24 gallons

6. 252 miles to 14 gallons

7. 550 miles in 11 hours

8. 3,000 miles to 160 hours

9. 640 customers to 280 toys

10. $45,000 to 800 hours

11. 4,500 books to 1,050 readers

12. 5,100 pounds of rations to 300 soldiers

13. 9,000 miles to 450 gallons

14. 250 million people to 3,500,000 square miles

15. In 8 hours a machine can bind 1,880 books. Write the rate of books bound per hour.

16. In 21 seasons Babe Ruth hit 714 home runs. Write the rate of home runs per season.

17. The Fahrenheit scale has 180 degrees. The Celsius scale has 100 degrees. Write the ratio of Fahrenheit degrees to Celsius degrees.

18. 500 liters of a solution contains 20 liters of acid. Write the ratio of liters of acid to liters of solution.

2 *Change the name in the rate as indicated. Then reduce.*

19. 30 miles to 1 hour: change hours to minutes

20. 1 foot to 2 minutes: change feet to inches

21. $90 to 150 pounds: change dollars to cents

22. $320 to 7 pounds: change pounds to ounces

23. 8 hours to 24 boxes: change hours to minutes

24. 6 spark plugs to 4 minutes: change minutes to seconds

25. $5 to 12 ounces: change dollars to cents

26. $80 to 5 dozen: change dozen to units

27. 40 dozen eggs cost a food store $24.
  a. Write the rate of dollars to dozens of eggs.

  b. Change dollars to cents and reduce.

  c. Rewrite the rate in part b using 1 dozen = 12 eggs.

28. 12 boxes of cigars cost $200.
  a. Write the rate of boxes to dollars.

  b. Rewrite this rate as cigars to dollars if there are 24 cigars in each box.

  c. Rewrite the rate in part b as cigars to cents.

29. A $17,500 car weighs 3,500 pounds.
  a. Write the rate of dollars to pounds.

  b. Rewrite this rate as cents to pounds.

  c. Rewrite the rate in part b as cents to ounces.

30. Mike lost 20 pounds in 48 days.
  a. Write the rate of pounds to days.

  b. Rewrite this rate as ounces to days.

  c. Rewrite the rate in part b as ounces to hours.

31. A man paid $720 in taxes on a 2-acre tract of land.
  a. Write the rate of tax dollars to acres.

  b. 1 acre = 43,560 square feet. Rewrite the rate of tax dollars to square feet.

  c. Rewrite the rate in part b by changing dollars to cents.

32. Greta can run 10 miles in 1 hour and 12 minutes.
  a. Write the rate of miles run to minutes.

  b. Rewrite the rate by changing miles to feet.

  c. Rewrite the rate in part b by changing minutes to seconds.

SKILLSFOCUS (Section 2.3) *Find the average.*

33. 14, 10, 20, 18, 8, 24, 11

34. 2.3, 4, 1.06, 2.09, 0.612

EXTEND YOUR THINKING ▶▶▶▶
▶TROUBLESHOOT IT

Find and correct the error.

35. 15 miles to 3 hours $= \dfrac{15 \text{ mi}}{3 \text{ hr}} = \dfrac{15 \text{ mi}}{180 \text{ min}} = \dfrac{1 \text{ mi}}{12 \text{ min}} = 12$ min per 1 mi

WRITING TO LEARN ▶▶▶▶

36. Explain the difference between a ratio and a rate.

37. Write a word problem whose answer in lowest terms is $10.50/hr. The problem asks you to make and reduce a rate.

6.3 REWRITING RATIOS INVOLVING FRACTIONS AND DECIMALS AS RATIOS OF WHOLE NUMBERS

OBJECTIVES

1 Write a ratio as a division.

2 Write a ratio whose terms are fractions as a ratio of whole numbers (Sections 3.4, 3.5).

3 Write a ratio of decimals as a ratio of whole numbers (Sections 1.8, 5.5).

---

1 Writing a Ratio as a Division

Any ratio can be written in fraction form. A fraction is a division. Therefore, a ratio is a division. For example,

$$8 \text{ to } 5 = \frac{8}{5} = 8 \div 5.$$

> A ratio is a division.
>
> $$a \text{ to } b = a \div b \qquad \frac{a}{b} = a \div b$$
>
> $$a{:}b = a \div b \qquad a \text{ per } b = a \div b$$

EXAMPLE 1 Write each ratio as a division.

a. $50 \text{ to } 15 = 50 \div 15$
b. $6{:}11 = 6 \div 11$
c. $3/10 = 3 \div 10$
d. $\$20 \text{ per } 8 \text{ hours} = \$20 \div 8 \text{ hours}$ ■

▶ **You Try It** Write each ratio as a division.

1. $40 \text{ to } 7$ **2.** $25{:}10$ **3.** $3/16$ **4.** $40 \text{ miles per } 3 \text{ gallons}$

---

2 Writing a Ratio Whose Terms Are Fractions as a Ratio of Whole Numbers

Write the ratio $\frac{1}{4}$ to $\frac{3}{8}$ as a ratio of two whole numbers.

Write the ratio as a division.

$$\frac{1}{4} \text{ to } \frac{3}{8} = \frac{1}{4} \div \frac{3}{8} = \frac{1}{\overset{}{\underset{1}{4}}} \times \frac{\overset{2}{\cancel{8}}}{3} = \frac{2}{3} = 2 \text{ to } 3$$

The ratio $\frac{1}{4}$ to $\frac{3}{8}$ is equivalent to 2 to 3. A ratio is easier to understand when written in terms of whole numbers. For instance, a profit to cost ratio of $\$\frac{1}{4}$ to $\$\frac{3}{8}$ means there are $2 profit to every $3 cost.

> If one or both terms in a ratio are fractions, rewrite the ratio in terms of whole numbers by dividing the first term by the second.

EXAMPLE 2   Write 2 to $\frac{1}{6}$ as a ratio of whole numbers.

$$2 \text{ to } \frac{1}{6} = \frac{2}{1} \div \frac{1}{6} = \frac{2}{1} \times \frac{6}{1} = \frac{12}{1} = 12 \text{ to } 1$$

The ratio 2 to $\frac{1}{6}$ is a 12 to 1 ratio. ■

▶ **You Try It**   **5.** Write the ratio 8 to $\frac{3}{4}$ as a ratio of whole numbers.

EXAMPLE 3   Write $2\frac{1}{4}$ to $3\frac{3}{5}$ as a ratio of whole numbers.

$$2\frac{1}{4} \text{ to } 3\frac{3}{5} = 2\frac{1}{4} \div 3\frac{3}{5} = \frac{9}{4} \div \frac{18}{5} = \frac{\overset{1}{\cancel{9}}}{4} \times \frac{5}{\underset{2}{\cancel{18}}} = \frac{5}{8} = 5 \text{ to } 8 \quad ■$$

▶ **You Try It**   **6.** Write $3\frac{5}{8}$ to $4\frac{5}{6}$ as a ratio of whole numbers.

EXAMPLE 4   The old computer can process a payroll in $2\frac{3}{4}$ hours. The new computer can process the same payroll in $\frac{1}{4}$ hour. Compare the speed with which the two computers process a payroll.

The ratio of old computer processing time to new computer processing time is $2\frac{3}{4}$ hours to $\frac{1}{4}$ hour.

$$\overset{old}{2\frac{3}{4} \text{ hr}} \text{ to } \overset{new}{\frac{1}{4} \text{ hr}} = \frac{11}{4} \div \frac{1}{4} = \frac{11}{\underset{1}{\cancel{4}}} \times \frac{\overset{1}{\cancel{4}}}{1} = \frac{11}{1} = 11 \text{ to } 1$$

The ratio 11 to 1 means the work the old computer could do in 11 minutes can be done by the new computer in 1 minute. The old computer took 11 times longer to process the same payroll. ■

▶ **You Try It**   **7.** Before training, Chris ran a sprint in $16\frac{1}{2}$ seconds. After training, she ran it in 11 seconds. What is the ratio of her pretraining to posttraining times?

3 Writing a Ratio Involving Decimals as a Ratio of Whole Numbers

Write 9.6 to 0.48 as a ratio of whole numbers.

$$9.6 \text{ to } 0.48 = \frac{9.6}{0.48} = \frac{9.6 \times 100}{0.48 \times 100} = \frac{960}{48} = \frac{\overset{20}{\cancel{960}}}{\underset{1}{\cancel{48}}} = \frac{20}{1} = 20 \text{ to } 1$$

Multiply numerator and denominator by the power of 10 that changes both decimals into whole numbers.

The ratio 9.6 to 0.48 is a 20 to 1 ratio.

> **To write a ratio of decimal numbers as a ratio of two whole numbers**
>
> 1. Multiply both terms of the ratio by the power of 10 having as many zeros as the decimal with the greater number of decimal places.
> 2. Reduce.

EXAMPLE 5 Write 2.4 to 0.036 as a ratio of whole numbers.

$$2.4 \text{ to } 0.036 = \frac{2.4}{0.036} = \frac{2.4 \times 1{,}000}{0.036 \times 1{,}000} = \frac{\overset{200}{\cancel{2{,}400}}}{\underset{3}{\cancel{36}}} = 200 \text{ to } 3 \quad \blacksquare$$

▶ **You Try It**
8. Write 5.4 to 0.042 as a ratio of whole numbers.
9. Write 18 to 3.6 as a ratio of whole numbers.

▶ **Answers to You Try It** 1. 40 ÷ 7   2. 25 ÷ 10   3. 3 ÷ 16
4. 40 mi ÷ 3 gal   5. 32 to 3   6. 3 to 4   7. 3 to 2   8. 900 to 7   9. 5 to 1

1 **2** *Rewrite each ratio as a ratio of whole numbers.*

1. $\frac{1}{2}$ to $\frac{1}{4}$

2. $\frac{9}{24}$ to $\frac{6}{8}$

3. $\frac{7}{5}$ to $\frac{21}{10}$

4. $\frac{9}{4}$ to $\frac{7}{6}$

5. $\frac{5}{11}$ to 6

6. $3\frac{3}{4}$ to 9

7. 5 to $7\frac{1}{5}$

8. 10 to $\frac{5}{8}$

9. $\frac{4}{7} : \frac{3}{5}$

10. $\frac{3}{10} : \frac{5}{6}$

11. $6\frac{2}{3}$ to $1\frac{1}{9}$

12. $4\frac{5}{7}$ to $1\frac{6}{13}$

13. 12 to $2\frac{2}{5}$

14. 7 to $10\frac{1}{2}$

15. $\frac{5}{12}$ to $\frac{1}{12}$

16. $\frac{2}{6}$ to $\frac{5}{6}$

17. $\frac{1}{9}$ to $\frac{1}{10}$

18. $\frac{1}{20}$ to $\frac{1}{200}$

19. $2\frac{1}{4}$ to $1\frac{1}{8}$

20. $4\frac{1}{3}$ to $8\frac{2}{3}$

21. $4\frac{1}{2}$ to 1

22. 1 to $6\frac{3}{4}$

23. $\frac{1}{8}$ to $\frac{1}{8}$

24. $7\frac{4}{9}$ to $7\frac{4}{9}$

---

25. A share of XYZ stock gained $\$\frac{5}{8}$. A share of ABC gained $\$1\frac{1}{4}$. Find the whole number ratio of ABC gain to XYZ gain.

26. In an engine, wheel V turns 14 times when wheel W turns $2\frac{1}{3}$ times. Find the whole number ratio of turns in wheel V to turns in wheel W.

27. Tom walks $5\frac{3}{4}$ miles in 2 hours. Write the rate of miles per hour in terms of whole numbers.

28. Write the rate of $\$3$ to $3\frac{3}{10}$ gallons of gas in terms of whole numbers.

29. Write 3 yards to $4\frac{1}{2}$ inches as a whole number ratio.

30. Write $2\frac{1}{4}$ minutes to $4\frac{1}{2}$ seconds as a whole number ratio.

3 *Rewrite each ratio as a ratio of whole numbers.*

31. 0.4 to 0.06

32. 0.08 to 0.36

33. 0.006 to 18

34. 0.1 to 1

35. 5 to 0.09

36. 20 to 0.004

37. 0.09 to 0.03

38. 0.4 to 0.6

39. 0.1 to 0.09

40. 6.05 to 2.5

41. 7.7 to 1.1

42. 0.8 to 0.008

43. 4.8 to 0.048

44. 5.5 to 0.44

45. 0.001 to 0.01

46. 0.0014 to 0.00007

47. 0.00009 to 0.027

48. 0.064 to 0.00032

49. 3.6 to 5

50. 12.5 to 20

51. 1 to 0.06

52. 4 to 8.4

---

53. The fuel for a gas saw consists of 1.5 parts oil to 31.5 parts gas. Write the whole number ratio of oil to gas.

55. Write the ratio of 7.25 gallons to $8\frac{3}{4}$ as a ratio of two whole numbers.

54. A widget takes 22.5 hours to construct and 4.5 hours to finish. Find the whole number ratio of construction time to finish time.

56. Write the ratio of $2\frac{3}{4}$ hours to 5.5 miles as a ratio of two whole numbers.

SKILLSFOCUS (Section 2.6) *Solve the equation.*

57. $4.5x = 27$

58. $7y = 30$

59. $\dfrac{t}{8} = 1.25$

60. $\dfrac{k}{5.48} = 2.35$

EXTEND YOUR THINKING ▶▶▶▶
▶TROUBLESHOOT IT

Find and correct the error.

61. $3\dfrac{1}{2}$ to $7 = 3\dfrac{1}{2} \div 7 = \dfrac{7}{2} \div \dfrac{7}{1} = \dfrac{2}{\overset{}{\underset{1}{7}}} \times \dfrac{\overset{1}{7}}{1} = \dfrac{2 \times 1}{1 \times 1} = 2$ to 1

62. 0.06 to $1.8 = \dfrac{0.06 \times 100}{1.8 \times 100} = \dfrac{\overset{1}{\overset{1}{6}}}{\underset{30}{180}} = 30$ to 1

WRITING TO LEARN ▶▶▶▶

63. Explain the difference between the ratios 2 to 1 and 1 to 2.

64. Explain each step you use to write 12.5 to 0.375 as a ratio of whole numbers.

6.4 UNIT RATIOS, UNIT PRICE, AND THE ONE-NUMBER RATIO

OBJECTIVES

1. Change a ratio or rate to a unit ratio or unit rate (Sections 3.5, 4.2, 5.3, 5.8).
2. Calculate unit price.
3. Define the one-number ratio.

NEW VOCABULARY

unit ratio

unit rate

unit price

one-number ratio

---

1 Unit Ratio and Unit Rate

A **unit ratio** or **unit rate** is a ratio or rate with a second term of 1.

EXAMPLE 1 Examples of unit ratios.

a. 20 to 1   **b.** 7:1   **c.** 8/1   **d.** 100 feet to 1 foot   ■

EXAMPLE 2 Examples of unit rates.

a. 5 men to 1 car
c. $2 per 1 pound

b. 30 mpg = 30 miles per 1 gallon
d. 50 mph = 50 miles/1 hour   ■

▶ **You Try It** Label each as a unit ratio, unit rate, or neither.

1. 5 feet to 1 second   **2.** $9:$4   **3.** 8 inches/1 inch

4. 40 miles:3 gallons   **5.** $12 to 1 hour   **6.** $\dfrac{6}{1}$

Write the ratio 5 to 2 as a unit ratio.

Write the ratio in fraction form.   Divide both terms of the fraction by the denominator 2.

$$5 \text{ to } 2 = \frac{5}{2} = \frac{5 \div 2}{2 \div 2} = \frac{2.5}{1} = 2.5 \text{ to } 1$$

> To write any ratio or rate as a unit ratio or unit rate, divide both terms by the second term.

EXAMPLE 3 Write 7 feet to 4 feet as a unit ratio.

Divide each term by the second term, 4.

$$7 \text{ feet to } 4 \text{ feet} = \frac{7 \text{ feet}}{4 \text{ feet}} = \frac{7 \div 4}{4 \div 4} = \frac{1.75}{1} = 1.75 \text{ to } 1 \quad ■$$

EXAMPLE 4 Pat lost 15 pounds in 12 days. Write the unit rate.

$$15 \text{ pounds to } 12 \text{ days} = \frac{15 \text{ lb} \div 12}{12 \text{ days} \div 12} = \frac{1.25 \text{ lb}}{1 \text{ day}} = 1.25 \text{ lb/day}$$

Pat lost weight at a rate of 1.25 pounds per day.   ■

▶ **You Try It** **7.** Write 8 hours to 5 hours as a unit ratio.
8. Ricardo drove 800 miles in 2 hours. Write the unit rate.

EXAMPLE 5   A small town has 871 families with 1,643 children. Find children per family (round to tenths).

$$1{,}643 \text{ children} \quad \text{per} \quad 871 \text{ families}$$

$$\frac{1{,}643 \text{ children}}{871} \quad \text{per} \quad \frac{871 \text{ families}}{871}$$

$$1.9 \text{ children} \quad \text{per} \quad \text{family} \quad \blacksquare$$

▶ **You Try It**   **9.** A $60,000 prize was split among 7 winners. Find dollars per winner. Round to dollars.

2 Unit Pricing   **Unit price** is the price for one unit of an item. Unit price is a unit rate.

EXAMPLE 6   Examples of unit prices.

a. 50 cents per foot $= \dfrac{50 \text{ cents}}{1 \text{ foot}}$ **b.** $3.49 per pound $= \dfrac{\$3.49}{1 \text{ lb}}$

c. $1.19 per gallon $= \dfrac{\$1.19}{1 \text{ gal}}$ ∎

▶ **You Try It**   Write each unit price in fraction form.

10. $3.89 per yard   **11.** $17 per hour   **12.** 5¢ per pretzel

> To find unit price, divide the price paid by the quantity purchased.
> $$\text{unit price} = \frac{\text{price}}{\text{quantity}}$$

EXAMPLE 7   Peaches cost $7.25 for 5 pounds. Find the unit price.

$$\text{unit price} = \frac{\text{price}}{\text{quantity}} = \frac{\$7.25}{5 \text{ lb}} = \$1.45 \text{ per lb} \quad \blacksquare$$

▶ **You Try It**   **13.** Alicia paid $42.80 for 8 pounds of seed. Find the unit price.

EXAMPLE 8   A 28-ounce jar of honey costs $2.29. Find the unit price rounded to the nearest tenth of a cent.

$$\text{unit price} = \frac{\text{price}}{\text{quantity}} = \frac{\$2.29}{28 \text{ oz}} = \frac{229¢}{28 \text{ oz}} \doteq 8.2¢/\text{oz} \quad \blacksquare$$

EXAMPLE 9   7.62 acres of land sold for $165,900. Find the unit price. Round to the nearest hundred dollars.

$$\text{unit price} = \frac{\text{price}}{\text{quantity}} = \frac{\$165{,}900}{7.62 \text{ acres}} \doteq \$21{,}800 \text{ per acre} \quad \blacksquare$$

14. An 18-ounce box of cereal costs $3.19. Find the unit price in cents per ounce. Round to the nearest tenth of a cent.

15. 450 acres of land were sold for $750,000. Find the unit price. Round to the nearest ten dollars.

3 The One-Number Ratio

Occasionally you will see a ratio expressed as one number, not two. To see how this is done, write the ratio 3 to 1 in fraction form.

$$3 \text{ to } 1 = \frac{3}{1} = 3$$

3 is a **one-number ratio**. It means 3 to 1. 3 is a shortcut way of writing 3 to 1.

CAUTION Though the ratio 3 is written as one number, it is *understood* that the second term of this ratio is 1.

> The ratio r to 1 can be written as r because r to $1 = r/1 = r$.
> A one-number ratio r means r to 1 because $r = r/1 = r$ to 1.

EXAMPLE 10 **a.** 5 to 1 can be written as the one-number ratio 5, because 5 to $1 = 5/1 = 5$.
b. 2.7 to 1 can be written as the one-number ratio 2.7, because 2.7 to $1 = 2.7/1 = 2.7$. ∎

► **You Try It** Write each ratio as a one-number ratio.   **16.** 14 to 1   **17.** 3.8 to 1

EXAMPLE 11 **a.** The one-number ratio 16 is the ratio 16 to 1, because $16 = 16/1 = 16$ to 1.
b. The one-number ratio 4.5 is the ratio 4.5 to 1, because $4.5 = 4.5/1 = 4.5$ to 1. ∎

► **You Try It** Write each one-number ratio using two terms.   **18.** 7   **19.** 2.65

The housing table at the beginning of this chapter uses one-number ratios.

EXAMPLE 12 **Chapter Problem** Read the problem at the beginning of this chapter.

The price to income ratio of 9.9 in 1900 means 9.9 to 1.

$$9.9 = 9.9/1 = 9.9 \text{ to } 1 = \$9.90 \text{ to } \$1$$

This means in 1900 there was, on the average, $9.90 in home price to every $1 in family income. Similarly, the ratio 3.1 in 1980 means there was $3.10 in home price to every $1 in family income. ∎

► **You Try It** **20.** Explain what the price to income ratio for 1930 means in the table at the beginning of this chapter.

> To change any ratio a to b to a one-number ratio, divide the first term a by the second term b.
> $$a \text{ to } b = a \div b$$

EXAMPLE 13   Convert each ratio to a one-number ratio.

     a. $7 to $2 = $7 ÷ $2 = 3.5 (meaning $3.5 to $1)

     b. 0.06 to 0.015 = 0.06 ÷ 0.015 = 4 (meaning 4 to 1)   ■

▶ **You Try It**   Change each ratio to a one-number ratio.

     21. 2.4 ounces to 0.16 ounces   **22.** 90 to 2.4

SOMETHING MORE

Loss Ratio   In the insurance industry, a loss ratio is the ratio of total losses paid out by an insurance company to total premiums collected for a given time period.

$$\text{loss ratio} = \text{losses paid to premiums collected} = \frac{\text{losses paid}}{\text{premiums collected}}$$

The loss ratio is a one-number ratio. For instance, in one month the XYZ insurance company paid losses of $6,200,000 and collected premiums of $10,472,000. Their loss ratio for that month was

$$\text{loss ratio} = \frac{\$6,200,000}{\$10,472,000} = 0.59 \quad \text{(rounded to the nearest cent).}$$

Since 0.59 means $0.59 to $1, $0.59 in losses were paid out for every $1 collected in premiums for the month.

a. In one year, $460,000,000 in losses were paid and $616,000,000 in premiums were collected. Compute the loss ratio rounded to the nearest cent.

b. In one quarter, $52,968,000 in losses were paid and $60,800,000 in premiums were collected. Compute the loss ratio rounded to the nearest cent.

c. Explain what a loss ratio of 1 means.

d. Explain what a loss ratio greater than 1 means.

e. **You Be the Judge**   Is it true that an insurance company wants to keep its loss ratio small? Explain your decision. What do you think insurance companies do to keep this ratio small?

▶ **Answers to You Try It**   **1.** unit rate   **2.** neither   **3.** unit ratio
4. neither   **5.** unit rate   **6.** unit ratio   **7.** 1.6 to 1   **8.** 400 mi/hr

9. $8,571/winner   **10.** $\dfrac{\$3.89}{1 \text{ yd}}$ **11.** $\dfrac{\$17}{1 \text{ hr}}$ **12.** $\dfrac{5¢}{1 \text{ pretzel}}$ **13.** $5.35/lb

14. 17.7¢/oz   **15.** $1,670/acre   **16.** 14   **17.** 3.8   **18.** 7 to 1   **19.** 2.65 to 1
20. $5.20 to $1, meaning $5.20 in home price to every $1 in income
21. 15   **22.** 37.5

▶ **Answers to Something More**   **a.** 0.75 to 1, or 0.75   **b.** 0.87 to 1, or 0.87
c. $1 paid out in losses for every $1 collected in premiums   **d.** More than $1 paid out in losses for every $1 collected in premiums

SECTION 6.4 EXERCISES

1 *Find the unit ratio or unit rate.*

1. 60 to 16

2. 20 to 25

3. 300 to 20

4. 72 to 24

5. 84 children to 40 families

6. $180 to 50 hours

7. 80 dozen eggs to 12 chickens

8. $420 to 12 women

9. According to a Census Bureau report, in 1980 there were 600,000 millionaires out of a population of 226,000,000 Americans. Find the ratio of population to 1 millionaire.

11. The ratio of nitrogen to oxygen is 0.075 grams to 0.0003 grams. What is the ratio of nitrogen to 1 unit of oxygen?

10. The ratio of Chinese to Americans is 1,000,000,000 to 250,000,000. Find the ratio of Chinese to 1 American.

12. The ratio of cost to profit is 0.72 to 0.012. What is the ratio of dollars of cost to 1 dollar of profit?

2 *Calculate the unit price.*

13. $48 is paid for 16 pounds of butter.

14. $440 is charged for 50 hours of bookkeeping.

15. You pay $7.28 for 5.2 gallons of gas.

16. 75 cents is paid for 5 marbles.

17. $90 was paid for 18 plastic models.

18. $70 is the cost for 28 pens.

19. The cost for $\frac{1}{4}$ pound of lunch meat was $1.27.

20. The cost for $16\frac{9}{10}$ gallons of gas is $20.28.

21. 2.945 acres of land sold for $84,000. Round to hundreds.

22. 0.28 acres of land costs $34,000. Round to thousands.

23. 20 ounces of chips cost $2.19. Round to tenth of a cent.

24. 6.37 pounds of chicken cost $10.13. Round to nearest cent.

25. A dozen eggs cost 89 cents. Find the price per egg to the nearest tenth of a cent.

27. 2 pounds 8 ounces of bacon cost $4.79. What is the price per ounce to the nearest tenth of a cent (1 lb = 16 oz)?

26. A pack of 20 cigarettes costs $1.75. What is the price per cigarette to the nearest tenth of a cent?

28. A board 5 feet 6 inches long costs $7.31. What is the price per 1 foot of board to the nearest cent $\left(6 \text{ in} = \frac{1}{2} \text{ ft}\right)$?

3 *Write each ratio as a one-number ratio.*

29. 7 to 1

30. 400 to 1

31. 2.6 to 1

32. 7.8 to 1

33. 10 to 2

34. 20 to 5

35. 7 to 1.4

36. 64 to 3.2

37. 27 to 8 (round to tenths)

38. 16 to 9 (round to hundredths)

39. $4\frac{1}{2} : 3$

40. $\frac{3}{5}$ to 12

SKILLSFOCUS (Section 3.4) *Simplify by reducing.*

41. $\dfrac{25 \times 10}{40}$

42. $\dfrac{16 \times 9}{72}$

43. $\dfrac{7 \times 5}{15 \times 2}$

44. $\dfrac{18 \times 7.5}{2.5 \times 3}$

EXTEND YOUR THINKING ▶▶▶▶

45. In one year an insurance company paid out $63,600,000 in losses, and collected $108,900,000 in premiums. Calculate the loss ratio. Round to the nearest cent. Explain what this ratio means.

46. In one quarter, $450,000 in losses were paid out and $400,000 in premiums were collected. Calculate the loss ratio. Round to the nearest cent. Explain what this ratio means.

▶TROUBLESHOOT IT

Find and correct the error.

47. Write 12 feet to 15 feet as a unit ratio.

$$12 \text{ ft to } 15 \text{ ft} = \frac{12 \text{ ft} \div 12}{15 \text{ ft} \div 12} = \frac{1}{1.25} = 1 \text{ to } 1.25$$

WRITING TO LEARN ▶▶▶▶

48. Jockey Bill Shoemaker ran 40,350 races and won 8,833 in his 40-year career.

   a. Write the ratio of races to wins as a unit ratio rounded to tenths. Explain what this ratio means.

   b. Mr. Shoemaker finished second 6,136 times and finished third 4,987 times. Write the unit ratio of total top three finishes, to years. Round to units. Explain what this ratio means.

49. Explain how to calculate unit price.

50. Explain how unit ratio and unit rate are different.

51. Explain what a one-number ratio means. Is it really a ratio? Use your own examples in your explanation.

6.5 APPLICATIONS

OBJECTIVES

1 Split an amount according to a given ratio.

2 Compare ratios to solve problems.

---

1 Splitting an Amount According to a Given Ratio

Al and Bob are in business together. They made a $4,000 profit this year. They agreed to split any profit they make according to a 3:5 ratio. How much money does each man get?

If Al and Bob split their profit according to a 3:5 ratio, then Al gets 3 parts of the profit, and Bob gets 5 parts. This means the profit must be split into $3 + 5 = 8$ equal parts.

$$\text{Al:Bob} \quad \text{parts for Al} \quad \text{parts for Bob}$$
$$3 : 5 \longrightarrow \quad 3 \quad + \quad 5 \quad = 8 \text{ equal parts}$$

Divide the $4,000 profit by 8 to get one part. Then give 3 of these parts to Al and 5 to Bob.

$$\frac{\$500}{8 \,)\, \$4,000} = \text{one part} \longrightarrow \begin{array}{l} \text{Al gets } 3 \times \$500 = \$1,500 \\ \text{Bob gets } 5 \times \$500 = \underline{\$2,500} \\ \text{Add to check} \longrightarrow \$4,000 \end{array}$$

To check your answer, add Al's share of the profit to Bob's share. You get the original profit of $4,000.

To split an amount according to a given ratio

1. Add the terms of the ratio together.
2. Divide the amount to be split by this sum. This gives one part.
3. Multiply one part by each term in the ratio. The answers represent the original amount split according to the given ratio.

EXAMPLE 1

300 women are split into two teams A and B according to an 11:4 ratio. How many women are on each team?

$$\begin{array}{l} \text{A : B} \\ 11 : 4 \longrightarrow 11 + 4 = 15 \longrightarrow 15 \,)\, 300 \end{array} \qquad \frac{20 \text{ women}}{} = \text{one part}$$

Multiply 20 women by each term in the ratio.

$$\begin{array}{l} \text{Team A} = 11 \times 20 = 220 \text{ women} \\ \text{Team B} = 4 \times 20 = \underline{80 \text{ women}} \\ \text{Add to check} \longrightarrow 300 \text{ women} \quad \blacksquare \end{array}$$

▶ **You Try It** **1.** A dog and pony show consisted of 80 dogs and ponies in a 13:3 ratio. How many of each animal participated in the show?

EXAMPLE 2 A mother is preparing medicine for her child. Instructions say mix the medicine with apple juice in a 1:6 ratio. How much of each ingredient does she use to prepare 28 ounces of the mixture?

$$\text{medicine} : \text{apple juice}$$
$$1 : 6 \longrightarrow 1 + 6 = 7 \longrightarrow 7\overline{)28}\quad 4\text{ oz} = \text{one part}$$

$$1 \times 4\text{ oz} = 4\text{ oz of medicine}$$
$$6 \times 4\text{ oz} = 24\text{ oz of juice}$$
$$\text{Add to check} \longrightarrow \overline{28\text{ oz of mixture}} \quad \blacksquare$$

▶ **You Try It**   **2.** A college bookstore ordered 1,200 ballpoint pens. The pens had black ink or red ink in a 9 to 7 ratio. How many of each color pen was ordered?

Occasionally you will see a relationship between three or more numbers expressed using ratio notation. An example is 5:3:2. Though some texts call this a ratio, it is not a standard ratio because it involves more than two numbers. However, it can be used to solve problems like the ones in this section.

EXAMPLE 3 The total purse for a woman's golf tournament is $120,000. It will be split among the top three finishers according to 5:3:2. How much will the first, second, and third place finishers win?

$$\text{first} : \text{second} : \text{third}$$
$$5 : 3 : 2 \longrightarrow 5 + 3 + 2 = 10\text{ parts}$$

$$\$12,000 = \text{one part}$$
$$10\overline{)\$120,000}$$

$$\text{first place wins}\quad 5 \times \$12,000 = \$60,000$$
$$\text{second place wins}\quad 3 \times \$12,000 = \$36,000$$
$$\text{third place wins}\quad 2 \times \$12,000 = \$24,000$$
$$\text{Add to check} \longrightarrow \overline{\$120,000\text{ total purse}} \quad \blacksquare$$

▶ **You Try It**   **3.** Ann, Marie, and Betty split a $4,200 profit according to 8:5:1. How much profit did each woman get?

---

2 Comparing Ratios to Solve Problems

Unit price gives the consumer a tool for comparing prices.

EXAMPLE 4 **Regular vs Jumbo** Donna bought an 18-ounce jar of jam for $1.59. Her friend Chris spent $2.79 on the 30-ounce jumbo size. Who made the better buy?

The person who paid less per ounce made the better buy. To decide who this is, compute the unit prices and compare them.

$$\text{Donna's unit price} = \frac{\text{price}}{\text{quantity}} = \frac{\$1.59}{18 \text{ oz}} = \frac{159\cancel{c}}{18 \text{ oz}} \doteq 8.8\cancel{c} \text{ per oz}$$

$$\text{Chris's unit price} = \frac{\text{price}}{\text{quantity}} = \frac{\$2.79}{30 \text{ oz}} = \frac{279\cancel{c}}{30 \text{ oz}} \doteq 9.3\cancel{c} \text{ per oz}$$

Donna paid 8.8¢ per ounce. Chris paid 9.3¢ per ounce. Donna made the better buy. She saved $9.3¢ - 8.8¢ = 0.5¢$ or $\frac{1}{2}$ cent per ounce by purchasing the smaller 18-ounce jar. ■

▶ **You Try It** **4.** Benjamin bought 144 baseball cards for $29.50. Sebastian bought 240 for $53.75. Assuming each card has the same value, who made the better buy? Round to cents.

To compare two ratios or rates

1. The first term of each ratio or rate must have the same name. The second terms must also have the same name.
2. Rewrite each ratio or rate as a unit ratio or unit rate. Do this by dividing both terms of each ratio or rate by the second term.
3. The larger ratio or rate will have the larger first term. The smaller will have the smaller first term.

EXAMPLE 5   **Regular vs Family Pack**   A 14.2-ounce bag of chips costs $1.89. An eight-pack of chips costs $1.29 and consists of 8 bags of chips, each bag weighing $\frac{3}{4}$ ounce. Compare the prices.

$$\text{unit price for 14.2-ounce bag} = \frac{\text{price}}{\text{quantity}} = \frac{\$1.89}{14.2 \text{ oz}} = \frac{189\cancel{c}}{14.2 \text{ oz}} \doteq 13.3\cancel{c} \text{ per oz}$$

The eight-pack consists of 8 bags of chips. Each bag weighs $\frac{3}{4}$ ounce.

$$\text{quantity in eight-pack} = 8 \times \frac{3}{4} \text{ oz} = \frac{\overset{2}{\cancel{8}}}{1} \times \frac{3}{\cancel{4}} \text{ oz} = 6 \text{ oz of chips in the eight-pack}$$

$$\text{unit price for eight-pack} = \frac{\text{price}}{\text{quantity}} = \frac{\$1.29}{6 \text{ oz}} = \frac{129\cancel{c}}{6 \text{ oz}} = 21.5\cancel{c} \text{ per oz}$$

You are paying $21.5¢ - 13.3¢ = 8.2¢$ more per ounce for the eight-pack for the convenience and for the extra packaging. ■

▶ **You Try It** **5.** A 22.8-ounce box of pretzels costs $2.79. A ten-pack of pretzels costs $1.87 and consists of 10 bags, each weighing $1\frac{4}{5}$ ounces. Compare the prices. Round to the nearest tenth of a cent.

EXAMPLE 6 **Tune-ups and Miles per Gallon** Ella drove 361 miles on 15.7 gallons of gas. After a tune-up, she drove 312 miles on 11.2 gallons of gas under the same driving conditions. Use ratios to decide if the tune-up gave Ella better gas mileage.

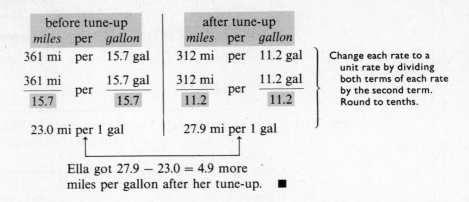

Ella got $27.9 - 23.0 = 4.9$ more miles per gallon after her tune-up. ■

▶ **You Try It** 6. Vince drove 186 miles in the city and used 9.6 gallons of gas. He drove 397 miles on the highway and used 17.8 gallons. Use ratios to compare Vince's city mileage to his highway mileage. Round to tenths.

▶ **Answers to You Try It** 1. 65 dogs and 15 ponies 2. 675 have black ink, 525 have red ink 3. Ann gets $2,400, Marie gets $1,500, and Betty gets $300 4. Benjamin paid 20¢ per card, Sebastian paid 22¢ per card; Sebastian paid 2¢ more per card 5. The unit price for the box is 12.2¢/oz; for the ten-pack it is 10.4¢/oz; you are paying 1.8¢ more per ounce with the box 6. 19.4 mpg in the city, and 22.3 mpg on the highway; Vince got 2.9 more mpg on the highway

I *Split each amount according to the given ratio.*

1. Jim and Sue split $5,000 according to a 7:3 ratio.

2. Mike and Debbie split $8,000 according to a 9:7 ratio.

3. Groups W and Z share 560 gallons according to a 13:1 ratio.

4. Buildings J and K share 320 gallons of oil according to a 1:7 ratio.

5. Meg and Millie split $7,200 according to a 5:4 ratio.

6. Zeke and Zack split 430 pounds according to a 30:13 ratio.

7. Cal and Don split 600 marbles according to $3\frac{1}{2}:2\frac{1}{2}$.

8. Fred and Ed split 50 acres according to $\frac{1}{2}:\frac{1}{3}$.

9. Lynn and Lou split $7.50 according to 3.3 to 1.7.

10. Ben and Chris split $1,100 according to a 4.9 to 0.6 ratio.

11. Alice, Betty, and Cath split $2,400 according to 3:2:1. How much does each woman get?

12. Mitch, Nick, and Otto split $4,560 according to 13:5:2. How much does each man receive?

13. Companies A, B, and C employ 3,627 people according to 4:3:2. How many people does each company employ?

14. 390 deliveries were made to cities F, G, and H according to 5:5:3. How many deliveries did each city receive?

15. Saul wants to mix 32 pints of fuel. The fuel is a mixture of gasoline and oil in the ratio 13:3. How many pints each of gasoline and oil are needed to get 32 pints of fuel?

16. The ratio of staff to inmates at County Prison is 7:6. If total staff plus inmates equals 2,327 people, how many people are in each category?

17. A team of construction workers are available for 160 hours. The team must split this time between working on a bridge and working on a building. They are being paid $51,750 to work on the bridge, and $20,250 to work on the building.

  a. The team wants to split its total time between the two jobs according to the ratio of bridge salary to building salary. What is this lowest terms ratio?

  b. Split the total time of 160 hours according to this ratio. How much time will the team spend on each job?

18. Amy and Beth have a jar of 279 jelly beans. They agree to split them according to how much time each spent doing yard work. Amy spent 270 minutes in the yard. Beth spent 216 minutes.

  a. Express Amy's time to Beth's time as a ratio in lowest terms.

2 *Solve each problem by comparing ratios.*

23. Meg planted 176 bulbs in 8 hours. Susan planted 144 bulbs in 6 hours. Who planted more bulbs per hour?

24. Zack painted 75 figurines in 15 hours. Zeke painted 48 in 12 hours. Who painted more figurines per hour?

25. Last year Gail scored 437 points in 23 basketball games. This year she scored 378 points in 18 games. In which year was her point per game average better?

  b. Split the 279 jelly beans according to this ratio. How many does each girl get?

19. Three accountants, Al, Bob, and Cy, form a tax service and agree to split all profits according to 7:5:4.

  a. What is each person's share of a $7,926.40 profit?

  b. Suppose the three are inefficient and lose $3,412. If the three agree to pay losses according to the same ratio, how much does each accountant pay?

20. Mary, Pat, and Cathy agree to spend a total of 820 hours restoring an old building. How much time will each woman spend on the building if they split their total time according to 9:7:4?

21. The first, second, and third place finishers in an auto race will split $180,000 according to 7:4:1. How much money will each place receive?

22. The first six finishers in a golf tournament will share $420,000 according to 10:4:3:2:1:1. How much money will each place receive?

26. Last year Johnny passed for 2,688 yards in 16 football games. This year he passed for 1,794 yards in 13 games. In which season did he pass for more yards per game?

27. Jayne purchased 12.6 gallons of unleaded gas for $14.52. Her friend paid $20.95 for 16.8 gallons of unleaded.

  a. Determine who paid the better price per gallon. Round to the nearest cent.

  b. How much less was paid per gallon at the better price?

28. The jumbo 64-ounce box of detergent costs $4.59. The regular 40-ounce box sells for $3.19.

  a. Find the better buy. Round to the nearest tenth of a cent.

  b. How much less per ounce was paid at the better price?

29. Mike bought a 12-ounce can of tomato sauce for 79¢. Debbie bought the 28-ounce size for $1.69.

  a. Who made the better buy? Round to tenths of a cent.

  b. How much less per ounce was paid at the better price?

30. Paul bought a 32-ounce can of fruit for $1.19. Kim bought the 14-ounce size for 53¢.

  a. Find the better buy. Round to tenths of a cent.

  b. How much less per ounce was paid at the better price?

31. Six 12-ounce cans of soda sell for $1.79. The 2-liter bottle (67.6 ounces) sells for $0.99.

  a. Find the better buy by computing the unit price per ounce of soda. Round to the nearest tenth of a cent.

  b. How much less per ounce was paid at the better price?

32. The 12-pack of bar soap costs $2.79. The 5-pack costs $1.09.

  a. Find the better buy by computing the unit price per bar of soap. Round to the nearest tenth of a cent.

  b. How much less per bar was paid at the better price?

33. Recently the population of the former Soviet Union was estimated to be 290,178,000. Its land area was 8,650,000 square miles. At the same time the population of China was estimated to be 1,117,700,000. China's land area is 3,678,000 square miles.

  a. Find the number of people per square mile in each country. Round to the nearest person

  b. About how many times more populated is China per square mile than the former Soviet Union?

34. Recently the population of New York state was estimated at 18,142,000. Its land area is 49,108 square miles. At the same time the population of California was estimated to be 25,481,000. Its land area is 158,706 square miles.

  a. Find the number of people per square mile in each state. Round to the nearest person.

  b. About how many times more populated is New York per square mile than California?

SKILLSFOCUS (Section 4.5) *Subtract.*

35. $\dfrac{5}{6} - \dfrac{2}{5}$

36. $\dfrac{7}{10} - \dfrac{3}{8}$

37. $4\dfrac{1}{3} - \dfrac{8}{9}$

38. $7\dfrac{1}{12} - 4\dfrac{7}{8}$

EXTEND YOUR THINKING ▶▶▶▶

▶SOMETHING MORE

39. More on Tune-ups   Before her tune-up, Carla drove 248 miles on 13.6 gallons of gas. After her tune-up she drove 334 miles on 14.1 gallons of the same grade of gas, under the same driving conditions.

   a. Compute Carla's mileage before her tune-up and after her tune-up. Round to the tenth of a mile.

   b. How much did her gas mileage improve after the tune-up?

   c. The tune-up cost her $49.95. Assume a tune-up lasts 10,000 miles (meaning you can drive 10,000 miles before gas mileage per gallon begins to drop). If the average price for a gallon of gas is $1.259, how much money did the tune-up save her over the 10,000 miles? Round to the nearest ten dollars.

40. The Keg   A market charges $48.50 for $\frac{1}{2}$ keg of soda, which is equivalent to 216 drinks of 12 ounces each. It charges $29.50 for $\frac{1}{4}$ keg, which gives 108 drinks of 12 ounces each. 6 12-ounce cans of soda can be purchased for $1.89.

   a. Determine the best buy by comparing the unit prices of each of the three items.

   b. How many 6-packs of 12-ounce cans must be purchased to have as much soda as in $\frac{1}{2}$ keg?

   c. What would the cost be for the cans in part b?

WRITING TO LEARN ▶▶▶▶

41. Write a word problem that splits a profit among two women. The women receive $3,500 and $2,000 in profit, respectively. State the ratio you use. Then solve the problem.

OBJECTIVE

Define proportions and cross product.

NEW VOCABULARY

proportion   false proportion   cross product

true proportion   cross product property

Defining Proportion and Cross Product

A **proportion** is a statement that two ratios are equal. For example,

$$\frac{1}{3} = \frac{2}{6}$$

is a proportion. This is read "1 is to 3 as 2 is to 6." It is a statement that the ratio $\frac{1}{3}$ equals the ratio $\frac{2}{6}$. You can write this proportion in *colon form* as 1:3::2:6. 1 and 6 are called the *extremes*. 3 and 2 are called the *means*.

A proportion is either true or false. A **true proportion** has equal cross products. **Cross products** are found by multiplying in a crosswise fashion.

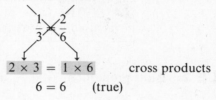

$$\boxed{2 \times 3} = \boxed{1 \times 6} \quad \text{cross products}$$
$$6 = 6 \quad \text{(true)}$$

The cross products are 2×3 and 1×6. $\frac{1}{3} = \frac{2}{6}$ is a true proportion because both cross products equal 6. You also say it is true because the product of the means (2×3) equals the product of the extremes (1×6). A **false proportion** has unequal cross products.

The Cross Product Property

A true proportion has equal cross products. This is called the **cross product property**, and is pictured as follows.

$$\frac{a}{b} = \frac{c}{d}$$
$$b \cdot c = a \cdot d$$

EXAMPLE I   Determine if the proportion is true or false.

a. $\frac{2}{5} \overset{?}{=} \frac{4}{10}$

$$\boxed{4 \times 5} \overset{?}{=} \boxed{2 \times 10}$$
$$20 = 20 \quad \text{(true)}$$

b. $\frac{6}{16} \overset{?}{=} \frac{9}{24}$

$$\boxed{9 \times 16} \overset{?}{=} \boxed{6 \times 24}$$
$$144 = 144 \quad \text{(true)}$$

c. $\frac{1}{2} \overset{?}{=} \frac{6}{8}$

$$\boxed{6 \times 2} \overset{?}{=} \boxed{1 \times 8}$$
$$12 \neq 8 \quad \text{(false)}$$

Since the cross products are unequal, write $\frac{1}{2} \neq \frac{6}{8}$. ■

▶ **You Try It** Determine if each proportion is true or false.

1. $\dfrac{7}{10} \overset{?}{=} \dfrac{3}{5}$ 2. $\dfrac{8}{20} \overset{?}{=} \dfrac{6}{15}$ 3. $\dfrac{21}{56} \overset{?}{=} \dfrac{39}{104}$ 4. $\dfrac{64}{80} \overset{?}{=} \dfrac{16}{24}$

OBSERVE A proportion is true if both of its ratios reduce to the same lowest terms answer. The proportion in Example 1a is true because $\dfrac{4}{10}$ can be reduced to $\dfrac{2}{5}$. The proportion in Example 1b is true because $\dfrac{6}{16}$ and $\dfrac{9}{24}$ both reduce to $\dfrac{3}{8}$. In Example 1c, the proportion is false because $\dfrac{6}{8}$ reduces to $\dfrac{3}{4}$, and $\dfrac{1}{2} \neq \dfrac{3}{4}$.

CAUTION Cross multiplication is only used with proportions. It is never used when multiplying fractions.

EXAMPLE 2 Is the proportion $\dfrac{4.5}{6} \overset{?}{=} \dfrac{0.9}{1.2}$ true or false?

$$\dfrac{4.5}{6} \overset{?}{=} \dfrac{0.9}{1.2}$$

$$6 \times 0.9 \overset{?}{=} 4.5 \times 1.2$$

$$5.4 = 5.4 \qquad \text{The proportion is true.} \ \blacksquare$$

EXAMPLE 3 Is the proportion $\dfrac{2\frac{1}{2}}{\frac{3}{5}} \overset{?}{=} \dfrac{5}{\frac{1}{3}}$ true or false?

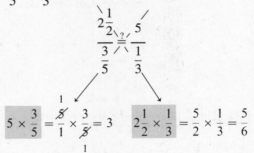

$$5 \times \dfrac{3}{5} = \dfrac{\cancel{5}}{1} \times \dfrac{3}{\cancel{5}} = 3 \qquad 2\dfrac{1}{2} \times \dfrac{1}{3} = \dfrac{5}{2} \times \dfrac{1}{3} = \dfrac{5}{6}$$

$$3 \neq \dfrac{5}{6} \qquad \text{The proportion is false.} \ \blacksquare$$

▶ **You Try It** Determine if each proportion is true or false.

5. $\dfrac{0.8}{3.5} \overset{?}{=} \dfrac{4.2}{9.25}$ 6. $\dfrac{4\frac{4}{5}}{6} \overset{?}{=} \dfrac{1\frac{1}{2}}{1\frac{7}{8}}$

▶ **Answers to You Try It** 1. $35 \neq 30$, false   2. $120 = 120$, true
3. $2{,}184 = 2{,}184$, true   4. $1{,}536 \neq 1{,}280$, false   5. $7.4 \neq 14.7$, false
6. $9 = 9$, true

SECTION 6.6 EXERCISES

Cross multiply to determine if each proportion is true or false.

1. $\frac{1}{2} \stackrel{?}{=} \frac{3}{6}$

2. $\frac{3}{4} \stackrel{?}{=} \frac{9}{12}$

3. $\frac{2}{5} \stackrel{?}{=} \frac{8}{20}$

4. $\frac{4}{8} \stackrel{?}{=} \frac{2}{6}$

5. $\frac{5}{7} \stackrel{?}{=} \frac{2}{3}$

6. $\frac{1}{5} \stackrel{?}{=} \frac{1}{6}$

7. $\frac{7}{8} \stackrel{?}{=} \frac{21}{24}$

8. $\frac{12}{27} \stackrel{?}{=} \frac{20}{45}$

9. $\frac{7}{10} \stackrel{?}{=} \frac{11}{15}$

10. $\frac{3}{5} \stackrel{?}{=} \frac{5}{8}$

11. $\frac{14}{35} \stackrel{?}{=} \frac{6}{15}$

12. $\frac{10}{18} \stackrel{?}{=} \frac{30}{54}$

13. $\frac{60}{72} \stackrel{?}{=} \frac{45}{54}$

14. $\frac{200}{650} \stackrel{?}{=} \frac{80}{260}$

15. $\frac{24}{77} \stackrel{?}{=} \frac{16}{44}$

16. $\frac{32}{45} \stackrel{?}{=} \frac{132}{145}$

17. $\frac{8}{13} \stackrel{?}{=} \frac{72}{117}$

18. $\frac{28}{84} \stackrel{?}{=} \frac{33}{99}$

19. $\frac{7}{5} \stackrel{?}{=} \frac{10}{7}$

20. $\frac{12}{17} \stackrel{?}{=} \frac{15}{19}$

🖩 21. $\frac{3.6}{6} \stackrel{?}{=} \frac{6}{10}$

🖩 22. $\frac{2}{0.4} \stackrel{?}{=} \frac{0.5}{0.1}$

🖩 23. $\frac{100}{2.4} \stackrel{?}{=} \frac{25}{0.8}$

🖩 24. $\frac{9.6}{8} \stackrel{?}{=} \frac{3.6}{2.8}$

25. $\frac{\frac{1}{2}}{0.2} \stackrel{?}{=} \frac{8}{5}$

26. $\frac{\frac{1}{4}}{0.6} \stackrel{?}{=} \frac{1}{2}$

27. $\frac{\frac{5}{8}}{\frac{2}{5}} \stackrel{?}{=} \frac{4}{\frac{8}{9}}$

28. $\frac{\frac{9}{4}}{\frac{3}{4}} \stackrel{?}{=} \frac{6}{\frac{2}{3}}$

29. $\frac{\frac{3}{5}}{2} \stackrel{?}{=} \frac{6}{20}$

30. $\frac{\frac{16}{4}}{\frac{9}} \stackrel{?}{=} \frac{9}{\frac{1}{4}}$

31. $\frac{100}{\frac{1}{2}} \stackrel{?}{=} \frac{50}{\frac{1}{4}}$

32. $\frac{\frac{1}{8}}{\frac{1}{4}} \stackrel{?}{=} \frac{\frac{1}{6}}{\frac{1}{3}}$

SKILLSFOCUS (Section 5.5) *Multiply.*

33. 2.4×0.68

34. 4.5×8.2

35. 0.7×1.9

36. 20×4.3

EXTEND YOUR THINKING ▶▶▶▶
▶SOMETHING MORE

37. A blueprint is drawn according to the scale

    $\frac{1}{2}$ inch: 6 feet.

    A wall $3\frac{5}{8}$ inches long on the blueprint is
    actually 38 feet long. Was the wall constructed
    according to blueprint specifications? (Hint:
    Make two ratios, and see if they form a true
    proportion.)

38. A blueprint is drawn according to the scale

    1 inch: 8 feet.

    A floor $7\frac{1}{4}$ inches long on the blueprint is
    actually 58 feet long. Was the floor constructed
    according to blueprint specifications?

Find and correct the error.

39. $\dfrac{5}{8} \times \dfrac{3}{10} = \dfrac{5}{8} \times \dfrac{3}{10} = \dfrac{24}{50} = \dfrac{12}{25}$

40. $\dfrac{12}{21} \overset{?}{=} \dfrac{7}{4} \rightarrow 12 \times 7 \overset{?}{=} 21 \times 4 \rightarrow 84 = 84$. Therefore, $\dfrac{12}{21} = \dfrac{7}{4}$ is true.

WRITING TO LEARN ▶▶▶▶

41. What is a proportion? Create an example for your explanation.

42. Make up your own examples of a true proportion and a false proportion. Then explain why each proportion is true or false.

6.7 SOLVING PROPORTIONS

OBJECTIVE

Solve proportions using cross multiplication (Sections 2.6, 3.3).

Solving Proportions

A proportion has four numbers. If you are only given three of the numbers, you can solve for the fourth. For example, solve for x.

$$\frac{2}{3} = \frac{x}{6}$$

The letter x represents the number that will make this a true proportion. Even though x is unknown, if this is to be a true proportion the cross products must be equal.

$$\frac{2}{3} = \frac{x}{6}$$

$$3 \cdot x = 2 \cdot 6$$
$$3 \cdot x = 12$$

Divide both sides of the equation by 3, the number multiplied by x.

$$\frac{\overset{1}{\cancel{3}} \cdot x}{\underset{1}{\cancel{3}}} = \frac{12}{3}$$

$$x = 4$$

The answer is $x = 4$. To check, replace x with 4 in the original proportion. The proportion is true if the cross products are equal.

Check:
$$\frac{2}{3} \overset{?}{=} \frac{4}{6}$$

$$3 \cdot 4 \overset{?}{=} 2 \cdot 6$$
$$12 = 12 \qquad \text{(true)}$$

To solve a proportion for the missing number x

1. Cross multiply. Set the two cross products equal to each other.
2. Solve this equation for x. Do this by dividing both sides of the equation by the number multiplied by x.
3. Check your answer. In the original proportion, replace x with the number found in Step 2, and cross multiply. If the cross products are equal, your answer is correct.

EXAMPLE 1   Solve for x. $\dfrac{3}{8} = \dfrac{x}{32}$

$$\frac{3}{8} = \frac{x}{32}$$

Cross multiply. Set the two cross products equal to each other.

$$8 \cdot x = 3 \cdot 32$$

$$\frac{\overset{1}{\cancel{8}} \cdot x}{\cancel{8}} = \frac{3 \cdot \overset{4}{\cancel{32}}}{\cancel{8}}$$

Divide both cross products by 8, the number multiplied by x. Reduce.

$$x = 12 \qquad \text{The solution.}$$

Check: Replace x with 12 in the original proportion.

$$\frac{3}{8} \overset{?}{=} \frac{12}{32}$$

$$8 \cdot 12 \overset{?}{=} 3 \cdot 32$$

$$96 = 96 \qquad (\text{true}) \quad \blacksquare$$

▶ **You Try It**   **1.** Solve for y. $\dfrac{y}{12} = \dfrac{14}{21}$

EXAMPLE 2   Solve for x. $\dfrac{15}{x} = \dfrac{24}{40}$

$$\frac{15}{x} \diagup \frac{24}{40}$$

Cross multiply. Set the two cross products equal to each other.

$$24 \cdot x = 15 \cdot 40$$

$$\frac{\overset{1}{\cancel{24}} \cdot x}{\cancel{24}} = \frac{\overset{5}{\cancel{15}} \cdot \overset{5}{\cancel{40}}}{\cancel{24}}$$

Divide both sides of the equation by 24, the number multiplied by x. Reduce.

$$x = 25 \qquad \text{The solution.}$$

Check: Replace x with 25 in the original proportion

$$\frac{15}{25} \overset{?}{=} \frac{24}{40}$$

$$25 \cdot 24 \overset{?}{=} 15 \cdot 40$$

$$600 = 600 \qquad (\text{true}) \quad \blacksquare$$

▶ **You Try It**   Solve.

2. $\dfrac{18}{x} = \dfrac{63}{49}$

3. $\dfrac{27}{2} = \dfrac{45}{y}$

OBSERVE   You may reduce one of the ratios in a proportion before solving for x. In Example 2, reduce $\dfrac{24}{40}$. Then cross multiply.

$$\frac{15}{x} = \frac{\overset{3}{\cancel{24}}}{\underset{5}{\cancel{40}}} \longrightarrow \frac{15}{x} \diagup \frac{3}{5}$$

Cross multiply. Set the cross products equal to each other.

$$3 \cdot x = 15 \cdot 5$$

$$\frac{\cancel{8}^{1} \cdot x}{\cancel{8}_{1}} = \frac{\cancel{15}^{5} \cdot 5}{\cancel{8}_{1}}$$

$$x = 25 \qquad \text{Same answer as in Example 2.}$$

The check for each of the following examples is left as an exercise for the student.

EXAMPLE 3   Solve for n. $\dfrac{n}{10} = \dfrac{16}{28}$

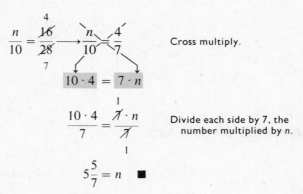

$$\frac{n}{10} = \frac{\cancel{16}^{4}}{\cancel{28}_{7}} \longrightarrow \frac{n}{10} = \frac{4}{7} \qquad \text{Cross multiply.}$$

$$10 \cdot 4 = 7 \cdot n$$

$$\frac{10 \cdot 4}{7} = \frac{\cancel{7}^{1} \cdot n}{\cancel{7}_{1}} \qquad \begin{array}{l}\text{Divide each side by 7, the}\\ \text{number multiplied by } n.\end{array}$$

$$5\frac{5}{7} = n \quad \blacksquare$$

▶ **You Try It**   **4.** Solve. $\dfrac{x}{7} = \dfrac{4}{9}$

EXAMPLE 4   Solve for y and round to hundredths. $\dfrac{17}{11} = \dfrac{9}{y}$

$$\frac{17}{11} = \frac{9}{y} \qquad \begin{array}{l}\text{Cross multiply. Set the cross products}\\ \text{equal to each other.}\end{array}$$

$$17 \cdot y = 11 \cdot 9$$

$$\frac{\cancel{17}^{1} \cdot y}{\cancel{17}_{1}} = \frac{99}{17} \qquad \text{Divide each side by 17.} \quad 17\overline{)99.000}^{\,5.823\;\doteq\;5.82}$$

$$y \doteq 5.82 \text{ rounded to the nearest hundredth} \quad \blacksquare$$

CAUTION   When you round your answer, your check will be approximately correct. Checking Example 4 gives the following.

$$\frac{17}{11} \stackrel{?}{=} \frac{9}{5.82} \longrightarrow 11 \cdot 9 \stackrel{?}{=} 17 \cdot 5.82 \longrightarrow 99 \doteq 98.94$$

▶ **You Try It**   **5.** Solve and round to tenths. $\dfrac{15}{s} = \dfrac{22}{17}$

EXAMPLE 5   Solve for x. $\dfrac{x}{6} = \dfrac{3\frac{1}{2}}{10}$

$$\dfrac{x}{6} = \dfrac{3\frac{1}{2}}{10} \qquad \text{Cross multiply.}$$

$$10 \cdot x = 6 \cdot 3\frac{1}{2} \qquad \text{Multiply: } 6 \cdot 3\frac{1}{2} = \dfrac{6}{1} \cdot \dfrac{7}{2} = 21.$$

$$10 \cdot x = 21$$

$$\dfrac{\cancel{10} \cdot x}{\cancel{10}} = \dfrac{21}{10} \qquad \text{Divide each side by 10.}$$

$$x = 2\frac{1}{10} \quad \text{or} \quad 2.1 \quad \blacksquare$$

EXAMPLE 6   Solve for t. $\dfrac{4}{t} = \dfrac{5.2}{3.51}$

$$\dfrac{4}{t} = \dfrac{5.2}{3.51} \qquad \text{Cross multiply.}$$

$$5.2 \cdot t = 4 \cdot 3.51$$

$$\dfrac{\cancel{5.2} \cdot t}{\cancel{5.2}} = \dfrac{14.04}{5.2} \qquad \text{Divide both sides by 5.2.} \quad 5.2\overline{)14.04} \longrightarrow 52\overline{)140.4}^{\,2.7}$$

$$t = 2.7 \quad \blacksquare$$

▶ **You Try It**   Solve.

6. $\dfrac{m}{5\frac{1}{4}} = \dfrac{8}{30}$

7. $\dfrac{7.5}{0.18} = \dfrac{w}{0.045}$

▶ **Answers to You Try It**   1. $y = 8$ 2. $x = 14$ 3. $y = 20$ 4. $x = 3\frac{1}{9}$

5. $s = 11.6$ 6. $m = 1\frac{2}{5}$ or 1.4   7. $w = 1.875$

SECTION 6.7 EXERCISES

Solve each proportion, and check your answer.

1. $\dfrac{4}{9} = \dfrac{x}{36}$

2. $\dfrac{y}{24} = \dfrac{7}{8}$

3. $\dfrac{20}{t} = \dfrac{4}{3}$

4. $\dfrac{7}{10} = \dfrac{42}{n}$

5. $\dfrac{8}{h} = \dfrac{12}{9}$

6. $\dfrac{x}{25} = \dfrac{40}{50}$

7. $\dfrac{27}{40} = \dfrac{y}{120}$

8. $\dfrac{55}{72} = \dfrac{165}{z}$

9. $\dfrac{4}{21} = \dfrac{s}{168}$

10. $\dfrac{x}{6} = \dfrac{63}{42}$

11. $\dfrac{14}{18} = \dfrac{49}{x}$

12. $\dfrac{y}{56} = \dfrac{10}{35}$

13. $\dfrac{2}{3} = \dfrac{x}{27}$

14. $\dfrac{30}{x} = \dfrac{12}{20}$

15. $\dfrac{75}{120} = \dfrac{5}{x}$

16. $\dfrac{x}{385} = \dfrac{200}{350}$

17. $\dfrac{x}{15} = \dfrac{16}{12}$

18. $\dfrac{35}{21} = \dfrac{x}{36}$

19. $\dfrac{88}{33} = \dfrac{56}{y}$

20. $\dfrac{60}{x} = \dfrac{50}{20}$

21. $\dfrac{x}{12} = \dfrac{10}{14}$

22. $\dfrac{55}{44} = \dfrac{z}{22}$

23. $\dfrac{17}{51} = \dfrac{100}{v}$

24. $\dfrac{120}{t} = \dfrac{74}{37}$

25. $\dfrac{60}{92} = \dfrac{x}{138}$

26. $\dfrac{64}{40} = \dfrac{96}{x}$

27. $\dfrac{45}{35} = \dfrac{x}{14}$

28. $\dfrac{3}{25} = \dfrac{60}{x}$

29. $\dfrac{x}{8} = \dfrac{132}{24}$

30. $\dfrac{4}{x} = \dfrac{28}{63}$

31. $\dfrac{9}{h} = \dfrac{60}{200}$

32. $\dfrac{500}{800} = \dfrac{x}{240}$

33. $\dfrac{40}{9} = \dfrac{x}{5}$

34. $\dfrac{18}{y} = \dfrac{4}{7}$

35. $\dfrac{\frac{1}{2}}{4} = \dfrac{x}{6}$

36. $\dfrac{x}{\frac{3}{4}} = \dfrac{10}{4}$

37. $\dfrac{x}{8} = \dfrac{3\frac{1}{2}}{7}$

38. $\dfrac{2\frac{1}{4}}{4} = \dfrac{y}{10}$

39. $\dfrac{5}{3\frac{1}{3}} = \dfrac{y}{4}$

40. $\dfrac{12\frac{1}{2}}{100} = \dfrac{5}{y}$

41. $\dfrac{2.4}{6} = \dfrac{x}{10}$

42. $\dfrac{y}{2.5} = \dfrac{4}{10}$

43. $\dfrac{t}{3.6} = \dfrac{4.5}{6}$

44. $\dfrac{x}{12} = \dfrac{4.5}{2.25}$

45. $\dfrac{1.4}{0.4} = \dfrac{0.7}{x}$ **46.** $\dfrac{0.4}{t} = \dfrac{0.29}{2.03}$ **47.** $\dfrac{0.2}{0.8} = \dfrac{s}{0.5}$ **48.** $\dfrac{p}{0.75} = \dfrac{0.7}{1}$

49. $\dfrac{2.07}{k} = \dfrac{15.9}{4.93}$ (round to thousandths) **50.** $\dfrac{7.84}{5} = \dfrac{4.2}{h}$ (round to tenths)

SKILLSFOCUS (Section 4.7) Order the fractions from smallest to largest.

51. $\dfrac{1}{4}, \dfrac{5}{16}, \dfrac{3}{8}$ **52.** $\dfrac{7}{10}, \dfrac{19}{30}, \dfrac{13}{20}$ **53.** $\dfrac{5}{6}, \dfrac{11}{12}, \dfrac{7}{9}, \dfrac{2}{3}$ **54.** $\dfrac{4}{7}, \dfrac{5}{14}, \dfrac{1}{2}, \dfrac{15}{28}$

EXTEND YOUR THINKING ▶▶▶▶
▶ TROUBLESHOOT IT

Find and correct the error.

55. $\dfrac{x}{16} = \dfrac{12}{5}$

$$\dfrac{x}{\overset{}{\underset{4}{\cancel{16}}}} = \dfrac{\overset{3}{\cancel{12}}}{5}$$

$$5 \cdot x = 3 \cdot 4$$
$$x = 2.4$$

56. $\dfrac{10}{y} = \dfrac{50}{80}$

$$\dfrac{10}{y} = \dfrac{\overset{5}{\cancel{50}}}{\underset{8}{\cancel{80}}}$$

$$8 \cdot y = 5 \cdot 10$$
$$y = 6.25$$

WRITING TO LEARN ▶▶▶▶

57. Explain in writing each step used to solve $\dfrac{6}{10} = \dfrac{x}{45}$.

58. Explain your answer to each problem. Use your own examples to demonstrate your point.

   a. (True/False) Every ratio can be expressed as a fraction.

   b. (True/False) A false proportion has unequal ratios.

   c. (True/False) A mixed number cannot be written as a ratio.

6.8 APPLICATIONS

OBJECTIVE	NEW VOCABULARY
Solve applications using proportions.	proportional

Solving Applications Proportions are used to solve word problems whose quantities are **proportional**. Two quantities are proportional if doubling one means the other will double, tripling one means the other will triple, halving one means the other will halve, and so on.

For example, gallons of gas and miles are proportional. If you get 25 miles on 1 gallon, you can expect to get 50 miles on 2 gallons driving at the same speed under the same conditions. Ingredients in a recipe are proportional. If you halve one ingredient, the rest are halved to maintain the same taste.

Not all units are proportional. A person's age and height are not proportional. If a 20-year-old woman is 5 feet tall, at double the age (40 years) she will not be double the height (10 feet).

When deciding to use proportions to solve a problem, ask the following question. If one quantity is increased or decreased, will the other increase or decrease by the same factor? If the answer is yes, use proportions to solve the problem.

EXAMPLE 1 Ron drove 120 miles on 6 gallons of gas. How many gallons will he use on a 300-mile trip?

Let x = number of gallons used on a 300-mile trip.

Ron drove 120 miles on 6 gallons of gas. How many gallons will he use on a 300-mile trip?

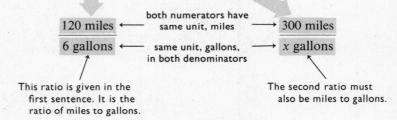

both numerators have same unit, miles

same unit, gallons, in both denominators

This ratio is given in the first sentence. It is the ratio of miles to gallons.

The second ratio must also be miles to gallons.

Set the two ratios equal to make a proportion.

$$\frac{120 \text{ miles}}{6 \text{ gallons}} = \frac{300 \text{ miles}}{x \text{ gallons}}$$

You may drop the units to solve the proportion.

$$\frac{120}{6} = \frac{300}{x}$$

$$120 \cdot x = 6 \cdot 300 \qquad \text{Cross multiply.}$$

$$\frac{\overset{1}{\cancel{120}} \cdot x}{\cancel{120}} = \frac{\overset{1}{\cancel{6}} \cdot \overset{15}{\cancel{300}}}{\cancel{120}} \qquad \text{Divide both sides by 120. Reduce.}$$

$$x = 15 \text{ gallons}$$

Ron will use 15 gallons of gas on the 300-mile trip. ■

OBSERVE Either of the following proportions could have been used to solve Example 1. Both give the same cross products as in Example 1.

$$\frac{6 \text{ gallons}}{120 \text{ miles}} = \frac{x \text{ gallons}}{300 \text{ miles}} \qquad \frac{120 \text{ miles}}{300 \text{ miles}} = \frac{6 \text{ gallons}}{x \text{ gallons}}$$

$$120 \cdot x = 6 \cdot 300 \qquad\qquad 6 \cdot 300 = 120 \cdot x$$

CAUTION The next proportion could not be used to solve Example 1, because the ratios are unlike. A ratio of gallons to miles is unequal to a ratio of miles to gallons. In addition, the cross products are not the same as in Example 1.

$$\frac{6 \text{ gallons}}{120 \text{ miles}} \neq \frac{300 \text{ miles}}{x \text{ gallons}}$$

$$120 \cdot 300 \neq 6 \cdot x$$

▶ **You Try It**   1. Donald paid $98 for 7 shirts. What will he pay for 18 shirts?

To solve a word problem using proportions

1. Let x represent the unknown amount.
2. Use the information in the problem to make two ratios.
3. The first ratio is often given in the statement of the word problem. Write it in fraction form. Write the unit name next to each number in the ratio.
4. Write the second ratio so both numerators have the same name, and both denominators have the same name.

$$\frac{\text{miles}}{\text{gallons}} \xleftarrow[\text{denominators have same name}]{\text{numerators have same name}} \frac{\text{miles}}{\text{gallons}}$$

5. Make a proportion by setting the ratios equal. Solve the proportion for x.

EXAMPLE 2   5 pens cost $8. How much will 12 pens cost?

Let x = the cost of 12 pens.

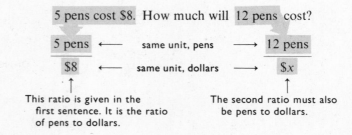

5 pens cost $8. How much will 12 pens cost?

| 5 pens | ← same unit, pens → | 12 pens |
| $8 | ← same unit, dollars → | $x |

This ratio is given in the first sentence. It is the ratio of pens to dollars.

The second ratio must also be pens to dollars.

Write the proportion by setting the two ratios equal.

$$\frac{5 \text{ pens}}{\$8} = \frac{12 \text{ pens}}{\$x}$$

$5 \cdot x = 8 \cdot 12$ Cross multiply.

$$\frac{\overset{1}{\cancel{5}} \cdot x}{\underset{1}{\cancel{5}}} = \frac{96}{5}$$ Divide each side by 5.

$x = \$19.20$

The cost of 12 pens is $19.20. ■

▶ **You Try It** **2.** Melissa made $49.26 for 3 hours work. How many hours must she work to make $410.50?

EXAMPLE 3 The scale for a blueprint is 1 inch represents 6 feet. What is the actual length of a wall that measures $4\frac{3}{8}$ inches on the blueprint?

Let x = actual length of the wall.

$$\frac{1 \text{ in (blueprint)}}{6 \text{ ft (actual)}} = \frac{4\frac{3}{8} \text{ in (blueprint)}}{x \text{ ft (actual)}}$$

Write this ratio from the first sentence: "The scale for a blueprint is 1 inch represents 6 feet."

Also write this ratio as blueprint inches to actual feet.

$$\frac{1}{6} = \frac{4\frac{3}{8}}{x}$$

$1 \cdot x = 6 \cdot 4\frac{3}{8}$ Cross multiply.

$$x = \frac{\overset{3}{\cancel{6}}}{1} \cdot \frac{35}{\underset{4}{\cancel{8}}} = \frac{105}{4} = 26\frac{1}{4}$$

$x = 26\frac{1}{4} \text{ ft}$

The actual wall length is $26\frac{1}{4}$ feet. ■

▶ **You Try It** **3.** The scale for a blueprint is $\frac{1}{8}$ inch represents 4 feet. What is the actual width of a room that measures $1\frac{3}{4}$ inches on the blueprint?

EXAMPLE 4 A rancher owns 6,000 cows. He randomly tests 160 for a disease and finds that 36 are infected. How many of his 6,000 cows can the rancher expect are infected?

36 infected out of 160 cows is this ratio.

Let x = the expected number of infected cows in the herd of 6,000.

$$\frac{36 \text{ infected}}{160 \text{ cows}} = \frac{x \text{ infected}}{6{,}000 \text{ cows}}$$

$$160 \cdot x = 36 \cdot 6{,}000 \qquad \text{Cross multiply.}$$

$$\frac{\overset{1}{\cancel{160}} \cdot x}{\cancel{160}} = \frac{\overset{9}{\cancel{36}} \cdot \overset{150}{\cancel{6{,}000}}}{\underset{\underset{1}{\cancel{40}}}{\cancel{160}}} \qquad \text{Divide each side by 160. Reduce.}$$

$$x = 1{,}350$$

The rancher expects about 1,350 of his 6,000 cows are infected. ■

▶ **You Try It** **4.** The population of a city is 250,000 people. 400 people are randomly tested and 24 are found to have a rare virus. How many city residents do you expect have the virus?

EXAMPLE 5 Diane reads 25 pages of a novel in 40 minutes. How long will it take her to read a 325-page novel if she reads at the same rate?

Let x = number of minutes to read 325 pages.

Reading 25 pages in 40 minutes is this ratio.

Reading 325 pages in x minutes is this ratio.

$$\frac{25 \text{ pages}}{40 \text{ minutes}} = \frac{325 \text{ pages}}{x \text{ minutes}}$$

$$40 \cdot 325 = 25 \cdot x \qquad \text{Cross multiply.}$$

$$\frac{40 \cdot \overset{13}{\cancel{325}}}{\underset{1}{\cancel{25}}} = \frac{\overset{1}{\cancel{25}} \cdot x}{\underset{1}{\cancel{25}}} \qquad \text{Divide each side by 25. Reduce.}$$

$$520 \text{ minutes} = x$$

Diane will read the novel in about 520 minutes, or 8 hours 40 minutes. ■

▶ **You Try It** **5.** A child reads 9 pages of a book in 6 minutes. How long will it take to finish the book if it has 51 pages?

EXAMPLE 6 **Unit Price** A 15-ounce box of raisins costs $1.39. Find the unit price per ounce.

Let x = the price per ounce.

15-ounce box costs   1 ounce costs
$1.39 (= 139¢) x¢

$$\frac{15 \text{ oz}}{139¢} = \frac{1 \text{ oz}}{x¢}$$

$$15 \cdot x = 1 \cdot 139$$

$$x = \frac{139}{15} \doteq 9.3¢ \text{ per oz} \quad \blacksquare$$

▶ **You Try It**   **6.** A 36-ounce jar of peanut butter costs $2.79. Find the unit price per ounce. Round to tenths.

Proportions can be used to find parts per hundred, parts per thousand, and parts per million (or ppm).

EXAMPLE 7 Of 263,000 people infected with a virus, 18,410 died. How many people died per 100 infected?

x = number dead per 100 infected

$$\frac{18,410 \text{ dead}}{263,000 \text{ infected}} = \frac{x \text{ dead}}{100 \text{ infected}}$$

$$263,000x = 100 \cdot 18,410$$

$$x = \frac{1,841,000}{263,000}$$

$$x = 7$$

There were 7 deaths per 100 people infected. $\blacksquare$

> **OBSERVE** Parts per 100 is called percent. 7 deaths per 100 means 7% died. Percents are studied in the next chapter.

▶ **You Try It**   **7.** Of 42,600 eligible voters, 30,672 voted. How many voters per 100 voted?

SOMETHING MORE

Shadows On a sunny day, the length of an object's shadow is proportional to its height. If a pole is twice as tall as you, its shadow will be twice as long as yours. If your pet is one-third your height, its shadow will be one-third as long as yours. You can use this fact to find the heights of buildings, towers, trees, and so on.

To find the height of a building, use a yardstick to measure the length of its shadow. Immediately afterward (why?), hold the yardstick vertically on the ground. Mark the end of its shadow.

Then measure the shadow's length. Use the proportion below to find the height of the building.

$$\frac{\text{yardstick height}}{\text{yardstick shadow length}} = \frac{\text{building height}}{\text{building shadow length}}$$

a. A building casts a shadow 136 feet long. A yardstick casts a shadow 2 feet long. How tall is the building? (Let x = the height of the building.)

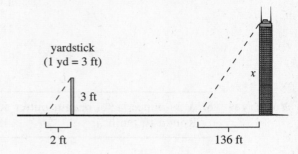

yardstick
(1 yd = 3 ft)

3 ft

2 ft

136 ft

x

b. A flagpole casts a shadow 60 feet long. At the same time, a yardstick casts a 4-foot shadow. How high is the pole?

c. A tree casts a shadow 55 feet long. At the same time, a yardstick has a 2 foot 6 inch shadow. How tall is the tree?

d. You Be the Judge Susan claims the procedure used in this problem only gives accurate answers when the ground over which the shadows lie is perfectly level. Do you agree? Explain your decision.

▶ **Answers to You Try It** 1. $252 2. 25 hr 3. 56 ft 4. 15,000 people
5. 34 min 6. 7.8¢/oz 7. 72 per 100 voted

▶ **Answers to Something More** a. 204 ft b. 45 ft c. 66 ft

$\left(\text{Note: since 6 in} = \frac{6}{12} \text{ ft} = \frac{1}{2} \text{ ft, write 2 ft 6 in as } 2\frac{1}{2} \text{ ft}\right)$

SECTION 6.8 EXERCISES

Use proportions to solve each word problem.

1. Suzanne drove 140 miles on 5 gallons of gas. How far can she drive on 12 gallons?

2. Tom drove 57 miles on 3 gallons of gas. How far will 8 gallons take him?

3. If 8 favors cost $28.48, how many can you buy for $74.76?

4. If 6 burgers cost $7.74, how many can you buy for $19.35?

5. The scale for a blueprint is 1 inch represents 20 feet. What is the actual length of a kitchen wall measuring $\frac{5}{8}$ inch on the blueprint?

6. The scale for a blueprint is $\frac{1}{2}$ inch represents 4 feet. What is the actual length of a living room floor measuring $2\frac{3}{8}$ inches?

7. Of 480 people who voted today, 250 voted for Candidate A. If 30,240 people will vote today, how many will vote for Candidate A if voting patterns do not change?

8. 270 geese were randomly examined. 95 were contaminated with crude oil. How many geese in a population of 8,100 can you expect to be contaminated by oil?

9. If 2 pounds of beef are needed to feed 6 adults, how many pounds will be needed for a cookout with 27 adults?

10. If 9 tablespoons of coffee are needed to serve 6 people, how many people can be served with 24 tablespoons of coffee?

11. If a car travels 140 miles in 3 hours, at the same speed how long will it take to travel 560 miles?

12. An express train travels 195 miles in two hours. At that rate, how long will it take to go 1,755 miles?

13. The scale for a map is 1 inch equals 13.5 miles. What is the actual distance between two cities that are $8\frac{3}{4}$ inches apart on the map?

14. The scale for a globe is 660 miles equals 1 inch. Find the actual distance between New York and Hawaii if they are $7\frac{3}{4}$ inches apart on the globe.

15. Ed drove 264 miles on 14.1 gallons of gas. If he plans to drive 9,300 miles on a cross-country vacation, how many gallons of gas will he need, to the nearest gallon?

16. Tammy gets 16.3 miles on each gallon of gas. How many gallons will she need to drive to her mother's house 450 miles away? Round to the nearest gallon.

17. A baseball team won 16 of their first 27 games. If they play 162 games in one season, how many games will the team win if they keep winning at the same rate?

18. A mining company extracts 30 pounds of iron from 100 pounds of iron ore. How many pounds of iron can be extracted from 3,600 pounds of iron ore?

19. If an average family of 4 produces 58 pounds of garbage a week, about how much garbage is produced weekly in a city of 1,200,000 people?

20. A doctor found that 3 out of every 200 people have contracted a virus. About how many of the 256,000,000 people in the United States are infected with the virus?

21. A 20-ounce jar of honey costs $1.69. What is the unit price per pound rounded to the nearest tenth of a cent?

22. A 54-ounce can of meat costs $8.49. What is the unit price per pound rounded to the nearest tenth of a cent?

23. 3 ounces of sirloin steak contain 75 milligrams of cholesterol. How much cholesterol is in an 8-ounce cut of sirloin?

24. 55 milligrams of cholesterol are contained in $3\frac{1}{2}$ ounces of brook trout. How much cholesterol is in an 8-ounce fillet of trout?

25. The recipe for fixing two servings of oatmeal calls for $1\frac{1}{2}$ cups of water, $\frac{1}{2}$ teaspoon salt, and $\frac{2}{3}$ cup of oats.

    a. If you use $1\frac{2}{3}$ cups of oats, how much salt and water must be used to keep the recipe in proportion?

    b. How many servings will you now be making?

26. To make 1 quart of instant milk you need $1\frac{1}{3}$ cups of instant milk mix and $3\frac{3}{4}$ cups of water.

    a. If you use 4 cups of instant milk mix, how much water will you need?

    b. How many quarts will you now be making?

27. Candidate A received 71,592 votes out of 125,600 votes cast. How many votes per 100 did Candidate A receive?

28. In a population of 255 million people, 765 were infected with a rare disease. How many people per million were infected?

SKILLSFOCUS (Section 5.8) *Change each fraction to a decimal.*

29. $\frac{3}{5}$

30. $\frac{7}{10}$

31. $\frac{5}{8}$

32. $\frac{8}{11}$ (round to thousandths)

EXTEND YOUR THINKING ▶▶▶▶
▶ SOMETHING MORE

33. The Capitol Building in Washington, DC, casts a 500-foot shadow. A yardstick has a 5-foot shadow at the same time of day. Calculate the height of the Capitol.

34. At the moment a plane is directly overhead, its shadow is 1,200 feet away. Calculate the altitude of the plane if a yardstick casts a 1 foot 9 inch shadow at the same time of day.

35. **Constructing a Proportion Slide** A proportion slide is a ruler, yardstick, meterstick, or other straight measuring device with a thick slide that moves back and forth. The slide does not have to be attached to the ruler, and can be something as simple as an eraser or building block.

    If you know the distance to an object, you can use the proportion slide to calculate its height. (Or, if you know the height you can calculate the distance.) Place the ruler under your eye. Point it toward the object whose height you are measuring. Move the slide back and forth until the height of the slide matches the height of the object being measured. Use the following proportion to calculate the object's height.

$$\frac{\text{distance from eye to slide}}{\text{height of slide}}$$
$$= \frac{\text{distance from eye to object}}{\text{height of object}}$$

a. You point your proportion slide at a lamp 18 feet away. Move the slide until its height matches the lamp's height. At that point, the distance from your eye to the slide is 6 inches. The slide is $1\frac{1}{4}$ inches high. How tall is the lamp? (Let $x =$ the height of the lamp.)

slide is $1\frac{1}{4}$ in high

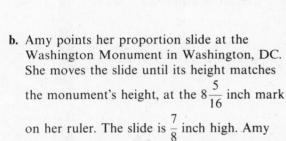

←6 in→

←—— 18 ft ——→

b. Amy points her proportion slide at the Washington Monument in Washington, DC. She moves the slide until its height matches the monument's height, at the $8\frac{5}{16}$ inch mark on her ruler. The slide is $\frac{7}{8}$ inch high. Amy is exactly 1 mile (= 5,280 feet) away from the monument. Calculate the height of the Washington Monument to the nearest ten feet.

c. Nicole knows a tower is 200 feet high. She sights the tower with her proportion slide. She finds the $\frac{1}{2}$-inch slide sitting on the $8\frac{1}{4}$ inch mark on her ruler. Calculate how far Nicole is from the tower.

36. A dining room measures $5\frac{1}{2}$ inches by $3\frac{7}{8}$ inches on a blueprint. If the scale is $\frac{1}{4}$ inch equals 1 foot, what are the actual dimensions of the room?

37. a. Immediately after jogging, Sam's heart beats 26 times in 10 seconds. What is his heart rate per minute?

b. After 1 minute, he counts 19 heartbeats in 10 seconds. What is his heart rate per minute?

c. What is his recovery rate in the first minute?

38. *Voyager II* The *Voyager II* spacecraft is expected to be returning data to Earth until the year 2020, when its nuclear power source is expected to be exhausted. It was launched in 1977. By 1989 it had traveled a total of 4,400,000,000 miles. How far from Earth can we expect *Voyager II* to be by 2020? Round to hundred millions.

WRITING TO LEARN ▶▶▶▶

39. Write a proportion word problem whose answer is $3.60, rounded to the nearest cent. The problem is about candy and a child named Ted. Solve the problem, explaining each step.

▶ YOU BE THE JUDGE

40. Is this proportion valid? Explain your decision.

$$\frac{7 \text{ pens}}{\$4.95} = \frac{\$x}{12 \text{ pens}}$$

CALCULATOR TIPS

Solving Proportion Problems You can solve proportion problems using your calculator and the following procedure.

> ### Quick 'N' Dirty Proportion Solver
> 1. Cross multiply where you can.
> 2. Divide this product by the number left over.

The next example demonstrates why this procedure works.

EXAMPLE 1 $\dfrac{x}{5} = \dfrac{21}{15}$

$15 \cdot x = 5 \cdot 21$ Cross multiply. Then divide each side by 15.

$x = \dfrac{5 \cdot 21}{15}$ In short, cross multiply $5 \cdot 21$. Then divide this product by 15, the number left over.

$x = 5 \boxed{\times} 21 \boxed{\div} 15 \boxed{=} 7$

Therefore, $\dfrac{7}{5} = \dfrac{21}{15}$.

To check, find both cross products, and subtract them in memory. If the answer in memory is 0, the cross products are equal, and the answer is correct. (In addition, if your answer is correct, the "M" indicating memory is in use will disappear from your display.)

Check: $7 \boxed{\times} 15 \boxed{=} \boxed{M+} 5 \boxed{\times} 21 \boxed{=} \boxed{M-} \boxed{MR} 0$ (correct) ∎

EXAMPLE 2 $\dfrac{3.125}{5} = \dfrac{2.8}{y}$

$y = 5 \boxed{\times} 2.8 \boxed{\div} 3.125 \boxed{=} 4.48$

Therefore, $\dfrac{3.125}{5} = \dfrac{2.8}{4.48}$.

Check: $5 \boxed{\times} 2.8 \boxed{=} \boxed{M+} 4.48 \boxed{\times} 3.125 \boxed{=} \boxed{M-} \boxed{MR} 0$ ∎

Use your calculator to solve each proportion. Check.

1. $\dfrac{9}{10} = \dfrac{x}{25}$

2. $\dfrac{x}{42} = \dfrac{17}{28}$

3. $\dfrac{3.78}{0.63} = \dfrac{1.92}{y}$

4. $\dfrac{31}{17} = \dfrac{x}{58}$ (round to hundredths)

5. $\dfrac{2.68}{w} = \dfrac{11.3}{15.4}$ (round to thousandths)

CHAPTER 6 REVIEW

VOCABULARY AND MATCHING

New words and phrases introduced in this chapter are shown in the left-hand column. Match each term on the left with the phrase or sentence on the right that best describes it.

A. ratio   _____ also written 4:6 and 4/6

B. order in a ratio   _____ a statement that two ratios are equal

C. 4 to 6   _____ price for 1 unit of an item

D. terms of the ratio   _____ ratio or rate whose second term is 1

E. reducing ratios   _____ $\frac{24}{40} = \frac{35}{60}$, for example

F. rate   _____ also means $6 \div 4$

G. per   _____ increase one quantity, and you increase the other by the same factor

H. 6 to 4   _____ cross products are equal for a true proportion

I. unit ratio or unit rate   _____ numbers are written in the same order as the names

J. unit price   _____ for $\frac{6}{4} = \frac{3}{2}$, they are $6 \cdot 2$ and $3 \cdot 4$

K. comparing ratios   _____ word often used when writing rates, as in mpg or mph

L. proportion   _____ a comparison of two quantities

M. true proportion   _____ when 6 to 4 becomes 3 to 2, or 100:30 becomes 10:3

N. false proportion   _____ requires rewriting both ratios with the same second term

O. cross products   _____ $\frac{15}{24} = \frac{35}{56}$, for example

P. cross product property   _____ what 4 and 6 are in the ratio 4 to 6

Q. proportional   _____ a ratio whose names cannot be eliminated

REVIEW EXERCISES

6.1 The Meaning of Ratio

1. A school has 163 students, 17 teachers, and 4 administrators. Write the following ratios.
   a. students to teachers

   b. teachers to administrators

   c. teachers and administrators to students

2. There are 39 rafters, 9 guides, and 5 kayakers on the river. Write the following ratios.
   a. rafters to kayakers

   b. guides and kayakers to rafters

   c. kayakers to people

3. There are 27 cars and 61 bikes on a ferryboat. Write the ratio of bikes to cars in three ways.

4. There are $30,615 of investment to $12,059 of profit. Write the ratio of investment to profit in three ways.

Reduce each ratio to lowest terms.

5. 50 to 35

6. 26:40

7. 8/10

8. 2,000:6,000

9. 75/125

10. 480 to 320

11. Joan reads a 640-page novel in the same time it takes her to read 160 pages of a science text. What is the lowest terms ratio of novel pages to science pages?

c. 2 minutes to 40 seconds

d. 3 feet 9 inches to 2 feet

12. $450 in profit was made on a $3,000 investment. What is the lowest terms ratio of profit to investment?

14. Mike is 6 feet tall. His left to right arm extension measures 72 inches. What is Mike's height to arm extension ratio in lowest terms?

13. Cancel the names in each ratio. Reduce to lowest terms.
a. 10 miles to 13 miles

15. A board 12 feet long is 8 inches wide. What is the lowest terms ratio of length to width?

b. 4 feet to 10 inches

6.2 Ratio as a Rate

Reduce each rate to lowest terms.

16. 450 miles to 100 hours

17. $36 per 12 toys

18. 240 miles per 15 gallons

19. $87 per 9 hours

20. In 21 working days a salesman sold 30 cars. Write the rate of cars sold per day in lowest terms.

21. In 56 minutes Greg can run 8 miles. What is the lowest terms rate of minutes per mile?

Change the name in each rate as indicated. Reduce.

22. 45 miles to 2 hours: change hours to minutes

23. \$6 per 8 gallons: change dollars to cents

24. \$336 per 2 weeks: change weeks to days

25. 10 hours/24 boxes: change hours to minutes

6.3 Rewriting Ratios Involving Fractions and Decimals as Ratios of Whole Numbers

Rewrite each ratio as a ratio of whole numbers.

26. $\frac{5}{8}$ to 4

27. 3 to $2\frac{1}{4}$

28. $5\frac{1}{3}$ to $2\frac{3}{4}$

29. 0.8 to 0.05

30. 6.04 to 3.2

31. 0.024 to 0.0003

32. The ratio of cups of water to cups of flour is $2\frac{1}{2}$ to $4\frac{1}{3}$. Rewrite this as a ratio of whole numbers.

34. A table takes 4.5 hours to make and 7.75 hours to finish. Write the ratio of make time to finish time in terms of whole numbers.

33. The rate of feet per minute is 12 feet to $\frac{3}{4}$ minute. Write this rate in terms of whole numbers.

6.4 Unit Ratios, Unit Price, and the One-Number Ratio

35. Write \$280 to \$14 as a unit ratio.

36. Write 18 pounds lost in 10 days as a unit rate.

37. Find the unit price if 6 pounds of apples cost \$2.34.

38. Find the unit price if 22 ounces of jam cost \$1.76.

39. A 7.41-pound turkey breast costs \$10.27. Find the unit price to the nearest tenth of a cent.

40. 2.945 acres of land sells for \$82,500. Find the unit price per acre to the nearest \$100.

6.5 Applications

41. Fran bought a 40-ounce jar of honey for \$2.89. Bob bought the 24-ounce size for \$1.89. Who made the better buy?

42. 2 liters (67.6 fluid ounces) of soda costs \$1.09. A six-pack of 12-ounce cans is selling for \$1.69. Which is the better buy?

43. David drove 278 miles on 16.5 gallons of gas A. He drove 312 miles on 17.2 gallons of gas B. Which gas gave the better gas mileage?

44. The jumbo 96-ounce box of detergent sells for $7.69. The regular 52-ounce size sells for $3.89. Which box of detergent is the better buy?

45. Melissa and Jason split $460 according to a 3:1 ratio. How much money did each get?

46. Buildings K and L split 960 gallons of oil according to a 7 to 5 ratio. How much oil does building K get?

6.6 Introduction to Proportions

Determine if each proportion is true or false.

47. $\dfrac{5}{9} \overset{?}{=} \dfrac{4}{7}$

48. $\dfrac{15}{40} \overset{?}{=} \dfrac{21}{56}$

49. $\dfrac{0.9}{1.5} \overset{?}{=} \dfrac{0.21}{0.35}$

50. $\dfrac{\frac{2}{5}}{4} \overset{?}{=} \dfrac{1}{15}$

6.7 Solving Proportions

Solve the proportion. Check each answer.

51. $\dfrac{x}{6} = \dfrac{6}{9}$

52. $\dfrac{10}{16} = \dfrac{x}{24}$

53. $\dfrac{8}{10} = \dfrac{6}{x}$

54. $\dfrac{5}{x} = \dfrac{0.7}{1.4}$

55. $\dfrac{x}{5} = \dfrac{8}{7}$ (round to tenths)

56. $\dfrac{4.62}{0.9} = \dfrac{3.8}{x}$ (round to hundredths)

57. $\dfrac{5\frac{2}{3}}{4} = \dfrac{x}{1\frac{7}{8}}$

58. $\dfrac{48}{x} = \dfrac{7\frac{1}{2}}{5\frac{3}{4}}$

6.8 Applications

59. The daily dosage for a certain medicine is 3 ounces for each 60 pounds of body weight. What is the daily dosage for a man weighing 200 pounds?

60. Four tablespoons of weed killer are mixed with 3 quarts of water. How many quarts of water must be mixed with 10 tablespoons of weed killer?

61. The property tax on a $50,000 home is $720. At the same rate, what is the property tax on a home worth $130,000?

62. The fuel mixture for a gas saw is 1 ounce of oil for every $2\frac{1}{2}$ pints of gas. How much oil must be mixed with 4 pints of gas?

63. The scale for a blueprint is $\frac{1}{4}$ inch equals 2 feet. What is the actual length of a wall that measures $2\frac{5}{8}$ inches on the blueprint?

64. The scale for a map is 1 inch equals 7.2 miles. What is the distance between two cities $4\frac{1}{2}$ inches apart on the map?

65. If 1 gallon of paint covers 450 square feet of wall space, how many gallons are needed for 2,950 square feet?

66. 52 out of 800 computers tested were found to be defective. At this rate, how many of the 22,000 computers assembled at the same time are expected to be defective?

This test will measure your understanding of ratio and proportion. Allow 50 minutes to complete it. Show the work for each problem. When done, check your answers. Rework each problem solved incorrectly.

1. There are 280 students, 35 teachers, and 10 administrators in a small school. Write the following ratios in lowest terms.

   a. students to teachers

   b. teachers to administrators

   c. teachers plus administrators to students

2. There are 6 men and 8 children on a boat. Write the ratio of men to children in three ways.

3. Write the ratio of $500 profit to $2,100 cost in lowest terms.

4. A machine can wrap 5,000 packages per 8-hour period. Write this rate in lowest terms.

5. The rate of consumption is 24 pounds per hour. Rewrite this rate by changing hours to minutes.

6. Write $4\frac{5}{6}$ to $1\frac{3}{8}$ as a ratio of whole numbers.

7. Write 4.08 to 0.6 as a ratio of whole numbers.

8. Find the unit price if 11.2 ounces cost $1.59. Round to the nearest tenth of a cent.

9. A 340-acre tract of land sold for $1,650,000. Find the unit price per acre rounded to the nearest $100.

10. A 40-ounce box of brand A pretzels sells for $4.69. The 28-ounce brand B box sells for $3.19. Find the better buy.

11. Millie, Nick, and David agree to split $80 according to 9:5:2. How much does each receive?

12. Decide if the proportion is true or false. $\frac{60}{36} \stackrel{?}{=} \frac{35}{21}$

13. Find x. $\dfrac{4}{9} = \dfrac{x}{8}$

14. Find y. $\dfrac{13}{y} = \dfrac{26}{3}$

15. Find n. $\dfrac{n}{\frac{5}{7}} = \dfrac{14}{\frac{1}{2}}$

16. Find h. $\dfrac{0.8}{4} = \dfrac{6.25}{h}$

---

17. The scale for a blueprint is $\dfrac{1}{2}$ inch represents 6 feet. If a wall is actually 46 feet long, what is its length on the blueprint?

18. The dosage for a certain medicine is $1\dfrac{1}{2}$ pills per 80 pounds of body weight. What should the body weight be to the nearest 10 pounds to get a 4-pill dosage?

Explain the meaning of each term. Use your own examples.

19. rate

20. cross product

21. unit price

NAME   DATE   HOUR

1. Evaluate 3^2.

2. Evaluate 2^4.

3. Solve $5w = 60$.

4. Evaluate $\sqrt{100}$.

5. Simplify $4\sqrt{81} + 6 \cdot 2^3$.

6. Simplify $3(6 - 4)$.

7. Add $4\frac{2}{3} + 1\frac{3}{8}$.

8. Subtract $\frac{4}{7} - \frac{1}{6}$.

9. Multiply $3\frac{3}{10} \times 2\frac{2}{11}$.

10. Divide $\frac{13}{20} \div \frac{2}{5}$.

11. One share of stock is selling for $\$12\frac{3}{8}$. In one hour, it lost $\frac{1}{3}$ of its value.

    a. How much value did one share lose?

    b. What is one share now worth?

12. Add $7.6 + 90 + 4.073 + 8.06$.

13. Subtract $0.6 - 0.097$.

14. Multiply 4.73×10.8.

15. Divide $4.2 \div 7.05$ (round to thousandths).

16. Simplify the ratio. 5 feet to 10 inches

17. Solve $\frac{w}{21} = \frac{25}{35}$.

18. Write the ratio 5.6 to 0.21 as a ratio of whole numbers.

19. Split $840 according to a $4:3$ ratio.

20. If 2.57 pounds of lunch meat costs $11.80, what will 0.75 pound cost? Round to the nearest cent.

Explain the meaning of each term. Use your own examples.

21. composite number

22. repeating decimal

Percents

OIL

The following table presents world crude oil production and consumption by major producing region.

Region	World Production (57.6 million barrels per day)	World Consumption (55.2 million barrels per day)	World Reserves (541.4 million barrels per day)
Middle East	37.0%	2.0%	58.4%
The former Soviet Union and Eastern Europe	16.8%	15.8%	11.9%
Africa	9.9%	1.9%	9.7%
United States	18.9%	31.2%	6.5%
Western Europe	0.7%	26.2%	3.09%
Other Western Hemisphere	6.4%	8.0%	4.1%
Caribbean	6.4%	0.9%	2.9%
Japan	0.03%	9.1%	0.01%
Other Eastern Hemisphere	3.87%	4.9%	3.4%

SOURCE: Federal Energy Administration, Project Independence Blueprint.

From the table, the United States consumes 31.2% of the 55.2 million barrels used each day in the world. How many barrels is this? The United States has about 6% of the world's population. Estimate how many times more oil the United States consumes when compared to its size. These questions will be answered in this chapter.

Percents are widely used in business, science, and everyday living to make comparisons. It is easier to make comparisons using percents because percents are based on 100.

For example, suppose 3,700 votes were cast in an election. 1,887 votes were for candidate Paul. This does not immediately tell you if Paul won. But if you know that 1,887 out of 3,700 is 51 out of every 100 votes cast, then Paul was victorious. 51 votes out of every 100 is 51% of the votes.

In this chapter you will learn to make percent conversions and solve percent problems. You will then study some of the many practical applications that use percents, including commission, discount, advertising, interest, and finding a monthly payment.

Work through each section in this chapter. Use this test to identify topics you are not familiar with. These topics may require additional study. You may *not* use a calculator.

7.1

1. 42% means _____ parts out of _____ equal parts.

2. _____ parts out of _____ equal parts is $33\frac{1}{3}\%$.

3. The complement of 24% is _____.

7.2

4. Change 0.48 to a percent.

5. Change 9.2% to a decimal.

7.3

6. Change $\frac{7}{8}$ to a percent.

7. Change 96% to a fraction in lowest terms.

7.4

8. Identify the base, rate, and amount: 47% of $300 is $141.

9. Identify the base, rate, and amount: 13.5 is 27% of what number?

7.5

10. What is 60% of $45?

11. 13 is what percent of 65?

12. 48 is 120% of what number?

7.6

13. 25 pounds of lettuce contain 23.7 pounds of water by weight. What percent of the weight of lettuce is water?

14. In a country of 80 million people, 1.5% are employed by the government. How many people are employed by the government?

15. If $4\frac{1}{2}\%$ of a cow's milk is butterfat, how many gallons of milk are needed to get $3\frac{3}{8}$ gallons of butterfat?

7.7

16. Eve works on an 8% commission. How much merchandise must she sell to make $1,440 in commissions?

17. A $450 dryer is on sale for 30% off.
    a. What is the discount?

    b. What is the sale price?

18. Georgia made a $45,000 profit last year. This year her profit was $63,000. What was the percent gain in her profit?

7.8

19. How much simple interest does $420 earn at 16% for 6 months?

20. What principal will earn $968 over 2 years at 11% simple interest?

VOCABULARY *Explain the meaning of each term. Use your own examples.*

21. percent

22. discount

7.1 THE MEANING OF A PERCENT

OBJECTIVES	NEW VOCABULARY
1 Define percent.	percent
2 Define complement of a percent.	base
3 Define percent greater than 100%.	complement

1 Defining Percent

Percent means parts per 100. A **percent** is a number of parts out of 100 equal parts.

A percent is written as a number followed by the percent symbol, %. 36% means 36 parts out of 100 equal parts.

FIGURE 7.1
36% shaded.

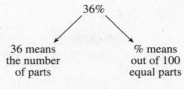

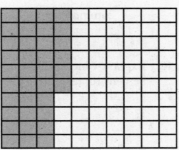

The **base** is the entire population being considered. The base represents 100%, or 100 parts out of a total of 100 equal parts. In Figure 7.1, the base is represented by a large rectangle. It is divided into 100 equal parts, or boxes, of which 36 are shaded. 36% of the figure is shaded.

> **OBSERVE** Since percent means parts per 100, a percent is a ratio of parts to 100 equal parts. 36% is the ratio of 36 parts to 100 parts.

EXAMPLE 1

Explain what each percent means. Represent each percent as a shaded region in Figure 7.2.

a. 50% means 50 parts out of 100 equal parts. This is half of the boxes in Figure 7.2.

b. 1% means 1 part out of 100 equal parts. This is 1 of the 100 boxes in Figure 7.2.

c. $6\frac{1}{4}\%$ means $6\frac{1}{4}$ parts out of 100 equal parts. This is 6 boxes plus $\frac{1}{4}$ of a seventh box in Figure 7.2.

d. $\frac{1}{2}\%$ means $\frac{1}{2}$ of 1 part out of 100 equal parts. This is $\frac{1}{2}$ of 1 box in Figure 7.2.

e. 7.3% means 7.3 parts out of 100 equal parts. This is 7 boxes plus $0.3 = \frac{3}{10}$ of an eighth box in Figure 7.2. That is, divide the eighth box into 10 equal pieces, and take 3. ■

FIGURE 7.2

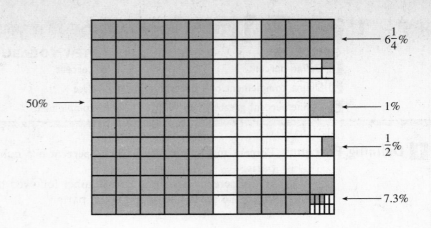

50% →

← $6\frac{1}{4}\%$

← 1%

← $\frac{1}{2}\%$

← 7.3%

OBSERVE To understand percents, it may be helpful to use one dollar (100¢) to represent 100%. Then 36% is 36¢ out of one dollar. 1% is one cent. What is $\frac{1}{2}\%$? It is $\frac{1}{2}$ of one penny. $\frac{1}{2}\%$ is not 50¢, a common wrong answer.

▶ **You Try It** Explain what each percent means. How many boxes in Figure 7.2 does each percent represent?

1. 70%   **2.** 3%   **3.** $16\frac{2}{3}\%$

4. $\frac{1}{5}\%$ **5.** 2.8%

2 Complement of a Percent

36% of the boxes in Figure 7.1 are shaded. Subtract 36% from 100%.

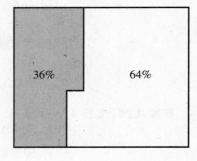

$$100\% - 36\% = 64\%$$
whole   shaded   unshaded

The remaining 64% of the boxes are unshaded. 64% is called the complement of 36%. It is the rest of the whole. A percent plus its complement equals 100%: 36% + 64% = 100%.

The **complement** of a percent is found by subtracting that percent from 100%.

EXAMPLE 2 Find the complement of each percent.

a. The complement of 80% is 100% − 80% = 20%.

b. The complement of $37\frac{1}{2}\%$ is $100\% - 37\frac{1}{2}\% = 62\frac{1}{2}\%$.

c. The complement of 4.7% is 100% − 4.7% = 95.3%. ∎

▶ **You Try It**   Find the complement of each percent.

6. 3%   **7.** 60%   **8.** $12\frac{3}{4}\%$

9. $\frac{1}{2}\%$ **10.** 72.4%

---

3 **Percents Greater than 100%**

A percent greater than 100% means one or more wholes of 100 equal parts each, plus a portion of a final 100 equal parts.

EXAMPLE 3   Explain what 150% means.

150% means 100% plus 50%. This is one whole of 100 equal parts, plus 50 parts of a second whole.

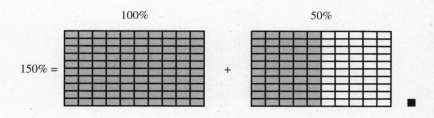

> **CAUTION**   150% makes sense in some contexts, but not in others. To say *150% of the students in your class passed the exam* does not make sense. If 100% pass, then everyone passes. No one is left.
>   However, suppose it costs you $10 to build a chair. *You sell it for a 150% profit.* What do you sell it for? You sell it for $25. This includes the $10 it cost you, plus 100% of $10 ($10), plus another 50% of $10 ($5), or $25.

EXAMPLE 4   Explain what 325% means.

325% means 100% + 100% + 100% + 25%. This is 3 wholes of 100 equal parts each, plus 25 parts out of a fourth whole.

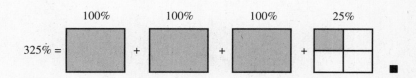

▶ **You Try It**   Explain what each percent means.

11. 175%   **12.** 200%   **13.** 430%   **14.** 101%

▶ **Answers to You Try It** 1. 70 parts out of 100 equal parts (70 boxes in Figure 7.2) 2. 3 parts out of 100 equal parts (3 boxes) 3. $16\frac{2}{3}$ parts out of 100 equal parts $\left(16 \text{ boxes plus } \frac{2}{3} \text{ of a 17th box}\right)$ 4. $\frac{1}{5}$ of 1 part out of 100 equal parts $\left(\frac{1}{5} \text{ of 1 box}\right)$ 5. 2.8 parts out of 100 equal parts $\left(2 \text{ boxes plus } \frac{8}{10} \text{ of a third box}\right)$ 6. 97% 7. 40% 8. $87\frac{1}{4}\%$ 9. $99\frac{1}{2}\%$ 10. 27.6% 11. 1 whole of 100 equal parts plus 75 parts of a second whole. 12. 2 wholes of 100 equal parts each 13. 4 wholes of 100 equal parts each, plus 30 parts of a fifth whole 14. 1 whole of 100 equal parts, plus 1 part out of a second whole

SECTION 7.1 EXERCISES

1 *Fill in the blanks. Draw and shade a diagram to represent each percent.*

1. 40% represents _____ parts out of _____ equal parts.

2. 5% means _____ parts out of _____ equal parts.

3. _____% represents 85 parts out of 100 equal parts.

4. _____% means 72 parts out of 100 equal parts.

5. _____% represents 12.5 parts out of 100 equal parts.

6. 6.25% means _____ parts out of _____ equal parts.

7. $\frac{1}{4}$% represents _____ of one part out of _____ equal parts.

8. _____% means $33\frac{1}{3}$ parts out of _____ equal parts.

2 *Find the complement of the percent.*

9. 70%   10. 23%   11. 61%   12. 20%   13. 80%   14. 0%

15. $83\frac{1}{3}$%   16. $37\frac{1}{2}$%   17. $\frac{1}{2}$%   18. 10.6%   19. 0.08%   20. 65.7%

21. If 10% of the politicians smoke cigars, then _____% do not smoke cigars.

22. If 63% of the cattle are branded, then _____% are not branded.

23. If 96.1% of the weight of a cucumber is water, then _____% of the cucumber is not water.

24. If 3% of the bacteria die in one hour, then _____% will still be alive.

25. If _____% of a paycheck goes toward purchasing food, then 82% goes toward nonfood items.

26. If _____% of a solution is acid, then $66\frac{2}{3}\%$ is not acid.

Draw and shade a diagram to represent each percent.

27. 125%

28. 250%

29. 400%

30. 320%

31. 190%

32. 675%

SKILLSFOCUS (Section 5.8) *Change each decimal to a fraction and reduce.*

33. 0.64

34. 0.9

35. 0.06

36. 1.25

EXTEND YOUR THINKING ▶▶▶▶

▶SOMETHING MORE

37. A soap commercial claims its product is 99 and 44 one hundredths percent pure. What percent is not pure?

▶TROUBLESHOOT IT

Find and correct the error.

38. $\frac{3}{4}\%$ is 75 parts out of 100 equal parts.

39. The complement of 0.6% is 100% − 0.6% = 94%.

WRITING TO LEARN ▶▶▶▶

40. Explain how a percent is a ratio. Use your own example.

41. A coach said his team gave 150% in a championship game.
   a. Explain why this does not make sense.

   b. What is the coach trying to say?

42. It costs a pharmacy 10¢ to make a pill. They sell the pill for a 400% profit. What is the cost of one pill? Explain.

▶YOU BE THE JUDGE

43. Tomas claims 120% has a complement? Is he correct? Explain your decision.

7.2 DECIMALS AND PERCENTS

OBJECTIVES

1 Change a decimal to a percent (Section 5.8).

2 Change a percent to a decimal (Section 5.6).

Any number can be written in three ways: as a fraction, as a decimal, and as a percent. In this section and the next, you will learn how to change a percent to a fraction or decimal. You do this before you use a percent to solve a word problem. You also learn how to change a fraction or decimal to a percent. You do this in problems that ask you to find percent.

1 Changing a Decimal to a Percent

Percent means hundredths, or parts per 100. To change a decimal to a percent, express the decimal as parts per 100. For example, change 0.42 to a percent.

$$0.42 = 42 \text{ hundredths} = 42 \text{ parts out of 100 equal parts} = 42\%$$

decimal point moved two places to the right (note, $42\% = 42.\%$)

42 parts out of 100 equal parts means 42%, so $0.42 = 42\%$. Notice that the decimal point moved two places to the right.

EXAMPLE 1   Change each decimal to a percent.

a. $0.24 = \dfrac{24}{100} = 24 \text{ hundredths} = 24\%$

decimal point moved two places to the right

b. $0.5 = \dfrac{5}{10} = \dfrac{5 \times 10}{10 \times 10} = \dfrac{50}{100} = 50 \text{ hundredths} = 50\%$

decimal point moved two places to the right

c. $0.742 = \dfrac{742}{1,000} = \dfrac{742 \div 10}{1,000 \div 10} = \dfrac{74.2}{100} = 74.2 \text{ hundredths} = 74.2\%$

decimal point moved two places to the right

d. $3.68 = 3\dfrac{68}{100} = \dfrac{368}{100} = 368 \text{ hundredths} = 368\%$

decimal point moved two places to the right ∎

> **To change a decimal to a percent**
>
> **1.** Move the decimal point exactly two places to the right.
> **2.** Add the percent symbol, %.

EXAMPLE 2   **a.** $0.28 = 0\ 28.\% \text{ or } 28\%$ **b.** $0.0638 = 0\ 06.38\% \text{ or } 6.38\%$

c. $0.7 = 0.70 = 0\ 70.\% \text{ or } 70\%$ **d.** $0.3\dfrac{1}{2} = 0.35 = 0\ 35.\% \text{ or } 35\%$

e. $1.27 = 1\ 27.\% = 127\%$ **f.** $1 = 1.00 = 1\ 00.\% \text{ or } 100\%$ ∎

From Example 2f, 1 equals 100%. Therefore,

$$2 = 2 \times 1 = 2 \times 100\% = 200\%;$$
$$5 = 5 \times 1 = 5 \times 100\% = 500\%;$$
$$24 = 24 \times 1 = 24 \times 100\% = 2,400\%; \text{ etc.}$$

Change each decimal to a percent.

1. 0.19   **2.** 0.2   **3.** 0.84   **4.** 0.06   **5.** 2.58

6. 0.2153   **7.** 4.6   **8.** 0.007   **9.** 3   **10.** $0.9\frac{1}{2}$

---

2 Changing a Percent to a Decimal

Percent means hundredths, or parts per 100. 63% means 63 hundredths.

$$63\% = 63 \text{ hundredths} = \frac{63}{100} = 0.63$$

<u>decimal point moved two places left</u>

Division by 100 moves a decimal point two places to the left (Section 5.6). Since 63 divided by 100 is 0.63, 63% = 0.63.

EXAMPLE 3   Change each percent to a decimal.

a. $25\% = 25 \text{ hundredths} = \frac{25}{100} = 0.25$ **b.** $4\% = 4 \text{ hundredths} = \frac{4}{100} = 0.04$

decimal point moved two places left   decimal point moved two places left

c. $106\% = 106 \text{ hundredths} = \frac{106}{100} = 1.06$ **d.** $8.8\% = 8.8 \text{ hundredths} = \frac{8.8}{100} = 0.088$

decimal point moved two places left   decimal point moved two places left   ∎

> **To change a percent to a decimal**
>
> **1.** Move the decimal point exactly two places to the left.
> **2.** Drop the % symbol.

EXAMPLE 4   Change each percent to a decimal.

a. $28\% = 28.\% = .28\ \% = 0.28$ **b.** $0.05\% = 00.05\%$
$$= .00\ 05\% = 0.0005$$

c. $260\% = 260.\% = 2.60\ \% = 2.60$ **d.** $3\frac{1}{4}\% = 3.25\%$
$$= 03.25\% = .03\ 25\% = 0.0325 \quad ∎$$

▶ **You Try It**   Change each percent to a decimal.

11. 42%   **12.** 75%   **13.** 9%   **14.** 265%   **15.** 600%

16. 2.5%   **17.** 0.003%   **18.** 19.37%   **19.** $12\frac{1}{2}\%$ **20.** 0.4%

▶ **Answers to You Try It**   **1.** 19%   **2.** 20%   **3.** 84%   **4.** 6%   **5.** 258%
6. 21.53%   **7.** 460%   **8.** 0.7%   **9.** 300%   **10.** 95%   **11.** 0.42   **12.** 0.75
13. 0.09   **14.** 2.65   **15.** 6   **16.** 0.025   **17.** 0.00003   **18.** 0.1937   **19.** 0.125
20. 0.004

1 *Change the following decimals to percents.*

1. 0.62	**2.** 0.33	**3.** 0.05	**4.** 0.01
5. 0.212	**6.** 0.683	**7.** 0.018	**8.** 0.007
9. 0.0608	**10.** 0.0009	**11.** 0.1004	**12.** 0.0908
13. 0.3	**14.** 0.8	**15.** 0.50001	**16.** 0.00702
17. 4	**18.** 7	**19.** 2.5	**20.** 9.8
21. 12	**22.** 50	**23.** 750	**24.** 4400
25. 3.0007	**26.** 5.732	**27.** 12.55	**28.** 60.6
29. $0.62\frac{1}{2}$	**30.** $0.6\frac{1}{4}$	**31.** $4\frac{1}{2}$	**32.** $3.72\frac{1}{4}$

2 *Change the following percents to decimals.*

33. 48%	**34.** 62%	**35.** 5%	**36.** 2%
37. 72.2%	**38.** 60.9%	**39.** 200%	**40.** 350%
41. 234.7%	**42.** 600.4%	**43.** 4,000%	**44.** 1,245%
45. 0.04%	**46.** 0.25%	**47.** 0.0006%	**48.** 0.0012%
49. 80.91%	**50.** 69.331%	**51.** 2.3%	**52.** 5.007%
53. 290%	**54.** 385%	**55.** 2,263.7%	**56.** 100.01%
57. $6\frac{1}{2}\%$	**58.** $22\frac{1}{2}\%$	**59.** $35\frac{1}{4}\%$	**60.** $4\frac{1}{4}\%$
61. $12\frac{3}{8}\%$	**62.** $7\frac{1}{10}\%$		

SKILLSFOCUS (Section 5.8) *Change each fraction to a decimal.*

63. $\dfrac{4}{5}$ **64.** $\dfrac{3}{8}$ **65.** $\dfrac{7}{10}$ **66.** $\dfrac{1}{6}$ (round to thousandths)

EXTEND YOUR THINKING ▶▶▶▶

▶ **TROUBLESHOOT IT**

Find and correct the error.

67. $0.004\% = 0\ 00.4\% = 0.4$

68. $250 = 250. = 2.50\ \% = 2.5\%$

WRITING TO LEARN ▶▶▶▶

69. Change 0.84 to a percent. Write an explanation for each step.

70. Change 240% to a decimal. Write an explanation for each step.

7.3 FRACTIONS AND PERCENTS

OBJECTIVES

1 Change a fraction to a percent (Sections 3.4, 5.8).

2 Change a percent to a fraction (Sections 3.5, 3.6).

3 Change a decimal percent to a fraction (Section 5.8).

---

1 Changing a Fraction to a Percent

To change a fraction to a percent, first change the fraction to a decimal (Section 5.8). Then change the decimal to a percent (Section 7.2). For example, change $\frac{3}{4}$ to a percent.

$$\frac{3}{4} \quad \boxed{\text{change to a decimal}} \longrightarrow \quad 4\overline{)3.00} \begin{array}{r} 0.75 \\ \underline{2\,8} \\ 20 \\ \underline{20} \\ 0 \end{array} \quad \boxed{\text{change to a percent}} \longrightarrow \quad 0.75 = 0\,75.\% = 75\%$$

> **To change a fraction to a percent**
>
> 1. Change the fraction to a decimal by dividing the numerator by the denominator.
> 2. Change the decimal to a percent.

EXAMPLE 1 Change each fraction to a percent.

a. $\frac{3}{5}$ $\boxed{\text{change to a decimal}} \longrightarrow$ $5\overline{)3.0} \begin{array}{r} 0.6 \\ \underline{3\,0} \\ 0 \end{array}$ $\boxed{\text{change to a percent}} \longrightarrow$ $0.6 = 0\,60.\% = 60\%$

Therefore, $\frac{3}{5} = 60\%$.

b. $\frac{1}{8}$ $\boxed{\text{change to a decimal}} \longrightarrow$ $8\overline{)1.000} \begin{array}{r} 0.125 \\ \underline{8} \\ 20 \\ \underline{16} \\ 40 \\ \underline{40} \\ 0 \end{array}$ $\boxed{\text{change to a percent}} \longrightarrow$ $0.125 = 0\,12.5\%$
$= 12.5\%$

Therefore, $\frac{1}{8} = 12.5\%$. $\left(\text{NOTE, } 12.5\% = 12\frac{1}{2}\%\right)$ ■

▶ **You Try It** Change each fraction to a percent.

1. $\frac{2}{5}$ **2.** $\frac{3}{8}$ **3.** $\frac{9}{20}$

When changing a fraction to a decimal, the decimal may repeat. In this case, you can carry the division to two decimal places. Write any remainder over the divisor. Then change to a percent.

EXAMPLE 2   Change $\frac{2}{3}$ to a percent.

$$
\begin{array}{r}
0.66\frac{2}{3} \\
3 \overline{)\,2.00} \\
1\,8 \\
\overline{20} \\
18 \\
\overline{2}
\end{array}
$$

$0.66\frac{2}{3} = 66\frac{2}{3}\%$

Therefore, $\frac{2}{3} = 66\frac{2}{3}\%$ ■

▶ **You Try It**   Change each fraction to a percent.

4. $\frac{1}{3}$

5. $\frac{7}{11}$

When changing a fraction to a decimal, round the decimal to any place value you choose. Then change to a percent.

EXAMPLE 3   Change $\frac{3}{7}$ to a percent. Round to thousandths.

$$
\begin{array}{r}
0.4285 \doteq 0.429 = 42.9\% \\
7 \overline{)\,3.0000} \\
2\,8 \\
\overline{20} \\
14 \\
\overline{60} \\
56 \\
\overline{40} \\
35 \\
\overline{5}
\end{array}
$$

To round to thousandths, add zeros to the ten-thousandths place.

Therefore, $\frac{3}{7} = 42.9\%$.   ■

▶ **You Try It**   Change each fraction to a percent. Round each decimal to the indicated place before changing to a percent.

6. $\frac{8}{9}$ (thousandths)

7. $\frac{5}{24}$ (ten-thousandths)

To change a mixed number to a percent, first change the mixed number to an improper fraction.

EXAMPLE 4 Change $3\frac{17}{20}$ to a percent.

$$3\frac{17}{20} = \frac{77}{20} \longrightarrow 20\overline{)77.00} \qquad 3.85 = 3\ 85.\% \quad \text{or} \quad 385\%$$
$$\underline{60}$$
$$170$$
$$\underline{16\,0}$$
$$1\,00$$
$$\underline{1\,00}$$
$$0$$

Therefore, $3\frac{17}{20} = 385\%$. ∎

▶ **You Try It** **8.** Change $1\frac{4}{5}$ to a percent.

2 **Changing a Percent to a Fraction**

Percent means hundredths, or parts per 100. To change a percent to a fraction, write the percent as hundredths. Then simplify.

$$72\% = 72 \text{ hundredths} = \frac{72}{100} = \frac{\overset{18}{\cancel{72}}}{\underset{25}{\cancel{100}}} = \frac{18}{25}$$

Drop % symbol, write number that remains over 100.

EXAMPLE 5 Change each percent to a fraction.

a. $25\% = 25 \text{ hundredths} = \frac{25}{100} = \frac{\overset{1}{\cancel{25}}}{\underset{4}{\cancel{100}}} = \frac{1}{4}$ **b.** $9\% = 9 \text{ hundredths} = \frac{9}{100}$

c. $300\% = 300 \text{ hundredths} = \frac{300}{100} = \frac{\overset{3}{\cancel{300}}}{\underset{1}{\cancel{100}}} = \frac{3}{1} = 3$ ∎

> **To change a percent to a fraction**
>
> **1.** Drop the % symbol.
> **2.** Write the number that remains over 100.
> **3.** Simplify.

EXAMPLE 6 Change each percent to a fraction.

a. $79\% = \frac{79}{100}$ **b.** $60\% = \frac{60}{100} = \frac{\overset{3}{\cancel{60}}}{\underset{5}{\cancel{100}}} = \frac{3}{5}$

c. $2\% = \frac{2}{100} = \frac{\overset{1}{\cancel{2}}}{\underset{50}{\cancel{100}}} = \frac{1}{50}$ **d.** $100\% = \frac{100}{100} = 1$

e. $160\% = \dfrac{160}{100} = \dfrac{\overset{8}{\cancel{160}}}{\underset{5}{\cancel{100}}} = \dfrac{8}{5}$ or $1\dfrac{3}{5}$ ■

▶ **You Try It** Change each percent to a fraction.

9. 30% **10.** 75% **11.** 80% **12.** 6%

13. 200% **14.** 650% **15.** 21% **16.** 1%

EXAMPLE 7 Change $16\dfrac{2}{3}\%$ to a fraction.

$$16\dfrac{2}{3}\% = \dfrac{16\dfrac{2}{3}}{100} = 16\dfrac{2}{3} \div 100 = \dfrac{50}{3} \div \dfrac{100}{1} = \dfrac{\overset{1}{\cancel{50}}}{3} \times \dfrac{1}{\underset{2}{\cancel{100}}} = \dfrac{1}{6}$$

Fraction bar means divide $16\dfrac{2}{3}$ by 100.

To divide fractions, invert the divisor and multiply.

Suppose $16\dfrac{2}{3}\%$ of a paycheck is tax. Since $16\dfrac{2}{3}\% = \dfrac{1}{6}$, this means $\dfrac{1}{6}$ of the paycheck, or \$1 out of every \$6 earned is tax. In this example the fraction form of the percent is easier to understand. ■

EXAMPLE 8 Change $\dfrac{1}{4}\%$ to a fraction. Interpret the answer.

$$\dfrac{1}{4}\% = \dfrac{\dfrac{1}{4}}{100} = \dfrac{1}{4} \div \dfrac{100}{1} = \dfrac{1}{4} \times \dfrac{1}{100} = \dfrac{1}{400}$$

$\dfrac{1}{4}\% = \dfrac{1}{400}$ and can be interpreted to mean 1 part out of 400 equal parts. If $\dfrac{1}{4}\%$ of a population lives to be 100 years old, then 1 person in 400 lives to that age. ■

▶ **You Try It** Change each percent to a fraction.

17. $4\dfrac{1}{6}\%$ **18.** $\dfrac{5}{8}\%$

---

3 Changing a Decimal Percent to a Fraction

Change 12.7% to a fraction.

$$12.7\% = .12\ 7\% = .127 = 127 \text{ thousandths} = \dfrac{127}{1,000}$$

First change the percent to a decimal.

Then change the decimal to a fraction.

> To change a decimal percent to a fraction
>
> 1. Change the percent to a decimal.
> 2. Change the decimal to a fraction.
> 3. Reduce.

EXAMPLE 9   Change each percent to a fraction.

a. $3.5\% = .03\ 5\% = .035 = 35\ \text{thousandths} = \dfrac{35}{1,000} = \dfrac{\overset{7}{\cancel{35}}}{\underset{200}{\cancel{1,000}}} = \dfrac{7}{200}$

b. $72.25\% = .72\ 25\% = .7225 = 7,225\ \text{ten-thousandths}$

$= \dfrac{7,225}{10,000} = \dfrac{\overset{289}{\cancel{7,225}}}{\underset{400}{\cancel{10,000}}} = \dfrac{289}{400}$ Reduce using 25.

c. $0.0008\% = .00\ 0008\% = .000008 = 8\ \text{millionths}$

$= \dfrac{8}{1,000,000} = \dfrac{\overset{1}{\cancel{8}}}{\underset{125,000}{\cancel{1,000,000}}} = \dfrac{1}{125,000}$

d. $1.2\dfrac{1}{6}\% = .01\ 2\dfrac{1}{6}\% = .012\dfrac{1}{6} = 12\dfrac{1}{6}\ \text{thousandths}$

$= \dfrac{12\dfrac{1}{6}}{1,000} = \dfrac{73}{6} \div \dfrac{1,000}{1} = \dfrac{73}{6} \times \dfrac{1}{1,000} = \dfrac{73}{6,000}$ ■

EXAMPLE 10   In some states a blood alcohol level of 0.1% or higher is considered proof of intoxication. Express this level as a fraction.

$$0.1\% = 0.00\ 1\% = 0.001 = \dfrac{1}{1,000}$$

Intoxicated means $\dfrac{1}{1,000}$ (or more) of your blood volume is alcohol. ■

▶ **You Try It**   Change each percent to a fraction.

19. 7.5%   **20.** 6.25%   **21.** 0.02%   **22.** $2.4\dfrac{1}{3}\%$

23. A machine can produce parts with measurements accurate to 2.5% of a centimeter. Express this accuracy as a fraction.

▶ **Answers to You Try It**   **1.** 40%   **2.** 37.5%   **3.** 45%   **4.** $33\dfrac{1}{3}\%$ **5.** $63\dfrac{7}{11}\%$

6. 88.9%   **7.** 20.83%   **8.** 180%   **9.** $\dfrac{3}{10}$ **10.** $\dfrac{3}{4}$ **11.** $\dfrac{4}{5}$ **12.** $\dfrac{3}{50}$ **13.** 2

14. $6\dfrac{1}{2}$ **15.** $\dfrac{21}{100}$ **16.** $\dfrac{1}{100}$ **17.** $\dfrac{1}{24}$ **18.** $\dfrac{1}{160}$ **19.** $\dfrac{3}{40}$ **20.** $\dfrac{1}{16}$ **21.** $\dfrac{1}{5,000}$

22. $\dfrac{73}{3,000}$ **23.** accurate to $\dfrac{1}{40}$ of a centimeter

SOMETHING MORE

Deception with Percents? Advertisements have been known to lead the unwary into thinking a product is good based on research the consumer seldom sees. A TV ad used some time ago to sell shaving equipment claimed:

> 57% of the people who use Smoothe say they like it as well as or better than their current blade.

This boast makes one feel that more people prefer Smoothe. Further thought, however, could lead to something quite different.

Determining what the claim means depends on how you interpret "as well as or better than." When someone uses a Smoothe blade, there are three distinct possibilities.

1. like Smoothe better than their current blade
2. like Smoothe as well as their current blade
3. like Smoothe less than their current blade

The following table presents three possible ways to interpret 57%.

liked Smoothe "as well as or better than" their current blade	liked Smoothe better	as well as	liked Smoothe less	liked current blade "as well as or better than" Smoothe
57% (53% + 4%)	53%	4%	43%	47% (4% + 43%)
57% (7% + 50%)	7%	50%	43%	93% (50% + 43%)
57% (0% + 57%)	0%	57%	43%	100% (57% + 43%)

For instance, 57% liked Smoothe "as well as or better than" could mean 50% liked it "as well as" and 7% liked it "better than." Then 100% − 57% = 43% liked it "less." In other words, 43% liked the competing blade better. This means 50% + 43% = 93% liked their current blade "as well as or better than" Smoothe, an unexpected turn of the tables on the Smoothe ad.

The extreme case comes from the bottom row of the table. Assume 0% liked Smoothe "better" and 57% liked it "as well as." Then the remaining 43% liked it "less." From this you conclude 57% + 43% = 100% liked the other brand "as well as or better than" Smoothe!

You have no way of knowing how 57% was actually split between the "as well as" and the "better than" people when Smoothe consumer tested its blade. However, a company may very well take its test results and word them in an ad so that their product is viewed in the best possible light. 57% may give a false impression or mislead consumers, even though it may be a perfectly accurate number derived from consumer data.

a. 18% of the people tested liked CRX better than and 34% liked it as well as their current brand. What percent liked CRX "as well as or better than" their current brand?

b. What percent liked their current brand "as well as or better than" CRX?

▶ **Answers to Something More** a. 18% + 34% = 52% liked CRX as well as or better than their current brand. b. 100% − 52% = 48% liked their current brand better. Therefore, 34% + 48% = 82% liked their current brand as well as or better than CRX.

1 *Change each fraction to a percent.*

1. $\dfrac{1}{4}$

2. $\dfrac{1}{3}$

3. $\dfrac{4}{5}$

4. $\dfrac{1}{7}$

5. $\dfrac{5}{8}$

6. $\dfrac{1}{6}$

7. $\dfrac{4}{11}$

8. $\dfrac{2}{9}$

9. $\dfrac{5}{12}$

10. $\dfrac{5}{6}$

11. $\dfrac{9}{10}$

12. $\dfrac{7}{9}$

13. $\dfrac{7}{5}$

14. $\dfrac{21}{10}$

15. $\dfrac{15}{8}$

16. $\dfrac{7}{2}$

17. $2\dfrac{1}{4}$

18. $5\dfrac{3}{8}$

19. $1\dfrac{1}{2}$

20. $3\dfrac{23}{25}$

21. $4\dfrac{1}{40}$

22. $9\dfrac{11}{50}$

23. $6\dfrac{1}{3}$

24. $2\dfrac{4}{15}$

25. $0.2\dfrac{1}{6}$

26. $0.4\dfrac{5}{9}$

27. $0.04\dfrac{2}{3}$

28. $0.36\dfrac{4}{11}$

29. Change $\dfrac{5}{16}$ into a percent.

30. Change $4\dfrac{2}{7}$ into a percent.

31. One-twentieth of the people in a village are hunters. What percent are hunters?

32. Twenty-four twenty-fifths of the weight of a head of lettuce is water. What percent of the lettuce is water?

33. Sam lost $\dfrac{2}{27}$ of his body weight playing football. What percent weight loss is this?

34. About $\frac{7}{10}$ of the Earth's surface is covered with water. What percent is this?

2 *Change each percent to a fraction. Reduce to lowest terms.*

35. 55%

36. 38%

37. 90%

38. 20%

39. 69%

40. 12%

41. 8%

42. 94%

43. 180%

44. 240%

45. 700%

46. 450%

47. $66\frac{2}{3}\%$

48. $12\frac{1}{2}\%$

49. $83\frac{1}{3}\%$

50. $28\frac{4}{7}\%$

51. $237\frac{1}{2}\%$

52. $444\frac{4}{9}\%$

53. $\frac{1}{5}\%$

54. $\frac{7}{10}\%$

55. $\frac{1}{7}\%$

56. $\frac{8}{11}\%$

---

57. $5\frac{1}{3}\%$ of the taxes collected in our city pays for sanitation. Change this percent to a fraction and interpret the answer.

58. $32\frac{3}{4}\%$ of a paycheck is tax. Change this percent to a fraction and interpret the answer.

59. $82\frac{1}{2}\%$ of your students study a foreign language. Change this percent to a fraction, and interpret the answer.

60. $2\frac{4}{5}\%$ of the bulbs are defective. Change this percent to a fraction and interpret the answer.

61. If 98% of a product is pure,
  a. what percent is not pure?

  b. what fraction of the product is not pure?

62. 42% of a family's income is spent on rent, and $20\frac{1}{2}\%$ is spent on food.

  a. What percent of the income is spent on rent and food?

  b. What percent is spent on items other than rent and food?

  c. What fraction is spent on items other than rent and food?

3 *Change the decimal percent to a fraction.*

63. 2.2%   **64.** 0.3%   **65.** 16.25%   **66.** 8.4%

67. 42.5%   **68.** 90.05%   **69.** 62.5%   **70.** 1.025%

71. 0.162%   **72.** 0.05%   **73.** 0.00002%   **74.** 0.0006%

75. 140.5%   **76.** 330.75%   **77.** $5.2\frac{1}{3}$%   **78.** $0.4\frac{1}{7}$%

79. 51.3% of the babies born in the United States are female. What fraction of the babies are female?

82. 71.4% of the Earth is covered with water. What fraction of the Earth is this?

80. 5.26% of each dollar made is profit. What fraction of each dollar is this?

83. 8.125% of the budget is spent on defense. What fraction of the budget is spent on defense?

81. 0.9% of a substance is niacin. What fraction is niacin?

84. Inflation rose 0.75% this month. By what fraction did inflation rise this month?

Any number can be written in three ways: as a fraction, as a decimal, and as a percent. Write the given number in the other two forms. Round decimals to the ten-thousandths place.

	Fraction	Decimal	Percent
85.		0.36	
86.			72%
87.	$\frac{5}{4}$		
88.			45%
89.		1.5	
90.	$\frac{9}{11}$		
91.	$\frac{6}{10}$		

	Fraction	Decimal	Percent
92.			$2\frac{1}{6}$%
93.		0.008	
94.	$1\frac{6}{7}$		
95.			$166\frac{2}{3}$%
96.			375%
97.		0.6225	
98.	$\frac{17}{20}$		

Solve each equation. Round to hundredths.

99. $8b = 28$ **100.** $120r = 45$ **101.** $0.72y = 400$ **102.** $57.3m = 9.6$

EXTEND YOUR THINKING ▶▶▶▶

▶SOMETHING MORE

103. 12% of the women tested liked MBQ better than and 51% liked it as well as their current brand. What percent liked MBQ "as well as or better than" their current brand? What percent liked their current brand "as well as or better than" MBQ?

104. **Using a Calculator to Change a Fraction to a Percent**   You can change a fraction to a percent by using the ⟨%⟩ key on your calculator. For example, to change $\frac{3}{4}$ to a percent, enter the following sequence of keypresses.

$$3 \div 4 \, \%$$

The calculator displays 75, meaning 75%. Use your calculator to change the following fractions to percents.

a. $\dfrac{1}{2}$ b. $\dfrac{5}{8}$

c. $\dfrac{3}{5}$ d. $\dfrac{2}{3}$

e. $5\dfrac{1}{4}$

▶TROUBLESHOOT IT

Find and correct the error.

105. $\dfrac{2}{25} \longrightarrow 25 \overline{)2.00} = 0 \; 80.\% \text{ or } 80\%. \text{ So, } \dfrac{2}{25} = 80\%.$
$\dfrac{2\,00}{0}$

106. $0.6\% = \dfrac{0.6}{100} = \dfrac{6}{10} = \dfrac{3}{5}$

WRITING TO LEARN ▶▶▶▶

107. Explain how to change $\dfrac{7}{20}$ to a percent, as you would explain it to a classmate. Write a reason for each step.

108. Explain how to change 0.8 to a fraction, as you would explain it to a classmate. Write a reason for each step.

7.4 IDENTIFYING THE THREE NUMBERS IN A PERCENT WORD PROBLEM

OBJECTIVE	NEW VOCABULARY
Identify the three numbers in a percent word problem.	rate
	amount

Percent Word Problems

Every percent word problem involves three numbers.

R% = the rate (or percent)
B = the base
A = the amount (also called percentage)

> **BASE = the whole population; it represents 100%**
> You take a percent of the whole population, or base. Hence the base usually follows the words "percent of" or follows a numerical percent (such as "25% of").
>
> **RATE = a portion of the base, expressed as a percent**
> It is the number written with the word "percent" or with the percent symbol, %.
>
> **AMOUNT = a portion of the base**
> It is the remaining number after the base and rate have been identified.

EXAMPLE 1

25% of 8 cars is 2 cars.

rate R%	base B	amount A
Has the % symbol.	The whole population. It follows "25% of."	Equals 25% of 8 cars. ∎

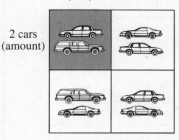

25% of
the cars
(rate)

2 cars
(amount)

whole rectangle is 8 cars
(base, or 100%)

The diagram shows how the amounts in Example 1 are related. The whole rectangle equals the base, or 100%. It represents 8 cars. 25% of the cars are represented by the shaded region. The shaded region also represents 2 cars. As you see, *the rate (25%) and the amount (2 cars) represent the same portion of the base*.

The first step to solve any percent word problem is to identify which numbers represent the rate, the base, and the amount. The rate is easiest to identify. It is written with the word "percent" or with the percent symbol, %. Identify the rate first. Then identify the base. The base follows the words "percent of." The amount is the remaining number after rate and base have been identified.

EXAMPLE 2

75% of $400 is $300.

rate	R% = 75%	(number written with percent symbol, %)
base	B = $400	(follows the words "75% of")
amount	A = $300	(number remaining after rate and base are identified) ∎

EXAMPLE 3

0.00024 is 0.3% of 0.8.

rate	R% = 0.3%	
base	B = 0.8	(follows "0.3% of")
amount	A = 0.00024	(remaining number) ∎

▶ **You Try It**   Identify the base, rate, and amount.

1. 60% of 30 is 18.   **2.** 0.0232 is 2.9% of 0.8.   **3.** 15 is $33\frac{1}{3}$% of 45.

When solving a percent word problem, two of the three numbers will be given. You must identify the unknown third number.

EXAMPLE 4   40% of what number is $20?

rate   R% = 40%
base   B = ?   (the base is unknown because the words
                         "what number" follow "40% of")
amount   A = $20   (remaining number) ■

EXAMPLE 5   What percent of 60 men is 36 men?

rate   R% = ?   (the rate is unknown because no number is
                         with the word "percent" in the problem)
base   B = 60 men   (follows "percent of")
amount   A = 36 men   (remaining number) ■

EXAMPLE 6   What number is $37\frac{1}{2}$% of 64?

rate   R% = $37\frac{1}{2}$%

base   B = 64   (follows "$37\frac{1}{2}$% of")

amount   A = ?   (the remaining number is unknown) ■

▶ **You Try It**   Identify the base, rate, and amount. Do not solve.

4. What percent of 40 is 25?   **5.** $3\frac{1}{2}$% of $50 is how much?

6. What is 10% of 65?   **7.** 17% of what number is 34?

EXAMPLE 7   40 is what percent of 30?

rate   R% = ?
base   B = 30   (follows "percent of")
amount   A = 40   (remaining number) ■

CAUTION   Just because 40 is larger than 30 does not make 40 the base. 40 is not the base because it does not follow "percent of."

▶ **You Try It**   Identify the base, rate, and amount.

8. What percent of 8 is 20?   **9.** $60 is 150% of what?

▶ **Answers to You Try It**   **1.** R% = 60%, B = 30, A = 18
2. R% = 2.9%, B = 0.8, A = 0.0232   **3.** R% = $33\frac{1}{3}$%, B = 45, A = 15

4. R% = ?, B = 40, A = 25   **5.** R% = $3\frac{1}{2}$%, B = $50, A = ?

6. R% = 10%, B = 65, A = ?   **7.** R% = 17%, B = ?, A = 34
8. R% = ?, B = 8, A = 20   **9.** R% = 150%, B = ?, A = 60

SECTION 7.4 EXERCISES

Identify the rate R%, base B, and amount A.

1. 65% of 300 cars is 195 cars.

2. 19% of 400 women is 76 women.

3. 8% of $35 is $2.80.

4. 12.6% of 500 men is 63 men.

5. 0.07% of 6,500 kilograms is 4.55 kilograms.

6. 32 cats is 25% of 128 cats.

7. 0.648 ounces is 120% of 0.540 ounces.

8. 900 cattle is 10% of 9,000 cattle.

9. $300 is 600% of $50.

10. 93 doctors is 150% of 62 doctors.

Identify the rate R%, base B, and the amount A. Denote the unknown number with a question mark, ?. Do not solve.

11. 44% of what number is 35?

12. $83\frac{1}{3}$% of what number is 75?

13. 12 is 20% of what number?

14. 6.8 is 87% of what number?

15. What is 38% of 65 pounds?

16. What is 37.5% of 1,200 geese?

17. 90% of $550 is how much?

18. 170% of 1,000 people is how many people?

19. What percent of 63.2 is 21.8?

20. What percent of 0.35 is 0.07?

21. 55 is what percent of 88?

22. 100 dogs is what percent of 150 dogs?

23. What percent of 40 is 128?

24. What percent of 2 is 3.5?

25. 450 is what percent of 180?

26. $120 is what percent of $75?

27. Find 34% of $60.50.

28. Calculate 190% of $26.

29. $\dfrac{2}{3} + \dfrac{4}{5}$ **30.** $\dfrac{7}{8} + \dfrac{3}{4}$ **31.** $1\dfrac{7}{9} + 3\dfrac{5}{6}$ **32.** $5\dfrac{3}{7} + \dfrac{7}{8}$

EXTEND YOUR THINKING ▶▶▶▶

▶ TROUBLESHOOT IT

Find and correct the error.

33. $50 is what percent of $36?

     R% = ?
     B   = $50
     A   = $36 (remaining number)

WRITING TO LEARN ▶▶▶▶

34. In your own words, define rate, base, and amount.

35. In your own words, explain how the word *rate* used in Section 6.2 is like *rate* used in this chapter. Explain how they are different.

7.5 SOLVING THE SHORT PERCENT WORD PROBLEM: R% OF B IS A

OBJECTIVES

1 Define the short percent problem R% of B is A.

2 Find the amount, A (Section 3.7).

3 Find the base, B (Section 2.6).

4 Find the rate, R% (Section 2.6).

5 Solve percent problems using the proportion method (optional) (Sections 6.6, 6.7).

NEW VOCABULARY

short form

proportion method

---

1 Defining the Short Percent Problem R% of B is A

Every percent word problem involves taking a percent (R%) of the base (B) to get the amount (A). This is written

$$R\% \text{ of B is A}$$

or, reversing it,

$$A \text{ is } R\% \text{ of B.}$$

This is called the **short form** for a percent word problem. An example is

$$25\% \text{ of 8 is 2}$$

or, reversing it,

$$2 \text{ is } 25\% \text{ of 8.}$$

This example does not qualify as a problem because the rate, base, and amount are known. An actual percent problem will give you two of these numbers and ask you to find the third.

Three Types of Percent Problems

To find the:	You must be given the:	For example:
1. rate (R%) ⟶	base (B) & amount (A) ⟶	What percent of 8 is 2?
2. base (B) ⟶	rate (R%) & amount (A) ⟶	25% of what is 2?
3. amount (A) ⟶	rate (R%) & base (B) ⟶	25% of 8 is what?

2 Finding the Amount, A

The short percent word problem A is R% of B can be rewritten as an equation.

$$\left. \begin{array}{l} A \text{ is } R\% \text{ of B} \\ A = R\% \times B \end{array} \right\}$$ Recall, "of" indicates multiplication (Section 3.7).

To find the amount A, this equation says multiply the rate times the base.

$$\text{amount} = \text{rate} \times \text{base}$$
$$A = R\% \times B$$

To solve percent word problems, first identify the three numbers R%, B, and A. Change the rate R% to a fraction or decimal before calculating with it.

EXAMPLE 1   42% of 250 is what number?

R% = 42% (= 0.42)
B = 250 (follows "42% of")
A = ?

$$A = R\% \times B$$
$$A = 0.42 \times 250$$
$$A = 105 \quad \blacksquare$$

EXAMPLE 2   What is $37\frac{1}{2}\%$ of 64?

R% = $37\frac{1}{2}\%$ (= 37.5% or 0.375)

B = 64 $\left(\text{follows "}37\frac{1}{2}\% \text{ of"}\right)$

A = ?

$$A = R\% \times B$$
$$A = 0.375 \times 64$$
$$A = 24 \quad \blacksquare$$

You can also solve Example 2 by changing $37\frac{1}{2}\%$ to a fraction instead of a decimal. Since $37\frac{1}{2}\% = \frac{3}{8}$,

$$A = R\% \times B \quad = \frac{3}{8} \times 64 = \frac{3}{\cancel{8}} \times \frac{\cancel{64}^{\,8}}{1} = 24.$$

▶ **You Try It**

1. 30% of $180 is how much?
2. What is 4.5% of 80?

---

3 Finding the Base, B   The next example demonstrates how to find the base.

EXAMPLE 3   15 is 75% of what number?

R% = 75% (= 0.75)
B = ? (the words "what number" follow "75% of")
A = 15 (remaining number)

$$A = R\% \times B$$
$$15 = 0.75 \times B$$
$$\frac{15}{0.75} = \frac{0.75}{0.75} \times B \qquad \left\{ \begin{array}{l} \text{Divide each side by} \\ \text{0.75 to solve for B.} \end{array} \right.$$

amount A

$$\frac{15}{0.75} = B \qquad \left\{ \begin{array}{l} \text{As seen at this step, the base B} \\ \text{equals the amount 15 divided} \\ \text{by the rate 0.75.} \end{array} \right.$$

rate R%

$$B = 20 \quad \blacksquare$$

From Example 3, the base was found by dividing the amount by the rate.

$$\text{base} = \frac{\text{amount}}{\text{rate}}$$

$$B = \frac{A}{R\%}$$

EXAMPLE 4   32% of what number is 4.8?

$R\% = 32\% \ (= 0.32)$
$B \quad = ?$ (the words "what number" follow "32% of")
$A \quad = 4.8$

$$B = \frac{A}{R\%}$$

$$B = \frac{4.8}{0.32}$$

$$B = 15 \quad \blacksquare$$

▶ **You Try It**   **3.** 24 is 60% of what number?   **4.** 76% of what number is 38?

EXAMPLE 5 $16\frac{2}{3}\%$ of what number is 10?

$R\% = 16\frac{2}{3}\% \left(= \frac{1}{6}, \text{ see Example 7, Section 7.3, page 420} \right)$

$B \quad = ? \left(\text{the words "what number" follow "}16\frac{2}{3}\% \text{ of"} \right)$

$A \quad = 10$

$$B = \frac{A}{R\%}$$

$$B = \frac{10}{\frac{1}{6}} = 10 \div \frac{1}{6} = \frac{10}{1} \times \frac{6}{1} = 60 \quad \blacksquare$$

▶ **You Try It**   **5.** $87\frac{1}{2}\%$ of what number is 63?

4 Finding the Rate, R%

The next example demonstrates how to find the rate.

EXAMPLE 6   What percent of 25 is 6?

$R\% = ?$
$B \quad = 25$ (follows "percent of")
$A \quad = 6$ (the remaining number)

$$A = R\% \times B$$

$$6 = R\% \times 25$$

$$\frac{6}{25} = R\% \times \frac{25}{25} \quad \left\{ \begin{array}{l} \text{Divide each side by} \\ \text{25 to solve for R\%.} \end{array} \right.$$

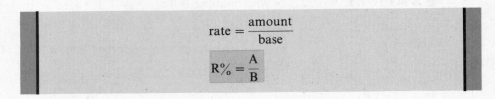

amount A $\longrightarrow \dfrac{6}{25} = $ R% $\quad$ { As seen at this step, the rate R% equals the amount 6 divided by the base 25.

base B

$$R\% = 0.24 \qquad \text{Change 0.24 to a percent.}$$
$$R\% = 24\% \quad \blacksquare$$

From Example 6, the rate equals the amount divided by the base.

$$\text{rate} = \frac{\text{amount}}{\text{base}}$$

$$R\% = \frac{A}{B}$$

EXAMPLE 7   0.03 is what percent of 0.6?

R% = ?
B = 0.6 (follows "percent of")
A = 0.03

$$R\% = \frac{A}{B}$$

$$R\% = \frac{0.03}{0.6}$$

$$R\% = 0.05 \qquad \text{Change 0.05 to a percent.}$$
$$R\% = 5\% \quad \blacksquare$$

▶ **You Try It**   **6.** What percent of 12 is 9?
   7. 1.6 is what percent of 40?

EXAMPLE 8   What percent of 280 is 120?

R% = ?
B = 280 (follows "percent of")
A = 120

$$R\% = \frac{A}{B}$$

$$R\% = \frac{120}{280} \qquad \text{Reduce } \frac{\overset{3}{\cancel{120}}}{\underset{7}{\cancel{280}}} = \frac{3}{7}.$$

$$R\% = \frac{3}{7} \qquad \text{Change } \frac{3}{7} \text{ to a percent.}$$

$$R\% = 42\frac{6}{7}\% \quad \blacksquare$$

▶ **You Try It**   **8.** What percent of 126 is 98?

The Percent Triangle

The following triangle summarizes the three percent formulas discussed in this section. To use it, cover the number you are looking for with a finger. The triangle then shows the formula you need to find that number.

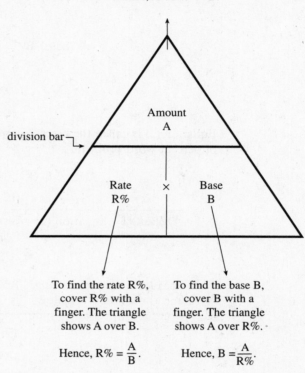

To find the amount A,
cover A with a finger.
The triangle shows
R% × B.
Hence, A = R% × B.

division bar

Amount
A

Rate
R%

×

Base
B

To find the rate R%,
cover R% with a
finger. The triangle
shows A over B.

Hence, R% = $\dfrac{A}{B}$.

To find the base B,
cover B with a
finger. The triangle
shows A over R%.

Hence, B = $\dfrac{A}{R\%}$.

Just remember that the amount, A, always belongs in the top of the triangle (try to remember *amount, A, always above*). The percent triangle is a memory aid. Use it to help you remember the percent equations.

CONNECTIONS

The percent triangle can be used to help you remember formulas involving three unknowns from many different areas. Examples are distance, speed, and time problems (D = S · T), Ohm's Law (E = I · R), and the area of a rectangle (A = L · W). The letter by itself on one side of the equal sign is written in the top of the triangle. The letters in the product are written below.

---

5 Using the Proportion Method to Solve Percent Word Problems (Optional)

Example 1, Section 7.4, said that 25% of 8 cars is 2 cars. 25% is the ratio $\dfrac{25}{100}$.

2 cars out of 8 cars is the ratio $\dfrac{2}{8}$. Set these ratios equal. You get a true proportion.

$$\frac{2}{8} = \frac{25}{100} \tag{1}$$

Recall from Section 6.6 that a proportion is true if both cross products are equal. Both cross products, 8×25 and 2×100, equal 200.

If 25% of 8 cars is 2 cars, then you know the following.

$$R\% = 25\% \text{ (so } R = 25)$$
$$B = 8 \quad \text{(follows "25% of")}$$
$$A = 2$$

Substitute these letters for the numbers in Equation (1).

(amount) A $\longrightarrow$ $\qquad$ $\longleftarrow$ R (rate)

$$\frac{2}{8} = \frac{25}{100}$$

(base) B $\longrightarrow$

or

$$\frac{A}{B} = \frac{R}{100}. \tag{2}$$

Equation (2) is called the **percent proportion** and can be used to solve any kind of percent word problem.

OBSERVE Equation (2) is also written

$$\frac{\text{is}}{\text{of}} = \frac{\%}{100}$$

because in the short form, R% of B is A, the base B follows "of," and the amount A follows "is."

The Proportion Method

To use proportions to solve percent word problems

1. Identify R%, B, and A as shown in Section 7.4.

2. Substitute these values into the percent proportion. Represent the unknown quantity with its own letter.

$$\frac{A}{B} = \frac{R}{100}$$

With the proportion method, there is no need to change R% to a fraction or decimal before using it. Just drop the percent symbol, %, and substitute the number that remains for R. For example, if R% = 36%, then R = 36. Replace R with 36. (In fact, the right side of the percent proportion already has the rate written in fraction form, with a denominator of 100.)

3. Solve the proportion.

EXAMPLE 9   30% of what number is 24?

R% = 30% (so R = 30)
B   = ? ("what number" follows "30% of")
A   = 24

$$\frac{A}{B} = \frac{R}{100}$$

$$\frac{24}{B} = \frac{30}{100}$$

$$B = \frac{24 \times 100}{30}$$

$$B = 80 \blacksquare$$

EXAMPLE 10   4 is $12\frac{1}{2}$% of what number?

$$R\% = 12\frac{1}{2}\% \left(\text{so } R = 12\frac{1}{2} = 12.5 \right)$$

$$B = ? \left(\text{"what number" follows "}12\frac{1}{2}\%\text{ of"} \right)$$

$$A = 4$$

$$\frac{A}{B} = \frac{R}{100}$$

$$\frac{4}{B} = \frac{12.5}{100}$$

$$B = \frac{4 \times 100}{12.5}$$

$$B = 32 \blacksquare$$

▶ **You Try It**   Solve using the proportion method.

9. 70% of what number is 56?

10. 121 is $31\frac{1}{4}$% of what number?

EXAMPLE 11   What is 4.8% of 7.5?

R% = 4.8% (so R = 4.8)
B   = 7.5 (follows "4.8% of")
A   = ?

$$\frac{A}{B} = \frac{R}{100}$$

$$\frac{A}{7.5} = \frac{4.8}{100}$$

$$A = \frac{4.8 \times 7.5}{100}$$

$$A = 0.36 \blacksquare$$

EXAMPLE 12   18 is what percent of 60?

R% = ?
B = 60 (follows "percent of")
A = 18

$$\frac{A}{B} = \frac{R}{100}$$

$$\frac{18}{60} = \frac{R}{100}$$

$$R = \frac{18 \times 100}{60}$$

$$R = 30$$

Attach the percent symbol to 30. The answer is 30%. ■

▶ **You Try It**   Solve using the proportion method.
11. Find 2.3% of $150.
12. 34 is what percent of 40?

Use the proportion method to solve problems 1–48 in Exercises 7.5.

▶ **Answers to You Try It**   1. A = $54   2. A = 3.6   3. B = 40   4. B = 50
5. B = 72   6. R% = 75%   7. R% = 4%   8. R% = $77\frac{7}{9}$%   9. B = 80
10. B = 400   11. A = 3.45   12. R% = 85%

■ ② ③ ④ ⑤

1. 42% of 600 is what number?

2. 6% of 40 is what number?

3. 87% of 380 is what amount?

4. 2.2% of 750 is what amount?

5. What number is 7.2% of $65?

6. What number is 100% of 25.3 pounds?

7. $12\frac{1}{2}$% of 640 men is how many men?

8. $83\frac{1}{3}$% of $420 is how much?

9. What is 0.06% of 22,000 pounds?

10. Find 0.3% of 0.85 grams?

11. 240% of $550 is how much money?

12. 338% of 62.5 is what number?

13. 55% of what number is 44?

14. 27% of what number is 162?

15. 83% of what number is 41.5?

16. $37\frac{1}{2}$% of what number is 15?

17. 20 is $22\frac{2}{9}$% of what number?

18. 50 is $83\frac{1}{3}$% of what quantity?

19. 60 people is 0.3% of how many people?

20. $126 is 140% of how much money?

21. 275% of what number is 330?

22. 103% of what number is 49.44?

23. 0.1% of how many liters is 0.01 liter?

24. 38% of how many cars is 171 cars?

25. What percent of 300 is 90?

26. What percent of 20 is 12?

27. What percent of $65 is $19.50?

28. What percent of 680 children is 51 children?

29. 667 men is what percent of 2,900 men?

30. 14 is what percent of 30?

31. 45 senators is what percent of 75 senators?

32. 13.2 is what percent of 13.2?

33. 0.8 is what percent of 0.5?

34. 230 is what percent of 57.5?

35. What percent of $5.10 is $1.70?

36. What percent of 0.4 is 0.016?

37. 2.9% of what number is 10.15?

38. 38 is 20% of what number?

39. What is $18\frac{2}{11}\%$ of 33?

40. 185% of $340 is how much?

41. 54 is what percent of 72?

42. What percent of 8 is 3?

43. $\frac{1}{3}$ is 20% of what number?

44. 140% of what number is $\frac{7}{8}$?

45. 7.2 is what percent of 9?

46. $\frac{3}{7}$ is what percent of $\frac{4}{7}$?

47. What is 80% of $80?

48. 150% of $25 is how much?

SKILLSFOCUS (Section 5.3) *Add.*

49. $2.7 + 30 + 5.06 + 0.91$

50. $600.3 + 25.08 + 4 + 0.6$

51. $0.07 + 0.008 + 1 + 0.95$

52. $200.003 + 20.03 + 2.3 + 2,000$

WRITING TO LEARN ▶▶▶▶

53. Draw the percent triangle from memory. Explain how to use it. Show how to get the three percent formulas from it.

▶ YOU BE THE JUDGE

54. Paula paid $200 for a painting. She sold it for 150% of what she paid for it. What did she sell the painting for? Explain how you arrived at your answer.

55. Suppose 200 people are in a train station. Phil says 150% of them are holding train tickets to New York. Is Phil making sense? Explain your decision.

7.6 SOLVING PERCENT WORD PROBLEMS

OBJECTIVE	NEW VOCABULARY
Identify R%, B, and A in a long percent word problem, then solve it.	long form

Long Percent Word Problems

A percent problem not stated in the concise short form of A is R% of B is written in **long form**. In long form, the problem is often stated in two or more sentences. These story problems give additional descriptive information that may make it more difficult to identify the rate, base, and amount.

Problems stated in long form use the exact same rules to identify R%, B, and A as used with the short form.

> To solve a percent word problem
>
> 1. **Identify the rate R%.** It is the number with the percent symbol, %, or it is represented by the word *percent*.
> 2. **Identify the base B.** It is the number following the words "percent of" or following a numerical percent (such as "25% of").
> 3. **Identify the amount A.** It is the number remaining after the rate and base have been identified.
> 4. **Solve for the unknown number.** Represent the unknown number with its own letter. Solve for it by using the percent triangle in Section 7.5 on page 435.

EXAMPLE 1 Joe said that 60% of the cars he repaired last year needed brake work. He repaired 540 cars last year. How many needed brake work?

Step #1: Identify R%, B, and A.

$$R\% \qquad\qquad B$$
Joe said that 60% of the cars he repaired last year
A   needed brake work. He repaired 540 cars last year.
How many needed brake work?

$R\% = 60\%$
$B \ \ = 540$ (The base follows "60% of," and refers to the number of cars he repaired last year. The second sentence says he repaired 540 cars last year.)
$A \ \ = ?$ (Unknown number of cars needing brake work. It is a part of the 540 cars repaired.)

Step #2: Solve for A using the percent triangle.

$$A = R\% \times B$$
$$A = 0.60 \times 540$$
$$A = 324 \quad \text{cars needed brake work} \quad \blacksquare$$

▶ **You Try It** 1. The USDA claims that 67.1% of the weight of a chicken is water. How many pounds of water are contained in 5 pounds of chicken? Round to the nearest tenth of a pound.

The unknown number in a long percent problem is often identified in the question at the end of the problem. In Example 1, the question "How many needed brake work?" tells you to solve for A.

EXAMPLE 2   The TNT Construction Company employs 90 skilled technicians. If this represents 30% of their employees, how many people work for TNT?

Step #1: Identify R%, B, and A.

The TNT Construction Company employs 90 skilled technicians. If this represents 30% of their employees, how many people work for TNT?

$$R\% = 30\%$$
B = ?   (the words their employees follow "30% of," and refer to the total number of people working for TNT. The last sentence asks you to find how many people work for TNT.)

A = 90   (90 skilled technicians is a portion of the total number of TNT employees.)

Step #2: Solve for B using the percent triangle.

$$B = \frac{A}{R\%} = \frac{90}{0.30}$$

$$B = 300 \text{ people work for TNT} \quad \blacksquare$$

▶ **You Try It**   2. 88 gallons of oil are left in a tank. This is 40% of the tank's capacity. What is the capacity of the tank?

EXAMPLE 3   Hank's salary last year was $24,000. He paid $6,720 in taxes. What percent of Hank's salary was taxes?

Step #1: Identify R%, B, and A.

Hank's salary last year was $24,000. He paid $6,720 in taxes. What percent of Hank's salary was taxes?

$$R\% = ?$$ (The last sentence asks you to find percent.)
B   = $24,000   (Hank's salary is the base because it follows " percent of." The first sentence says his salary was $24,000.)

A   = $6,720   ($6,720 paid in taxes is a part of Hank's total salary.)

Step #2: Solve for R%.

$$R\% = \frac{A}{B} = \frac{\$6,720}{\$24,000}$$

$$R\% = 0.28$$

$$R\% = 28\% \text{ of his salary was paid as tax} \quad \blacksquare$$

▶ **You Try It**   3. A Chevy van sells for $28,600. To finance the van, the buyer must put down $4,290. What percent of the selling price is the buyer being asked to put down?

The next problem does not include the phrase "percent of." The base follows the two words "percent of" and will not necessarily follow the word "percent" by itself or the word "of" by itself. Reword these problems to include the phrase "percent of" before solving them.

EXAMPLE 4 A group of students took an exam today. 80% passed. If 24 passed, how many students took the exam?

Step #1: Identify R%, B, and A. Though "passed" follows 80%, the number who passed is *not* the base. "80% passed" means 80% of *the students who took the exam* passed. The base is the total number of students who took the exam. Reword the problem as follows.

A group of students took an exam today. 80% of the students passed it. If 24 passed, how many students took the exam?

$R\% = 80\%$
$B = ?$ (The number of students is the base because it follows "80% of." The last sentence asks you to find how many students took the exam.)
$A = 24$ (number who passed)

Step #2: Solve for B using the percent triangle.

$$B = \frac{A}{R\%} = \frac{24}{0.80}$$

$B = 30$ students took the exam ∎

▶ **You Try It** **4.** A development has 350 homes. 217 participate in a neighborhood watch program. What percent participate?

EXAMPLE 5 **Chapter Problem** Read the problem at the beginning of this chapter.

a. The United States consumes 31.2% of the 55.2 million barrels used worldwide each day. How many barrels is this?

$R\% = 31.2\%$
$B = 55.2$ million barrels (follows "31.2% of")
$A = ?$

$A = R\% \times B$
$A = 0.312 \times 55.2$
$A \doteq 17.2$ million barrels consumed by the United States each day

b. The United States has about 6% of the world's population. It consumes 31.2% of the oil used in the world each day. About how many times more oil does the United States consume daily when compared to its population size?

Find how many times 6% is contained in 31.2%.

$$\frac{31.2\%}{6\%} = \frac{0.312}{0.06} \doteq 5$$

The United States consumes about 5 times its share of oil. ∎

---

5. Refer to the table at the beginning of this chapter.
   a. The Middle East produces 37% of the 57.6 million barrels of oil produced daily in the world. How many barrels is this?
   b. About how many times more oil does the Middle East produce daily compared to the United States?

▶ **Answers to You Try It**   **1.** A = 3.4 lb of water   **2.** B = 220 gal
3. R% = 15%   **4.** R% = 62%   **5. a.** A = 21.3 million barrels
b. $\dfrac{37\%}{18.9\%} = \dfrac{0.37}{0.189} \doteq 2$. The Middle East produces about twice the oil the
United States produces daily.

Identify the three numbers R%, B, and A in each problem, then solve.

1. Thirty-eight percent of the employees for the Mutual Insurance Company are salespeople. Mutual employs 5,500 people. How many are in sales?

2. There are 40 pizza shops in town. Thirty percent of them advertise in the local newspaper. How many advertise in the local paper?

3. Maria claims that 10% of the women who belong to her country club are doctors. If 23 women are doctors, how many women belong to the club?

4. In the school, 760 students are science majors. This is 95% of all the students attending the school. How many students attend the school?

5. In a survey of 300 children, 213 preferred playing with toy widgets. What percent of the children preferred the widget?

6. Sam received a check for $615.25. He used $516.81 to purchase an oak desk. What percent of his check was spent on the desk?

7. The tax on the purchase of a stereo is $44.00. If this is 5% of the purchase price, what was the price of the stereo?

8. A football team won 10 games, which was 62.5% of the games the team played. How many games did the team play?

9. A congresswoman polled 1,500 of her constituents on the question of a tax increase. 645 favored a tax increase.
   a. What percent of those polled favored the increase?

   b. What percent opposed it?

10. In a class of 30 students, 6 received a grade of A.
   a. What percent of the class received an A grade?

   b. What percent received a different grade?

11. $\frac{2}{3}$% of an ore is gold. If 600 pounds of the ore is mined, how many pounds of gold can you expect?

12. 85% of a donation of $4,200 will be used to clothe the needy. How much of the donation will be used to clothe the needy?

13. A builder constructed a home and sold it for $145,000. If 16% of the sale price of the home is the builder's profit, how much profit did the builder make?

14. A rat weighing 36 ounces ate 16 ounces of food today. What percent of the rat's body weight does this food represent?

15. In a class of 35 students, 14 students received a grade of B. What percent of the class received a grade of B?

16. John lost 10 pounds on a fat free diet. His goal is to lose 25 pounds. What percent of his goal has John reached?

17. Sal uses 38% of his net salary to pay his home mortgage. If Sal's net salary is $1,450, how large is his mortgage payment?

18. Alberta sold her home for $66,500. If 40% of this is profit, how much profit has she made on the sale of her home?

19. George weighs 92% of what his doctor says is his ideal weight. If George weighs 138 pounds, what is his ideal weight?

20. If 0.17% of the weight of the human body is chlorine, what would a body weigh that contained 0.306 pound of chlorine?

21. A 3-ounce piece of lean beef contains 77 milligrams of cholesterol. A 3-ounce piece of beef liver contains 483% of the cholesterol in lean beef. How much cholesterol is contained in 3 ounces of beef liver?

22. One frankfurter contains 27 milligrams of cholesterol. If one egg contains 1,000% of the cholesterol in a frankfurter, how many milligrams of cholesterol are in one egg?

27 mg   ? mg

Reword each problem to contain the phrase "percent of," then solve.

23. Ed won $180 on a game show. He spent $65 on a fancy dinner afterward. What percent did he spend on the dinner?

24. There are 4,500 cats in this city. 750 are strays. What percent are strays?

25. Last week 1,500 men were unemployed. 70% found work. How many found a job?

26. Gary purchased 175 pounds of bananas to sell to his customers. 8% spoiled before they were sold. How many pounds spoiled?

27. Cathy sold her cabin cruiser for an unknown sum of money. She used 20%, or $35,000, to buy a catamaran. She banked the rest. How much did Cathy get for the cabin cruiser?

30. 4,500 people work for Munks Stores. If 3.2% will retire this year, how many Munk employees will retire this year?

28. A solution is 12% acid. If there are 15 liters of acid in the solution, how many liters of solution are there?

31. Paul and Kim dined out this evening. Their dinner bill was $37.60. They left a 15% tip. How large was the tip?

29. Lucy purchased 80 flowering plants and 5% died. How many died?

32. Mike and Debbie want to leave a 20% tip for their waiter. If the bill was $62.40, how much of a tip do they leave?

SKILLSFOCUS (Section 5.6) *Divide. Round to hundredths.*

33. $7.6 \div 0.19$ **34.** $48.9 \div 3$ **35.** $0.8 \div 1.7$ **36.** $4 \div 6.3$

EXTEND YOUR THINKING ▶▶▶▶
▶SOMETHING MORE

37. Your medical insurance covers 80% of the cost of your surgery. You must pay the rest, which amounts to $450. What is the total charge for the surgery?

39. A farmer sold 50 pounds of broccoli for $20 to a neighbor. The neighbor sold it to a wholesaler for $30. The wholesaler increased the price by 25% and sold it to a retailer. The retailer marked the price up by 40% to sell to his customers. What price per pound will the retailer charge?

38. In a group, 20% are adults, and the rest are children. Seventy percent of the adults and forty percent of the children will be attending a seminar. What percent of the group will attend the seminar?

40. Constructing a Percent Solver The following diagram can be used to estimate solutions to percent word problems. You need a flat piece of cardboard or plywood to which a piece of graph paper is attached. Also needed is a ruler, pencil, string, and two tacks used to serve as pivots for the string. (Note: A ruler or straightedge pivoted at 0% can be used in place of string and tacks.)

a. To find R% or A, adjust the movable tack pivot so that the string passes directly over the base on the amount axis. To find A, locate R% on the rate axis. Move horizontally to the right. When you reach the string move vertically downward to A on the amount axis. To find R%, locate A on the amount axis. Move vertically up. When you reach the string move horizontally left to R% on the rate axis.

b. To find the base, locate R% on the rate axis and A on the amount axis. Draw a rectangle. Adjust the movable pivot until the string touches the vertex point of the rectangle. The base is the intersection of the string with the amount axis.

In this diagram, the base is 80. Use this diagram to estimate answers to the following problems.

c. 50% of 80 is what?

d. 110% of 80 is what?

e. What is 35% of 80?

f. 80 is what percent of 80?

g. 12 is what percent of 80?

h. 56 is what percent of 80?

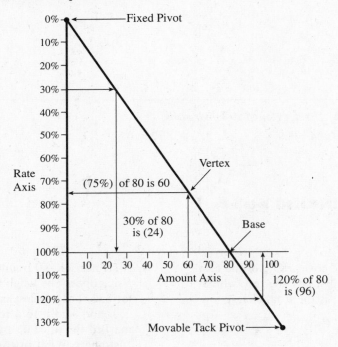

WRITING TO LEARN ▶▶▶▶

41. Write a word problem whose answer is 40%. The problem must be about birds. Then write a complete solution for the problem.

42. Write a word problem about purchasing clothes that gives the rate and amount, and asks you to find the base. The base is four times the amount. Then write the complete solution.

7.7 COMMISSION, DISCOUNT, AND PERCENT INCREASE AND DECREASE

OBJECTIVES
1. Solve commission problems.
2. Solve discount problems.
3. Solve percent increase and decrease problems.

NEW VOCABULARY
commission	discount
straight commission	percent increase
salary plus commission	percent decrease

1 Commission

Some salespersons earn salaries that are partially or totally based on commissions. This means the salesperson receives a part of the selling price of each item he or she sells. This part is called the **commission**. It is a percent of the selling price, called the *commission rate*.

> R% = commission rate (expressed as a percent of the selling price)
> B = selling price
> A = amount of commission

The commission triangle is used the same way as the percent triangle introduced in Section 7.5. Cover the number you are looking for with a finger. The triangle then shows the formula needed to find that number.

Commission Triangle

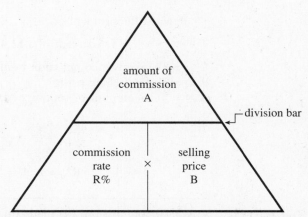

Working **straight commission** means one's entire salary comes from commission.

EXAMPLE 1

The Century Real Estate Agency makes a 7% commission on each home it sells. What commission does Century make on the sale of a $132,000 home?

This problem can be reworded.

7% of the $132,000 selling price is what commission?

> R% = 7% (commission rate)
> B = $132,000 (selling price of the home)
> A = ? (unknown amount of commission made on the sale)

$$A = R\% \times B$$

$$A = 0.07 \times \$132,000$$

$$A = \$9,240 \text{ commission} \quad ■$$

▶ **You Try It**

1. Agnes makes a 9% commission when she sells a piece of unimproved land. What commission does she make on a 3.72-acre plot of land that she sells for $58,500?

EXAMPLE 2 Trina sells major appliances. She makes a straight 14% commission. To support her family, she must make $2,310 in commissions per month. How many dollars in major appliances must she sell each month to meet her financial needs?

This problem can be reworded.

14% of what sales total gives $2,310 in commissions?

R% = 14% (commission rate)
B = ? (unknown total monthly sales)
A = $2,310 (monthly commission needed)

$$B = \frac{A}{R\%} = \frac{\$2,310}{0.14}$$

B = $16,500 total sales Trina must make monthly ∎

▶ **You Try It** **2.** Miguel works on a 12.5% straight commission. How much merchandise must he sell to make $1,781.25 in commissions?

EXAMPLE 3 A commission of $116 was made on the sale of a $1,450 dining room set. What was the commission rate?

Reword the problem.

What percent of a $1,450 sale is a $116 commission?

R% = ? (unknown commission rate)
B = $1,450 (selling price)
A = $116 (amount of commission)

$$R\% = \frac{A}{B} = \frac{\$116}{\$1,450}$$

R% = 0.08 = 8% commission rate ∎

▶ **You Try It** **3.** A $300 commission was made on the sale of a $5,000 bedroom set. Find the rate of commission.

Working **salary plus commission** means one is paid a fixed salary plus a commission on the sales made.

EXAMPLE 4 Jan sells power tools. She is paid a fixed salary of $80 per week plus a 5.5% commission on sales. If she sold $680 in power tools this week, find her salary.

Step #1: Find Jan's commission. Reword the problem.

5.5% of $680 in sales is what commission?

R% = 5.5% (commission rate)
B = $680 (total weekly sales)
A = ? (amount of commission)

$$A = R\% \times B$$

A = 0.055 × $680

A = $37.40 earned in commission

Step #2: Jan's total weekly salary is her fixed weekly salary plus commission.

$$\begin{array}{ll} \$37.40 & \text{commission earned} \\ +\,\$80.00 & \text{weekly salary} \\ \hline \$117.40 & \text{total salary} \end{array} \qquad \blacksquare$$

> **CAUTION** Commission is not figured on $80, because $80 is not sales.

▶ **You Try It**   **4.** Ellen is paid a fixed salary of $110.50 a week, plus a 3% commission on sales. Her sales for the week were $1,680. What is her salary for the week?

2 Discount

When an item is put on sale, the original selling price is reduced. The amount of this reduction is called the **discount**. The discount is a percent of the original selling price, called the *discount rate*.

R% = discount rate (expressed as a percent of original selling price)
B = original selling price
A = amount of discount

sale price = original selling price − discount

To use the discount triangle that follows, cover the number you are looking for with a finger. The triangle then shows the formula needed to find that number.

Discount Triangle

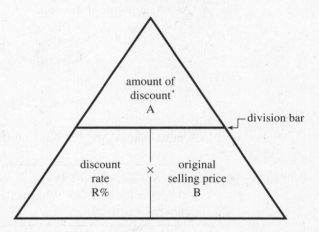

EXAMPLE 5   A winter coat regularly sells for $90. For today only, it is being sold for a 30% discount.

a. What is the amount of discount?
Reword this problem as follows.

30% of the $90 original selling price is what discount?

R% = 30% (discount rate)
B = $90 (original selling price)
A = ? (unknown amount of discount)

$$A = R\% \times B$$

$$A = 0.3 \times \$90$$

$$A = \$27 \text{ discount}$$

b. What is the sale price of the coat?

$$\begin{array}{rl} \$90 & \text{original selling price} \\ -27 & \text{discount} \\ \hline \$63 & \text{sale price} \qquad \blacksquare \end{array}$$

> **OBSERVE** In Example 5, you can find the sale price by using complements. A 30% discount means you get 30% off of the $90 price. The complement of 30% is 70%. If you get 30% off, you must pay the other 70%. You pay 70% of $90 = 0.70 × $90 = $63.

▶ **You Try It**   **5.** A suit originally priced at $240 is being sold at a 15% discount.
   a. What is the amount of the discount?
   b. What is the sale price of the suit?

EXAMPLE 6   Jack's Electronics sells a TV for $350. His competition sells the same TV for $280. What percent discount must Jack offer on the TV to meet his competitor's price?

Jack's competition sells the TV for $350 − $280 = $70 less. Jack must discount his $350 price tag by $70. Reword the problem as follows.

What percent of the $350 original selling price gives a $70 discount?

R% = ? (unknown discount rate)
B = $350 (Jack's original selling price)
A = $70 (amount of discount)

$$R\% = \frac{A}{B} = \frac{\$70}{\$350} \qquad \text{Reduce } \frac{\overset{1}{\cancel{\$70}}}{\underset{5}{\cancel{\$350}}} = \frac{1}{5}.$$

$$R\% = \frac{1}{5} = 20\% \text{ discount}$$

Jack must offer a 20% discount on the TV to meet his competitor's price.   ■

▶ **You Try It**   **6.** Stan's Stereos sells a CD player for $240. His competition sells the same CD player for $211.20. What percent discount must Stan offer to meet his competitor's price?

EXAMPLE 7   Eddie purchased a train set at a 15% discount. This discount amounted to a savings of $5.40.

a. What was the original selling price of the set?
   Reword the problem as follows.

15% of what original selling price gives a $5.40 discount?

R% = 15% (discount rate)
B   = ? (original selling price)
A   = $5.40 (amount of discount)

$$B = \frac{A}{R\%} = \frac{\$5.40}{0.15}$$

B = $36 original selling price of the train set

b. What was the sale price of the train set?

$36.00   original selling price
− 5.40   discount
─────────────────
$30.60   sale price   ■

▶ You Try It 7. Kelly bought a mink stole at a 20% discount.
    a. The discount was $290. Find the original price of the stole.
    b. What was the sale price?

3 **Percent Increase and Decrease**

Percent increase and **percent decrease** are also called *percent of change*. Percent increase is sometimes called *percent gain*. Percent decrease is sometimes called *percent loss*.

EXAMPLE 8

Dave earned $1,000 last month. This month he earned $1,500. What is the percent increase in Dave's wages?
    The amount of increase in wages is $1,500 − $1,000 = $500. This $500 is an increase over the original earnings of $1,000. To find the percent increase, find the percent that $500 is of $1,000.

R% = ? (unknown percent increase in wages)
B = $1,000 (the original earnings)
A = $500 (the amount of increase, found by subtracting
        $1,500 − $1,000 = $500)

$$R\% = \frac{A}{B} = \frac{\$500}{\$1,000} = \frac{1}{2}$$

R% = 50% increase in wages   ■

> To find percent increase or decrease
>
> **1.** Subtract to find the amount of increase or decrease, A.
> **2.** The base B is the original number.
> **3.** Solve for the rate
>
> $$R\% \text{ (percent increase/decrease)} = \frac{A \text{ (amount increase/decrease)}}{B \text{ (original number)}}$$

To use the increase/decrease triangle that follows, cover the number you are looking for with a finger. The triangle then shows the formula needed to find that number.

Increase/Decrease Triangle

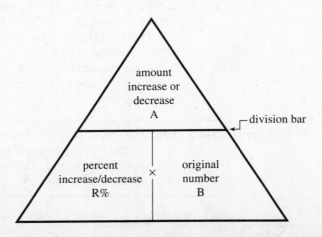

EXAMPLE 9   The price of gold today is $326.40 per ounce. One month ago the price was $340 per ounce. By what percent has the price of gold decreased during the month?

R% = ? (unknown percent decrease)
B = $340 (original price per ounce)
A = $13.60 (subtract $340 − $326.40 = $13.60 to get the amount decrease per ounce)

$$R\% = \frac{A}{B} = \frac{\$13.60}{\$340}$$

$$R\% = 0.04 = 4\% \text{ decrease} \quad \blacksquare$$

▶ **You Try It**   8. Nathan earned $2,100 last month. This month he earned $1,680. Find the percent decrease in Nathan's wages.

9. A loaf of bread selling for $0.35 ten years ago, sells for $0.98 today. What percent increase in price is this?

EXAMPLE 10   **a.** Last year Buddy weighed 200 pounds. This year he weighs 250 pounds. What is the percent increase in his weight?

R% = ? (unknown percent increase in weight)
B = 200 pounds (Buddy's original weight)
A = 50 pounds increase in weight (= 250 − 200)

$$R\% = \frac{A}{B} = \frac{50}{200} = \frac{1}{4} = 25\% \text{ increase in weight}$$

b. This year Buddy weighs 250 pounds. Next year he wants to drop his weight to 200 pounds. What percent decrease is this?

R% = ? (unknown percent decrease in weight)
B = 250 pounds (original weight)
A = 50 pounds decrease in weight (= 250 − 200)

$$R\% = \frac{A}{B} = \frac{50}{250} = \frac{1}{5} = 20\% \text{ decrease in weight} \quad \blacksquare$$

OBSERVE   In Example 10a, Buddy went from 200 to 250 pounds, an increase of 50 pounds. In Example 10b, he went from 250 to 200 pounds, a decrease of 50 pounds. But in the first problem, a 50-pound increase in weight is a 25% increase. In the second problem, a 50-pound decrease is a 20% decrease. How can 50 pounds represent one percent in one case, and a different percent in a second case?

The percent that 50 pounds represents changes in each case because the base changes in each case. In the first example it was 200 pounds. In the second example it was 250 pounds. *Whenever the base changes, the percent represented by a number will also change.*

▶ **You Try It**   10. **a.** Last year Jason weighed 140 pounds. This year he weighs 154 pounds. What is the percent increase in his weight?

b. Next year Jason wants to drop his weight back to 140 pounds. What percent decrease is this?

In the next two examples, you will find the amount increase or decrease.

EXAMPLE 11 Greg's Gas & Go made a profit of $29,000 last year. This year there was a 16% decrease in profits.

a. By how much did profits decrease this year?

The decrease in profit is 16% of $29,000.

R% = 16% (percent decrease in profit)
B = $29,000 (the original profit)
A = ? (unknown amount of decrease in profit)

$$A = R\% \times B = 0.16 \times \$29,000$$

$$A = \$4,640 \quad \text{decrease in profit}$$

b. What were this year's profits?

$29,000	profit last year
−$4,640	decrease in profit
$24,360	profit this year ∎

▶ **You Try It** **11.** A small market made a profit of $8,400 last year. This year profits were down 7%.
a. By how much did profits decrease this year?
b. What were this year's profits?

EXAMPLE 12 The Ablix Company is giving all of its employees a 5% across-the-board salary increase. This means every employee's salary is increased by 5%. Elaine makes $24,000 a year. What is her new salary?

Step #1: Find Elaine's increase in salary. Find 5% of $24,000.

R% = 5% (rate of increase in Elaine's salary)
B = $24,000 (Elaine's current salary)
A = ? (unknown amount of increase in Elaine's salary)

$$A = R\% \times B = 0.05 \times \$24,000$$

$$A = \$1,200 \quad \text{increase in Elaine's salary}$$

Step #2: Elaine's new salary is

$24,000	current salary
+ 1,200	increase in salary
$25,200	new salary ∎

OBSERVE Elaine's current salary is the base, or 100%. Her increase in salary is 5%. Therefore her new salary is 100% + 5% = 105% of her current salary. Her current salary is $24,000, so her new salary is 105% of $24,000 = 1.05 × $24,000 = $25,200.

▶ **You Try It** **12.** Susan's company gave each employee a 3.8% across-the-board salary increase. Susan makes $47,500 a year. What is her new salary?

SOMETHING MORE

Real Estate Commissions   Sam works for the Sage Real Estate Agency. He has agreed to sell his next door neighbor's house. Sam then becomes the *listing agent* for the home, and Sage is the *listing agency*. If Sam sells this home, the agency makes a 7% commission on the selling price of the property. Sam and the agency split this commission, each getting $3\frac{1}{2}\%$.

a. Sam sells the home for $88,500. What commission is made?

b. How much of this commission does Sam get?

c. How much does Sage get?

William works for the Winter Real Estate Agency, a strong Sage competitor. Because Sage and Winter both belong to a *Multiple List Service*, the two agents can sell homes listed with either agency. Suppose William sells Sam's neighbor's home for $95,000. A 7% commission is made on this sale. But because Winter sold a home listed for sale by the Sage agency, the two agencies split the 7% commission, each getting $3\frac{1}{2}\%$. Because Sam is the listing agent, Sam and Sage evenly split this $3\frac{1}{2}\%$, each getting $1\frac{3}{4}\%$. Because William sold the home, he and Winter split their $3\frac{1}{2}\%$, each getting $1\frac{3}{4}\%$.

d. How much commission is made on the sale of this home?

e. How much do Sam and Sage each receive?

f. How much do William and Winter each receive?

▶ **Answers to You Try It**   **1.** A = $5,265 in commission   **2.** B = $14,250 in sales   **3.** R% = 6%   **4.** A = $50.40 in commission; salary for the week = $160.90   **5. a.** A = $36 discount   **b.** sale price = $204   **6.** R% = 12%   **7. a.** B = $1,450 original selling price   **b.** sale price = $1,160   **8.** R% = 20% decrease   **9.** R% = 180% increase   **10. a.** R% = 10% increase   **b.** R% = $9\frac{1}{11}\%$ decrease   **11. a.** A = $588 decrease in profit   **b.** this year's profit = $7,812   **12.** A = $1,805 increase; new salary = $49,305

▶ **Answers to Something More**   **a.** $6,195   **b.** $3,097.50   **c.** $3,097.50   **d.** $6,650   **e.** $1,662.50   **f.** $1,662.50

1

1. Jack sells boats. He makes a 4.7% commission on each sale. What is his commission on the sale of a $92,000 powerboat?

2. A townhouse sold for $56,600. If the agency who sold the home made a 6% commission on the sale, what commission did they earn?

3. Arnold is paid a 15% commission on everything he sells. If he needs $1,200 per month to meet his financial obligations, how much merchandise must he sell monthly?

4. What sales must Janet make in one week to earn $1,170 in commission? Her commission rate is 7.8%.

5. Barb works for a marina. She works on an 8% commission. This month she sold a $28,000 catamaran and a $46,000 speedboat. What was Barb's total commission for the month?

6. Sally made a 3.5% commission on the sale of a $349,000 mansion. What commission did Sally make?

7. A real estate agency sold a 3-acre piece of land. The agency made a $5,450 commission on the sale. If the rate of commission was 10%, what did the land sell for?

8. A 20% commission was made on the sale of a diamond ring. If the commission was $255, what did the ring sell for?

9. A $2.78 commission was made on a $55.60 sale. What was the commission rate?

10. What is the commission rate if a $6,507 commission is made on the sale of a $144,600 home?

11. Ed is paid a fixed salary of $200 per month plus a 12% commission on sales. What is Ed's July salary if his sales for July totaled $5,620?

12. Irma sold $2,250 in jewelry this week. If her salary is $125 per week plus a 5.4% commission on sales, what did Irma earn this week?

13. Jerry made $920 this week selling cars. If he is paid a fixed salary of $200 per week plus a 4% commission on sales, what were his total car sales for the week?

14. Howard earns a fixed salary of $40 per week plus a 16% commission on sales. What were his sales for the week if his salary was $200?

2

15. A $900 sofa is being sold at a 35% discount.
    a. What is the amount of discount?

    b. What is the sale price of the sofa?

16. A $450 pair of skis is on sale at a 40% discount.
    a. What is the amount of discount?

    b. What is the sale price of the skis?

17. The Charles Men's Shop is having a storewide clearance sale. Every item is being discounted 40%. For what sale price can you purchase the following items?
    a. A pair of slacks regularly selling for $32.

    b. A tie that regularly sells for $11.50.

    c. A suit priced at $249.99 (round to the nearest cent).

18. To celebrate customer appreciation day, a department store is offering 20% off on every item it stocks. For what price can you purchase each item?

   a. A $40 pair of slacks.

   b. A watch that regularly sells for $29.95.

   c. A basketball net for $3.69 (round to the nearest cent).

19. Frances works for a local department store. She wants to buy a TV regularly priced at $438. The TV is on sale for 20% off. Frances gets another 20% off of the sale price because she is an employee of the store. What will she pay for the TV set?

20. Jason works for a department store. A coat regularly selling for $180 is on sale for 25% off. Jason gets an additional 15% off of the sale price because he is a store employee. What does Jason pay for the coat?

21. Fred puts a microwave oven on sale for $238. If the regular price is $340, what percent discount is Fred offering?

22. Alice sells a boom box for $180. Her competition sells the same one for $135. What discount rate must Alice offer to meet her competitor's price?

23. A stereo is discounted 20%, giving a savings of $173.

   a. What is the original selling price of the stereo?

   b. What is the sale price?

24. A set of china is discounted 60%, resulting in a savings of $330.

   a. What was the original price of the china?

   b. What is the sale price?

3

25. If George pays his electric bill before November 2, he must pay $55.60. After that date he must pay $58.38. What percent increase is this?

26. A wool sweater regularly selling for $48 was reduced to $16 because of a flaw. What percent reduction is this?

27. A home selling for $130,000 was reduced to $117,000 for a quick sale. By what percent was the home reduced in price?

28. A book originally selling for $14 was sold in a clearance sale for $10.50. What percent decrease in price is this?

29. George made $800 last month. He estimates he will make $1,000 this month.

   a. What is the expected percent increase in earnings?

   b. George only made $940 this month. What percent increase did he actually make?

30. Bonnie made $50 last summer selling lemonade. This summer she expects to make $80.

   a. What percent increase in sales does Bonnie expect to make?

   b. This summer Bonnie only made $70. What percent increase in sales did she actually make over last summer?

31. **Depreciation** Depreciation is the loss in value of an item due to old age or obsolescence. Suppose a new car depreciates 30% in the first year.

   a. How much will a new $23,400 car depreciate in its first year?

   b. What will it then be worth?

32. Appreciation Appreciation is the increase in value of an item over time. A $180,000 home appreciates 16% in one year.

a. How much has the home appreciated?

b. What is the home then worth?

33. Wolfe's Bakery purchased a $3,600 oven. It is expected to depreciate 20% in the first year.

a. How much will the oven depreciate in the first year?

b. What will it then be worth?

34. A car dealer pays the manufacturer $14,000 for a new car. The dealer increases this price by 22% to get the base sticker price.

a. How much of an increase is this?

b. What is the base sticker price?

35. A house sold for $18,000 nine years ago. Today it sells for $44,000. What percent increase in value is this?

36. According to the U.S. Bureau of Statistics, a certain basket of groceries cost $10 in 1967. In 1987, the same groceries cost $29.60. What is the percent increase in cost?

37. a. Willy weighed 144 pounds in the spring. Due to a hectic summer, his weight dropped to 120 pounds. What percent decrease in weight is this?

b. Willy weighed 120 pounds in the fall. During the winter, his weight rose to 144 pounds. What was Willy's percent increase in weight?

38. a. A $250 investment grew to $400 in one year. What was the percent gain?

b. A $400 investment was worth $250 in one year. What was the percent loss?

39. If George makes $20,500 a year, and receives a 5.5% increase in salary, what is his new yearly salary?

40. A company pays its employees $9.40 per hour and decides to increase the hourly wage by 5%.

a. What is the new hourly wage?

b. If an employee works a 40-hour week, how much more money will the employee be making per week?

SKILLSFOCUS (Section 2.5) *Evaluate each formula.*

41. Find M if $M = 6S$ and $S = 7$.

42. Find Y if $Y = 2L + 2W$ and $L = 3$ and $W = 8$.

43. Find I if $I = PRT$ and $P = 100$, $R = 0.06$, and $T = 2$.

44. Find D if $D = 16T^2$ and $T = 5$.

EXTEND YOUR THINKING ▶▶▶▶
▶ **SOMETHING MORE**

45. Alida sold a home for $175,000. She was the listing agent for the home, and her company was the listing agency. She earned a 7% commission on the sale.

a. How much commission was made on the sale?

b. How much did Alida get? How much did her company get?

46. Connie, who works for CDE Real Estate, sold a townhome for $72,450. An 8% commission was made on the sale. The listing agent for the home was Tara, who works for TUV, the listing agency.

a. What commission was made on the sale?

b. How much did Connie, CDE, Tara, and TUV each receive?

47. Lots for Sale Anita owned two building lots. She sold them for $24,000 each. She sold one for 25% less than she originally paid for it. She sold the other for 25% more than she originally paid for it.

a. What did Anita originally pay for each lot?

b. Did she make a profit or loss on this transaction, or did she break even?

48. Baseball Averages This year Ben has a baseball batting average of .320. Since .320 = 32%, this means he got a hit 32% of the time he came to bat.

a. Ben came to bat 650 times. How many hits did he get?

b. Next season, Ben wants to increase his batting average to 0.380. If he comes to bat 650 times next season, how many more hits will he need?

c. A baseball player started his season with 5 hits in his first 20 at bats. His current batting average is $\frac{5}{20} = .250$. How many hits must he make in the next 30 times at bat to raise his average to .360?

49. Flogging In 1989, Hong Kong decided to abolish the use of flogging as a form of corporal punishment. The punishment, in which a male offender was beaten at most 18 times with a cane in the presence of a doctor, declined from 476 floggings in 1952 to only 2 in 1988. What was the percent decrease in floggings in this period of time?

50. Inflation Inflation refers to a time of rising prices for goods and services. Suppose an economic study concludes that a family of four needs $36,000 a year to maintain a moderate standard of living. Over the next year, the inflation rate is 4.2%. This means the overall price of goods and services in the economy increases 4.2%.

a. How much more money will the family of four need per year to maintain their standard of living?

b. What yearly income would they then have?

c. The inflation rate for the following year is 5.5%. What increase in annual income is needed for the family to continue to maintain their standard of living?

WRITING TO LEARN ▶▶▶▶

51. Write a commission word problem about Mary who makes an 8.6% commission on each yacht she sells. Then write the solution.

▶YOU BE THE JUDGE

52. Which is more, $\frac{1}{2}$ of a million dollars, or $\frac{1}{2}$% of a million dollars? Explain your decision.

53. Ron earns $100 per week. His employer increased his salary by 10%. A month later, Ron's employer decreased his salary by 10%. Ron claims this is unfair. Is he right? Explain your decision.

54. Farmer Prints His Own Money Recently an evening news program reported that a farmer, unable to get a loan from a bank, decided to print his own money. Actually, to raise money, the farmer printed a note valued at $9, and sold it for $9. Later, when the farmer harvested his crops, the owner of the note could use it to purchase $10 worth of produce from the farmer. The farmer raised money to stay in business, and at the same time paid out a profit in food to his investors.

a. The reporter for this story said that the farmer paid a 10% profit in food on each one of his notes. Is this correct? If not, what do you think is the actual profit?

b. Give an example of a sale price and a redemption value for this note that results in a 10% profit being made.

7.8 SIMPLE AND COMPOUND INTEREST

OBJECTIVES

1. Calculate simple interest (Sections 7.2, 7.3).
2. Application: calculate a monthly payment.
3. Calculate principal, rate, and time (Sections 2.5, 2.6).
4. Calculate compound interest.

NEW VOCABULARY

interest	maturity value
principal	installment buying
time	finance charge
simple interest	compound interest

1 Calculating Simple Interest

Interest is what you pay to use someone else's money. You pay interest on money you borrow to buy a car, and on purchases made with a credit card.

You earn interest for letting other people use your money. You earn interest on bank deposits, certificates of deposit, and bonds.

Principal is the amount on which interest is calculated. It is the amount borrowed or invested. The rate is the percent at which interest is calculated. The **time**, or term, is the total time over which interest is calculated. Interest calculated only on the principal is called **simple interest**.

---

simple interest = principal × rate × time

$$I = P \times R \times T$$

I = interest (amount earned or charged)
P = principal (amount borrowed or deposited)
R = rate of interest (per year)
T = time (in years)

The *rate R* and *time T* must be expressed in the *same time unit.*

EXAMPLE 1

How much will $2,000 earn at 7% simple interest for 3 years?

I = interest earned
P = $2,000
R = 7%
T = 3 years

$I = P \times R \times T$
$I = \$2,000 \times 0.07 \times 3$
$I = \$420$ earned in interest ■

▶ **You Try It**

1. How much will $800 earn at 6% simple interest for 2 years?

EXAMPLE 2

Lucy borrowed $846 at 18% for 7 months. How much interest will she pay?

I = ?
P = $846
R = 18%
$T = \dfrac{7}{12}$ years

{ The rate is 18% per year. The time, 7 months, must also be expressed in years. Since 12 months = 1 year, 7 months = $\dfrac{7}{12}$ years.

$I = P \times R \times T$

$I = \$846 \times 0.18 \times \dfrac{7}{12} = \dfrac{\$846}{1} \times \dfrac{0.18}{1} \times \dfrac{7}{12} = \dfrac{\$1065.96}{12}$

$I = \$88.83$

Lucy will pay $88.83 in interest to borrow $846 for 7 months. ■

2. Tomas borrowed $1,248 at 14% simple interest for 5 months. How much interest will he pay?

2 Application: Calculating a Monthly Payment

Maturity value, S, is total principal plus interest.

> maturity value = principal + interest
>
> $$S = P + I$$

In Example 1, the principal is $2,000. The interest earned is $420. The maturity value is S = P + I = $2,000 + $420 = $2,420.

With many purchases, such as a car or stereo, you can go home with it now and pay for it later. This is called **installment buying**. You often pay for the item in equal monthly payments. This payment includes an interest charge, called a **finance charge**.

> To figure a monthly payment.
>
> 1. Subtract the down payment or trade-in from the total cost. This gives the principal to be borrowed.
> 2. Compute the interest on this principal. For the time T, use the *total length of time* over which payments will be made.
> 3. Add principal to interest to get maturity value.
> 4. To find the monthly payment, divide maturity value by the number of months over which payments will be made.
>
> $$\text{Monthly Payment} = \frac{\text{Maturity Value}}{\text{Number of Months}}$$

EXAMPLE 3

Alberto borrows $600 to finance the purchase of a microwave oven. The simple interest charge is 12% over 24 months. Find the monthly payment.

Step #1: Calculate the interest charge on $600 at 12% for 24 months. The time for the loan is 24 months = 24 ÷ 12 = 2 years.

$$I = ?$$
$$P = \$600$$
$$R = 12\%$$
$$T = 2 \text{ years}$$

$$I = P \times R \times T$$
$$I = \$600 \times 0.12 \times 2$$
$$I = \$144$$

Step #2: maturity value = P + I = $600 + $144 = $744

Step #3: monthly payment = $\dfrac{\text{maturity value}}{\text{number of payments}} = \dfrac{\$744}{24} = \$31$

Alberto must pay $31 per month for 24 months. ∎

3. Barbara borrowed $1,800 to finance the purchase of a large screen TV. The simple interest charge is 10% for 36 months. Find the monthly payment

EXAMPLE 4 Maria purchased a stereo system priced at $1,620. She made a $180 down payment and financed the rest at 14.5% for 18 months.

a. What is her monthly payment?

Step #1: Find the amount to be borrowed.

$$\begin{array}{rl} \$1,620 & \text{cost} \\ - \ \$180 & \text{down payment} \\ \hline \$1,440 & \text{principal to be borrowed} \end{array}$$

Step #2: Calculate the interest on $1,440 at 14.5% for 18 months.

$$\begin{array}{l} I = ? \\ P = \$1,440 \\ R = 14.5\% = 0.145 \\ T = 1.5 \text{ yrs} \left(18 \text{ months} = \dfrac{18}{12} = 1.5 \text{ years} \right) \end{array}$$

$$I = P \times R \times T$$
$$I = \$1,440 \times 0.145 \times 1.5$$
$$I = \$313.20$$

Step #3: Find the maturity value.

maturity value = P + I = $1,440 + $313.20 = $1,753.20

Step #4: Find the monthly payment.

$$\text{monthly payment} = \frac{\text{maturity value}}{\text{number of months}} = \frac{\$1,753.20}{18}$$
$$= \$97.40$$

Maria must pay $97.40 per month for 18 months.

b. What is the total cost of the stereo on the installment plan?

The total cost of the $1,620 stereo on the installment plan is

$$\begin{array}{rl} \$1,753.20 & \text{P + I} \\ + \quad 180.00 & \text{down} \\ \hline \$1,933.20 & \quad \blacksquare \end{array}$$

▶ **You Try It** **4.** Carl bought a riding mower for $3,200. He put $200 down, and financed the rest over 15 months at 9%.
a. What is his monthly payment?
b. What is the total cost of the mower on the installment plan?

---

3 Calculating the Principal, Rate, and Time

To find interest I, multiply $P \cdot R \cdot T$. But how do you find P, R, or T? For example, find the principal P that will earn $50 in interest at 5% over 2 years. First, identify the givens.

$$\begin{array}{l} I = \$50 \\ P = \text{unknown} \\ R = 5\% = 0.05 \\ T = 2 \end{array}$$

Next, substitute into I = PRT, and solve the equation for P.

$$I = P \cdot R \cdot T$$

$$50 = P \cdot \underline{0.05 \cdot 2} \qquad \text{Multiply rate by time.}$$

$$50 = P \cdot \quad 0.1 \qquad 0.05 \cdot 2 = 0.1$$

$$\frac{50}{0.1} = \frac{P \cdot \overset{1}{\cancel{0.1}}}{\underset{1}{\cancel{0.1}}} \qquad \begin{array}{l}\text{Divide each side by 0.1, the}\\ \text{number multiplied by P.}\end{array}$$

$$\$500 = P$$

> To find principal, rate, or time
>
> 1. Identify the letter you are to find (P, R, or T).
> 2. Multiply the other two together.
> 3. Divide the interest by this product.

This procedure is summarized in the simple interest triangle.

Simple Interest Triangle
$$I = P \times R \times T$$

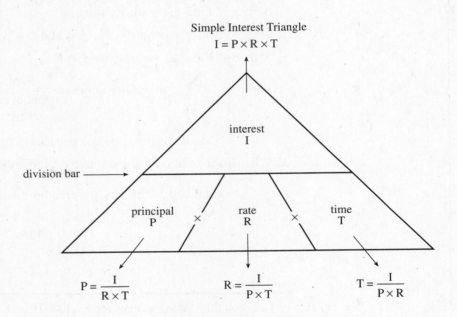

$$P = \frac{I}{R \times T} \qquad\qquad R = \frac{I}{P \times T} \qquad\qquad T = \frac{I}{P \times R}$$

This triangle is similar to the percent triangle used in Section 7.5. Cover the letter you are looking for with your finger. The triangle then shows the formula to find that amount.

For example, find the principal P. Cover P with your finger. The triangle shows the interest I divided by the product R × T. This is the formula for finding principal.

EXAMPLE 5 How long does it take for $4,000 to earn $960 at 12% simple interest?

Step #1: Identify the amounts given and the amount to be found.

$$\begin{array}{l} I = \$960 \text{ (the amount earned is the interest)} \\ P = \$4,000 \\ R = 12\% = 0.12 \\ T = ? \end{array}$$

Step #2: Use the triangle. You want to find time T. Cover T with your finger. The triangle shows the following formula.

$$T = \frac{I}{P \times R} = \frac{\$960}{\$4,000 \times 0.12} = \frac{\$960}{\$480}$$

$$T = 2 \text{ years}$$

Check: $I = P \times R \times T = \$4,000 \times 0.12 \times 2 = \$960$ ■

▶ **You Try It** 5. How long does it take $750 to earn $240 at 8% simple interest? Check your answer.

EXAMPLE 6 How much money must be deposited for 6 months at 8.75% to earn $280?

I = $280 (amount earned is interest)
P = ? (unknown principal to be deposited)
R = 8.75% = 0.0875
T = 0.5 years (6 months = $\frac{6}{12} = \frac{1}{2}$ or 0.5 year)

$$P = \frac{I}{R \times T} = \frac{\$280}{0.0875 \times 0.5} = \frac{\$280}{0.04375}$$

$$P = \$6,400$$

Check: $I = P \times R \times T = \$6,400 \times 0.0875 \times 0.5 = \$280$ ■

▶ **You Try It** 6. How much money must be deposited at 21% simple interest for 3 months to earn $78.75? Check your answer.

EXAMPLE 7 Ed wants to invest $15,000 over 5 years and make $6,000 in interest income. What rate of interest must he get?

I = $6,000 (amount of interest income to be made)
P = $15,000 (amount to invest)
R = ?
T = 5 years

$$R = \frac{I}{P \times T} = \frac{\$6,000}{\$15,000 \times 5} = \frac{\$6,000}{\$75,000}$$

R = 0.08 ⟵ change this decimal to a percent
R = 8% simple interest

Check: $I = P \times R \times T = \$15,000 \times 0.08 \times 5 = \$6,000$ ■

▶ **You Try It** 7. What rate of interest is needed to earn $1,008 on $4,200 over 6 years? Check your answer.

4 Calculating Compound Interest

Simple interest is calculated one time on the original principal for the entire term of the loan.

EXAMPLE 8 Calculate simple interest on $1,000 at 10% for 2 years.

$I = ?$
$P = \$1,000$
$R = 10\%$
$T = 2$ years

$$I = P \times R \times T$$
$$I = \$1,000 \times 0.10 \times 2$$
$$I = \$200$$

The total interest earned is $200. ■

Compound interest allows you to earn interest on previously earned interest. This is possible because interest is calculated more than once. Savings accounts often earn compound interest. Solve Example 8 again using compound interest.

EXAMPLE 9 Calculate the compound interest earned on $1,000 at 10% compounded annually for 2 years.

Compounded annually means calculate the interest once a year. Then add this interest to the principal before calculating interest again.

First year: $I = ?$
$P = \$1,000$
$R = 10\%$
$T = 1$ year

$$I = P \times R \times T$$
$$I = \$1,000 \times 0.10 \times 1$$
$$I = \$100 \text{ earned interest}$$

The interest earned in the first year is $100. It is added to the old principal $1,000 to get the new principal $1,100.

Second year: $I = ?$
$P = \$1,100$ (new principal)
$R = 0.10$
$T = 1$ year

$$I = P \times R \times T$$
$$I = \$1,100 \times 0.10 \times 1$$
$$I = \$110 \text{ earned interest}$$

The interest earned in the second year is $110. After 2 years the total interest earned is $100 + $110 = $210. ■

In Example 8, $1,000 earned $200 in 2 years at 10% simple interest. In Example 9, $1,000 earned $210 in 2 years at 10% compounded annually. $10 more was earned with compound interest. The reason is that interest was earned on interest. The extra $10 in interest was earned on the $100 in interest made in the first year.

▶ **You Try It**   **8.** Calculate simple interest on $100 at 20% for 2 years.

9. Calculate compound interest on $100 at 20% compounded annually for 2 years. How much more interest is earned here when compared to Problem 8?

---

compounded annually—means interest is calculated 1 time a year
$$(T = 1 \text{ year})$$
compounded semiannually—interest is calculated every 6 months
$$\left(T = \frac{6}{12} = 0.5 \text{ year}\right)$$
compounded quarterly—interest is calculated every 3 months
$$\left(T = \frac{3}{12} = 0.25 \text{ year}\right)$$
compounded monthly—interest is calculated each month
$$\left(T = \frac{1}{12} \text{ year}\right)$$
compounded daily—interest is calculated every day
$$\left(T = \frac{1}{365} \text{ year}\right)$$

EXAMPLE 10   How much interest will $15,000 earn at 8% compounded quarterly for 1 year.

Quarterly means interest is calculated every 3 months. In 1 year, interest will be calculated 4 times. 3 months = $\frac{3}{12}$ year = 0.25 year.

first 3 months:
$$I = P \times R \times T = \$15,000 \times 0.08 \times 0.25 = \$300$$
$$\text{new P} = \$15,000 + \$300 = \$15,300$$

second 3 months:
$$I = P \times R \times T = \$15,300 \times 0.08 \times 0.25 = \$306$$
$$\text{new P} = \$15,300 + \$306 = \$15,606$$

third 3 months:
$$I = P \times R \times T = \$15,606 \times 0.08 \times 0.25 = \$312.12$$
$$\text{new P} = \$15,606 + \$312.12 = \$15,918.12$$

fourth 3 months:
$$I = P \times R \times T = \$15,918.12 \times 0.08 \times 0.25 \doteq \$318.36$$
$$\text{maturity value} = \$15,918.12 + \$318.36 = \$16,236.48$$

Find the total interest earned as follows.

$16,236.48	maturity value
−$15,000.00	original principal
$1,236.48	compound interest earned ∎

total compound interest earned	=	maturity value − original principal

▶ **You Try It** **10.** How much interest will $12,000 earn at 15% compounded semiannually for 2 years?

SOMETHING MORE

Annual Yield You invest $1,000 at 8% compounded monthly. A bank tells you the *annual yield* at this rate is 8.2995%. This means to find the annual interest, just take 8.2995% of $1,000.

$$I = P \times \text{Annual Yield} = \$1,000 \times 0.082995 \doteq \$83.00$$

8.2995% compounded annually earns the same amount of interest in one year as 8% compounded monthly. They are equivalent rates. The annual yield is given to consumers so they can easily and quickly figure the yearly interest earned on an investment.

Solve each problem.
a. A $5,000 CD earns 9% compounded daily, with an annual yield of 9.4162%. What interest is earned in the first year?
b. Suppose $2,500 is invested at 12% compounded monthly with an annual yield of 12.6825%. What interest is earned in the first year?

▶ **Answers to You Try It**   1. I = $96   2. $T = \dfrac{5}{12}$; I = $72.80

3. $T = \dfrac{36}{12} = 3$; I = $540; S = $2,340; monthly payment = $65

4. a. $T = \dfrac{15}{12} = 1.25$;   I = $337.50; maturity value = $3,337.50; monthly payment = $222.50   b. total cost = $3,537.50   5. T = 4 yr

6. $T = \dfrac{3}{12} = 0.25$; P = $1,500   7. R% = 4%   8. I = $40

9. first year: I = $20; new P = $120; second year: I = $24; total interest earned = $44; $4 more in interest is earned
10. T = 0.5 yr; first 6 mo: I = $900; new P = $12,900; second 6 mo: I = $967.50; new P = $13,867.50; third 6 mo: I = $1,040.06; new P = $14,907.56; fourth 6 mo: I = $1,118.07; maturity value = $16,025.63; total interest earned = $4,025.63

▶ **Answers to Something More**   a. $470.81   b. $317.06

SECTION 7.8 EXERCISES

1 *Find the simple interest. Round answers to the nearest cent.*

1. P = $800, R = 10%, T = 1 year

2. P = $450, R = 11%, T = 3 years

3. P = $4,500, R = 8%, T = 4 years

4. P = $260, R = 14%, T = 2 years

5. P = $569.95, R = 7.5%, T = 2 years

6. P = $934.56, R = 10%, T = 2 years

7. P = $4.000. R = 16%, T = 12 months

8. P = $350. R = $9\frac{1}{2}$%, T = 24 months

9. P = $700, R = 6%, T = 18 months

10. P = $2,420, R = 18%, T = 42 months

11. What interest do you earn if you invest $5,000 at 15% for 2 years?

12. What interest is earned if you deposit $2,150 at 6.75% for 4 years?

13. What interest do you pay on a $600 loan at 13.5% for 1 year?

14. What interest do you pay on a $3,500 loan at 12% for 3 years?

🖩 15. $1,500 is borrowed at 6.5% for 30 months. Find the interest.

🖩 16. $780 is invested at 8.8% for 48 months. Find the interest.

---

2 *Find the monthly payment.*

17. Daniele borrowed $4,000 at 14% simple interest for 12 months to build a deck. Calculate her monthly payment.

18. Susan purchased a computer for $1,540. She financed it at 16% simple interest for 24 months. Find her monthly payment.

19. A couple borrowed $3,000 for a vacation to Disneyworld. Find their monthly payment if they are paying 20% simple interest for 6 months.

20. Ed borrowed $1,620 to buy office equipment. If he pays the money back at 8% simple interest over 9 months, find his monthly payment.

21. Jana purchased a car for $14,600. She put $2,600 down and financed the rest at 10% simple interest over 24 months. What is her monthly payment?

22. A $748 TV can be purchased for $148 down with the rest financed at 11% simple interest for 15 months. Find the monthly payment.

23. A motorboat costs $12,000. It may be purchased for 15% down, with the rest paid in 36 monthly payments at 16% simple interest.
  a. What is the finance charge?

  b. What is the monthly payment (round to the nearest cent)?

24. A jeweled watch priced at $2,400 is sold on the installment plan for 20% down with the rest paid in 18 monthly payments at 16% simple interest.
  a. What is the finance charge?

  b. What is the monthly payment (round to the nearest cent)?

25. A lot is on sale for $46,000. The owner will finance the sale for $6,000 down, with the remainder paid in 60 monthly payments at 10% simple interest.
  a. What is the finance charge?

  b. What is the monthly payment?

  c. What is the total cost of the lot on the installment plan?

26. A necklace is on sale for $640. The store owner will finance the purchase for $\frac{1}{4}$ down with the balance paid in 16 equal monthly installments at 12% simple interest.
  a. What is the finance charge?

  b. What is the monthly payment?

  c. What is the total cost on the installment plan?

---

3 *Use the simple interest triangle to solve each of the following problems for the unknown amount.*

	Interest	Principal	Rate	Time
27.	$200	?	10%	1 year
28.	$480	?	8%	2 years
29.	$90	$300	7.5%	?
30.	$432	$3,600	6%	?
31.	$75	$1,250	?	1.5 years
32.	$1,200	$8,000	?	3 years
33.	?	$700	9%	2.75 years
34.	?	$4,200	5.25%	4 years
35.	$26.65	$820	?	6 months
36.	$33.12	$184	?	2 years 3 months
37.	$537.60	?	8.4%	1 year
38.	$9.45	?	4.5%	4 years
39.	?	$212	6.25%	0.4 year
40.	?	$26.80	12.5%	0.6 year
41.	$8.10	$90	4.5%	?
42.	$450.20	$10,000	9.004%	?

43. What rate earns $60 on $480 over 1 year?

44. What rate is needed to charge $120 interest on $400 over 3 years?

45. How long will it take for $750 to earn $180 at 6% simple interest?

46. In how many years will $1,600 earn $400 interest at 2.5%?

47. What principal must be invested at 16% to earn $3,200 in 2 years?

48. What amount will make $500 over 6 months at 8%?

49. How long will it take $5,000 to double at 6.25% simple interest?

50. How long will it take $1,200 to triple at 8% simple interest?

51. Nancy and Ken borrowed $14,000 for 2 years to finance a sailboat. The maturity value for the loan was $16,520. What rate of interest did they get on the loan?

52. Lee Ann borrowed $18,500 to buy a BMW. The maturity value for the loan was $23,772.50. If the rate was 9.5%, find the term.

4 *Find the maturity value and the total interest earned.*

53. $800 at 6% compounded annually for 2 years

54. $500 at 9% compounded annually for 3 years

55. $2,400 at 12% compounded quarterly for 1 year

56. $250 at 8% compounded quarterly for 1 year

57. $3,750 at 12% compounded monthly for 5 months

58. $600 at 24% compounded monthly for 3 months

59. $40,000 at 9% compounded semiannually for 30 months

60. $25,000 at 7% compounded semiannually for 2 years

61. Ed deposits $5,000 in an account earning 8% compounded quarterly.

  a. Find his maturity value after one and one-half years.

  b. What total interest did he earn?

62. Simone invested $3,000 in a CD earning 12% compounded semiannually for 4 years.

  a. How much will she have when her CD matures?

  b. What total interest will she have earned?

63. John invested $2,000 in a bond earning 7.2% compounded annually.

  a. What will the bond be worth when it matures in 7 years?

  b. How much interest will have been earned?

64. Regina deposited $45,000 at 10.91% compounded annually.

  a. How much will she have in 4 years?

  b. How much interest will she have earned?

SKILLSFOCUS (Section 3.6) *Divide.*

65. $\dfrac{8}{9} \div \dfrac{2}{9}$ **66.** $\dfrac{6}{15} \div \dfrac{7}{10}$ **67.** $4 \div \dfrac{5}{8}$ **68.** $3\dfrac{4}{7} \div 2$

EXTEND YOUR THINKING ▶▶▶▶
▶ SOMETHING MORE

69. $4,500 is invested at 8% compounded monthly. The annual yield is 8.2995%. Find the first year's interest.

70. A certificate of deposit for $17,200 earns 12% compounded monthly. The annual yield is 12.6825%. What interest is earned in the first year?

71. Find the maturity value and the amount of interest earned in each case if $1,000 is deposited for 1 year at 12% compounded
  a. annually.

  b. semiannually.

  c. quarterly.

  d. monthly.

  e. What conclusion do you draw about the amount of interest earned when you compound more often?

72. Find the maturity value and amount of interest earned if each sum of money is deposited at 10% compounded annually for 4 years.

  a. $1,000

  b. $2,000

  c. $4,000

  d. $8,000

  e. If the principal is doubled, does the interest earned also double?

73. Find the maturity value and total interest earned if you deposit $1,000 at 12% compounded quarterly for each period of time.

  a. 3 months

  b. 6 months

  c. 9 months

  d. 12 months

  e. Does adding 3 months to the term each time increase the interest earned by a fixed amount each time?

74. Compute the maturity value and interest earned if you invest $1,000 for 5 years at each compounded annual rate.

  a. 2%

  b. 4%

  c. 8%

  d. 16%

  e. If the rate doubles, does the interest earned also double?

WRITING TO LEARN ▶▶▶▶

75. In your own words, explain the meaning of the terms *simple interest*, *principal*, *rate*, *time*, and *maturity value*.

76. Draw the simple interest triangle. In your own words, explain how to use it, and how to get four formulas from it.

77. Explain the difference between simple interest and compound interest.

CALCULATOR TIPS

Using Your Calculator's Percent Key, %, to Solve Word Problems
Your calculator's percent key, %, can be used to solve percent word problems. To enter a rate into your calculator, enter the number followed by the % key. There is no need to first change the rate to a fraction or decimal before using it. The calculator takes care of that. For consistency, always enter the rate into the calculator last. Therefore, in each example below, the % key is the last key pressed.

Find the AMOUNT using the formula $A = B \times R\%$.

Enter these keypresses: B × R %

EXAMPLE 1   42% of 250 is what number?

$R\% = 42\%$
$B = 250$
$A = ?$

$A = B \times R\% = 250 \times 42\%$
Enter: 250 × 42 %
Calculator displays: 105 ∎

CAUTION   Some calculators may require you to press the = key after the % key to get the final answer.

Find the BASE using the formula $B = \dfrac{A}{R\%}$.

Enter these keypresses: A ÷ R %

EXAMPLE 2 $62\frac{1}{2}\%$ of what number is 25?

$R\% = 62\frac{1}{2}\% = 62.5\%$ $B = \dfrac{A}{R\%} = \dfrac{25}{62.5\%}$

$B = ?$
$A = 25$

Enter: 25 ÷ 62.5 %
Calculator displays: 40 ∎

Find the RATE using the formula $R\% = \dfrac{A}{B}$.

Enter these key presses: A ÷ B %

EXAMPLE 3   What percent of 30 is 21?

$R\% = ?$
$B = 30$
$A = 21$

$R\% = \dfrac{A}{B} = \dfrac{21}{30}$

Enter: 21 ÷ 30 %
Calculator displays: 70
(meaning 70%) ∎

Use your calculator to solve these problems.
1. 27% of \$350 is what?
2. What is 8.7% of 40 pounds?
3. 65% of what is 156?
4. 0.6 is 75% of what?
5. What percent of 20 is 19?
6. $\dfrac{3}{4}$ is what percent of 2?

CHAPTER 7 REVIEW

VOCABULARY and MATCHING

New words and phrases introduced in this chapter are shown in the left-hand column in the order they appeared. Match each term on the left with the phrase or sentence on the right that best describes it.

A. percent

B. base

C. complement

D. rate

E. amount

F. short form for a percent word problem

G. long form for a percent word problem

H. commission

I. straight commission

J. salary plus commission

K. discount

L. sale price

M. percent increase or percent gain

N. percent decrease or percent loss

O. interest

P. principal

Q. time, or term

R. simple interest

S. maturity value

T. compound interest

_____ remaining number after R% and B are identified

_____ salary that is made up entirely of commissions

_____ story problems often having two or more sentences

_____ paid to borrow money; earned when you invest it

_____ interest earns interest

_____ income earned, expressed as a percent of what you sell

_____ number of parts out of 100 equal parts

_____ quantity on which interest is calculated

_____ interest calculated only on the principal for the entire term of the loan

_____ found by subtracting a percent from 100%

_____ original principal plus all interest earned

_____ difference between two numbers divided by the larger number

_____ original selling price minus discount

_____ whole amount, follows the words "percent of"

_____ earn a fixed salary plus commission on sales made

_____ number with the percent symbol

_____ difference between two numbers divided by the smaller number

_____ A is R% of B, or R% of B is A

_____ original selling price minus sale price

_____ entire time over which simple interest is calculated

REVIEW EXERCISES

7.1 The Meaning of a Percent

1. 30% means _____ parts out of _____ equal parts.

2. _____ parts out of _____ equal parts is 7.5%.

3. 99 parts out of 100 equal parts is _____ %.

4. _____ % means 70 parts out of 100 equal parts.

5. The complement of 28% is _____ %.

6. The complement of 3.7% is _____ %.

7. _____ % is the complement of $22\frac{2}{9}$%.

8. _____ % is the complement of $\frac{1}{4}$%.

7.2 Decimals and Percents

Change each decimal to a percent.

9. 0.37   **10.** 0.009   **11.** 1.4   **12.** 2.0137   **13.** 4   **14.** 0.875

Change each percent to a decimal.

15. 91%   **16.** 5%   **17.** 215%   **18.** 0.0635%   **19.** 7.2%   **20.** $6\frac{1}{2}\%$

7.3 Fractions and Percents

Change each fraction to a percent.

21. $\dfrac{3}{5}$ **22.** $\dfrac{5}{8}$ **23.** $\dfrac{7}{3}$ **24.** $\dfrac{9}{11}$ **25.** $\dfrac{9}{10}$ **26.** $\dfrac{11}{12}$

Change each percent to a fraction, and reduce.

27. 45%   **28.** 17%   **29.** $88\frac{8}{9}\%$ **30.** $27\frac{3}{11}\%$ **31.** 137.5%

32. 14.4%   **33.** 8.25%   **34.** 0.6%   **35.** $\frac{1}{3}\%$ **36.** $1\frac{5}{8}\%$

7.4 Identifying the Three Numbers in a Percent Word Problem

37. 35% of 400 cars is 140 cars.

38. 115% of $240 is $276.

39. 38 is 10% of 380.

40. 173.6 grams is 155% of 112 grams.

41. What percent of 60 is 50?

42. 30 is what percent of 24?

43. 20 is 40% of what number?

44. 74% of what number is 333?

45. 106% of 45 is what number?

46. What is 0.75% of $40?

7.5 Solving the Short Percent Word Problem

47. 42% of what number is $147?

48. 114% of what number is 627?

49. What percent of 70 is 50?

50. 36 is what percent of 24?

51. 120% of 45 is what number?

52. What is 1.25% of $40?

7.6 Solving Percent Word Problems

53. Suppose 64 people belong to the country club. If 24 are senior citizens, what percent are seniors?

54. A new car is advertised to get 40 miles to a gallon of gas. George bought the car and found he got 28 miles to the gallon. What percent of the advertised miles per gallon did George actually get?

55. There are 750 yachts in a marina. If 38% of them exceed 36 feet in length, how many exceed 36 feet?

56. Dale had $48.50 in his wallet. He spent 16% of his money on lunch. How much did he spend on lunch?

57. Sol's mortgage payment is $650. If this is 40% of his monthly net pay, what is Sol's monthly net pay?

58. Suppose 17% of a small town's yearly budget is spent on education. If $391,000 is spent on education, how large is the town's annual budget?

59. Suppose 1,650 plants were grown in an experimental soil. If 56% died before maturity, how many died before maturity?

60. The museum had 4,000 visitors today. Of those, 1,600 reside out of state. What percent reside out of state?

7.7 Commission, Discount, and Percent Increase and Decrease

61. Harry works for an 11% commission on sales. If he sells $6,000 in merchandise, what will his commission be?

62. Jack made $870 in commissions this month. If he is paid a 15% straight commission, what were Jack's total sales for the month?

63. Al is paid a monthly salary of $440 plus a 7.5% commission on sales. If he sold $5,400 in merchandise this month, what was his salary?

64. Edna works on salary plus commission. She made $1,160 this week. If her weekly salary is $260, and she earns a 9% commission, what were her total sales for the week?

65. A ten-speed bike usually sells for $180. It is being sold this week at a 30% discount. Find the sale price.

66. A motorcycle is discounted 30%, giving a savings of $1,440.

    a. What was the original price of the motorcycle?

    b. What is the sale price?

67. Dave sold $1,340 in books yesterday, and $1,400.30 in books today. Find Dave's percent increase in sales?

68. Cal's Car Wash served 450 cars last week, but only 414 cars this week. What percent decrease is this?

69. Isadore makes $34,560 per year. He is getting a 4.2% increase in salary.
  a. How much of an increase will Isadore get?

  b. What will his new annual salary be?

70. A man weighing 465 pounds had a 60% decrease in weight over a two-year period.
  a. How much weight did he lose?

  b. What was his new weight?

7.8 Simple Interest

71. Find simple interest if $4,000 earns 12% over 3 years.

72. What interest does $360 earn at 8.5% simple interest over 6 years?

73. How much interest will Gail pay if she borrows $1,450 at 5% simple interest for 18 months?

74. What will $66,000 earn at 4.8% simple interest for 9 months?

75. What monthly payment will be made if $1,200 is borrowed at 16% simple interest for 12 months?

76. Lucy is planning to borrow $2,800 for a trip to Hawaii. She plans to pay the money back at 9% simple interest over 5 months. Find her monthly payment.

77. What principal can earn $400 at 10% for 2 years?

78. What rate will earn $700 on $3,500 over 4 years?

79. How long must $5,000 be deposited to earn $1,200 at 6%?

80. What principal will earn $4,000 at 12.5% over 8 years?

81. What rate will earn $88.20 on $784 over 30 months?

82. How long will it take $800 to earn $70 at 7% simple interest?

83. What will $250 earn at 8% compounded annually for 3 years?

84. What will $40,000 earn at 6% compounded semiannually for 2 years?

85. What will $9,400 earn at 9% compounded monthly for half a year?

86. Ida deposits $15,000 in an account earning 12% compounded quarterly.
  a. Find her maturity value after one and one-half years.

  b. What total interest did she earn?

This test will measure your understanding of percent. Allow yourself 50 minutes to complete it. Show the work for each problem. When done, check your answers. Rework each problem solved incorrectly.

1. 7.2% means _____ parts out of _____ equal parts.

2. If 29% of a solution is water, then _____ percent is not water.

3. Change 0.07 to a percent.

4. Change 1.5 to a percent.

5. Change 5.8% to a decimal.

6. Change 0.06% to a decimal.

7. Change $\frac{5}{12}$ to a percent.

8. Change $2\frac{7}{8}$ to a percent.

9. Change 84% to a fraction in lowest terms.

10. Change 3.6% to a fraction in lowest terms.

11. 40% of how much money is $52?

12. 120% of 2,500 bushels is how many bushels?

13. 28 feet is what percent of 35 feet?

14. Justin works on a 7% commission. If he earned an $840 commission last month, what were his total sales for that month?

15. Phil made $2,460 last month selling brushes. If $1,045.50 was reinvested in purchasing more brushes, what percent did Phil reinvest?

16. Thirty-eight percent of the children in Alma's school are boys. If 1,500 children are enrolled in Alma's school, how many are boys?

17. Polly is a golfer. Yesterday she shot a 75. Today she shot a 69. What percent decrease did Polly realize in her golf score?

18. The original selling price of a coat was $125. The coat was put on sale for a 40% discount. What was the sale price of the coat?

19. This year Catherine's company earned $75,600. This is 126% of last year's earnings. What did her company earn last year?

20. The company will allow 0.09% of its annual budget to be used for recreation. How much money does this amount to if the budget this year is $3,600,000?

21. What simple interest will you pay if you borrow $1,100 at 8% for 2 years?

22. What principal will earn $450 at 7.5% simple interest for 2 years?

23. What simple interest rate is needed to earn $600 on $2,000 in 36 months?

24. How long will it take for $25,000 to earn $3,187.50 at 8.5% simple interest?

25. Find the maturity value if $6,400 earns 3% over 4 years.

26. Gus borrows $2,100 to buy a lawn tractor. If he finances his loan at 14% for 24 months, what is his monthly payment?

27. Find the interest earned on $3,500 at 16% compounded quarterly for 1 year.

28. What is the maturity value in 18 months if $6,000 earns 20% compounded semiannually?

Explain the meaning of each term. Use your own examples.

29. base

30. principal

31. commission

1. Write 2.603 in words.

2. Write five thousand and five ten-thousandths in decimal notation.

3. Round each number to the indicated place.

   a. 508.36 tens   **b.** 0.0528 hundredths   **c.** 6.99975 tenths

4. $5.6 + 23.564 + 7 + 0.083 + 5.76$

5. $45.06 - 7.814$ **6.** 7.05×12.8

7. $75.361 \div 5.27$ **8.** $378.48 \div 8.6$ (round to hundredths)

9. A yacht owner purchased 68.3 gallons of gas at $1.239 per gallon. What was the total cost rounded to the nearest cent?

10. How many $0.79 bags of licorice can be purchased for $16?

11. Change $\dfrac{15}{8}$ to an exact decimal.   **12.** Change $\dfrac{7}{11}$ to a decimal rounded to three places.

13. Change 0.46 to a fraction in lowest terms.   **14.** Change 2.075 to a fraction in lowest terms.

15. Simplify $0.6 + 1.4 \times 2.85$.

16. Arrange the following numbers in order from smallest to largest.
7.07, 7.087, 7, 7.107, 7.0077

17. There are 12 yachts and 44 people at the marina. Write the ratio of people to yachts in lowest terms.

18. A mobile home traveled 2,112 miles on 264 gallons of gas. Write the rate of miles per gallon in lowest terms.

19. Write $7\dfrac{1}{3}$ to $2\dfrac{3}{4}$ as a ratio of whole numbers.   **20.** Write 0.16 to 12.8 as a ratio of whole numbers.

21. Find the unit price if 48 ounces costs $7.29. Round to the nearest tenth of a cent.

22. Find the price per egg if 12 eggs cost $0.93. Round to the nearest tenth of a cent.

23. Tom, Dick, and Harry split $72,522 according to 8:3:7. How much money does each man receive?

24. Find x. $\dfrac{x}{24} = \dfrac{45}{72}$

25. Find k. $\dfrac{150}{360} = \dfrac{70}{k}$

26. If 38 shirts cost a retailer $579.50, how much will 54 shirts cost?

27. The property tax on a piece of land worth $85,000 is $340. At the same rate, what is the property tax on a parcel of land worth $64,000?

28. Change 70% to a fraction in lowest terms.

29. Change 0.2 to a percent.

30. Change 15.6% to a decimal.

31. Change $\dfrac{4}{15}$ to a percent.

32. Change 265% to a decimal.

33. What is 80% of $450?

34. $\dfrac{3}{4}$% of what number is $6?

35. 9.4 is what percent of 188?

36. Joan works on a 12% commission. She sells a sailboat for $58,900. What commission does she earn on the sale?

37. A company earned a profit of $420,000 last quarter. They invested $260,400 of it in real estate. What percent of the profit was invested in real estate?

38. Last year Angela swam the 50 meters in 48.2 seconds. This year she swam it in 43.6 seconds. By what percent did Angela improve her time? Round to the nearest tenth of a percent.

39. A $240 leather coat was discounted $33\dfrac{1}{3}$%. What is the sale price of the coat?

40. What percent of 25 is 40?

41. What will $4,600 earn at 8% simple interest for 3 years?

42. What principal can earn $72.60 at 11% simple interest for 9 months?

Explain the meaning of each term. Use your own examples.

43. hundredths

44. proportion

45. simple interest

Basic Statistics
and Graphs

THE CIRCLE GRAPH AND WORLD POPULATION

Suppose you could shrink the world down to 100 people, maintaining all existing human ratios. The world would be populated as shown in the circle graph.

100 People

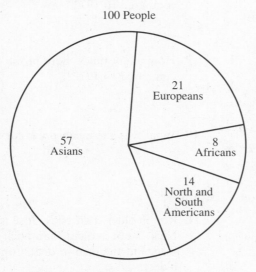

Other facts:

• 1 would have a university education

• 70 would be unable to read

• 6 would have 50% of the world's wealth

Source: United Nations Demographic Data

If there are 5 billion people on Earth, about how many are Europeans? This question will be answered in this chapter.

Information, or data, is used to help you make decisions. This is true in the business world and in your personal life. What data tells you sometimes depends on what numbers you calculate from it, or how you display it. Mean, median, mode, and range are calculated from data, and each tells you something different. On the other hand, a graph allows you to see a picture of your data. This is helpful for discovering trends and making comparisons.

Work through each section in this chapter. Use this test to identify topics you are not familiar with. These topics may require additional study. You may *not* use a calculator.

8.1

1. Given the set of scores 2, 5, 8, 10, and 50, find the

  a. mean.   **b.** median.

  c. mode.   **d.** range.

2. Given the set of scores 12, 15, 25, 30, 20, and 42, find the

  a. mean.   **b.** median.

  c. mode.   **d.** range.

8.2

Use the circle graph to answer the questions. Round answers to tenths.

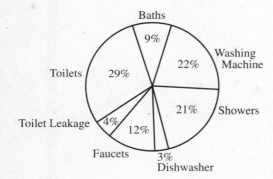

Average Family of 4 Uses
280 Gallons of Water a Day

3. How many gallons of water are used in showers each day?

4. How many more gallons of water are used a day with faucets than with baths?

5. About how many times more water is used in the washing machine than in the dishwasher?

6. How many gallons are lost each day through leakage?

8.3

Use the bar graph to answer the questions.

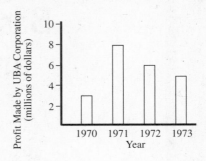

7. How much profit did the company make in 1972?

8. How much less profit did UBA make in 1973 compared to 1972?

9. How many times more profit was made in 1971 compared to 1970?

10. What was the mean profit for the four years?

8.4

11. Ron is older than Nick. José is younger than Nick. Steve is older than Nick but younger than Ron. Put the men in order from youngest to oldest.

VOCABULARY *Explain the meaning of each term. Use your own examples.*

12. mean

13. range

14. mode

8.1 MEAN, MEDIAN, MODE, and RANGE

OBJECTIVES

1 Find the mean (Section 2.3, Objective 2).
2 Find the median.
3 Find the mode.
4 Find the range.
5 Define percentile (Section 7.1, Objectives 1, 2).

NEW VOCABULARY

statistics	mean	range
data	median	percentile
population	mode	raw score
sample	tally	

Suppose 50 students took a math exam. You want an estimate of the class average. You randomly ask five students for their test scores. They are 90, 80, 86, 69, and 75. You then calculate the average of the five scores.

$$\text{average} = \frac{90 + 80 + 86 + 69 + 75}{5} = \frac{400}{5} = 80$$

Can you then predict that the whole class averaged about 80?

That is the kind of question statistics attempts to answer. **Statistics** is the study of information, or **data**. The whole class of 50 students is called the **population**. The five scores 90, 80, 86, 69, and 75 are called **sample data**. Statistics uses sample data to attempt to make predictions about the whole population. This is how you learn about a population when you do not know, or cannot get, all the scores in it.

In this chapter you will learn to calculate the mean, median, mode, and range. Mean, median, and mode are called *measures of central tendency*. They are measures of the center of a set of data.

1 Finding the Mean

The **mean** for a group of numbers is the same as the average discussed in Section 2.3.

$$\text{mean} = \frac{\text{sum of the numbers}}{\text{how many numbers were summed}}$$

EXAMPLE 1 Find the mean for the scores: 15, 20, 17, 20, 15, and 18.

$$\text{mean} = \frac{15 + 20 + 17 + 20 + 15 + 18}{6} \quad \longleftarrow \text{ Sum the scores.}$$
$$\longleftarrow \text{ Divide by 6, the number of scores.}$$
$$= \frac{105}{6}$$
$$= 17.5 \quad \blacksquare$$

▶ **You Try It** 1. Find the mean for 10, 7, 12, 8, 9, 4, and 6.

2 Finding the Median

The mean is a measure of the center of data calculated from all the given scores. Therefore, it is affected by very large or very small numbers.

Another measure of central tendency is the median. The **median** is the middle score. You use it when you want a measure of the center of data that is not affected by very large or very small numbers. To find the median, first arrange all your data in order.

> Given a group of numbers ordered from smallest to largest (or largest to smallest), the **median** is the middle number. If there are two middle numbers, the median is the mean of the two middle numbers.

EXAMPLE 2   The number of goals scored by the Lake High lacrosse team in the last seven games were 5, 9, 7, 2, 2, 6, and 3. Find the median number of goals scored.

First, arrange the numbers in order.

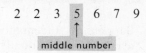

2   2   3   **5**   6   7   9

↑
middle number

The median is 5. ∎

The median is the middle number. Half the numbers are smaller than the median, and half are larger. In Example 2, three numbers, 2, 2, and 3, are smaller than the median, 5. Three numbers, 6, 7, and 9, are larger.

▶ **You Try It**   **2.** Find the median for 2, 6, 9, 20, 4, 7, and 3.

EXAMPLE 3   Find the median for the test scores

87, 94, 82, 100, 76, and 90.

First, arrange the scores in order.

76   82   87   90   94   100

The median is the mean of two middle numbers.

$$\text{median} = \frac{87 + 90}{2} = 88.5$$

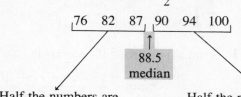

76   82   87   90   94   100

↑
88.5
median

Half the numbers are less than the median.   Half the numbers are greater than the median.

OBSERVE   The median 88.5 is not one of the six original numbers. ∎

▶ **You Try It**   **3.** Find the median for 19, 40, 9, 32, 25, 19, 21, and 38.

3 Finding the Mode   The **mode** is the number that occurs most often in a group of numbers. Like mean and median, the mode is also called an average.

EXAMPLE 4   Find the mode for each set of scores.

a. 2, 3, 5, 5, 5, 7, 9, 9   The mode is 5, because 5 occurs most often.

b. 2, 4, 4, 4, 6, 7, 7, 7, 8   There are two modes, 4 and 7. Each occurs three times. Thus the data are *bimodal*.

c. 2, 3, 6, 7, 8   There is no mode. No number occurs more often than any other. ∎

▶ **You Try It**   Find the mode.

4. 7, 2, 9, 2, 4

5. 1, 6, 7, 3, 0, 5, 8

6. 7, 2, 5, 7, 5, 2, 5, 1, 3

EXAMPLE 5   A conservationist counted the number of eggs in ten nests of bald eagles. She found 2, 1, 2, 3, 2, 2, 4, 3, 2, and 2 eggs, respectively. Find the mode.

Create a **tally** to see how often each number of eggs occurred.

	number of eggs	frequency
	1	/
mode	2	ЖЖ /
	3	//
	4	/

The mode is 2. It is the number that occurs most often. 2 eggs were found in bald eagle nests more often than any other number of eggs. ∎

The mode may be an indicator of a popular item.

EXAMPLE 6   A store sells three kinds of shoes, flats (#204), pumps (#317), and heels (#165), for the same price. Each time a pair of shoes is sold, its number is recorded on a sales sheet in tally form as shown.

shoe	number sold
flat (#204)	////
pump (#317)	ЖЖ //
heel (#165)	///

The sales mode is shoe #317, the pump. It is the shoe that sold most often. The mode may tell the store owner that the pump is the most popular of the three shoes. As an inventory item, the sales mode may also indicate the shoe in most need of restocking. ∎

▶ **You Try It**   **7.** Fourteen fifth graders have ages 10, 11, 11, 10, 11, 11, 12, 10, 11, 11, 10, 11, 10, and 10. Find the mode.

8. The number of children in seventeen families is 2, 1, 1, 3, 1, 2, 1, 0, 0, 2, 1, 2, 2, 3, 4, 0, and 1. Find the mode.

---

4 Finding the Range   The **range** for a group of numbers is the largest number minus the smallest.

EXAMPLE 7   Find the range for each group of numbers.

a. 2, 5, 6, 6, 8, 9:   range = largest − smallest = 9 − 2 = 7
b. 2, 3, 2, 2, 3, 2, 2:   range = largest − smallest = 3 − 2 = 1
c. 54, 6, 700:   range = 700 − 6 = 694   ∎

Find the range.

    9. 2, 1, 7, 5, 3

  10. 16, 20, 17, 16, 33, 16, 25, 15, 30, 28, 18

  11. 63, 0, 7, 95

Mean, median, and mode measure the center of data. Range is called a *measure of dispersion*. It measures how spread out a set of scores is. Two sets of scores can have the same mean, median, and mode, but different ranges.

EXAMPLE 8

Test scores for two students are shown here. Calculate the mean, median, mode, and range for each student.

Student #1	*Student #2*
test scores: 40, 70, 100	test scores: 69, 70, 71
mean $= \dfrac{40 + 70 + 100}{3} = \dfrac{210}{3} = \boxed{70}$	mean $= \dfrac{69 + 70 + 71}{3} = \dfrac{210}{3} = \boxed{70}$
median $= \boxed{70}$	median $= \boxed{70}$
mode $= \boxed{\text{none}}$	mode $= \boxed{\text{none}}$
range $= 100 - 40 = \boxed{60}$	range $= 71 - 69 = \boxed{2}$

Range is sometimes used to measure consistency in data. Student #2 has a much smaller range in test scores than Student #1. The scores of Student #2 are more consistent. Could the large range in Student #1's scores indicate inconsistent study habits? What else could they indicate? ■

▶ **You Try It**

  12. Student A had test scores 40, 60, 70, and 90. Student B had test scores 63, 64, 66, and 67. Find the mean, median, mode, and range for each student.

5 Defining Percentile

Percentile is a value below which a given percent of scores lie. For example, Diane took the SAT exam. She scored in the 85th percentile. This means 85% of the scores were lower than Diane's score, and 15% were higher.

EXAMPLE 9

a. Susan scored 710 out of a possible 800 points on a national exam. Susan's actual score of 710 is called her **raw score**. She was told her raw score was in the 93rd percentile. Interpret what this means.

    93rd percentile means 93% of the scores on the exam were lower than Susan's score of 710. The complement, or 7%, were higher than 710.

b. Suppose 40,000 people took the exam. How many scored lower than Susan? How many scored higher?

    If Susan scored in the 93rd percentile, then

$$93\% \text{ of } 40{,}000 \text{ people} = .93 \times 40{,}000 = 37{,}200 \text{ people}$$

had a score lower than Susan's. $40{,}000 - 37{,}200 = 2{,}800$ people scored higher than Susan. ■

▶ **You Try It**

  13. Maria scored 312 out of 400 on an exam. Her score is in the 78th percentile. What percent of the scores were lower than Maria's? What percent were higher?

  14. Suppose 56,000 people took the exam in problem 13. How many people scored lower than Maria? How many scored higher?

SOMETHING MORE

Deception with Averages The mean and median are both called averages. However, the word *average* is sometimes used without stating whether it is the mean or median. This can result in misleading data. For example, in a small town of 5 people, the annual salaries are $12,000, $14,250, $14,750, $21,000, and $129,000.

The mayor tells state officials he is in need of state aid because the average income in his town is only

$$\text{average (median)} = \$14{,}750.$$

Later, the mayor tells a group of voters he should be reelected. The city is doing well under his leadership. He says the average income in his city is a state high of

$$\text{average (mean)} = \frac{\$12{,}000 + \$14{,}250 + \$14{,}750 + \$21{,}000 + \$129{,}000}{5}$$

$$= \$38{,}200.$$

Both statements are correct.

The mean is not a good measure of average income. It is strongly affected by very large (or very small) numbers. The mean, $38,200, gives a distorted picture of average income. The very large $129,000 salary pushes the mean so high that it is no longer representative of city salaries. The median, $14,750, gives a more realistic picture of average income. It is the middle income. It is not affected by very large (or very small) numbers. You can replace $129,000 by $1,000,000, and the median will not change. $14,750 will still be the middle number. The next example shows how increasing one number increases the mean, but not the median.

numbers	median	mean
5, 10, 15	10	10
5, 10, 150	10	55
5, 10, 1,500	10	505

The mean is a more realistic average than the median when dieting. For example, you must average 1,000 calories a day for 7 days a week on a new diet. You decide to use the median as your average. You plan to consume the following number of calories each day of the week.

MON	TUES	WED	THUR	FRI	SAT	SUN
1,000	1,000	1,000	1,000	7,000	8,000	9,000

The average (median) calories per day is 1,000. But the mean number of calories per day is (1,000 + 1,000 + 1,000 + 1,000 + 7,000 + 8,000 + 9,000)/7 = 4,000! The median does not provide a realistic measure if your intention is to lose weight because it is not changed by very large (or very small) numbers.

Find the mean and median for each set of scores.

a. 10, 20, 30   **b.** 10, 20, 3,000   **c.** 10, 20, 300,000

▶ **Answers to You Try It** 1. 8   2. 6   3. 23   4. 2   5. none   6. 5   7. 11
8. 1   9. 6   10. 18   11. 95   12. A: 65, 65, none, 50; B: 65, 65, none, 4
13. 78%; 22%   14. 43,680; 12,320

▶ **Answers to Something More**   a. mean = 20, median = 20
b. mean = 1,010, median = 20   c. mean = 100,010, median = 20

1 2 3 4 *Find the mean, median, mode, and range.*

1. 2, 3, 5, 8

2. 1, 1, 4, 5, 7, 12

3. 25, 32, 40, 45, 50

4. 5, 5, 6, 8, 12

5. 1, 0, 3, 2, 0, 6

6. 15, 15, 15

7. 2, 2, 5, 5, 5, 5

8. 7, 8, 9, 10, 11, 12, 13

9. 24, 10, 51, 134, 17, 10

10. 1, 5, 3, 7, 4, 2, 3, 8, 5, 0

11. 2, 3, 2, 2, 3, 3

12. 88, 73, 90, 42, 100, 84, 77, 90

13. 10, 20, 30, 40

14. 10, 20, 30, 400

15. 6, 6, 6, 6

16. 1, 0, 1, 1, 1, 0, 0, 1, 0, 1

17. 0.1, 2.7, 0.33, 0.2, 0.1

18. 1.4, 20, 0.830, 0.007, 7, 100

19. Compute the mean, median, mode, and range for each set of numbers.
   a. 90, 100, 110   **b.** 80, 100, 120   **c.** 50, 100, 150

   d. Which measures changed, and which were the same?

20. Compute the mean, median, mode, and range for each set of numbers.
   a. 20, 40, 60   **b.** 20, 40, 600   **c.** 20, 40, 6,000

   d. Which measures changed, and which were the same?

---

21. In a one-year period, a prime stock had a range in price of $18.25. If the low price for the stock was $57.75, what was the high price for the year?

22. In four years of high school, Sam's weight had a range of 42 pounds. If his heaviest weight was 203 pounds, what did Sam weigh at his slimmest?

23. Alice must take five history exams. Her scores on the first four are 99, 88, 92, and 93. What must she score on her last exam to have a mean of 90? (Hint: See Section 2.3, Example 10, page 113.)

24. Eva bowled 157 and 180 in two games. What must she bowl in her third game to have a mean score of 175?

25. A Census Survey A market researcher surveyed 40 families. The number of children in each family is given below.

```
2  3  1  0  3  2  4  0  2  2
3  2  0  1  3  3  6  2  0  1
1  0  0  3  1  2  1  2  5  2
3  3  1  2  0  1  2  2  1  4
```

a. Construct a tally for the 40 numbers.

b. Compute the mean number of children per family.

c. Compute the median, mode, and range.

26. Quality Control A quality control engineer had to determine if a soda machine is properly adjusted to give 8 fluid ounces of soda. The engineer had the machine fill 24 cups. She then measured the volume of soda in each cup. The volumes are shown here.

```
7.8  8.2  8.1  7.9  7.7  8.1  7.9  7.8
8.1  7.7  7.9  8.0  8.0  7.6  8.1  7.8
8.1  8.0  7.8  7.6  7.9  7.7  8.1  8.0
```

a. Construct a tally for the 24 volumes.

b. Compute the mean volume of soda per cup. Round to tenths.

c. The machine is considered to be properly dispensing soda if this mean is within 0.2 ounces of 8 ounces. That is, within the range 8 − 0.2 to 8 + 0.2 ounces. Is this machine dispensing soda in proper amounts?

d. Compute the median, mode, and range for the 24 volumes.

5

27. Phil scored in the 76th percentile on an exam. His raw score was 518.

a. What percent of the test scores were lower than Phil's?

b. What percent of the test scores were higher than 518?

c. If 80,000 people took the exam, about how many scored lower than Phil?

28. Amanda's exam score of 785 put her in the 97th percentile.

a. What percent of the test scores were below 785?

b. What percent scored higher than Amanda?

c. If 350,000 people took the exam, about how many scored lower than Amanda?

SKILLSFOCUS (Section 6.7) *Solve for x.*

29. $\dfrac{12}{x} = \dfrac{18}{12}$

30. $\dfrac{16}{25} = \dfrac{x}{40}$

31. $\dfrac{10}{7} = \dfrac{15}{x}$

32. $\dfrac{x}{28} = \dfrac{31.5}{42}$

EXTEND YOUR THINKING ▶▶▶▶
▶ SOMETHING MORE

33. Find the mean and median for each set of scores.
   a. 7, 8, 9   **b.** 7, 8, 90   **c.** 7, 8, 900

a. Find the mean and median number of rooms occupied at each hotel.

34. A truck carried ten sacks weighing 120 pounds each, fifteen weighing 135 pounds each, and five weighing 160 pounds each. Find the mean, median, mode, and range for the weights.

b. Which hotel did better business for the week, and why?

c. Compute the range for each hotel. What does this tell you?

35. A company employs ten accountants at $28,000 each, eight finance officers at $24,000 each, five programmers at $32,000 each, two vice presidents at $42,000 each, and pays the president $54,000. Find the mean, median, mode, and range for the salaries. Round to tens.

37. Six numbers are bimodal with three numbers in each mode. The range is 4. The median is 7. What are the six numbers?

36. Teresa owns two identical hotels. The number of rooms occupied at each hotel in one seven-day period were as follows.

Hotel East:	28	24	27	30	20	17	22
Hotel West:	25	28	26	29	31	24	26

38. Fifteen numbers are trimodal (has three modes), with 5 numbers in each of the 3 modes. The median is 4. The range is 5. The largest number is 7. What are the fifteen numbers?

▶ TROUBLESHOOT IT *Find and correct the error.*

39. The mean for 6, 2, 7, 0, and 5 $= \dfrac{6 + 2 + 7 + 0 + 5}{4} = \dfrac{20}{4} = 5.$

40. The median for 6, 4, 8, and 3 is $\dfrac{4 + 8}{2} = \dfrac{12}{2} = 6.$

41. Explain in your own words the difference between mean and median.

42. Twenty-four students took a test with the following results. Half scored 80. A few scored 100. The rest scored 90. Answer each question, and explain in your own words why you chose that answer.

   a. What is the mode for the test scores?

   b. What is the median?

   c. Why can't you calculate the mean from the information given?

   d. Is the mean greater than the mode?

   e. Is the mean less than the median?

43. Thirty students take a quiz. Half scored 70. From one to five scored 80. And the rest scored 60. Answer each question, and explain in your own words why you chose that answer.

   a. Does the median equal the mode?

   b. Is the mean less than the median?

   c. Is the mean greater than the mode?

   d. Is it possible for the mean to be 65 or lower?

▶ YOU BE THE JUDGE

44. Elaine claims the 50th percentile is the median. Is she correct? Explain your decision.

45. The mean age for a group of people is 40.

   a. Does this mean there is someone in the group who is 40 years old? Explain, using your own examples.

   b. Does this mean half the people are older than 40 and half are younger? Explain, using your own examples.

8.2 PICTOGRAPHS, TABLES, AND CIRCLE GRAPHS

OBJECTIVES

1. Read and interpret a pictograph.
2. Read and interpret a table.
3. Read and interpret a circle graph (Section 7.5, Objectives 1, 2).

NEW VOCABULARY

graph	table
pictograph	interpolation
key	circle graph

A **graph** is a picture of data. It conveys a lot of information at a glance. A graph helps you see trends and relationships. It helps you compare, interpret, and analyze data. The next two sections show you how to read and draw conclusions from graphs.

1 Reading a Pictograph

A **pictograph** uses a picture or symbol to represent a fixed quantity. Two symbols represent twice the quantity. One-half of a symbol represents half the quantity, and so on. A **key** is used to show the quantity represented by one symbol. In Figure 8.1, one dollar symbol, $, represents $100.

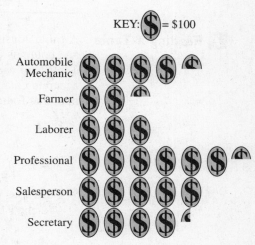

FIGURE 8.1
Median Weekly Earnings of Full-Time Occupations

SOURCE: U. S. Department of Labor, Bureau of Labor Statistics, 1989.

EXAMPLE I   Use the pictograph in Figure 8.1 to answer each question.

a. What are the approximate weekly earnings of an auto mechanic?

The weekly salary of an auto mechanic is represented by four and one-half $ symbols. Since one $ represents $100,

$$\text{auto mechanic} = 4.5 \times \$100$$
$$= \$450 \text{ per week.}$$

b. What is the approximate annual salary of a salesperson?

A salesperson has five $. This is approximately $5 \times \$100 = \$500$ per week. One year has 52 weeks. The approximate annual salary of a salesperson is $52 \times \$500 = \$26,000$.

c. Which occupation is closest to having a median weekly income of $661?

$661 is closer to $650 than to $600 or $700. It would be represented by 6 $ symbols plus one-half of a symbol in the pictograph. The professional is closest to having a weekly median income of $661.

d. How much more does a secretary make per week than a farmer?

secretary = four and one-quarter $	farmer = two and one-half $
$= 4.25 \times \$100$	$= 2.5 \times \$100$
$= \$425$	$= \$250$

A secretary makes about $425 − $250 = $175 more per week than a farmer.

e. Which occupation has an annual salary closest to $15,964?

$$\$15,964 \text{ per year} \div 52 \text{ weeks per year} = \$307 \text{ per week}$$

This would be represented by three $\$$ in the pictograph. $15,964 is closest to the salary of a laborer. ∎

▶ **You Try It**   Answer the following questions using Figure 8.1.

1. What are the approximate weekly earnings of a professional?
2. What is the approximate annual salary of a farmer?
3. Which occupation has a median weekly income closest to $429?
4. How much more does a salesperson make per week than a farmer?
5. Which occupation has an annual salary closest to $23,400?

---

2 Reading a Table

A **table** consists of rows and columns of numbers. Tables are used when more accurate data is wanted than that found in a graph.

Height		Weight	
in	ft in	Men (lb)	Women (lb)
58	4′10″	—	102 (92–119)
60	5′ 0″	—	107 (96–125)
62	5′ 2″	123 (112–141)	113 (102–131)
64	5′ 4″	130 (118–148)	120 (108–138)
66	5′ 6″	136 (124–156)	128 (114–146)
68	5′ 8″	145 (132–166)	136 (122–154)
70	5′10″	154 (140–174)	144 (130–163)
72	6′ 0″	162 (148–184)	152 (138–173)
74	6′ 2″	171 (156–194)	—
76	6′ 4″	181 (164–204)	—

FIGURE 8.2
Desirable Weights and Weight Ranges Versus Height

SOURCE: *Recommended Dietary Allowances*, 9th ed., 1980.

EXAMPLE 2   Use Figure 8.2 to answer each question.

a. According to the National Center for Health Statistics, the average American male is 5′10″ tall and the average female is 5′4″ tall. What are their desirable weights?

5′10″ male: Desirable weight is 154 pounds; desirable weight range is from 140 pounds to 174 pounds.

5′4″ female: Desirable weight is 120 pounds; desirable weight range is from 108 pounds to 138 pounds.

b. What is the highest desirable weight for a 6′2″ man?

The weight range for a 6′2″ man is 156 pounds to 194 pounds. Therefore, his highest desirable weight is 194 pounds.

c. What is the desirable range of heights for a 135-pound woman?

A woman's weight of 135 pounds is contained in the weight ranges corresponding to heights of 5′4″ through 5′10″.

d. Estimate the desirable weight for a 5′5″ adult female.

5′5″ is not in Figure 8.2. It is between 5′4″ and 5′6″. To estimate the desirable weight for a 5′5″ female, use the mean of the weights for 5′4″ and 5′6″.

$$\text{desirable weight for 5′5″ female} = (120 + 128)/2 = 124 \text{ pounds}$$

$$\text{desirable weight range} \quad = (108 + 114)/2 \text{ to } (138 + 146)/2$$

$$= 111 \text{ pounds to } 142 \text{ pounds} \quad ∎$$

The process of estimating values between two given values is called **interpolation**. In Example 2d, the desirable weight for 5'5" was interpolated from the data given for 5'4" and 5'6".

▶ **You Try It**

Answer the following questions using Figure 8.2.

6. Find the desirable weight and weight range for a 5'8" woman.

7. What is the lowest desirable weight for a 5'6" man?

8. What is the desirable range in heights for a 158-pound man?

9. Estimate the desirable weight for a 6'1" man.

3 **Reading a Circle Graph**

A **circle graph**, or **pie chart**, is a circle cut into pie shaped wedges. The wedges display the portion of the whole represented by different quantities.

The circle represents the whole amount. In the percent circle graph in Figure 8.3, the entire circle represents 80,100 deaths. Each wedge is a certain percent of 80,100 deaths. For example, deaths due to choking in 1987 are 4% of 80,100 = 0.04 × 80,100 = 3,204. The sum of the percents in a circle graph is 100%.

FIGURE 8.3
Accidental Deaths by Principal Types, 1987

Total Accidental Deaths = 80,100

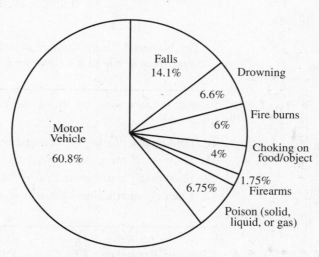

SOURCE: National Safety Council.

EXAMPLE 3

Use the circle graph in Figure 8.3 to answer each question. In each case, round the number of deaths to hundreds.

a. How many accidental deaths were caused by motor vehicles?

60.8% of the deaths were caused by motor vehicles. The whole circle represents 80,100 deaths. Find 60.8% of 80,100.

$$60.8\% \text{ of } 80,100 = 0.608 \times 80,100$$
$$\doteq 48,700 \text{ deaths caused by motor vehicles}$$

b. How many more accidental deaths were caused by falls than drowning?

falls: 14.1% of 80,100 = 0.141 × 80,100 ≐ 11,300 deaths

drowning: 6.6% of 80,100 = 0.066 × 80,100 ≐ 5,300 deaths

About 11,300 − 5,300 = 6,000 more deaths were caused by falls than drowning.

c. How many times more deaths were caused by motor vehicles than poisoning?

60.8% of the deaths were caused by motor vehicles. This is 48,700 from part a. 6.75% of the deaths were caused by poison. This is 6.75% of 80,100 = 0.0675 × 80,100 ≐ 5,400 deaths. Find how many times 5,400 divides into 48,700.

$$\frac{48,700 \text{ motor vehicle deaths}}{5,400 \text{ poisoning deaths}} \doteq 9$$

There were about 9 times more motor vehicle deaths than poisoning deaths.

> **OBSERVE** Dividing percents gives the same answer:
>
> $$\frac{60.8\%}{6.75\%} = \frac{0.608}{0.0675} \doteq 9.$$

d. Sam claims that 3,200 accidental deaths were caused by firearms in 1987. Does the National Safety Council support Sam's claim?

The graph states that 1.75% of the deaths were due to firearms.

$$1.75\% \text{ of } 80,100 = 0.0175 \times 80,100$$
$$\doteq 1,400 \text{ deaths}$$

The National Safety Council says 1,400 accidental deaths were due to firearms in 1987. Sam's claim of 3,200 deaths is too high. ∎

▶ **You Try It**   Answer the following questions using Figure 8.3. Round all answers to hundreds.

10. How many deaths were caused by fire burns?

11. How many more deaths were caused by poison than choking?

12. How many times more deaths were caused by falls than firearms?

13. Ellen claims that 5,437 deaths were due to poison. Is she approximately correct?

EXAMPLE 4   **Chapter Problem**   Read the problem at the beginning of this chapter.

From the graph, 21 out of 100 people, or 21%, are European. If there are 5 billion people on Earth, then 21% of 5 billion are Europeans.

R% = 21%	A = R% × B
B = 5 billion	= 0.21 × 5 billion
A = ? (number of Europeans)	= 1.05 billion Europeans ∎

▶ **You Try It**   **14.** Using the same graph as in Example 4, how many people on Earth are of Asian descent?

▶ **Answers to You Try It**   1. $650   2. $13,000   3. secretary   4. $250
5. auto mechanic   6. 136 lb; 122 lb to 154 lb   7. 124 lb   8. 5' 8" to 6' 2"
9. 166.5 lb; 152 lb to 189 lb   10. 4,800   11. 2,200   12. 8 times   13. yes
14. 2.85 billion

SECTION 8.2 EXERCISES

1 *Use Figure 8.4 to answer questions 1–6.*

FIGURE 8.4
Land Area of the 7 Continents

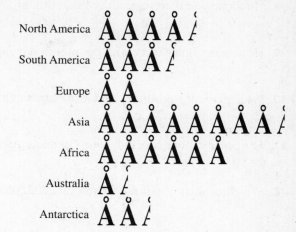

KEY:

$\overset{\circ}{A}$ = 2,000,000 square miles

North America
South America
Europe
Asia
Africa
Australia
Antarctica

1. a. What is the approximate land area of North America?

   b. There were about 430,000,000 people inhabiting North America in 1990. How many people is this per square mile, to the nearest person?

2. a. What is the approximate land area of Asia?

   b. There were about 3,200,000,000 people inhabiting Asia in 1990. How many people is this per square mile, to the nearest person?

3. Which continent has a land area closest to 7,000,000 square miles?

4. Approximately how many more square miles does Antarctica have than Australia?

5. Which continents are closest in area, and by how much?

6. Which continents are farthest apart in area, and by how much?

Use Figure 8.5 to answer questions 7–12.

FIGURE 8.5
Worldwide Passenger Car Production, 1987
SOURCE: United Nations, *Monthly Bulletin of Statistics*, Jan 1989.

KEY:

🚗 = 50,000 cars per month

France
West Germany
Italy
Japan
Spain
United Kingdom
United States
All Other Countries

7. How many cars were produced monthly by the United Kingdom in 1987?

8. How many cars were produced monthly in 1987 in the entire world?

9. How many times more cars per month did Japan produce in 1987 versus Spain?

10. Which country produced about the same number of cars per month as Italy?

11. How many more cars did West Germany produce per month than France?

12. Which country produced closest to 7,000,000 cars in 1987?

2 *Use the table in Figure 8.6 to answer questions 13–18.*

FIGURE 8.6
World Population Growth
from 1650 to 1989

	Population in Millions				
Continent	1650	1750	1850	1950	1989
North America	5	5	39	219	420
South America	8	7	20	111	287
Europe	100	140	265	530	685
Asia	335	476	754	1,418	3,131
Africa	100	95	95	199	642
Australia/Pacific Islands	2	2	2	13	26
Antarctica	. . . . .uninhabited.				
World Total	550	725	1,175	2,490	5,192

SOURCE: Rand McNally & Co.

13. What was the population of Europe in 1850?

14. Name the only two continents to experience a drop in population. Give the dates over which these drops occurred.

15. Which continent had a 1989 population about 7 times its 1650 population?

16. Which continent more than tripled in population from 1950 to 1989?

17. Which continent had the smallest increase in population from 1850 to 1950, and by how much?

18. By what percent did the world's total population increase from
  a. 1650 to 1850?   **b.** 1850 to 1950?   **c.** 1950 to 1989?

Use Figure 8.7 to answer questions 19–24.

FIGURE 8.7
Income Needed to Get a
30-Year Mortgage

Interest Rate	Amount Borrowed			
	$50,000	$75,000	$100,000	$150,000
	Annual Income Needed			
9%	$17,242	$25,863	$34,484	$51,726
9.5%	18,018	27,028	36,037	54,055
10%	18,085	28,208	37,611	56,415
10.5%	19,602	29,403	39,203	58,805
11%	20,407	30,611	40,814	61,221
11.5%	21,221	31,831	42,441	63,662
12%	22,042	33,063	44,084	66,125

SOURCE: National Association of Realtors.

19. What annual income is needed to borrow $75,000 at 10.5% for 30 years?

20. About how much can be borrowed at 9.5% if your annual income is $36,150?

21. What additional annual income is needed to borrow $75,000 versus $50,000 at 10%?

22. What additional annual income is needed to borrow $100,000 at 10% versus $75,000 at 12%?

23. A family has two annual incomes of $28,950 and $27,896. What rate must they get on a loan to borrow $150,000?

24. Your annual income is $39,850. What is the maximum rate allowed for you to borrow $100,000?

3 *Use Figure 8.8 to answer questions 25–29.*

FIGURE 8.8
Family Budget for a
Monthly Income of $2,800

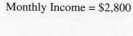

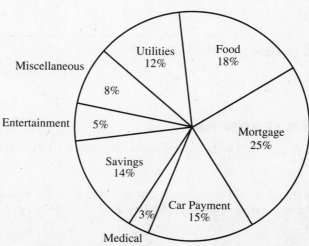

25. How much is budgeted for utilities each month?

26. What is the ratio of money budgeted for entertainment to money budgeted for the mortgage?

27. How much more money is budget for food than for utilities each month?

28. What fraction of the budget is the mortgage?

29. Suppose the monthly income increases to $3,200. If the percents in Figure 8.8 do not change, how much additional money will go into savings each month?

Use Figure 8.9 for problems 30–34.

FIGURE 8.9

Activities for a 24-Hour Weekday

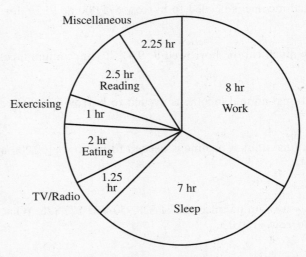

Circle = 24 hours

30. What fraction of the day is spent eating?

31. What is the ratio of time spent sleeping to time spent exercising?

32. What percent of the day is spent working?

33. About how many total hours are spent reading Monday through Friday?

34. Name each activity that, in one 5-day work-week, does not amount to the time spent in one night's sleep.

SKILLSFOCUS (Section 5.5) *Multiply.*

35. 2.5 × 0.68

36. 36.2 × 5.1

37. 0.6 × 0.09

38. 30 × 0.15

WRITING TO LEARN ▶▶▶▶

39. Explain in your own words the difference between a circle graph and a pictograph?

40. Explain how you would construct a pictograph out of the circle graph in Figure 8.8. How would you define your key? What would each wedge in the circle become in the pictograph?

OBJECTIVES
1 Read and interpret bar graphs.
2 Read and interpret line graphs.

NEW VOCABULARY

bar graph	double bar graph
ticks	line graph
trend	double line graph
upward trend	downward trend
forecast	

1 Reading a Bar Graph

A **bar graph** is used to compare quantities. A bar graph has a horizontal axis and a vertical axis. In Figure 8.10, the horizontal axis stretches left to right and is labeled in years. The vertical axis stretches up and down and is length of life. The vertical axis is marked at intervals with short lines called **ticks**. In Figure 8.10, the distance between two consecutive ticks represents 5 years.

FIGURE 8.10
Life Expectancy in the United States for Females Born in a Given Year, 1850–1990

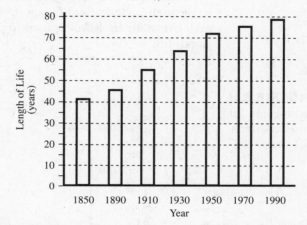

SOURCE: Department of Health and Human Services, National Center for Health Statistics.

EXAMPLE 1 Use Figure 8.10 to answer each question.

a. How long was a woman born in 1850 expected to live?

Align a ruler with the top of the bar at 1850. Move right to left until you reach the vertical axis. A woman born in 1850 was expected to live about 41 years.

b. Estimate the female life expectancy in 1990, and compare it to 1850.

A woman born in 1990 is expected to live about 79 years. In 1850 she was expected to live 41 years. From 1850 to 1990, the life expectancy of a woman almost doubled, rising $79 - 41 = 38$ years.

c. When was the average life expectancy 62 years?

Align your ruler with 62 years on the vertical axis. The bar closest to this height is at 1930. The life expectancy for a woman was 62 years if she was born about 1930.

d. Find the percent increase in life expectancy from 1890 to 1950.

In 1890, life expectancy was about 45 years. In 1950, it was about 72 years. From 1890 to 1950, life expectancy for a woman rose $72 - 45 = 27$ years. This is a $\frac{27}{45} = 60\%$ increase. ■

▶ **You Try It** Answer the following questions using Figure 8.10.

1. How long was a woman born in 1910 expected to live?
2. Estimate female life expectancy in 1950 and compare it to 1910.
3. When was the average life expectancy 76 years?
4. Find the percent increase in life expectancy from 1850 to 1910.

A bar graph is also used to show **trends** or patterns. In Figure 8.10, the life expectancy for a female rose continuously from 1850 to 1990. This is called an **upward trend**. It rose the fastest from 1890 to 1950, since the height of the bars increased the most in that period. It began to level off from 1950 to 1990.

A bar graph can also be used to **forecast** or make predictions about future trends based on recent trends. The leveling off from 1950 to 1990 may cause one to forecast a slight decrease in life expectancy in the near future.

Figure 8.11 is a **double bar graph**. It is used to compare two quantities to themselves, and to each other. Different shadings are used to distinguish each bar in a pair. A key is used to display what each differently shaded bar stands for.

FIGURE 8.11
Median Annual Income for Different Educational Levels, 1987

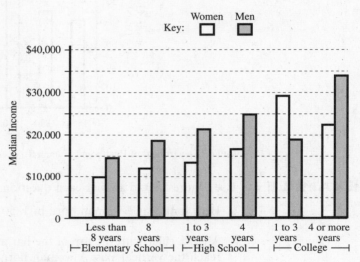

SOURCE: Department of Commerce, Bureau of the Census.

EXAMPLE 2   Use Figure 8.11 to answer each question.

a. What is the median income for a man with 1 to 3 years of high school?

About $21,000 a year. (See third shaded bar from the left.)

b. What is the median income for a woman with 4 or more years of college?

About $23,000 a year. (See unshaded bar farthest right.)

c. How much more does a woman make than a man with 1 to 3 years of college?

A woman makes about $29,000 a year with 1 to 3 years of college.
A man makes about $19,000 a year with 1 to 3 years of college.
A woman makes about $29,000 − $19,000 = $10,000 more per year.

d. How much more does a man make with 4 or more years of college versus 4 years of high school? What percent more is this?

A man makes about $34,000 a year with 4 or more years of college. He makes about $25,000 a year with 4 years of high school. The difference is $34,000 − $25,000 = $9,000 more per year. A man makes about $9,000 ÷ $25,000 = 36% more income per year with 4 or more years of college. ■

▶ **You Try It**   Answer the following questions using Figure 8.11.

5. What is the median income for a man with 8 years of elementary school?

6. What is the median income for a woman with 4 years of high school?

7. How much more does a man make than a woman when each has less than 8 years of elementary school?

8. How much more does a woman make with 4 years of high school compared to 8 years of elementary school?

---

2 Reading a Line Graph

A **line graph** is used to show trends. One common trend is showing how a quantity changes over time. Each dot on the graph represents a certain quantity at a given time. The dots are connected with straight lines, called *line segments*. The line segments show how a quantity increases or decreases over a given time period.

A line graph has a horizontal axis and a vertical axis. In Figure 8.12, birthrate in the United States (vertical axis) is graphed against time measured in 10-year intervals (horizontal axis). There is a "squiggle" at the bottom of the vertical axis. This tells you there are no birthrates from 0 through 15 on this graph.

FIGURE 8.12

Birthrates in the United States, 1910–1990

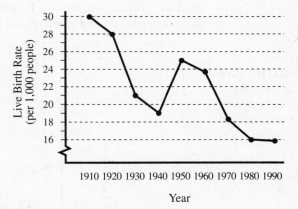

SOURCE: Department of Health and Human Resources.

EXAMPLE 3   Use Figure 8.12 to answer each question.

a. How many live births were there in 1920 per 1,000 people?

Locate 1920 on the horizontal axis. More upward until you reach the graph. Then move left. You reach 28 on the vertical axis. There were about 28 live births per 1,000 people in 1920.

b. When were there 18 live births per 1,000 people?

Locate 18 on the vertical axis. Move right until you reach the graph. Then move down. In 1970 there were 18 live births per 1,000 population.

---

c. Interpolate the birthrate for 1945. (See Section 8.2, Example 2d).

1945 is midway between 1940 and 1950. Mark this spot with your pencil. Using a ruler, move upward from 1945 to the graph. Then move left. There were approximately 22 live births per 1,000 people in 1945.

d. What was the only 10-year period to have an increase in birthrate?

The only increase in birthrate occurred from 1940 to 1950. This period is represented by the only upward sloping line segment on the graph. It increased from 19 per 1,000 in 1940 to 25 per 1,000 in 1950.

e. What upward and downward trends do you see in this graph?

Where the graph slants upward indicates an upward trend. This means the birthrate is rising. Where the graph slants downward indicates a **downward trend**. This means the birthrate is falling.
From 1910 to 1940 there was a downward trend in birthrate.
From 1940 to 1950 there was an upward trend.
From 1950 to 1990 there was a downward trend. ■

▶ **You Try It** Answer the following questions using Figure 8.12.

9. How many live births were there in 1960 per 1,000 people?
10. When were there 28 live births per 1,000 people?
11. Interpolate the birthrate in 1975.
12. What was the drop in birthrate from 1960 to 1970?
13. Between which two dates did the birthrate decline for 40 years?

A **double line graph** shows the change in two quantities over time. It is often used to compare two trends. Figure 8.13 shows the percent of the labor force in farm occupations and nonfarm occupations from 1830 to 1990.

FIGURE 8.13
Percent of Persons in Farm Versus Nonfarm Occupations in the United States, 1830–1990

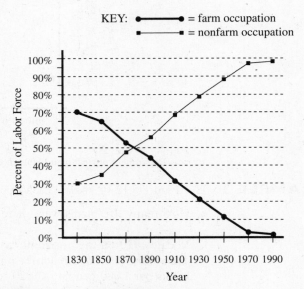

SOURCE: Department of Commerce, Census Bureau.

EXAMPLE 4   Use Figure 8.13 to answer each question.

a. What percent of the labor force was employed in farm occupations in 1830? About 70%. Start at 1830 on the time axis. Move vertically upward to the thicker line. Move left to 70%.

OBSERVE   The two graphs in Figure 8.13 are complements of each other. The sum of the two percents above any date is 100%. For example, above 1830, percent farm + percent nonfarm occupations = 70% + 30% = 100% of the labor force.

---

b. What percent of the labor force was employed in farm occupations in 1990?

About 2%.

c. Estimate when 80% of the labor force was employed in nonfarm occupations.

About 1930. Locate 80% on the percent axis. Move right to the thinner line. Move down to the time axis to 1930.

d. Estimate when the percent of farm occupations and nonfarm occupations were equal.

About 1875. This occurs where the two graphs intersect. Using a ruler, move downward from the point of intersection to 1875. Here, 50% of the labor force was employed in farm occupations, and 50% had other employment.

e. What has been the trend in farm occupations from 1830 to 1990?

The graph for farm occupations slants constantly downward. There has been a downward trend in farm occupations from 1830 to 1990. ■

▶ **You Try It**   Answer the following questions using Figure 8.13.

14. What percent of the labor force was employed in farm occupations in 1910?

15. What percent of the labor force was employed in nonfarm occupations in 1950?

16. Estimate when 10% of the labor force was employed in farm occupations.

17. What was the trend in nonfarm occupations from 1830 to 1990?

▶ **Answers to You Try It   1.** 54 yr   **2.** 72 yr; live 18 years longer if born in 1950   **3.** 1970   **4.** about 32%   **5.** $19,000   **6.** $16,000   **7.** $5,000 more **8.** $4,000 more   **9.** 23   **10.** 1920   **11.** 17   **12.** 12.5 per 1,000 **13.** 1950 to 1990   **14.** 32%   **15.** 89%   **16.** about 1955   **17.** increasing

SECTION **8.3** EXERCISES

1 *Use Figure 8.14 to answer questions 1–8.*

FIGURE 8.14
U.S. Unemployment Rate
for Selected Years

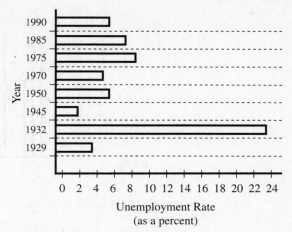

SOURCE: Department of Labor, Bureau of Labor Statistics.

1. What was the unemployment rate in 1970?

2. What was the unemployment rate in 1945?

3. In what year was the unemployment rate closest to the 1950 rate?

4. In what years were the unemployment rates under 4%?

5. In what year was the unemployment rate the lowest, and what was that rate?

6. In what year was the unemployment rate the highest, and what was that rate?

7. **a.** What were the unemployment rates in 1929 and 1932?

   b. About how many times greater was the 1932 rate?

8. About how many times smaller was the 1945 unemployment rate compared to 1932?

Use Figure 8.15 to answer questions 9–16.

FIGURE 8.15
Daily Recommended
Calories for Men and
Women by Age

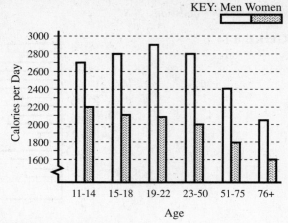

SOURCE: National Academy of Sciences.

9. How many calories are recommended daily for a 21-year-old man?

10. What age range recommends about 1,800 calories daily for a woman?

11. **a.** In what age range are men expected to have the most calories per day?

   b. How many calories is this?

   c. Describe the trend in calories needed by men from ages 11 through 76+.

12. **a.** In what two age ranges do women have the same number of calories recommended daily?

   b. How many calories is this?

   c. Describe the trend in calories needed by women from ages 11 through 76+.

13. **a.** In what age range is the difference between calories recommended for men and women the greatest?

   b. What is this difference?

14. **a.** In what age range is the difference between calories recommended for men and women the smallest?

   b. What is this difference?

15. What total number of calories should two eighty-year-old grandparents consume daily?

16. A woman nursing a baby needs 500 additional calories per day. If the woman is 26 years old, how many calories should she consume daily?

2 *Use Figure 8.16 to answer questions 17–24.*

FIGURE 8.16
Percent of College
Graduates that Are
Women in Selected Years

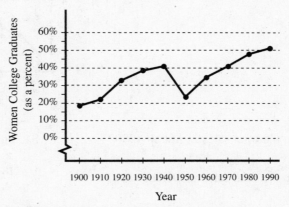

SOURCE: Department of Education, Center for Educational Statistics.

17. About what percent of college graduates in 1990 were women?

18. About what percent of college graduates in 1990 were men?

19. Between which two dates did the percent of women college graduates drop? By how many percentage points did it drop?

20. On which two dates did women make up about 41% of all college graduates?

21. a. What percent of college graduates were women in 1970?

 b. 877,676 people graduated from college in 1970. To the nearest thousand, how many were women?

22. a. What percent of college graduates were women in 1900?

 b. 27,410 people graduated from college in 1900. To the nearest hundred, how many were women?

23. Estimate the ratio of male to female college graduates in 1900.

24. When were the upward trends in Figure 8.16?

FIGURE 8.17
Comparison of Sales in
1991 and 1990

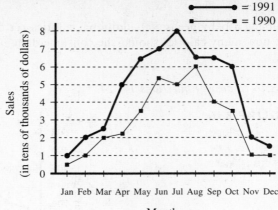

25. What were the total sales in March 1990?

26. What were the total sales in October 1991?

27. a. In which month in 1990 were sales the greatest?

   b. What were the sales for that month?

28. a. In which month in 1991 were sales the greatest?

   b. What were the sales for that month?

29. How much more in sales were made in July 1991 compared to July 1990?

30. In which months in 1990 were sales about $10,000?

31. In what time periods in 1990 were the trends
   a. upward?
   b. downward?
   c. steady (no change)?

32. In what time periods in 1991 were the trends
   a. upward?
   b. downward?
   c. steady?

33. If 30% of sales is profit for the business in Figure 8.17, about how much profit was made in September 1991?

34. If 30% of sales is profit in both 1990 and 1991, estimate how much more profit was made in May 1991 compared to May 1990.

SKILLSFOCUS (Section 7.2) *Change each decimal to a percent.*

35. 0.58   **36.** 0.3   **37.** 2.075   **38.** 0.008   **39.** 4   **40.** $0.4\frac{1}{2}$

WRITING TO LEARN ▶▶▶▶

41. Explain how a line graph and a bar graph are alike in two ways. Explain how they are different in two ways.

▶ YOU BE THE JUDGE

42. What kind of product could a business be selling whose sales curves look like Figure 8.17? Explain why.

8.4 DRAWING GRAPHS TO HELP YOU REASON THROUGH PROBLEMS

OBJECTIVES

1 Draw a bar graph to help you reason through a problem.

2 Build a table to help you reason through a problem.

1 Drawing a Bar Graph to Help You Solve a Problem

Pencil, paper, and drawing can be used to help you think through problems. The key is to work with one sentence at a time.

EXAMPLE 1 Bill is younger than Sam. Chuck is older than Sam. Carl is younger than Bill. Is Sam older or younger than Carl?

To solve this problem using a bar graph, draw a bar for each person. Label each bar with a name. The length of each bar corresponds to the age of the person. Work with one sentence at a time.

Bill is younger than Sam.

Bill < Sam (Draw Bill's bar lower than Sam's.)

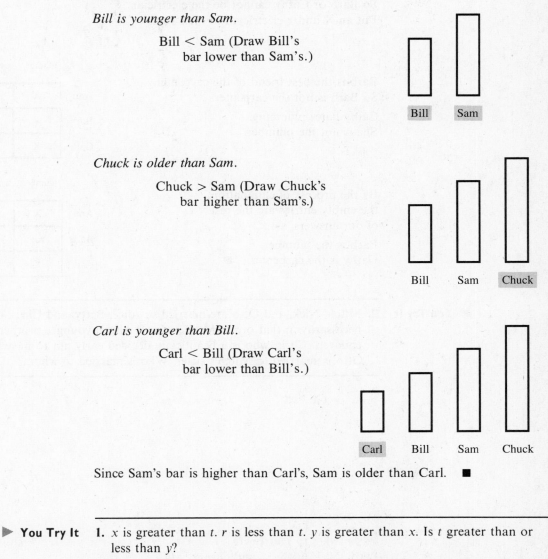

Chuck is older than Sam.

Chuck > Sam (Draw Chuck's bar higher than Sam's.)

Carl is younger than Bill.

Carl < Bill (Draw Carl's bar lower than Bill's.)

Since Sam's bar is higher than Carl's, Sam is older than Carl. ∎

▶ **You Try It** 1. x is greater than t. r is less than t. y is greater than x. Is t greater than or less than y?

Building a Table to Help You Solve a Problem

The following procedure uses a table and the process of elimination to solve problems.

EXAMPLE 2

Ann, Barb, and Cathy are a plumber, a carpenter, and an electrician, but not necessarily in that order. Ann can't do carpentry work and is married to the brother of the plumber. Barb is the best friend of the carpenter. Cathy has hated plumbing since she was a child, and has avoided it ever since. What is each woman's occupation?

Draw a table. Make each row a name.
Make each column an occupation.

Ann can't do carpentry work.
Therefore, Ann is not a carpenter.
Put an X in Ann's row under carpenter.

Ann is married to the plumber's brother.
So Ann is not the plumber.
Therefore, Ann is the electrician.

So Barb or Cathy cannot be the electrician.
Put an X under electrician in their rows.

	plumber	carpenter	electrician
Ann	X	X	Yes
Barb			X
Cathy			X

Barb is the best friend of the carpenter.
So Barb is not the carpenter.

Cathy hates plumbing.
She is not the plumber.

	plumber	carpenter	electrician
Ann	X	X	Yes
Barb		X	X
Cathy	X		X

By the process of elimination, the empty entries are the rest of our answers.

Barb is the plumber.
Cathy is the carpenter. ■

	plumber	carpenter	electrician
Ann	X	X	Yes
Barb	Yes	X	X
Cathy	X	Yes	X

▶ **You Try It**

2. Mitch, Nick, and Otto are married to Alice, Betty, and Chris, but not necessarily in that order. Mitch, who is Alice's younger brother, has 3 children. Chris, who is a beautician, decided early not to have any children. Otto is married to Mitch's sister. Who is married to whom?

▶ **Answers to You Try It** 1. $r < t < x < y$ so $t < y$ 2. Mitch is married to Betty, Nick to Chris, and Otto to Alice.

1

1. Cathy is taller than Mary. Pat is shorter than Mary. Laura is an inch taller than Cathy. Put the women in order from shortest to tallest.

2. Jimmy is faster than Nina. Chris is faster than Jimmy. Nina is faster than Bob. Who is the slowest and who is the fastest?

3. Bill and Frank are both older than Tom. Hank is older than Bill but younger than Frank. Order the men from youngest to oldest.

4. *a* is greater than *c*. *b* is less than *c*. *g* is greater than *a*. Which letter is greatest and which is least?

2

5. Nancy, Pat, and Chris are a secretary, a singer, and a teacher, but not necessarily in that order. Pat can't sing and is married to the brother of the teacher. Nancy is good friends with the singer. What is each woman's occupation?

6. Donna, Linda, and Jayne are married to Greg, Brian, and Christopher, but not necessarily in that order. Jayne, who is best friends with Christopher's wife, has never met Brian. Donna's father-in-law is Christopher's dad. Who is married to whom?

7. At Dail Chemical, the foreman, the stockclerk, the salesperson, and the chemist are named Bob, Henry, George, and Lou, but not necessarily in that order. The supervisor, Ms. Ferget, can't remember who is who. However, she does know the following from casual conversation. Lou goes rafting with the foreman and the stockclerk. Henry rides to work with the foreman and the salesperson. Bob can't stand the stockclerk. The stockclerk golfs with Henry. Match each man with his job.

8. Susan, Carol, and Gail own between them a total of 42 cars. They are compacts, midsize cars, and luxury cars. Carol has 7 luxury cars and 4 compacts. Susan, who has a total of 15 cars, has 9 compacts, and no luxury cars. There are 25 compact cars all together. Gail, who has a total of 14 cars, has as many luxury cars as midsize. How many cars of each type does each woman own? Display all of your numbers in a table.

SKILLSFOCUS (Section 3.5) *Multiply.*

9. $\dfrac{3}{8} \times \dfrac{4}{5}$

10. $\dfrac{15}{12} \times \dfrac{4}{3}$

11. $\dfrac{11}{20} \times \dfrac{15}{22} \times \dfrac{2}{3}$

12. $\dfrac{36}{60} \times \dfrac{80}{100}$

CALCULATOR TIPS

The Percent Circle Graph and Memory The circle in a circle graph represents the whole. In this circle graph, the circle represents $360,000.

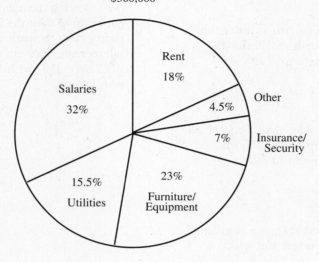

Business Expenses for
Lisa's Fashion Shop

$360,000

You can use memory to store a number you will use over and over again in your calculations. For example, how much did Lisa spend in each of the six categories in the circle graph?

Multiply each percent times $360,000. Since you will use $360,000 repeatedly, store it in memory by pressing

(CM) 360000 (M+)

Now whenever you press (MR), you will get 360000. To figure how much is spent in a category, enter the following.

Salaries: 32% of $360,000 = (MR) (×) 32 (%) $115,200
Utilities: 15.5% of $360,000 = (MR) (×) 15.5 (%) $55,800

Use memory to find Lisa's expenses for each of the remaining categories.
a. Furniture/Equipment
b. Insurance/Security
c. Other
d. Rent

CHAPTER **8** REVIEW

VOCABULARY AND MATCHING

New words and phrases introduced in this chapter are shown in the left-hand column. Match each term on the left with the phrase or sentence on the right that best describes it.

A. data _____ number that occurs most often

B. mean _____ uses symbols representing fixed amounts

C. median _____ displays rows and columns of data

D. mode _____ the actual numbers that make up the data before being graphed

E. range _____ pattern or tendency in a graph

F. percentile _____ the length of each rectangle is proportional to the size of the quantity represented

G. raw scores _____ displays portions of a whole using pie shaped wedges

H. graph _____ quantities, represented by dots, connected by line segments; often used for displaying trends

 I. pictograph _____ information, such as a set of numbers or attributes

J. key _____ sum of the numbers divided by the number of numbers

K. table _____ largest minus smallest

L. circle graph _____ displays the value of a symbol in a graph

M. bar graph _____ middle number

N. trend _____ value below which a given percent of the scores lie

O. line graph _____ a picture of data

REVIEW EXERCISES

8.1 Mean, Median, Mode, and Range

Find the mean, median, mode, and range.

1. 4, 8, 9, 14, 15

2. 65, 70, 100, 84, 65

3. 1, 4, 4, 7, 10, 10, 10, 18

4. 6, 1, 20, 0, 200, 7, 0, 2, 6, 8

5. 2, 2, 2, 2, 3, 3, 3, 23

6. 1.6, 4.3, 2.2, 5, 8.3, 2.6

7. Ron's range in weight was 24 pounds during his four years in college. His heaviest weight was 207 pounds. What was his lightest?

8. Donna will take six tests in her music class. The scores on her first five are 84, 91, 77, 80, and 73. What must she score on her sixth test to have an 80 average?

9. Find the mean, median, mode, and range for each set of numbers.

a. 0, 10, 20

b. 0, 10, 100

c. 0, 10, 1,000

10. A company employs five keypunch operators at $16,000 each, three programmers at $24,000 each, and two systems analysts at $31,000. Find the mean, median, mode, and range for the salaries.

8.2 Pictographs, Tables, and Circle Graphs

Use Figure 8.18 to solve problems 11–15.

FIGURE 8.18
The Annual Budget for
Frederic County

KEY: $ = $1,000,000

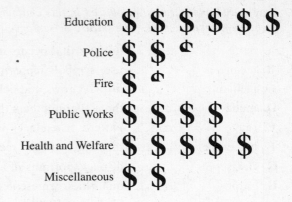

11. How much of the Frederic County budget is for education?

12. How many dollars are allotted to public works?

13. How many times more money goes to education than to fire protection?

14. What percent of the total budget is education?

15. Which category is closest to having a $2,100,000 budget?

Use Figure 8.19 to answer questions 16–20.

FIGURE 8.19
U.S. Deaths in Major
Wars

War	Number Engaged	Total Deaths
Civil War	2,213,363	364,511
World War I	4,734,991	116,516
World War II	16,112,566	405,399
Korean War	5,720,000	54,246
Vietnam War	8,744,000	58,135

SOURCE: Department of Defense.

16. What are the mean, median, and range for the column titled "Number Engaged"?

17. How many more died in the Civil War than in World War I?

18. What percent of Americans engaged in World War II died?

19. About how many times more Americans were engaged in the Vietnam War than the Civil War?

20. Figure 8.19 has five ratios of Number Engaged to Total Deaths. In each ratio, divide both terms by Total Deaths. Round to units. This gives five unit ratios of number engaged per 1 death.

    a. Which war had the least number of deaths when compared to number engaged?

    b. Which war had the most deaths when compared to number engaged?

Use Figure 8.20 to answer questions 21–25.

FIGURE 8.20
Grade Distribution for Credit Courses in the Fall Semester at Norwood College

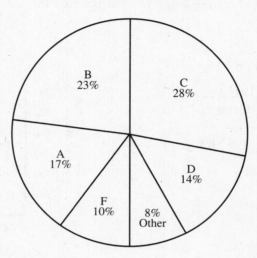

Circle = 14,291 grades assigned

21. About how many C grades were assigned? Round to tens.

22. What is the ratio of C grades to D grades?

23. How many more F grades were assigned than "Other" grades? Round to tens.

24. What percent of the grades were C or higher?

25. What fraction of the grades were C, D, or "Other"?

8.3 Bar Graphs and Line Graphs

Use Figure 8.21 to answer questions 26–30.

FIGURE 8.21
Death Rates per 100,000 from Selected Causes, 1900 and 1990

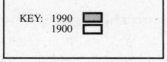

KEY: 1990
        1900

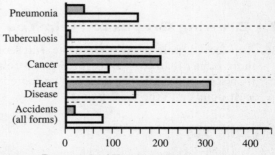

SOURCE: Department of Health and Human Services.

26. **a.** About how many people per 100,000 died from heart disease in 1900?

   b. In 1990?

   c. About how many times more people died from heart disease in 1990 compared to 1900?

27. Which cause of death was almost eliminated from 1900 to 1990?

28. Which causes resulted in more deaths per 100,000 in 1990 than 1900?

29. In 1900, which disease was closest to causing the same number of deaths as pneumonia?

30. In 1990, which of the five categories was the median?

Use Figure 8.22 in problems 31–35.

FIGURE 8.22
Mean Price of 1 Share of XYZ Stock for Each of Six Months, March through August

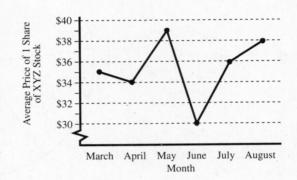

31. What was the average cost of one share of XYZ stock in May?

32. Between which two successive months did the value of one share rise the least?

33. **a.** Between which two successive months did the average value of one share drop the most?

   b. How much was this drop?

34. Estimate the average price per share for each of the six months shown. Using these numbers, find their
   a. mean.   **b.** median.   **c.** mode.   **d.** range.

35. Teresa purchased 240 shares of XYZ in June and sold them in July. Using the average price per share, estimate how much profit she made.

8.4 Drawing Graphs to Help You Reason through Problems

36. Ron scored more points than Mike. Jim scored more points than Ron. Bill scored fewer points than Ron but more than Mike. Who scored the least points and the most points?

37. Sally, Pam, and Terry own a boat, a plane, and a hot air balloon, but not necessarily in that order. Pam took the plane owner out to lunch, and confessed she did not like the owner of the balloon. Terry took the plane owner for a ride. Match each woman with what she owns.

Allow yourself 50 minutes to complete this test. Show the work for each problem. When done, check your answers. Rework each problem solved incorrectly.

1. Given the eight test scores 84, 91, 77, 100, 80, 84, 60, and 89, find the
   a. mean.　　　　**b.** median.　　　　**c.** mode.　　　　**d.** range.

2. The lengths in inches of 20 trout caught in the Chesapeake Bay are given below.

    | 6 | 7.5 | 8 | 7 | 5.5 | 9.5 | 8 | 10 | 8.5 | 7 |
    | 4 | 9 | 4.5 | 8 | 7 | 8 | 9.5 | 9 | 6.5 | 8 |

    Compute the mean, median, mode, and range for the lengths.

3. Given the quiz scores 17, 0, 15, 20, 18, 19, 0, and 11, find the
   a. mean.　　　　**b.** median.　　　　**c.** mode.　　　　**d.** range.

4. Mike will take 4 tests in his biology class. He scored 67, 84, and 75 on his first three. What must he score on his last test to average 80?

Use the table in Figure 8.23 to answer questions 5–10.

FIGURE 8.23
Median Age at First
Marriage

Year	Males	Females	Year	Males	Females
1900	25.9	21.9	1950	22.8	20.3
1910	25.1	21.6	1960	22.8	20.3
1920	24.6	21.2	1970	23.2	20.8
1930	24.3	21.3	1980	24.7	22.0
1940	24.3	21.5	1990	25.9	23.6

SOURCE: Department of Commerce.

5. In which years were both males and females the youngest at their first marriage?

6. What is the range in the male ages from 1900 to 1990?

7. What is the mode of the female ages from 1900 to 1990?

8. In which years were males the oldest at first marriage?

9. **a.** Is it true since 1960 that females have been waiting longer before getting married?

   b. Can the same be said of males?

10. When was the difference between male and female ages the greatest at first marriage?

Use Figure 8.24 to answer questions 11–16.

FIGURE 8.24
Number of Homes Sold by a Real Estate Agency in the First Six Months of 1992

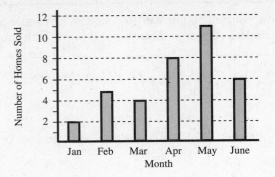

11. How many homes were sold in April?

12. How many fewer homes were sold in June than May?

13. What percent more homes were sold in April than March?

14. Find the mean, median, mode, and range for the number of homes sold over the six-month period.

15. How many more homes were sold in the second three months compared to the first three months of the year?

16. What was the trend in home sales from March through May?

Explain the meaning of each term. Use your own examples.

17. median

18. trend

NAME   DATE   HOUR

1. Subtract 6,000 − 4,209.

2. Simplify 4 + 6 · 10.

3. a. Estimate 480 · 237.

4. Simplify 30 ÷ 10 · 3.

b. What is the exact answer?

5. Add $\dfrac{5}{12} + \dfrac{7}{10}$.

6. Find $\dfrac{2}{3}$ of $144.

7. Subtract $3\dfrac{1}{4} - 2\dfrac{1}{10}$.

8. Change $\dfrac{4}{7}$ to a decimal. Round to thousandths.

9. Multiply 42.6 · 0.08.

10. Add 55.72 + 0.8 + 4 + 0.09.

11. Solve $s + 6.3 = 110$.

12. Solve $4.2w = 63$.

13. Simplify 2.5 + 3.5 · 4.

14. Order the decimals from smallest to largest.

0.707, 0.077, 0.7, 0.0079

15. Write the ratio $5\frac{3}{4}$ to $1\frac{7}{8}$ as a ratio of two whole numbers.

16. Split $5,274 according to a 7:2 ratio.

17. Solve $\frac{y}{8} = \frac{22.5}{30}$.

18. Change 0.07 to a percent.

19. Change 182% to a decimal.

20. What is 34% of $620?

21. 180% of what number is 9?

22. This month 750 teachers signed up for a workshop. 450 of them live out of state. What percent live out of state?

23. Thirteen sparrows' nests were examined. The following number of eggs were found in each nest: 2, 1, 2, 3, 2, 2, 1, 2, 2, 0, 3, 2, 4. Find the
a. mean.   **b.** median.   **c.** mode.   **d.** range.

Explain the meaning of each term. Use your own examples.

24. tenths   **25.** expand a fraction   **26.** simple interest

Measurement

DRIVING

Millions of Americans visit Canada each year by car. Canada uses the metric system of measurement on its road signs. Suppose you were traveling at 45 miles per hour on a Canadian road. The speed limit is posted as 60 kilometers per hour. Are you exceeding the speed limit?

To solve this problem, change kilometers per hour to miles per hour.

$$60 \text{ km/hr} = ? \text{ mi/hr}$$

This problem will be solved in this chapter.

Units of measurement give a number meaning by telling you how much. The football player weighs 270 pounds. You bought $14\frac{7}{10}$ gallons of gas. You need a board 5 feet 10 inches long for a shelf. Pounds, gallons, feet, and inches are English units of measurement. Since these measurements are used predominantly in the United States, you will study them in this chapter.

A can of string beans weighs 454 grams. The footrace is 10 kilometers long. You purchased 47 liters of gas. Grams, kilometers, and liters are metric units of measurement. The metric system is used by every country in the world, except the United States and a few others. One reason the United States is moving toward the metric system is to be more competitive in world trade. For these reasons, you will also study the metric system in this chapter.

Work through each section in this chapter. Use this test to identify topics you are not familiar with. These topics may require additional study. You may *not* use a calculator.

9.2

1. 18 pt = ? gal

2. 4.5 hr = ? min

3. 12 lb = ? oz

4. 110 in = ? ft

9.3

5. 3 ℓ = ? ml

6. 8.6 kg = ? g

7. 5,400 cl = ? ℓ

8. 250 mm = ? cm

9. 2,409 m = ? km

10. 690 mg = ? g

11. A jar of preserves weighs 1,062 g. How many kg is this?

9.4

12. 4 gal = ? ℓ (round to tenths)

13. 28 cm = ? in (round to tenths)

14. 118 lb = ? kg (round to units)

15. 36.7 m = ? ft (round to units)

VOCABULARY *Explain the meaning of each term. Use your own examples.*

16. meter

17. unit fraction

9.1 ABSTRACT AND DENOMINATE NUMBERS

OBJECTIVES

1 Define abstract and denominate numbers.

2 Add and subtract denominate numbers.

3 Multiply and divide denominate numbers.

NEW VOCABULARY

abstract number

denominate number

1 Abstract and Denominate Numbers

Abstract numbers are numbers without units. 5, 8, and 0.2 are called abstract numbers. **Denominate numbers** are numbers with units. 5 feet, 8 seconds, and 0.2 pounds are denominate numbers.

Units give practical meaning to numbers. $3 > 5$ is false. So you write $3 \not> 5$. But 3 feet > 5 inches is true. By attaching units, you can change the relationship between two numbers. $2 > 1$ is true. But 2 ounces $\not>$ 1 pound.

Can $1 = 12$? Yes, 1 foot = 12 inches. Though $1 = 12$ is false, by proper selection of units it can be made into a true statement.

EXAMPLE 1 Each statement below is false. Attach units to the abstract numbers to make each statement true.

a. $1 = 4$:   1 dollar = 4 quarters

b. $60 + 60 = 2$:   60 seconds + 60 seconds = 2 minutes

c. $\frac{1}{2} = 6$: $\frac{1}{2}$ dozen eggs = 6 eggs

d. $100 < 1$:   100 feet < 1 mile ∎

EXAMPLE 2 Each statement below is true. Attach units to the abstract numbers to make each statement false.

a. $4 = 4$:   4 pounds ≠ 4 ounces

b. $10 > 1$:   10 inches $\not>$ 1 foot ∎

▶ **You Try It**   Attach units to make each statement true.

1. $1 = 5$ **2.** $24 = 1$ **3.** $1\frac{1}{2} = 18$ **4.** $7 < 3$

Attach units to make each statement false.

5. $6 = 6$ **6.** $5 > 4$

2 Adding and Subtracting Denominate Numbers

You can add or subtract like denominate numbers. For example, $\$3 + \$5 = \$8$. You can add or subtract abstract numbers. For example, $3 + 5 = 8$.

But you cannot add or subtract a denominate number and an abstract number. The addition

$$\$3 + 5$$

cannot be done. You cannot assume 5 means \$5. It could just as well mean 5¢, 5 dimes, or 5 quarters. Since you do not know, you cannot add. Also, 3 feet + 5 pounds cannot be added. The units measure different quantities.

EXAMPLE 3 Find the answer, or label the problem as "cannot do."

a. 4 inches + 6 inches = 10 inches   **b.** 2 + 4 inches (cannot do)

c. 1 pound + 3 (cannot do)   **d.** 11 feet − 6 feet = 5 feet

e. $8 - 3 = 5$ ∎

Find the answer, or label the problem as "cannot do."

7. 5 yards + 8 yards   **8.** 4 + 3 tons   **9.** 12 − 7

10. 8 miles − 5 miles   **11.** 5 ounces + 3 days

You can write two different units together if they are like measures.

EXAMPLE 4
  a. 3 feet + 7 inches = 3 feet 7 inches
  b. 8 pounds + 3 ounces = 8 pounds 3 ounces
  c. 4 minutes + 48 seconds = 4 minutes 48 seconds ■

▶ **You Try It**   Write without the addition sign.

12. 6 hours + 20 minutes   **13.** 4 yards + 1 foot

14. 7 meters + 82 centimeters

CONNECTIONS

You can only add and subtract numbers with like units. For example, 5 inches + 3 inches = 8 inches. In algebra, you can only add and subtract like terms. $5n$ and $3n$ are like terms because they have the same letter name, n. Therefore, $5n + 3n = 8n$.

Arithmetic		*Algebra*
5 inches + 3 inches = 8 inches	⟶	$5n + 3n = 8n$
7 feet − 4 feet = 3 feet	⟶	$7p − 4p = 3p$
6 gallons + 4 (cannot do)	⟶	$6h + 4$ (cannot be added because the terms are unlike)
9 hours − 2 pounds (cannot do)	⟶	$9y − 2c$ (cannot be subtracted because the terms are unlike)

---

3 Multiplying and Dividing Denominate Numbers

You can multiply an abstract number and a denominate number. To see why, write the multiplication as a repeated addition. For example, the multiplication 4(2 feet) can be written as 2 feet added to itself four times.

$$4(2 \text{ feet}) = 2 \text{ feet} + 2 \text{ feet} + 2 \text{ feet} + 2 \text{ feet} = 8 \text{ feet}$$

In short, 4(2 feet) = (4 × 2) feet = 8 feet.

> To multiply an abstract number and a denominate number.
>
> **1.** Multiply the two numbers.
> **2.** Give the product the same unit name as the denominate number.

EXAMPLE 5   Multiply.

  a. 8 (4 gallons) = (8 × 4) gallons = 32 gallons

  b. 6 (10 cents) = (6 × 10) cents = 60 cents

  c. (5 ounces) 4.8 = (5 × 4.8) ounces = 24 ounces ■

> To divide a denominate number by an abstract number
>
> 1. Divide the two numbers.
> 2. Give the answer the same unit name as the denominate number.

EXAMPLE 6 Divide.

a. $\dfrac{80 \text{ men}}{2} = 40 \text{ men}$

b. $\dfrac{\$30}{5} = \$6$

c. $\dfrac{12 \text{ feet}}{3} = 4 \text{ feet}$ ■

▶ **You Try It** Simplify.

15. 4 (3 inches)

16. (7 minutes) 6

17. 2.5 ($2)

18. $\dfrac{35 \text{ minutes}}{7}$

19. $\dfrac{60 \text{ yards}}{20}$

20. $\dfrac{45 \text{ trucks}}{9}$

CONNECTIONS

The rules for multiplication and division of abstract and denominate numbers carry over into algebra.

Arithmetic		*Algebra*
$4(3 \text{ inches}) = (4 \cdot 3) \text{ inches}$ $= 12 \text{ inches}$	⟶	$4(3y) = (4 \cdot 3)y$ $= 12y$
$6(5 \text{ seconds}) = (6 \cdot 5) \text{ seconds}$ $= 30 \text{ seconds}$	⟶	$6(5t) = (6 \cdot 5)t$ $= 30t$
$\dfrac{32 \text{ gallons}}{8} = 4 \text{ gallons}$	⟶	$\dfrac{32p}{8} = 4p$

▶ **Answers to You Try It**   **1.** 1 nickel = 5 pennies   **2.** 24 hours = 1 day
3. $1\frac{1}{2}$ dozen eggs = 18 eggs   **4.** $7 > 3 quarters   **5.** 6 men ≠ 6 cars
6. 5 in > 4 yd   **7.** 13 yd   **8.** cannot do   **9.** 5   **10.** 3 mi   **11.** cannot do
12. 6 hr 20 min   **13.** 4 yd 1 ft   **14.** 7 m 82 cm   **15.** 12 in   **16.** 42 min
17. $5   **18.** 5 min   **19.** 3 yards   **20.** 5 trucks

1 *Each statement is false. Make it true by adding appropriate units to each abstract number.*

1. $7 < 4$ **2.** $5 > 12$ **3.** $3 < \dfrac{1}{2}$ **4.** $1 > 30$

5. $1 = 2$ pints   **6.** $1 = 100\text{¢}$ **7.** $6 = 3$ **8.** $6 < 2.5$

Each statement is true. Make it false by adding appropriate units to each abstract number.

9. $5 > 1$ **10.** $2 < 100$ **11.** $8 = 8$ **12.** $4 > 3$

2 **3** *Write the answer, or write "cannot do."*

13. 6 feet + 2 feet   **14.** $6 + 2$

15. $6 - 2$ gallons   **16.** $\$10 - 8$

17. 4(5 cents)   **18.** 12 yards + 5 quarts

19. $\dfrac{40 \text{ men}}{8}$ **20.** (6 pints)7

21. 3 pounds + 2 miles   **22.** $\dfrac{600 \text{ seats}}{50}$

23. $9 + 4$ **24.** 8(12)

25. 4(3.5 miles)   **26.** 6 acres − 4

27. (18 feet) ÷ 6   **28.** 8 yards − 3 yards

29. 6 feet + 2(4 feet)   **30.** (4 pounds)5 − 8(2)

31. 36% of $250 is what?

32. What percent of 20 is 16?

33. 50% of what is 30?

34. What is 75% of $36?

35. 270 is what percent of 450?

36. 138 is 92% of what?

EXTEND YOUR THINKING ▶▶▶▶

▶TROUBLESHOOT IT

Find and correct the error.

37. 5 qt − 2 qt = 5 q̶t̶ − 2 q̶t̶ = 5 − 2 = 3

38. 8 feet − 3 = (8 − 3) feet = (5) feet = 5 feet

▶CONNECTIONS

Perform each operation, or write "cannot do."

39. $6p + 8p$

40. $7m + 2g$

41. $5x + 9x$

42. $7t − 4t$

43. $8q − 2$

44. $10s − 4s$

45. $9(2y)$

46. $(3r)7$

47. $\dfrac{8d}{4}$

48. $\dfrac{35y}{7}$

49. $(8r)10$

50. $0(7y)$

WRITING TO LEARN ▶▶▶▶

51. Explain in your own words why you cannot add 5 + $3.

52. Explain in your own words why 1 ≠ 3, but 1 yard = 3 feet.

53. Explain the concept the graffiti artist did not understand when marking up this sign.

9.2 THE ENGLISH SYSTEM OF MEASUREMENT

OBJECTIVES

1. Define the English system of measurement.
2. Change from one unit to another using unit fractions.

NEW VOCABULARY

English system of measurement

unit fraction

---

1 The English System of Measurement

The United States is one of the few countries that still uses the customary or **English system of measurement**. England, where the English system developed about 1200 AD, now uses the metric system.

FIGURE 9.1 Common English Measures and Their Equivalents

Length:	
	1 foot (ft) = 12 inches (in)
	1 yard (yd) = 3 ft
	1 mile (mi) = 5,280 ft
	1 knot = 1.15 mi/hr
	1 mi/hr = 1.467 ft/sec

Time:	
	1 minute (min) = 60 seconds (sec)
	1 hour (hr) = 60 min
	1 day (dy) = 24 hr
	1 week (wk) = 7 dy

Weight:	
	1 pound (lb) = 16 ounces (oz)
	1 ton (t) = 2,000 lb

Volume:	
	1 pint (pt) = 16 oz
	1 quart (qt) = 2 pt
	1 gallon (gal) = 4 qt

Household Measures:
1 tablespoon (T or tbsp) = 3 teaspoons (t or tsp)
1 cup (c) = 8 oz = 16 tbsp
1 pt = 2c

---

2 Changing from One Unit to Another Using Unit Fractions

To change from one unit to another, multiply by a **unit fraction**. A unit fraction is a fraction equal to 1. Each equivalence in Figure 9.1 can be written as a unit fraction. Just divide one side of the equivalence by the other. For example, 1 foot = 12 inches. 1 foot and 12 inches are the same length. Therefore, 1 foot divided by 12 inches, or 12 inches divided by 1 foot, equals 1.

$$\frac{1 \text{ ft}}{12 \text{ in}} = \frac{12 \text{ in}}{1 \text{ ft}} = 1$$

Both are unit fractions. Each equals 1 because 1 ft = 12 in.

A unit fraction equals 1. Multiplying by 1 does not change the value of a number. Therefore, you can change one unit to another by multiplying by the appropriate unit fraction. The process of changing units by multiplying by a unit fraction is sometimes called *dimensional analysis*.

EXAMPLE 1   Change 48 inches to feet.

Find the equivalence in Figure 9.1 equating inches to feet. It is 1 ft = 12 in.

This fraction equals 1. ——— Put the unit for the answer, feet, in the numerator.

$$48 \text{ in} = \frac{\overset{4}{\cancel{48 \text{ in}}}}{1} \times \frac{1 \text{ ft}}{\underset{1}{\cancel{12 \text{ in}}}} = 4 \text{ ft}$$

Put the unit to be changed, inches, in the denominator. Reduce units like you reduce numbers when you multiply fractions. Inches are reduced. This leaves feet for the answer. ■

▶ **You Try It**   1. Change 66 inches to feet.

> To change from one unit to another
>
> 1. Select the equivalence relating these same units from Figure 9.1.
> 2. Write this equivalence as a unit fraction with
>   a. the unit for the answer in the numerator.
>   b. the unit to be changed in the denominator.
> 3. Multiply the unit to be changed by this unit fraction.
> 4. Reduce numbers and units.

EXAMPLE 2 Change 7.5 yards to feet.

From Figure 9.1, 1 yd = 3 ft.

Write 1 yd = 3 ft as this unit fraction. Since you are changing to feet, keep feet in the numerator.

$$7.5 \text{ yd} = \frac{7.5 \ \cancel{\text{yd}}}{1} \times \frac{3 \text{ ft}}{1 \ \cancel{\text{yd}}} = 22.5 \text{ ft} \quad \blacksquare$$

EXAMPLE 3 Change 40 ounces to pounds.

1 lb = 16 oz. You are changing to lb. Write lb in the numerator.

$$40 \text{ oz} = \frac{\overset{5}{\cancel{40 \ \text{oz}}}}{1} \times \frac{1 \text{ lb}}{\underset{2}{\cancel{16 \ \text{oz}}}} = \frac{5}{2} \text{ lb or } 2\frac{1}{2} \text{ lb} \quad \blacksquare$$

EXAMPLE 4 Change $5\frac{3}{4}$ gallons to quarts.

4 qt = 1 gal. To change to qt, write qt in the numerator.

$$5\frac{3}{4} \text{ gal} = \frac{23 \ \cancel{\text{gal}}}{\underset{1}{\cancel{4}}} \times \frac{\overset{1}{\cancel{4} \text{ qt}}}{1 \ \cancel{\text{gal}}} = 23 \text{ qt} \quad \blacksquare$$

▶ **You Try It**
2. Change 4.5 feet to inches.
3. Change 15 pints to quarts.
4. Change 60 ounces to pounds.

Sometimes you must multiply by two or more unit fractions to make a conversion.

EXAMPLE 5 Change $1\frac{1}{4}$ hr to sec.

Change hours to minutes, then change minutes to seconds.

$$1\frac{1}{4} \text{ hr} = \frac{5 \ \cancel{\text{hr}}}{\underset{1}{\cancel{4}}} \times \frac{\overset{15}{\cancel{60 \ \text{min}}}}{1 \ \cancel{\text{hr}}} \times \frac{60 \text{ sec}}{1 \ \cancel{\text{min}}} = 4,500 \text{ sec}$$

Changes hours to minutes.   Changes minutes to seconds.   ■

▶ **You Try It**
5. Change $2\frac{1}{2}$ days to minutes.

▶ **Answers to You Try It** 1. 5.5 ft 2. 54 in 3. 7.5 qt 4. 3.75 lb
5. 3,600 min

1 **2** *Use unit fractions to make the following conversions.*

1. 36 in = ? ft

2. 42 in = ? ft

3. 3 ft = ? in

4. 5 ft = ? in

5. 7 yd = ? ft

6. 12 yd = ? ft

7. 21 ft = ? yd

8. 36 ft = ? yd

9. 78 in = ? ft

10. 156 in = ? ft

11. 414 ft = ? yd

12. 54 ft = ? yd

13. 23,760 ft = ? mi

14. 3 mi = ? ft

15. 150 sec = ? min

16. 4 min = ? sec

17. $3\frac{1}{2}$ hr = ? min

18. 45 min = ? hr

19. 40 oz = ? lb

20. $2\frac{3}{4}$ lb = ? oz

21. 23 pt = ? qt

22. 1.25 gal = ? oz

23. 9 in = ? ft

24. $8\frac{1}{2}$ ft = ? in

25. $\frac{3}{4}$ lb = ? oz

26. 6.25 lb = ? oz

27. 14 oz = ? c

28. $1\frac{1}{4}$ c = ? tbsp

29. 72 oz = ? lb

30. 7,500 lb = ? t

31. $1\frac{2}{3}$ c = ? oz

32. 16 tsp = ? tbsp

33. 800 oz = ? gal

34. 4,400 sec = ? hr

35. $\frac{3}{4}$ hr = ? sec

36. $3\frac{1}{4}$ dy = ? min

37. 6 yd = ? in

38. 128 in = ? yd

39. The fuel mixture for a gas saw calls for 2.5 pt of gas to be mixed with 1 oz of oil. How many ounces of gas are needed per ounce of oil?

40. A recipe calls for $\frac{3}{4}$ c of oil. How many tablespoons is this?

41. Sally is worth her weight in gold. If Sally weighs 119 lb, and gold is worth \$348.72/oz, what salary should Sally demand?

42. A sprinter can run short distances at 21 mi/hr. How many ft/sec is this?

SKILLSFOCUS (Section 3.3) *Prime factor.*

43. 20   **44.** 15   **45.** 17   **46.** 72

EXTEND YOUR THINKING ▶▶▶▶

▶SOMETHING MORE

47. A sprinter ran 100 yd in 9.72 sec. How many mi/hr is this? Round to tenths. (Hint: Multiply the rate 100 yd/9.72 sec by unit fractions so that all units cancel except miles in the numerator and hours in the denominator.)

▶TROUBLESHOOT IT

Find and correct the error.

48. Change 8.25 yd to ft: $8.25 \text{ yd} = \dfrac{8.25 \text{ yd}}{1} \times \dfrac{1 \text{ yd}}{3 \text{ ft}} = 2.75 \text{ ft}$

WRITING TO LEARN ▶▶▶▶

49. In your own words, explain why a unit fraction equals 1. Choose one of the entries in Figure 9.1 to help make your point.

▶YOU BE THE JUDGE

50. Elaine wants to change 9.5 ft to in. She wants to do this by multiplying 9.5 ft by 1 ft/12 in. Is she correct? Explain your decision.

OBJECTIVES

1 Define the metric prefixes and the meter (Section 4.1).

2 Use the metric bar to change one prefix to another.

3 Define the liter and make conversions.

4 Define the gram and make conversions.

NEW VOCABULARY

meter

metric bar

liter

gram

The metric system has replaced the English system as the most widely used system of measurement in the world. There are two major reasons for this.

1. Like our decimal number system, the metric system is based on 10. Therefore, it is easy to learn and use.

2. The metric system has three commonly used measures, the meter, the liter, and the gram. Prefixes (such as kilo, centi, and milli) are used to say how much of each measurement you want.

The metric system was developed after the French Revolution, in the early 1790s.

The version of the metric system used today is called the International System of Units, or SI. The United States is using this system more and more. For this reason, it merits a closer look.

1 Metric Prefixes and the Meter

The metric system of measurement is entirely based on one unit of measure, the **meter**. The liter and gram are defined in terms of the meter, as you will see later.

The meter is the only measure of length in the metric system. Mile, yard, foot, and inch are used to measure length in the English system.

Different prefixes are attached to meter to say how many meters. For example, *kilo* means 1,000 times. Therefore, a kilometer means 1,000 times a meter, or 1,000 meters.

Memorize the prefixes in Figure 9.2.

FIGURE 9.2
Metric Prefixes and the Meter

*kilo (k)	$= 1,000$	1 kilometer (km)	$= 1,000$ meters
hecto (h)	$= 100$	1 hectometer (hm)	$= 100$ meters
deka (da)	$= 10$	1 dekameter (dam)	$= 10$ meters
no prefix	$= 1$	1 meter (m)	$= 1$ meter
deci (d)	$= \dfrac{1}{10}$	1 decimeter (dm)	$= \dfrac{1}{10}$ of a meter
*centi (c)	$= \dfrac{1}{100}$	1 centimeter (cm)	$= \dfrac{1}{100}$ of a meter
*milli (m)	$= \dfrac{1}{1,000}$	1 millimeter (mm)	$= \dfrac{1}{1,000}$ of a meter

* most commonly used prefixes

A meter is slightly longer than a yard.

1 yard   36 inches

1 meter   39.37 inches

A kilometer is 1,000 meters. It is about $\frac{6}{10}$ of a mile, or the length of six city blocks.

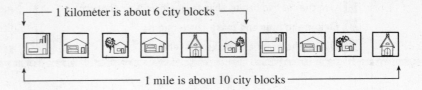

1 kilometer is about 6 city blocks

1 mile is about 10 city blocks

The kilometer is used in the metric system where the mile would be used in the English system.

In many countries, speed limits are given in kilometers per hour, not miles per hour. The speed 80 kilometers per hour is approximately equal to 50 miles per hour.

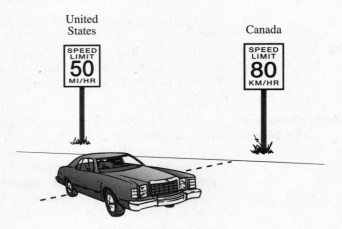

United States

SPEED LIMIT 50 MI/HR

Canada

SPEED LIMIT 80 KM/HR

A centimeter is $\frac{1}{100}$, or 1%, of one meter. 100 cm = 1 m. The centimeter is used in the metric system where the inch is used in the English system.

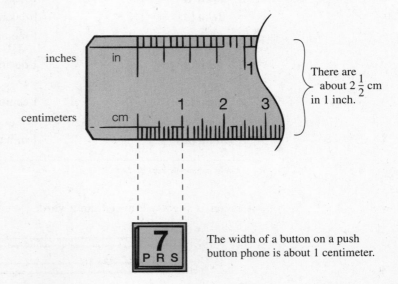

inches

centimeters

There are about $2\frac{1}{2}$ cm in 1 inch.

The width of a button on a push button phone is about 1 centimeter.

A millimeter is $\frac{1}{1,000}$ of a meter. 1,000 mm = 1 m. A millimeter is used to measure very small lengths.

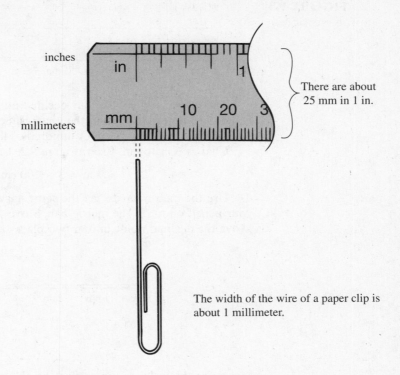

inches

There are about 25 mm in 1 in.

millimeters

The width of the wire of a paper clip is about 1 millimeter.

EXAMPLE I   Find the most reasonable metric measure.

a. distance from your house to the grocery store: 200 km, 2 km, 2 m

   Answer: 2 km. Recall, 1 km is about $\frac{6}{10}$ of a mile. So 2 km is a little more than a mile.

b. width of a light bulb: 3 m, 6 cm, 2 mm

   Answer: 6 cm. 6 cm is about the width of 6 buttons from a push-button phone.

c. height of a math teacher: $\frac{1}{2}$ km, 2 m, 6 cm

   Answer: 2 m. 1 m is about a yard, or 3 feet, so 2 m is about 6 feet.   ■

▶ **You Try It**   Find the most reasonable measure.

1. width of a pencil: 8 mm, $\frac{1}{4}$ km, 3 cm

2. length of a living room: 10 dm, 6 m, 100 cm

3. height of an oak tree: 5 cm, 3 km, 40 m

4. altitude of a jet plane: 8 km, 3 mm, $\frac{1}{2}$ m

2 Using the Metric Bar to Change One Prefix to Another

Use the **metric bar** to change from one metric prefix to another.

FIGURE 9.3
The Metric Bar

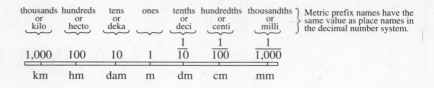

thousands or kilo	hundreds or hecto	tens or deka	ones	tenths or deci	hundredths or centi	thousandths or milli
1,000	100	10	1	$\frac{1}{10}$	$\frac{1}{100}$	$\frac{1}{1,000}$
km	hm	dam	m	dm	cm	mm

Metric prefix names have the same value as place names in the decimal number system.

The metric system, like our decimal number system, is based on 10. A place value is 10 times larger than the place value name to its right. In Figure 9.3, just as thousands are 10 times hundreds, kilometers are 10 times hectometers.

Change 3 m to cm. From Figure 9.2, 1 m = 100 cm. Therefore,

$$3 \text{ m} = 3 \times 100 \text{ cm} = 300 \text{ cm}.$$

Use the metric bar to get the same answer. To change from m to cm, put your pencil on m on the metric bar. Move to cm. You moved two places right. Move the decimal point in 3 m two places right.

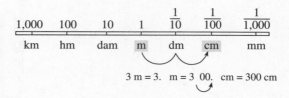

3 m = 3. m = 3 00. cm = 300 cm

Memorize the order of the prefixes on the metric bar. **You can change from one metric unit to another by just moving the decimal point.**

> To change from one metric unit to another
>
> 1. Locate the old metric prefix on the metric bar.
> 2. Draw an arrow from the old prefix to the new prefix. Count the number of places moved.
> 3. Move the decimal point in the number to be converted the **same number of places** in the **same direction**.
> 4. Replace the old unit with the new unit.

EXAMPLE 2 Change 54 cm to dm.

Step #1: Locate centimeters on the metric bar.
Step #2: Draw an arrow from cm to dm. The arrow moved one place left.
Step #3: Move the decimal point in 54 cm one place left. Replace the old unit, cm, with the new unit, dm.

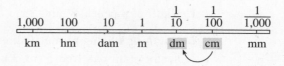

Move decimal point one place left.

54 cm = 54. cm = 5.4 dm or 5.4 dm

Therefore, 54 cm = 5.4 dm. ■

EXAMPLE 3   Change 4,000 m to km.

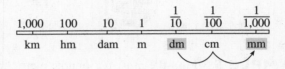

Draw an arrow from m to km.
You moved three places left.
Move the decimal point three places left.
4,000 m = 4,000. m = 4.000 km or 4 km ■

EXAMPLE 4   Change 0.73 dm to mm.

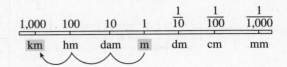

Draw an arrow from dm to mm.
You moved two places right.
Move the decimal point two places right.
0.73 dm = 073. mm or 73 mm ■

► **You Try It**   **6.** Change 4,700 mm to m.   **7.** Change 2.9 m to cm.

3 The Liter, the Metric Measure of Volume   The ounce, pint, quart, and gallon are used to measure volume or capacity in the English system. The only measure of volume in the metric system is the **liter**. It is a measure of how much space a substance occupies.

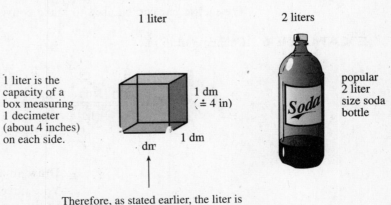

1 liter is the capacity of a box measuring 1 decimeter (about 4 inches) on each side.

1 liter

2 liters

1 dm (≐ 4 in)
1 dm
dm

popular 2 liter size soda bottle

Therefore, as stated earlier, the liter is defined in terms of the meter.

Units of volume use the same prefixes as units of length.

FIGURE 9.4
Metric Prefixes and
the Liter

1 kiloliter (kl) = 1,000 liters	1 deciliter (dl) = $\frac{1}{10}$ of a liter
1 hectoliter (hl) = 100 liters	1 centiliter (cl) = $\frac{1}{100}$ of a liter
1 dekaliter (dal) = 10 liters	*1 milliliter (ml) = $\frac{1}{1,000}$ of a liter
*1 liter (ℓ) = 1 liter	

*most commonly used prefixes

A liter is slightly larger than a quart. A milliliter is a common unit of measure in medicine. (It is called 1 cubic centimeter, or 1 cc.) Dosages of drugs are often measured in milliliters. 1 ml is the capacity of a box 1 cm on each side.

1 liter ≐

1 milliliter =   1 cm ←— about the size of
or 1 cc   1 cm   a sugar cube
      1 cm

EXAMPLE 5   Find the most reasonable metric measure in each case.

a. the capacity of a large coffee mug: 1 kl, $\frac{1}{2}$ ℓ, 1 ml

   Answer: $\frac{1}{2}$ ℓ.   1 liter is about 1 qt. $\frac{1}{2}$ ℓ is about 1 pt.

b. a marble: 3 dal, 14 ℓ, 2 ml

   Answer: 2 ml. 3 dal and 14 ℓ are both larger than 1 liter. ■

▶ **You Try It**   Find the most reasonable metric measure.

8. the capacity of a cooking pot: 5 cl, 1 kl, 2 ℓ

9. the capacity of a bathtub: 500 ml, $\frac{1}{2}$ kl, 5 ℓ

The metric bar can be used to make conversions involving capacity.

EXAMPLE 6   Change 3 dl to ml.

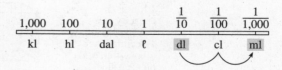

1,000	100	10	1	$\frac{1}{10}$	$\frac{1}{100}$	$\frac{1}{1,000}$
kl	hl	dal	ℓ	dl	cl	ml

Draw an arrow from dl to ml.
You moved two places right.
Move the decimal point two places right.
3 dl = 3. dl = 3 00. ml or 300 ml ■

10. Change 6.4 kl to ℓ.

EXAMPLE 7 Change 450 ml to ℓ.

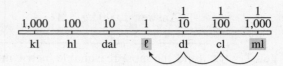

1,000	100	10	1	$\frac{1}{10}$	$\frac{1}{100}$	$\frac{1}{1,000}$
kl	hl	dal	ℓ	dl	cl	ml

Draw an arrow from ml to ℓ.
You moved three places left.
Move the decimal point three places left.
450 ml = 450. ml = 0.450 ℓ or 0.45 ℓ ■

▶ **You Try It** **11.** Change 6,500 ℓ to kl. **12.** Change 35 cl to ml.

4 The Gram, the Metric Measure of Weight

Your weight on the moon is about $\frac{1}{6}$ of your weight on Earth. A 60-pound child would weigh 10 pounds on the moon. However, the child's mass in both places is the same. Mass is a measure of how much matter an object has. On Earth, the words *mass* and *weight* are often used interchangeably.

The **gram** is the only measure of mass, or weight, in the metric system. The ounce, pound, and ton are used to measure weight in the English system. There are about 28 grams in one ounce. A gram is the weight of water that fills a box 1 cm on a side.

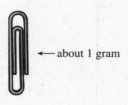

← about 1 gram

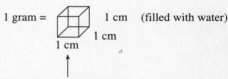

1 gram = 1 cm (filled with water)
1 cm
1 cm

Therefore, as stated earlier,
the gram is defined in terms of the meter.

Units of weight use the same prefixes as units of length and volume.

FIGURE 9.5
Metric Prefixes and the Gram

*1 kilogram (kg) = 1,000 grams
1 hectogram (hg) = 100 grams
1 dekagram (dag) = 10 grams
*1 gram (g) = 1 gram
1 decigram (dg) = $\frac{1}{10}$ of a gram
1 centigram (cg) = $\frac{1}{100}$ of a gram
*1 milligram (mg) = $\frac{1}{1,000}$ of a gram

* most commonly used prefixes

1 kg ≐

1 kilogram is a little more than 2 pounds. A small textbook weighs about 1 kilogram.

EXAMPLE 8   Find the most reasonable metric measure in each case.

    a. a piece of typing paper: 5 kg, 3 g, 1 mg

       Answer: 3 g

    b. a small car: 1,000 g, 1,000 kg, 100 mg

       Answer: 1,000 kg   ■

▶ **You Try It**   Find the most reasonable metric measure.

    13. a pocket calculator: 50 g, 4 mg, 2 kg
    14. a brick: 2 kg, 30 g, 500 mg

The metric bar can be used to make conversions involving the gram.

EXAMPLE 9   Change 250 mg to g.

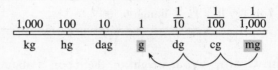

    Draw an arrow from mg to g.
    You moved three places left.
    Move the decimal point three places left.
    250 mg = 250. mg = 0.250 g = 0.25 g   ■

EXAMPLE 10   Change 0.82 kg to g.

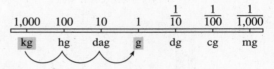

    Draw an arrow from kg to g.
    You moved three places right.
    Move the decimal point three places right.
    0.82 kg = 0 820. g = 820 g   ■

▶ **You Try It**   **15.** Change 1.09 g to mg.　　　　**16.** Change 375 g to hg.

The relationship between the meter, gram, and liter is summarized as follows.

      gram　　　　　meter　　　　　liter

$$\begin{array}{l}1\text{ g of}\\ \text{water}\end{array} = \boxed{}\quad 1\text{ cm} = 1\text{ ml of water}$$

        1 cm

(filled with water)

▶ **Answers to You Try It**　**1.** 8 mm　**2.** 6 m　**3.** 40 m　**4.** 8 km　**5.** 6 hm

6. 4.7 m　**7.** 290 cm　**8.** 2 ℓ　**9.** $\frac{1}{2}$ kl　**10.** 6,400 ℓ　**11.** 6.5 kl　**12.** 350 ml

13. 50 g　**14.** 2 kg　**15.** 1,090 mg　**16.** 3.75 hg

SECTION 9.3 EXERCISES

1 **2**

1. 4 m = ? cm

2. 50 cm = ? m

3. 580 mm = ? m

4. 0.72 m = ? mm

5. 6,000 m = ? km

6. 40 km = ? m

7. 1,520 cm = ? m

8. 0.88 km = ? m

9. 42 hm = ? km

10. 0.6 dam = ? cm

11. 780 m = ? km

12. 4 m = ? mm

13. 250 dm = ? dam

14. 500 m = ? hm

15. 8.2 cm = ? mm

16. 36,000 cm = ? km

17. 145 km = ? m

18. 3,600 cm = ? m

19. The speed of light is about 300,000 km per sec. How many meters per second is this?

20. How many km long is a 5,000 m foot race?

21. A man is 225 cm tall. How many meters tall is he?

22. The bullet missed its target by 0.25 cm. How many mm is this?

3

23. 3 ℓ = ? ml

24. 70 cl = ? ℓ

25. 580 ml = ? ℓ

26. 0.72 ℓ = ? ml

27. 8,000 ℓ = ? hl

28. 30 kl = ? ℓ

29. 2,590 cl = ? ℓ

30. 0.52 kl = ? ℓ

31. 75 dl = ? ml

32. 0.1 dal = ? cl

33. 724 ℓ = ? kl

34. 2.2 ℓ = ? ml

35. 700 ℓ = ? ml

36. 8.2 cl = ? ℓ

37. 36,000 ml = ? ℓ

38. 510 kl = ? ℓ

39. 9,200 cl = ? ℓ

40. 1.85 ml = ? cl

41. How many milliliters are in a 2-liter bottle of soda?

42. How many liters are in 420 dl of acid?

43. 4 ml of antibiotic are needed to inoculate one person.
   a. How many ml are needed to inoculate 60,000 people?
   b. How many liters is this?
   c. What is the cost of the antibiotic if one liter costs $250?

44. 2.5 ml of antibiotic are needed to inoculate one person.
   a. How many ml are needed to inoculate a town of 2,680 people?
   b. How many liters is this?
   c. What is the cost if one liter of antibiotic costs $870?

4

45. 4 g = ? mg

46. 900 cg = ? g

47. 510 mg = ? g

48. 0.42 g = ? mg

49. 3,000 g = ? hg

50. 60 kg = ? g

51. 3,800 cg = ? g

52. 1.52 kg = ? g

53. 175 dg = ? mg

54. 4.3 dag = ? cg

55. 354 g = ? kg

56. 14.2 g = ? kg

57. 718 g = ? dag

58. 80.4 g = ? mg

59. 9.6 cg = ? g

60. 16,500 mg = ? g

61. 890 kg = ? g

62. 12,506 g = ? kg

63. A pharmacy purchases 2.5 kg of a drug. How many grams is this?

64. A box of cereal weighs 454 g. How many kilograms is this?

65. One pill has 250 mg of a drug.
   a. How many mg of the drug are in a bottle of 30 pills?
   b. How many grams in this?

66. One dose of medicine contains 35 mg of a drug.
   a. How many mg are needed to make 4,000 doses?
   b. How many grams in this?

SKILLSFOCUS (Section 4.7) *Arrange the fractions in order from smallest to largest.*

67. $\dfrac{3}{5}, \dfrac{2}{3}, \dfrac{7}{10}$

68. $\dfrac{5}{8}, \dfrac{3}{5}, \dfrac{3}{4}$

69. $\dfrac{7}{12}, \dfrac{5}{6}, \dfrac{2}{3}, \dfrac{3}{4}$

70. $\dfrac{3}{10}, \dfrac{1}{5}, \dfrac{4}{15}, \dfrac{7}{30}$

WRITING TO LEARN ▶▶▶▶

71. Explain each step used to change 500 cm to meters using the metric bar.

72. Explain how the meter, liter, and gram are related.

9.4 CONVERSIONS BETWEEN THE ENGLISH AND METRIC SYSTEMS, AND TEMPERATURE

OBJECTIVE

Change English units to metric units or metric units to English units using unit fractions (Section 9.2), and conversions between Celsius and Fahrenheit temperatures.

Conversions between Systems, and Temperature

• Driving through Canada, you see a sign that reads 120 km to Toronto. How many miles is this?

• You weigh 85 kg on a metric scale. How many pounds is this?

• You purchase 40.1 ℓ of gasoline. How many gallons did you buy?

To answer these questions, you need a set of equivalents between the English and metric systems of measurement.

FIGURE 9.6
English-Metric Equivalents

Length:	1 inch (in) = 2.54 centimeters (cm)
	39.37 inches (in) = 1 meter (m)
	1 foot (ft) = 0.3048 meters (m)
	1 mile (mi) = 1.61 kilometers (km)
Weight:	1 ounce (oz) = 28.35 grams (g)
	1 pound (lb) = 453.6 grams (g)
	2.2 pounds (lb) = 1 kilogram (kg)
Volume:	1.06 quarts (qt) = 1 liter (ℓ)
	1 gallon (gal) = 3.78 liters (ℓ)

EXAMPLE 1

Change 120 kilometers to miles. Round to tenths.

Step #1: The equivalence relating km and mi is 1 mi = 1.61 km.

Step #2: Write this equivalence as the unit fraction $\dfrac{1\ \text{mi}}{1.61\ \text{km}}$. Since you are looking for miles, write miles in the numerator.

Step #3: Multiply 120 km by $\dfrac{1\ \text{mi}}{1.61\ \text{km}}$.

$$120\ \text{km} = \frac{120\ \text{km}}{1} \times \frac{1\ \text{mi}}{1.61\ \text{km}} = \frac{120\ \text{mi}}{1.61} \doteq 74.5\ \text{mi} \quad \blacksquare$$

▶ **You Try It**   **1.** Change 540 m to ft. Round to units.

> To change from an English unit to a metric unit, or from a metric unit to an English unit
>
> 1. Select the equivalence from Figure 9.6 having the same two units.
> 2. Write this equivalence as a unit fraction with
>   a. the unit for the answer in the numerator.
>   b. the unit to be changed in the denominator.
> 3. Multiply the unit to be changed by this unit fraction.
> 4. Simplify. The units reduce. The only unit left should be the unit for the answer. If not, recheck your work.

EXAMPLE 2 Tom weighs 85 kg on a metric scale. How many pounds is this? Round to the nearest pound.

The equivalence relating kg and lb is 2.2 lb = 1 kg.

Write 2.2 lb = 1 kg as this unit fraction. Since you are finding lb, keep lb in the numerator.

$$85 \text{ kg} = \frac{85 \text{ kg}}{1} \times \frac{2.2 \text{ lb}}{1 \text{ kg}} = \frac{85 \times 2.2 \text{ lb}}{1} \doteq 187 \text{ lb}$$

Therefore, Tom weighs 187 lb, rounded to the nearest pound. ∎

EXAMPLE 3 How many liters is 10.6 gallons of gas? Round to tenths.

The equivalence relating liters and gallons is 1 gal = 3.78 ℓ.

Write 1 gal = 3.78 ℓ as this unit fraction. Since you are finding liters, keep liters in the numerator.

$$10.6 \text{ gal} = \frac{10.6 \text{ gal}}{1} \times \frac{3.78 \text{ ℓ}}{1 \text{ gal}} = \frac{10.6 \times 3.78 \text{ ℓ}}{1} \doteq 40.1 \text{ ℓ} \quad \blacksquare$$

▶ **You Try It**
2. Change 20 cm to in. Round to tenths.
3. How many g are in 4.3 lb of antibiotic? Round to tens.

EXAMPLE 4 **Chapter Problem** Read the problem on the opening page of this chapter.

Change the 60 km/hr speed limit into miles per hour.

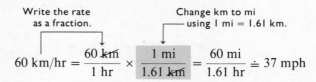

Write the rate as a fraction.   Change km to mi using 1 mi = 1.61 km.

$$60 \text{ km/hr} = \frac{60 \text{ km}}{1 \text{ hr}} \times \frac{1 \text{ mi}}{1.61 \text{ km}} = \frac{60 \text{ mi}}{1.61 \text{ hr}} \doteq 37 \text{ mph}$$

At 45 mph, you are traveling about 8 mph faster than the posted Canadian speed limit. ∎

▶ **You Try It**
4. How many g are in a 7.5 oz bag of chips? Round to units.
5. You are traveling 85 km/hr on an American road. The speed limit is 65 mph. Are you exceeding the speed limit? Round to units.

EXAMPLE 5   How many yards long is the 100 m dash? Round to tenths.

There is no conversion from m to ft in Figure 9.6. Therefore, you must make two conversions.

Change meters to feet using 1 ft = 0.3048 m. ──── Change feet to yards using 1 yd = 3 ft.

$$100 \text{ m} = \frac{100 \text{ m}}{1} \times \frac{1 \text{ ft}}{0.3048 \text{ m}} \times \frac{1 \text{ yd}}{3 \text{ ft}} = \frac{100 \text{ yd}}{0.3048 \times 3} = \frac{100 \text{ yd}}{0.9144}$$

$$\doteq 109.4 \text{ yd}$$

The 100 m dash is about 109.4 yd long.   ■

> **OBSERVE**   To change 0.4 yard to feet and inches, 0.4 yd × 3 ft/yd = 1.2 ft. Then 0.2 ft × 12 in/ft = 2.4 in. Therefore, 109.4 yd = 109 yd 1 ft 2.4 in.

▶ **You Try It**   **6.** How many meters long is the 1-mile run?

SOMETHING MORE

Temperature   The temperature outside is 30° Celsius. Do you wear a coat? The Fahrenheit and Celsius temperature scales are compared in the figure.

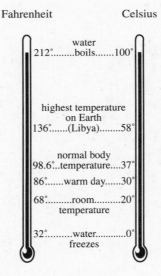

Celsius is the name for the metric temperature scale. (The name Celsius replaced the original name for this scale, centigrade, in 1948.)
   30°C (read "30 degrees Celsius") is equivalent to 86°F. This is a warm day. A coat is not needed.

To change Celsius (C) to Fahrenheit (F) use	To change Fahrenheit (F) to Celsius (C) use
$F = \dfrac{9}{5}C + 32.$	$C = \dfrac{5}{9}(F - 32).$

EXAMPLE I   Change 30°C to °F.

$$F = \frac{9}{5}C + 32 = \frac{9}{5} \cdot 30 + 32 = \frac{9}{\cancel{5}} \cdot \frac{\overset{6}{\cancel{30}}}{1} + 32 = 54 + 32 = 86°F \quad \blacksquare$$

EXAMPLE 2   Change 68°F to °C.

$$C = \frac{5}{9}(F - 32) = \frac{5}{9} \cdot (68 - 32) = \frac{5}{\cancel{9}} \cdot \frac{\overset{4}{\cancel{36}}}{1} = 20°C \quad \blacksquare$$

Make each conversion. Round answers to the nearest degree.

a. 35°C = ?°F   **b.** 80°C = ?°F   **c.** 41°F = ?°C   **d.** 122°F = ?°C

e. Dee is sick. Her temperature on a Celsius thermometer is 40°C. What is her Fahrenheit temperature?

f. The highest temperature recorded in the United States was 134°F in Death Valley, California. Change this to Celsius.

▶ **Answers to You Try It**   **1.** 1,772 ft   **2.** 7.9 in   **3.** 1,950 g   **4.** 213 g
5. No, 85 km/hr ≐ 53 mph   **6.** 1,610 m

▶ **Answers to Something More**   **a.** 95°F   **b.** 176°F   **c.** 5°C   **d.** 50°C
e. 104°F   **f.** 57°C

Make each conversion. Round all answers to two decimal places.

1. 14 in = ? cm

2. 40 km = ? mi

3. 100 lb = ? kg

4. 6 ℓ = ? qt

5. 3.2 oz = ? g

6. 140 cm = ? in

7. 8 mi = ? km

8. 27 ℓ = ? gal

9. 50 in = ? m

10. 920 g = ? lb

11. 90 ft = ? m

12. 200 m = ? ft

13. 106 qt = ? ℓ

14. 454 g = ? oz

15. 3.61 lb = ? g

16. 500 kg = ? lb

17. 40 qt = ? ℓ

18. 7.25 m = ? in

19. 400 lb = ? kg

20. 8 oz = ? g

21. 15 ℓ = ? qt

22. 8 gal = ? ℓ

23. 80 m = ? ft

24. 330 g = ? oz

25. 36 ft = ? m

26. 3.7 m = ? in

27. 289 km = ? mi

28. 72 in = ? cm

29. 935 cm = ? ft

30. 12 ℓ = ? pt

---

31. A marathon is a 26.2 mile foot race. How many kilometers is this?

32. How many miles long is a 10 km race?

33. A standard Olympic event is the 800 m run. How many yards is this?

34. The speed limit on some U.S. interstates is 55 mph. Change this to km/hr. Round to the nearest unit.

35. Susan bought a car in France. The gas tank has a 52 ℓ capacity. How many gallons of gas will the tank hold?

36. Gas costs 39.2¢ per liter. What is the cost per gallon, to the nearest tenth of a cent?

37. Gas costs $1.369 per gallon. What is the cost per liter, to the nearest tenth of a cent?

38. Which is longer, a 400 m race or a 440 yd race, and by how much?

Simplify each ratio.

39. 5 ft to 10 in

40. 3 lb to 18 oz

41. 3 m to 25 cm

42. 400 g to 2 kg

EXTEND YOUR THINKING ▶▶▶▶

▶ SOMETHING MORE

43. Change 16°C to Fahrenheit. Round to the nearest degree.

44. Change 451°F to Celsius. Round to the nearest degree.

45. The escape velocity from the Earth is 7 mi/sec. This means a projectile must be fired at a speed of 7 mi/sec to escape the Earth's gravitational pull, and not fall back. Change this to km/sec. Round to the nearest km.

46. 35 mm film is 35 mm wide. How many inches wide is this to the nearest tenth of an inch?

47. A standard 4-wall handball court is 6 m high, 6 m wide, and 12 m long. Express these dimensions in feet. Round to tenths.

48. Given kilometers, to estimate how many miles this is
    1. Multiply kilometers by 6.
    2. Round to the tens place.
    3. Drop the zero in the ones place.

For example, change 12 km to mi.

    Multiply by 6.   ⟶ 6 × 12 = 72
    Round to tens.   ⟶   72 ≐ 70
    Drop the ones digit. ⟶ 7Ø or 7 mi

Therefore, 12 km is about 7 mi. Change the following to miles by estimating.
a. 15 km   **b.** 6 km
c. 40 km   **d.** 175 km

49. **Hair** The average hair on the human scalp grows about 12 mm per month. The average scalp has about 100,000 hairs. How many miles of new hair growth does the average person have in one year?

▶ TROUBLESHOOT IT

Find and correct the error.

50. An advertising brochure for a popular amusement park in the eastern United States says the following.
"The (park name) is located 150 miles/93 kilometers from (city A), 400 miles/249 kilometers from (city B), and 276 miles/171 kilometers from (city C)."
a. What errors were made in this sentence?

b. Suppose the three distances given in miles are accurate. Explain in your own words how the conversion errors were made. Correct and rewrite the sentence.

WRITING TO LEARN ▶▶▶▶

51. To change 7 in to cm, explain why you multiply 7 in by the unit fraction $\dfrac{2.54 \text{ cm}}{1 \text{ in}}$ and

not by $\dfrac{1 \text{ in}}{2.54 \text{ cm}}$. Then solve the problem, explaining each step.

CALCULATOR TIPS

Using the Calculator to Make English/Metric Conversions Figure 9.7 shows the factors needed to make English to metric or metric to English conversions. To make the conversion, use your calculator to multiply by the appropriate factor.

EXAMPLE 1 Use the calculator to make each conversion.

a. 2.71 oz to g: 2.71 $\times$ 28.3495 $=$ 76.827145 $\doteq$ 76.8 g

b. 13.1 mi to km: 13.1 $\times$ 1.6093 $=$ 21.08183 $\doteq$ 21.1 km

c. 20 cm to in: 20 $\times$ 0.3937 $=$ 7.874 $\doteq$ 7.9 in

d. 7.5 kg to lb: 7.5 $\times$ 2.2046 $=$ 16.5345 $\doteq$ 16.5 lb ∎

FIGURE 9.7

English-Metric Conversion Table

To change	to	Multiply by
centimeters	inches	0.3937
centimeters	feet	0.03281
feet	centimeters	30.48
feet	meters	0.3048
gallons	liters	3.7853
grams	ounces	0.0353
grams	pounds	0.002205
inches	millimeters	25.4
inches	centimeters	2.54
kilograms	pounds	2.2046
kilometers	miles	0.6214
liters	gallons	0.2642
liters	pints	2.1134
liters	quarts	1.0567
meters	feet	3.2808
meters	miles	0.0006214
meters	yards	1.0936
miles	kilometers	1.6093
millimeters	inches	0.0394
ounces	grams	28.3495
pints	liters	0.4732
pounds	grams	453.6
pounds	kilograms	0.4536
quarts	liters	0.9463
yards	meters	0.9144

Use your calculator and Figure 9.7 to make each conversion. Round all answers to two decimal places.

1. 88 cm = ? ft

2. 4 kg = ? lb

3. 26 mi = ? km

4. 11.5 qt = ? ℓ

5. 62 mm = ? in

6. 8 in = ? cm

7. 14.9 gal = ? ℓ

8. 128 yd = ? m

9. 540 g = ? lb

10. 72 m = ? ft

11. 56.3 ℓ = ? gal

12. 90 km = ? mi

CHAPTER 9 REVIEW

VOCABULARY AND MATCHING

New words and phrases introduced in this chapter are shown in the left-hand column. Match each term on the left with the phrase or sentence on the right that best describes it.

A. abstract number   _____ examples are $\dfrac{4 \text{ qt}}{1 \text{ gal}}$, $\dfrac{1 \text{ yd}}{3 \text{ ft}}$, and $\dfrac{1 \text{ in}}{2.54 \text{ cm}}$

B. denominate number   _____ means 10 times

C. English system of measurement   _____ basic measure of length in the metric system

D. unit fraction   _____ means $\dfrac{1}{10}$ of

E. metric system of measurement   _____ slightly larger than a quart; measures capacity in the metric system

F. meter   _____ means $\dfrac{1}{1,000}$ of

G. deci-   _____ a tool to help make conversions from one metric unit to another

H. hecto-   _____ 1 gal, 3 ft, and 39.37 in are examples

I. centi-   _____ uses different prefixes on meter, liter, and gram to indicate how much

J. kilo-   _____ measure of mass or weight

K. milli-   _____ units include the pound, foot, and quart

L. deka-   _____ means 100 times

M. metric bar   _____ means $\dfrac{1}{100}$ of

N. liter   _____ means 1,000 times

O. gram   _____ $7, \dfrac{3}{4}$, and 0.5 are examples

REVIEW EXERCISES

9.1 Abstract and Denominate Numbers

Write the number or write "cannot do."

1. 6 gal + 4 gal   **2.** 4 ft + 2   **3.** 4 mi × 5

4. $8 − 6   **5.** (12 pounds)/4   **6.** 9 ft − 4 ft

7. 10 × 10 pt   **8.** 4 + 2 pens

9.2 The English System of Measurement

9. 42 in = ? ft   **10.** 5 gal = ? pt   **11.** 60 oz = ? lb

12. 16 ft = ? yd   **13.** 36,960 ft = ? mi   **14.** 45 qt = ? gal

15. $\dfrac{2}{3}$ hr = ? sec

16. 80 mi = ? ft

17. 240 min = ? hr

18. A recipe calls for $\dfrac{1}{3}$ c of oil. How many tsp is this?

19. A plane is flying at 240 knots. How many mi/hr is this?

9.3 The Metric System of Measurement

20. 2 m = ? cm

21. 1,357 mm = ? m

22. 28 cm = ? mm

23. 2.85 kg = ? g

24. 424 mg = ? cg

25. 5,800 g = ? kg

26. 38 ml = ? cl

27. 8 ℓ = ? ml

28. 470 cl = ? ℓ

29. A pharmacy purchases 7 dg of a drug. How many mg is this?

30. 25 ml of antibiotic are needed to inoculate one person.
   a. How many ml are needed to inoculate 16,500 people?

   b. How many liters is this?

   c. What is the cost of the antibiotic if 1 liter costs $98?

31. A marksman missed the bull's-eye by 5.3 cm. How many mm is this?

9.4 Conversions between the English and Metric Systems, and Temperature

32. 2 ft = ? cm

33. 3.5 m = ? ft

34. 690 g = ? lb

35. 8 ℓ = ? qt

36. 235 g = ? oz

37. 7.2 kg = ? lb

38. 7 yd = ? m

39. 43 ℓ = ? gal

40. 6,450 ft = ? km

41. How many meters long is a 44-ft yacht?

42. A foreign car has a 38 ℓ gas tank. How many gallons can it hold?

Allow yourself 50 minutes to complete this test. Show the work for each problem you solve. When done, check your answers. Rework each problem solved incorrectly.

1. 3.5 gal = ? qt

2. 9.75 min = ? sec

3. 75 in = ? ft

4. 270 oz = ? lb

5. 16 mm = ? cm

6. 5.2 kg = ? g

7. 140 cl = ? ℓ

8. 0.04 ℓ = ? ml

9. 1.85 km = ? m

10. 450 mg = ? g

11. A box of cereal weighs 640 g. How many kg is this?

Convert and round your answers to tenths.

12. 16 in = ? cm

13. 6 ℓ = ? pt

14. 7.2 kg = ? lb

15. 9.4 m = ? ft

Explain the meaning of each term. Use your own examples.

16. liter

17. gram

18. metric bar

NAME   DATE   HOUR

1. Multiply $3,506 \cdot 1,000$.

2. Simplify $10 + 10 \div 10$.

3. Write 483 in expanded notation.

4. Subtract $6,000 - 384$.

5. Find $4\frac{3}{10} + 7\frac{5}{6}$.

6. Find $10\frac{3}{5} \div 2\frac{1}{4}$.

7. Divide $28.3 \div 0.72$. Round to hundredths.

8. Change 0.648 to a fraction in lowest terms.

9. Add $0.26 + 5.079 + 3 + 10.002$.

10. Multiply $0.056 \cdot 0.98$.

11. Teresa traveled 362 miles on 16.7 gallons of gas.
    a. How many gallons will she need to drive 2,140 miles? Round to the nearest gallon.

    b. What will the gas cost at $1.19 per gallon? Round to the nearest dollar.

12. Change 64% to a fraction.

13. Change 0.085 to a percent.

14. 13 is 65% of what number?

15. A coat originally selling for $120 was discounted 30%. Find the sale price of the coat.

16. Given the eight ages 45, 36, 40, 41, 40, 33, 42, and 35, find the
    a. mean.   **b.** median.   **c.** mode.   **d.** range.

Use the circle graph to answer questions 17–21.

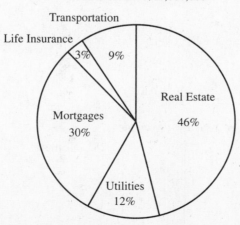

Total Invested: $80,000,000

17. How much money is invested in utilities?

18. What is the ratio of money invested in transportation to money invested in life insurance?

19. What fraction of the total investment is mortgages?

20. How much more is invested in real estate than mortgages?

21. How many more dollars must be invested in transportation to bring the total amount invested in transportation to $10 million?

22. 90 in = ? ft

23. 148 oz = ? lb

Convert and round your answer to tenths.

24. 32 in = ? cm

25. 120 m = ? ft

26. A piece of wire is 6 meters long.
   a. How many 23-cm pieces can be cut from the wire?

   b. How many centimeters of wire are left over?

Explain the meaning of each term. Use your own examples.

27. proportion

28. kilometer

10

Introduction to Geometry

Can the human race, about 5 billion people, fit in 1 cubic mile?

The average human is less than 2 feet wide, less than 2 feet deep, and less than 6 feet tall. Therefore, he or she can fit in a "phone booth" 2 feet wide, 2 feet deep, and 6 feet high.

A cubic mile is a box measuring 1 mile on each side.

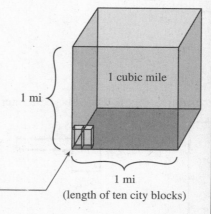

1 mi

1 cubic mile

1 mi

1 mi
(length of ten city blocks)

2 ft

2 ft

6 ft

A 747 jet flies about 5 miles up. A cubic mile is $\frac{1}{5}$ that high.

Put one person in each "phone booth." To find how many people fit in 1 cubic mile, find how many 2 ft × 2 ft × 6 ft phone booths will fit.

$$\frac{\text{number of people}}{\text{in 1 cubic mile}} = \frac{\text{volume of 1 cubic mile}}{\text{volume of "phone booth"}}$$

You will solve this problem in this chapter.

Geometry (*Geo* means Earth, *metry* means to measure) was used by the ancient Egyptians to solve problems on measuring the Earth's surface. Land bordering the Nile River was very good for farming. Each year the Nile overflowed, covering the land with rich soil. In the process, farm boundaries were washed away or buried. The ancient Egyptians used geometry to restore the boundaries to their farms. They also used geometry to build the pyramids.

Geometry is used by architects to draw blueprints for homes. Ships need it to navigate the globe. You use it at home to figure how many square yards of carpet to buy. Geometry is used in surveying, city planning, construction, horticulture, art, science, and business.

Work through each section in this chapter. Use this test to identify topics you are not familiar with. These topics may require additional study. You may *not* use a calculator.

10.1 *Classify each angle as acute, right, straight, or obtuse.*

1. 70°

2.

3.

4. 178°

10.2

5. Find the perimeter.
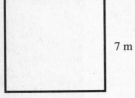

7 m

7 m

6. Find the perimeter.
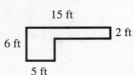

15 ft

6 ft

2 ft

5 ft

10.3

7. Find the area.
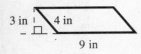

3 in 4 in

9 in

8. Find the area.
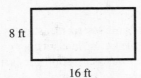

8 ft

16 ft

9. Carpet is on sale for $21.50 per square yard. What will it cost to carpet a den 24 ft long by 12 ft wide?

10.4

10. Find the area.
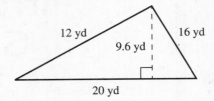

12 yd

9.6 yd

16 yd

20 yd

10.5

11. Find the area and circumference.
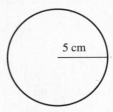

5 cm

10.6

12. Find the volume.
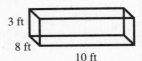

3 ft

8 ft

10 ft

10.7

13. How many cubic feet of oil can the tank hold?
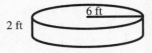

6 ft

2 ft

VOCABULARY *Explain the meaning of each term. Use your own examples.*

14. perpendicular lines

15. radius

16. volume

OBJECTIVE
Define basic geometric terms.

NEW VOCABULARY

point	horizontal line	angle	obtuse angle
line	vertical line	vertex point	straight angle
line segment	intersect	right angle	perpendicular
ray	parallel	acute angle	

Defining Basic Terms

A **point** is a position. It is represented by a dot and has no length or width. Capital letters name points.

point
• P

A **line** is a collection of points extending indefinitely in two directions.

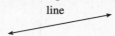

line

A **line segment** is a part of a line. It has two distinct endpoints plus all the points in between.

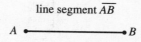

line segment $\overline{AB}$

A •————————• B

A **ray** is a part of a line. It starts at a point and extends indefinitely in one direction.

ray

A **horizontal line** extends left to right. A **vertical line** extends up and down.

horizontal line

vertical line

Parallel and Intersecting Lines

Two lines **intersect** if they cross at one point, or would cross if extended.

intersecting lines

Two lines are **parallel** if they do not intersect. The distance between parallel lines is always the same. The rails of a train track are parallel.

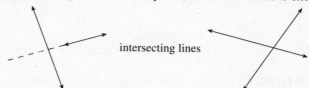

parallel lines

L_1 L_2

$L_1 \mathbin{/\!/} L_2$ is read "line L_1 is parallel to line L_2"

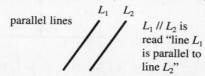

Angles

An **angle** is formed when two rays are drawn from a common point. The point is called the **vertex point**.

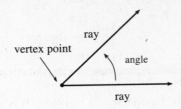

Angles are measured in degrees. The symbol for degrees is °. A circle has 360°. It represents one complete revolution.

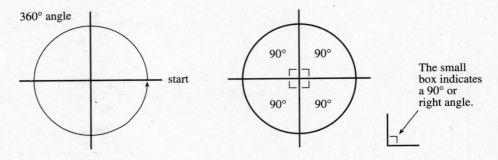

Divide the circle into four equal pieces. Each piece has 360° ÷ 4 = 90°. A 90° angle is called a **right angle**. A circle has four right angles. A right angle is the angle a vertical wall makes with a level floor. Other angles that you will see in this chapter are described as follows.

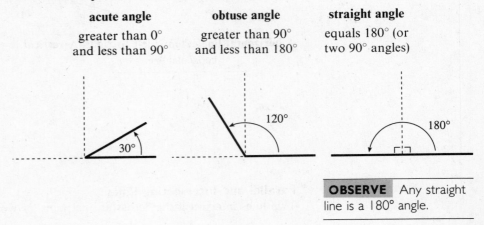

acute angle	**obtuse angle**	**straight angle**
greater than 0° and less than 90°	greater than 90° and less than 180°	equals 180° (or two 90° angles)

OBSERVE Any straight line is a 180° angle.

Perpendicular

Two lines are **perpendicular** if they meet at a right angle.

EXAMPLE I   Examples of perpendicular lines.

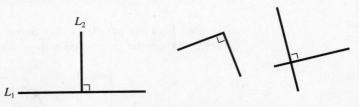

$L_1 \perp L_2$ is read "line L_1 is perpendicular to line L_2." ∎

Classify the lines as parallel, perpendicular, or neither.

1.
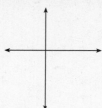

2.
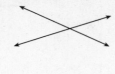

3.

4.
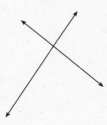

5.
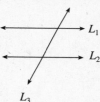

6.
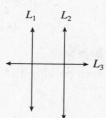

State which lines are parallel, perpendicular, horizontal, and vertical.

7.

8.
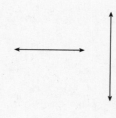

9.
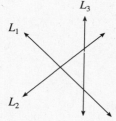

10.
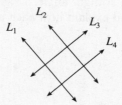

Classify each angle as acute, right, obtuse, or straight.

11. $47°$

12. $2°$

13. $165°$

14. $90°$

15. $100°$

16. $180°$

17. $(160°) \div 2$

18. $4 \times (45°)$

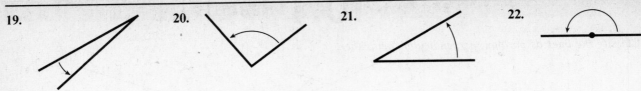

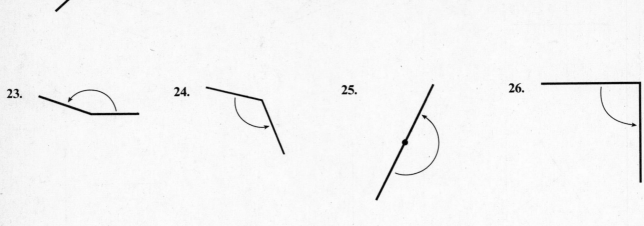

SKILLSFOCUS (Section 1.9) *Evaluate each power.*

27. 5^3 **28.** 6^2 **29.** 7^0 **30.** 2^5

WRITING TO LEARN ▶▶▶▶

31. Explain the difference between parallel and perpendicular lines. Give an example of each on a car.

32. Draw an acute angle, a straight angle, a right angle, and an obtuse angle. Label each angle with its name, and explain how each angle is different. Give an example of each in a kitchen.

33. Explain the difference between a horizontal line and a vertical line. Give an example of each in a forest.

▶ YOU BE THE JUDGE

34. Ellen claims a closed textbook has less than 20 right angles at all its corners combined. Is she correct? Explain your decision.

10.2 PERIMETER

1 Defining Polygon and Perimeter

A **polygon** is a closed geometric figure made up of line segments joined end to end.

EXAMPLE 1 Examples of polygons.

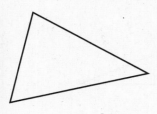 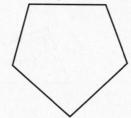  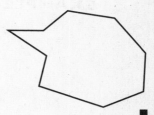

The **perimeter** of a polygon is the sum of the lengths of its sides.

EXAMPLE 2 Find the perimeter of the polygon.

Perimeter = 5 ft + 3 ft + 8 ft + 4 ft
= 20 ft

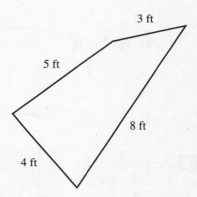

Perimeter is the distance around a polygon. It is a length, and so is measured in **linear units** such as inches, feet, and meters.

▶ **You Try It**   **1.** Find the perimeter of the polygon at the right.

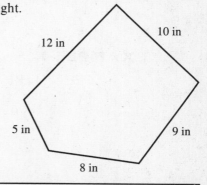

---

2 Perimeter of a Triangle

The **triangle** (*tri* means three) is a polygon with three sides and three angles.

The perimeter of a triangle is the sum of the lengths of the three sides.

$$P = s_1 + s_2 + s_3$$

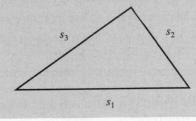

EXAMPLE 3   Find the perimeter of the triangle.

$P = \boxed{13 \text{ cm}} + \boxed{12 \text{ cm}} + \boxed{5 \text{ cm}}$

$P = 30$ cm

The perimeter is 30 cm.

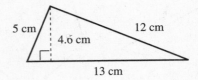

OBSERVE   Do not add 4.6 cm into the perimeter because 4.6 cm is not the length of a side of the triangle. ■

▶ **You Try It**   2. Find the perimeter of the triangle at the right.

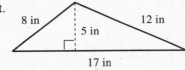

---

3 Perimeter of a Rectangle

A **rectangle** is a four-sided polygon with

1. opposite sides equal.
2. opposite sides parallel.
3. four right angles.

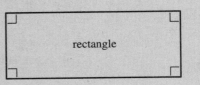

rectangle

EXAMPLE 4   Find the perimeter of the rectangle.

$P = \boxed{10 \text{ m}} + \boxed{10 \text{ m}} + \boxed{6 \text{ m}} + \boxed{6 \text{ m}}$

$= 32$ m

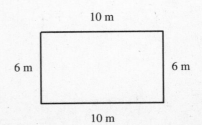

■

Since two sides are 10 m long, and the other two are 6 m long, you can write the perimeter in Example 4 as

$$P = 2(10 \text{ m}) + 2(6 \text{ m})$$
$$= 20 \text{ m} + 12 \text{ m}$$
$$= 32 \text{ m}$$

The perimeter of a rectangle is two times the length L plus two times the width W.

$$P = L + L + W + W$$
$$P = 2L + 2W$$

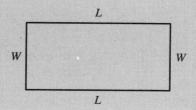

EXAMPLE 5 Find the perimeter of a 16 ft by 12.5 ft rectangular room.

First, draw and label the room.

$$P = 2L + 2W$$
$$= 2(16 \text{ ft}) + 2(12.5 \text{ ft})$$
$$= 32 \text{ ft} + 25 \text{ ft}$$
$$= 57 \text{ ft}$$

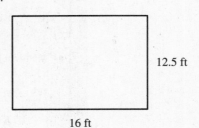

12.5 ft

16 ft

▶ **You Try It** Find the perimeter.

3.

3 ft

7 ft

4.

6 in

8.4 in

4 Perimeter of a Square

A **square** is a rectangle with

1. four equal sides.
2. opposite sides parallel.
3. four right angles.

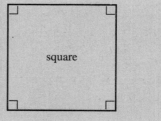

square

EXAMPLE 6 Find the perimeter of the square.

$P = \boxed{5\text{ cm}} + \boxed{5\text{ cm}} + \boxed{5\text{ cm}} + \boxed{5\text{ cm}}$

  $= 20$ cm

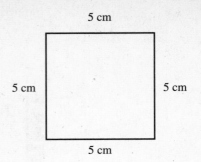

5 cm

5 cm   5 cm

5 cm

■

In Example 6, each side is 5 cm long. Therefore, you can write the perimeter as

$$P = 4(\boxed{5\text{ cm}})$$
$$P = 20\text{ cm}$$

The perimeter of a square is 4 times the length of one side.

$P = s + s + s + s$
$P = 4s$

s

s   s

s

EXAMPLE 7 Find the perimeter of the square.

$P = 4s$

  $= 4(\boxed{8\text{ ft 6 in}})$

  $= 34$ ft

$\left\{ \begin{array}{l} \text{8 ft 6 in} \\ \underline{\times \quad\quad 4} \\ \text{32 ft 24 in} = \text{34 ft} \end{array} \right.$

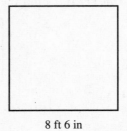

8 ft 6 in

■

▶ **You Try It** Find the perimeter of each square.

5.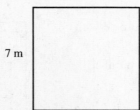

7 m

6.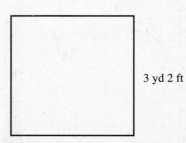

3 yd 2 ft

5 Applications   You find perimeter when you want to fence a piece of land. The perimeter of the land tells you how much fence to buy.

EXAMPLE 8   How much fence is needed to enclose the rectangular garden shown in the figure?

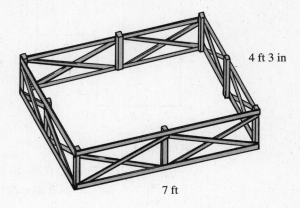

$$P = 2L + 2W$$
$$= 2 \cdot \boxed{7 \text{ ft}} + 2 \cdot (\boxed{4 \text{ ft 3 in}})$$
$$= 14 \text{ ft} + 8 \text{ ft 6 in}$$
$$= 22 \text{ ft 6 in}$$

You need 22 ft 6 in of fence to enclose the garden.   ■

▶ **You Try It**   **7.** How much fence is needed to enclose the yard below?

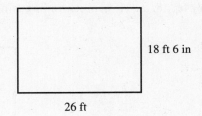

Wood, including shelving boards and molding, is sold by the linear foot. One linear foot is a length of wood 1 ft long.

EXAMPLE 9   Wood molding will be installed around the perimeter of the ceiling shown in the figure.

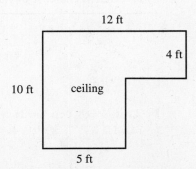

a. How much molding is needed?

First find the two missing sides. Let x and y represent their unknown lengths.

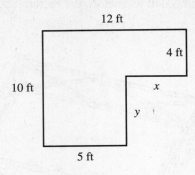

Side x is 7 ft long because:

$$5 \text{ ft} + x = 12 \text{ ft}$$
$$x = 12 \text{ ft} - 5 \text{ ft}$$
$$\boxed{x = 7 \text{ ft}}$$

Side y is 6 ft long because:

$$y + 4 \text{ ft} = 10 \text{ ft}$$
$$y = 10 \text{ ft} - 4 \text{ ft}$$
$$\boxed{y = 6 \text{ ft}}$$

Therefore,

$$P = 12 \text{ ft} + 10 \text{ ft} + 5 \text{ ft} + 6 \text{ ft} + 7 \text{ ft} + 4 \text{ ft}$$
$$= 44 \text{ ft of molding needed.}$$

b. What is the cost of the molding at $0.79 per linear foot. Include a sales tax of 5¢ on the dollar.

```
        44 ft needed
    × $0.79 per linear ft
      $34.76
    ×    1.05 sales tax factor (see Section 5.7)
      $36.498 ≐ $36.50 to the nearest cent
```

The cost of 44 ft of molding including tax is $36.50. ■

▶ **You Try It**   **8. a.** How much railing is needed to enclose the courtyard?

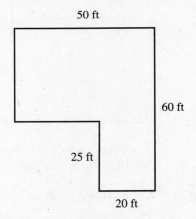

b. What is the cost of the railing at $4.69 a foot? Include a 4¢ sales tax.

▶ **Answers to You Try It**   **1.** 44 in   **2.** 37 in   **3.** 20 ft   **4.** 28.8 in   **5.** 28 m
6. 14 yd 2 ft   **7.** 89 ft   **8. a.** 220 ft   **b.** $1,073.07

SECTION 10.2 EXERCISES

1 2 3 4 *Find the perimeter of each figure.*

1.
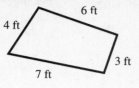
6 ft, 4 ft, 3 ft, 7 ft

2.
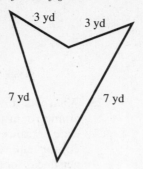
3 yd, 3 yd, 7 yd, 7 yd

3.
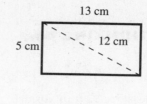
13 cm, 12 cm, 5 cm

4.
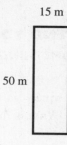
15 m, 50 m

5.
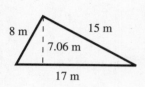
8 m, 15 m, 7.06 m, 17 m

6.
10 ft, 10 ft, 10 ft, 10 ft (square)

7.
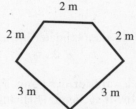
2 m, 2 m, 2 m, 3 m, 3 m

8.
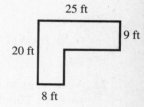
25 ft, 9 ft, 20 ft, 8 ft

5

9.
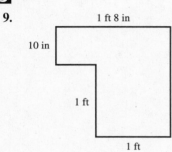
1 ft 8 in, 10 in, 1 ft, 1 ft

10.
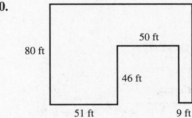
80 ft, 50 ft, 46 ft, 51 ft, 9 ft

Draw and label a figure, then solve the problem.

11. Find the perimeter of a 20 ft by 40 ft rectangular garden.

12. Find the perimeter of a rectangular room 14 ft 3 in long by 10 ft 8 in wide.

13. What is the perimeter of a square room with 15 ft 8 in on a side?

14. Find the perimeter of a square room with 8.37 m on each side.

15. You plan to put a border around the walls of a 15 ft by 11 ft rectangular bedroom. The border costs $0.39 per linear foot. What will the border cost?

16. A rectangular yard 105 ft by 317 ft is fenced for a pony. Fence costs $2.90 per foot. What will it cost to fence the yard?

17. A 32 in by 32 in square painting needs a frame. Framing material costs $3.75 per foot. What will it cost to frame the painting?

18. Janice plans to install crown style molding around the perimeter of her dining room ceiling. Her dining room is 12 ft 6 in by 13 ft. What will the molding cost at $1.35 per linear foot?

19. $m + 30 = 180$ **20.** $2.4 + t = 3.8$ **21.** $s - 50 = 40$ **22.** $r - 1.9 = 2.63$

EXTEND YOUR THINKING ▶▶▶▶
▶ SOMETHING MORE

23. A square garden measures 12 ft on a side.
  a. What is the perimeter of the garden?

  b. Suppose you double the length of each side of the garden. What is the new perimeter?

  c. By what percent did the perimeter increase?

24. Alice plans to fence a yard for her pony. The yard is a 200-ft by 160-ft rectangle. Fencing costs $14.50 per 8-ft section. What is the cost of the fence? (Hint: You can use proportions to find the cost.)

25. What will it cost to buy floor molding for the dining room in the figure? There are two 4-ft wide entranceways into the dining room that do not require molding. Molding costs $1.19 per linear foot.

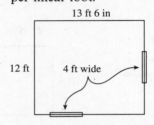

13 ft 6 in

12 ft   4 ft wide

26. The perimeter of a rectangular room is 60 ft. The length is 12 ft. Find the width.

▶ TROUBLESHOOT IT
Find and correct the error.

27. Find the perimeter.

3 ft

8 ft

$P = 8 \text{ ft} \times 3 \text{ ft}$
$\quad = 24 \text{ ft}$

WRITING TO LEARN ▶▶▶▶

28. Use the properties of rectangles and squares discussed in this section to answer each question. Justify each answer.
  a. Is a rectangle a square?

  b. Is a square a rectangle?

▶ YOU BE THE JUDGE

29. Don said the figure shown contains 5 rectangles. Fran said it had more than 10. Who is right? Explain your decision.

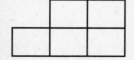

10.3 AREA

OBJECTIVES

1 Compute the area of a rectangle (Sections 2.5, 3.5).
2 Compute the area of a square.
3 Compute the area of a parallelogram.
4 Compute the area of a composite geometric figure.
5 Solve applications.

NEW VOCABULARY

area

parallelogram

composite geometric figure

A line segment is a one-dimensional figure. It has length only. Length is measured in linear units such as inches, feet, miles, and meters.

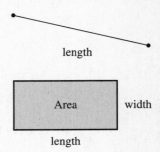

A rectangle is a two-dimensional figure. It has length and width. It is made of four one-dimensional line segments arranged to enclose an area.

1 Area of a Rectangle

Area is the amount of surface within a two-dimensional figure. Area is measured in square units, such as square inches (in^2), square feet (ft^2), or square centimeters (cm^2).

1 square inch = a square, 1 inch on each side. It is about the size of a postage stamp.

1 square foot = a square, 1 foot on each side. It is about the size of a standard floor tile.

The rectangle below is 4 feet long by 3 feet wide. Mark off 3 feet on the left side. Mark off 4 feet across the top. Then divide the rectangle into squares.

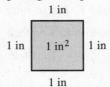

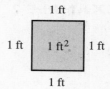

Each small square is 1 foot by 1 foot, or 1 square foot. By counting, the rectangle has 12 square feet. Note that length times width gives 12 square feet.

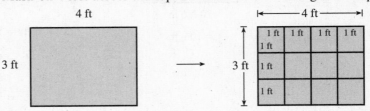

$$\text{length} \times \text{width} = \boxed{4 \text{ ft}} \times \boxed{3 \text{ ft}}$$
$$= 12 \text{ sq ft or } 12 \text{ ft}^2$$

The area of a rectangle is length times width.

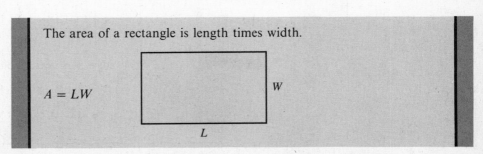

$$A = LW$$

EXAMPLE 1   Find the area of a 12 ft by 20 ft rectangle.

First, draw and label the figure.

$A = L \times W$

$\quad = \boxed{12 \text{ ft}} \times \boxed{20 \text{ ft}}$

$\quad = 240 \text{ ft}^2$

20 ft

12 ft

■

▶ **You Try It**   1. Find the area of the rectangle.

12 yd

7 yd

When measurements are given in mixed units, change all units to the same unit before finding area.

EXAMPLE 2   Find the area of a floor 10 ft 6 in long by 9 ft 4 in wide.

Solution #1: Change all units to feet.

9 ft 4 in
$= 9\frac{1}{3}$ ft

10 ft 6 in
$= 10\frac{1}{2}$ ft

10 ft 6 in $= 10$ ft $+ \boxed{6 \text{ in}}$

$\qquad\qquad = 10$ ft $+ \boxed{\frac{1}{2} \text{ ft}}$

$\qquad\qquad = 10\frac{1}{2}$ ft

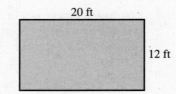

9 ft 4 in $= 9$ ft $+ \boxed{4 \text{ in}}$

$\qquad\qquad = 9$ ft $+ \boxed{\frac{1}{3} \text{ ft}}$

$\qquad\qquad = 9\frac{1}{3}$ ft

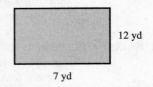

$A = L \times W$

$= \boxed{10\frac{1}{2}} \times \boxed{9\frac{1}{3}} = \overset{7}{\cancel{\frac{21}{2}}} \times \overset{14}{\underset{1}{\cancel{\frac{28}{3}}}}$

$= 98 \text{ ft}^2$

Solution #2: Change all units to inches.

$$10 \text{ ft } 6 \text{ in} = \boxed{10 \text{ ft}} + 6 \text{ in}$$
$$= \boxed{10 \times 12 \text{ in}} + 6 \text{ in} \longleftarrow 1 \text{ ft} = 12 \text{ in}$$
$$= 126 \text{ in}$$

$$9 \text{ ft } 4 \text{ in} = \boxed{9 \text{ ft}} + 4 \text{ in}$$
$$= \boxed{9 \times 12 \text{ in}} + 4 \text{ in}$$
$$= 112 \text{ in}$$

$$A = L \times W = 126 \text{ in} \times 112 \text{ in} = 14{,}112 \text{ in}^2 \quad \blacksquare$$

OBSERVE In Example 2, 98 ft² and 14,112 in² are the same area expressed in different units. To verify this, you know 1 ft × 1 ft = 1 ft². Also, 1 ft × 1 ft = 12 in × 12 in = 144 in². Therefore, 1 ft² = 144 in². Now change 14,112 in² to square feet.

$$14{,}112 \text{ in}^2 = \frac{14{,}112 \text{ in}^2}{1} \times \frac{1 \text{ ft}^2}{144 \text{ in}^2} = 98 \text{ ft}^2$$

CAUTION 10 ft 6 in ≠ 10.6 ft. As shown in solution #1, 10 ft 6 in = $10\frac{1}{2}$ ft = 10.5 ft because $\frac{1}{2} = 0.5$.

▶ **You Try It** **2.** A dining room is 12 ft 9 in long by 8 ft 8 in wide.
    a. Find its area in square feet.
    b. Find its area in square inches.

---

2 Area of a Square The square in the figure has 5 cm on a side. Its area is length times width.

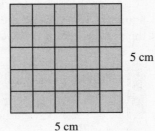

$$A = \boxed{5 \text{ cm}} \times \boxed{5 \text{ cm}}$$
$$= 25 \text{ cm}^2$$

5 cm

5 cm

Since 5 cm × 5 cm = (5 cm)², write the area as

$$A = (\boxed{5 \text{ cm}})^2$$

The area of a square is side times side.

$$A = s \cdot s$$
or, $$A = s^2$$

EXAMPLE 3   Find the area of a square 7.2 m on a side.

$$A = s^2$$
$$= (7.2\text{ m})^2$$
$$= 7.2\text{ m} \cdot 7.2\text{ m}$$
$$= 51.84\text{ m}^2$$

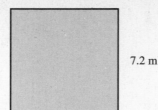

7.2 m

∎

▶ **You Try It**   **3.** Find the area of a square 15.6 ft on a side.

3 Area of a Parallelogram

A **parallelogram** is a four-sided polygon with

1. opposite sides equal.
2. opposite sides parallel.
3. opposite angles equal.

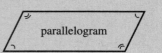

parallelogram

A parallelogram has a base B and a height H. Remove the triangle on the left side of the parallelogram below and attach it to its right side. You get a rectangle with the same base B and height H.

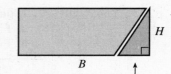

Since the parallelogram has just been rearranged to get the rectangle, both figures have the same area. The area of the rectangle is $A = BH$. The area of the parallelogram is the same.

The area of a parallelogram is the base times the height.

$$A = BH$$

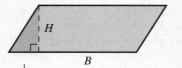

The height H is perpendicular to the base B.

EXAMPLE 4   Find the area of the parallelogram.

$$A = BH$$
$$= 10\text{ ft} \cdot 4\text{ ft}$$
$$= 40\text{ ft}^2$$

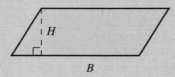

4 ft   5 ft

10 ft

OBSERVE   5 ft is not the height because that side is not perpendicular to the base.

∎

▶ **You Try It**   **4.** Find the area of the parallelogram.

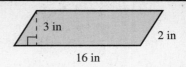

4 **Area of a Composite Geometric Figure**

A **composite geometric figure** consists of two or more other geometric figures joined together. Three examples are shown below. Dashed lines are included to show how the figure can be divided into rectangles and squares.

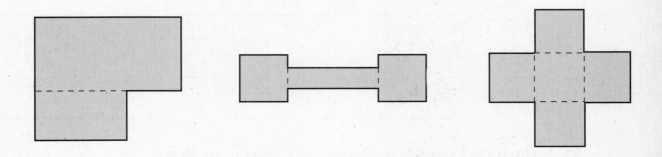

The area of each figure is the sum of the areas of the rectangles and squares into which it is subdivided.

To find the area of a composite geometric figure

1. Subdivide it into smaller figures whose area formulas you know. (Note: There may be more than one way to do this. In each case, the total area will be the same.)

2. Calculate the area of each smaller figure.

3. Sum the areas found in Step 2.

EXAMPLE 5   Find the area.

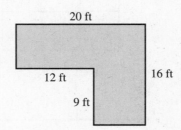

Solution #1: Divide the figure into two smaller rectangles. Label them I and II.

This side is
16 ft − 9 ft = 7 ft

This side is
20 ft − 12 ft = 8 ft

$$\text{total area} = \text{area of I} + \text{area of II}$$
$$= \underbrace{20\ \text{ft} \times 7\ \text{ft}} + \underbrace{9\ \text{ft} \times 8\ \text{ft}}$$
$$= 140\ \text{ft}^2 + 72\ \text{ft}^2$$
$$= 212\ \text{ft}^2$$

Solution #2:

$$\text{Area} = \frac{\text{area of large}}{\text{rectangle}} - \frac{\text{area of cutout}}{\text{rectangle}}$$

$$= \boxed{20 \text{ ft}} \times \boxed{16 \text{ ft}} - \boxed{12 \text{ ft}} \times \boxed{9 \text{ ft}}$$

$$= \quad 320 \text{ ft}^2 \quad - \quad 108 \text{ ft}^2$$

$$= 212 \text{ ft}^2 \quad \blacksquare$$

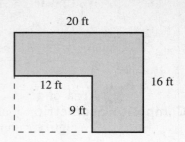

▶ **You Try It**   **5.** Find the area.

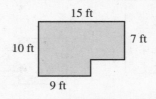

5 Applications   You often calculate floor area in square feet. Carpet is sold by the square yard. To find how much carpet you need to cover a floor, you must know how to change the floor area from square feet to square yards.

EXAMPLE 6   How many square feet are in 1 square yard?

Draw 1 square yard. This is a square 1 yd on each side. Since 1 yd = 3 ft, the length of each side is also 3 ft.

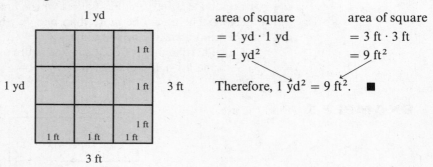

area of square
$$= 1 \text{ yd} \cdot 1 \text{ yd}$$
$$= 1 \text{ yd}^2$$

area of square
$$= 3 \text{ ft} \cdot 3 \text{ ft}$$
$$= 9 \text{ ft}^2$$

Therefore, $1 \text{ yd}^2 = 9 \text{ ft}^2$. $\blacksquare$

▶ **You Try It**   **6.** How many square inches are in 1 sq ft?

EXAMPLE 7   **The Carpet Problem**   Carpet costs $24 per sq yd. What will it cost to carpet the living room shown below?

Step #1: Find the area of the living room.

$$A = L \cdot W$$
$$= \boxed{18 \text{ ft}} \cdot \boxed{15 \text{ ft}}$$
$$= 270 \text{ ft}^2$$

Step #2: Change square feet into square yards. From Example 6, you know that $9 \text{ ft}^2 = 1 \text{ yd}^2$.

$$270 \text{ ft}^2 = \frac{\overset{30}{\cancel{270 \text{ ft}^2}}}{1} \cdot \frac{1 \text{ yd}^2}{\underset{1}{\cancel{9 \text{ ft}^2}}} = 30 \text{ yd}^2 \text{ (area of living room)}$$

OBSERVE To change square feet to square yards, you just divide square feet by 9.

Step #3: Cost of carpet $= 30 \text{ yd}^2 \cdot \$24 \text{ per yd}^2 = \frac{30 \text{ yd}^2}{1} \cdot \frac{\$24}{1 \text{ yd}^2}$

$= \$720$ ∎

▶ **You Try It**   **7.** Carpet costs \$18.50 per yd^2. What will it cost to carpet the den below?

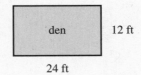

den   12 ft

24 ft

Land area is measured in acres.

1 acre (ac) = 43,560 square feet (ft^2)

1 acre is about the area of a football field from the 0 yard line down to the 9 yard line at the other end of the field.

1 acre is about the area of this rectangle.

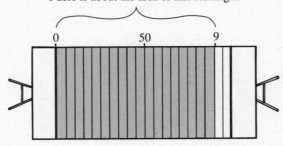

EXAMPLE 8   Tom purchased a rectangular piece of land 100 ft wide by 317 ft deep. How many acres is this? Round to hundredths.

Step #1: Find the area of the land in square feet.

$$A = L \cdot W$$
$$= 100 \text{ ft} \cdot 317 \text{ ft}$$
$$= 31{,}700 \text{ ft}^2$$

Step #2: Change square feet to acres. Since 1 ac = 43,560 ft^2,

$$31,700 \text{ ft}^2 = \frac{31,700 \text{ ft}^2}{1} \cdot \frac{1 \text{ ac}}{43,560 \text{ ft}^2} = \frac{31,700}{43,560} \text{ ac}$$

$$\doteq 0.73 \text{ ac (rounded to hundredths).} \quad \blacksquare$$

▶ **You Try It**   **8.** Amy bought a rectangular lot 124 ft wide by 457 ft deep. How many acres is this? Round to hundredths.

Area Conversion Summary

1 sq ft = 144 sq in
1 sq yd = 9 sq ft
1 acre = 43,560 sq ft

▶ **Answers to You Try It**   **1.** 84 yd^2 **2. a.** 110.5 ft^2 **b.** 12,168 in^2
3. 243.36 ft^2 **4.** 48 in^2 **5.** 132 ft^2 **6.** 144 in^2 **7.** $592   **8.** 1.30 ac

SECTION 10.3 EXERCISES

 2 **3** **4** *Find the area and perimeter of each figure.*

1.
15 ft
10 ft

2.
4 m
6 m

3.
9 cm
9 cm

4.
5 yd
5 yd

5.
4 ft 6 in
2 ft

6.
3 ft 9 in
5 ft 4 in

7.
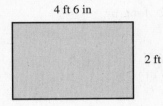
7 cm   6 cm
8 cm

8.
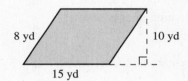
8 yd   10 yd
15 yd

9.
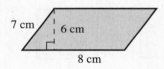
30 ft
12 ft   24 ft
14 ft

10. Each side is 4 ft long.
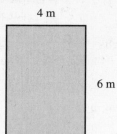
4 ft

11.
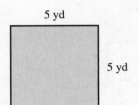
14 in
10 in   3 in
5 in
8 in

12.
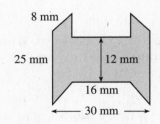
8 mm
25 mm   12 mm
16 mm
30 mm

13. Estimate the area of a square 42.7 m on each side.

14. Estimate the area of a square 89.2 ft on each side.

15. Estimate the area of a rectangle 6 ft 3 in long by 9 ft 10 in wide.

16. Estimate the area of a rectangle 67′2″ long by 82′9″ wide.

17. A rectangle is 4 ft 6 in long by 2 ft 9 in wide.
   a. What is the area in square feet?

   b. What is the area in square inches?

18. A square is 5 ft 3 in on each side.
   a. What is the area in square feet?

   b. What is the area in square inches?

19. A kitchen floor measures 12 ft by 9 ft. What is the cost to tile this floor at $0.39 per tile? Each tile is 1 sq ft.

20. A basement measures 24 ft by 36 ft. What is the cost to tile it if tiles cost $0.62 each? Each tile is 1 sq ft.

21. A living room is 24 ft by 18 ft. What will be the cost to carpet this room at $23 per square yard?

22. A high-grade carpet costs $38 per sq yd. What will be the cost to carpet a family room 21 ft long by 15 ft wide?

23. A dining room is 12 ft by 9 ft. What will be the cost to carpet this room at $21.50 per sq yd?

24. A porch measures 30 ft by 15 ft. What will be the cost to cover this porch with indoor/outdoor carpet at $6.98 per sq yd?

25. What will it cost to carpet the living room/dining room floor plan in the figure? Carpet costs $26.75 per sq yd.

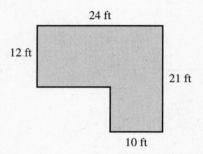

26. What will it cost to carpet both rooms and the hallway pictured in the floor plan? Carpet costs $27 per sq yd.

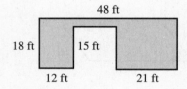

27. How many acres is an 8,000-sq ft lot (round to hundredths)?

28. How many acres is a 173,550-sq ft lot (round to hundredths)?

29. How many acres are in a rectangular lot 135 ft long by 68 ft wide (round to hundredths)?

30. How many acres are in a rectangular lot 285 ft long by 790 ft wide (round to hundredths)?

31. An American football field is 120 yd by $53\frac{1}{3}$ yd. How many acres is this (round to hundredths)?

32. A lacrosse field is 110 yd by 60 yd. How many acres is this (round to hundredths)?

SKILLSFOCUS (Section 7.8) *Find the missing value at simple interest.*

33. I = ?
P = $400
R = 10%
T = 2 yr

34. I = ?
P = $312
R = 7.5%
T = 6 mo

35. I = $1,218
P = $5,800
R = 6%
T = ?

36. I = $947.60
P = ?
R = 8%
T = 15 mo

EXTEND YOUR THINKING ▶▶▶▶

▶ SOMETHING MORE

37. The picture shown has a 3 in border around it. What is the area of the border?

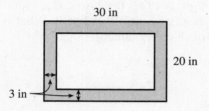

30 in

20 in

3 in

38. A *rhombus* is a parallelogram with four equal sides.

a. Find the perimeter of the rhombus shown here.

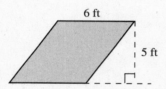

6 ft

5 ft

b. Find the area of the rhombus.

39. A *trapezoid* is a polygon with one pair of parallel sides. The area of a trapezoid is

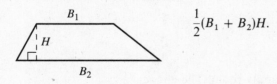

B_1

H

B_2

$\frac{1}{2}(B_1 + B_2)H.$

Find the area of the trapezoid in the figure.

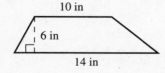

10 in

6 in

14 in

40. The MISL Answer these questions using the excerpt from an article taken from a national newspaper and the figure shown.

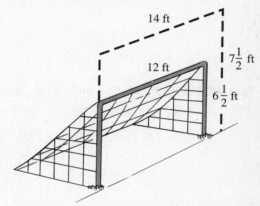

14 ft

12 ft

$7\frac{1}{2}$ ft

$6\frac{1}{2}$ ft

MISL owners will vote to increase size of goals

Major Indoor Soccer League team owners will vote today on proposals that would increase the size of the goals by a total of 6 feet.

The proposed rule

Rule Changes

1. Increase the size of goals by 2 ft in length (12 to 14 ft) and 1 ft in height (6 1/2 to 7 1/2 ft).

2.

a. By how many square feet will the plane of the goal increase?

b. What percent increase is this?

c. Troubleshoot It The article claims the size of the goals will increase by a total of 6 ft. Is this an accurate statement? Explain.

41. An imported floor tile is a square 9 in on each side. It costs $1.26 per tile. What will it cost to tile a kitchen 12'9" by 14'3"?

42. How many acres are in 1 sq mi?

43. A standard basketball court is 94 ft long by 50 ft wide.

a. What is its area?

b. How many acres is this?

c. To the nearest court, how many courts are in 1 acre.

44. Malls A large shopping mall is designed to have 960,000 sq ft of floor space.

a. How many acres is this?

b. 28% of the mall is nonstore space. This area will be tiled at a cost of $3.60 per sq yd of linoleum. Find the total cost.

c. This cost will be spread among the shop owners in the mall. Each owner will pay an amount proportional to the square footage of his/her store, relative to the total square footage of all the stores. Greg's furniture store occupies 10,192 sq ft. What is his share of the linoleum cost?

▶**TROUBLESHOOT IT** *Find and correct the error.*

45. Find the area of a rectangle 5 ft 6 in by 8 ft 9 in.

$$A = L \times W = (5 \text{ ft } 6 \text{ in}) \times (8 \text{ ft } 9 \text{ in})$$
$$= (5.6 \text{ ft}) \times (8.9 \text{ ft})$$
$$= 49.84 \text{ ft}^2$$

46. The area of a square pendant 1 cm on a side is

$$A = s^2 = (1 \text{ cm})^2 = 2 \text{ cm}^2$$

WRITING TO LEARN ▶▶▶▶

47. Explain why every rectangle is also a parallelogram.

48. Are all parallelograms rectangles? Explain.

49. Explain, using a diagram, why 1 sq yd equals 9 sq ft.

50. Write a word problem to find the total price to carpet a family room. The answer for the total price is $600. In your problem give the dimensions of the room in feet. Also give the price per sq yd of carpet. Then write the step-by-step solution to the problem. (Hint: Work backward starting with $600.)

▶**YOU BE THE JUDGE**

51. What is wrong with the statement, "There are 6 square yards in 18 feet"?

10.4 TRIANGLES

OBJECTIVES

1 Find the missing angle in a triangle (Section 2.4).

2 Calculate the area of a triangle (Section 2.5).

NEW VOCABULARY

right triangle

1 Finding the Missing Angle in a Triangle

A triangle is a polygon having three angles and three sides. The sum of the three angles in a triangle is a straight angle, or 180°. To show this, cut the three angles off of any triangle. Then rearrange them in a straight angle, as shown.

> The sum of the three angles in a triangle is a straight angle, or 180°. To find the missing angle in a triangle, subtract the sum of the two given angles from 180°.

EXAMPLE 1 How many degrees are in angle C?

$$\text{angle } C = 180° - (25° + 110°)$$
$$= 180° - 135°$$
$$= 45°$$

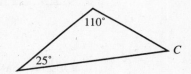

EXAMPLE 2 How many degrees are in angle B?

$$\text{angle } B = 180° - (30° + 90°)$$
$$= 180° - 120°$$
$$= 60°$$

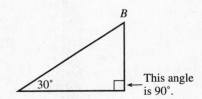

This angle is 90°.

The triangle in Example 2 has a 90°, or right angle. Therefore, it is called a **right triangle**.

▶ **You Try It** Find the missing angle.

1.

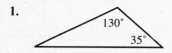

2.

2 Area of a Triangle

In Figure 10.1, two identical triangles are arranged to form a rectangle (or parallelogram) with the same base and height. Therefore, the area of the triangle is half the area of the rectangle (or parallelogram) with the same base and height.

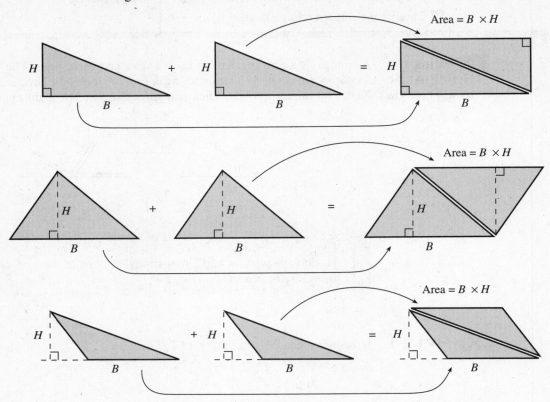

Observe that the height H of a triangle is always perpendicular to the base B, even if the base has to be extended.

The area of a triangle is the base times the height, divided by 2.

$$A = \frac{B \cdot H}{2} \quad \text{or} \quad \frac{1}{2}BH$$

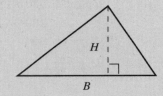

The height H is always perpendicular to the base B.

EXAMPLE 3 Find the area of the triangle.

$$A = \frac{B \cdot H}{2} = \frac{6 \text{ ft} \cdot 4 \text{ ft}}{2} = \frac{24 \text{ ft}^2}{2}$$
$$A = 12 \text{ ft}^2$$

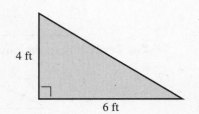

4 ft

6 ft

▶ **You Try It**   **3.** Find the area.

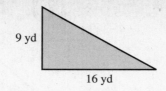

EXAMPLE 4   Find the area of the triangle.

$$A = \frac{B \cdot H}{2} = \frac{8 \text{ cm} \cdot 5.2 \text{ cm}}{2} = \frac{41.6 \text{ cm}^2}{2}$$

$$A = 20.8 \text{ cm}^2$$

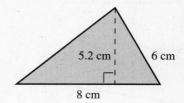

> **CAUTION**   6 cm is not the height of the triangle, because that side is not perpendicular to the base. It is just the length of one of the sides. ■

EXAMPLE 5   Find the area and perimeter of the triangle.

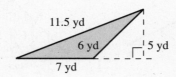

a. Area: $A = \dfrac{B \cdot H}{2} = \dfrac{7 \text{ yd} \cdot 5 \text{ yd}}{2} = \dfrac{35 \text{ yd}^2}{2}$

   $A = 17.5 \text{ yd}^2$

> **CAUTION**   6 yd is not the height, because that side is not perpendicular to the base. Extend the base of the triangle to draw the height.

b. Perimeter: $P = 11.5 \text{ yd} + 6 \text{ yd} + 7 \text{ yd} = 24.5 \text{ yd}$

> **CAUTION**   5 yd is not a side of the triangle, so it is not added into the perimeter. ■

▶ **You Try It**   Find the area and the perimeter.

4.

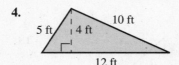

5.

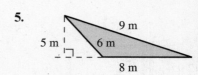

▶ **Answers to You Try It**   **1.** 15°   **2.** 53°   **3.** 72 yd²   **4.** $A = 24 \text{ ft}^2$, $P = 27 \text{ ft}$ **5.** $A = 20 \text{ m}^2$, $P = 23 \text{ m}$

SECTION 10.4 EXERCISES

1 *Find the missing angle.*

1.

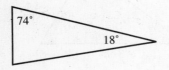

50°

2.

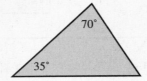

70°

35°

3.

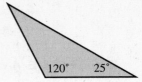

120° 25°

4.

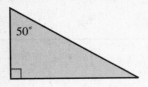

74°

18°

5.

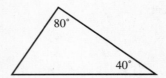

80°

40°

6.

45°

2 *Find the area.*

7.

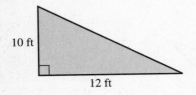

10 ft

12 ft

8.

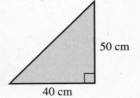

50 cm

40 cm

9.

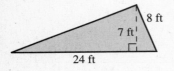

5 yd 4 yd

6.5 yd

10.

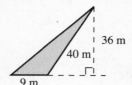

8 ft

7 ft

24 ft

11.

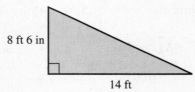

36 m

40 m

9 m

12.

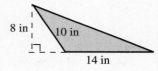

8 in 10 in

14 in

13.

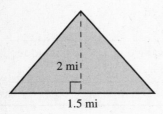

2 mi

1.5 mi

14.

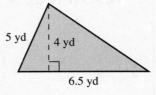

8 ft 6 in

14 ft

15.

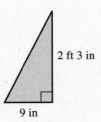

2 ft 3 in

9 in

16.

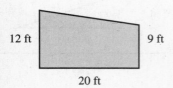

12 ft 9 ft

20 ft

17.

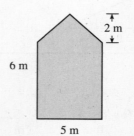

2 m

6 m

5 m

18.

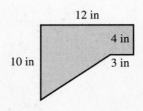

12 in

4 in

10 in

3 in

19. Find the area and perimeter.

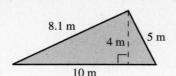

20. Find the area and perimeter.

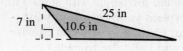

21. Find the area in sq ft and sq in.

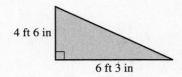

22. Find the area in sq ft and sq in.

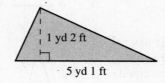

SKILLSFOCUS (Section 2.5) *Evaluate each formula.*

23. Find T when $T = 4P - 2Q$ and $P = 3$ and $Q = 5$.

24. Find M when $M = A + BC$ and $A = 2.5$, $B = 3.4$, and $C = 7$.

25. Find A when $A = 2\pi R$ and $\pi = 3.14$ and $R = 6$.

26. Find V when $V = LWH$ and $L = 5$, $W = 2$, and $H = 3.5$.

EXTEND YOUR THINKING ▶▶▶▶
▶SOMETHING MORE

27. Using pencil, paper, ruler, and scissors, carefully draw a triangle. Cut out the triangle. Cut off the three angles. Rearrange the angles to form a straight angle. State in words what this shows.

28. Using pencil, paper, ruler, and scissors, carefully draw each pair of triangles shown in Figure 10.1. Give each triangle a 6-in base and a 4-in height. Cut out the triangles. Use a pencil to label the base and the height on each triangle.

Now rearrange each pair of triangles into a rectangle or parallelogram.

a. What is the area of each triangle?

b. What is the area of each rectangle or parallelogram?

c. How are these areas related?

29. Is it possible to build a triangle with sides 3 in, 9 in, and 5 in? Explain your answer in your own words. (Hint: Cut pieces of wood or wire to those lengths and build the triangle.)

30. Explain why a triangle cannot have two right angles.

31. **Triangles and Their Properties**

An equilateral triangle has:
   3 equal sides.
   3 equal angles.

$A + A + A = 180°$
or, $3A = 180°$

An isosceles triangle has:
   2 sides that are equal.
   2 angles that are equal.

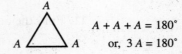

$A + A + B = 180°$
or, $2A + B = 180°$

A scalene triangle has:
   all sides unequal.
   all angles unequal.

$A + B + C = 180°$

A right triangle has:
   1 right angle.
   2 acute angles.

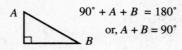

$90° + A + B = 180°$
or, $A + B = 90°$

Answer each question. Give an explanation with reasons for each answer. Begin by drawing and labeling the triangle.

a. How large is each angle in an equilateral triangle?

b. If the unequal angle in an isosceles triangle is 40°, how large are the other two angles?

c. Can a scalene triangle have a 100° angle?

d. Is every equilateral triangle an isosceles triangle?

e. Can an equilateral triangle have a 90° angle?

f. If one of the equal angles in an isosceles triangle is 30°, how large are the other two angles?

g. Sum the three angles in a scalene triangle. Sum the three angles in an equilateral triangle. What is the difference between the two answers?

h. Does every right triangle have two acute angles?

10.5 CIRCLES

OBJECTIVES

1 Calculate the circumference of a circle.

2 Applications of circumference.

3 Calculate the area of a circle.

4 Applications of area.

NEW VOCABULARY

circle	diameter
center	circumference
radius	pi

---

1 Circumference of a Circle

Place each person in your math class exactly 5 feet away from you. What geometric figure will they form? They will form a circle. 5 feet is called the radius of the circle.

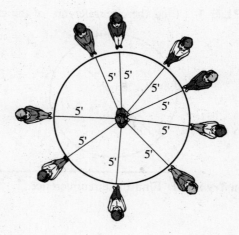

A **circle** is the set of points the same distance from a fixed point. The fixed point is the **center** of the circle. The **radius** R is the distance from the center to the circle. The **diameter** D is the distance from one point on the circle, through the center, to another point. The diameter is two times the radius, $D = 2R$.

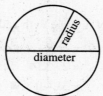

R = radius

D = diameter

$D = 2 \cdot R$

The **circumference** (*circum* means around) of a circle is the distance around the circle. It is similar to perimeter. If you broke the circle at one point and stretched it out, that length is the circumference.

Divide the circumference of any circle by its diameter. In every case, your answer is about 3.14. This number is so common in mathematics, it is given its own name. It is called π, or **pi**.

$$\pi = \frac{\text{circumference}}{\text{diameter}} \doteq 3.14$$

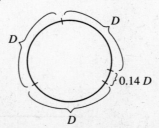

This means there are about 3.14 diameters in one circumference. That is, $C = 3.14 \cdot D$ or $C = \pi D$. Since $D = 2R$,

$$
\begin{aligned}
C &= \pi \cdot \boxed{D} \\
&= \pi \cdot \boxed{2R} \quad \longleftarrow \text{Substitute 2R for D.} \\
&= 2\pi R.
\end{aligned}
$$

> The circumference of a circle is
>
> $$C = \pi D \qquad \text{or} \qquad C = 2\pi R$$
>
> where $\quad D = $ diameter
> $\qquad\quad R = $ radius.

EXAMPLE 1 Find the circumference of the circle with a 7-in diameter.

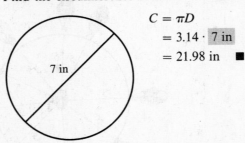

$$
\begin{aligned}
C &= \pi D \\
&= 3.14 \cdot \boxed{7 \text{ in}} \\
&= 21.98 \text{ in} \quad \blacksquare
\end{aligned}
$$

▶ **You Try It** 1. Find the circumference.

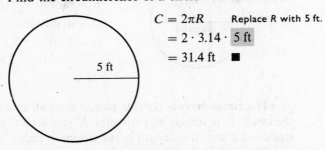

EXAMPLE 2 Find the circumference of a circle with radius 5 ft.

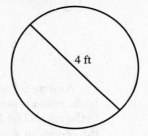

$$
\begin{aligned}
C &= 2\pi R \qquad \text{Replace R with 5 ft.} \\
&= 2 \cdot 3.14 \cdot \boxed{5 \text{ ft}} \\
&= 31.4 \text{ ft} \quad \blacksquare
\end{aligned}
$$

▶ **You Try It** 2. Find the circumference.

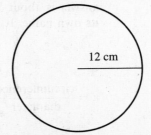

2 Applications of Circumference

EXAMPLE 3 A decorative metal guardrail will be placed around a circular fountain. The rail costs $17.58 per foot installed. The radius of the fountain is 18 ft. What is the total cost?

Step #1: To find how much rail is needed, find the circumference of the fountain.

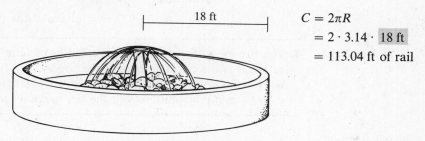

$$C = 2\pi R$$
$$= 2 \cdot 3.14 \cdot \boxed{18 \text{ ft}}$$
$$= 113.04 \text{ ft of rail}$$

Step #2: Cost = 113.04 ft of rail · $17.58 per ft
$$\doteq \$1{,}987.24 \text{ installed} \quad \blacksquare$$

▶ **You Try It**   **3.** A curtain will be installed around a circular stage. The radius of the stage is 14 ft. What is the total cost of the curtain at $8.29 per foot installed?

EXAMPLE 4 An automobile tire has a 15-in radius.

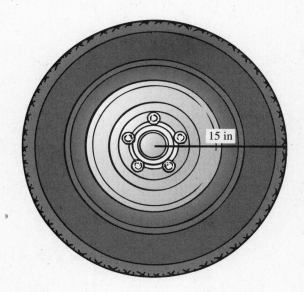

a. What is the circumference of the tire?

$$C = 2\pi R$$
$$= 2 \cdot 3.14 \times \boxed{15 \text{ in}}$$
$$= 94.2 \text{ in (or 7 ft 10.2 in)}$$

b. How far will the car go when the wheel turns one revolution?

94.2 inches. The distance traveled in one revolution is the circumference.

c. How many revolutions will the tire make when the car travels 1 mile?

$$1 \text{ mile} = 5{,}280 \text{ ft} = \frac{5{,}280 \text{ ft}}{1} \cdot \frac{12 \text{ in}}{1 \text{ ft}} = 63{,}360 \text{ in}$$

The car travels 94.2 inches in one tire revolution. To find the number of tire revolutions per mile, find how many times 94.2 in divides into 63,360 in.

$$\frac{\text{number of}}{\text{revolutions}} = \frac{63{,}360 \text{ in}}{94.2 \text{ in}}$$

$$\doteq 673 \text{ revolutions of the tire in 1 mile} \quad \blacksquare$$

▶ **You Try It**   **4.** The tire for a 10-speed bike has a radius of 12 in.
  a. What is the circumference of the tire?
  b. How far will the bike go when the tire turns one revolution?
  c. How many revolutions will the tire make when the bike travels 1 mile?

---

3 **Area of a Circle**   Cut a circle into equal slices as shown below left. Rearrange the slices into the approximate figure of a parallelogram. The base of the parallelogram is approximately half the circumference, $\left(\dfrac{1}{2}\right) \cdot 2\pi R$. The height is the radius R. The area is base times height.

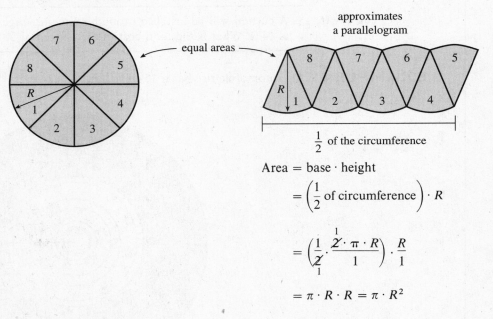

approximates a parallelogram

equal areas

$\dfrac{1}{2}$ of the circumference

Area = base · height

$$= \left(\frac{1}{2} \text{ of circumference}\right) \cdot R$$

$$= \left(\frac{1}{\cancel{2}} \cdot \frac{\overset{1}{\cancel{2}} \cdot \pi \cdot R}{1}\right) \cdot \frac{R}{1}$$

$$= \pi \cdot R \cdot R = \pi \cdot R^2$$

The area of a circle is π times the square of the radius R.

$$A = \pi \cdot R^2$$

or, $$A = \pi \cdot R \cdot R$$

EXAMPLE 5   Find the area of the circle with radius 6 yd.

$A = \pi \cdot R^2$

  $= 3.14 \cdot (6 \text{ yd})^2$

  $= 3.14 \cdot 36 \text{ yd}^2$

  $= 113.04 \text{ yd}^2$

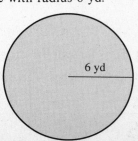

6 yd

$\blacksquare$

► **You Try It**   **5.** Find the area.

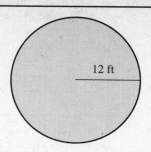

12 ft

EXAMPLE 6   Find the area of a 16-inch circle.

16-inch circle

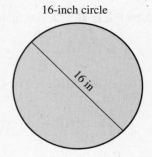

16 in

A 16-inch circle is a circle with diameter 16 in. The radius is half the
diameter, or $R = 16$ in $\div 2 = $ 8 in .

$$A = \pi \cdot R^2$$
$$= 3.14 \cdot (8 \text{ in})^2$$
$$= 3.14 \cdot 64 \text{ in}^2$$
$$= 200.96 \text{ in}^2 \quad \blacksquare$$

► **You Try It**   **6.** Find the area of an 18-yard circle. Draw and label a figure.

**4 Applications
of Area**

EXAMPLE 7   All residents within 5 miles of a toxic spill will be evacuated. How many square
miles is this?

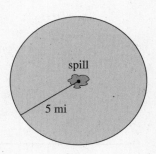

spill

5 mi

The region within 5 miles of the spill forms a circle. The spill is the center.
5 miles is the radius.

$$A = \pi \cdot R^2$$
$$= 3.14 \cdot (5 \text{ mi})^2$$
$$= 3.14 \cdot 25 \text{ mi}^2$$
$$= 78.5 \text{ mi}^2 \text{ of land to be evacuated} \quad \blacksquare$$

▶ **You Try It**   **7.** A rotating sprinkler system will water the lawn within a radius of 30 feet. How many square feet of lawn is this?

EXAMPLE 8   In the figure, a circle is inscribed within a square. Find the area of the shaded region.

40 ft

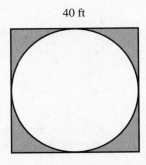

From the figure, the length of the side of the square is 40 ft. The diameter of the circle is 40 ft. So the radius of the circle is 20 ft.

$$
\begin{aligned}
\begin{array}{c}\text{area of}\\\text{shaded}\\\text{region}\end{array} &= \begin{array}{c}\text{area}\\\text{of}\\\text{square}\end{array} - \begin{array}{c}\text{area}\\\text{of}\\\text{circle}\end{array}\\
A &= S^2 - \pi R^2\\
&= (40\text{ ft})^2 - 3.14 \cdot (20\text{ ft})^2\\
&= 1{,}600\text{ ft}^2 - 3.14 \cdot 400\text{ ft}^2\\
&= 344\text{ ft}^2 \quad \blacksquare
\end{aligned}
$$

▶ **You Try It**   **8.** Given the circle and square arranged as shown, find the area of the shaded region.

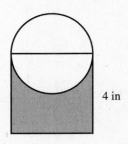

4 in

▶ **Answers to You Try It**   **1.** 12.56 ft   **2.** 75.36 cm   **3.** $728.86
4. a. 75.36 in   **b.** 75.36 in   **c.** about 841 revolutions   **5.** 452.16 ft^2
6. 254.34 yd^2 **7.** 2,826 ft^2 **8.** 9.72 in^2

1 *Find the circumference. If not provided, draw and label a diagram.*

1.

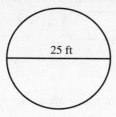

25 ft

2.

8 m

3.

9 cm

4.

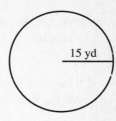

15 yd

5. Find the circumference of a circle with diameter 4 ft.

6. Find the circumference of a circle with diameter 1 mile.

7. What is the circumference of a circle with radius 32 ft?

8. What is the circumference of a circle with radius 80 cm?

9. A circle has a 4 ft 6 in diameter. Find its circumference in
  a. inches.

  b. feet.

10. A circle has a 6 yd 1 ft radius. What is its circumference in
  a. feet?

  b. inches?

2

11. A figure skater must make the figure eight shown. It consists of two identical circles. Each has radius 10 ft. How far will the skater go one time around the eight?

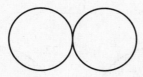

12. How many linear feet of wood are needed to build the window in the figure? Each line in the figure is made of wood. The figure at the top of the window is a semicircle (half a circle).

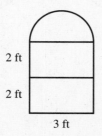

2 ft

2 ft

3 ft

13. The outfield wall in a baseball park is in the shape of a quarter circle.

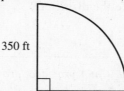

350 ft

  a. How long is the wall?

  b. What will it cost to paint the wall at $24.80 per foot?

14. Refer to the indoor, circular racetrack shown in the figure.

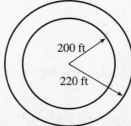

200 ft

220 ft

a. What is the distance around the inside of the track?

b. What is the distance around the outside of the track?

c. How much farther will a runner have to run on the outside of the track versus the inside?

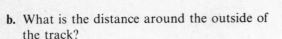 **15.** Calculate the circumference of each wheel with the given radius.

    a. 2 in   **b.** 4 in

    c. 8 in   **d.** 16 in

    e. What happens to the circumference when you double the radius?

16. Calculate the circumference of each wheel with the given diameter.

    a. 1 ft   **b.** 2 ft

    c. 4 ft   **d.** 8 ft

    e. What happens to the circumference when you double the diameter?

17. a. Find the distance around the track shown in the figure.

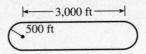

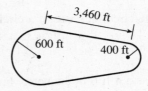

b. Express this distance in miles (round to tenths).

18. a. Find the distance around the track shown in the figure.

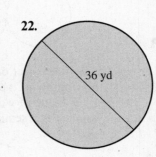

b. Express this distance in miles (round to tenths).

---

❸ *Find the area. If not provided, draw and label a diagram.*

19.

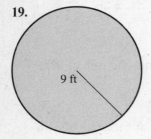

20.

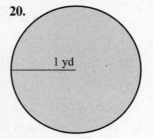

21.

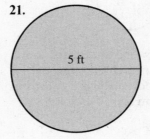

22.

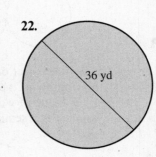

23. Find the area of a circle with a radius of 2 in.

24. What is the area of a circle with a 7-ft radius?

25. What is the area of a 30-yard circle?

26. Find the area of a 3-mile circle?

27. Find the area of a circle with a 2 ft 6 in radius
   a. in square feet.

   b. in square inches.

28. Find the area of a circle with a 10 yd 2 ft radius
   a. in square feet.

   b. in square inches.

4

29. How many sq ft of glass are needed in the construction of the window shown in the figure? The top of the window is a semicircle (half a circle).

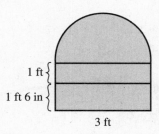

1 ft

1 ft 6 in

3 ft

30. Find the area of the shaded region. Each circle has a radius of 1 in.

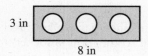

3 in

8 in

31. a. Find the area of the circular racetrack.

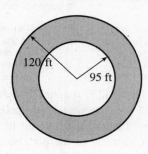

120 ft

95 ft

   b. Find the cost of asphalting the track at $70 per 100 sq ft.

32. The driving circle at a busy intersection will be landscaped at a cost of $1.75 per sq ft. Find the total cost.

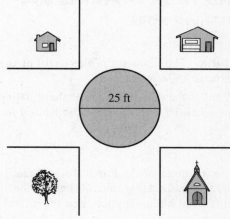

25 ft

33. A radio station can be heard 600 miles away on a clear broadcasting day. How many square miles are in the station's listening area on such a day?

34. A giant sequoia has a base diameter of 18 ft.
   a. Find the approximate circumference at the base of the tree.

   b. Find the approximate area of the tree's footprint (ground area covered by the base of the tree).

35. a. Find the area of the ice-skating rink in the figure.

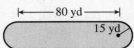

|← 80 yd →|

15 yd

b. In one week it costs $0.075 per sq yd to maintain the rink. What is the weekly maintenance cost?

36. Find the area of the shaded portion of the figure. The small circle has a 1 mm radius.

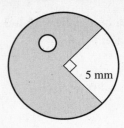

5 mm

SKILLSFOCUS (Section 5.2) *Round to hundredths.*

37. 5.609   **38.** 17.9958   **39.** 0.0729   **40.** 9.0909

EXTEND YOUR THINKING ▶▶▶▶
▶SOMETHING MORE

41. The Earth The diameter of the Earth at the equator is 7,926 miles.

   a. What is the circumference of the Earth at the equator? Use $\pi = 3.14159$. Round to units.

   b. The diameter of the Earth through the poles is 7,900 miles. How much longer is the equatorial circumference than the polar circumference?

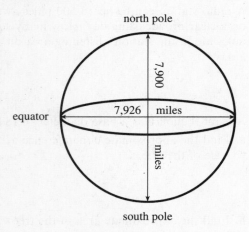

north pole

7,900

equator   7,926 | miles

miles

south pole

42. Circles and the 10-Speed Bike Most 10-speed bikes have 2 circular sprockets connected to the front pedals (sprockets have teeth, and hold the chain). They have 5 circular sprockets connected to the rear wheel. This gives a total of $2 \times 5 = 10$ possible gears. A chain connects a front sprocket to a rear one.

front sprocket (pedals)

rear sprocket (rear wheel)

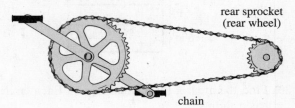

chain

To determine how far a bike travels when the pedals (and so the front sprocket) make 1 revolution, evaluate the following formula. (See Section 2.4, 2.5.)

$$\text{distance traveled on 1 revolution of the pedals} = \frac{(\text{number of teeth in front sprocket})}{(\text{number of teeth in rear sprocket})} \cdot \frac{\pi D}{12} \text{ ft}$$

This is the *gear ratio.* It tells you how many times the rear wheel rotates with one revolution of the pedals.

πD is the circumference of the rear wheel. *D* is the diameter in inches. Divide by 12 to change inches to feet.

With a 27″ bike (27″ means 27 inches), the diameter of the rear wheel is approximately 27 inches. How far will the bike go in 10th gear with one revolution of the pedals? In 10th gear, the chain is around the larger sprocket in the front, and the smallest in the rear. If the larger front sprocket has 52 teeth, and the smallest rear sprocket has 14 teeth, the bike will go

$$\frac{52}{14} \cdot \frac{3.14 \cdot 27}{12} = \frac{52 \cdot 3.14 \cdot 27}{14 \cdot 12}$$

$$\doteq 26 \text{ ft with one revolution of the pedals in 10th gear.}$$

Note, in 10th gear, the gear ratio is $\frac{52}{14} \doteq 3.7$.

This means the rear wheel rotates about 3.7 times with one revolution of the pedals.

a. In first gear, the chain is around the smaller sprocket in the front (39 teeth) and the largest in the rear (28 teeth). How far will the 27″ bike go in 1st gear with one revolution of the pedals?

b. About how many times farther will this bike go in 10th gear versus 1st gear on one revolution of the pedals?

c. Suppose you ride 1 mile on your 27″ bike. How many revolutions of the pedals do you make in 10th gear? 1st gear? Round to tens.

43. Pizza Complete the chart below for three different sizes of pepperoni pizza sold by a national pizza chain. Compute percent increase based on the previous amount given in the chart. Round to units.

Size of Pizza	Cost	% Increase in Cost	Area	% Increase in Area	Cost per Square Inch
9″	$5.60	xxxxx		xxxxx	
12″	$8.50				
15″	$11.40				

a. Which increases faster, percent increase in cost or percent increase in area?

b. Which pizza is the best buy per square inch?

44. Compute π Compute π using a ruler, a circular object such as a dish, and a calculator. Mark the dish on its edge. Place the dish on a table-top with the mark down. Mark this spot. Roll the dish exactly one revolution (be careful not to slide it). This occurs when the mark again touches the table. Mark this spot. Measure the distance between the two marks. This distance is the circumference of the plate. Why?

Next, measure the diameter of the plate. Fix one end of the ruler on the mark at the edge of the plate. Move the ruler back and forth on the opposite side of the plate. The longest measurement you can get is the diameter of the plate. Why?

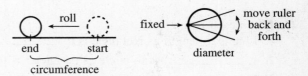

Estimate π by dividing the circumference by the diameter.

$$\pi = \frac{\text{circumference}}{\text{diameter}}$$

Use your calculator. Round to two places. How far is your value from 3.14?

45. Find the area of the shaded region bounded by the four quarter circles.

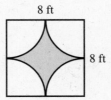

46. Explain what π means. Relate it to a circular pie pan.

47. Explain the difference between area and circumference.

48. Explain how circumference and perimeter are alike. Explain how they are different.

49. Fill in each blank.
  1. Line is to length as rectangle is to _____.
  2. _____ is to circle as perimeter is to square.
  3. Four is to rectangle as _____ is to triangle.
  4. 40° is to acute as 120° is to _____ .
  5. Right angle is to straight angle as _____ is to 180°.

▶ **YOU BE THE JUDGE**

50. Suppose a car tire makes 673 revolutions in 1 mile traveling 30 mph. How many revolutions will the same tire make in 1 mile traveling 60 mph?

10.6 VOLUME: THE RECTANGULAR SOLID

OBJECTIVES

1 Find the volume of a rectangular solid (Sections 3.5, 9.2).

2 Applications.

NEW VOCABULARY

volume   rectangular solid

cubic units   cube

A box is a 3-dimensional figure. It has length, width, and height. Its sides are six 2-dimensional rectangles arranged to enclose a space. This space is called volume. Volume is measured in cubic units such as cubic inches (in³), cubic feet (ft³), or cubic centimeters (cm³).

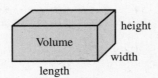

1 Volume of a Rectangular Solid

Volume is the amount of space within a solid, 3-dimensional figure. It is measured in **cubic units**.

1 cubic inch is the volume of a box with 1 inch on all edges.

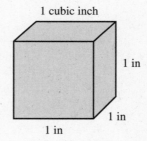

OBSERVE Cylinder displacement in some gasoline engines is measured in cubic inches.

The volume of a **rectangular solid**, or box, can be found by filling it with cubic inches.

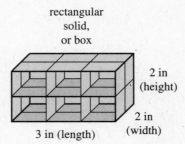

12 cubic inches fit in the figure. Therefore, its volume is 12 cubic inches, or 12 in³. The product of length, width, and height also equals 12 in³.

$$3 \text{ in} \cdot 2 \text{ in} \cdot 2 \text{ in} = 12 \text{ in}^3$$

The volume of a rectangular solid, or box, is length L times width W times height H.

$$V = L \cdot W \cdot H$$

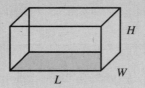

OBSERVE 3 in · 2 in · 2 in = 3 · 2 · 2 · in · in · in = 12 · in³ = 12 in³ or 12 cu in. Cubic inch can be written cu in or in³. The unit in³ comes from power notation. Since $4 \cdot 4 \cdot 4 = 4^3$, write in · in · in = in³.

EXAMPLE 1   Find the volume of a rectangular solid 8 in long, 5 in wide, and 4 in high.

Draw and label a picture.

$V = L \cdot W \cdot H$
$= 8\ \text{in} \cdot 5\ \text{in} \cdot 4\ \text{in}$
$= 160\ \text{in}^3$

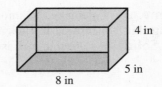

■

▶ **You Try It**   **1.** Find the volume.

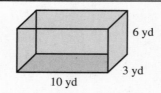

---

2 Applications

EXAMPLE 2   Find the volume of a refrigerator with inside dimensions 2 ft 3 in long, 1 ft 6 in deep, and 4 ft 8 in high. Round to tenths.

Change all dimensions to feet.

$\text{length} = 2\ \text{ft}\ 3\ \text{in} = 2\dfrac{3}{12}\ \text{ft} = 2\dfrac{1}{4}\ \text{ft}$

$\text{width} = 1\ \text{ft}\ 6\ \text{in} = 1\dfrac{6}{12}\ \text{ft} = 1\dfrac{1}{2}\ \text{ft}$

$\text{height} = 4\ \text{ft}\ 8\ \text{in} = 4\dfrac{8}{12}\ \text{ft} = 4\dfrac{2}{3}\ \text{ft}$

$V = L \cdot W \cdot H$

$= 2\dfrac{1}{4} \cdot 1\dfrac{1}{2} \cdot 4\dfrac{2}{3} = \dfrac{9}{4} \cdot \dfrac{3}{2} \cdot \dfrac{14}{3} = \dfrac{63}{4}$

$\doteq 15.8\ \text{ft}^3$ ■

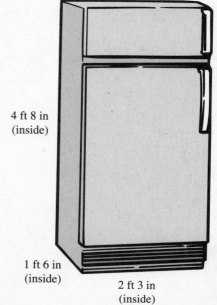

4 ft 8 in
(inside)

1 ft 6 in
(inside)

2 ft 3 in
(inside)

▶ **You Try It** 2. Find the volume of a freezer with inside dimensions 4'6" long, 1'9" deep, and 2'4" high. Round to tenths.

EXAMPLE 3   How many cubic feet are in 1 cubic yard?

1 cubic yard is the volume of a box with 1 yd on each side. Since 1 yd = 3 ft, it is also a box with 3 ft on each side.

$V = LWH$
$\quad = 1 \text{ yd} \cdot 1 \text{ yd} \cdot 1 \text{ yd}$
$\quad = 1 \text{ yd}^3$
$V = LWH$
$\quad = 3 \text{ ft} \cdot 3 \text{ ft} \cdot 3 \text{ ft}$
$\quad = 27 \text{ ft}^3$

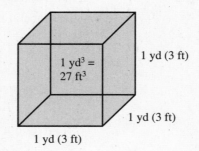

Therefore, $1 \text{ yd}^3 = 27 \text{ ft}^3$.   ■

▶ **You Try It** 3. How many cubic inches are in 1 cubic foot?

> **OBSERVE**   Natural gas and water are sold by the cubic foot. Concrete and stone are sold by the cubic yard.

EXAMPLE 4   **The Patio Problem**   You plan to build a concrete patio 24 ft long, 20 ft wide, and 4 in deep. What is the cost for the concrete if concrete sells for $80 per cubic yard?

Step #1: The patio is a rectangular solid. Find its volume.

$$V = L \cdot W \cdot H$$

$$= \frac{\overset{8}{24}}{1} \text{ ft} \cdot \frac{20}{1} \text{ ft} \cdot \frac{1}{\underset{1}{3}} \text{ ft}$$

$$= 160 \text{ ft}^3$$

Step #2: Change cubic feet to cubic yards. Recall, $1 \text{ yd}^3 = 27 \text{ ft}^3$. Round to tenths.

$$160 \text{ ft}^3 = \frac{160 \text{ ft}^3}{1} \cdot \frac{1 \text{ yd}^3}{27 \text{ ft}^3} = \frac{160}{27} \text{ yd}^3 \doteq 5.9 \text{ yd}^3 \text{ of concrete}$$

Step #3: Find the cost at $80 per cubic yard.

$$\text{cost} = 5.9 \text{ yd}^3 \cdot \$80 \text{ per yd}^3 = \$472 \quad ■$$

▶ **You Try It** 4. You plan to build a concrete patio 12 ft long, 18 ft wide, and 6 in deep. What is the cost for the concrete if concrete sells for $75 per cubic yard?

A **cube** is a rectangular box where length, width, and height are equal. If the length of each edge of a cube is E, its volume is

$$V = L \cdot W \cdot H$$
$$V = E \cdot E \cdot E$$
$$V = E^3.$$

EXAMPLE 5   Find the volume of a cube with 6 inches on each side.

Draw and label a picture.

$V = E^3$
$V = (\,6 \text{ in}\,)^3$
$V = 216 \text{ in}^3$

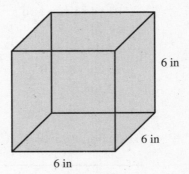

6 in

6 in

6 in

▶ **You Try It**   **5.** Find the volume of a cube with 10 ft on each edge.

EXAMPLE 6   **Chapter Problem**   Reread the problem at the beginning of this chapter.

$$\frac{\text{number of people}}{\text{in 1 cubic mile}} = \frac{\text{volume of 1 cubic mile}}{\text{volume of "phone booth"}}$$

$$= \frac{E^3}{LWH}$$

$$= \frac{(5{,}280 \text{ ft})^3}{2 \text{ ft} \cdot 2 \text{ ft} \cdot 6 \text{ ft}} \qquad \text{Note, } E = 1 \text{ mi or } 5{,}280 \text{ ft.}$$

$$= \frac{147{,}197{,}952{,}000 \text{ ft}^3}{24 \text{ ft}^3}$$

$$= 6{,}133{,}248{,}000 \text{ people can fit in 1 cubic mile}$$

Therefore, the human race, about 5 billion people, will fit.   ■

▶ **You Try It**   **6.** A storage bin is a cube with 12 ft on each side. How many crates can be packed into the bin if each crate is 1′ by 3′ by 4′?

▶ **Answers to You Try It**   **1.** 180 yd^3 **2.** 18.4 ft^3 **3.** 1,728 in^3
4. $300   **5.** 1,000 ft^3 **6.** 144 crates

SOMETHING MORE
Understanding Your Water and Sewer Bill

1. You are billed for water used between these two dates.

2. There were 81 days between the two dates.

3. Meter readings are in units of 100 ft³.

1,265	present reading (6/26/90)
−1,253	previous reading (4/6/90)
12	units of water used

Since 1 unit = 100 ft³,
12 units = 12 × 100 ft³
= 1,200 ft³ of water used (to nearest hundred).

4. 1,200 ft³ ÷ 81 days $\doteq$ 14 ft³ of water used each day (rounded down)

5. 1 ft³ $\doteq$ 7.48 gal
So, 1,200 ft³ = 1,200 × 7.48 gal
= 8,976 gal.
To the nearest thousand, this is 9,000, or 9 thousand gallons of water.

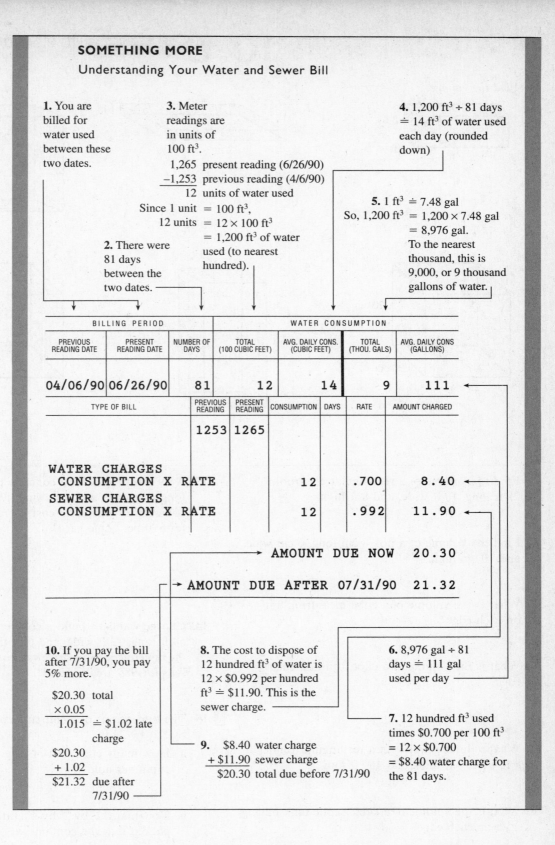

BILLING PERIOD			WATER CONSUMPTION			
PREVIOUS READING DATE	PRESENT READING DATE	NUMBER OF DAYS	TOTAL (100 CUBIC FEET)	AVG. DAILY CONS. (CUBIC FEET)	TOTAL (THOU. GALS)	AVG. DAILY CONS (GALLONS)
04/06/90	06/26/90	81	12	14	9	111

TYPE OF BILL	PREVIOUS READING	PRESENT READING	CONSUMPTION	DAYS	RATE	AMOUNT CHARGED
	1253	1265				
WATER CHARGES CONSUMPTION X RATE			12		.700	8.40
SEWER CHARGES CONSUMPTION X RATE			12		.992	11.90

AMOUNT DUE NOW   20.30

AMOUNT DUE AFTER 07/31/90   21.32

10. If you pay the bill after 7/31/90, you pay 5% more.

$20.30 total
× 0.05
1.015 $\doteq$ $1.02 late charge

$20.30
+ 1.02
$21.32 due after 7/31/90

8. The cost to dispose of 12 hundred ft³ of water is 12 × $0.992 per hundred ft³ $\doteq$ $11.90. This is the sewer charge.

9. $8.40 water charge
+ $11.90 sewer charge
$20.30 total due before 7/31/90

6. 8,976 gal ÷ 81 days $\doteq$ 111 gal used per day

7. 12 hundred ft³ used times $0.700 per 100 ft³
= 12 × $0.700
= $8.40 water charge for the 81 days.

SECTION 10.6 EXERCISES

1 *Find the volume.*

1.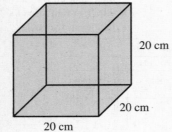
3 ft
9 ft
5 ft

2.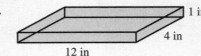
1 in
4 in
12 in

3.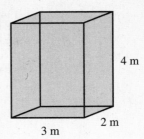
4 m
3 m
2 m

4.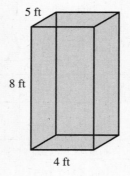
20 cm
20 cm
20 cm

5.
5 ft
8 ft
4 ft

6.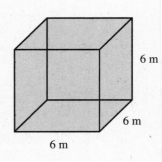
6 m
6 m
6 m

---

7. Find the volume of a rectangular classroom 25 ft long, 18 ft wide, and 8 ft high.

8. Find the volume of a box 8 cm long, 6 cm wide, and 10 cm high.

9. What is the volume of a cube measuring 8 ft on each edge?

10. What is the volume of a cube 10 in on each edge?

2

11. What volume of air is in a rectangular tunnel 8 ft wide, 8 ft high, and 150 ft long?

12. What is the volume of a cubic septic tank having 2 yd on each edge?

13. Amy wants to build a concrete patio 18 ft long, 9 ft wide, and 4 in thick. What will the concrete for this patio cost her if concrete sells for $90 per cubic yard?

14. George wants to build a concrete driveway 54 ft long, 12 ft wide, and 6 in thick. What will the concrete for this driveway cost him if concrete sells for $85 a cubic yard?

15. The dimensions of a shipping container are 8 ft by 8 ft by 24 ft.
   a. How many cubic feet of cargo can this container hold?

   b. How many 2′ by 2′ by 4′ crates can be packed into the container?

16. A refrigerator has inside dimensions 1′2″ by 1′6″ by 4′6″.

    a. What is its volume in cubic inches?

    b. How many 2″ by 4″ by 9″ egg cartons can be packed into it?

17. Stella plans to cover her front yard with 4 inches of topsoil. The yard is 81 ft long by 40 ft wide.

    a. How many cubic yards of topsoil must Stella order?

b. What is the cost at $26 per cubic yard?

18. You purchased a small tub of cement. It can fill 47 cu in of space.

    a. Do you have enough cement to fill a sidewalk gap 6 ft long, 2 in deep, and $\frac{1}{2}$ in wide?

    b. If so, how much cement will be left over? If not, how much more cement do you need?

SKILLSFOCUS (Section 3.4) *Multiply.*

19. $\dfrac{2}{3} \times \dfrac{9}{10}$ **20.** $\dfrac{4}{3} \times 6$ **21.** $\dfrac{4}{9} \times 9 \times 9$ **22.** $\dfrac{8}{3} \times \dfrac{3}{5} \times \dfrac{15}{16}$

EXTEND YOUR THINKING ▶▶▶▶
▶SOMETHING MORE

23. A Cord of Wood Firewood is often sold in a stack 4 ft wide, 4 ft high, and 8 ft long. This stack is called a *cord* of wood.

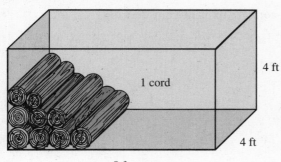

    a. How many cubic feet are in 1 cord?

    b. Assume the average piece of firewood occupies a space 8 in by 8 in by 24 in. About how many pieces will be in 1 cord?

    c. A *face cord* is a stack of wood 4 ft by 4 ft by 2 ft. How many cubic feet are in a face cord?

    d. What fraction of a cord is a face cord?

    e. If 1 cord costs $100, what is a fair price for a face cord?

24. A cube measures 4 in on each side. You double the length of each side of the cube. By how many times will the volume increase?

25. a. How many cubic feet of water can the aquarium in the figure hold?

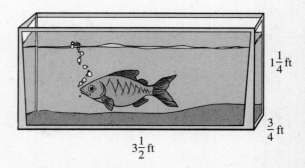

---

b. If water weighs 62.5 lb per cubic foot, how many pounds of water is this?

26. a. How many 1-oz gold ingots must be melted to form the gold bar in the figure?

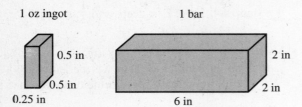

1 oz ingot   1 bar

0.5 in
0.5 in
0.25 in

2 in
2 in
6 in

b. What will the bar weigh?

27. A rectangular sheet of metal is shown in the figure. A 4-in square is cut from each corner. The metal edges are then folded up along the dotted lines to form a rectangular container.

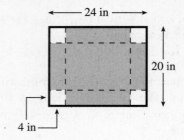

24 in

20 in

4 in

a. What are its dimensions?

b. What is its volume?

WRITING TO LEARN ▶▶▶▶

28. Explain the difference between area and volume.

29. Write a word problem that involves finding the volume of a rectangular solid. The answer is 60 cu in. Draw and label a diagram. Then write the solution.

▶ YOU BE THE JUDGE

30. Carl claims that a cube is not a rectangular solid. Is he correct? Explain your decision.

10.7 VOLUME: THE CYLINDER AND THE SPHERE

OBJECTIVES

1 Find the volume of a cylinder (Sections 2.4, 2.5).

2 Applications (Section 9.2).

3 Find the volume of a sphere (Section 3.5).

4 Applications.

NEW VOCABULARY

right circular cylinder

sphere

hemisphere

---

1 Volume of a Cylinder

The **right circular cylinder** is a familiar figure. A soup can is in the shape of a right circular cylinder. It is called *circular* because the top and bottom are circles. It is called *right* because the side is at right angles to the top and bottom.

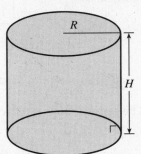

Right circular cylinder

R = radius of top and bottom

H = height

OBSERVE The height is always perpendicular to the base.

The area of the circle at the top of the cylinder is πR^2. Move the circle down a distance H. You generate the shape of a cylinder. Its volume is the area of the top circle times the distance moved, or $\pi R^2 \cdot H$.

OBSERVE Moving an area through a distance generates volume.

> The volume of a cylinder = area of top $(\pi R^2) \cdot$ height (H).
>
> $$V = \pi R^2 H$$

EXAMPLE 1 Find the volume of the cookie tin. Round to the nearest unit.

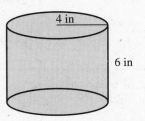

$V = \pi R^2 H$

$\quad = 3.14 \cdot (4\text{ in})^2 \cdot (6\text{ in})$

$\quad = 3.14 \cdot 4\text{ in} \cdot 4\text{ in} \cdot 6\text{ in}$

$\quad = 301\text{ in}^3$ ∎

▶ **You Try It**　**1.** Find the volume. Round to units.

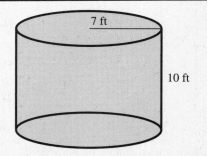

---

2 Applications

EXAMPLE 2 **The Swimming Pool Problem** The swimming pool in the figure has a 20-ft diameter. It is filled with water to a depth of 4 ft.

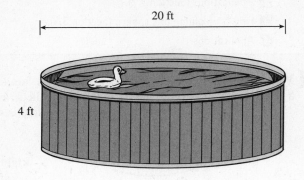

20 ft

4 ft

a. How many cubic feet of water can the pool hold?

The pool is a cylinder. The radius of the top is 20 ft ÷ 2 = 10 ft.

$$V = \pi R^2 H$$
$$= 3.14 \cdot (\,10\,\text{ft}\,)^2 \cdot 4\,\text{ft}$$
$$= 1,256\,\text{ft}^3$$

The pool holds 1,256 ft³ of water.

b. How many gallons of water is this? Round to units.

$$1,256\,\text{ft}^3 = \frac{1,256\,\text{ft}^3}{1} \cdot \frac{7.48\,\text{gal}}{1\,\text{ft}^3} \qquad \text{(Note, 1 cu ft } \doteq 7.48\,\text{gal)}$$
$$\doteq 9,395\,\text{gal}$$

The pool holds about 9,395 gal of water.

c. How long will it take to fill the pool using a hose that yields 5 gal/min?

$$\text{time to fill} = \frac{9,395\,\text{gal}}{5\,\text{gal/min}}$$
$$= 1,879\,\text{min} \qquad \blacksquare$$

It will take 1,879 min, or 31 hr 19 min, to fill the pool.

▶ **You Try It** **2. a.** How many cubic feet of water can the cylindrical cistern hold? Round to units.

b. How many gallons is this?

c. How long will it take to fill the cistern using a pump that yields 13 gal/min? Round to units.

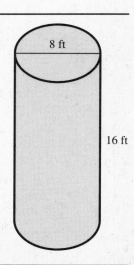

8 ft

16 ft

---

3 Volume of a Sphere

A **sphere** is pictured below. A sphere is the set of points the same distance from a fixed point in 3-dimensional space. A basketball and a marble are examples of spheres. The radius R of a sphere is the distance from the center to any point on its surface.

sphere with radius R

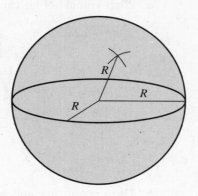

The volume of a sphere with radius R is

$$V = \frac{4}{3}\pi R^3.$$

EXAMPLE 3   Find the volume of a 10-inch sphere. Round to units.

A 10-inch sphere has a diameter of 10 in. Its radius is 10 in $\div$ 2 = 5 in.

$$V = \frac{4}{3}\pi R^3$$

$$= \frac{4}{3} \cdot 3.14 \cdot (\boxed{5 \text{ in}})^3$$

$$= \frac{4}{3} \cdot \frac{3.14}{1} \cdot \frac{125 \text{ in}^3}{1}$$

$$= \frac{1{,}570}{3} \text{ in}^3$$

$$\doteq 523 \text{ in}^3$$

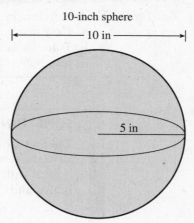

10-inch sphere

◄ **You Try It**   **3.** Find the volume of the 4-ft sphere at the right. Round to tenths.

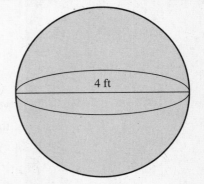

4 Applications A **hemisphere** is half of a sphere. Therefore, its volume is half the volume of a sphere.

EXAMPLE 4 A city in the form of a hemisphere is constructed under the ocean. The city has a 50 ft radius.

a. What volume of air can the city hold? Round to thousands.

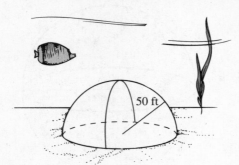

$$V = \frac{1}{2} \cdot \left(\frac{4}{3} \pi R^3 \right)$$

The volume of a hemisphere is half the volume of a sphere.

$$= \frac{1}{\overset{1}{2}} \cdot \frac{\overset{2}{4}}{3} \cdot 3.14 \cdot (50 \text{ ft})^3$$

$$= \frac{2}{3} \cdot \frac{3.14}{1} \cdot \frac{125,000 \text{ ft}^3}{1}$$

$$\doteq 262,000 \text{ ft}^3 \text{ of air.}$$

b. Three people live and work in this city. How long will the air last if the average person uses 1 cubic foot of air per minute?

Three people will use 3 cubic feet of air per minute. The air will last about

$$262,000 \text{ ft}^3 \div 3 \text{ ft}^3/\text{min} \doteq 87,333 \text{ min.}$$

This is about 61 days, or 2 months. This information means that the air replenishment system must be set to replace the complete volume of air in the city once every 2 months. ■

▶ **You Try It** **4.** A container in the shape of a hemisphere has an 8-ft radius.
 a. How many cubic feet of water can it hold?
 b. Water is used at a rate of 16 cubic feet per hour. To the nearest hour, how long will it take to empty the container?

SOMETHING MORE

The Soft Drink Problem Two cups are shown in the figure. They have the same height H and radius R. The cylindrical cup on the left is filled with soda. It costs 90¢. The cup on the right is a cone. It is also filled with soda. What is a fair price for it?

$$V = \pi R^2 H$$

Cylinder 90¢

H

Cone ?¢

$$V = \frac{1}{3} \pi R^2 H$$

A common answer is 45¢. But this is too much. The formulas for volume are shown below each figure. Compare them. The formula for the volume of a cone is $\frac{1}{3}$ of the formula for the volume of the cylinder. So the cone holds $\frac{1}{3}$ the soda that the cylinder holds.

If the cylindrical drink costs 90¢, a fair price for the conical drink is $\frac{1}{3}$ of 90¢, or 30¢.

▶ **Answers to You Try It**   **1.** 1,539 ft³   **2. a.** 804 ft³   **b.** 6,014 gal
c. 463 min or 7 hr 43 min   **3.** 33.5 ft³   **4. a.** 1,072 ft³   **b.** 67 hr or 2 dy 19 hr

SECTION 10.7 EXERCISES

1 *Find the volume. Draw and label a diagram for each word problem.*

1.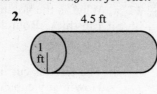

4 ft

8 ft

2.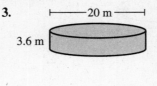

4.5 ft

1 ft

3.

├── 20 m ──┤

3.6 m

4.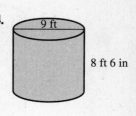

9 ft

8 ft 6 in

5. Find the volume of a cylinder with a 4″ radius and an 11″ height.

6. Find the volume of a cylinder with a 14 cm radius and a 10 cm height.

2

7. A cylindrical water tank is pictured in the figure.

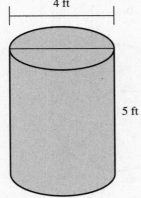

4 ft

5 ft

  a. How many cubic feet of water can it hold?

  b. How many gallons of water is this? Round to tenths.

8. A cylindrical oilcan is shown in the figure.

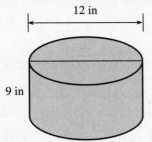

12 in

9 in

  a. How many cubic inches of oil can it hold?

  b. How many gallons of oil is this? Round to tenths. (1 gal ≐ 231 cu in)

9. The foundation for a circular fountain is a cylinder. The cylinder is 20 ft in diameter and 4 in high. How many cubic yards of concrete are needed to build the foundation?

10. A cylindrical manhole cover is 24″ in diameter and 1″ thick. How many cubic inches of steel are needed to make the cover?

11. A segment of oil pipeline is shown in the figure.

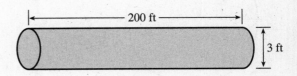

├──────── 200 ft ────────┤

3 ft

  a. How many cubic feet of oil can it hold?

  b. How many gallons of oil is this? (1 ft³ ≐ 7.48 gal)

12. A cylindrical hot water heater is shown in the figure.

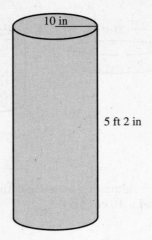

10 in

5 ft 2 in

a. What is its volume in cubic inches? (Hint: Change all measurements to inches.)

b. How many gallons is this? (See Problem 8b.)

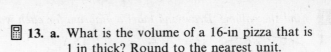 **13. a.** What is the volume of a 16-in pizza that is 1 in thick? Round to the nearest unit.

b. The pizza is cut into 8 equal pieces. You eat 3. How many cubic inches of pizza did you eat? Round to the nearest unit.

14. a. How many cubic yards of rock are needed to build a circular rock garden 12 ft in diameter and 6 in deep? Round to units.

b. What is the cost of the rock at $30 per cubic yard?

---

3 *Compute the volume. Round to hundredths, unless otherwise specified. Draw and label a diagram for each word problem.*

15.

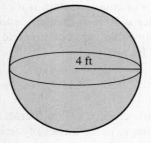

4 ft

16.

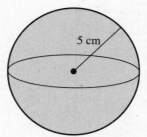

5 cm

17.

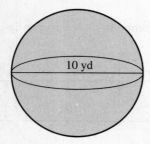

10 yd

18.

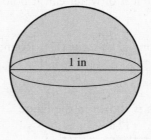

1 in

19. Find the volume of a sphere with a 1-ft radius. Round to thousandths.

20. Find the volume of a sphere with a radius of 7 in. Round to units.

21. What is the volume of a 40-ft sphere? Round to units.

22. What is the volume of a 5-ft spherical weather balloon? Round to tenths.

---

4 **23.** The sphere in the figure has a 2 ft 3 in radius.

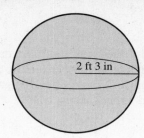

2 ft 3 in

a. Find its volume in cu ft. Round to tenths.

b. Find its volume in cu in. Round to hundreds.

24. Refer to the figure.

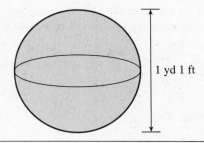

1 yd 1 ft

a. Find the volume in cu ft. Round to tenths.

b. Find the volume in cu in. Round to hundreds.

25. **a.** Find the volume of a ball bearing with a 0.125-in radius. Round to ten-thousandths.

b. How many such ball bearings can be made from 1 cu in of aluminum? Round to units.

26. A spherical space station is orbiting the Earth. It is 200 ft wide.

a. Find the volume of the station. Round to ten-thousands.

b. Assume the station can hold the volume of air found in part a. How long can 10 people live on this air if each uses 1 ft^3 of air per minute? Round to the nearest day.

SKILLSFOCUS (Section 5.10) *Simplify.*

27. $2.6 + 4 \cdot 3.8$ **28.** $0.2 \cdot 4.1 + 0.6 \cdot 3.5$ **29.** $(1.2)^2 + (2.5)^2$ **30.** $4.8 \div 2.4 + 2.4$

EXTEND YOUR THINKING ▶▶▶▶
▶SOMETHING MORE

31. The Well Problem Homes living with well water have water pumped from the well into a holding tank in the home.

a. Find the volume of the holding tank in cubic feet. Round to hundredths.

b. How many gallons does this tank hold when full? Round to units.

c. Suppose the well supplies water to the house at a rate of 3 gal/min. A lawn sprinkler system is using water at a rate of 5 gal/min. How long will it take to empty a full tank?

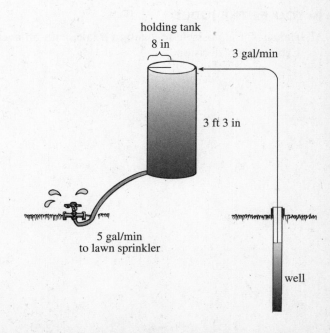

holding tank
8 in

3 gal/min

3 ft 3 in

5 gal/min
to lawn sprinkler

well

32. Water for a small town is stored in a spherical tank sitting on top of a pedestal.

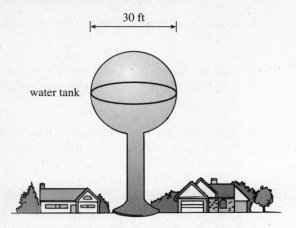

30 ft

water tank

a. How many cubic feet of water can the tank hold?

b. How many gallons is this? Round to hundreds.

c. The city uses 1 full tank per day. If the city must make $1,500 per tank to cover its costs, what must it charge per 100 gallons of water? Round to the nearest cent.

33. a. Find the volume of a sphere with a 3-ft radius.

b. Find the volume of a sphere with a 6-ft radius.

c. How many times does the volume of a sphere increase when you double the radius?

WRITING TO LEARN ▶▶▶▶

34. Write a word problem to find the volume of a cylindrical oil tank. The answer to the nearest unit must be 3,140 ft³. Then solve it.

35. How are a cylinder and a sphere alike? How are they different?

36. Write a word problem to find the volume of any kind of ball in your home. Identify the kind of ball you are using. Clearly explain how you will measure its diameter. Then solve the problem.

▶ YOU BE THE JUDGE

37. Hal says it is possible to swim in a lake with an average depth of 1 inch. Is he correct? Explain your decision.

CALCULATOR TIPS
Using Your Calculator to Find Area and Volume

EXAMPLE 1 Find the area of the rectangle in square feet.

4 ft 5 in

7 ft 8 in

$A = L \cdot W$
$= (\,7 \text{ ft } 8 \text{ in}\,) \cdot (\,4 \text{ ft } 5 \text{ in}\,)$

Now use your calculator. Change inches to feet by dividing inches by 12.

$8 \,÷\, 12 \,+\, 7 \,=\, \boxed{\text{M+}}$ ⟵ Change the 7 ft 8 in length to feet, and store in memory.

$5 \,÷\, 12 \,+\, 4 \,=$ ⟵ Change the 4 ft 5 in width to feet.

$\boxed{×}\,\boxed{\text{MR}}\,\boxed{=}\, 33.861111 \text{ ft}^2$ ⟵ Multiply the width on the display times the length in memory. ■

EXAMPLE 2 Find the volume of the sphere.

8.5 cm

$V = \frac{4}{3}\pi R^3$

$= \frac{4}{3} \cdot 3.14 \cdot (8.5 \text{ cm})^3$

$= 4 \,÷\, 3 \,×\, 3.14 \,×\, 8.5 \,×\, 8.5 \,×\, 8.5 \,=\, 2{,}571.1365 \text{ cm}^3$

OBSERVE By writing (8.5 cm)³ as 8.5 × 8.5 × 8.5, you can apply the order of operations and evaluate all multiplications and divisions in order from left to right. See Section 2.4. ■

Solve each problem using your calculator.

1. Find the area of each rectangle in square feet. Round to tenths.

a.

4 ft 1 in

13 ft 11 in

b.

3 ft 10 in

5 ft 2 in

2. Find the volume of each sphere and round to tenths.

a.

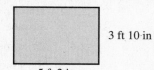

2.6 ft

b.

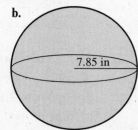

7.85 in

CHAPTER **10** REVIEW

VOCABULARY AND MATCHING

New words and phrases introduced in this chapter are shown in the left-hand column. Match each term on the left with the phrase or sentence on the right that best describes it.

Angles and Lines

A. point   _____ perpendicular to the horizontal

B. line   _____ 90°

C. line segment   _____ between 90° and 180°

D. ray   _____ always the same distance apart

E. horizontal line   _____ starts at one point, extends indefinitely in one direction

F. vertical line   _____ an endpoint shared by two rays or two line segments

G. intersecting lines   _____ intersect at a right angle

H. parallel lines   _____ extends indefinitely in two directions

I. angle   _____ 180°

J. vertex point   _____ less than 90°

K. right angle   _____ any two lines that cross at one point

L. acute angle   _____ two rays with a common starting point

M. obtuse angle   _____ two endpoints plus all the points in between

N. straight angle   _____ a position; has no width

O. perpendicular lines   _____ perfectly level; the ocean's horizon looks like this

Geometric Figures

A. polygon   _____ a rectangle with four equal sides

B. perimeter   _____ half the diameter

C. triangle   _____ amount of space within a 3-dimensional figure

D. rectangle   _____ made up of simpler figures whose area and perimeter formulas are known

E. square   _____ a geometric solid whose length, width, and height are all equal

F. area   _____ the "perimeter" of a circle

G. parallelogram   _____ the set of points equally distant from a fixed point in 3-dimensional space

H. composite geometric figure   _____ the distance around a polygon

I. right triangle   _____ set of points equally distant from the same point on a 2-dimensional surface

J. circle   _____ circumference ÷ diameter

K. radius   _____ the amount of surface within a 2-dimensional figure

L. diameter   _____ a box or brick, for example

M. circumference   _____ opposite sides are equal and parallel; has 4 angles; opposite angles are equal

N. pi or π   _____ a 2-dimensional geometric figure with three or more line segments for sides

O. volume   _____ half of a baseball, for example

P. rectangular solid   _____ a soup can is shaped like this

Q. cube   _____ opposite sides are equal and parallel; has four right angles

R. cylinder   _____ one of its three angles is a 90° angle

S. sphere   _____ has three angles; their sum is 180°

T. hemisphere   _____ two times the radius

REVIEW EXERCISES

10.1 Basic Concepts

State which pairs of lines are parallel or perpendicular; state which lines are horizontal or vertical.

1.
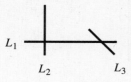

2.
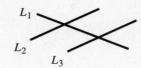

3.
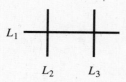

4.
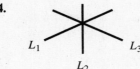

Classify each angle as acute, right, obtuse, or straight.

5. 91°   **6.** 180°   **7.** 90°   **8.** 29°

9.

10.

11.
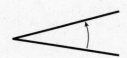

12.

10.2 Perimeter

Find the perimeter of each figure.

13.
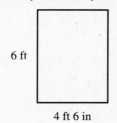
6 ft
4 ft 6 in

14.
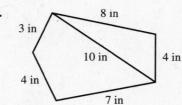
3 in 8 in 10 in 4 in 4 in 7 in

15.
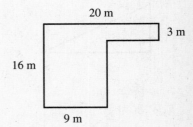
20 m 3 m 16 m 9 m

16. Find the perimeter of a rectangular garden 12 ft 6 in long by 15 ft wide.

17. Ceiling molding is $1.29 per linear foot. What will it cost to purchase molding for a 12 ft by 13 ft dining room?

18. Find the perimeter of a square boxing ring with 14 ft 9 in on each side.

10.3 Area

Find the area and perimeter.

19.

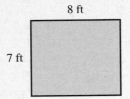

8 ft

7 ft

20.

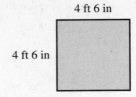

4 ft 6 in

4 ft 6 in

21.

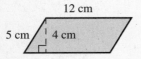

12 cm

5 cm   4 cm

22.

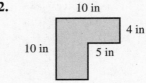

10 in

4 in

10 in

5 in

23. What is the area of a 15-in square?

24. What is the area of a 9 ft by 13 ft rectangle?

25. A 12 ft by 15 ft rectangular kitchen floor is tiled at a cost of $0.79 per tile. Each tile covers 1 ft². What is the total cost for tile?

27. A rectangular lot measures 125 ft by 400 ft. How many acres is this to the nearest hundredth? (1 ac = 43,560 ft²)

26. What will it cost to carpet a square family room 18 ft on a side, if carpet is $22.50 per sq yd?

28. How many sq ft are in a 0.31 ac lot? Round to tens.

10.4 Triangles

29. Find the missing angle.

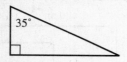

35°

30. Find the missing angle.

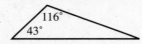

116°

43°

Find the area.

31.

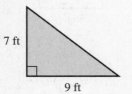

7 ft

9 ft

32.

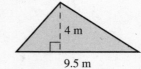

4 m

9.5 m

33.

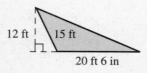

12 ft   15 ft

20 ft 6 in

34.

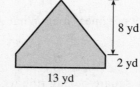

8 yd

2 yd

13 yd

35. Find the area of a triangular garden with a base of 40 ft and a height of 9 ft.

36. A triangular lot has a 50-yd base and a 40-yd height.

    a. Find the area in square yards.

    b. Find the area in square feet.

10.5 Circles

Find the circumference of each circle.

37.

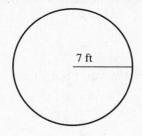

7 ft

38.

9.6 mm

Find the area of each circle.

39.

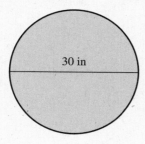

30 in

40.

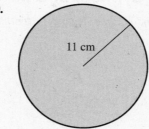

11 cm

41. Find the circumference of a circle with a 50-ft radius.

42. Find the area of a circle with a 6.8-yd radius. Round to tenths.

43. Find the area of the circle if the square has a perimeter of 20 in.

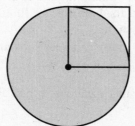

44. The shaded wedge in the circular arena is to be covered with oak flooring.

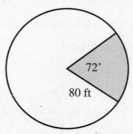

72°

80 ft

    a. What is the area of the wedge? (Hint: The wedge is $\frac{72}{360}$ of a circle.)

    b. What is the cost at $48.50 per square yard installed?

10.6 Volume: The Rectangular Solid

Find the volume.

45.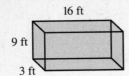

16 ft
9 ft
3 ft

46.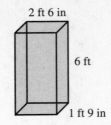

2 ft 6 in
6 ft
1 ft 9 in

47. How many cu in of lead are needed to make a rectangular anchor 8.5 in long, 6.2 in wide, and 10 in high?

a. How many cubic yards of concrete will he need?

48. Tom is building a concrete patio 40 ft long, 24 ft wide, and 6 in deep.

b. What will the concrete cost at $68.50 per cu yd?

10.7 Volume: The Cylinder and the Sphere

In problems 49–52, find the volume. Round to tenths.

49.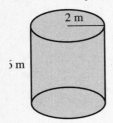

2 m
5 m

50.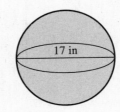

17 in

51.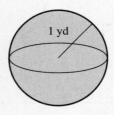

1 yd

52.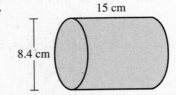

15 cm
8.4 cm

53. A cylindrical oil tank is 4 m long and has a radius of $\frac{1}{2}$ m. How many cubic meters of oil can the tank hold?

54. How many cu ft of air can a spherical diving bell hold if its diameter is 10 ft? Round to tenths.

55. A spherical ball bearing has a radius of 0.4 cm. Find its volume? Round to thousandths.

56. A child's swimming pool is in the shape of a cylinder. Its diameter is 8 ft, and its height is 1 ft 6 in.

a. How many cubic feet of water can it hold?

b. How many gallons is this? Round to tenths.

c. How long will it take to fill the pool from a hose that yields 6 gal of water per minute?

Allow yourself 50 minutes to complete this test. Write the work for each problem you solve. When done, check your answers. Rework each problem solved incorrectly.

Classify each angle as acute, right, obtuse, or straight.

1. 155°

2. 88°

3.

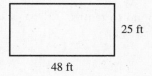

4.

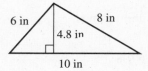

5. Find the perimeter.

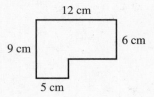

25 ft
48 ft

6. Find the perimeter.

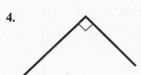

6 in   8 in
4.8 in
10 in

7. Find the perimeter.

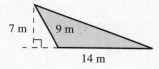

12 cm
9 cm   6 cm
5 cm

8. Find the area.

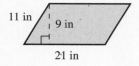

11 in   9 in
21 in

9. Find the area.

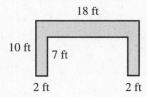

7 m   9 m
14 m

10. Find the area.

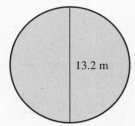

18 ft
10 ft   7 ft
2 ft   2 ft

11. Find the circumference.

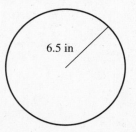

6.5 in

12. Find the area. Round to tenths.

13.2 m

13. Find the area of a square 8 ft 6 in on each side.

14. Ceiling molding is installed in a rectangular room 12 ft long by 14 ft wide. What will the molding cost at $1.79 per linear foot?

15. Carpet is on sale for $18.75 a sq yd. What will it cost to carpet a den 21 ft long by 15 ft wide?

16. The stage in the figure will be covered with hardwood flooring at a cost of $2.60 per sq ft. Find the total cost. The front of the stage is a semicircle.

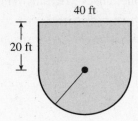

40 ft

20 ft

17. Find the volume. Round to units.

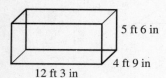

5 ft 6 in

4 ft 9 in

12 ft 3 in

18. Find the volume. Round to tenths.

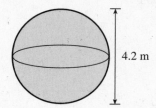

4.2 m

19. How many cu ft of oil can the tank in the figure hold?

10 ft

7 ft

20. A concrete landing pad for a helicopter will be 60 ft square and 18 in deep. What is the cost for the concrete at $67 per cu yd?

Explain the meaning of each term. Use your own examples.

21. perimeter

22. right angle

23. radius

24. circumference

NAME   DATE   HOUR

1. What is the quotient and remainder when you divide $4{,}286 \div 73$?

2. Myra wrote checks for $246, $38, $793, $2,809, and $7. Find the total.

3. Subtract $4\dfrac{3}{8} - 1\dfrac{5}{6}$.

4. Multiply $\dfrac{2}{7} \cdot 5\dfrac{1}{4}$.

5. You have $16\dfrac{2}{3}$ oz of an antibiotic.

   a. How many $\dfrac{4}{5}$-oz dosages can be made from the antibiotic?

   b. What fraction of a dose will be left over?

6. Find $4.017 - 3.8$.

7. Divide $105.69 \div 26$.

8. Change 0.88 to a fraction.

9. Simplify $0.6 + 0.4 \cdot 5$.

10. Express 3.6 to 0.072 as a ratio of two whole numbers.

11. Split $4,590 according to a $13:5$ ratio.

12. Solve $\dfrac{9}{24} = \dfrac{7}{w}$. Round to the nearest hundredth.

13. What is 56% of $295?

14. 36 oz is 75% of what?

15. Last year Emma made $24,215. This year she made $26,152.20. What was the percent increase in her earnings?

16. How much will $600 earn at 9% simple interest over 3 years?

17. The number of children in 16 randomly sampled families is given here.

$$2, 0, 1, 2, 3, 0, 1, 5$$
$$2, 1, 0, 3, 2, 2, 2, 4$$

  a. Construct a tally for the 16 numbers.

  b. What is the mode?

  c. Is the median equal to the mode?

  d. What is the mean number of children per family in the sample?

18. 6,300 mg = ? g   **19.** 0.045 kl = ? cl

20. 500 g = ? kg   **21.** 68 mm = ? cm

22. What is the perimeter of a rectangular yard 50 m long by 75 m wide?

23. What is the area of a room 12 ft long by 30 ft wide?

24. Find the circumference.   **25.** Find the area.

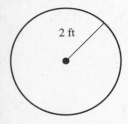
2 ft

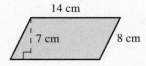
14 cm
7 cm
8 cm

Explain the meaning of each term. Use your own examples.

26. mean   **27.** polygon

The Integers

Have you heard of pinochle or rummy? They are examples of games where you may at times have a negative score. When you keep score in such games, you are adding positive and negative numbers, called integers.

What is each team's score in the game below?
Who has the higher score?

WE	THEY
10	−15
+15	+10
25	
−30	
−5	
−15	
−20	
+10	

The last two chapters of this text are devoted to algebra. In this chapter, you will study operations on a new set of numbers called integers. The integers are positive and negative whole numbers, including 0. You already have seen and used integers. A negative temperature, such as $-10°$, is an example of an integer. A checking account that is overdrawn by fifteen dollars can have its balance represented by the negative integer $-\$15$. Have you ever kept score in a game and found yourself with a negative score? If so, then you have already performed addition of integers. In this chapter you will learn to add, subtract, multiply, and divide integers, with applications.

Work through each section in this chapter. Use this test to identify topics you are not familiar with. These topics may require additional study. You may *not* use a calculator.

11.1

1. $|-26|$

2. $-(-3)$

3. Circle the larger integer. $-6 \quad -8$

11.2

4. $(-7) + (-4)$

5. $(-6) + 6 + (+3) + (-7)$

6. Miguel had a score of -36 in a game. An hour later, he had gained 68 points. What was his new score?

11.3

7. $(-6) - (-10)$

8. $9 - 12 + 6 - (-2)$

11.4

9. $(-4)(5)$

10. $6(-10)$

11. $(-7)(-5)$

12. $14 \div (-7)$

13. $\dfrac{-100}{-25}$

14. $(-6)(+4)(-1)$

15. Find the mean and range for the temperatures.
$14°, -20°, 0°, -12°, 8°, -20°, 2°$

11.5

16. $(-5)^2$

17. -8^2

18. $60 \div (-30)(-2)$

19. $6 + 6(-3)$

20. Evaluate each expression using $X = -4$ and $Y = 7$.
   a. $2Y - 3X$
   b. $4(X^2 + Y)$

11.6

21. Change to decimal notation.
   a. 5.31×10^5
   b. 2.07×10^{-3}

22. Change to scientific notation.
   a. 0.000604
   b. $2,870,000,000$

VOCABULARY *Explain the meaning of each term. Use your own examples.*

23. absolute value

24. negative integer

11.1 INTEGERS AND THE NUMBER LINE

OBJECTIVES	NEW VOCABULARY	
1 Define the integers (Section 1.1).	negative integers	negative sign
2 Compare integers.	integers	absolute value
3 Find the absolute value of an integer.	positive sign	opposite
4 Find the opposite of an integer.		

1 The Integers   The number line was introduced in Chapter 1.

However, there are numbers you cannot represent on this line. For example, a winter temperature of 5 degrees below zero. Or a checkbook balance that is $40 overdrawn.

These numbers are less than 0. To represent them on the number line, extend the number line to the left of 0. These new numbers are called **negative integers**.

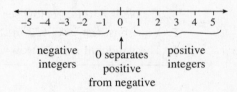

A **positive sign** (+) means a number is to the right of 0. A number without a sign is understood to be positive.

$$4 = +4$$
$$7 = +7$$

A **negative sign** (−) means a number is to the left of 0.

The numbers . . . , −5, −4, −3, −2, −1, 0, 1, 2, 3, 4, 5 , . . . are called the **integers**. Integers are also called signed numbers. The integer 0 is neither positive nor negative.

EXAMPLE 1   3 is read three, or positive three.
+9 is read positive nine, or just nine.
−7 is read negative 7.   ■

▶ **You Try It**   Read each integer, and write it in words.

1. 8   **2.** +4   **3.** −12

EXAMPLE 2   Negative integers are used in everyday life.

a. The lowest point in Death Valley, California, is 282 ft below sea level, written −282 ft.
b. A checkbook balance overdrawn $40 is written −$40.

$20	A positive $20 balance means you are *in the black*.
−$60	You write a check for $60.
−$40	A negative balance means you are *in the red*.

c. A temperature 5 degrees below zero is written $-5°$.

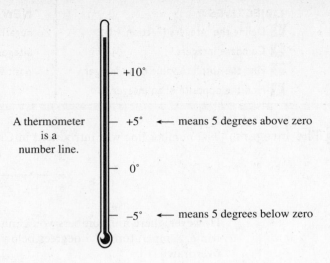

A thermometer
is a
number line.

+10°

+5° ⟵ means 5 degrees above zero

0°

−5° ⟵ means 5 degrees below zero

▶ **You Try It**   Write an integer for each statement.

4. A cave is 410 feet below sea level.

5. Your business is $3,522 in the red.

6. The windchill is 28° below zero.

2 Comparing Integers   Given two integers, the smaller one is farther to the left on the number line. The larger one is farther to the right.

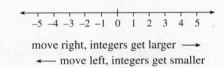

−5 −4 −3 −2 −1 0 1 2 3 4 5

move right, integers get larger ⟶
⟵ move left, integers get smaller

EXAMPLE 3   Comparing integers.

a. $-4 < -2$: -4 is smaller because it is farther to the left on the number line. (Which temperature is lower, $-4°$ or $-2°$? Since $-4°$ is lower, write $-4° < -2°$.)

b. $-1 > -5$: -1 is larger because it is farther to the right on the number line. (Which is the higher checkbook balance, $-\$1$ or $-\$5$? Since $-\$1$ is higher, write $-\$1 > -\$5$.)

c. $0 > -3$:   0 is larger because it is farther to the right on the number line. (A score of 0 in a game is higher than a score of -3.)   ■

▶ **You Try It**   Write < or > between the integers.

7. -6 -8 **8.** -4 0 **9.** -9 -3 **10.** 1 -12

3 Absolute Value

The **absolute value** of an integer n is written $|n|$. It is the distance n is from 0 on the number line. Since distances are positive, the absolute value of any number is always positive.

For example, $|-5|$ is read "the absolute value of -5."

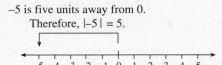

-5 is five units away from 0.
Therefore, $|-5| = 5$.

EXAMPLE 4
a. $|-3| = 3$ because -3 is 3 units from 0 on the number line
b. $|+7| = 7$ because $+7$ is 7 units from 0 on the number line
c. $4 + |-6| = 4 + 6 = 10$ ■

▶ **You Try It**   Evaluate each absolute value.

11. $|-8|$ **12.** $|-2|$ **13.** $|9|$ **14.** $|+3|$

4 The Opposite of a Number

The **opposite** of a number is a second number the same distance from 0 on the number line, but on the opposite side. The opposite of $+4$ is -4. -4 and $+4$ are both 4 units from 0 on the number line, but on opposite sides.

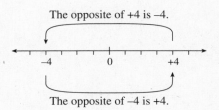

The opposite of +4 is –4.

The opposite of –4 is +4.

Opposite numbers have the same absolute value. For example, $|-4| = 4$ and $|+4| = 4$.

EXAMPLE 5
The opposite of $+7$ is -7.⎫
The opposite of 5 is -5.⎭   The opposite of a positive number is negative.

The opposite of -2 is $+2$, or just 2.⎫
The opposite of -100 is $+100$, or 100.⎭   The opposite of a negative number is positive.

The opposite of 0 is 0.⎭   0 is its own opposite.
Note, $0 = -0 = +0$. ■

▶ **You Try It**   Find the opposite of each integer.

15. $+20$ **16.** -14 **17.** -1 **18.** 300

A negative sign, $-$, can be read "the opposite of." For example, -5 is read "the opposite of 5." Then,

$$-(-5) = \text{the opposite of } -5 = +5$$

or,

$$-(-5) = +5.$$

EXAMPLE 6   **a.** $-(+9) =$ the opposite of $+9 = -9$ So, $-(+9) = -9.$

b. $-(+15) = -15$

c. $-(-7) =$ the opposite of $-7 = +7$ So, $-(-7) = 7.$

d. $-(-36) = 36$ ∎

▶ **You Try It**   Simplify.

19. $-(+4)$ **20.** $-(-8)$ **21.** $-(-48)$ **22.** $-(+2)$

▶ **Answers to You Try It** **1.** eight, or positive eight **2.** positive four, or four
3. negative 12 **4.** -410 ft **5.** $-\$3,522$ **6.** $-28°$ **7.** > **8.** < **9.** <
10. > **11.** 8 **12.** 2 **13.** 9 **14.** 3 **15.** -20 **16.** 14 **17.** 1 **18.** -300
19. -4 **20.** 8 **21.** 48 **22.** -2

SECTION 11.1 EXERCISES

1 *Write a positive or negative integer for each statement.*

1. The temperature is 40 degrees above zero.

2. Today's low temperature in Nome, Alaska, was 54 degrees below zero.

3. The submarine is 35 ft below sea level.

4. The plane is 33,000 ft above sea level.

5. Tom's checkbook balance is $110 in the black.

6. Alice scored 28 points in the last hand.

7. George lost 23 points this hand.

8. Deduct a check for $29.50 from your checkbook balance.

9. As of today, his business is operating $4,000 in the red.

10. The elephant's weight dropped 36 pounds in one week.

2 **3** *Find the absolute value, and simplify.*

11. $|-13|$ **12.** $|7|$ **13.** $|+15|$ **14.** $|-25|$

15. $5 + |-3|$ **16.** $|+7| + 7$ **17.** $8 + |-8|$ **18.** $|-50| + 50$

19. $|-1| + |-3| + |-6|$ **20.** $|-10| + 10 + |10|$

Insert the correct symbol $(<, >, =)$ between the integers.

21. $-3 \quad -5$ **22.** $-6 \quad -2$ **23.** $-4 \quad +3$ **24.** $-7 \quad +10$

25. $-30 \quad -25$ **26.** $-50 \quad +1$ **27.** $|-3| \quad -3$ **28.** $|0| \quad 0$

29. $|-9| \quad 9$ **30.** $2 \quad |-5|$ **31.** $0 \quad -1$ **32.** $17 \quad -62$

33. $-3 \quad -4$ **34.** $-23 \quad -19$ **35.** $|6| \quad 8$ **36.** $|-2| \quad -8$

4

37. The opposite of 8 is _____.

38. The opposite of -8 is _____.

39. The opposite of -45 is _____.

40. The opposite of $+19$ is _____.

41. $-(+6)$

42. $-(+10)$

43. $-(-6)$

44. $-(-38)$

45. $-(-(-7))$

46. $-(+(-(+8)))$

SKILLSFOCUS (Section 7.3) *Change each fraction to a percent.*

47. $\dfrac{3}{4}$

48. $\dfrac{1}{8}$

49. $\dfrac{5}{6}$

50. $3\dfrac{1}{2}$

EXTEND YOUR THINKING ▶▶▶▶
▶ TROUBLESHOOT IT
Find and correct the error.

51. $|-9| + |-6| = 9 - 6 = 3$

52. $-(-9) = -(+9) = -9$

WRITING TO LEARN ▶▶▶▶

53. If two numbers are opposites, explain why their absolute values are equal.

54. Explain how the number line is used in the launch of the space shuttle. Let 0 be the moment of lift-off. Then what will negative numbers, such as -5 and -3, represent? What will positive numbers, such as $+1$ and $+60$, represent?

11.2 ADDITION OF INTEGERS

OBJECTIVES

1 Add integers with like signs (Section 1.3).

2 Add integers with unlike signs (Section 1.4).

NEW VOCABULARY

additive inverse

Have you ever had a negative score in a game? Have you ever found a negative balance in your checkbook? If so, then you have added integers.

For example, check the scores shown here. Based on this, can you state some rules for adding integers?

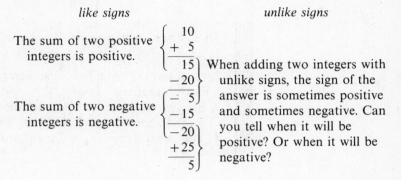

like signs

The sum of two positive integers is positive.
$$\begin{array}{r} 10 \\ +\ 5 \\ \hline 15 \end{array}$$

The sum of two negative integers is negative.
$$\begin{array}{r} -20 \\ -\ 5 \\ \hline -15 \end{array}$$

$$\begin{array}{r} -20 \\ +25 \\ \hline 5 \end{array}$$

unlike signs

When adding two integers with unlike signs, the sign of the answer is sometimes positive and sometimes negative. Can you tell when it will be positive? Or when it will be negative?

1 Adding Integers with Like Signs

EXAMPLE 1 Different ways to write addition of signed numbers.

a. $(+7) + (+5)$ is read positive 7 added to positive 5

- $(+7)$ — is read positive 7
- $+$ — means addition
- $(+5)$ — is read positive 5

OBSERVE $(+7) + (+5)$ is another way to write $7 + 5$.

b. $(-2) + (-9)$ is read negative 2 added to negative 9

- (-2) — negative 2
- $+$ — addition
- (-9) — negative 9

c. $6 + (-8)$ is read 6 added to negative 8

- 6 — six, or positive 6
- $+$ — addition
- (-8) — negative 8

d. $-4 + 9 = (-4) + (+9)$ ← You may rewrite the problem this way. This is read negative 4 added to 9.

- -4 — negative 4
- $+$ — addition
- 9 — positive 9 ■

▶ **You Try It** Write each addition in words.

1. $(+2) + (+6)$ **2.** $(-3) + (-4)$ **3.** $1 + (-9)$ **4.** $-12 + 4$

EXAMPLE 2   Add $(+7) + (+5)$ using the number line.

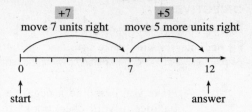

Therefore, $(+7) + (+5) = +12$. ∎

> To add two integers using the number line
>
> 1. Start at 0.
> 2. Count to the right if the first addend is positive. Count to the left if it is negative.
> 3. Starting where you left off with the first addend, count right if the second addend is positive. Count left if it is negative.
> 4. Your final stopping point is the answer.

EXAMPLE 3   Add $-4 + (-2)$ using the number line.

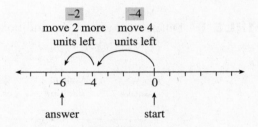

Therefore, $-4 + (-2) = -6$. ∎

▶ **You Try It**   Add using the number line.

5. $(+3) + (+5)$ **6.** $+1 + (+5)$ **7.** $(-8) + (-3)$ **8.** $-6 + (-7)$

In Examples 2 and 3, the sum has the same sign as the addends.

> To add two integers with like signs
>
> 1. Add the absolute values of the two integers.
> 2. If both integers are positive, the sum is positive.
>   If both integers are negative, the sum is negative.

EXAMPLE 4   Add $(+7) + (+9)$.

Signs are the same.
Add the absolute values.

$$|+7| + |+9| = 7 + 9 = 16$$

$$(+7) + (+9) = +16$$

Both integers are positive,
so the sum is positive. ∎

EXAMPLE 5   Add $-6 + (-8)$.

The signs are the same.
Add the absolute values.

$$|-6| + |-8| = 6 + 8 = 14$$

$$(-6) + (-8) = -14$$

Both integers are negative,
so the sum is negative. ∎

▶ **You Try It**   Add the integers.

9. $(+4) + (+11)$ **10.** $(+5) + (+5)$ **11.** $-7 + (-9)$ **12.** $(-8) + (-5)$

2 Adding Integers with Unlike Signs

EXAMPLE 6   Add $(-5) + (+8)$ using the number line.

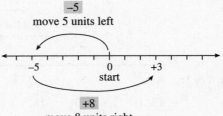

Therefore, $(-5) + (+8) = +3$. ∎

EXAMPLE 7   Add $6 + (-10)$ using the number line.

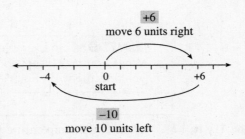

Therefore, $6 + (-10) = -4$. ∎

Add using the number line.

13. $(+2) + (-6)$ **14.** $(-3) + (+4)$ **15.** $-1 + (+7)$ **16.** $-12 + 4$

In Examples 6 and 7, the sum has the sign of the integer with the larger absolute value.

> **To add two integers with unlike signs**
> 1. Subtract the smaller absolute value from the larger.
> 2. Give this difference the sign of the integer with the larger absolute value.

EXAMPLE 8   Add $(-6) + (+5)$.

The signs are unlike.
Subtract the absolute values.

$$|-6| = 6, |+5| = 5, \text{ and } 6 - 5 = 1$$

$$(-6) + (+5) = -1$$

Since $6 > 5$, -6 has the larger absolute value.
Use the negative sign for the answer.

EXAMPLE 9   Add $8 + (-3)$.

The signs are unlike.
Subtract the absolute values.

$$|8| = 8, |-3| = 3, \text{ and } 8 - 3 = 5$$

$$8 + (-3) = +5$$

Since $8 > 3$, 8 has the greater absolute value.
Use the positive sign for the answer. ■

▶ **You Try It**   Add.

17. $(+2) + (-9)$ **18.** $(+13) + (-4)$ **19.** $14 + (-6)$ **20.** $(-18) + (+15)$

EXAMPLE 10   Rita's score in a game was -6. She just made 10 points. What is her new score? Rita's new score is her old score, -6, plus 10.

$$\text{new score} = -6 + 10 = (-6) + (+10)$$
$$= 4 \text{ points} ■$$

> **CAUTION**   In Example 10, it is an error to write $-6 + 10 = -16$. Why? You may insert your own parentheses around integers to help you add them.

▶ **You Try It**   **21.** At 6 A.M., the temperature was $-15°$. By noon the temperature rose $9°$. What was the noon temperature?

Add an integer to its opposite and you get 0. The opposite of -8 is $+8$, and

$$(-8) + (+8) = 0.$$

> The sum of an integer and its opposite is 0.
> The opposite of an integer is called its **additive inverse**.

EXAMPLE 11
a. The opposite of 6 is (-6) and $6 + (-6) = 0$.
-6 is also called the additive inverse of 6.

b. The opposite of -1 is $+1$ and $-1 + (+1) = 0$.
$+1$, or 1, is the additive inverse of -1. ■

▶ **You Try It**   Find the opposite of each integer. Then show that the sum of each integer and its opposite is 0.

22. 5   **23.** -8 **24.** -2 **25.** $+7$

Adding Three or More Integers
To add three or more integers together, follow the order of operations. Add the two integers on the left. Then add the next integer to this sum. Continue this way until all integers have been added.

EXAMPLE 12   Add $(+6) + (-11) + (+8)$.

Add $+6$ to -11.
You get -5.

Add -5 to $+8$.
You get $+3$.

$$(+6) + (-11) + (+8) = +3$$
$$-5$$
$$+3 \text{ (answer)} ■$$

EXAMPLE 13   Add $-12 + 3 + (-6) + 1$.

$$-12 + 3 + (-6) + 1 = -14$$
$$-9$$
$$-15$$
$$-14 ■$$

▶ **You Try It**   Add.

26. $-4 + 7 + (-8)$ **27.** $8 + (-10) + (-5) + 9$

EXAMPLE 14   **Chapter Problem**   Read the problem at the beginning of this chapter. The score is:

WE	THEY
-20	-15
$+10$	$+10$
-10	-5

Since $-10 < -5$, "THEY" are leading. ■

Summary

To add two integers

1. If the integers have the same sign, add the absolute values, keep the sign.
2. If the integers have different signs, subtract the smaller absolute value from the larger. Use the sign of the integer with the larger absolute value.

▶ **Answers to You Try It**   1. positive 2 added to positive 6
2. negative 3 added to negative 4   3. 1 added to negative 9
4. negative 12 added to 4   5. +8   6. +6   7. −11   8. −13   9. +15
10. +10   11. −16   12. −13   13. −4   14. +1   15. +6   16. −8
17. −7   18. +9   19. +8   20. −3   21. −6°   22. −5; 5 + (−5) = 0
23. +8; −8 + (+8) = 0   24. +2; −2 + (+2) = 0   25. −7; +7 + (−7) = 0
26. −5   27. +2

1 **2** *Write the addition in words, then add.*

1. $8 + 5$ **2.** $7 + 12$ **3.** $-6 + 4$ **4.** $-10 + 2$

5. $3 + (-4)$ **6.** $8 + (-11)$ **7.** $-5 + 16$ **8.** $-2 + 3$

9. $6 + (-4)$ **10.** $5 + (-1)$ **11.** $8 + (-8)$ **12.** $1 + (-1)$

13. $-7 + (-13)$ **14.** $-8 + (-2)$ **15.** $(-9) + (+5)$ **16.** $(4) + (-14)$

Add.

17. $-5 + (+5)$ **18.** $(-5) + (5)$ **19.** $-6 + 0$ **20.** $-2 + 0$

21. $0 + (-12)$ **22.** $0 + (+6)$ **23.** $(7) + (-12)$ **24.** $(-3) + (5)$

25. $-16 + 16$ **26.** $-4 + 4$ **27.** $12 + (-5)$ **28.** $13 + (-14)$

29. $-3 + 2$ **30.** $-12 + 3$ **31.** $(+2) + (+9)$ **32.** $(-8) + (-8)$

33. $-10 + (-12)$ **34.** $-1 + (-1)$ **35.** $15 + (-9)$ **36.** $-9 + (15)$

37. $7 + (-4)$ **38.** $-4 + (7)$ **39.** $-7 + 2$ **40.** $-8 + 12$

41. $-6 + (-12)$ **42.** $-2 + (-2)$ **43.** $+9 + 7$ **44.** $+4 + 9$

45. $-121 + 84$ **46.** $-82 + 106$ **47.** $-80 + 70$ **48.** $-80 + 90$

49. $-9 + (-4)$ **50.** $-2 + (-5)$ **51.** $20 + (-14)$ **52.** $10 + (-14)$

53. $40 + (-40)$ **54.** $80 + (-20)$ **55.** $(-65) + 52$ **56.** $(-33) + 50$

57. $-6 + (-3) + (-6)$ **58.** $7 + (-4) + 5$

59. $7 + (-10) + (4) + (-2)$ **60.** $-8 + (-5) + 6 + (-6)$

61. $-8 + 8 + 9 + (-9)$ **62.** $7 + 8 + (-7) + (-8)$

63. $-1 + (-1) + 1$

64. $6 + 9 + (-12)$

65. $-10 + (-12) + (-21) + (-15)$

66. $-23 + (-40) + 34 + (-18)$

67. The temperature in St. Helena at 6 A.M. was $-19°$. It is expected to rise $31°$ by noon. What is the expected noon temperature?

68. The midnight temperature in Nome, Alaska, was $-41°$. By noon the next day, it had risen $27°$. What was the noon temperature?

69. Tom's score in a game of cards is -16. He just won 25 points. What is his new score?

70. Maria's score in a game is -29. She just made 13 points. What is her new score?

71. Edna's checking account balance is $-\$42$. She deposits a \$27 check in her account. What is her new balance?

72. The county payroll office has a balance of $-\$250,680$. Tax revenues recorded this morning added \$1,078,000 to this account. What is the new balance?

SKILLSFOCUS (Section 9.5)

73. $12 \text{ cm} = ? \text{ mm}$

74. $8,200 \text{ g} = ? \text{ kg}$

75. $50 \text{ kl} = ? \ell$

76. $52,000 \text{ mg} = ? \text{ g}$

EXTEND YOUR THINKING ▶▶▶▶
▶ TROUBLESHOOT IT
Find and correct the error.

77. $-5 + 10 = -15$

78. $-4 + (-2) + (-8) = (-6) + (-8) = -2$

WRITING TO LEARN ▶▶▶▶

79. a. Explain how to add two integers with like signs.

b. Explain how to add two integers with unlike signs.

80. Use the number line, and the addition $(+5) + (-8)$, to show that addition of integers is commutative.

81. Explain why you start at 0 when adding integers using the number line.

82. In the CAUTION following Example 10, explain why $-6 + 10 = -16$ is an error.

11.3 SUBTRACTION OF INTEGERS

OBJECTIVES

1 Subtract integers (Section 1.4).

2 Applications.

1 Subtracting Integers

Addition of integer problems are written:

$$4 + 7$$
$$(-3) + (-5)$$
$$-6 + (+2)$$

addition sign ⌐

Subtraction of integer problems are written:

$$4 - 7$$
$$(-3) - (-5)$$
$$-6 - (+2)$$

subtraction sign ⌐

EXAMPLE 1 Write each subtraction problem in words.

a. $4 - 7$ is read **4 minus 7.** You can also write

$$4 - 7 = (+4) - (+7)$$
or $4 - 7 = (4) - (7)$

↑ means subtraction

b. $(-3) - (-5)$ is read **negative 3 minus negative 5**

negative 3   subtraction   negative 5

c. $-6 - (+2)$ is read **negative 6 minus positive 2**

negative 6   subtraction   positive 2 ∎

▶ **You Try It** Write each subtraction problem in words.

1. $4 - 9$ **2.** $(-10) - (-7)$ **3.** $-2 - (+5)$

You subtract integers by adding the opposite of the subtrahend.

For example, show that the subtraction $(+5) - (-3)$ gives the same answer as the addition $(+5) + (+3)$.

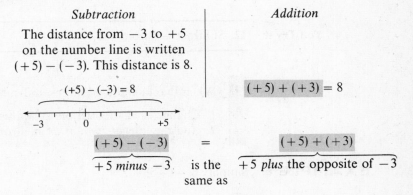

Subtraction

The distance from -3 to $+5$ on the number line is written $(+5) - (-3)$. This distance is 8.

$$(+5) - (-3) = 8$$

Addition

$$(+5) + (+3) = 8$$

$$\underbrace{(+5) - (-3)}_{+5 \text{ minus } -3} \quad = \quad \underbrace{(+5) + (+3)}_{+5 \text{ plus the opposite of } -3}$$

is the same as

To subtract two integers

1. Change the subtraction sign to addition.

2. Change the sign of the subtrahend to its opposite.

3. Add.

To subtract two integers, *add the opposite* of the subtrahend.

EXAMPLE 2 Find $-14 - (-9)$.

Change subtraction ⟶   ⟵ Change the sign of the
to addition.   subtrahend to its opposite.

$$-14 - (-9)$$
$$= -14 + (+9)$$
$$= \quad -5 \quad \blacksquare$$

EXAMPLE 3 Find $-20 - (+70)$.

Change subtraction ⟶   ⟵ Change the sign of the
to addition.   subtrahend to its opposite.

$$-20 - (+70)$$
$$= -20 + (-70)$$
$$= \quad -90 \quad \blacksquare$$

EXAMPLE 4 What is $10 - 18$?

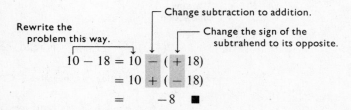

Rewrite the
problem this way.

Change subtraction to addition.

Change the sign of the
subtrahend to its opposite.

$$10 - 18 = 10 - (+18)$$
$$= 10 + (-18)$$
$$= \quad -8 \quad \blacksquare$$

▶ **You Try It**

4. $-18 - (-12)$	**5.** $-4 - (-7)$	**6.** $-15 - (+20)$
7. $-25 - (+16)$	**8.** $6 - 11$	**9.** $8 - 9$
10. $3 - (-10)$	**11.** $50 - (-20)$	

EXAMPLE 5 Subtract -16 from 12.

$$\text{subtract } -16 \text{ from } 12 = 12 - (-16)$$
$$= 12 + (+16)$$
$$= 28 \quad \blacksquare$$

▶ **You Try It**

12. Subtract -8 from -5.   **13.** Subtract -4 from 6.

> When two or more additions and subtractions occur together
>
> **1.** Change each subtraction to addition by adding the opposite.
> **2.** Add the integers two at a time from left to right.

EXAMPLE 6 Simplify $6 - (-5) + 2$.

Write the subtraction as addition of the opposite.

$$6 - (-5) + 2$$
$$= 6 + (+5) + 2 = 13$$
$$\underbrace{}_{+11}$$
$$\underbrace{}_{+13} \quad \blacksquare$$

EXAMPLE 7   Simplify $-5 - (-2) + 8 - (4) - 3$.

Write each of the three subtractions as addition of the opposite.

$$-5 - (-2) + 8 - (4) - 3$$
$$= -5 + (+2) + 8 + (-4) + (-3) = -2$$

-3

$+5$

$+1$

-2 ■

▶ **You Try It**   **14.** $8 - (-3) + 4$ **15.** $10 - (-7) - 5$

16. $-1 - (-2) + (-6) + 3$ **17.** $-8 - (-5) + 6 - 6 - (+5)$

2 Applications   What practical meaning can $5 - (-3)$ have? Where is such a subtraction used in real life? Suppose the temperature this morning was $-3°$. This afternoon the temperature was $5°$. How much did the temperature rise?

The rise in temperature is the higher temperature minus the lower.

rise $= 5° - (-3°)$
$= 5° + (+3°)$
$= 8°$

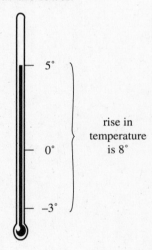

rise in temperature is 8°

To find the difference between two numbers on a ruler, thermometer, or any other form of number line, subtract the lower number from the higher.

Draw and label a number line to help you solve problems containing negative integers.

EXAMPLE 8   Ed's bank balance was $-\$27$ yesterday. After a deposit, the balance was $49. How large was the deposit?

Draw a number line for the bank balances. The amount of the deposit is the difference between $49 and $-\$27$.

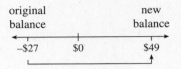

$$\text{deposit} = \$49 - (-\$27)$$
$$= \$49 + (+\$27)$$
$$= \$76$$

Check: $-\$27 + \$76 = \$49$ ∎

▶ **You Try It**   **18.** Tom's score in a game was -34. An hour later his score was 27. By how many points did his score rise?

SECTION 11.3 EXERCISES

I *Subtract by adding the opposite of the subtrahend.*

1. $8 - 5$ **2.** $12 - 9$ **3.** $3 - 7$ **4.** $1 - 8$

5. $-5 - 5$ **6.** $0 - 3$ **7.** $0 - 10$ **8.** $-12 - 12$

9. $7 - (-4)$ **10.** $5 - (-8)$ **11.** $-9 - (-4)$ **12.** $-1 - (-4)$

13. $-10 - (10)$ **14.** $-2 - (2)$ **15.** $-8 - (-8)$ **16.** $-3 - (-3)$

17. $16 - 21$ **18.** $26 - 28$ **19.** $0 - (-1)$ **20.** $0 - (-15)$

21. $-12 - (-4)$ **22.** $-25 - (-20)$ **23.** $16 - (-2)$ **24.** $2 - (-7)$

25. $-80 - (-80)$ **26.** $-1 - (-1)$ **27.** $-100 - (-3)$ **28.** $-9 - (-4)$

29. $-40 - (-140)$ **30.** $-9 - (-79)$ **31.** $16 - (-14)$ **32.** $4 - (-24)$

33. $8 - 0$ **34.** $-5 - 0$ **35.** $3 - (-3)$ **36.** $9 - (-9)$

37. $0 - (-50)$ **38.** $0 - (-12)$ **39.** $-8 + 9$ **40.** $-12 + 8$

41. $-10 + 10$ **42.** $-23 + (23)$ **43.** $-30 - (-45)$ **44.** $6 - (-40)$

45. $-9 + 4 - (-5)$ **46.** $-2 + 6 - (-4)$

47. $8 - (-3) - 8$ **48.** $20 - (-5) - 20$

49. $4 - (-7) - 1$ **50.** $3 - (-5) - 9$

51. $-4 + 7 + 2 - 8 - 3$ **52.** $+5 - 1 - 12 + 5 - 9$

53. $8 - (-12) - 8 + 3 - 7$ **54.** $16 + (-12) - (-14) - 2$

2 *Draw and label a number line for each problem.*

55. Al's checkbook balance yesterday was −$57. After a deposit his balance was $140. How large was the deposit?

56. George had a score of −38 in a game of cards. His score is now 14. By how many points did his score rise?

57. Zelda had a checkbook balance of $8. She wrote checks for $26 and $41. She then made a $60 deposit. What is her new balance?

58. A business had a bank balance of −$400. The owner wrote checks for $120, $67, and $240 and made a $900 deposit. What was the new balance?

59. At 2 A.M. the temperature was −21°. By noon, it was 17°. By how many degrees did the temperature rise?

60. The temperature at an Arctic station at midnight was −56°. By noon the next day the temperature was −18°. By how many degrees did the temperature rise?

61. The ocean surface is at 0 ft elevation. A diver is underwater at the base of a rock formation. Her elevation is −178 ft. The top of the formation has an elevation of −113 ft. What is the height of the rock formation?

62. An underwater missile was launched from a −251 ft elevation. It reached a height of 1,375 ft above the surface of the water. What vertical distance did the missile travel?

SKILLSFOCUS (Section 1.9) *Simplify.*

63. 2^6 **64.** 6^3 **65.** 5^0 **66.** 1^{25} **67.** 10^1

WRITING TO LEARN ▶▶▶▶

68. Write a short essay to explain how to subtract $(-8) - (-3)$.

69. Write a word problem whose answer is given by the expression $21 - (-6)$. Then solve it.

70. What is $(-9) - (-14)$? Reverse the integers. What is $(-14) - (-9)$?
  a. What is similar about the two answers?

  b. How are the two answers different?

  c. What happens when you reverse the order of a subtraction?

  d. Is subtraction of integers commutative?

OBJECTIVES

1 Multiply two integers with unlike signs (Sections 1.7, 1.8).
2 Multiply two integers with like signs.
3 Multiply by 1, −1, and 0.
4 Divide two integers (Sections 2.1, 2.2).

NEW VOCABULARY

multiplicative identity

---

1 Multiplying Two Integers with Unlike Signs

The product of two positive integers is positive. What sign do you give the product of a negative integer and a positive integer? Find the answer by looking for the pattern in the following products.

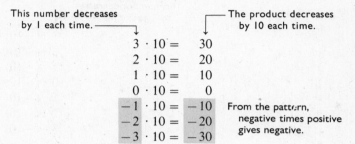

This number decreases by 1 each time.

The product decreases by 10 each time.

$$3 \cdot 10 = 30$$
$$2 \cdot 10 = 20$$
$$1 \cdot 10 = 10$$
$$0 \cdot 10 = 0$$
$$-1 \cdot 10 = -10$$
$$-2 \cdot 10 = -20$$
$$-3 \cdot 10 = -30$$

From the pattern, negative times positive gives negative.

To multiply a negative integer times a positive integer

1. Multiply the absolute values.
2. Attach a negative sign to the product.

EXAMPLE 1 Multiply $(-5) \cdot (4)$.

$(-5) \cdot (4) = -20$ Multiply the absolute values: $5 \cdot 4 = 20$. Since one integer is negative and the other positive, attach a negative sign to the product. ∎

EXAMPLE 2 $7 \cdot (-3) = -21$ Multiply the absolute values: $7 \cdot 3 = 21$. Attach a negative sign. ∎

EXAMPLE 3 $-6 \cdot 8 = -48$ ∎

EXAMPLE 4 $(+10)(-10) = -100$
└─────── Multiplication is implied. ∎

EXAMPLE 5 $5(-2) = -10$
└─────── Multiplication is implied. ∎

OBSERVE In Example 3, the multiplication $-6 \cdot 8$ can be rewritten in any of the following equivalent ways.

$-6 \cdot 8$	$(-6) \cdot (+8)$	$-6(8)$	$-6(+8)$
$-6 \cdot (8)$	$(-6) \cdot (8)$	$(-6)(8)$	$(-6)(+8)$
$-6 \cdot (+8)$	$(-6) \cdot 8$	$(-6)8$	

▶ **You Try It**

1. $(-3) \cdot (7)$ 2. $(-6) \cdot (+2)$ 3. $5 \cdot (-9)$
4. $8 \cdot (-8)$ 5. $(+1)(-1)$ 6. $(-10)(+5)$
7. $(9)(-6)$ 8. $(-4) \cdot 0$ 9. $(-7)5$

EXAMPLE 6   What is -4 added to itself 42 times?

This is repeated addition. Multiply for the answer. (See Section 1.7.)

-4 added to itself 42 times $= -4 \cdot 42 = -168$ ∎

► **You Try It**   **10.** What is -7 added to itself 36 times?

2 **Multiplying Two Integers with Like Signs**

What sign do you give the product of a negative integer and a negative integer? Find the answer by looking for the pattern in the following products.

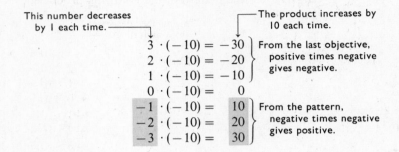

This number decreases by 1 each time.

The product increases by 10 each time.

$3 \cdot (-10) = -30$ } From the last objective, positive times negative gives negative.
$2 \cdot (-10) = -20$
$1 \cdot (-10) = -10$
$0 \cdot (-10) = 0$
$-1 \cdot (-10) = 10$ } From the pattern, negative times negative gives positive.
$-2 \cdot (-10) = 20$
$-3 \cdot (-10) = 30$

To multiply a negative integer times a negative integer

1. Multiply the absolute values.
2. Attach a positive sign to the product.

EXAMPLE 7   Multiply $-6 \cdot (-5)$.

$-6 \cdot (-5) = +30$ or 30 Multiply the absolute values: $6 \cdot 5 = 30$. Since both integers are negative, attach a positive sign to the product. ∎

EXAMPLE 8 $(-4) \cdot (-7) = +28$ or 28 Multiply the absolute values: $4 \cdot 7 = 28$. Attach a positive sign to the product. ∎

EXAMPLE 9 $-3(-4) = 12$

↑_____ The multiplication is implied. ∎

EXAMPLE 10 $(-8)(-10) = 80$ ∎

► **You Try It**   **11.** $-8 \cdot (-3)$ **12.** $-6 \cdot (-2)$ **13.** $(-4) \cdot (-1)$
14. $(-10)(-10)$ **15.** $(-9)(-7)$ **16.** $-20(-2)$

To multiply three or more integers, multiply the two on the left. Multiply this product by the next integer on the right, and so on.

EXAMPLE 11   Multiply $-4(-2)(3)(-1)$.

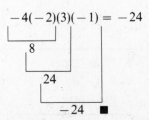

$-4(-2)(3)(-1) = -24$

8

24

-24 ∎

17. $-6(-4)(-2)$ 18. $-8(5)(-1) \cdot 3$

3 Multiplying by 1, −1, and 0

The **multiplicative identity** for integers is the number 1. Multiplying 1 times any integer gives the same integer.

EXAMPLE 12   **a.** $1 \cdot (-4) = -4$ **b.** $(-4) \cdot 1 = -4$
c. $(-5)1 = -5$ **d.** $1(-5) = -5$
e. $1 \cdot x = x$ **f.** $x \cdot 1 = x$
g. $1y = y$ **h.** $y1 = y$ ■

Multiplying an integer by -1 gives the opposite of the integer.

EXAMPLE 13   **a.** $-1 \cdot (+6) = -6$ **b.** $-1 \cdot (-6) = +6$ or 6
c. $-1(-15) = 15$ **d.** $4(-1) = -4$
e. $(-7)(-1) = 7$ **f.** $-1 \cdot p = -p$ ■

0 times any integer equals 0.

EXAMPLE 14   **a.** $0 \cdot (-6) = 0$ **b.** $(0)(+5) = 0$
c. $(-8)0 = 0$ ■

▶ **You Try It**   **19.** $1 \cdot (-8)$ **20.** $(-6)1$ **21.** $1 \cdot m$
22. $-1 \cdot (3)$ **23.** $-1 \cdot (-5)$ **24.** $(-9)(-1)$
25. $0 \cdot (+2)$ **26.** $(-25)0$

4 Dividing Two Integers

Recall from Section 2.1, every division has a related multiplication. This fact can be used to show that the rules for dividing integers are the same as the rules for multiplying integers.

Division		*Related Multiplication*	

$\dfrac{10}{2} = 5$ because $2 \cdot 5 = 10$ Normal check for division.

$\dfrac{-10}{2} = -5$ because $2 \cdot (-5) = -10$ ⎫

$\dfrac{10}{-2} = -5$ because $-2 \cdot (-5) = 10$ ⎬ Dividing a positive and a negative gives a negative.

$\dfrac{-10}{-2} = +5$ because $-2 \cdot (+5) = -10$ ⎭ Dividing a negative by a negative gives a positive.

> **To divide (or multiply) two integers**
>
> 1. Divide (or multiply) the absolute values.
>   The sign of the answer is
>   **a.** positive if both integers are positive or both are negative.
>   **b.** negative if one integer is negative and the other is positive.

EXAMPLE 15   Divide $\dfrac{-40}{5}$.

$$\frac{-40}{5} = -8$$ Divide the absolute values: $\frac{40}{5} = 8$. The signs are unlike, so the answer is negative.

Check: $5 \times (-8) = -40$ ∎

▶ **You Try It**   **27.** $\dfrac{-64}{8}$ **28.** $\dfrac{+30}{-6}$

EXAMPLE 16 $\dfrac{-24}{-4} = +6$ or 6 Divide the absolute values: $\frac{24}{4} = 6$.

The signs are both negative, so the answer is positive.

Check: $(-4)(6) = -24$ ∎

EXAMPLE 17 $-36 \div (-9) = 4$

Check: $(-9) \cdot 4 = -36$ ∎

EXAMPLE 18 $\dfrac{63}{-7} = -9$

Check: $(-7) \cdot (-9) = 63$ ∎

▶ **You Try It**   **29.** $\dfrac{-27}{-9}$ **30.** $\dfrac{30}{-2}$ **31.** $-28 \div (-7)$ **32.** $\dfrac{-10}{10}$

EXAMPLE 19   How many -2 are in -38?

To find how many a are in b, write $b \div a$. (See Section 2.3.)

The number of -2 in $-38 = -38 \div -2$
$$= 19 \quad ∎$$

▶ **You Try It**   **33.** How many -3 are in -24?

Summary

To multiply or divide two integers

1. Multiply or divide the two absolute values.
2. The sign of this answer is
   a. positive if both integers are positive or both are negative.
   b. negative if one integer is negative and the other is positive.

▶ **Answers to You Try It**   1. -21 2. -12 3. -45 4. -64
5. -1 6. -50 7. -54 8. 0   9. -35 10. -252 11. 24   12. 12
13. 4   14. 100   15. 63   16. 40   17. -48 18. 120   19. -8 20. -6
21. m 22. -3 23. 5   24. 9   25. 0   26. 0   27. -8 28. -5 29. 3
30. -15 31. 4   32. -1 33. 8

SECTION 11.4 EXERCISES

1 **2** **3** *Perform each multiplication.*

1. $5 \cdot (+7)$ **2.** $6 \cdot (+4)$ **3.** $(3)(2)$ **4.** $(+5)(+5)$

5. $-6 \cdot (+5)$ **6.** $-4 \cdot (+7)$ **7.** $-8 \cdot 2$ **8.** $-5 \cdot 3$

9. $4(-5)$ **10.** $7(-2)$ **11.** $-1 \cdot (9)$ **12.** $-1 \cdot (1)$

13. $(-10) \cdot 0$ **14.** $(-3) \cdot 0$ **15.** $-8(-7)$ **16.** $-10(-4)$

17. $-15(-1)$ **18.** $-3(-1)$ **19.** $(-6) \cdot (-6)$ **20.** $(-4)(-3)$

21. $(8) \cdot (-3)$ **22.** $(5) \cdot (-6)$ **23.** $-9(3)$ **24.** $-2(7)$

25. $0(-1)$ **26.** $0(-120)$ **27.** $(-3) \cdot (-9)$ **28.** $(-10) \cdot (-10)$

29. $6(-12)$ **30.** $8(-6)$ **31.** $(-12)(-2)$ **32.** $(-3)(-10)$

33. $-2 \cdot (-4)(-2)$ **34.** $-3 \cdot (-1)(-5)$ **35.** $-1 \cdot (-6)(2) \cdot (4)$ **36.** $-9 \cdot (-8)(7)(0)$

37. $5(-2)(-3)(-1)$ **38.** $2(-3)(-3)(-2)$ **39.** $-15(-12)(0)$ **40.** $-8(-2)(2)$

41. What is -7 added to itself 60 times?   **42.** What is -3 added to itself 45 times?

43. What is $6 \cdot t$ if $t = -8$?   **44.** What is $-3 \cdot v$ if $v = -4$?

45. What is $-5 \cdot n$ if $n = -7$?   **46.** What is $-10 \cdot y$ if $y = +6$?

4 *Divide. Check using the related multiplication.*

47. $\dfrac{-30}{5}$ **48.** $\dfrac{-18}{6}$ **49.** $\dfrac{24}{-3}$ **50.** $\dfrac{40}{-8}$

51. $\dfrac{-45}{-9}$ **52.** $\dfrac{-30}{-6}$ **53.** $\dfrac{8}{-1}$ **54.** $\dfrac{6}{-1}$

55. $\dfrac{60}{-5}$ **56.** $\dfrac{100}{-10}$ **57.** $\dfrac{-90}{0}$ **58.** $\dfrac{0}{-25}$

59. $\dfrac{-80}{10}$

60. $\dfrac{-65}{5}$

61. $\dfrac{200}{-40}$

62. $\dfrac{300}{-3}$

63. $\dfrac{-72}{-8}$

64. $\dfrac{-84}{-7}$

65. $\dfrac{0}{7}$

66. $\dfrac{+64}{0}$

67. $\dfrac{-35}{-1}$

68. $\dfrac{-7}{-1}$

69. $\dfrac{-480}{60}$

70. $\dfrac{-600}{-50}$

71. $12 \div (-2)$

72. $30 \div (0)$

73. $-36 \div 0$

74. $-70 \div (-7)$

75. $(-8) \div (-1)$

76. $-90 \div (-1)$

77. $0 \div (-2)$

78. $0 \div (-5)$

79. Find $\dfrac{k}{-4}$ when $k = 28$.

80. Find $\dfrac{p}{-12}$ when $p = -36$.

81. What is $\dfrac{-35}{t}$ when $t = -5$?

82. What is $\dfrac{-88}{y}$ when $y = 11$?

83. How many -3 are in 60?

84. How many 8 are in -72?

85. How many -1 are in -15?

86. How many 10 are in -300?

SKILLSFOCUS (Section 3.4) *Reduce.*

87. $\dfrac{12}{18}$

88. $\dfrac{21}{36}$

89. $\dfrac{72}{27}$

90. $\dfrac{84}{240}$

EXTEND YOUR THINKING ▶▶▶▶

▶ TROUBLESHOOT IT

Find and correct the error.

91. $(-2)(-3)(-7) = (-5)(-7) = +35$

92. $\dfrac{-100}{+50} = -50$

WRITING TO LEARN ▶▶▶▶

93. In your own words, explain how to multiply -4 by -5.

94. Explain the difference between the two divisions: $\dfrac{0}{-4}$ and $\dfrac{-4}{0}$. Write your answer for each with a reason why.

▶ YOU BE THE JUDGE

95. Is division of integers commutative? Explain your decision using your own examples.

96. If $3 \cdot \$10 = \$30$ can be interpreted as selling 3 tickets at $10 each, for a total of $30, how can you interpret $3 \cdot (-\$10) = -\$30$?

11.5 ORDER OF OPERATIONS

OBJECTIVES

1 Evaluate powers of integers (Section 1.9, Objectives 1, 2, 3).

2 Apply the order of operations to integers (Section 2.4).

3 Evaluate algebraic expressions involving integers (Section 2.5).

---

1 Powers of Integers

Recall from Section 1.9 that a power is a short way to write a repeated multiplication.

EXAMPLE 1 Simplify the powers.

a. $5^2 = 5 \cdot 5 = 25$

b. $(+4)^3 = (+4)(+4)(+4) = +64$

$+16$

$+64$

c. $(-6)^2 = (-6)(-6) = 36$

d. $(-2)^5 = (-2)(-2)(-2)(-2)(-2) = -32$

$+4$

-8

$+16$

-32

e. $(-3)^4 = (-3)(-3)(-3)(-3) = 81$

$+9$

-27

$+81$

f. $(-10)^3 = (-10)(-10)(-10) = -1{,}000$

$+100$

$-1{,}000$ ■

> A negative integer raised to an even power is positive (see Examples 1c and 1e).
> A negative integer raised to an odd power is negative (see Examples 1d and 1f).

▶ **You Try It** Evaluate.

1. 8^2 **2.** $(+5)^3$ **3.** $(-9)^2$

4. $(-1)^9$ **5.** $(-3)^3$ **6.** $(-1)^8$

7. $(-10)^5$ **8.** $(7)^3$ **9.** $(-2)^7$

EXAMPLE 2 Show that $(-3)^2 \neq -3^2$.

$$(-3)^2 = (-3) \cdot (-3) = 9$$

The negative sign is in the parentheses with the 3.
So -3 is multiplied twice.

$$-3^2 = -(3 \cdot 3) = -(9) = -9$$

The negative sign is not written in parentheses
with the 3. So only the 3 is multiplied twice.

Therefore, $(-3)^2 \neq -3^2$. ■

OBSERVE You may also write $-3^2 = -1 \cdot 3^2 = -1 \cdot 9 = -9$.

EXAMPLE 3 **a.** $(-6)^2 = (-6) \cdot (-6) = 36$

b. $-6^2 = -(6 \cdot 6) = -36$ ■

▶ **You Try It** Evaluate.

10. $(-8)^2$ **11.** -8^2 **12.** $(-1)^2$ **13.** -1^2

14. -7^2 **15.** $(-7)^2$ **16.** -2^2 **17.** $(-2)^2$

---

2 Order of Operations

The order of operations for integers is the same as for whole numbers. Review the procedure in Section 2.4.

EXAMPLE 4 Simplify $-40 \div 8(-4)$.

$$-40 \div 8(-4) = \underbrace{-40 \div 8} \cdot (-4) \quad \text{Insert the multiplication symbol.}$$

$$= \quad -5 \quad \cdot (-4) \quad \text{Multiply and divide in order from left to right. Divide first.}$$

$$= 20 \quad ■$$

EXAMPLE 5 Evaluate $-6 + 6(-2)$.

$$-6 + 6(-2) = -6 + \underbrace{6 \cdot (-2)}$$

$$= -6 + \quad (-12) \quad \text{Do multiplication first. Then add.}$$

$$= -18 \quad ■$$

▶ **You Try It** **18.** $-60 \div 10(-2)$ **19.** $-8 + 8(-3)$

EXAMPLE 6 Find $7 - 5 - 3(-4)$.

$$7 - 5 - 3(-4) = 7 - 5 - \underbrace{3 \cdot (-4)}$$

$$= 7 - 5 - (-12) \quad \text{Multiply first: } 3 \cdot (-4) = -12.$$

$$= 2 - (-12) \quad \text{Subtract from left to right.}$$

$$= 2 + (+12)$$

$$= 14 \quad ■$$

▶ **You Try It**   **20.** $10 - 6 - 4(-7)$

EXAMPLE 7   Evaluate $(-6)^2(2) + \sqrt{49}(-10)$.

$$
\begin{aligned}
(-6)^2(2) + \sqrt{49}(-10) &= \underbrace{(-6)^2}_{} \cdot (2) + \underbrace{\sqrt{49}}_{} \cdot (-10) && \text{Powers and roots} \\
&= \underbrace{(+36) \cdot (2)}_{} + \underbrace{7 \cdot (-10)}_{} && \text{Multiply.} \\
&= \quad 72 \quad + \quad (-70) && \text{Add.} \\
&= 2 \quad \blacksquare
\end{aligned}
$$

▶ **You Try It**   **21.** $(-4)^2(3) + \sqrt{81}(-6)$

3 Evaluating Algebraic Expressions

Algebraic expressions can also be evaluated using negative numbers. To avoid confusion with signs, use parentheses when replacing a letter with a negative number.

EXAMPLE 8   Evaluate $5 - y$ when $y = -8$.

subtraction sign ⟍   ⟋ negative sign

$$
\begin{aligned}
5 - y &= 5 - (-8) && \text{Replace } y \text{ with } -8. \text{ Enclose } -8 \text{ in} \\
& && \text{parentheses to keep the subtraction} \\
& && \text{sign and the negative sign separate.} \\
&= 5 + (+8) \\
&= 13 \quad \blacksquare
\end{aligned}
$$

▶ **You Try It**   **22.** Evaluate $7 - x$ when $x = -9$.

EXAMPLE 9   Find $5k + 2m$ when $k = -2$ and $m = 4$.

$$
\begin{aligned}
5k + 2m &= 5 \cdot k + 2 \cdot m \\
&= \underbrace{5 \cdot (-2)}_{} + \underbrace{2 \cdot (4)}_{} \\
&= \quad -10 \quad + \quad (8) \\
&= -2 \quad \blacksquare
\end{aligned}
$$

▶ **You Try It**   **23.** Evaluate $3p + 7w$ when $p = -4$ and $w = 3$.

EXAMPLE 10   Evaluate $-7x^2$ when $x = -3$.

$$
\begin{aligned}
-7x^2 &= -7 \cdot x^2 \\
&= -7 \cdot (-3)^2 && \text{Replace } x \text{ with } -3. \\
&= -7 \cdot (+9) && \text{Evaluate the exponent before multiplying.} \\
&= -63 \quad \blacksquare
\end{aligned}
$$

▶ **You Try It**   **24.** Evaluate $-5w^2$ when $w = -4$.

EXAMPLE 11 Evaluate $b^2 - 4ac$ when $b = 3$, $a = 1$, and $c = -4$.

$$b^2 - 4ac = b^2 - 4 \cdot a \cdot c$$
$$= \underbrace{3^2} - 4 \cdot 1 \cdot (-4) \qquad \text{Evaluate the exponent first.}$$
$$= 9 - \underbrace{(4 \cdot 1)} \cdot (-4)$$
$$= 9 - \underbrace{(4) \cdot (-4)}$$
$$= 9 - \quad (-16)$$
$$= 9 + \quad (+16)$$
$$= 25 \quad \blacksquare$$

▶ **You Try It** **25.** Evaluate $b^2 - 4ac$ when $b = 6$, $a = 2$, and $c = -5$.

▶ **Answers to You Try It** 1. 64 2. 125 3. 81 4. -1 5. -27 6. 1
7. $-100,000$ 8. 343 9. -128 10. 64 11. -64 12. 1 13. -1
14. -49 15. 49 16. -4 17. 4 18. 12 19. -32 20. 32 21. -6
22. 16 23. 9 24. -80 25. 76

SECTION 11.5 EXERCISES

1 *Evaluate each expression.*

1. $(+5)^2$

2. $(+8)^2$

3. $(-6)^2$

4. $(-7)^2$

5. $(-10)^2$

6. $(-1)^2$

7. -6^2

8. -7^2

9. $(-3)^3$

10. $(-2)^4$

11. $(-4)^2$

12. $(-1)^7$

13. -3^3

14. -2^4

15. -4^2

16. -1^7

2

17. $6(3-9)$

18. $4(6-8)$

19. $-3(1-4+2)$

20. $-7(8-5-3)$

21. $6 \cdot (-3)^2$

22. $4 \cdot (-5)^2$

23. $10 \div (-2) \cdot 5$

24. $-60 \div (6) \cdot (-10)$

25. $4 + 6(-2)$

26. $6 + 3(-7)$

27. $-8 + 4(2)$

28. $-1 + 1(1)$

29. $-9 + 3 \cdot (-4)$

30. $-4 + 4 \cdot (-5)$

31. $9 - 9(-4)$

32. $6 - 8(-1)$

33. $-3 - 3(-3)$

34. $-8 - 5(-2)$

35. $-4 \cdot (6) \div (-3)$

36. $8 \cdot (-9) \div (-6)$

37. $-100 \div (50)(2)$

38. $-80 \div (20) \cdot (-4)$

39. $-6 + 5^2$

40. $-4 + (-2)^2$

41. $(-2)(-4) + (-3)(-8)$

42. $(-1)(-2) + (-4)(5)$

43. $60 \div (-10) + 55 \div (11)$

44. $(-36) \div (-9) + (-2)(9)$

45. $(-6)(-6) - (-4)(-4)$

46. $(-1)(-1) - (-2)(-3)$

47. $\dfrac{-40}{(-5)(8)}$

48. $\dfrac{(-9)(-8)}{6(-6)}$

49. $\dfrac{(-6) - (-2)}{(-8) - (-4)}$

50. $\dfrac{(10) - (-10)}{(-2) - (3)}$

51. $\dfrac{(-8)(-6)}{(-4)(-3)}$

52. $\dfrac{(-40)(60)}{(20)(-30)}$

3 *Evaluate each expression using $x = -2$, $y = 3$, $a = -4$, and $b = -1$.*

53. $8 - x$

54. $3 - b$

55. $2y - 7b$

56. $-4a + 1$

57. $4 - a - y$

58. $12 - y - 1$

59. $6a^2$

60. $-4x^2$

61. $b^2 - 4ay$	62. $y^2 - 4xb$	63. $1 - 3x$	64. $-8 - 4b$
65. axy	66. abx	67. $b - 9$	68. $x + a$
69. $-7bx$	70. $3ab$	71. $b - 1$	72. $x - 2y$
73. $(x + y)^2$	74. $(b - a)^3$	75. $x^2 - a^2$	76. $(ax)/(2b)$

SKILLSFOCUS (Section 8.1) *Find the mean, median, mode, and range for each set of numbers.*

77. 4, 6, 7, 7, 11

78. 4, 0, 3, 5, 1, 0, 1

79. 25, 25, 30, 25, 30, 30

80. 98, 88, 90, 100, 80, 87, 95, 70

EXTEND YOUR THINKING ▶▶▶▶

▶**SOMETHING MORE**

81. Simplify $(20 - (19 - (18 - (\ldots(3 - (2 - 1))\ldots))))$.

82. Given $a = 4$, $b = -5$, $x = 3$, and $y = -2$, do the following.
   a. Write three expressions which, when evaluated, give 13. (Hint: $5x + y$ is an example because $5x + y = 5 \cdot 3 + (-2) = 15 + (-2) = 13$.)

   b. Write an expression that evaluates to 0.

   c. Write an expression that evaluates to -20.

   d. Write an expression that evaluates to -37.

▶**TROUBLESHOOT IT**

Find and correct the error.

83. $(-4)^3 = (-4)(-4)(-4) = -12$

84. $-5^2 = (-5)(-5) = 25$

85. $(-100) \div (10) \cdot (-10) = (-100) \div (-100) = 1$

WRITING TO LEARN ▶▶▶▶

86. Evaluate $-w$ when $w = -6$. Explain why the sign of the answer is positive.

▶**YOU BE THE JUDGE**

87. Maria claims the product of an even number of negative integers always gives a positive answer. Is she correct? Explain your decision using your own examples.

88. Kim claims the product of an odd number of negative integers always gives a negative answer. Is she correct? Explain your decision using your own examples.

11.6 SCIENTIFIC NOTATION

OBJECTIVES

1 Define scientific notation (Section 1.8; Section 1.9, Objectives 1, 2, 3; Section 2.4).

2 Write numbers greater than 1 in scientific notation (Section 5.6, Objective 3).

3 Write numbers between 0 and 1 in scientific notation (Section 5.5, Objective 2).

NEW VOCABULARY

scientific notation

---

1 Scientific Notation

Scientific notation is a convenient way to express very large or very small numbers. Computers and some calculators display answers in scientific notation. It is used extensively in science. For example, in chemistry, Avogadro's number (the number of atoms in 16 grams of oxygen) is

$$602,000,000,000,000,000,000,000.$$

It is much more convenient to express this number in scientific notation as

$$6.02 \times 10^{23}.$$

must be a number 1 or larger, but less than 10   times   a power of 10

> A number is written in scientific notation if it is written in the form
> $$a \times 10^n.$$
> *a* is greater than (or equal to) 1 and less than 10.   The power *n* is an integer.

The number 5.2×10^3 is written in scientific notation. To change it to decimal notation, apply the order of operations.

$$5.2 \times 10^3 = 5.2 \times 1,000 \qquad \text{Evaluate the power.}$$
$$= 5,200 \qquad \text{Multiply.}$$

$10^3 = 1,000$ has 3 zeros. To multiply 5.2 by 1,000, just move the decimal point 3 places right (see Section 5.5).

$$5.2 \times 1,000 = 5\ 200.\ \text{or } 5,200$$

3 zeros   3 places right

> To multiply by a power of 10 when the exponent on 10 is positive, move the decimal point to the right the number of places equal to the power on 10.

EXAMPLE 1 Change from scientific notation to decimal notation.

a. $6 \times 10^2 = 6\ 00. = 600$

2 zeros   2 places right

b. $1.076 \times 10^5 = 1\ 07600. = 107,600$

5 zeros   5 places right   ∎

▶ **You Try It** Change to decimal notation.

1. 8.93×10^4 **2.** 9×10^7

---

2 Writing Numbers Greater Than 1 in Scientific Notation

Write 6,800 in scientific notation.

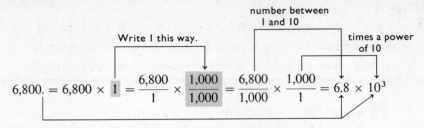

$$6{,}800. = 6{,}800 \times 1 = \frac{6{,}800}{1} \times \frac{1{,}000}{1{,}000} = \frac{6{,}800}{1{,}000} \times \frac{1{,}000}{1} = 6.8 \times 10^3$$

In short, move the decimal point 3 places left.
Multiply by 10 raised to the power of 3.

In summary,

$$6\ 800. = 6.8 \times 10^3.$$

Move the decimal point to the left until you get a number between 1 and 10. 6.8 is between 1 and 10.

The power on 10 equals 3, the number of places the decimal point moved left.

Check: $6.8 \times 10^3 = 6.8 \times 1{,}000 = 6{,}800$

> **To write a decimal number greater than 1 in scientific notation**
>
> 1. Move the decimal point to the left until you get a number 1 or larger and less than 10. (At this point, there will be one digit to the left of the decimal point.)
> 2. Count the number of places the decimal point moved left. Write this number as the power on 10.
> 3. Multiply the number in Step 1 by the power of 10 in Step 2.

EXAMPLE 2   Write each number in scientific notation.

a. $45.3 = 4\ 5.3 = 4.53 \times 10^1$

Move the decimal point 1 place left to get 4.53, a number between 1 and 10.

The power on 10 is 1, because the decimal point moved 1 place left.

b. $300 = 3\ 00. = 3 \times 10^2$

number between 1 and 10

2 places left

c. $62{,}490{,}000 = 6\ 2\ 490\ 000. = 6.249 \times 10^7$

number between 1 and 10

7 places left   ∎

▶ **You Try It**   Write each number in scientific notation.

3. 73.2   **4.** 506   **5.** 810,000   **6.** 6,073.4

3 Writing Numbers between 0 and 1 in Scientific Notation

10^{-1}, 10^{-2}, 10^{-3}, and so on, are needed to write numbers between 0 and 1 in scientific notation. To understand what these numbers mean, try to find the pattern with the following powers of 10.

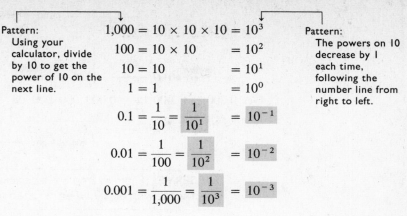

Pattern: Using your calculator, divide by 10 to get the power of 10 on the next line.

$$1,000 = 10 \times 10 \times 10 = 10^3$$
$$100 = 10 \times 10 = 10^2$$
$$10 = 10 = 10^1$$
$$1 = 1 = 10^0$$

Pattern: The powers on 10 decrease by 1 each time, following the number line from right to left.

$$0.1 = \frac{1}{10} = \frac{1}{10^1} = 10^{-1}$$
$$0.01 = \frac{1}{100} = \frac{1}{10^2} = 10^{-2}$$
$$0.001 = \frac{1}{1,000} = \frac{1}{10^3} = 10^{-3}$$

Therefore, $10^{-1} = \frac{1}{10^1}$, $10^{-2} = \frac{1}{10^2}$, $10^{-3} = \frac{1}{10^3}$, and so on. As a result, you can write

$$\frac{4.3}{10^2} = \frac{4.3}{1} \times \frac{1}{10^2} = 4.3 \times 10^{-2}.$$

Dividing by 10^2 is the same as multiplying by 10^{-2}.

Write 0.043 in scientific notation. To do this, write 0.043 as a number between 1 and 10, times a power of 10.

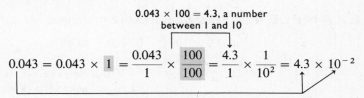

$0.043 \times 100 = 4.3$, a number between 1 and 10

$$0.043 = 0.043 \times 1 = \frac{0.043}{1} \times \frac{100}{100} = \frac{4.3}{1} \times \frac{1}{10^2} = 4.3 \times 10^{-2}$$

Move the decimal point 2 places right.
Multiply by 10 raised to the −2 power.

This is summarized as follows.

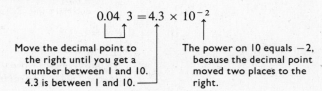

$$0.04\ 3 = 4.3 \times 10^{-2}$$

Move the decimal point to the right until you get a number between 1 and 10. 4.3 is between 1 and 10.

The power on 10 equals −2, because the decimal point moved two places to the right.

> **To write a decimal number between 0 and 1 in scientific notation**
>
> 1. Move the decimal point to the right until you get a number between 1 and 10. (At this point, there will be one nonzero digit to the left of the decimal point.)
> 2. Count the number of places the decimal point moved right. Write this number as a negative power on 10.
> 3. Multiply the number in Step 1 by the power of 10 in Step 2.

EXAMPLE 3 Write each number in scientific notation.

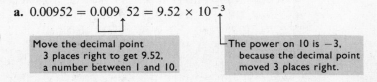

a. $0.00952 = 0.009\ 52 = 9.52 \times 10^{-3}$

Move the decimal point 3 places right to get 9.52, a number between 1 and 10.

The power on 10 is −3, because the decimal point moved 3 places right.

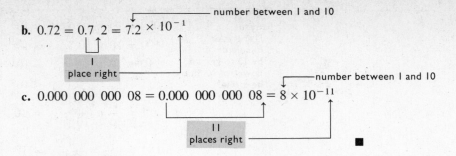

b. $0.72 = 0.7\ 2 = 7.2 \times 10^{-1}$

number between 1 and 10

1 place right

c. $0.000\ 000\ 000\ 08 = 0.000\ 000\ 000\ 08 = 8 \times 10^{-11}$

number between 1 and 10

11 places right

∎

> **OBSERVE** From Example 2, a decimal number between 0 and 1 has a negative power on 10 when written in scientific notation.

▶ **You Try It**   Write each number in scientific notation.

7. 0.00006 **8.** 0.904 **9.** 0.000 000 027 **10.** 0.013

> To change a number from scientific notation to decimal notation when the power on 10 is negative, move the decimal point to the left the number of places indicated by the power on 10.

EXAMPLE 4   Change from scientific to decimal notation.

a. $4.9 \times 10^{-2} = 0.04\ 9 = 0.049$

−2 power means move the decimal point 2 places left

b. $3.706 \times 10^{-5} = 0.00003\ 706 = 0.00003706$

−5 power means move the decimal point 5 places left

c. $8 \times 10^{-3} = 0.008 = 0.008$

−3 power means move the decimal point 3 places left

▶ **You Try It**   Write each number in decimal notation.

11. 6.02×10^{-4} **12.** 1.73×10^{-1} **13.** 3×10^{-7} **14.** 9.043×10^{-9}

> **OBSERVE** From Example 3, a negative power on 10 means the number in scientific notation is between 0 and 1 when changed to decimal notation.

▶ **Answers to You Try It**   1. 89,300   2. 90,000,000   3. 7.32×10^{1}
4. 5.06×10^{2} 5. 8.1×10^{5} 6. 6.0734×10^{3} 7. 6×10^{-5} 8. 9.04×10^{-1}
9. 2.7×10^{-8} 10. 1.3×10^{-2} 11. 0.000602   12. 0.173   13. 0.0000003
14. 0.000000009043

SECTION 11.6 EXERCISES

1 *Write each number in decimal notation.*

1. 4.6×10^2 **2.** 3.7×10^1 **3.** 5.92×10^2 **4.** 6.30×10^3

5. 8×10^4 **6.** 1×10^6 **7.** 9.932×10^5 **8.** 5.768×10^4

9. 6.002×10^7 **10.** 5.514×10^1 **11.** 7.3×10^9 **12.** 4.021×10^0

13. The circumference of the Earth is 2.5×10^4 miles.

14. The diameter of the Earth is 1.3×10^7 meters.

2 *Write each number in scientific notation.*

15. 56   **16.** 33   **17.** 902   **18.** 526

19. 5,800   **20.** 6,040   **21.** 38.6   **22.** 923.7

23. 60,000   **24.** 4,000,000   **25.** 7.8   **26.** 1.2

27. 465,000   **28.** 14,700   **29.** 255,000,000   **30.** 71,380,000

31. The speed of light is 186,000 miles per second.

32. There are 31,600,000 seconds in one year.

33. The Earth is 4,500,000,000 years old.

34. Light travels 5,878,000,000,000 miles in one year.

35. The average distance between the Earth and the Sun is 92,600,000 miles.

36. The Earth weighs 6,000,000,000,000,000,000,000 metric tons.

3 *Write each number in scientific notation.*

37. 0.43   **38.** 0.88   **39.** 0.063   **40.** 0.051

41. 0.00953   **42.** 0.00812   **43.** 0.000052   **44.** 0.0000803

45. 0.1   **46.** 0.9   **47.** 0.000007   **48.** 0.004

49. 0.006045   **50.** 0.000006824   **51.** 0.00056   **52.** 0.000011

53. 0.0000061   **54.** 0.00000702   **55.** 0.0925   **56.** 0.06251

57. One inch equals 0.0000157 miles.

58. One hour equals 0.0001141 years.

59. A 150-pound man weighs 0.075 tons.

60. The fraction of radon gas present is 0.00000045.

61. Planck's constant is 0.0000000000000000000000000006624.

62. The mass of a hydrogen atom is 0.0000000000000000000000016617 g.

Write each number in decimal notation.

63. 3.6×10^{-2}

64. 2.7×10^{-1}

65. 7.03×10^{-3}

66. 9.405×10^{-2}

67. 1.026×10^{-1}

68. 6.6×10^{-3}

69. 9.04×10^{-5}

70. 3.85×10^{-4}

71. 4.8×10^{-2}

72. 9.93×10^{-5}

73. 8.507×10^{-4}

74. 6.204×10^{-3}

75. 5.4×10^{-12}

76. 2.2×10^{-7}

77. 1.077×10^{-6}

78. 6×10^{-9}

79. The mass of the oxygen atom is 2.6372×10^{-23}

80. The Leaning Tower of Pisa leans another 2.2×10^{-5} inches each day.

SKILLSFOCUS (Section 2.4) *Simplify.*

81. $7 + 4 \cdot 3$

82. $5 \cdot 2 + 6 \cdot 4$

83. $90 \div 30 \cdot 3$

84. $4 + 3(9 - 4)$

WRITING TO LEARN ▶▶▶▶

85. Explain each step you would use to change 45,300,000 into scientific notation.

86. Explain each step you would use to change 0.00048 into scientific notation.

▶ YOU BE THE JUDGE

87. Nicole claims when you change a number from scientific to decimal notation, the number of zeros you attach is the absolute value of the power, minus 1. Is she right? Explain your decision with your own examples.

CALCULATOR TIPS

The Opposite Key, $\boxed{+/-}$, and Integers   Some calculators have an opposite key, $\boxed{+/-}$. Each time you press $\boxed{+/-}$, you change the sign of the number on the display to its opposite. Such a key is called a *toggle*.

$\qquad\qquad$ 4 $\boxed{+/-}$ Displays -4 on the calculator.

$\qquad\qquad$ -4 $\boxed{+/-}$ Displays $\quad$ 4 on the calculator.

EXAMPLE 1   Simplify $-(-7)$ using the $\boxed{+/-}$ key.

$-(-7)$:$\qquad$ Enter $\quad$ 7 $\qquad$ Display shows $\quad$ 7.

$\qquad\qquad$ Press $\boxed{+/-}$ $\qquad$ Display shows -7.

$\qquad\qquad$ Press $\boxed{+/-}$ $\qquad$ Display shows $\quad$ 7.

$\qquad\qquad\qquad\qquad\qquad\qquad$ Therefore, $-(-7) = 7$. $\blacksquare$

EXAMPLE 2   Add $-6 + (+2)$.

$\qquad\qquad$ 6 $\boxed{+/-}$ $\boxed{+}$ 2 $\boxed{=}$ -4 $\blacksquare$

EXAMPLE 3   Add $-432 + (-68) + 140 + (-275) + 600$.

432 $\boxed{+/-}$ $\boxed{+}$ 68 $\boxed{+/-}$ $\boxed{+}$ 140 $\boxed{+}$ 275 $\boxed{+/-}$ $\boxed{+}$ 600 $\boxed{=}$ -35 $\blacksquare$

EXAMPLE 4   Use your calculator to subtract $-8 - (-3)$.

$\qquad\qquad$ $-8 - (-3) =$ 8 $\boxed{+/-}$ $\boxed{-}$ 3 $\boxed{+/-}$ $\boxed{=}$ -5 $\blacksquare$

EXAMPLE 5   Multiply and divide.

a. $-8(-4) =$ 8 $\boxed{+/-}$ $\boxed{\times}$ 4 $\boxed{+/-}$ $\boxed{=}$ 32

b. $(4)(-5)(-2) =$ 4 $\boxed{\times}$ 5 $\boxed{+/-}$ $\boxed{\times}$ 2 $\boxed{+/-}$ $\boxed{=}$ 40

c. $\dfrac{-120}{8} =$ 120 $\boxed{+/-}$ $\boxed{\div}$ 8 $\boxed{=}$ 15 $\blacksquare$

Use your calculator to simplify each problem.

1. $-(-9)$ $\qquad\qquad\qquad\qquad$ **2.** $-(+4)$

3. $(-6) + (-5)$ $\qquad\qquad\qquad$ **4.** $(+8) + (-17)$

5. $(-260) + (-427) + 216 + (-73) + (-92)$

6. $12 - (-3)$ $\qquad\qquad\qquad\quad$ **7.** $(-7) - (-7)$

8. $(+10)(-8)$ $\qquad\qquad\qquad\quad$ **9.** $(-2)(+6)(-3)(-5)$

10. $\dfrac{-60}{-4}$ $\qquad\qquad\qquad\qquad\quad$ **11.** $\dfrac{-267}{89}$

12. Explain why 0 $\boxed{-}$ 9 $\boxed{=}$ gives the same result as 9 $\boxed{+/-}$.

Square Roots of Negative Integers   You know $\sqrt{9} = 3$ because $3^2 = 3 \cdot 3 = 9$ (see Section 1.9). But what is $\sqrt{-9}$? Is the answer $+3$ or -3? Now $(+3)^2 = (+3)(+3) = 9$, not -9. And $(-3)^2 = (-3)(-3) = 9$, not -9. So, $\sqrt{-9}$ has no integer answer. You can demonstrate this on your calculator. To find $\sqrt{-9}$, enter

$$9 \; \boxed{+/-} \; \boxed{\sqrt{}}.$$

You see E on your calculator display. E means error. Your calculator is telling you $\sqrt{-9}$ has no integer answer. The square root of any negative integer is not an integer.

Calculators and Scientific Notation   Use your calculator to multiply $5{,}000{,}000 \times 7{,}000{,}000$. If your calculator displays the letter E (for error), then your calculator cannot express answers in scientific notation. The E means the answer is too big to fit on the calculator's display.

If instead your answer was displayed as

$$\boxed{3.5 \quad\quad 13}$$

then your calculator can express answers in scientific notation.

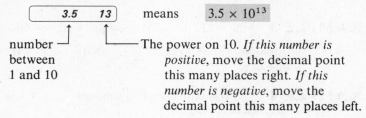

$$\boxed{3.5 \quad\quad 13} \quad \text{means} \quad 3.5 \times 10^{13}$$

number between 1 and 10 ⟶ | The power on 10. *If this number is positive*, move the decimal point this many places right. *If this number is negative*, move the decimal point this many places left.

So, $\boxed{3.5 \quad\quad 13} = 3.5000000000000 = 35{,}000{,}000{,}000{,}000$

13 places

EXAMPLE   Write the calculator display in decimal notation.

a. $\boxed{3.426 \quad\quad 8} = 3.42600000 = 342{,}600{,}000$

8 places

b. $\boxed{7.03 \quad\quad -11} = .00000000007\,03 = 0.0000000000703$ ∎

11 places

Write each calculator display in decimal notation.

1. $\boxed{4.6 \quad\quad 10}$

2. $\boxed{2.07 \quad\quad -12}$

3. $\boxed{8 \quad\quad -9}$

CHAPTER **11** REVIEW

VOCABULARY AND MATCHING

New words and phrases introduced in this chapter are shown in the left-hand column. Match each term on the left with the phrase or sentence on the right that best describes it.

A. integers   _____ add the opposite of the subtrahend

B. positive integers   _____ attached to each number on the right side of 0 on the number line

C. negative integers   _____ -25

D. 0   _____ have positive integers as powers on 10 when written in scientific notation

E. positive sign   _____ $+1$

F. negative sign   _____ positive and negative counting numbers and 0

G. absolute value   _____ product of two like signs is positive; product of two unlike signs is negative

H. opposite of a number   _____ lie to the left of 0 on the number line

I. adding integers   _____ have negative integers as powers on 10 when written in scientific notation

J. subtracting integers   _____ distance between a number and 0 on the number line

K. multiplicative identity   _____ numbers in the form $a \times 10^n$; a is 1 or larger and less than 10; n is an integer

L. multiply integers   _____ $+25$

M. divide integers   _____ attached to each number on the left side of 0 on the number line

N. $(-5)^2$ _____ this number is the same distance from 0 on the number line, but on the other side

O. -5^2 _____ the only integer that is neither positive nor negative

P. scientific notation   _____ same sign rules as for multiplying two integers

Q. numbers greater than 1   _____ two like signs, keep the sign; two unlike signs, use the sign of the larger absolute value

R. numbers between 0 and 1   _____ same as the counting numbers

REVIEW EXERCISES

11.1 Integers and the Number Line

1. Write an integer for $37°$ below zero.

2. Write an integer for a checkbook balance $25 below zero.

3. $|-3|$ **4.** $|+32|$ **5.** $|-16|$

6. Insert the proper symbol $<$, $=$, or $>$.
   a. -7 ___ -10 **b.** -20 ___ -15 **c.** $|-6|$ ___ -6 **d.** 12 ___ $|-12|$

7. The opposite of -6 is ___.   **8.** The opposite of $+15$ is ___.

9. $-(-22)$ **10.** $-(+30)$

11.2 Addition of Integers

11. $5 + (-3)$

12. $6 + (-13)$

13. $-3 + (-7)$

14. $-10 + (-5)$

15. $-7 + (-7)$

16. $-15 + (+9)$

17. $-20 + (-25)$

18. $-8 + (8)$

19. $-10 + (-12)$

20. $(30) + (-25)$

21. $-4 + (-7)$

22. $-11 + (20)$

23. $-3 + (-18) + (32) + (+11)$

24. $-40 + (-36) + (-100) + (50)$

25. The temperature in Nome, Alaska, was $-47°$ this morning. It rose $31°$ by noon. What was the noon temperature?

26. Tom has a score of -25. He made 42 points. What is his new score?

11.3 Subtraction of Integers

27. $7 - 6$

28. $9 - 12$

29. $-16 - (-16)$

30. $-9 - (+4)$

31. $+13 - (-17)$

32. $60 - (-45)$

33. $-5 - (+12)$

34. $-10 - (-10)$

35. $-100 - (-90)$

36. $-17 - (-27)$

37. $-21 - (30)$

38. $-1 - (+3)$

39. $-5 - (-10) - (+9) + (-3)$

40. $-10 - (-36) + (-5) - (+12)$

41. The temperature at midnight was $-56°$. At noon it was $-13°$. By how many degrees did the temperature rise?

42. Sally had a checkbook balance of $-\$136$. After a deposit her new balance was $\$48$. How much did she deposit?

11.4 Multiplying and Dividing Integers

43. $(+5)(+5)$

44. $(+8)(-8)$

45. $(-7)(-3)$

46. $(-2)(-9)$

47. $-8(+10)$

48. $-12(+7)$

49. $-6(-6)$

50. $-15(-4)$

51. $+20(-3)$

52. $+30(-6)$

53. $-1(+1)$

54. $-7(0)$

55. $-2(-5)(-2)$

56. $-4(2)(-2)$

57. $(-1)(-2)(-3)(-4)$

58. What is -5 added to itself 60 times?

59. What is -25 added to itself 18 times?

60. $-6 \div (+2)$

61. $-80 \div (-10)$

62. $(20) \div (-4)$

63. $30 \div (+6)$

64. $\dfrac{-45}{-9}$

65. $\dfrac{+50}{-25}$

66. $\dfrac{-24}{+24}$

67. $\dfrac{-8}{-1}$

68. $\dfrac{-90}{-3}$

69. $\dfrac{-40}{5}$

70. $\dfrac{150}{-30}$

71. $\dfrac{-54}{-6}$

72. What is -350 divided by 7?

73. How many -8 are in -96?

11.5 Order of Operations

74. $(-4)^3$

75. $(-9)^2$

76. -6^2

77. -10^3

78. $-4 + 7 - 9 - 3$

79. $-7 + 4(-3)$

80. $-5 - 5(-5)$

81. $-3(-5) + (-2)(6)$

82. $(-40) \div 8 \cdot (-5)$

83. $-3(-4)^2$

84. Evaluate using $x = -4$, $y = -2$, $a = 3$, and $b = 10$.

    a. $5xy$

    b. $-4y$

    c. $4y - 3b$

    d. $2b - x + (-3)$

    e. $-x$

    f. $\dfrac{b^2}{x}$

11.6 Scientific Notation

Change each number to decimal notation.

85. 7.05×10^4

86. 4×10^{-2}

87. 1.8×10^{-5}

88. 9.88×10^7

89. 6.025×10^1

90. 7.6003×10^0

91. 9.4×10^3

92. 3.24×10^{-6}

Write each number in scientific notation.

93. 15,000

94. 477

95. 0.9

96. 0.0023

97. 507,000

98. 0.00308

99. 0.0000048

100. 1,060,000,000

101. 6,400

102. 40.7

103. 0.00011

104. 0.000000000041

Allow yourself 50 minutes to complete this test. Show the work for each problem.
When done, check your answers. Rework each problem solved incorrectly.

1. $|-150|$ **2.** $|-28| + (-2)$ **3.** $-(-70)$ **4.** $|-5| + |-8|$

5. Insert the correct symbol $<$, $>$, or $=$.
    a. -10 -8 **b.** -3 -4 **c.** $|-4|$ 4 **d.** 5 $|-12|$

6. What is the opposite of -5?

7. $-10 + (-7)$ **8.** $-40 - (-13)$ **9.** $16 - (-11)$

10. $-12 + (-16) + (-22)$ **11.** $-5 - (-3) + (-2)$

12. $(-10)(+10)$ **13.** $12(-1)$ **14.** $-6(-20)$

15. $(-70) \div (-14)$ **16.** $-48 \div (6)$ **17.** $120 \div (-24)$

18. $-3(-1)(-5) \cdot 4$ **19.** $-30 \div (-10) \cdot (-3) \cdot (-1)$

20. The coldest temperature ever recorded on Earth was $-127°$F in Vostok Station, Antarctica, in August 1960. If the average winter temperature in Antarctica is $-58°$F, how many degrees below average was this?

21. Yana's balance was \$321.50. If she writes a check for her mortgage, her new balance will be $-\$507.93$. How much is her mortgage payment?

22. $(-6)^3$

23. -4^2

24. $-(-7)^1$

25. $(-20)(-4) + 6(-5)^0$

26. $(100) \div (-25) \cdot 4$

27. $30 - 8(-2)$

28. Evaluate each expression for $A = -4$, $B = 2$, and $C = -5$.
   a. $3A + 2B$
   b. $-6BC$

   c. $-2(B - A)$
   d. $A + B - C$

29. Find the mean and range for the following numbers.

$$-8, -12, 5, -2, 0, 0, -15, 10, 10, -1, 0, 1$$

Write the following in decimal notation.

30. 4.6×10^3

31. 7.93×10^{-4}

32. 1.006×10^5

33. 8×10^{-1}

34. 3.55×10^{-6}

35. 5×10^4

Write the following in scientific notation.

36. 0.053

37. 90,000

38. 423

39. 0.7

40. 23,500,000

41. 0.000504

Explain the meaning of each term. Use your own examples.

42. absolute value

43. negative integer

NAME   DATE   HOUR

1. Multiply $386 \cdot 419$.

2. Simplify $7 \cdot 2 + 4 \cdot 5$.

3. Solve $4w = 36$.

4. Divide $4{,}500 \div 100$.

5. Evaluate $a + 4y$ when $a = 3$ and $y = 10$.

6. Add $\dfrac{5}{12} + \dfrac{11}{16}$.

7. Subtract $8 - 5\dfrac{2}{7}$.

8. Divide $11\dfrac{7}{18} \div 5$.

9. Multiply $48 \cdot \dfrac{7}{8}$.

10. Add $4.093 + 28 + 13.054 + 0.09$.

11. Multiply $0.084 \cdot 50$.

12. You have \$28.52. A toy costs \$3.11. That includes tax.

   a. How many of the toys can you buy with your money?

   b. How much money is left over?

13. Solve $\dfrac{16}{45} = \dfrac{x}{100}$. Round to the nearest hundredth.

14. A recreation council spent 14% of its budget on administrative expenses. If it spent \$11,564 on administrative expenses, how large is its budget?

15. Dana earns a 9% commission on sales. She sold a bedroom set for \$3,450. What was her commission?

Use the line graph to answer questions 16–20.

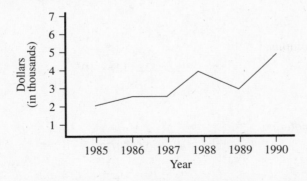

Average Weekly Sales for Eldridge Music, Inc., 1985–1990

16. What were the average weekly sales in 1988?

17. About how much more was sold per week in 1989 compared to 1985?

18. When did average weekly sales drop? By how much?

19. In what two years were average weekly sales the same?

20. Estimate the total sales for all of 1986.

21. 3.2 m = ? cm

22. 3,200 dm = ? km

23. Find the perimeter.

18 ft

12 ft

24. Find the area.

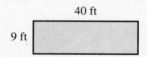
40 ft

9 ft

25. Add $(-6) + (5)$.

26. Add $-3 + 8 + 2 - 4 - 10$.

27. Subtract $(-12) - (-8)$.

28. Multiply $(-5)(-7)$.

29. Divide $(60) \div (-6)$.

30. Simplify $(-30) \div (5)(-6)$.

Write the following in decimal notation.

31. 4×10^{-2}

32. 1.8×10^{3}

Write the following in scientific notation.

33. 405,000

34. 0.072

Explain the meaning of each term. Use your own examples.

35. area

36. commission

Solving Equations with Applications

USING EQUATIONS TO HELP YOU SOLVE
PRACTICAL PROBLEMS

A coat is on sale for $24. This is a 40% discount off the original price. What is the original price of the coat?

Let p = original price of coat. The discount is 40% of $p = 0.4 \cdot p$. The sale price is

$$\text{original price} - \text{discount} = \text{sale price}$$

$$p - 0.4p = 24.$$

To find the original price, solve this equation for p. You will learn how to solve this equation in this chapter.

Equations are a tool that you can use to solve many real-life problems. In science, equations are used to find the distance a projectile travels. In business, equations are used in computer programs to calculate mortgage payments and compute compound interest. In electronics, equations are used to design the circuits in the television and microwave oven used in the home. Equations are a fundamental concept in many areas. In this chapter, you will learn how to solve linear equations in one variable.

Work through each section in this chapter. Use this test to identify topics you are not familiar with. These topics may require additional study. You may *not* use a calculator.

12.1 *Identify each term and the coefficient of each term.*

1. $5x - 3y$

2. $8x - y + 6$

3. Which number, if any, is a solution for $5x - 9 = 7x + 3$?
   a. $x = 6$ b. $x = -6$ c. $x = 5$

4. Which expressions, if any, are equivalent to $5x - 7y$?
   a. $5y - 7x$ b. $7y - 5x$ c. $-7y + 5x$

12.2

5. Solve $x - 7 = -10$.

6. Solve $12 = -5 + x$.

12.3

7. Solve $-6x = 24$.

8. Solve $7x + 8 = -6$.

12.4 *Remove the parentheses.*

9. $4(2x + 9)$ 10. $-3(5 - 7x)$

11. $6(4 - x + 7y)$

Combine like terms.

12. $5x + 12y$ 13. $4y - y + 6y$

14. $7p - 9 - p + 13$

15. $9x + 3x = 36$ 16. $5x + 9 = 16 - 2x$

17. $x + 7 - 4x = 1 + 3x - 12$

12.5 *Translate each word phrase into an algebraic expression.*

18. The product of 7 and p.

19. Eight less than v.

20. y more than the product of 7 and t.

21. 14 less than 6 times an unknown number is -2. Find the number.

12.6

22. Find the missing side of the triangle.

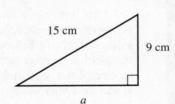

VOCABULARY *Explain the meaning of each term. Use your own examples.*

23. variable

24. equivalent equations

680

12.1 VARIABLES, TERMS, EXPRESSIONS, AND EQUATIONS

OBJECTIVES

1 Define algebraic terminology (Section 2.5).

2 Determine when a variable has a coefficient of $+1$ or -1 (Section 1.7).

3 Change the order of the terms in an expression.

4 Define an equation (Section 2.6).

NEW VOCABULARY

variable	constant
algebraic expression	equation
term	solution
coefficient	equivalent equation

1 Algebraic Terminology

Throughout this text you have seen letters represent numbers. These letters are called **variables**.

An **algebraic expression** is any collection of numbers, variables, and operations. For example,

$$4x - 7y + 2$$

is an algebraic expression. The variables are x and y.

An algebraic expression consists of one or more **terms**. Terms are separated by plus and minus signs. The expression $4x - 7y + 2$ has three terms: $+4x$, $-7y$, and $+2$.

OBSERVE $4x - 7y + 2$ is also written $4x + (-7y) + 2$.

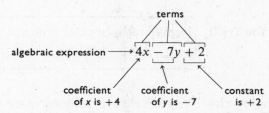

A **coefficient** is the number in a term that is multiplied by the variable. The coefficient of the term $+4x$ is $+4$. The coefficient of the term $-7y$ is -7. $+2$ is not a coefficient. It has no letter. It is called a **constant** term.

CAUTION A coefficient consists of a *number and its sign*.

EXAMPLE 1 Identify each variable, term, and coefficient.

a. $3x - 8w$
  - variables: x and w
  - terms: $3x$ and $-8w$
  - coefficients: $+3$ is the coefficient of x
  - -8 is the coefficient of w

b. $-4 + 6x - 3y + 1.2z$
  - variables: x, y, and z
  - terms: -4 (constant term), $6x$, $-3y$, and $1.2z$
  - coefficients: $+6$ is the coefficient of x
  - -3 is the coefficient of y
  - $+1.2$ is the coefficient of z ∎

▶ **You Try It** Identify each variable, term, and coefficient.

1. $5x - 3w$ **2.** $-6s + 3t - 8$ **3.** $-5x - 4 + 2y - 1.5z$

---

2 Coefficients of +1 and −1

The variable x has a coefficient of $+1$.

$$x = \boxed{+1} \cdot x = 1x$$

1 is the multiplicative identity (Section 1.7). 1 times any number or variable gives the same number or variable as an answer. A negative variable, such as $-x$, has a coefficient of -1.

$$-x = \boxed{-1} \cdot x = -1x$$

EXAMPLE 2

$y = \boxed{+1} \cdot y$ or $1y$ $-y = \boxed{-1} \cdot y$ or $-1y$

$p = \boxed{+1} \cdot p$ or $1p$ $-p = \boxed{-1} \cdot p$ or $-1p$ ■

▶ **You Try It**

What is the coefficient of each term?

4. t **5.** $+g$ **6.** $-m$ **7.** $-1p$ **8.** $1w$

EXAMPLE 3

Identify each variable, term, and coefficient in $1 - x + y$.

Observe, $1 - x + y = 1 - 1x + 1y$.

variables: x and y
terms: $+1$ (constant term), $-x$, and $+y$
coefficients: -1 is the coefficient of x
                $+1$ is the coefficient of y ■

▶ **You Try It**

Identify each variable, term, and coefficient.

9. $a - b - 1$

---

3 Changing the Order of Terms in an Expression

You can change the order of the terms in an algebraic expression. This gives an equivalent expression. Just be sure each term keeps the same coefficient (sign *and* number) and variable.

$$\left.\begin{array}{l} 4x - 7y + 2 \\ -7y + 2 + 4x \\ +2 + 4x - 7y \end{array}\right\}$$ Three different ways to write the same expression. Notice that each term kept its coefficient and variable.

EXAMPLE 4

a. $3y - 7z$ is equivalent to $-7z + 3y$
b. $-x + 3t$ is equivalent to $3t - x$
c. $2x + y - 3$ can be rewritten in five equivalent ways.

$$2x - 3 + y, \quad y + 2x - 3, \quad y - 3 + 2x,$$
$$-3 + 2x + y, \quad \text{and} \quad -3 + y + 2x \quad ■$$

▶ **You Try It**

Rewrite each expression in one equivalent way.

10. $5m - 2n$ **11.** $-f + 6g$

Rewrite the expression in five equivalent ways.

12. $-4x + 7y - 2$

---

4 Equations

An **equation** is a statement that two expressions are equal. If an equation contains only numbers, then it is either true or false.

$$5 + 3 = 8 \text{ is a true equation}$$
$$4 \cdot 5 = 12 \text{ is false}$$
$$8 - 10 = -2 \text{ is true}$$

Many equations contain a variable.

EXAMPLE 5   Example of an equation containing one variable x.

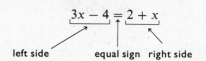

left side   equal sign   right side   ∎

To solve the equation in Example 5 means to find the number to replace x so that the left side $3x - 4$ equals the right side $2 + x$. This number is called the **solution** to the equation.

> A solution to an equation is a number that replaces the variable and makes the equation a true statement.

EXAMPLE 6   Is $x = 2$ a solution to $3x - 4 = 2 + x$?

Replace x with 2.

$$3 \cdot 2 - 4 \overset{?}{=} 2 + 2 \Bigg\}$$
$$6 - 4 \overset{?}{=} 2 + 2 \Bigg\}$$

To simplify, apply the order of operations.

$$2 \neq 4 \qquad \text{false}$$

Since the two sides are unequal, $x = 2$ is not a solution.   ∎

▶ **You Try It**   **13.** Is $x = 5$ a solution to $3x - 4 = 2x$?

EXAMPLE 7   Is $x = 3$ a solution to $3x - 4 = 2 + x$?

Replace x with 3.

$$3 \cdot 3 - 4 \overset{?}{=} 2 + 3$$
$$9 - 4 \overset{?}{=} 2 + 3$$
$$5 = 5 \qquad \text{true}$$

Since the two sides are equal, $x = 3$ is a solution of the equation. You also say $x = 3$ *satisfies* the equation.   ∎

▶ **You Try It**   **14.** Is $y = 2$ a solution to $9 + 2y = 8y - 3$?

Two equations are **equivalent equations** if they have the same solution. For example, $5x + 3 = 2x + 9$ and $3x = 6$ are equivalent equations because both have the same solution, $x = 2$.

Replace x with 2:	*Replace x with 2:*
$5x + 3 = 2x + 9$	$3x = 6$
$5 \cdot 2 + 3 = 2 \cdot 2 + 9$	$3 \cdot 2 = 6$
$13 = 13$ true	$6 = 6$ true

OBSERVE The solution, $x = 2$, is also an equation. It is its own solution because replacing x with 2 gives $2 = 2$, a true statement. Therefore, $x = 2$, $5x + 3 = 2x = 9$, and $3x = 6$ are all equivalent.

Equivalent equations are equations that have the same solution.

OBSERVE Some equations have more than one solution. The equation $x^2 = 16$ has two solutions, $x = 4$ and $x = -4$.

$$(4)^2 = 16 \quad \text{and} \quad (-4)^2 = 16$$

OBSERVE In the world of computer programming, words can be used as variables. Recall, the area of a rectangle is given by $A = L \cdot W$. You can also write this formula using words.

AREA = LENGTH · WIDTH

Here the variables A, L, and W are replaced with the variables AREA, LENGTH, and WIDTH. If LENGTH = 5 and WIDTH = 6, the computer replaces each word with its number, and evaluates the formula. The answer, 30, is stored in the variable AREA.

▶ **Answers to You Try It**   1. x, w; $+5x$, $-3w$; $+5$ is the coefficient of x, -3 is the coefficient of w 2. s, t; $-6s$, $+3t$, -8; -6 is the coefficient of s, 3 is the coefficient of t, -8 is the constant term   3. x, y, z; $-5x$, -4, $+2y$, $-1.5z$; -5 is the coefficient of x, -4 is the constant term, $+2$ is the coefficient of y, -1.5 is the coefficient of z 4. 1   5. 1   6. -1 7. -1 8. 1   9. a, b; $+a$, $-b$, -1; $+1$ is the coefficient of a, -1 is the coefficient of b, -1 is the constant term   10. $-2n + 5m$ 11. $6g - f$ 12. $-4x - 2 + 7y$, $7y - 4x - 2$, $7y - 2 - 4x$, $-2 - 4x + 7y$, $-2 + 7y - 4x$ 13. no, $11 \neq 10$ 14. yes, $13 = 13$

SECTION 12.1 EXERCISES

1 2 *List each term, coefficient, and constant.*

1. $6x + 2$ **2.** $9 + 10y$ **3.** $5w - 4$ **4.** $-5 + 6w$

5. $-8x + 4s$ **6.** $4m - 2n$ **7.** $3v$ **8.** $-4x$

9. $x + 1$ **10.** $1 - y$ **11.** $-2x + 4c - 2$

12. $12x - 4r - 1$ **13.** $8 - g + y$ **14.** $-h - 6 + s$

15. $x - 2y - 3z + 1$ **16.** $-7w + 4x - y - 8$

3 *Which of the three expressions, if any, are equivalent to the one on the left?*

17. $3x + 5y$ **a.** $5x + 3y$ **b.** $5y + 3x$ **c.** $3x - 5y$

18. $7x - 2$ **a.** $-2 + 7x$ **b.** $2x - 7$ **c.** $2 - 7x$

19. $2x - 3y + 1$ **a.** $1 + 2x - 3y$ **b.** $-3y + 1 + 2x$ **c.** $3y - 2x + 1$

20. $8 + y - z$ **a.** $y - z + 8$ **b.** $-y + 8 + z$ **c.** $z - y + 8$

21. $-7x + 5 + 3y$ **a.** $5 + 3y - 7x$ **b.** $-7x + 3y + 5$ **c.** $3y - 7x + 5$

22. $x - 2y + 3z$ **a.** $2y - x + 3z$ **b.** $3z - 2y + x$ **c.** $x - 2y + 3$

4 *Determine which number solves the given equation, if any.*

23. $x + 3 = 5$ **a.** $x = +8$ **b.** $x = -2$ **c.** $x = +2$

24. $6 + y = 2$ **a.** $y = -4$ **b.** $y = +2$ **c.** $y = +4$

25. $3x + 8 = 2$ **a.** $x = -1$ **b.** $x = -2$ **c.** $x = -3$

26. $-7 + 2w = 5$ **a.** $w = -1$ **b.** $w = 0$ **c.** $w = 6$

27. $7 - 4v = 3$ **a.** $v = -1$ **b.** $v = 2.5$ **c.** $v = 3$

28. $4x - 1 = -9$ **a.** $x = -2$ **b.** $x = -3$ **c.** $x = 5$

29. $3x + 7 = 5x - 1$ **a.** $x = 3$ **b.** $x = -4$ **c.** $x = 4$

30. $6 - 2y = 3y - 19$ **a.** $y = 5$ **b.** $y = 0$ **c.** $y = -5$

31. $-10 = 14 - 4z$ **a.** $z = 1$ **b.** $z = 6$ **c.** $z = -1$

32. $4x + 5 = 7x + 5$ **a.** $x = 2$ **b.** $x = -1$ **c.** $x = 1$

SKILLSFOCUS (Section 6.7) *Solve each proportion.*

33. $\dfrac{x}{14} = \dfrac{18}{21}$ **34.** $\dfrac{x}{24} = \dfrac{25}{40}$ **35.** $\dfrac{y}{7} = \dfrac{5}{3}$ **36.** $\dfrac{x}{9} = \dfrac{6}{13}$

WRITING TO LEARN ▶▶▶▶

37. In your own words, explain the difference between a coefficient and a constant. Give an example of each.

38. In your own words, explain why $5 - 3x$ is not the same as $3x - 5$.

39. Explain what equivalent equations are. Give examples.

▶ YOU BE THE JUDGE

40. Angela said the constant term for the expression $7x + 2y$ is 0. Is she correct? Explain your decision.

12.2 SOLVING EQUATIONS: THE ADDITION RULE

OBJECTIVES
1. Define the addition rule (Section 2.6).
2. Solve equations of the form $x + a = b$.

NEW VOCABULARY
addition rule

---

1 The Addition Rule

The simplest equation is a number set equal to itself. For example,

$$5 = 5$$

is an equation. Add the same amount, say 3, to each side. You still get a true equation.

EXAMPLE 1

$5 = 5$	Start with a true equation.
$5 + \boxed{3} = 5 + \boxed{3}$	Add 3 to each side.
$8 = 8$	You get a true equation. ∎

If you add the same negative number to each side, you get a true equation.

EXAMPLE 2

$5 = 5$	Start with a true equation.
$5 + \boxed{(-3)} = 5 + \boxed{(-3)}$	Add -3 to each side.
$2 = 2$	You get a true equation. ∎

Now suppose you are given an equation with a variable x in it.

$$x + 4 = 9$$

To solve an equation for x means find $x = +1x = 1x$. To do this, change the equation into the equivalent equation

$$x = a$$

where a is the solution. Note: $x = a$ can also be written

$$+1 \cdot x + 0 = a.$$

means coefficient of x must be $+1$ means the only other term allowed to be on the same side of the equation as x is 0

EXAMPLE 3   Solve for x: $x + 4 = 9$

$x + 4 = 9$	Adding the same number to each side gives a true equation. To solve for x, add the opposite of 4, or -4, to each side.
$x + \underbrace{4 + (-4)} = \underbrace{9 + (-4)}$	
$x + \quad 0 \quad = \quad 5$	A number plus its opposite equals 0.
$x = 5$	By the additive identity property of 0, $x + 0 = x$. ∎

$x = 5$ is the solution to $x + 4 = 9$. To check, replace x with 5 in the equation.

Check: $x + 4 = 9$
$\boxed{5} + 4 \overset{?}{=} 9$
$9 = 9$ true

▶ **You Try It**   **1.** Solve for x and check: $x + 6 = 14$

> **The Addition Rule**
>
> If you add or subtract the same quantity on each side of an equation, you get an equivalent equation.

2 Solving Equations of the Form
$x + a = b$

> To solve the equation $x + a = b$ for x
>
> 1. Locate the constant term on the same side of the equation as x.
> 2. Add the opposite of this term to both sides (addition rule).
> 3. Simplify. The variable $x = 1x$ will be on one side of the equation, by itself. The solution is on the other side.

EXAMPLE 4   Solve for x: $x + 7 = 2$

$$x + 7 = 2$$

The term on the same side as x is $+7$.

$$x + \underbrace{7 + (-7)} = \underbrace{2 + (-7)}$$

Add the opposite of $+7$, or -7, to each side.

$$x + \quad 0 \quad = \quad -5$$

Simplify. The sum of a number and its opposite is 0.

$$x = -5$$

By the zero property of addition, $x + 0 = x$.

The solution is $x = -5$.

Check: $x + 7 = 2$
$$-5 + 7 \overset{?}{=} 2$$
$$2 = 2 \quad \blacksquare$$

▶ **You Try It**   **2.** Solve for x and check: $x + 4 = 1$

Write each subtraction in an equation as addition of the opposite.

EXAMPLE 5   Solve for x: $x - 7 = -4$

$$x - 7 = -4$$
$$x + (-7) = -4$$

Write the subtraction $x - 7$ as addition of the opposite, $x + (-7)$.

$$x + \underbrace{(-7) + 7} = \underbrace{-4 + 7}$$
$$x + \quad 0 \quad = \quad 3$$
$$x = 3$$

The term on the same side as x is -7. Add the opposite of -7, or 7, to each side.

The solution is $x = 3$.

Check: $x - 7 = -4$
$$3 - 7 \overset{?}{=} -4$$
$$-4 = -4 \quad \blacksquare$$

▶ **You Try It**   **3.** Solve for x and check: $x - 9 = -1$

> **Summary**
>
> Properties used to solve equations are:
>
> *The sum of a number and its opposite is 0:* $a + (-a) = 0$
> *The additive identity property of 0:* $a + 0 = a$ and $0 + a = a$
> *The multiplicative identity property of 1:* $1x = x$ and $x1 = x$

▶ Answers to You Try It   1. 8   2. −3   3. 8

SECTION 12.2 EXERCISES

1 **2** *Solve each equation. Check all answers.*

1. $x + 5 = 12$

2. $x + 3 = 8$

3. $4 + x = -7$

4. $-3 + x = 8$

5. $x - 3 = -2$

6. $x - 10 = 7$

7. $-5 = x + 8$

8. $10 = x + 10$

9. $4 = x - 9$

10. $-7 = x - 7$

11. $y + 6 = 3$

12. $y + 12 = -8$

13. $-6 + w = 1$

14. $13 + w = 7$

15. $z - 8 = 0$

16. $z - 12 = 12$

17. $20 = v + 30$

18. $-12 = v + 6$

19. $-1 = t - 1$

20. $2 = t - 5$

21. $a + 6 = -6$

22. $a - 6 = 6$

23. $b + 5 = -7$

24. $-3 + h = -12$

25. $k - 9 = -10$

26. $x - 40 = 32$

27. $-15 = x + 9$

28. $-23 = x + 12$

29. $-3 = x - 14$

30. $100 = x - 50$

31. $x + \dfrac{1}{4} = \dfrac{3}{4}$

32. $x + \dfrac{1}{6} = \dfrac{5}{6}$

33. $\dfrac{7}{10} + t = -\dfrac{3}{10}$

34. $x - \dfrac{16}{25} = -\dfrac{9}{25}$

35. $-\dfrac{5}{7} = d + \dfrac{5}{7}$

36. $-\dfrac{3}{7} = x - \dfrac{3}{7}$

37. $x + 3.05 = 2$

38. $x + 0.8 = -1.4$

39. $-8.35 + x = 20$

40. $0.05 + v = 1$

41. $-0.48 = d - 0.6$

42. $2.04 = x + 5.5$

Perform each operation.

43. $(-3)(+5)$ **44.** $(-8)(-6)$ **45.** $-1 \cdot (+3)$ **46.** $-4 \cdot (-9)$

47. $\dfrac{-12}{-4}$ **48.** $\dfrac{42}{-6}$ **49.** $\dfrac{-7}{-1}$ **50.** $\dfrac{-72}{+72}$

EXTEND YOUR THINKING ▶▶▶▶

▶ TROUBLESHOOT IT

Find and correct the error.

51. Solve $x - 5 = 8$ for x.
$$x - 5 = 8$$
$$x - 5 + (-5) = 8 + (-5)$$
$$x + 0 = 3$$
$$x = 3$$

52. Solve $y + 8 = 3$ for y.
$$y + 8 = 3$$
$$y + 8 + (-3) = 3 + (-3)$$
$$y + 5 = 0$$
$$y = 5$$

WRITING TO LEARN ▶▶▶▶

53. In a short paragraph, explain to a friend how to solve $x + 3 = -8$. Give a written reason for each step you take.

54. Find another equation equivalent to $w = -5$. Then explain why the two equations are equivalent.

55. **a.** Write three equations, each having $+4$ for a solution.

     b. Write three equations, each having -7 for a solution.

     c. Write an equation with a solution of 0.

12.3 SOLVING EQUATIONS: THE MULTIPLICATION RULE

OBJECTIVES

1 Define the multiplication rule (Section 2.7, 3.5).

2 Solve equations of the form $ax = b$.

3 Use both the addition and multiplication rules to solve equations of the form $ax + b = c$.

NEW VOCABULARY

multiplication rule

---

1 The Multiplication Rule

Multiply each side of the equation

$$6 = 6$$

by the same number, say 3. Your answer is another true equation.

EXAMPLE 1

$6 = 6$ Start with a true equation.

$3 \cdot 6 = 3 \cdot 6$ Multiply each side by 3.

$18 = 18$ You get a true equation. ■

Divide each side by the same number. You also get a true equation.

EXAMPLE 2

$6 = 6$ Start with a true equation.

$\dfrac{6}{3} = \dfrac{6}{3}$ Divide each side by 3.

$2 = 2$ You get a true equation. ■

Division and multiplication are related operations. Dividing by 3 gives the same answer as multiplying by its reciprocal, $\dfrac{1}{3}$.

EXAMPLE 3

$6 = 6$ Start with a true equation.

$\dfrac{1}{\cancel{3}} \times \dfrac{\cancel{6}^{2}}{1} = \dfrac{1}{\cancel{3}} \times \dfrac{\cancel{6}^{2}}{1}$ Multiply each side by $\dfrac{1}{3}$, the reciprocal of 3.

$2 = 2$ You get a true equation. ■

▶ **You Try It**

1. Show that you get a true equation when you multiply each side of the equation $8 = 8$ by 5.

2. Show that you get a true equation when you divide each side of the equation $30 = 30$ by 6.

3. Show that you get a true equation when you multiply each side of the equation $30 = 30$ by the reciprocal of 6.

Now suppose you are given an equation with a variable x in it.

EXAMPLE 4   Solve for x: $3x = 12$

To solve for x, divide each side of the equation by 3, the coefficient of x. This will leave $1x = x$ on the left side.

$$3x = 12$$

$$\frac{3x}{3} = \frac{12}{3} \qquad \text{Divide each side by 3, the coefficient of } x.$$

$$\frac{\overset{1}{\cancel{3x}}}{\underset{1}{\cancel{3}}} = \frac{\overset{4}{\cancel{12}}}{\underset{1}{\cancel{3}}} \qquad \text{Note that } \frac{\overset{1}{\cancel{3x}}}{\cancel{3}} = \frac{1x}{1} = 1x.$$

$$1x = 4$$

$$x = 4 \qquad \text{By the multiplicative identity property of 1, } 1x = x. \quad \blacksquare$$

$x = 4$ is the solution to $3x = 12$. To check, replace x with 4 in the original equation.

$$
\begin{aligned}
\text{Check:} \quad 3x &= 12 \\
3 \cdot 4 &\overset{?}{=} 12 \\
12 &= 12 \qquad \text{true}
\end{aligned}
$$

▶ **You Try It**   **4.** Solve $5x = 20$ by dividing each side by the coefficient of x.

The Multiplication Rule

If you multiply or divide each side of an equation by the same number, you get an equivalent equation. Note: Dividing by a number n gives the same result as multiplying by its reciprocal, $\dfrac{1}{n}$.

2 **Solving Equations of the Form $ax = b$**

EXAMPLE 5   Solve $6x = -30$.

$$6x = -30 \qquad \text{Note that } 6x \text{ means } 6 \cdot x.$$

$$\frac{\overset{1}{\cancel{6}}x}{\underset{1}{\cancel{6}}} = \frac{-30}{6} \qquad \text{Divide each side by 6, the coefficient of } x.$$

$$
\begin{aligned}
1x &= -5 \qquad \text{Dividing unlike signs gives a negative answer.} \\
x &= -5
\end{aligned}
$$

The solution is $x = -5$.   Check:
$$
\begin{aligned}
6x &= -30 \\
6(-5) &\overset{?}{=} -30 \\
-30 &= -30 \quad \blacksquare
\end{aligned}
$$

▶ **You Try It**   **5.** Solve $7x = -42$.   **6.** Solve $4w = 36$.

EXAMPLE 6   Solve $16 = -8x$.

$$16 = -8x \qquad \text{Note, } -8x \text{ means } -8 \cdot x.$$

$$\frac{16}{-8} = \frac{\overset{1}{\cancel{-8}}x}{\underset{1}{\cancel{-8}}} \qquad \text{Divide each side by } -8, \text{ the coefficient of } x.$$

$$-2 = 1x$$
$$-2 = x$$

The solution is $x = -2$.   Check: $16 = -8x$

$$16 \overset{?}{=} -8(\boxed{-2})$$
$$16 = 16 \quad \blacksquare$$

▶ **You Try It**   **7.** Solve $15 = -5x$.   **8.** Solve $-24 = -6x$.

EXAMPLE 7   Solve $-x = 3$.

$$-x = 3$$
$$-1 \cdot x = 3 \qquad \text{Recall, } -x = -1 \cdot x.$$
$$\frac{-1 \cdot x}{-1} = \frac{3}{-1} \qquad \text{Divide each side by } -1, \text{ the coefficient of } x.$$
$$1x = -3$$
$$x = -3$$

The solution is $x = -3$.   Check: $-x = 3$

$$-1 \cdot x = 3$$
$$-1 \cdot (\boxed{-3}) \overset{?}{=} 3$$
$$3 = 3 \quad \blacksquare$$

▶ **You Try It**   Solve.

9. $-y = -8$ **10.** $18 = -w$

If the variable is divided by an integer, multiply both sides of the equation by the integer.

EXAMPLE 8   Solve $\dfrac{x}{7} = -4$.

$$\frac{x}{7} = -4 \qquad \begin{array}{l}\text{The variable } x \text{ is divided by 7. Multiply}\\ \text{both sides of the equation by } 7 = \dfrac{7}{1}.\end{array}$$

$$\frac{\overset{1}{\cancel{7}}}{1} \cdot \frac{x}{\underset{1}{\cancel{7}}} = \frac{7}{1} \cdot \frac{-4}{1}$$

$$1 \cdot x = 7(-4)$$
$$x = -28$$

The solution is $x = -28$.

Check: $\dfrac{x}{7} = -4$

$$\dfrac{-28}{7} \overset{?}{=} -4$$

$$-4 = -4 \qquad \blacksquare$$

▶ **You Try It** **11.** Solve $\dfrac{x}{3} = -8$.

12. Solve $\dfrac{y}{-6} = -5$.

---

3 **Solving Equations of the Form** $ax + b = c$

Some equations must be solved using both the addition rule and the multiplication rule.

EXAMPLE 9 Solve $3x + 2 = 14$.

$3x + 2 = 14$

Solve for $x = +1 \cdot x + 0$. This means change the equation so the term on the side of the equation with x is 0, and the coefficient of x is $+1$.

$3x + \underbrace{2 + (-2)}_{} = \underbrace{14 + (-2)}_{}$

First, apply the addition rule. Add the opposite of 2, or -2, to each side.

$3x + \qquad 0 \quad = \qquad 12$

This leaves the x term by itself on the left side.

$$\dfrac{3x}{3} = \dfrac{12}{3}$$

Second, apply the multiplication rule. Divide each side by 3, the coefficient of x.

$1x = 4$

$x = 4$

The solution is $x = 4$.

Check: $3x + 2 = 14$

$3(4) + 2 \overset{?}{=} 14$

$12 + 2 \overset{?}{=} 14$

$14 = 14 \qquad \blacksquare$

▶ **You Try It** **13.** Solve $5x + 3 = 13$.

14. Solve $6x + 1 = -11$.

To solve an equation of the form $ax + b = c$

1. First, *apply the addition rule* to remove the constant term being added to the x term. Add the opposite of this term to each side of the equation.
2. Second, *apply the multiplication rule* to make the coefficient of the x term $+1$.
3. *Check* your solution.

EXAMPLE 10   Solve $7x - 6 = -20$.

$$7x + (-6) = -20$$ Write subtraction as addition of the opposite.

$$7x + (-6) + 6 = -20 + 6$$ Add the opposite of -6, or 6, to each side.

$$7x + 0 = -14$$

$$\frac{7x}{7} = \frac{-14}{7}$$ Divide each side by 7, the coefficient of x.

$$1x = -2$$

$$x = -2$$

The solution is $x = -2$.

Check: $7x - 6 = -20$

$$7(-2) - 6 \overset{?}{=} -20$$

$$-14 - 6 \overset{?}{=} -20$$

$$-20 = -20 \quad\blacksquare$$

► **You Try It**   **15.** Solve $4x - 7 = 9$.   **16.** Solve $3x - 5 = -17$.

The variable term may be written to the right of the constant term in an equation. Follow the same procedure to solve it.

EXAMPLE 11   Solve $3 - 5x = -12$.

$$3 - 5x = -12$$

$$3 + (-3) - 5x = -12 + (-3)$$ Add the opposite of 3, or -3, to each side.

$$0 - 5x = -15$$

$$\frac{-5x}{-5} = \frac{-15}{-5}$$ Divide each side by -5, the coefficient of x.

$$+1x = 3$$ $-5 \div -5 = +1$ and $-15 \div -5 = +3$

$$x = 3$$

The solution is $x = 3$.

Check: $3 - 5x = -12$

$$3 - 5(3) \overset{?}{=} -12$$

$$3 - 15 \overset{?}{=} -12$$

$$-12 = -12 \quad\blacksquare$$

► **You Try It**   **17.** Solve $2 - 3x = -7$.   **18.** Solve $5 - 8x = 21$.

The variable term may be on the right side of the equation. Follow the same procedure to solve the equation.

EXAMPLE 12   Solve $-14 = -9 + 5x$.

$$-14 = -9 + 5x$$

$$-14 + 9 = -9 + 9 + 5x$$ Add the opposite of -9, or 9, to each side.

$$-5 = 0 + 5x$$ This leaves the x term by itself on the right.

$$\frac{-5}{5} = \frac{5x}{5}$$ Divide each side by 5, the coefficient of x.

$$-1 = x$$

The solution is $x = -1$.

$$\text{Check:} \quad -14 = -9 + 5x$$
$$-14 \overset{?}{=} -9 + 5(\boxed{-1})$$
$$-14 \overset{?}{=} -9 + (-5)$$
$$-14 = -14 \quad \blacksquare$$

▶ **You Try It**

19. Solve $45 = 6x - 3$.

20. Solve $-16 = 20 + 4x$.

SECTION 12.3 EXERCISES

1 **2** *Solve each equation. Check each answer.*

1. $4x = 20$

2. $2x = 12$

3. $5x = -35$

4. $3x = -18$

5. $-7x = 28$

6. $-9x = 63$

7. $-6x = -48$

8. $-12x = -60$

9. $-x = 8$

10. $-x = -4$

11. $16 = -4x$

12. $36 = -3x$

13. $\dfrac{x}{5} = 6$

14. $\dfrac{x}{8} = -2$

15. $\dfrac{x}{7} = -7$

16. $\dfrac{x}{6} = 30$

17. $2.4x = 7.2$

18. $6.5x = -45.5$

19. $-12.4 = 4x$

20. $-0.06 = -0.5x$

3

21. $3x + 6 = 21$

22. $6x + 8 = 20$

23. $2x - 9 = -1$

24. $7x - 12 = 9$

25. $8x + 6 = -2$

26. $5x + 10 = -5$

27. $8x - 20 = -4$

28. $12x - 12 = -60$

29. $9x + 27 = 0$

30. $-3x - 21 = 0$

31. $10 - 8x = 34$

32. $-20 - 6x = 16$

33. $-1 = 4 + 5x$

34. $50 = 200 + 3x$

35. $2.5x + 9 = -1$

36. $0.7x - 1.2 = 2.3$

SKILLSFOCUS (Section 3.5)

Multiply.

37. $\dfrac{6}{1} \cdot \dfrac{5}{8}$

38. $\dfrac{9}{1} \cdot \dfrac{7}{12}$

39. $4 \cdot \dfrac{5}{6}$

40. $15 \cdot \dfrac{3}{10}$

▶TROUBLESHOOT IT

Find and correct the error.

41. Solve for x: $5x = -20$

$$\frac{5x}{-5} = \frac{-20}{-5}$$

$$x = 4$$

42. Solve for x: $8 - 5x = 9$

$$3x = 9$$

$$x = 3$$

43. Solve for x: $\frac{x}{5} = -40$

$$x = \frac{-40}{5}$$

$$x = -8$$

WRITING TO LEARN ▶▶▶▶

44. In your own words, explain what the addition rule says. Make up and solve an example to accompany your explanation.

45. In your own words, explain what the multiplication rule says. Make up and solve an example to accompany your explanation.

46. Write three equations each with a solution of -5. Each equation must have a coefficient other than $+1$ on the variable term.

12.4 THE DISTRIBUTIVE PROPERTY AND ITS APPLICATION

OBJECTIVES	NEW VOCABULARY
1 Define the distributive property.	distributive property
2 Use the distributive property to solve equations.	combining like terms
3 Combine like terms (Section 9.1).	
4 Solve equations using a general procedure.	

1 The Distributive Property

The following two expressions give the same answer when simplified.

$$5(6 + 4) = 5(10) = 50$$
$$5 \cdot 6 + 5 \cdot 4 = 30 + 20 = 50$$

Therefore, the two expressions are equal.

$$5(6 + 4) = 5 \cdot 6 + 5 \cdot 4$$

multiply 5 times each
term in parentheses

Each term in the parentheses is multiplied by the number in front of the parentheses to get the expression on the right side. This is called the **distributive property** of multiplication over addition. This rule also applies when variables are used.

$$3(x + 6) = 3 \cdot x + 3 \cdot 6$$

multiply 3 times each
term in parentheses

or,

$$3(x + 6) = 3x + 18$$

The distributive property of multiplication over addition

$$a(b + c) = a \cdot b + a \cdot c \quad \text{and} \quad (b + c)a = b \cdot a + c \cdot a$$

for any numbers a, b, and c.

EXAMPLE 1 Apply the distributive property to remove the parentheses.

a. $4(2 + 5x) = 4 \cdot 2 + 4 \cdot 5x$
$$= 8 + 20x$$

b. $6(x - 7) = 6(x + (-7))$ Write the subtraction as addition of the opposite.

$$= 6 \cdot x + 6 \cdot (-7)$$
$$= 6x + (-42) \;\Big\}$$ Recall, adding the opposite is another way to write
$$= 6x - 42 \;\Big\}$$ subtraction.

c. $(8 + 3x)9 = 8 \cdot 9 + 3x \cdot 9$
$$= 72 + 27x$$

d. $-7(5 - 4x) = -7(5 + (-4x))$ Write the subtraction as addition of the opposite.

$$= (-7) \cdot 5 + (-7)(-4x)$$
$$= -35 + (+28x)$$
$$= -35 + 28x \quad \text{or} \quad 28x - 35 \quad \blacksquare$$

CAUTION A common error when applying the distributive property is shown here.

$$5(x + 2) = 5x + 2$$

This is **wrong**. The error is in forgetting to multiply 5 times the second term, 2. What should the answer be? The correct answer is $5x + 10$.

▶ **You Try It** Use the distributive property to remove parentheses.

1. $6(5 + 4x)$ 2. $2(5x + 9)$
3. $7(x - 2)$ 4. $9(2 - 3x)$
5. $(10x + 5)3$ 6. $(1 + 6x)7$
7. $-9(3 - 2x)$ 8. $-4(5 - 8x)$

2 **Using the Distributive Property to Solve Equations**

The distributive property is used to remove parentheses when solving equations.

EXAMPLE 2 Solve $3(2x + 5) = 21$.

$3(2x + 5) = 21$ Apply the distributive property to remove the parentheses.

$6x + 15 = 21$ Add the opposite of 15, or -15, to each side.

$6x + 15 + (-15) = 21 + (-15)$

$6x + 0 = 6$

$\dfrac{6x}{6} = \dfrac{6}{6}$

$x = 1$

Check: $3(2x + 5) = 21$

$3(2 \cdot 1 + 5) \stackrel{?}{=} 21$

$3(2 + 5) \stackrel{?}{=} 21$

$3(7) \stackrel{?}{=} 21$

$21 = 21$ ∎

▶ **You Try It** 9. Solve $4(3x + 1) = -32$.

Suppose an equation contains one or more fractions. All the denominators can be eliminated by multiplying both sides of the equation by the least common denominator. This is called *clearing the fractions*.

EXAMPLE 3 Solve $\dfrac{2x}{9} = \dfrac{8}{3}$.

$\dfrac{2x}{9} = \dfrac{8}{3}$ Multiply each side by $9 = \dfrac{9}{1}$, the least common denominator for 9 and 3.

$$\overset{1}{\underset{1}{\frac{\cancel{9}}{1}}} \cdot \frac{2x}{\underset{1}{\cancel{9}}} = \overset{3}{\underset{1}{\frac{\cancel{9}}{1}}} \cdot \frac{8}{\underset{1}{\cancel{3}}}$$ Reduce each side separately.

$$2x = 24$$

$$\frac{2x}{2} = \frac{24}{2}$$

$$x = 12$$

Check:
$$\frac{2x}{9} = \frac{8}{3}$$

$$\frac{2 \cdot 12}{9} \overset{?}{=} \frac{8}{3}$$

$$\underset{3}{\frac{\overset{8}{\cancel{24}}}{\cancel{9}}} = \frac{8}{3} \quad \blacksquare$$

▶ **You Try It**   **10.** Solve $\dfrac{3x}{8} = \dfrac{9}{4}$.

EXAMPLE 4   Solve $\dfrac{2x}{5} + 2 = \dfrac{9}{4}$.

$$\frac{2x}{5} + 2 = \frac{9}{4}$$ Clear the fractions. The least common denominator for 5 and 4 is 20.

$$\boxed{20}\left(\frac{2x}{5} + 2\right) = \boxed{20}\left(\frac{9}{4}\right)$$ Multiply each side of the equation by 20.

$$\overset{4}{\underset{1}{\frac{\cancel{20}}{1}}} \cdot \frac{2x}{\underset{1}{\cancel{5}}} + 20 \cdot 2 = \overset{5}{\underset{1}{\frac{\cancel{20}}{1}}} \cdot \frac{9}{\underset{1}{\cancel{4}}}$$ Apply the distributive property. Write $20 = \dfrac{20}{1}$. Reduce the fractions.

$$8x + 40 = 45$$

$$8x + 40 + \boxed{(-40)} = 45 + \boxed{(-40)}$$ Add the opposite of 40, or -40, to each side.

$$8x + 0 = 5$$

$$\frac{8x}{8} = \frac{5}{8}$$

$$x = \frac{5}{8} \quad \blacksquare$$

▶ **You Try It**   **11.** Solve $\dfrac{2x}{3} + 6 = \dfrac{7}{2}$.

If an equation has one or more decimals, you can eliminate all the decimals before solving. Just multiply each side of the equation by the power of ten with the number of zeros equal to the greatest number of decimal places in any number in the equation. This is called *clearing the decimals*.

EXAMPLE 5   Solve $2.5x + 1.75 = 11.75$.

$$2.5x + 1.75 = 11.75$$

Clear the decimals. The greatest number of decimal places is 2. Multiply each side by 100.

$$100(2.5x + 1.75) = 100(11.75)$$
$$250x + 175 = 1,175$$

Apply the distributive property.

$$250x + 175 + (-175) = 1,175 + (-175)$$
$$250x = 1,000$$
$$\frac{250x}{250} = \frac{1,000}{250}$$
$$x = 4$$

Check:
$$2.5x + 1.75 = 11.75$$
$$2.5(4) + 1.75 \stackrel{?}{=} 11.75$$
$$10 + 1.75 \stackrel{?}{=} 11.75$$
$$11.75 = 11.75 \quad \blacksquare$$

▶ **You Try It**   **12.** Solve $4.72x + 1.44 = 15.6$.

---

3 Combining Like Terms

Like terms have the same letter, or literal, part.

EXAMPLE 6   **a.** $3x$ and $5x$ are like terms because each term has the same letter, x.

b. $8y$ and $-3y$ are like terms because each term has the same letter, y.

c. $6x$ and $4y$ are unlike terms because they have different letters.

d. 3 and $7w$ are unlike terms because one term has a letter and the other does not.

e. $5x$ and $2x^2$ are unlike terms because x is not the same term as $x^2 = x \cdot x$. $\blacksquare$

▶ **You Try It**   Determine if the terms are like or unlike.

13. $8w$ and $2w$ **14.** $3p$ and -7 **15.** $6x$ and $6y$

16. $5m$ and $3m$ **17.** $4xy$ and $10xy$ **18.** $7x^2$ and $-5x^4$

To combine like units, you add the numbers and keep the unit.

$$5 \text{ ft} + 3 \text{ ft} = (5 + 3) \text{ ft}$$
$$= 8 \text{ ft}$$

Combine like terms in the same way. Add the coefficients and keep the variable.

$$5x + 3x = (5 + 3)x$$
$$= 8x$$

> **To combine like terms**
>
> **1.** Add the coefficients of the like terms together.
> **2.** Multiply this sum by the common letter.

Write each subtraction as addition of the opposite before combining terms.

EXAMPLE 7   Combine the like terms.

    a. $8x + 4x = (8 + 4)x$ Add the coefficients.
                 $= 12x$

    b. $9y - 5y = 9y + (-5y)$ Write subtraction as addition of the opposite.
                $= (9 + (-5))y$
                $= 4y$

    c. $-7x - x = -7x + (-1x)$
                  $= (-7 + (-1))x$
                  $= -8x$

    d. $4w - 6w + 3w = (4 + (-6) + 3)w$
                     $= 1w$ or w

    e. $7x + x - 5x + 4x - 3x = (7 + 1 + (-5) + 4 + (-3))x$
                          $= 4x$

    f. $-5x + 5x = (-5 + 5)x$
                $= (0)x$
                $= 0$

    g. $7 + 2x$ Cannot combine because the terms are unlike.

    h. $4x + 3y$ Cannot combine because the terms are unlike. ∎

▶ **You Try It**

19. $7x + 4x$	**20.** $8w - 6w$	**21.** $6x + 1x$
22. $-x - 4x$	**23.** $-6y + 10y$	**24.** $-8p + 8p$
25. $7x - 5x + 8x$	**26.** $-2x + 5x - 9x + x$	
27. $5 + 5w$	**28.** $9a + 3b$	

How do you combine like terms in the following expression?

$$5x + 7y + 2x - 4y$$

This expression contains x terms and y terms. First, write each subtraction as addition of the opposite.

$$5x + 7y + 2x + (-4y)$$

Then combine like terms as before.

$$5x + 7y + 2x + (-4y)$$
$$7x + 3y$$

OBSERVE It may help to identify like terms using circles and underlines.

To combine like terms in expressions having two or more different terms

1. Write each subtraction as addition of the opposite.
2. Combine terms that are alike.

EXAMPLE 8   Combine like terms.

a. $5x + 7 - 2x = \boxed{5x} + 7 + \boxed{(-2x)}$

$= 3x + 7$

b. $7 - 3x - 2 - 4x = \boxed{7} + \boxed{(-3x)} + \boxed{(-2)} + \boxed{(-4x)}$

$= 5 + (-7)x$

$= 5 - 7x$

c. $2x + 4z - 3x + 6 - 4z - 8 = \boxed{2x} + 4z + \boxed{(-3x)} + \boxed{6} + \boxed{(-4z)} + \boxed{(-8)}$

$= -1x + 0z + (-2)$

$= -x - 2$ ■

▶ **You Try It**   **29.** $7x + 6 - 5x$ **30.** $12 - 2x - 5 - 6x$

31. $3y + 7t - 10 - 8y + 6 + t$

---

4 **Solving Equations Using a General Procedure**

To solve equations

1. *Apply the distributive property* to remove parentheses, if there are any.
2. *Combine* like variable *terms* that are on the same side of the equal sign. Combine constants that are on the same side of the equal sign.
3. *Use the addition rule* to group variable terms on one side of the equal sign. Use it again to group constant terms on the other side. Combine like terms.
4. *Use the multiplication rule* to solve for $+1x$.
5. *Check* your answer.

If like terms are on the same side of the equal sign, combine like terms first. Then solve the equation.

EXAMPLE 9   Solve $5x + 7 - 3x = 19$.

$\boxed{5x} + 7 + \boxed{(-3x)} = 19$ Rewrite the subtraction as addition of the opposite.

$2x + 7 = 19$ Combine like terms on the left side.

$2x + 7 + (-7) = 19 + (-7)$ Now, solve as before.

$2x + 0 = 12$

$\dfrac{2x}{2} = \dfrac{12}{2}$

$x = 6$ ■

▶ **You Try It** **32.** Solve $8x + 2 - 4x = 22$.

Suppose variable and constant terms are on both sides of the equal sign. Use the addition rule to group the variable terms on one side. Use it again to group constants on the other side.

EXAMPLE 10 Solve $5x + 4 = 2x - 11$.

$$5x + 4 = 2x - 11$$

$$5x + \boxed{(-2x)} + 4 = \underbrace{2x + \boxed{(-2x)}} - 11 \qquad$$ Use the addition rule to group x terms on the left side. Add the opposite of $2x$ to each side.

$$3x \qquad + 4 = \qquad 0 \qquad - 11 \qquad$$ Combine like terms on each side.

$$3x + 4 + \boxed{(-4)} = -11 + \boxed{(-4)} \qquad$$ Use the addition rule to combine constant terms on the right side. Add the opposite of 4 to each side.

$$3x + 0 = -15$$

$$\frac{3x}{3} = \frac{-15}{3}$$

$$x = -5$$

Check:
$$5x + 4 = 2x - 11$$
$$5(\boxed{-5}) + 4 \stackrel{?}{=} 2(\boxed{-5}) - 11$$
$$-25 + 4 \stackrel{?}{=} -10 - 11$$
$$-21 = -21 \quad \blacksquare$$

▶ **You Try It** **33.** Solve $6x + 3 = 2x + 7$.

If an equation contains parentheses, first apply the distributive property to remove the parentheses.

EXAMPLE 11 Solve $10 - 4(3x - 5) = 3(1 - x)$.

$$10 - 4(3x - 5) = 3(1 - x)$$

$$10 + (-4)(3x + (-5)) = 3(1 + (-1x)) \qquad$$ Write each subtraction as addition of the opposite.

$$\boxed{10} + (-12x) + \boxed{20} = 3 + (-3x) \qquad$$ Apply the distributive property.

$$30 + (-12x) = 3 + (-3x) \qquad$$ Combine constants on the left.

$$30 + \underbrace{(-12x) + \boxed{3x}} = 3 + \underbrace{(-3x) + \boxed{3x}} \qquad$$ Add the opposite of $-3x$, or $3x$, to each side.

$$30 + \qquad (-9x) \qquad = 3 + \qquad 0$$

$$30 + \boxed{(-30)} + (-9x) = 3 + \boxed{(-30)}$$

$$0 + (-9x) = -27$$

$$\frac{-9x}{-9} = \frac{-27}{-9}$$

$$x = 3 \quad \blacksquare$$

▶ **You Try It** **34.** Solve $8 - 3(2x - 6) = 4(7 - x)$.

SOMETHING MORE

Equations with No Solutions   Solve the equation $x = x + 1$.

$$x = x + 1$$ Add the opposite of x, or $-x$, to each side.

$$\underbrace{x + (-x)}_{0} = \underbrace{x + (-x)}_{0} + 1$$

$$0 = 0 + 1$$ Combine like terms.

$$0 = 1$$ False. Therefore, there is no solution.

$0 = 1$ is a false statement. It tells you there is no solution to $x = x + 1$. In fact, $x = x + 1$ is impossible. Since x is a number, $x = x + 1$ is read *what number is equal to itself plus 1*. No number equals itself plus 1. Therefore, there is no solution.

When solving an equation, if the variables are eliminated, and you are left with a number equal to a different number, then there is no solution to the equation.

Solve each equation for x.

a. $3 + 5x = 5x + 7$

b. $6 - 3x = 4x + 1 - 7x$

c. $8 - 4x = 2(2x - 12)$

d. $-6(1 - x) = 8x - 2(x + 5)$

▶ **Answers to You Try It**   1. $30 + 24x$ 2. $10x + 18$ 3. $7x - 14$
4. $18 - 27x$ 5. $30x + 15$ 6. $7 + 42x$ 7. $-27 + 18x$ 8. $-20 + 32x$
9. $x = -3$ 10. $x = 6$ 11. $x = -\dfrac{15}{4}$ 12. $x = 3$ 13. like   14. unlike

15. unlike   16. like   17. like   18. unlike   19. $11x$ 20. $2w$ 21. $7x$
22. $-5x$ 23. $4y$ 24. 0 25. $10x$ 26. $-5x$ 27. cannot combine
28. cannot combine   29. $2x + 6$ 30. $7 - 8x$ 31. $-5y + rt - 4$
32. $x = 5$ 33. $x = 1$ 34. $x = -1$

▶ **Answers to Something More**   a. no solution   b. no solution   c. $x = 4$
d. no solution

SECTION 12.4 EXERCISES

1 *Remove the parentheses by applying the distributive property.*

1. $2(3x + 7)$ **2.** $8(5x + 9)$ **3.** $3(4x - 6)$

4. $9(3x - 5)$ **5.** $6(4 - 2x)$ **6.** $2(1 - x)$

7. $-3(6x + 1)$ **8.** $-2(-4 - 2x)$ **9.** $(7x - 3)7$

10. $(1 + 8x)(-1)$ **11.** $-(6 - 5x)$ **12.** $-(7x + 1)$

13. $8(-x - 3)$ **14.** $-5(1 + x)$ **15.** $2(3x + 2y + 4)$

16. $-5(1 - 3x + y)$ **17.** $7(9w - 4t + 2z)$ **18.** $0(x + y + z)$

2 *Solve by clearing the fractions.*

19. $\dfrac{3x}{4} = \dfrac{15}{2}$ **20.** $\dfrac{4x}{9} = \dfrac{5}{6}$ **21.** $\dfrac{2x}{5} - 10 = 6$

22. $\dfrac{3x}{4} + 12 = -9$ **23.** $\dfrac{4x}{5} + 3 = \dfrac{1}{6}$ **24.** $\dfrac{5x}{7} + 2 = \dfrac{3}{10}$

25. $\dfrac{3}{4} + \dfrac{5x}{8} = \dfrac{7}{2}$ **26.** $\dfrac{2x}{9} + \dfrac{2}{3} = \dfrac{5}{6}$

Solve by clearing the decimals.

27. $0.6x = 1.8$ **28.** $1.25x = 10$ **29.** $2.4x + 0.8 = -4$

30. $2.8x - 0.07 = 11.13$ **31.** $-1 = 0.69 + 1.3x$ **32.** $0.7x - 3.1 = -2.4$

3 *Combine like terms.*

33. $9x + 6x$ **34.** $7x + x$ **35.** $7w - 3w$

36. $10w - 4w$ **37.** $4x - 4x$ **38.** $y - y$

39. $2x - x$ **40.** $5z - z$ **41.** $-14x + 9x$

42. $-5y + 15y$ **43.** $-12x - 6x$ **44.** $-8x - 8x$

45. $-7x + 7x$ **46.** $-2z + 2z$ **47.** $-y - 6y$

48. $-x - 3x$ **49.** $7w + 2w + w$ **50.** $x + 2x + 3x$

51. $8x - 3x - 5x$ **52.** $9t - t - 8t$ **53.** $6x + 5 + 2x$

54. $3 + 4w + 8$ **55.** $2x + 7 - 5x$ **56.** $-4z + 2 + 4z$

57. $7x + 8y - 10x + 7y$

58. $-9x + 16 - 20 + 10x$

59. $8x - 5y + 3 - 11x + 5y - 1$

60. $3x - 7 + 4x + 5 - 7x + 2$

4 *Solve each equation.*

61. $6x + 4 - 4x = 12$

62. $5x - 7 - 8x = 5$

63. $1 - 8x + 7 = 16$

64. $1 - 3x + 3 = 7$

65. $-x = 2 + x$

66. $3x = 6 - 3x$

67. $10x - 1 = 5 + 4x$

68. $2x + 9 = 7x - 1$

69. $7y + 1 = 12 - 4y$

70. $8y + 25 = y - 3$

71. $4 + 2(6 + 3x) = 5x - 1$

72. $5 + 2(7x - 2) = 3x + 23$

73. $9 - 2(7 - 3x) = 4(2x - 5)$

74. $1 + 3(4 - x) = -3 - (8x - 5)$

SKILLSFOCUS (Section 2.4) *Evaluate each expression.*

75. Evaluate $x^2 + y^2$ when $x = 4$ and $y = 5$.

76. Find $5x + 6 - 2x$ when $x = 7$.

EXTEND YOUR THINKING ▶▶▶▶
▶TROUBLESHOOT IT

Find and correct the error.

77. $5 + 3x = (5 + 3)x = 8x$

78. $1 - x = 1 - 1 \cdot x = (1 - 1) \cdot x = 0 \cdot x = 0$

79. Solve. $4(2x + 5) = -3$
$$8x + 5 = -3$$
$$8x + 5 + (-5) = -3 + (-5)$$
$$8x = -8$$
$$x = -1$$

80. Combine like terms.
$$-3y + 7y - 6x + 6x$$
$$-3y + 7y + (-6)x + 6x$$
$$(-3 + 7)y + (-6 + 6)x$$
$$4y + x$$

81. $-3(y - x) = -3y - 3x$

82. Solve. $\dfrac{x}{2} = \dfrac{3}{5}$

$$\frac{\overset{1}{\cancel{5}}}{\cancel{10}} \cdot \frac{x}{\underset{1}{\cancel{2}}} = \frac{10}{1} \cdot \frac{3}{\underset{1}{\cancel{5}}}$$

$$x = 30$$

WRITING TO LEARN ▶▶▶▶

83. Explain the distributive property in your own words. Make up an example for your explanation.

84. Write a paragraph explaining how to combine like terms. Make up examples to use in your explanation. Include an example where terms cannot be combined.

12.5 APPLICATIONS

OBJECTIVES

1 Translate word phrases into algebraic expressions (Figure 1.3 and Figure 1.4 in Section 1.6, Figure 2.1 and Figure 2.2 in Section 2.3).

2 Solve applications using equations.

---

1 Translating Word Phrases into Algebraic Expressions

In order to use algebra to solve applications, you must be able to translate word phrases into algebraic expressions. For example, the sum of five and x is written

$$5 + x$$

because *sum* means add.

EXAMPLE 1 Translate the word phrases into algebra.

a. Addition: The sum of 2 and a number N is $2 + N$.

An unknown number x plus 1 is $x + 1$.

Y increased by 7 is $Y + 7$.

10 more than a number Q is $Q + 10$.

b. Subtraction: T minus 7 is $T - 7$.

20 decreased by W is $20 - W$.

4 less than x is $x - 4$.

Subtract x from 12 is $12 - x$.

OBSERVE With *less than* and *subtract* the terms are written in reverse order.

---

c. Multiplication: The product of 6 and x is $6 \cdot x$ or $6x$.

L times W is LW.

8 ties at D dollars each is $8D$ dollars.

$\frac{3}{4}$ of B is $\frac{3}{4} \cdot B$ or $\frac{3}{4}B$ or $\frac{3B}{4}$.

d. Division: The quotient of T and 4 is $\frac{T}{4}$.

X divided by A is $\frac{X}{A}$.

How many X's are in A is $\frac{A}{X}$. ∎

▶ **You Try It** Translate the word phrases into algebra.

1. the sum of L and -6
2. 11 more than $-3y$
3. m decreased by 80
4. 25 less than x
5. twice y
6. the product of 10 and $2x$
7. 8 divided by h
8. 6 divided into w

Some expressions have two or more different operations. Parentheses are sometimes used when one operation must be performed before another of higher precedence.

EXAMPLE 2   Translate expressions with two or more operations into algebra.

a. the product of y and 2 more than y

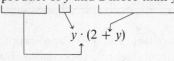

$$y \cdot (2 + y)$$

b. the quotient of z minus 6, and z

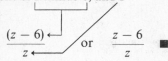

$$\frac{(z-6)}{z} \quad \text{or} \quad \frac{z-6}{z} \quad \blacksquare$$

▶ **You Try It**   Translate into algebra.

  9. the product of 3 and the sum of n and 5

  10. the quotient of 4, and the sum of 8 and k

---

2 **Solving Applications Using Equations**

An unknown number can be written using any letter. For example, the sum of 5 and a number can be written in any of the following ways.

$$\left.\begin{array}{l} \text{The sum of} \\ \text{5 and a number.} \end{array}\right\} \begin{array}{l} 5 + x \\ 5 + y \\ 5 + t \\ 5 + A \end{array} \left\{\begin{array}{l} \text{In each case,} \\ \text{the letter} \\ \text{represents the} \\ \text{unknown number.} \end{array}\right.$$

EXAMPLE 3   Seven times a number increased by 5 gives 33. Find the number.

Step 1: Read the problem carefully. You are to find the number.
Step 2: Let x = the unknown number.
Step 3: Translate the problem into an equation.

Seven times a number increased by 5 gives 33.

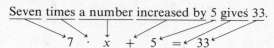

$$7 \cdot x + 5 = 33$$

Step 4: Solve the equation.

$$7x + 5 = 33$$
$$7x + 5 + (-5) = 33 + (-5)$$
$$\frac{7x}{7} = \frac{28}{7}$$
$$x = 4$$

The number x equals 4.

Step 5: Check: Test $x = 4$ in the original word problem.

7 times a number 4 is $7 \cdot 4 = 28$.
28 increased by 5 is $28 + 5 = 33$. $\blacksquare$

▶ **You Try It**   **11.** Six times a number increased by 4 results in 58. Find the number.

To solve applications using equations

1. Read and understand the problem. Determine what you are to find.
2. Choose a letter to represent the amount to be found in the problem. Represent any other amounts to be found in terms of this letter. Draw and label a picture if possible.
3. Translate the problem into an equation.
4. Solve the equation. Write a sentence stating your answer.
5. Check your solution in the original word problem.

EXAMPLE 4   Five times the current temperature decreased by 20 equals -35. What is the temperature?

Step 1: Find the current temperature.
Step 2: Let t = the unknown temperature.
Step 3: Translate the problem into an equation.

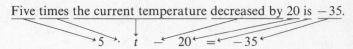

Five times the current temperature decreased by 20 is -35.

$$5 \cdot t - 20 = -35$$

Step 4: Solve the equation.

$$5t - 20 = -35$$
$$5t + (-20) + \boxed{(20)} = -35 + \boxed{20}$$
$$\frac{5t}{5} = \frac{-15}{5}$$
$$t = -3$$

The current temperature is $t = -3$.

Step 5: Check: 5 times the current temperature of -3 is $5 \cdot (-3) = -15$.
-15 decreased by 20 is $-15 - (20) = -15 + (-20) = -35$. ■

▶ **You Try It**   12. Three times Carmen's score increased by 7 gives -11. What is Carmen's score?

EXAMPLE 5   The length of a rectangle is 5 ft longer than the width. Its perimeter is 78 ft. Find the dimensions of the rectangle.

Step 1: Find the dimensions means find length and width.

Step 2:   Let W = the width of the rectangle.
$W + 5$ = the length of the rectangle, because the length is 5 ft more than the width.

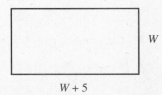

W

$W + 5$

Step 3: Perimeter = $2 \cdot$ length $+ 2 \cdot$ width   (See Section 10.2)
$$78 = 2(W + 5) + 2W$$

Step 4:
$$78 = 2(W + 5) + 2W$$
$$78 = 2W + 10 + 2W$$ Use the distributive property to eliminate parentheses.

$$78 = 4W + 10$$ Combine like terms.
$$78 + (-10) = 4W + 10 + (-10)$$
$$68 = 4W$$
$$\frac{68}{4} = \frac{4W}{4}$$
$$17 = W$$

The width is $W = 17$ ft. From Step 2, the length is $W + 5 = 17 + 5 = 22$ ft. The rectangle is 22 ft long by 17 ft wide.

Step 5: Check: The 22 ft length is 5 ft longer than the 17 ft width.

$$\text{Perimeter} = 2 \cdot \text{length} + 2 \cdot \text{width}$$
$$= 2 \cdot 22 \text{ ft} + 2 \cdot 17 \text{ ft}$$
$$= 44 \text{ ft} + 34 \text{ ft}$$
$$= 78 \text{ ft} \quad \blacksquare$$

▶ **You Try It**   **13.** The length of a rectangle is 8 inches longer than the width. The perimeter is 124 inches. Find its dimensions?

EXAMPLE 6   **Chapter Problem**   Read the problem at the beginning of this chapter.

Solve the equation $p - 0.4p = 24$. You may use a calculator.

$$\underbrace{1p - 0.4p}_{0.6p} = 24$$ Recall, $p = 1p$.

$$0.6p = 24$$ Combine like terms: $1 - 0.4 = 0.6$.

$$\frac{\overset{1}{\cancel{0.6}p}}{\underset{1}{0.6}} = \frac{24}{0.6}$$

$$p = \$40$$

The original cost of the coat was \$40.

Check: The discount is 40% off the original \$40 cost. 40% of \$40 = $0.40 \cdot \$40 = \$16$. The sale price is $\$40 - \$16 = \$24$.   ■

▶ **You Try It**   **14.** The price for a set of golf clubs was reduced 35%. If the sale price is \$312, use equations to find the original price.

▶ **Answers to You Try It**   **1.** $L + (-6)$ or $L - 6$ **2.** $-3y + 11$ or $11 - 3y$

3. $m - 80$ **4.** $x - 25$ **5.** $2y$ **6.** $10(2x)$ or $20x$ **7.** $\frac{8}{h}$ **8.** $\frac{w}{6}$ **9.** $3(n + 5)$

10. $\frac{4}{8 + k}$ **11.** 9   **12.** -6 **13.** width = 27 in, length = 35 in   **14.** \$480

SECTION 12.5 EXERCISES

1 *Translate the word phrase into an algebraic expression.*

1. the sum of p and q

2. w minus 17

3. 45 increased by x

4. 12 divided into Y

5. the product of A and B

6. 3 times x plus 5

7. $\frac{1}{2}$ less than b

8. the quotient of P and 4

9. the total of x, 3, and z

10. $\frac{1}{7}$ of the number n

11. t take away k

12. 25 more than w

13. k divided by 8

14. 6 plus z minus b

15. 24 subtract g

16. 4 times d plus k

17. w decreased by 9

18. a minus b minus c

19. the product of 5 and t

20. l more than $r + 3$

21. the total of $2x$, 4, and $3m$

22. 11 decreased by the product of s and t

23. 5 more than the sum of x and y

24. r less than the product of d and 2

25. subtract the sum of 2 and x from n

26. 5 more than a, divided by 5 less than a

---

2 *Use the 5-step procedure studied in this section*

27. Three times a number plus seven equals 31. Find the number.

28. Eight times a number take away 5 equals 43. Find the number.

29. The sum of six times a number and 4 is -2. Find the number.

30. Negative five added to 2 times a number is -5. Find the number.

31. A number increased by three times itself equals 48. Find the number.

32. Triple a number plus half of the number is -14. Find the number.

33. The length of a rectangle is 12 more than the width. If the perimeter of the rectangle is 240 ft, find its dimensions.

34. The width of a rectangle is 3 yd less than the length. If the perimeter of the rectangle is 48 yd, find its dimensions.

35. Lee Ann's age plus 6 equals twice Lee Ann's age less 20. How old is Lee Ann?

36. If three times Rita's age is subtracted from 100, the result is three times Rita's age plus 4. How old is Rita?

37. Dave is one year older than Greg. If the sum of their ages is 87, how old is each man?

38. Nona weighs 12 lb less than her older sister. If the sum of their weights is 108 lb, how much does each child weigh?

39. The length of a rectangle is twice its width. If the perimeter is 132 ft, find the dimensions of the rectangle.

40. The length of a rectangle is 8 more than four times the width. If the perimeter is 416 ft, find the dimensions of the rectangle.

41. A 70-in pipe is cut into two pieces. The smaller piece is 13 in shorter than the longer. How long is each piece?

42. An 84-in board is cut into two pieces. The smaller piece is 24 in shorter than the longer piece. How long is each piece?

43. Last year's price for a car was increased by 6% to get this year's price. If the car now costs $14,840,
   a. what did it cost last year?

   b. what was the increase in price?

44. After a 35% reduction, a jacket was put on sale for $55.25.
   a. What was the original price of the jacket?

   b. By how much was the jacket reduced?

SKILLSFOCUS (Section 2.4) *Evaluate using* $x = 4$, $y = 9$, *and* $w = -7$.

45. $x^2 + y$ **46.** $2xw$ **47.** $y^2 - x^2$ **48.** $10^2 + w^2 - 5^2$

EXTEND YOUR THINKING ▶▶▶▶

▶SOMETHING MORE

49. Using the diagram at the right, build an equation to solve for the angle labeled x. The sum of the 4 angles is 360°.

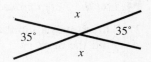

WRITING TO LEARN ▶▶▶▶

50. Write a word problem about two brothers. The sum of their weights is 260 lbs. The object of the problem is to figure out the weight of each brother by building and solving an equation. Then solve your problem using the 5-step procedure studied in this section.

▶YOU BE THE JUDGE

51. Romeo said of Juliet, "A rose by any other name would smell as sweet." Any letter can be used to represent the unknown in a word problem. Maria claims the last two sentences say the same thing. Is she correct? Explain your decision.

12.6 THE PYTHAGOREAN THEOREM

OBJECTIVES

1. Solve equations of the form $x^2 = a$ (Sections 1.9, 10.4).
2. Solve triangles using the Pythagorean Theorem.
3. Solve applications.

NEW VOCABULARY

hypotenuse

leg

Pythagorean Theorem

---

1 Solving Equations of the Form $x^2 = a$

Solve the equation

$$x^2 = 25.$$

This equation says *what number do you square to get 25*? There are two answers, one positive and one negative.

$$x = 5 \quad \text{is an answer because} \quad (5)^2 = 25$$

and,

$$x = -5 \quad \text{is an answer because} \quad (-5)^2 = 25$$

The answers, 5 and -5, equal $\sqrt{25}$ and $-\sqrt{25}$ respectively.

> The two solutions to $x^2 = a$ are
> $$x = \sqrt{a} \quad \text{and} \quad x = -\sqrt{a}.$$

EXAMPLE 1 Solve $x^2 = 81$.

$$x = \sqrt{81} \qquad \text{and} \qquad x = -\sqrt{81}$$
$$x = 9 \qquad\qquad\qquad\quad x = -9$$

Check: $x^2 = 81$ $\qquad$ Check: $x^2 = 81$

$\quad (9)^2 = 81$ $\qquad\qquad\quad (-9)^2 = (-9)(-9)$

$\qquad\qquad\qquad\qquad\qquad\qquad\qquad = 81$

The solutions to $x^2 = 81$ are $x = 9$ and $x = -9$. ■

▶ **You Try It** **1.** Solve $x^2 = 16$.

EXAMPLE 2 Solve $x^2 + 25 = 169$.

$$x^2 + 25 = 169 \qquad$$ Solve for x^2 the same way you solved for x earlier. Begin by adding the opposite of 25 to each side.

$$x^2 + 25 + (-25) = 169 + (-25)$$
$$x^2 = 144$$

$$x = \sqrt{144} \qquad \text{and} \qquad x = -\sqrt{144}$$
$$x = 12 \qquad\qquad\qquad\quad x = -12$$

Check: $x^2 + 25 = 169$ $\qquad$ Check: $x^2 + 25 = 169$

$\quad (12)^2 + 25 \overset{?}{=} 169$ $\qquad\qquad (-12)^2 + 25 \overset{?}{=} 169$

$\quad 144 + 25 \overset{?}{=} 169$ $\qquad\qquad\quad 144 + 25 \overset{?}{=} 169$

$\qquad\qquad 169 = 169$ $\qquad\qquad\qquad\qquad 169 = 169$

The solutions to $x^2 + 25 = 169$ are $x = 12$ and $x = -12$. ■

▶ **You Try It** **2.** Solve $x^2 + 64 = 100$.

---

2 Solving Triangles Using the Pythagorean Theorem

As you recall from Section 10.4, a right triangle has a 90° angle, called a right angle.

right triangle

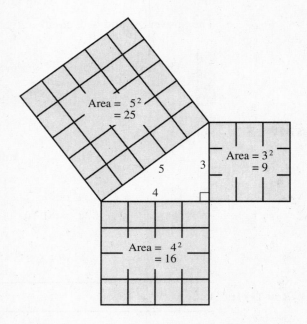

The side of the right triangle opposite the right angle is called the **hypotenuse**. It is the longest side of the triangle. The other two sides are called **legs**. They are perpendicular to each other.

A special property of all right triangles is that *the square of the hypotenuse equals the sum of the squares of the legs*. This is demonstrated with the triangle in the figure.

$$
\begin{aligned}
\text{hypotenuse squared} &= \text{sum of the squares of the legs} \\
5^2 &= 4^2 + 3^2 \\
25 &= 16 + 9 \\
25 &= 25
\end{aligned}
\tag{1}
$$

Another way to understand the relationship in equation (1) is to look at each term as the area of a square. Draw a square on each side of the triangle. The sum of the two smaller squares equals the larger in area.

This relationship is called the **Pythagorean Theorem**.

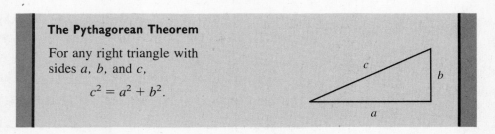

The Pythagorean Theorem

For any right triangle with sides a, b, and c,

$$c^2 = a^2 + b^2.$$

Given any two sides of a right triangle, you can use the Pythagorean Theorem to find the third side.

EXAMPLE 3 Find the missing side of the triangle.

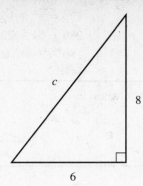

$$c^2 = a^2 + b^2$$ Write the Pythagorean Theorem.

$$c^2 = 6^2 + 8^2$$ Identify a, b, and c and substitute.
$$a = 6$$
$$b = 8$$
$$c = \text{unknown}$$

$$c^2 = 36 + 64$$ Solve for c.

$$c^2 = 100$$

$$c = \sqrt{100}$$

$$c = 10$$

OBSERVE The *answers to the equation* $c^2 = 100$ are $c = +10$ and $c = -10$. The *answer to the word problem* is $+10$. It cannot be -10 because the length of the side of a triangle cannot be negative.

Check: $$c^2 = a^2 + b^2$$

$$10^2 \overset{?}{=} 6^2 + 8^2$$

$$100 \overset{?}{=} 36 + 64$$

$$100 = 100 \quad \blacksquare$$

OBSERVE The role of a and b can be interchanged. In Example 3, if $a = 8$ and $b = 6$, the answer will be the same. See Exercise 41.

▶ **You Try It**   **3.** Find the missing side of the triangle.

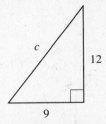

EXAMPLE 4 Find the missing side of the triangle.

$$c^2 = a^2 + b^2$$ Write the Pythagorean Theorem.

$$17^2 = a^2 + 15^2$$ Identify a, b, and c and substitute.
$$a = \text{unknown}$$
$$b = 15$$
$$c = 17$$

$$289 = a^2 + 225$$

$$289 + (-225) = a^2 + 225 + (-225)$$ Solve for a^2. Add the opposite of 225 to each side.

$$64 = a^2$$

$$\sqrt{64} = a$$

$$8 = a$$

Check:
$$c^2 = a^2 + b^2$$
$$17^2 \overset{?}{=} 8^2 + 15^2$$
$$289 \overset{?}{=} 64 + 225$$
$$289 = 289 \quad \blacksquare$$

▶ **You Try It** **4.** Find the missing side of the triangle.

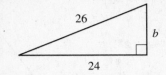

3 Solving Applications

EXAMPLE 5 How many feet of guy wire are needed to support the tower in the figure?

Assume the tower is perpendicular to the ground so a right triangle is formed. Then the guy wire is the hypotenuse of a right triangle with legs 30 ft and 16 ft.

$a = 16$

$b = 30$

$c = $ unknown length of guy wire

$c^2 = a^2 + b^2$

$c^2 = 16^2 + 30^2$

$c^2 = 256 + 900$

$c^2 = 1{,}156$

$c = \sqrt{1{,}156}$ **Use your calculator.**

$c = 34$ ft of guy wire

Check:
$$c^2 = a^2 + b^2$$
$$34^2 \overset{?}{=} 16^2 + 30^2$$
$$1{,}156 \overset{?}{=} 256 + 900$$
$$1{,}156 = 1{,}156 \quad \blacksquare$$

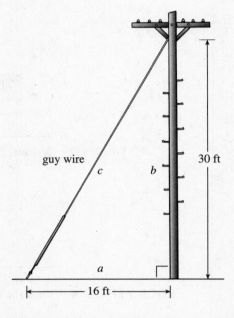

▶ **You Try It** **5.** Starting from home, Rose hiked north for 84 minutes and then east for 63 minutes. She then hikes directly home. In how many minutes will Rose be home?

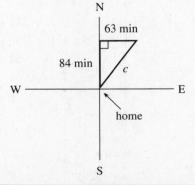

EXAMPLE 6 A plane flies 270 miles from Albuquerque to El Paso and 412 miles from El Paso to Amarillo. Assume the three cities form the right triangle shown in the figure. How far must the plane fly to return to Albuquerque?

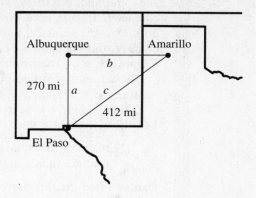

$a = 270$ mi

$b =$ unknown distance from Amarillo to Albuquerque

$c = 412$ mi (This is the hypotenuse because it is opposite the right angle.)

$$c^2 = a^2 + b^2$$
$$412^2 = 270^2 + b^2$$
$$169,744 = 72,900 + b^2$$
$$169,744 + (-72,900) = 72,900 + (-72,900) + b^2$$
$$96,844 = b^2$$
$$\sqrt{96,844} = b \qquad \text{Use your calculator.}$$
$$311 \text{ mi} \doteq b \qquad \text{rounded to the nearest mile}$$

The plane will fly about 311 miles from Amarillo to Albuquerque.

Check:
$$c^2 = a^2 + b^2$$
$$412^2 \stackrel{?}{=} 270^2 + 311^2$$
$$169,744 \stackrel{?}{=} 72,900 + 96,721$$
$$169,744 \doteq 169,621 \quad \blacksquare$$

OBSERVE Since the answer was rounded, the check will be approximately correct. ∎

▶ **You Try It** 6. How high above the ground does the ladder touch the building? Round to tenths.

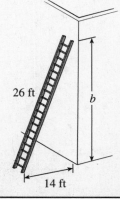

26 ft

b

14 ft

▶ **Answers to You Try It** 1. $x = +4$ and $x = -4$ 2. $x = +6$ and $x = -6$
3. $c = 15$ 4. $b = 10$ 5. $c = 105$ min from home 6. $b = 21.9$ ft above the ground

SECTION **12.6** EXERCISES

1 *Solve the equation. If needed, use a calculator. Round decimals to three places.*

1. $x^2 = 1$ **2.** $x^2 = 9$ **3.** $64 = x^2$ **4.** $4 = x^2$

5. $x^2 = 900$ **6.** $x^2 = 441$ **7.** $x^2 = 33$ **8.** $x^2 = 68$

2 *Solve for the missing side of the right triangle. Round all decimals to two places.*

9.
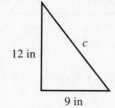
12 in c 9 in

10.
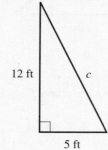
12 ft c 5 ft

11.

26 cm b 24 cm

12.
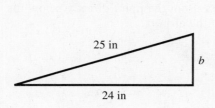
25 in b 24 in

13.
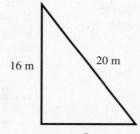
16 m   20 m a

14.
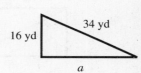
34 yd   16 yd a

15.
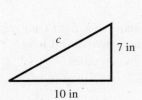
c 7 in   10 in

16.
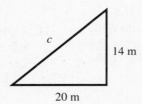
c 14 m   20 m

17.
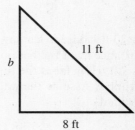
11 ft b 8 ft

18.
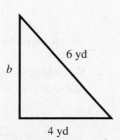
6 yd b 4 yd

19.
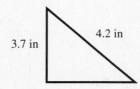
3.7 in   4.2 in a

20.
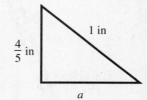
1 in $\frac{4}{5}$ in a

21.

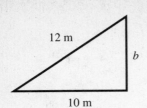

22.

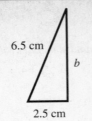

23.

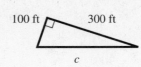

24.

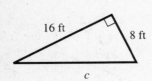

25.

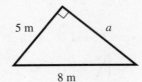

26.

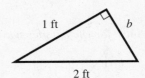

---

3 *Solve each problem by first drawing and labeling a picture. Round answers to tenths unless otherwise specified.*

27. Al drove 10 miles west and 9 miles south. How far is he from his starting point?

28. A football playing field is 300 ft long (excluding the end zones), and 160 ft wide. What is the diagonal distance *c* across the field?

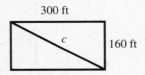

29. Find the length of a rectangle with a width of 20 ft and a diagonal of 40 ft.

30. A 20-ft ladder is leaning against a wall. The distance from the base of the ladder to the wall is 7 ft. How far is the top of the ladder from the ground? Assume the wall is perpendicular to the ground.

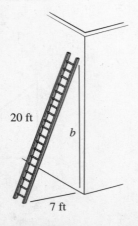

31. A pilot flies 300 miles east, 200 miles north, and 400 miles west. How far is she from her starting point?

32. A plane is 4 miles high. It is descending steadily to an airport 9 miles away. What is the ground distance *a* from the point directly below the plane to the airport?

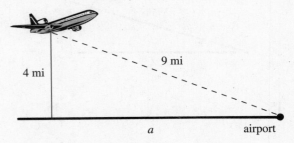

33. A garden is in the shape of a right triangle. The legs of the triangle are each 12 feet long.

a. How long is the hypotenuse?

b. How much will it cost to fence the garden at $5.40 a foot?

34. The two points on the plate are to be connected by a strip of platinum.

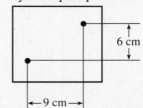

a. What length of platinum is needed?

b. What is the cost at $4.52 per cm?

SKILLSFOCUS (Section 5.8) *Change to a decimal. Round to thousandths.*

35. $\dfrac{3}{5}$ **36.** $\dfrac{7}{8}$ **37.** $\dfrac{1}{6}$ **38.** $\dfrac{9}{10}$

EXTEND YOUR THINKING ▶▶▶▶

▶SOMETHING MORE

39. TV Dimensions   A 25-inch TV measures 25 inches diagonally across the picture tube. To verify this at home, measure the length and width of your TV screen in inches using a yardstick. Then use the Pythagorean Theorem to compute the diagonal length of the TV. For the TV pictured, the length was measured to be 20 in, and the width 15 in.

$$c^2 = a^2 + b^2$$
$$c^2 = 20^2 + 15^2$$
$$c^2 = 400 + 225$$
$$c^2 = 625$$
$$c = \sqrt{625}$$
$$c = 25 \text{ inches, diagonal length}$$

This is a 25-inch TV.

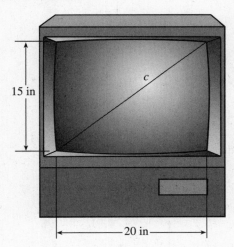

a. Measure the length and width of your home TV. Calculate the diagonal length to find the size of your TV. Round all numbers to the nearest unit.

b. A 20-inch TV measures 16 inches across the bottom of the picture tube. What is the height of the picture tube?

40. It costs $500 per linear foot to build a road. Round each answer to millions.

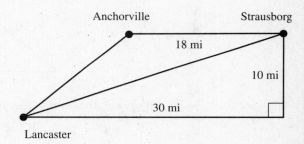

a. What is the cost of building a road from Lancaster to Strausborg by way of Anchorville?

b. How much cheaper would it be to build the road from Lancaster to Strausborg directly?

WRITING TO LEARN ▶▶▶▶

41. The letters a and b have been used to measure the lengths of the legs of a right triangle in this section. In Example 3, $a = 6$ miles and $b = 8$ miles. If you let $a = 8$ miles and $b = 6$ miles, you get the same answer. Resolve Example 3 to show this. In your own words, explain why reversing a and b does not change the answer.

CALCULATOR TIPS

You can use your calculator to evaluate the Pythagorean Theorem.

EXAMPLE 1 Solve for the missing side of the triangle.

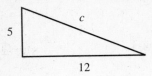

By the Pythagorean Theorem,

$$5^2 + 12^2 = c^2.$$

Find c using your calculator.

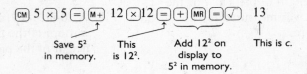

■

EXAMPLE 2 Solve for the missing side of the triangle.

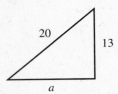

By the Pythagorean Theorem,

$$a^2 + 13^2 = 20^2.$$

Find a using your calculator.

CM 13 × 13 = M+ 20 × 20 = − MR = √ 15.2 (rounded to tenths) ■

Use your calculator to find the missing side of each triangle. Round to tenths.

1.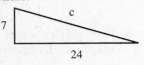

2.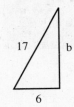

CHAPTER 12 REVIEW

VOCABULARY AND MATCHING

New words and phrases introduced in this chapter are shown in the left-hand column. Match each term on the left with the phrase or sentence on the right that best describes it.

A. variable

B. algebraic expression

C. terms

D. coefficient

E. constants

F. equation

G. solution

H. equivalent equations

I. addition rule

J. multiplication rule

K. distributive property

L. combining like terms

M. hypotenuse

N. leg

O. Pythagorean Theorem

_____ separated by $+$ and $-$ signs in an expression

_____ a statement that two expressions are equal

_____ $6(2 + x) = 6 \cdot 2 + 6 \cdot x$, for example

_____ multiply or divide each side of an equation by the same number and you get an equivalent equation

_____ the longest side of a right triangle

_____ 2, -7, and 5 are examples

_____ a letter that represents numbers

_____ can be done with $5x + 3x$ but not with $5x + 3y$

_____ a side of a right triangle adjacent to the right angle

_____ add or subtract the same term on each side of an equation and you get an equivalent equation

_____ the hypotenuse squared equals the sum of the squares of the two legs

_____ $5x = 10$ and $x = 2$ are; $3x = 9$ and $x = 2$ are not

_____ any collection of numbers, letters, and operations

_____ a number that makes an equation a true statement

_____ for $3x$ is 3, and for $-4y$ is -4

REVIEW EXERCISES

12.1 Variables, Terms, Expressions, and Equations

List each term, the coefficient of each variable, and the constant term.

1. $4 + 7t$

2. $15x - 7y - 3$

3. $x - y + 1$

Which number is a solution to the given equation, if any?

4. $4x - 6 = 2x + 8$ **a.** $x = -3$ **b.** $x = 7$ **c.** $x = 5$

5. $2y - 5 = 5 - 2y$ **a.** $y = -1$ **b.** $y = 0$ **c.** $y = 1$

6. $x^2 = 4x - 3$ **a.** $x = 2$ **b.** $x = 4$ **c.** $x = 3$

Which of the three expressions, if any, are equivalent to the one on the left?

7. $6x - 8y$ **a.** $6y - 8x$ **b.** $8y + 6x$ **c.** $-8y + 6x$

8. $-3x + 1 + 4y$ **a.** $3x - 1 + 4y$ **b.** $4y - 3x + 1$ **c.** $1 + 3y - 4x$

9. $2 - w + z$ **a.** $z - w + 2$ **b.** $2w - z$ **c.** $-z + 2 + w$

12.2 Solving Equations: The Addition Rule

Solve each equation and check.

10. $x + 6 = -4$

11. $x - 6 = -8$

12. $7 + x = 10$

13. $-9 + x = -9$

14. $x + 1 = -1$

15. $x - 4.5 = 6$

12.3 Solving Equations: The Multiplication Rule

Solve each equation and check.

16. $7x = 56$

17. $-4x = 18$

18. $-x = 4$

19. $-60 = 4w$

20. $1.75x = -5.25$

21. $-0.4z = -2.48$

22. $6x - 1 = 5$

23. $8 + 3x = 5$

24. $7x + 10 = 59$

25. $-21 = 3 - 8x$

26. $1.5x + 3 = 12$

12.4 The Distributive Property and Its Application

Apply the distributive property to remove the parentheses.

27. $4(3x + 5)$

28. $3(7x - 5 + 2y)$

29. $-5(x - 3y + 6)$

30. $-(-4x - 3y)$

Combine like terms.

31. $2x + 9x$

32. $5x - 8x$

33. $3x + x$

34. $4x - x$

35. $-2z + 2z$

36. $4c + c + 5c$

37. $9 + 2x - 1$

38. $12y + 9 - 8y$

39. $9x + 7z + 6x - 2z$

40. $x - 1 + x + 1$

Solve and check each equation.

41. $5x + 7x = 60$

42. $2x = 5 + 7x$

43. $3 + 2x - 9 = 6x$

44. $7x + 4 = 2x - 6$

45. $9 - 8x = 3 + 4x$

46. $-2 + x + 5 = 6x - 5 - x$

12.5 Applications

Translate each word phrase into an algebraic expression.

47. the product of 4 and v

48. the sum of eight and g

49. 3 times the sum of t and 2

50. x minus four plus y

51. y more than the product of 7 and z

52. 8 less than the sum of x and 8

Solve using the 5-step procedure studied in this section.

53. Nine times a number plus 4 is forty. Find the number.

54. Twenty less than 12 times a number is -32. Find the number.

55. A number decreased by eight times itself equals 42. What is the number?

56. The length of a rectangle is 5 ft more than the width. If the perimeter of the rectangle is 42 ft,
a. find its dimensions.

b. What is the area of the rectangle?

57. Nancy is three years older than Marie. The sum of their ages is 61. How old is each woman?

58. The length of a rectangle is twice the width. If the perimeter of the rectangle is 54, find its dimensions.

59. When five times Al's age is decreased by 36, the result is Al's age. How old is Al?

60. The cost of a car was reduced by 15%, dropping the price to $10,200.
a. What was the original price of the car?

b. What was the discount?

12.6 The Pythagorean Theorem

Solve each equation for x. Round decimals to two places.

61. $x^2 = 49$

62. $36 = x^2$

63. $x^2 = 75$

Solve for the missing side of the triangle. Round all decimals to two places.

64.

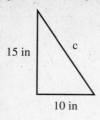

15 in   c

10 in

65.

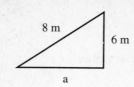

8 m

6 m

a

66.

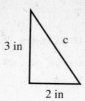

3 in   c

2 in

67.

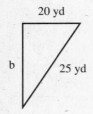

20 yd

b

25 yd

68.

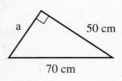

a

50 cm

70 cm

69.

13 ft   b

5 ft

---

70. Frances drove 27 miles west and 36 miles south. How far is she from her starting point?

71. A ball rests at the top of an inclined plane (shown below). How far will the ball roll before it hits the target?

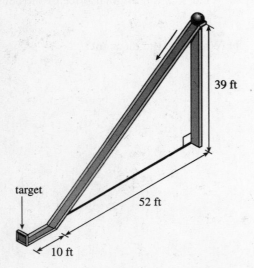

39 ft

52 ft

target

10 ft

72. A yard is in the shape of a right triangle. One leg is 80 ft long. The hypotenuse is 100 ft long.
  a. How long is the other leg?

  b. What is the perimeter of the yard?

  c. What is the cost to fence the yard at $9.90 a yard?

73. A wire is suspended between two poles. A 10-lb weight is attached to the center of the wire. The distance from the top of the weight to the ground is 18 ft. What is the total length of the wire?

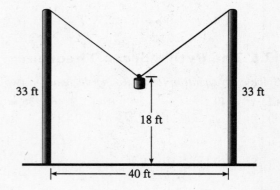

33 ft

33 ft

18 ft

40 ft

Allow 50 minutes to complete this test. Show the work for each problem you solve. When done, check your answers. Rework each problem solved incorrectly.

Solve for x.

1. $x - 7 = -10$

2. $4 + x = -3$

3. $x - 7 = -2$

4. $-9.5x = 4.75$

5. $10x = -70$

6. $-6 = -x$

7. $5x - 8 = 12$

8. $3x - 7 = -16$

9. $1.5 - 0.25x = 0$

10. $-1 = 20 - 3x$

Remove parentheses.

11. $4(6x + 5)$

12. $-7(3 - 4x + y)$

Combine like terms.

13. $3 + 7y - 8 + 4y - y$

14. $9x + 6 - 5x - 8 - 4x + 3$

Solve for x.

15. $7x - 10x = -15$

16. $4 - 5x = 8x + 17$

17. $3x + 1 - x = 10 + 7x - 14$

18. $6 + 2(4x - 1) = -12$

19. Six times a number increased by 25 gives 73. Find the number.

20. The width of a rectangle is 8 inches shorter than the length. If the perimeter is 100 inches,
   a. find the dimensions of the rectangle.

   b. What is the area of the rectangle?

21. Twice Rosser's age decreased by 40 equals Rosser's age plus five. How old is Rosser?

22. Dave is two years older than Ralph. The sum of their ages is 58 plus half of Dave's age. How old is each man?

23. Solve for the missing side of the right triangle.

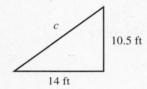

24. Solve for the missing side of the right triangle. Round to hundredths.

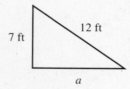

25. A baseball diamond is a square 90 ft on a side. What is the distance from home plate to second base? Round to tenths.

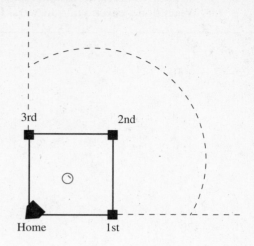

26. In the figure, two towers are connected by a cable. What is the distance between the towers if the cable is 200 ft long?

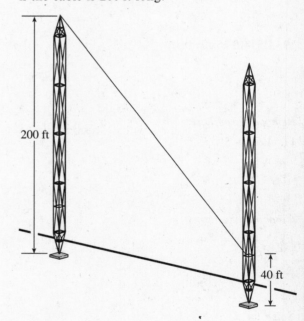

Explain the meaning of each word or phrase. Use your own examples.

27. variable

28. equivalent equations

29. distributive property

30. Pythagorean theorem

1. Simplify $4^2 \cdot 2^3$.

2. Simplify $6 + 2 \cdot 5$.

3. Find the answer, or write not possible.

    a. $12{,}768 \div 28$ **b.** $10 \div 0$ **c.** $1{,}000 \times 461$ **d.** $0 \div 7$

4. Multiply $\quad 4\frac{7}{8} \cdot 2\frac{4}{9}$.

5. Divide $\quad \frac{5}{16} \div \frac{3}{8}$.

6. Add $\quad 14\frac{3}{5} + 11\frac{7}{10}$.

7. Subtract $\quad 2\frac{3}{14} - \frac{20}{21}$.

8. Add $0.204 + 4.7821 + 6 + 0.79 + 49.956$.

9. Subtract $1 - 0.04$.

10. Multiply 800×0.072.

11. Divide $26.585 \div 6.5$.

12. Amy traveled 412 miles on 17.6 gallons of gas.

    a. How many gallons will she need to travel 2,305 miles? Round to tenths.

    b. What will this gas cost at $1.179 per gallon? Round to the nearest dollar.

13. What is 58% of $720?

14. What percent of 18 is 14.4?

15. Suppose 36% of the registered voters actually voted. If 1,710 actually voted, how many people were registered to vote?

16. A jogging suit originally selling for $46 is discounted 30%. What is the sale price?

17. Given the quiz scores 0, 6, 10, 10, 8, 3, 10, 0, and 7, find the
    a. mean.   **b.** median.   **c.** mode.   **d.** range.

18. 2.5 cm = ? mm

19. Find the area of a triangle with base and height both equal to 10 in.

20. Find the area of a circle with a radius of 35 ft.

21. Perform the indicated operations.
    a. $-6 + 4$ **b.** $(-7) - (-10)$ **c.** $-2 - 2$

    d. $(-9)(5)$ **e.** $(-40) \div (-8)$ **f.** $-5 + 5(-3)$

22. Write 0.000506 in scientific notation.

23. Write 350,000 in scientific notation.

24. Solve $-5w = 50$.   25. Solve $8t - 14 = -22$.

Explain the meaning of each word or phrase. Use your own examples.
26. percent   27. fraction

Appendix

TABLE OF FRACTION, DECIMAL, PERCENT EQUIVALENTS (DECIMAL FRACTIONS ARE ROUNDED TO TEN-THOUSANDTHS)

Fraction	Decimal	Percent	Fraction	Decimal	Percent
1	1	100%	$\frac{1}{9}$	0.1111	$11\frac{1}{9}\%$
$\frac{1}{2}$	0.5	50%	$\frac{1}{10}$	0.1	10%
$\frac{1}{3}$	0.3333	$33\frac{1}{3}\%$	$\frac{1}{12}$	0.0833	$8\frac{1}{3}\%$
$\frac{1}{4}$	0.25	25%	$\frac{1}{16}$	0.0625	$6\frac{1}{4}\%$
$\frac{1}{5}$	0.2	20%	$\frac{1}{20}$	0.05	5%
$\frac{1}{6}$	0.1667	$16\frac{2}{3}\%$	$\frac{1}{25}$	0.04	4%
$\frac{1}{7}$	0.1429	$14\frac{2}{7}\%$	$\frac{1}{100}$	0.01	1%
$\frac{1}{8}$	0.125	$12\frac{1}{2}\%$			

OBSERVE To change $\frac{4}{5}$ to a percent, $\frac{4}{5} = 4 \times \frac{1}{5} = 4 \times 20\% = 80\%$.

To change $\frac{5}{6}$ to a percent, $\frac{5}{6} = 5 \times \frac{1}{6} = 5 \times 16\frac{2}{3}\% = 80\frac{10}{3}\% = 83\frac{1}{3}\%$.

ANSWER SECTION

CHAPTER 1

Pretest (p. 2)

1. 2, 3, 4   **2. a.** >   **b.** <   **c.** =
3. 6,000 + 70 + 2   **4.** 4,028   **5.** $w = 2$ **6.** $y = 8$
7. 4,304   **8.** 100, 592   **9.** 7 + 2 = 9
10. 1,895   **11.** 29,878   **12.** 4,920   **13.** 63,000
14. 800 lb   **15.** $282   **16.** 9,568 mi   **17.** $y = 7$
18. $v = 9$ **19.** 37,062   **20.** 1,776,942
21. 251,000   **22.** 680,000   **23.** 14   **24.** 35,000
25. 7^4 **26.** 16   **27.** 125   **28.** 64   **29.** 9

Section 1.1 (p. 7)

1. 0   **3.** 10   **5.** 1   **7.** 0, 1, 2, 3   **9.** 0, 1, 2, 3,
4, 5, 6   **11.** 10, 12, 14   **13.** 7 is greater than 2
15. 4 equals 4   **17.** $\neq$, 10 is unequal to 11   **19.** $\not>$,
9 is not greater than 12   **21.** 6 is greater than 4   **23.** 8
is greater than 7   **25.** $\neq$, 8 is unequal to 10   **27.** >,
$\neq$, $\not<$ **29.** >, $\neq$, $\not<$ **31.** >, $\neq$, $\not<$ **33.** =, $\not<$, $\not>$
35. a. true   **b.** true   **c.** false   **37. a.** true
b. true   **c.** false   **39. a.** No   **b.** Yes
41. 7 = 7   **43.** >   **45.** <

Section 1.2 (p. 13)

1. forty-two   **3.** nine hundred four   **5.** five hundred
eleven   **7.** two thousand, three hundred forty-five
9. forty-five thousand, eight hundred twenty-four
11. sixty thousand, three hundred six
13. seven hundred ninety thousand, two hundred thirteen
15. twenty-seven million   **17.** 3,961   **19.** 1,310
21. 93,000,000   **23.** 6,000,000,000,000   **25.** 60 + 3
27. 700 + 10 + 9   **29.** 900 + 1
31. 1,000 + 700 + 30 + 9   **33.** 8,000 + 2
35. 70,000 + 1,000 + 300 + 40 + 8   **37.** 100,000 + 1
39. 3,000,000 + 400,000 + 6,000 + 500 + 50 + 6
41. >   **43.** <   **45.** 453 = 400 + 50 + 3

Section 1.3 (p. 21)

1. 8   **3.** 9   **5.** 9   **7.** 5   **9.** $x = 4$, a.
11. $n = 7$, b.   **13.** $c = 0$, a.   **15.** $p = 1$, c.
17. $s = 8$, c.   **19.** $x = 0$, b.   **21.** $x = 7$ **23.** $y = 7$
25. $k = 5$ **27.** $p = 0$ **29.** 68   **31.** 579
33. 5,289   **35.** 928   **37.** 9,818
39. 4,879   **41.** 36,888   **43.** $97   **45.** 80
47. 796   **49.** 1,026   **51.** 100   **53.** 615
55. 823   **57.** 1,222   **59.** 1,484   **61.** 6,317
63. 6,003   **65.** 4,237   **67.** 68,562   **69.** 3,793
71. 6,633   **73.** 9,963   **75.** 52,067   **77.** 11,288
79. 22,526,738   **81.** 400 + 20 + 6

83. 30,000 + 400 + 8   **85. a.** 435   **b.** 2,811
87. $t = 5$
89.
$$\begin{array}{r} \overset{2}{59} \\ 278 \\ +\ 14 \\ \hline 351 \end{array}$$

Section 1.4 (p. 31)

1. true, commutative   **3.** false   **5.** false   **7.** false
9. true, associative   **11.** true, zero property   **13.** true,
subtraction   **15.** true, addition   **17.** false
19. 9 − 5 = 4   **21.** 4 + 3 = 7   **23.** 1 + 0 = 1
25. 4 − 0 = 4   **27.** 212   **29.** 222   **31.** 432
33. 35   **35.** 521   **37.** 1,027   **39.** 30,832
41. $x = 9$ **43.** $m = 1$ **45.** $f = 0$ **47.** $z = 9$
49. 17   **51.** 19   **53.** 39   **55.** 52   **57.** 212
59. 697   **61.** 177   **63.** 3,187   **65.** 19   **67.** 387
69. 335   **71.** 5,491   **73.** 2,209   **75.** 3,214
77. 2,367   **79.** 2,428   **81.** 61,037   **83.** 1,203
85. 13,763   **87.** 3,000   **89.** $x = 8$ **91.** $f = 7$
93.
$$\begin{array}{r} \overset{9}{\overset{3\ 10\ 16}{40\!\!\!/6}} \\ -138 \\ \hline 268 \end{array}$$
95.
$$\begin{array}{r} \overset{5\ 99\ 10}{6,\!000\!\!\!/} \\ -2,\!194 \\ \hline 3,\!806 \end{array}$$

Section 1.5 (p. 39)

1. 60   **3.** 200   **5.** 1,780   **7.** 500   **9.** 4,400
11. 3,500   **13.** 0   **15.** 4,000   **17.** 5,000
19. 1,000   **21.** 60,000   **23.** 4,000,000
25. 99,990   **27.** 2,500,000   **29.** 70; 100; 0
31. 1,600; 1,700; 2,000   **33.** 7,290; 7,300; 7,000
35. 34,610; 34,600; 35,000   **37.** 262,910; 262,900;
263,000   **39.** 112,000   **41.** $12,500,000   **43.** 90
45. 700   **47.** 14,600   **49.** 20   **51.** 5,000
53. 700   **55.** 1,500   **57.** 290   **59.** 3,100
61. $2,100   **63.** 12,000 mi   **65.** $1,780
67. a. 270   **b.** 300   **69. a.** 2,900   **b.** 3,000
71. a. 7,000   **b.** 0   **73.** $x = 0$ **75.** $y = 18$
77. 650   **81. a.** 0   **b.** 8,000

Section 1.6 (p. 47)

1. $41   **3.** $33   **5.** 2,254 square miles   **7.** 304
calories   **9.** $59,000   **11.** 7,471 mi   **13.** $14,655
15. 57,254 mi   **17. a.** 7,449,803 square miles
b. 212,263 square miles   **19.** 1,320 lb   **21.** 363 ft
23. a. 6,845 lb   **b.** 1,655 lb   **c.** No   **25. a.** $444
b. $116   **27.** 1,218   **29.** $5   **31.** 78 ft
33. 407   **35.** 284,611   **37.** 25 rungs

Section 1.7 (p. 57)

1. a. product **b.** factor **c.** factor **d.** multiple
e. multiple **3. a.** factor **b.** multiple **c.** product
d. factor **e.** multiple **5. a.** factor **b.** product
c. multiple **7.** 54 **9.** 48 **11.** 0 **13.** 12
15. 18 **17.** 0 **19.** 9 **21.** 6 **23.** 0 **25.** 10
27. 8 **29.** 24 **31.** 0 **33.** 72 **35.** $a = 4$, c.
37. $a = 0$, a. **39.** $c = 2$, d. **41.** $f = 0$, a.
43. $c = 1$, b. **45.** $s = 0$, a. **47.** $w = 2$, d.
49. $z = 1$, b. **51.** $h = 1$, b. **53.** $x = 8$, c.
55. 7(9) **57.** 2×6 **59.** (8)4 **61.** $y8$
63. $3 \times (2 \times 5)$ **65.** $(8 \times 7) \times 2$ **67.** $(1 \cdot 2)3$
69. $5(n \cdot 3)$ **71.** $x = 5$ **73.** $t = 3$ **75.** $z = 5$
77. $w = 5$ **79.** $t = 0$ **81.** $v = 0$ **83.** 287
85. 318 **87. a.** $6 + 6 + 6 + 6 + 6 + 6 + 6$ **b.** $7 \cdot 6$
c. $6 \cdot 7$

Section 1.8 (p. 69)

1. 68 **3.** 3,213 **5.** 13,959 **7.** 289,416
9. 1,125 **11.** 810 **13.** 15,624 **15.** 21,793
17. 53,600 **19.** 121,680 **21.** 223,014
23. 309,372 **25.** 370,800 **27.** 3,036,376
29. 1,249,738 **31.** 2,015,018 **33.** 8,510,533
35. 2,430 **37.** 580,000 **39.** 360,500
41. 473,570 **43.** 560 **45.** 4,500 **47.** 30,000
49. 14,000 **51.** 4,900,000 **53.** $1,800
55. a. $400,000 **b.** No **57.** 6,000,000,000,000 mi
59. 13 **61.** 25 **63.** 16 **65.** 12 **67.** 15
69. 10 **71.** 7 **73.** 1 **75.** 14 **77.** 7 **79.** b.
81. c. **83.** b. **85.** $526 **87.** 1,830
89. a. $9 \times 10 = 90$ **b.** $50 \times 4 = 200$
c. $3 \times 30 = 90$ **d.** $28 \times 10 = 280$ **91.** 84,000
93. 12

Section 1.9 (p. 79)

1. 4^3 **3.** 7^2 **5.** 12^3 **7.** 6^1 **9.** 121^3 **11.** 3
13. 0 **15.** 1 **17.** 25 **19.** 1 **21.** 10,000
23. 15 **25.** 27,000 **27.** 2,500 **29.** 1 **31.** 243
33. 729 **35.** 1 **37.** 343 **39.** 1 **41.** 65,536
43. 4,096 **45.** 9,765,625 **47.** 5; $5^2 = 25$
49. 10; $10^2 = 100$ **51.** 15; $15^2 = 225$
53. 20; $20^2 = 400$ **55.** 16; $16^2 = 256$ **57.** 18;
$18^2 = 324$ **59.** 25 **61.** 349 **63.** 2.8284271
65. 3.1622777 **67.** 18 **69.** 9 **71.** 10^{100}
73. 1 **75.** t^3 **77.** y^6 **79.** $x \cdot x \cdot x \cdot x \cdot x \cdot x \cdot x$

Review (p. 83)

1. a. $>, \neq, \not<$ **b.** $=, \not>, \not<$ **c.** $<, \neq, \not>$ **3.** 8,
9, 10, 11, 12 **5. a.** True **b.** False **c.** False
7. a. forty-seven **b.** three thousand, five hundred two
c. four million, six hundred twenty-eight thousand, five
hundred thirty-one **9.** 4,629 **11.** 46,006
13. a. $t = 3$ **b.** $y = 4$ **c.** $f = 0$ **15.** 4,875
17. 131,195 **19.** 18,738 **21. a.** $3 + 4 = 7$
b. $2 + 7 = 9$ **c.** $0 + 5 = 0$ **23.** 32 **25.** 294
27. 54,242 **29.** 660 **31.** 7,000,000 **33.** 568,000
35. 1,500 **37.** 400 **39.** about $2,400 **41.** 178 lb
43. $132,500 **45.** 810 lb **47. a.** 0 **b.** 6
c. 56 **49. a.** $v = 5$ **b.** $t = 8$ **c.** $z = 9$

51. 47,772 **53.** 34,776 **55.** 8,173,202
57. 2,421,000 **59.** 240,000 **61.** 800,000
63. 8,000; low **65.** 17 **67.** 8 **69. a.** False
b. True **c.** False **71.** 9^3 **73.** 5^4 **75.** 11^2
77. 125 **79.** 1 **81.** 729 **83.** 8 **85.** 9
87. 1

Chapter Test (p. 89)

1. 4, 5, 6, 7 **2. a.** False **b.** False **c.** True
3. $800 + 5$ **4.** $40,000 + 7,000 + 600 + 90 + 2$
5. 70,423 **6.** $t = 0$ **7.** $t = 7$ **8.** $n = 7$
9. $y = 40$ **10.** 10,368 **11.** 68,791 **12.** 5,831
13. 68,311 **14.** $6 + 2 = 8$ **15.** $294 **16.** 9,410
17. 68,000 **18.** 5,600 **19.** 28,623
20. 2,614,920 **21.** 5,812,000 **22.** 95 **23.** Round
to hundreds: $100 + 400 + 0 + 1,100 + 800 = 2,400$
24. 720,000 **25.** 4^5 **26.** $5^2 \cdot 8^5$ **27.** 256
28. 343 **29.** 5 **30.** 12

CHAPTER 2

Pretest (p. 92)

1. $4 \cdot 9 = 36$ **2.** $6 \cdot 7 = 42$ **3.** $t = 24$ **4. a.** 1
b. 8 **c.** 0 **d.** not legal **5.** 38
6. 182 R 41 **7.** 49 **8.** 407 **9.** 6 **10.** about
2,000 people **11. a.** 18 **b.** $5 **12.** 80 **13.** 2
14. 16 **15.** 50 **16.** 18 **17.** 25
18. $4 + 6 \cdot 5 = 34$ **19.** $7 \cdot (9 + 2) = 77$ **20.** 18
21. 33 **22.** 38 **23.** $x = 341$ **24.** $k = 127$
25. $y = 18$ **26.** $x = 180$

Section 2.1 (p. 99)

1. $8 \div 2$, 8/2, $\frac{8}{2}$, $2\overline{)8}$ **3.** $6 \div 3$, 6/3, $\frac{6}{3}$, $3\overline{)6}$
5. $42 \div 7$, 42/7, $\frac{42}{7}$, $7\overline{)42}$

7. divisor $= 2$, dividend $= 10$, quotient $= 5$, $5 \cdot 2 = 10$
9. divisor $= 1$, dividend $= 6$, quotient $= 6$, $6 \cdot 1 = 6$
11. divisor $= 4$, dividend $= 4$, quotient $= 1$, $1 \cdot 4 = 4$
13. divisor $= 5$, dividend $= 25$, quotient $= 5$, $5 \cdot 5 = 25$
15. 2 **17.** 5 **19.** 9 **21.** 7 **23.** 6 **25.** 1
27. 8 **29.** 9 **31.** 0 **33.** 7 **35.** 8 **37.** NL
39. NL **41.** 7 **43.** 6, b and c **45.** 4, b
47. 5, none **49.** 4, c **51.** 3, b **53.** $s = 54$
55. $b = 80$ **57.** $f = 0$ **59.** $x = 63$ **61.** $y = 8$
63. $p = 7$ **65.** $0 \div 6 = 0$ **67.** $0 \div 0 =$ Not Legal

Section 2.2 (p. 107)

1. 4 R 2 **3.** 5 R 1 **5.** 3 R 2 **7.** 4 R 0 **9.** 4
R 2 **11.** 7 R 1 **13.** 381 R 4 **15.** 85 R 3
17. 900 R 4 **19.** 5,026 R 3 **21.** 8 R 0 **23.** 4
R 3 **25.** 13 R 2 **27.** 34 R 0 **29.** 65 R 67
31. 267 R 30 **33.** 1,288 R 45 **35.** 148 R 20
37. 34 R 0 **39.** 15 **41.** 7 **43.** 30 **45.** 173
47. 30 **49.** 20 **51.** 10 **53.** 30 **55.** 5
57. 200 **59.** estimate $= $2,000,000 \div 40$ or about
$50,000 each **61. a.** estimate $= $7,000,000 \div 70$ or
about $100,000 each **b.** No, it is too low.
63. 27,608 **65.** 5,319,000 **67.** about 40,000
69. 5 **71.** $\frac{60,200}{100} = 602$

Section 2.3 (p. 115)

1. $3,000   **3.** $23,556 ÷ 52 = $453 per week   **5.** 504 mi   **7.** 67 gal   **9.** Yes   **11.** about 250,000 words   **13.** $351   **15.** $12 × 40 = $480; $18 × 22 = $396; $480 + $396 = $876   **17.** 1,918 people can be seated   **19.** about 20 miles per second   **21.** $420   **23.** $8,162   **25.** $6,518   **27.** $684 − $120 down = $564, monthly payment = $564 ÷ 12 = $47   **29.** $250   **31.** 8   **33.** 10   **35.** $126,700   **37.** 5 × 45 = 225 points needed, he has 41 + 46 + 40 + 47 = 174 points, he still needs 225 − 174 = 51 on the fifth part   **39.** 16   **41.** 36   **43. a.** 15,000 seconds   **b.** 250   **c.** 4 hr 10 min   **45.** $7,280   **47.** about 250,000 people per day   **49.** 10:04 A.M.

Section 2.4 (p. 123)

1. 3   **3.** 16   **5.** 21   **7.** 6   **9.** 48   **11.** 11   **13.** 1   **15.** 1   **17.** 4   **19.** 54   **21.** 9   **23.** 75   **25.** 7   **27.** 21   **29.** 9   **31.** 18   **33.** 0   **35.** 36   **37.** 100   **39.** 44   **41.** 40   **43.** 35   **45.** 34   **47.** 250   **49.** 36   **51.** a and c   **53.** a and b   **55.** a and c   **57.** 26,432   **59.** 78,000   **61. a.** 13 = 5 + 12, false   **b.** 12 = 13 − 5, false   **c.** 20 = 4 · 5, true   **d.** $2 = \frac{6}{3}$, true   **63. A.** 22; C   **B.** 451; F   **C.** 2001; B   **D.** 109; A   **E.** 34; G   **F.** 911; E   **G.** 54; D   **65.** 7 + 2 · 3 = 7 + 6 = 13   **67.** 10 ÷ 5 · 2 = 2 · 2 = 4   **69.** 10 · 10 ÷ 10 · 10 = 100 ÷ 10 · 10 = 10 · 10 = 100   **71.** $2 \cdot 3^2 = 2 \cdot 9 = 18$ **73.** 3 · 2 + 4 · 7 = 6 + 28 = 34   **75.** 5(6 + 2) = 5(8) = 40   **77.** 4 + 6 ÷ 2 = 4 + 3 = 7

Section 2.5 (p. 131)

1. 7 + 10 = 17   **3.** 7 + 10 − 5 = 12   **5.** 12 · 10 = 120   **7.** 4 · 6 − 11 = 13   **9.** 40 ÷ 8 = 5   **11.** 3 + 4 · 7 = 31   **13.** 25 + 6 ÷ 2 = 28   **15.** $3^2 + 7 = 16$ **17.** 5 · 8 + 6 · 2 = 52   **19.** (12 + 8) · 3 = 60   **21.** 56 ÷ (4 · 2) = 7   **23.** 12 · 15 − 200 ÷ 4 = 130   **25.** $5\sqrt{49} = 35$ **27.** 15   **29.** 17   **31.** 29   **33.** 56   **35.** 96   **37.** 2   **39.** 80   **41.** 25   **43.** 32   **45.** 0   **47.** 30   **49.** 14   **51.** A = 48   **53.** P = 20   **55.** P = 32   **57.** F = 50   **59.** D = 12   **61.** L = 33   **63.** x = 7   **65.** t = 8   **67.** 4 + 3 · 5 = 4 + 15 = 19   **69.** N(P + 3) = 5(8 + 3) = 5 · 11 = 55

Section 2.6 (p. 137)

1. x = 1   **3.** x = 2   **5.** x = 7   **7.** x = 9   **9.** x = 8   **11.** w = 22   **13.** k = 45   **15.** s = 29   **17.** p = 288   **19.** x = 1,312   **21.** w = 1,600   **23.** x = 193   **25.** v = 766   **27.** z = 2,000   **29.** s = 0   **31.** t = 918   **33.** x = 2   **35.** x = 7   **37.** x = 7   **39.** x = 45   **41.** x = 225   **43.** x = 150   **45.** x = 360   **47.** x = 12   **49.** v = 28   **51.** x = 219   **53.** x = 180   **55.** n = 720   **57.** y = 1,152,000   **59.** x = 0   **61.** h = 0   **63.** x = 304   **65.** h = 418   **67.** x = 82   **69.** p = 980   **71.** x = 22   **73.** x = 31   **75.** t = 328   **77.** x = 20   **79.** t = 5,100   **81.** 3^3 **83.** 5^6 **85.** x = 8 · 8 = 64

Review (p. 141)

1. 3 · 2 = 6   **3.** 30 · 15 = 450   **5.** 1   **7.** 6   **9.** 8   **11.** 8   **13.** k = 32   **15.** h = 70   **17. a.** true   **b.** false   **c.** false   **19.** 4 R 2   **21.** 80 R 2   **23.** 62 R 61   **25.** 245 R 18   **27.** 63   **29.** 25   **31.** 30   **33.** about $300 per household   **35.** 84   **37.** 300 gal   **39.** $80   **41.** 77   **43.** 7   **45.** 14   **47.** 81   **49.** 19   **51.** 150   **53.** 18   **55.** 52   **57.** 207   **59.** 5 + 4 · 10 = 45   **61.** 3 + 12 ÷ 3 − 2 = 5   **63.** 300 ÷ (10 · 6) = 5   **65.** 70 − 20 · 2 = 30   **67.** $(6 \cdot 2)^2 = 144$ **69.** 37   **71.** 19   **73.** 25   **75.** S = 300   **77.** F = 86   **79.** x = 34   **81.** x = 165   **83.** v = 306   **85.** h = 5,732   **87.** x = 178   **89.** s = 0   **91.** x = 8   **93.** x = 204   **95.** v = 0   **97.** h = 13,840   **99.** x = 23   **101.** x = 336

Chapter Test (p. 145)

1. 27   **2.** 27 R 283   **3.** 300   **4.** 4 · 6 = 24   **5.** 25   **6.** 62   **7.** 49   **8. a.** 9 buses, 8 full and 1 partially full   **b.** 28   **9.** 4   **10.** 122   **11.** 60   **12.** 315   **13.** 98   **14.** L = 384   **15.** P = 330   **16.** x = 17   **17.** x = 329   **18.** w = 1,102   **19.** x = 2,016

Cumulative Test (p. 146)

1. 4,000 + 700 + 5   **2.** 7,482,026   **3.** 12,534   **4.** 3,382   **5.** 32,361   **6.** 452,000   **7. a.** 700 lb   **b.** 669 lb   **8.** $2,889   **9.** 3^5 **10.** 16   **11.** 308   **12.** 23 R 30   **13.** 8   **14.** 7   **15.** 17   **16. a.** 16 buses, 15 full and 1 partially full   **b.** 8   **17.** 38   **18.** d = 400   **19.** x = 17   **20.** n = 174   **21.** p = 18   **22.** w = 630

CHAPTER 3

Pretest (p. 148)

1. 9   **2.**

3. $\frac{25}{7}$ **4.** $\frac{51}{5}$ **5.** $5\frac{1}{6}$ **6.** $5\frac{8}{15}$ **7.** 1, 2, 3, 4, 6, 8, 12, 16, 24, 48   **8.** 2 · 3 · 5   **9.** 2 · 3 · 7   **10.** $\frac{5}{8}$ **11.** $4\frac{1}{5}$ **12.** $\frac{1}{5}$ **13.** $\frac{5}{18}$ **14.** 10   **15.** 11   **16.** $26\frac{2}{3}$ **17.** $1\frac{1}{3}$ **18.** 1   **19.** $3\frac{3}{14}$ **20.** $2\frac{5}{7}$ **21.** $w = \frac{5}{16}$ **22.** $42\frac{1}{4}$ gal   **23.** $8\frac{2}{5}$ **24.** $196   **25.** $11\frac{23}{35}$, or 11 wires

Section 3.1 (p. 153)

1. 3   **3.** 100   **5. a.** $\frac{1}{7}$ **b.** $\frac{7}{7} = 1$ whole pie   **7.** $\frac{2}{6}$ **9.** $\frac{14}{15}$ **11.** $\frac{4}{4}$ **13.** $\frac{5}{11}$

15. four-fifths   **17.** twelve-twelfths

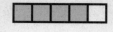

19. zero-sixths   **21.** one-tenth

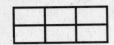

23. a. 7   **b.** $\frac{5}{7}$ **c.** $\frac{2}{7}$ **d.** $\frac{0}{7}$

25. a. $\frac{7}{10}$ **b.** $\frac{3}{10}$ **27. a.** $\frac{861}{2,145}$

b. $\frac{1,284}{2,145}$ **29.** 29   **31.** 8

Section 3.2 (p. 159)

1. improper   **3.** mixed   **5.** mixed   **7.** improper

9. mixed   **11.** $\frac{5}{2}$ **13.** $\frac{19}{7}$ **15.** $\frac{11}{3}$

17. $\frac{22}{3}$ **19.** $\frac{17}{2}$ **21.** $\frac{113}{9}$ **23.** $\frac{67}{21}$

25. $\frac{147}{10}$ **27.** $\frac{32}{8}$ **29.** $\frac{181}{6}$ **31.** $\frac{29}{4}$

33. $\frac{23}{7}$ **35.** $1\frac{1}{4}$ **37.** $2\frac{1}{2}$ **39.** $2\frac{3}{4}$

41. $6\frac{1}{2}$ **43.** $3\frac{5}{6}$ **45.** $9\frac{3}{4}$ **47.** $10\frac{1}{3}$

49. $7\frac{3}{10}$ **51.** 12   **53.** $5\frac{5}{11}$ **55.** $6\frac{1}{2}$

57. $5\frac{1}{8}$ **59.** $\$40\frac{5}{8}$ **61.** $7\frac{5}{6}$ **63.** $2\frac{4}{7}$ ft

65. $52\frac{5}{6}$ mi per gal   **67.** 470   **69.** 5,000

71. $7^5 \div 16 = 16,807 \div 16 = 1,050\frac{7}{16}$ lb

73. $\frac{17}{5} = 3\frac{2}{5}$

Section 3.3 (p. 165)

1. 1, 2, 3, 6 composite   **3.** 1, 2, 4, 8, 16 composite

5. 1, 2, 7, 14 composite   **7.** 1, 13 prime

9. 1, 2, 13, 26 composite   **11.** 1, 3, 11, 33 composite

13. 1, 2, 4, 5, 8, 10, 16, 20, 40, 80 composite

15. 1, 2, 3, 4, 6, 8, 12, 24 composite   **17.** 1, 79 prime

19. 1, 2, 71, 142 composite   **21.** 1, 3, 17, 51 composite

23. 1, 59 prime   **25.** 3   **27.** 11   **29.** 97   **31.** 4

33. 10   **35.** 9   **37.** False   **39.** 2^2 **41.** prime

43. $2 \cdot 7$ **45.** 2^4 **47.** $3 \cdot 11$ **49.** $2^4 \cdot 5$ **51.** 3^4

53. $2^4 \cdot 3^2$ **55.** $3 \cdot 17$ **57.** $2 \cdot 3 \cdot 11$ **59.** $7 \cdot 11$

61. $2 \cdot 3 \cdot 5 \cdot 7$ **63.** 11 R 1   **65.** 0 R 8

67. a. No   **b.** Yes   **c.** No   **d.** No   **e.** Yes, Yes

Section 3.4 (p. 171)

1. $\frac{3}{4}$ **3.** $\frac{2}{3}$ **5.** $\frac{3}{7}$ **7.** $\frac{3}{4}$ **9.** $\frac{3}{5}$

11. $\frac{3}{4}$ **13.** $\frac{1}{4}$ **15.** $\frac{3}{5}$ **17.** $\frac{4}{5}$ **19.** $\frac{3}{7}$

21. $\frac{3}{4}$ **23.** $\frac{3}{5}$ **25.** $\frac{4}{5}$ **27.** $\frac{2}{5}$ **29.** $\frac{3}{8}$

31. $\frac{17}{22}$ **33.** $\frac{19}{26}$ **35.** $\frac{3}{7}$ **37.** $\frac{3}{5}$ **39.** $\frac{1}{3}$

41. $\frac{1}{3}$ **43.** $\frac{1}{4}$ **45.** $\frac{1}{6}$ **47.** $\frac{7}{10}$

49. a. $\frac{6}{7}$ **b.** $\frac{1}{7}$ **51.** $\frac{18}{48} = \frac{3}{8}$ **53.** $4\frac{4}{5}$

55. $5\frac{5}{7}$ **57.** $1\frac{2}{9}$ **59.** $1\frac{8}{11}$ **61.** $1\frac{7}{8}$ **63.** $7\frac{1}{2}$

65. $11\frac{2}{9}$ **67.** $x = 12$ **69.** $y = 180$

71. $\frac{36}{54} = \frac{36 \div 9}{54 \div 9} = \frac{4}{6} = \frac{2}{3}$

Section 3.5 (p. 177)

1. $\frac{2}{5}$ **3.** $\frac{32}{63}$ **5.** $\frac{3}{7}$ **7.** $\frac{3}{7}$ **9.** $\frac{14}{25}$

11. 1   **13.** $\frac{5}{12}$ **15.** $\frac{15}{56}$ **17.** $\frac{6}{25}$

19. $\frac{4}{9}$ **21.** $\frac{5}{54}$ **23.** $\frac{1}{3}$ **25.** $3\frac{3}{4}$

27. $4\frac{2}{5}$ **29.** $7\frac{7}{8}$ **31.** $10\frac{1}{2}$ **33.** $13\frac{17}{27}$

35. $12\frac{4}{5}$ **37.** $4\frac{1}{2}$ **39.** 16   **41.** 28   **43.** $91\frac{2}{3}$

45. 33   **47.** 1   **49.** $\frac{9}{25}$ **51.** $8\frac{1}{36}$

53. $\frac{216}{343}$ **55.** $\frac{81}{10,000}$ **57.** $7\frac{58}{81}$ **59.** $11\frac{1}{9}$

61. $\frac{49}{64}$ **63.** 6   **65.** 1   **67.** $\frac{3}{8} \times \frac{1}{4} = \frac{3}{32}$

69. $3\frac{1}{4} \times \frac{2}{5} = 1\frac{3}{10}$

Section 3.6 (p. 183)

1. $\frac{7}{2}$ **3.** $\frac{1}{8}$ **5.** 10   **7.** $\frac{15}{11}$ **9.** $\frac{8}{41}$

11. $\frac{4}{7}$ **13.** $1\frac{1}{20}$ **15.** $1\frac{5}{7}$ **17.** $\frac{14}{15}$

19. $\frac{14}{15}$ **21.** $1\frac{1}{6}$ **23.** $\frac{10}{11}$ **25.** $1\frac{1}{3}$

27. $1\frac{1}{4}$ **29.** $\frac{4}{15}$ **31.** 9   **33.** 5   **35.** 1

37. $1\frac{17}{28}$ **39.** $1\frac{4}{5}$ **41.** 18   **43.** 3

45. $2\frac{3}{16}$ **47.** 5   **49.** $\frac{1}{4}$ **51.** 4

53. $\frac{1}{3}$ **55.** $\frac{306}{473}$ **57.** $1\frac{1}{8}$ **59.** $9\frac{1}{5}$

61. $s = \frac{6}{7}$ **63.** $k = 1\frac{11}{45}$ **65.** $s = 2\frac{4}{9}$

67. $m = 15$ **69.** $5 \cdot 8 = 40$

71. $\$252 \div \$7 = 36$ tapes

73. $\frac{7}{9} \div \frac{1}{6} = \frac{7}{\overset{3}{\cancel{9}}} \times \frac{\overset{2}{\cancel{6}}}{1} = \frac{14}{3} = 4\frac{2}{3}$

75. $\frac{1}{2} \div 2 = \frac{1}{2} \div \frac{2}{1} = \frac{1}{2} \times \frac{1}{2} = \frac{1}{4}$

Section 3.7 (p. 191)

1. 200   **3.** 200 men

5. $\dfrac{2}{9}$ of 12 $= \dfrac{2}{\cancel{9}} \times \dfrac{\cancel{12}^{4}}{1} = \dfrac{8}{3} = 2\dfrac{2}{3}$ gal

7. 8   **9.** 4   **11.** 20   **13.** 26 lb

15. 75 hamburgers   **17.** $15\dfrac{3}{4}$ **19.** $\dfrac{5}{7}$

21. $1\dfrac{3}{16}$ **23.** 48 lb   **25.** 32 tiles   **27.** $8\dfrac{3}{4}$

29. 5   **31.** $\dfrac{1}{4}$ of a pie each   **33.** $6\dfrac{2}{5}$ hr

35. $98\dfrac{1}{3} \div 7\dfrac{1}{2} = 13\dfrac{1}{9}$; you must make 14 trips.

37. 124 cups   **39. a.** 7 months   **b.** 17 months

41. Add $1\dfrac{1}{4}$ in to itself 24 times $= 24 \times 1\dfrac{1}{4} =$

$\dfrac{\cancel{24}^{6}}{1} \times \dfrac{5}{\cancel{4}_{1}} = 30$ in

43. 100 oz   **45.** $715

47. How many $\dfrac{3}{16}$ are in 24 $=$

$24 \div \dfrac{3}{16} = \dfrac{\cancel{24}^{8}}{1} \times \dfrac{16}{\cancel{3}_{1}} = 128$ ties

49. 3 out of 80 is $\dfrac{3}{80}$.

$\dfrac{3}{80}$ of 72,000 $= \dfrac{3}{80} \times \dfrac{72,000}{1} = 2,700$ will have the disease.   **51. a.** $2,136   **b.** $534

53. $x = 91$ **55.** $x = 356$

57. a. $\dfrac{5}{16}$ in   **b.** $1\dfrac{9}{16}$ in   **59. a.** No   **b.** 4 days

61. a. $\dfrac{1}{2}$ pint $\dfrac{1}{4}$ cup

     2 cherries $\dfrac{1}{6}$ cup

     $\dfrac{1}{9}$ cup $\dfrac{1}{3}$ package

b. 6 pints   3 cups

     24 cherries   2 cups

     $1\dfrac{1}{3}$ cups   4 packages

63. Tom, Terry, and Teresa each get $8,000. Dick gets $20,000, Dan and David each get $10,000. Harry gets $2,000. Helen, Hazel, and Harriet each get $10,000.

65. No   **67.** Split $\dfrac{5}{6}$ lb into 3 equal parts $= \dfrac{5}{6} \div 3 =$

$\dfrac{5}{6} \times \dfrac{1}{3} = \dfrac{5}{18}$ lb per part.

Review (p. 197)

1. 6   **3.** $\dfrac{2}{4}$ **5.** $\dfrac{0}{4}$ **7.**

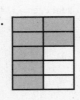

9. 2   **11.** 0   **13.** $\dfrac{21}{5}$ **15.** $\dfrac{38}{7}$

17. $4\dfrac{3}{4}$ **19.** $5\dfrac{5}{8}$ **21.** 1, 7, 49 composite

23. 1, 2, 4, 5, 7, 8, 10, 14, 20, 28, 35, 40, 56, 70, 140, 280 composite   **25.** 1, 167 prime   **27.** 1, 2, 47, 94 composite   **29.** $3^2 \cdot 13$ **31.** $2^3 \cdot 7$ **33.** $5 \cdot 73$

35. $3 \cdot 7 \cdot 13$ **37.** $\dfrac{2}{3}$ **39.** $\dfrac{7}{8}$ **41.** $1\dfrac{23}{45}$ **43.** $\dfrac{17}{19}$

45. $\dfrac{12}{35}$ **47.** $\dfrac{125}{324}$ **49.** $8\dfrac{1}{4}$ **51.** $30\dfrac{3}{4}$

53. $\dfrac{343}{729}$ **55.** $\dfrac{169}{400}$ **57.** $2\dfrac{1}{2}$ **59.** $\dfrac{5}{6}$

61. $4\dfrac{4}{5}$ **63.** 15   **65.** $x = \dfrac{9}{20}$ **67.** 112 quarts

69. $31\dfrac{1}{4}$ **71.** 34 plums   **73.** 28 wires

Chapter Test (p. 201)

1. $\dfrac{31}{6}$ **2.** $\dfrac{110}{9}$ **3.** $\dfrac{52}{15}$ **4.** $\dfrac{15}{8}$

5. $5\dfrac{1}{4}$ **6.** $2\dfrac{13}{16}$ **7.** $7\dfrac{4}{9}$ **8.** $33\dfrac{2}{3}$

9. 1, 2, 4, 5, 8, 10, 20, 25, 40, 50, 100, 200   **10.** 1, 2, 4, 8, 11, 22, 44, 88   **11.** $2 \cdot 3 \cdot 19$ **12.** $2 \cdot 5 \cdot 7$

13. $\dfrac{21}{80}$ **14.** 28   **15.** $37\dfrac{1}{2}$ **16.** $13\dfrac{1}{2}$

17. $1\dfrac{4}{45}$ **18.** $\dfrac{2}{3}$ **19.** 10   **20.** $5\dfrac{5}{21}$

21. $\dfrac{125}{512}$ **22.** $w = \dfrac{4}{7}$ **23.** 92 in   **24.** $1,500

25. 20 times   **26. a.** 33 half-acre lots

b. $\dfrac{1}{4}$ acre left over

Cumulative Test (p. 203)

1. 126,240   **2.** 47,758   **3.** 177,282   **4.** 140 R 2

5. 5,820   **6.** 62,900   **7.** 1   **8.** $P = 34$

9. $9 + 6 \cdot 4 = 33$ **10.** $7(10 + 30) = 280$ **11.** 1, 2, 3, 4, 6, 8, 12, 16, 24, 32, 48, 96   **12.** $3^2 \cdot 7$

13. $\dfrac{16}{25}$ **14.** $\dfrac{35}{36}$ **15.** $1\dfrac{1}{2}$ **16.** 6

17. $s = \dfrac{2}{3}$ **18.** $z = 600$ **19.** $1,060   **20.** 12 wires

CHAPTER 4

Pretest (p. 206)

1. $1\dfrac{1}{5}$ **2.** $9\dfrac{1}{8}$ **3.** $7\dfrac{4}{9}$ **4.** $\dfrac{2}{5}$ **5.** $3\dfrac{3}{7}$

6. Yes, $\dfrac{6}{14} = \dfrac{3}{7}$ and $\dfrac{15}{35} = \dfrac{3}{7}$

7. No, $\dfrac{21}{24} = \dfrac{7}{8}$ and $\dfrac{49}{64}$ cannot be reduced.

8. $\dfrac{12}{14}, \dfrac{18}{21}, \dfrac{24}{28}, \dfrac{30}{35}$ **9.** 18   **10.** 60   **11.** 14

12. 48   **13.** 90   **14.** $1\dfrac{1}{24}$ **15.** $3\dfrac{29}{48}$ **16.** $8\dfrac{23}{36}$

17. $\dfrac{11}{30}$ **18.** $2\dfrac{23}{40}$ **19.** $4\dfrac{11}{16}$ **20.** $48\dfrac{5}{8}$

21. $7\dfrac{5}{12}$ ft   **22.** $15\dfrac{1}{10}$ gal   **23.** $\dfrac{13}{15}$ **24.** $\dfrac{2}{3}, \dfrac{5}{6}, \dfrac{8}{9}$

Section 4.1 (p. 211)

1. $\frac{7}{9}$ **3.** $\frac{1}{2}$ **5.** $1\frac{3}{10}$ **7.** $\frac{1}{3}$ **9.** $\frac{5}{6}$ **11.** $\frac{17}{25}$

13. $1\frac{3}{4}$ **15.** 2   **17.** $2\frac{1}{5}$ **19.** 2   **21.** $7\frac{3}{5}$

23. 5   **25.** $12\frac{2}{3}$ **27.** $7\frac{1}{5}$ **29.** $18\frac{1}{2}$ **31.** 10

33. 49   **35.** $\frac{3}{7}$ **37.** $\frac{1}{4}$ **39.** $\frac{1}{2}$ **41.** $\frac{1}{10}$

43. $\frac{1}{3}$ **45.** $\frac{1}{3}$ **47.** $1\frac{2}{7}$ **49.** $\frac{1}{3}$

51. $2\frac{2}{3}$ **53.** $1\frac{1}{2}$ **55.** $8\frac{1}{2}$ **57.** $\frac{1}{5}$

59. $9\frac{5}{12}$ **61.** $s = \frac{5}{9}$ **63.** $k = \frac{5}{7}$

65. $x = 4\frac{1}{5}$ **67.** 49   **69.** 16   **71.** $\frac{5}{18}$

73. $\frac{5}{8} + \frac{7}{8} = \frac{12}{8} = 1\frac{1}{2}$

Section 4.2 (p. 217)

1. equivalent   **3.** not equivalent   **5.** not equivalent
7. equivalent   **9.** not equivalent   **11.** equivalent
13. not equivalent   **15.** not equivalent
17. not equivalent   **19.** equivalent

21. $\frac{18}{24}$ **23.** $\frac{20}{24}$ **25.** $\frac{56}{64}$ **27.** $\frac{18}{48}$

29. $\frac{16}{30}$ **31.** $\frac{18}{21}$ **33.** $\frac{2}{6}, \frac{3}{9}, \frac{4}{12}, \frac{5}{15}, \frac{6}{18}$

35. $\frac{14}{20}, \frac{21}{30}, \frac{28}{40}, \frac{35}{50}, \frac{42}{60}$ **37.** $\frac{6}{8}, \frac{9}{12}, \frac{12}{16}, \frac{15}{20}, \frac{18}{24}$ **39.** 4

41. 16   **43.** 16   **45.** 10   **47.** 6   **49.** 10
51. 36   **53.** 56   **55.** 175   **57.** 81   **59.** 66
61. 153   **63.** 21   **65.** 16   **67.** $2^3 \cdot 5$ **69.** $2 \cdot 3^2$
71. $3 \cdot 11$

Section 4.3 (p. 225)

1. 16   **3.** 60   **5.** 30   **7.** 24   **9.** 60   **11.** 80
13. 36   **15.** 210   **17.** 60   **19.** 18   **21.** 40
23. 28   **25.** 50   **27.** 90   **29.** 180   **31.** 126
33. 320   **35.** 1,200   **37.** 30   **39.** 140   **41.** 12
43. 60   **45.** 360   **47.** 1,020   **49.** 48
51. $\frac{25}{6}$ **53.** $\frac{27}{4}$ **55.** LCD for $\frac{3}{4}$ and $\frac{5}{8}$ is 8

Section 4.4 (p. 231)

1. $1\frac{1}{6}$ **3.** $\frac{17}{20}$ **5.** $\frac{5}{9}$ **7.** $1\frac{1}{12}$ **9.** $1\frac{1}{2}$

11. $\frac{11}{14}$ **13.** $1\frac{1}{4}$ **15.** $\frac{5}{6}$ **17.** $\frac{5}{36}$ **19.** $\frac{119}{160}$

21. $\frac{7}{144}$ **23.** $\frac{55}{108}$ **25.** $1\frac{2}{15}$ **27.** $\frac{57}{80}$

29. $\frac{13}{32}$ **31.** $\frac{11}{12}$ **33.** $1\frac{9}{16}$ **35.** $\frac{127}{150}$

37. $1\frac{65}{72}$ **39.** $1\frac{79}{120}$ **41.** $5\frac{5}{6}$ **43.** $9\frac{5}{21}$

45. $9\frac{5}{8}$ **47.** $6\frac{23}{24}$ **49.** $6\frac{1}{4}$ **51.** $18\frac{1}{8}$

53. $7\frac{5}{7}$ **55.** $11\frac{31}{40}$ **57.** $39\frac{1}{10}$ **59.** $12\frac{7}{9}$

61. $14\frac{47}{50}$ **63.** $13\frac{3}{14}$ **65.** $11\frac{11}{12}$ **67.** $8\frac{13}{24}$

69. $x = 32$ **71.** $y = 125$

73. Final addition should be $\frac{6}{15} + \frac{10}{15} = \frac{16}{15} = 1\frac{1}{15}$

Section 4.5 (p. 237)

1. $\frac{1}{4}$ **3.** $\frac{5}{8}$ **5.** $\frac{3}{8}$ **7.** $\frac{1}{6}$ **9.** $\frac{1}{24}$ **11.** $\frac{21}{40}$

13. $\frac{13}{24}$ **15.** $\frac{5}{36}$ **17.** $\frac{1}{12}$ **19.** $\frac{19}{36}$ **21.** $\frac{11}{35}$

23. $\frac{25}{54}$ **25.** $\frac{32}{105}$ **27.** $\frac{1}{4}$ **29.** $\frac{1}{5}$ **31.** $\frac{41}{180}$

33. $2\frac{1}{2}$ **35.** $3\frac{5}{21}$ **37.** $3\frac{3}{7}$ **39.** $2\frac{1}{8}$ **41.** $8\frac{1}{12}$

43. $1\frac{19}{24}$ **45.** $4\frac{3}{4}$ **47.** $6\frac{2}{9}$ **49.** $1\frac{1}{3}$ **51.** $14\frac{5}{8}$

53. $\frac{2}{7}$ **55.** $3\frac{1}{40}$ **57.** $11\frac{9}{10}$ **59.** $7\frac{7}{9}$ **61.** $\frac{17}{50}$

63. $7\frac{19}{42}$ **65.** $18\frac{5}{12}$ **67.** $2\frac{71}{84}$ **69.** 4

71. $35 + 26 = 61$ **73.** $52 - 37 = 15$

75. $\begin{aligned} 3\frac{8}{40} &= 2\frac{48}{40} \\ -1\frac{15}{40} &= 1\frac{15}{40} \\ \hline & 1\frac{33}{40} \end{aligned}$

Section 4.6 (p. 243)

1. $30\frac{1}{10}$ gal   **3.** $\frac{13}{24}$ in   **5.** $39\frac{5}{24}$ ft

7. a. $16\frac{7}{8}$ gal   **b.** $\$20\frac{1}{4}$

9. a. $\$46\frac{3}{4} + \$1\frac{3}{8} = \$48\frac{1}{8}$ **b.** $160 \times \$1\frac{3}{8} = \$220$ more

11. $1\frac{7}{12}$ cups of nuts   **13.** $1\frac{23}{30}$ **15.** $\frac{1}{12}$

17. $8\frac{3}{5}$ gal   **19.** $2\frac{1}{4}$ in $+ \frac{7}{8}$ in $+ \frac{7}{8}$ in $= 4$ in

21. $\frac{1}{3} + \frac{2}{5} = \frac{11}{15}$, Cy owns $1 - \frac{11}{15} = \frac{15}{15} - \frac{11}{15} = \frac{4}{15}$

23. $1\frac{11}{24}$ qt   **25.** $\frac{1}{8} + \frac{1}{10} = \frac{9}{40}$, $1 \div \frac{9}{40} = 4\frac{4}{9}$ hr

27. $a = 6\frac{3}{8}$ ft, $b = 1\frac{1}{4}$ ft

29. a. $\frac{2}{5} + \frac{4}{15} + \frac{1}{6} = \frac{5}{6}$ used. Therefore, $\frac{1}{6}$ are hanging.

b. $\frac{1}{6}$ of $60 = \frac{1}{6} \times \frac{60}{1} = 10$ balloons   **31.** 19   **33.** 28

35. a. Fill $\frac{1}{2}$ cup. Pour into $\frac{1}{3}$ cup until full. This

leaves $\frac{1}{6}$ of a cup of oil in the $\frac{1}{2}$ cup measure

because $\frac{1}{2} - \frac{1}{3} = \frac{1}{6}$.   **b.** $\frac{1}{3} - \frac{1}{4} = \frac{1}{12}$

37. a. $\frac{15}{16}$ of an acre each   **b.**

Section 4.7 (p. 253)

1. $1\frac{1}{10}$ **3.** 1 **5.** $1\frac{34}{45}$ **7.** $\frac{56}{75}$ **9.** $9\frac{1}{7}$ **11.** 5

13. $8\frac{1}{6}$ **15.** 72 **17.** $1\frac{1}{4}$ **19.** 7 **21.** $\frac{1}{8}$

23. 4 **25.** 11 **27.** $8\frac{1}{2}$ **29.** 170 **31.** $1\frac{1}{3}$

33. 15 **35.** $\frac{1}{80}$ **37.** $2\frac{3}{7}$ **39.** $6\frac{3}{4}$ **41.** $\frac{77}{96}$

43. $3\frac{1}{4} + 1\frac{5}{8} + 2 = 6\frac{7}{8}$ oz of drug,

$6\frac{7}{8} \div \frac{2}{3} = 10\frac{5}{16}$, 10 doses can be made

45. Each gets \$315 **47.** $\frac{3}{5} < \frac{5}{8}$ **49.** $\frac{9}{10} < \frac{11}{12}$

51. $\frac{3}{4} < \frac{4}{5} < \frac{5}{6}$ **53.** $\frac{1}{5} < \frac{7}{15} < \frac{2}{4}$ **55.** $\frac{1}{8} < \frac{5}{16} < \frac{8}{20}$

57. $\frac{1}{4} < \frac{3}{5} < \frac{5}{8} < \frac{2}{3}$ **59.** $\frac{7}{12} < \frac{5}{8} < \frac{3}{4} < \frac{5}{6}$ **61.** $\frac{5}{8}$ in

63. Company A; $\frac{7}{60}$ is higher **65.** 340 **67.** 64

69. a. midpoint = 5 **b.** midpoint = $7\frac{1}{2}$

c. midpoint = $\frac{7}{10}$ **d.** midpoint = $1\frac{1}{8}$

71. $\frac{1}{2} + \frac{1}{2} \cdot \frac{6}{1} = \frac{1}{2} + 3 = 3\frac{1}{2}$

Review (p. 256)

1. $1\frac{1}{4}$ **3.** 1 **5.** $5\frac{3}{5}$ **7.** $5\frac{7}{12}$ **9.** $\frac{2}{9}$ **11.** $\frac{1}{5}$

13. $3\frac{1}{2}$ **15.** $3\frac{1}{3}$ **17.** not equivalent

19. equivalent **21.** $\frac{12}{30}$ **23.** $\frac{21}{30}$

25. $\frac{2}{16}, \frac{3}{24}, \frac{4}{32}, \frac{5}{40}$ **27.** $\frac{6}{10}, \frac{9}{15}, \frac{12}{20}, \frac{15}{25}$

29. $\frac{25}{30}$ **31.** $\frac{63}{72}$ **33.** 36 **35.** 160 **37.** 72

39. 168 **41.** $1\frac{13}{60}$ **43.** $\frac{7}{10}$ **45.** $11\frac{13}{15}$ **47.** $9\frac{53}{72}$

49. $\frac{7}{36}$ **51.** $\frac{29}{84}$ **53.** $\frac{31}{45}$ **55.** $2\frac{7}{10}$ **57.** $3\frac{5}{12}$ ft

59. $23\frac{1}{20}$ mi **61.** \$3,050 **63.** 8 **65.** 9

67. 21 **69.** $\frac{2}{3} < \frac{11}{15} < \frac{7}{9}$ **71.** $\frac{5}{18} < \frac{9}{32} < \frac{7}{24}$

Chapter Test (p. 259)

1. $7\frac{23}{56}$ **2.** $\frac{52}{147}$ **3.** $10\frac{11}{18}$ **4.** $1\frac{37}{50}$

5. $\frac{47}{80}$ **6.** $3\frac{17}{24}$ **7.** $8\frac{19}{48}$ **8.** $\frac{1}{24}$

9. $12\frac{11}{12}$ yd **10.** $48\frac{1}{24}$ mi **11.** equivalent

12. equivalent **13. a.** $\frac{7}{8} - \frac{1}{4} = \frac{7}{8} - \frac{2}{8} = \frac{5}{8}$ of a tank

b. $\frac{5}{8}$ of 24 $= \frac{5}{8} \times \frac{\overset{3}{\cancel{24}}}{1} = 15$ gal
(with 1 below)

14. 19 **15.** $3\frac{23}{30}$ **16.** 2 **17.** $\frac{8}{45}$

18. $x = \frac{1}{10}$ **19.** $w = 1\frac{1}{2}$ **20.** $\frac{3}{40} < \frac{1}{12} < \frac{5}{48}$

Cumulative Test (p. 261)

1. $5,000 + 200 + 9$ **2.** 78,606

3. a. $y = 3$ **b.** $h = 6$ **c.** $n = 7$ **d.** $h = 12$

4. 4,872 **5.** 2,379 **6.** \$489 **7.** 540

8. 45,000 **9.** 33,096 **10.** 3,040,000 **11.** 903

12. 60 **13.** 34 **14.** 37 **15.** \$45 **16.** 84

17. $5^4 \cdot 7^2$ **18.** 81 **19.** 1 **20.** 76 **21.** 52

22. $Y = 47$ **23.** $x = 17$ **24.** $y = 536$ **25.** $h = 80$

26. $t = 360$ **27.** 1, 2, 4, 5, 8, 10, 16, 20, 40, 80

28. $2 \cdot 7^2$ **29.** $\frac{39}{7}$ **30.** $3\frac{3}{14}$ **31.** $\frac{18}{49}$ **32.** $1\frac{61}{72}$

33. \$7,025 **34.** $9\frac{1}{18}$ **35.** $3\frac{29}{40}$ **36.** $\$7\frac{5}{8}$

37. $1\frac{23}{27}$ **38.** 19 **39.** $\frac{16}{25}$

CHAPTER 5

Pretest (p. 264)

1. four **2.** $20 + 9 + \frac{7}{100} + \frac{4}{1,000}$

3. $200 + 80 + \frac{3}{10} + \frac{3}{100}$

4. four and sixty-two thousandths **5.** 290.0467

6. 1,675.10 **7.** 580 **8.** 919.38 **9.** 3.546

10. \$424.36 **11.** 146.50152 **12.** 0.369

13. 27.5 mpg **14.** \$5.84 **15.** 0.15

16. $\frac{1}{40}$ **17.** 9.78 **18.** $0.106 > 0.0660 > 0.06 > 0.0096$

Section 5.1 (p. 271)

1. tenths **3.** thousandths **5.** ten-thousandths

7. one **9.** two **11.** none

13. $1 + \frac{7}{10}$ **15.** $\frac{6}{10} + \frac{4}{1,000}$ **17.** $2 + \frac{6}{100}$

19. $\frac{2}{10} + \frac{8}{10,000}$ **21.** $1 + \frac{6}{10}$ **23.** $6,000 + \frac{6}{100}$

25. five and two tenths **27.** four hundredths

29. sixty-five and four tenths **31.** six tenths **33.** two hundred sixty-five and two tenths **35.** six and five hundredths **37.** 0.9 **39.** 6.07 **41.** 28.4

43. 0.312 **45.** 0.210 **47.** 200.010 **49.** 0.728

51. 75.100 **53.** 659.035 **55.** 800.4

57. 4,044,009.011 **59.** 0.0965 **61.** 50.50

63. 0.8 **65.** 0.173 **67.** $\frac{872}{100} = 8\frac{72}{100} = 8.72$

69. $\frac{6}{10}$ **71.** $\frac{2}{100}$ **73.** $4\frac{58}{100} = \frac{458}{100}$

75. 930 **77.** 1,000 **79.** 0.609

Section 5.2 (p. 277)

1. 4.7 **3.** 260 **5.** 4.48 **7.** 76 **9.** 90.027

11. 500 **13.** 6.5 **15.** 27 **17.** 500 **19.** 3.009

21. 0.3 **23.** 684 **25.** 0 **27.** 0.066 **29.** 427.0

31. 2.0048 **33.** 0.000571 **35.** 5.0 **37.** 0.01

39. 600.000 **41.** 35.1 **43.** 0.713 **45.** 35.0145

47. 0.0007 **49.** 25.0 **51.** 186,300 miles per second

53. \$252,500,000,000 or \$252.5 billion **55.** 1.9 children per family **57.** \$4,572.05 **59.** \$0.50 **61.** 261

63. 52 **65.** 78.⑨92 $\doteq$ 79.0

Section 5.3 (p.283)

1. 14.32 **3.** 696.44 **5.** 1.094 **7.** 14.068
9. 67.76 **11.** 613.2358 **13.** 605.3694 **15.** 486.9
17. 1108.99 **19.** 80 **21.** 3,100 **23.** 62.125 ft
25. 13.375 in **27.** 1.4 **29.** 9.65 **31.** 9.03
33. 3.887 **35.** 16.748 **37.** 0.001 **39.** 3.2
41. 6.528 **43.** 1.97 **45.** 334.69 **47.** 60.05
49. 5.4 **51.** 9.9 **53.** 4.7 **55.** $50 - 30 = 20$
57. $300 **59.** $x = 1.5$ **61.** $p = 4.8$
63. $k = 0.236$ **65.** $\frac{3}{4}$ **67.** $\frac{207}{1,000}$
69. $5.5 + 5.5 = 11$ **71.** $7x + 11y$ **73.** $5p + 2m$

Section 5.4 (p. 287)

1. $1,619.37 **3.** $2.64 **5.** 3.7°F **7.** 4.45 cm
9. a. 199.08 sec **b.** 1.65 sec **11.** 12.093
13. 29.321 **15. a.** needs about $110 more
b. $117.34 **17.** $1,355.57 **19. a.** By rounding to
tens, $670 - 30 - 250 - 170 - 20 + 40 + 100 = \$340$.
b. Total checks = $26.12 + $248.90 + $170 + $15.78 =
$460.80; total deposits = $35.56 + $100 = $135.56;
balance = $670.45 + $135.56 - $460.80 = $345.21
21. a. 52.3 gal **b.** $59.74 **23. a.** $900 **b.** $920
c. $920 **d.** $921.01 **25.** 36 **27.** 7 **29.** 1,000
31. 0.0000663 in **33.** 2 in

Section 5.5 (p. 293)

1. 0.28 **3.** 2.25 **5.** 18.395 **7.** 1,451.415
9. 0.024 **11.** 0.0000014 **13.** 29 **15.** 11.76
17. 18 **19.** 3.125 **21.** 10 **23.** 137.747
25. $50 \times 500 = 25,000$ **27.** 63; high **29.** $60
31. 23.4 **33.** 403.6 **35.** 52,300 **37.** 7,200
39. 460.3 **41.** 40,602 **43.** 38.5
45. $100 \times \$349.95 = \$34,995$ **47.** $1,290
49. 253,610,000 people **51.** 60,000,000,000 cells
53. 80 R 1 **55.** 61 R 9 **57. a.** $4.29 **b.** $0.89
c. $21.62 **d.** $507.25 **e.** $0.05 **f.** $2
59.

```
    0.02  ⟶   2
  ×  40  ⟶  +0
  ‾‾‾‾‾‾     ‾‾
   .80       2    2 places
```

Section 5.6 (p. 301)

1. 0.8 **3.** 1.4 **5.** 2.16 **7.** 78.5 **9.** 0.014
11. 0.875 **13.** 0.72 **15.** 0.1 **17.** 0.186
19. 4.9 **21.** 4.02 **23.** 0.0026 **25.** 0.22
27. 1.34 **29.** 4 **31.** 12 **33.** 0.7 **35.** 40
37. 6.4 **39.** 0.048 **41.** 1.58 **43.** 5.06
45. 300 **47.** 50 **49.** 6.86 **51.** 31.0 **53.** 560
55. 0.288 **57.** 1.26 **59.** 0.4175 **61.** 0.69301
63. 0.593016 **65.** 0.00608 **67.** 0.0000028
69. $2.56 **71.** $17.60 **73.** $12,503.41
75. $x = 2.5$ **77.** $h = 3.048$ **79.** $y = 0.86$
81. $m = 2002$ **83.** $v = 5.9$ **85.** 28 **87.** 12
89.

```
      .204
  4)0.816
    −8
    ‾‾
     16
    −16
    ‾‾‾
      0
```

Section 5.7 (p. 309)

1. $17.21 **3.** $5.26 **5. a.** about $10 \times 20 = \$200$
b. low **c.** $11 \times \$21.69 = \$238.59$ **7.** $547.36
9. $0.40 per brick **11.** $45.60
13. $3.15 \times 10 = \$31.50$, $\$0.92 \times 52 = \$47.84$
Total made = $31.50 + $47.84 = $79.34 **15.** 23.0 mpg
17. $2.57 **19.** $48 \times 1.06 = \$50.88$ **21. a.** First
minute = $0.67, next 16 minutes = $0.48 \times 16 = \$7.68$,
total charge = $0.67 + $7.68 = $8.35. **b.** 8.35×1.03
= $8.60 **23. a.** $34.5 \times \$10.68 = \$368.46$
b. $40 \times \$10.68 = \$427.20$ **c.** Wage for first 40 hours =
$427.20; wage for next 11 hours = $11 \times \$10.68 \times 1.5 =$
$176.22; total wage = $427.20 + $176.22 = $603.42.
25. 6 min 11 sec per mile **27.** $136.84
29. a. $42.16 \times 16 = \$674.56$ **b.** $674.56 - $499.95
= $174.61 **31. a.** $0.33 **b.** $0.30 **c.** save
$0.03 per bar **33. a.** $5.55 **b.** 40 ounces
35. $411.95 **37.** Savings per gallon = $1.269 - $1.209
= $0.06; total savings = $16.3 \times \$0.06 = \$0.98$
39. $7,620 **41.** $4,070 per person **43.** $269.76
45. a. $126.19 \div \$8.62 = 14.639211$, or 14 toys **b.** 14
toys $\times \$8.62 = \$120.68$; Change = $126.19 - $120.68 =
$5.51 **47.** 85.9 **49.** $2,400 **51.** $158.95
53. $2.95 **55.** 15 **57.** 17 **59.** 2.4 **61.** about
42,000 mph **63.** $1,159.11 **65.** 7 days
67. a. 0.2 in **b.** 125 years

Section 5.8 (p. 321)

1. 0.5 **3.** 0.7 **5.** 0.72 **7.** 0.25 **9.** 1.75
11. 4.4 **13.** 9.075 **15.** 15.5 **17.** 0.15625
19. 0.8 **21.** 0.7778 **23.** 0.727 **25.** 0.33
27. 2.857 **29.** 4.28 **31.** 2.6 **33.** 8.1333
35. 0.053
37. $\frac{1}{5}$ **39.** $\frac{9}{50}$ **41.** $\frac{9}{40}$ **43.** $\frac{1}{25}$
45. $\frac{3}{500}$ **47.** $\frac{2}{25}$ **49.** $5\frac{7}{10}$ **51.** $6\frac{2}{25}$
53. $10\frac{1}{2}$ **55.** $20\frac{18}{25}$ **57.** $\frac{1}{2,500}$ **59.** $37\frac{1}{2}$
61. $\frac{3}{8}$ **63.** $\frac{3}{800}$ **65.** $64\frac{17}{100}$ **67.** $36\frac{9}{25}$
69. $\frac{1}{8}$ **71.** $\frac{1}{3}$ **73.** $\frac{59}{60}$ **75.** 4
77. 1 **79. a.** $\frac{5}{9}$ **b.** $\frac{127}{999}$ **c.** $\frac{32}{33}$
81. a. all are equivalent
b. $\frac{24}{28} = \frac{138}{161} = \frac{96}{112} = \frac{54}{63} = \frac{30}{35}$ and $\frac{16}{21} = \frac{144}{189}$
83. $\frac{5}{8} = 8)\overline{5.000}^{\ .625}$. So $\frac{5}{8} = 0.625$ **85.** $0.28 = \frac{28}{100} = \frac{7}{25}$
89 a. $660,000 each **b.** $680,000 **c.** $666,666.66\frac{2}{3}$
d. $13,333.33\frac{1}{3}$

Section 5.9 (p. 329)

1. 34.16 **3.** 17.22 **5.** 400 **7.** 25 **9.** 30

11. 4.0205　　**13.** 10　　**15.** 45　　**17.** 25.48
19. 236　　**21.** 25.5　　**23.** 4017.6　　**25.** 1,923
27. 1,060.9　　**29.** 0　　**31.** $1.90 + $0.80 × 11 = $10.70
33. 3 × $24 + 6 × $14.95 = $161.70　　**35.** $0.83 +
$0.52(12 − 1) = $6.55　　**37.** 1.25　　**39.** 1.4
41. 0.31　　**43.** 30.1　　**45.** 1.5　　**47.** 3.09
49. $10.71　　**51. a.** $164.75　　**b.** $494.25
53. 0.089 < 0.8　　**55.** 0.0971 < 0.103
57. 0.00091 < 0.002　　**59.** 0.6 < 0.603 < 0.6104
61. 3.0063 < 3.026 < 3.03
63. 0.2042 < 0.205 < 0.254 < 0.26
65. 4 < 4.0069 < 4.03 < 4.1
67. 0.0061 < 0.019 < 0.02 < 0.1
69. 7.018 < 7.07 < 7.089 < 7.09 < 7.2
71. $\frac{3}{5}$　　**73.** $\frac{4}{9}$　　**75. a.** 0.9 + 0.9 ÷ 0.9 = 1.9
b. 2.5(2.5 − 2.5) = 0　　**c.** 1.8 + 0.2 × 1 = 2
d. 5 ÷ (5 + 5) = 0.5　　**e.** 0.08 + 0.05 + 0.07 = 0.2
f. 10 ÷ 0.1 ÷ 0.1 = 1,000　　**77. a.** $6\frac{2}{3}$ × $0.024 = $0.16
b. 30 × $0.16 = $4.80　　**c.** 365 × $0.16 = $58.40
79. 2.4 ÷ 1.2 · 2 = 2 · 2 = 4

Review (p.335)

1. a. hundredths　　**b.** tens　　**c.** tenths
d. ten-thousandths　　**3. a.** $1 + \frac{6}{10}$
b. $54 + \frac{9}{100} + \frac{7}{1,000}$　　**c.** $5 + \frac{6}{1,000} + \frac{2}{10,000}$
d. $\frac{8}{100}$　　**5. a.** 0.4　　**b.** 6.024　　**c.** 0.18
d. 0.624　　**e.** 0.005　　**f.** 20.1209　　**7.** 201
9. 28.29　　**11.** 0.0　　**13.** 1.805　　**15.** 33.413
17. 38.9802　　**19.** 337.372　　**21.** 66.44　　**23.** 9.61
25. 5,495.536　　**27.** 0.053　　**29.** $46.72
31. $545.49　　**33.** 2.128　　**35.** 0.0416　　**37.** 4,089.3
39. 0.065　　**41.** $45.72　　**43.** 15.2　　**45.** 12.63
47. 7.2　　**49.** 0.1704　　**51.** 16.36　　**53.** 0.0076
55. about 563 pounds of water　　**57.** $15.24
59. 24 months　　**61.** 8.2 mpg　　**63.** $17.50　　**65.** No
67. You save about $0.65　　**69.** Total cost = 180 mg ×
$0.048/mg × 200 = $1,728　　**71.** 0.45　　**73.** 0.9
75. 12.44　　**77.** 0.235　　**79.** 0.7917　　**81.** $\frac{81}{100}$
83. $2\frac{3}{100}$　　**85.** $20\frac{1}{2}$　　**87.** $100\frac{5}{16}$　　**89.** 10
91. 750,200　　**93.** 27.352　　**95.** 0.628　　**97.** 26.37
99 0.0619 < 0.062
101. 0.03 < 0.0359 < 0.038 < 0.0402

Chapter Test (p. 339)

1. fifty-two thousandths　　**2.** one hundred and seven
hundred ten ten-thousandths　　**3.** 3.03　　**4.** 0.0809
5. 4,060.2　　**6.** 53　　**7.** 50.9　　**8.** 0.900　　**9.** 5,800
10. 11.192　　**11.** 174.9566　　**12.** $16.86
13. 38.7847　　**14.** 15.19°C　　**15.** 29,095.476
16. 630　　**17.** 3.8　　**18.** 0.082　　**19.** 14 trips
20. $5.62　　**21.** about $0.18 per shot　　**22.** $13.40
23. 0.4375　　**24.** 4.429　　**25.** $\frac{9}{20}$　　**26.** $4\frac{41}{500}$
27. 5.57　　**28.** 80.3　　**29.** 10.19　　**30.** w = 7.25

31. y = 1.9　　**32.** z = 24.42
33. 2.06 < 2.0609 < 2.0614 < 2.065

Cumulative Test (p. 341)

1. 1,400　　**2.** 269,250　　**3.** 274　　**4.** 16
5. x = 20　　**6.** x = 48　　**7.** 4(12 + 8) − 6 = 74
8. $1\frac{4}{9}$　　**9.** $1\frac{29}{40}$　　**10.** $\frac{3}{10}$　　**11.** $3\frac{9}{11}$
12. $73\frac{4}{5}$ gallons　　**13.** $3\frac{52}{75}$　　**14.** 2.87　　**15.** 64.01
16. 2.71　　**17.** 0.3498　　**18.** 4.2　　**19.** 18.5
20. 0.001

CHAPTER 6

Pretest (p. 344)

1. a. 1,043 to 74　　**b.** 74 to 11　　**c.** 85 to 1,043
2. 8 to 5　　**3.** 15 to 2　　**4.** 15 mi to 1 hr, or 15 mph
5. $10 to 7 ft　　**6.** 10 to 3　　**7.** 12 to 1　　**8.** 2.5 to 1
9. 150 mi to 1 hr, or 150 mph　　**10.** 4.7¢ per ounce
11. Tom pays $20,306.25, Hal pays $12,183.75
12. Gas M gave 23.3 miles per gallon. Gas Q gave 21.7
miles per gallon. Gas M gave Ellen the better mileage.
13. False　　**14.** True　　**15.** x = 8　　**16.** y = 10.2
17. 188.2 gallons

Section 6.1 (p. 349)

1. a. 21 to 23　　**b.** 23 to 21　　**c.** 23 to 51
d. 51 to 44　　**3. a.** 12 to 17　　**b.** 23 to 24
c. 29 to 47　　**d.** 24 to 17　　**5.** 40 to 27, 40:27, $\frac{40}{27}$
7. 49 to 15, 49:15, $\frac{49}{15}$　　**9.** 80 to 21, 80:21, $\frac{80}{21}$
11. a. $\frac{24}{5}$　　**b.** 5:24　　**c.** 24 to 29, or twenty-four to
twenty-nine　　**13.** 5:3, $\frac{5}{3}$　　**15.** 6 to 5, 6:5
17. 40 to 1, $\frac{40}{1}$　　**19.** 1 to 64, 1:64　　**21.** 1:1, $\frac{1}{1}$
23. 8 to 9, $\frac{8}{9}$　　**25.** 11 to 8　　**27.** 19 to 10　　**29.** 4:1
31. 1:3　　**33.** $\frac{10}{3}$　　**35.** $\frac{4}{5}$　　**37.** 7 to 4
39. 25 to 7　　**41.** 18 to 5　　**43.** 1:5　　**45.** 6 to 1
47. 5 to 12　　**49.** 9 to 10　　**51.** 16 to 3　　**53.** 5:2
55. 1 to 25　　**57.** 5/12　　**59.** 7 to 2　　**61.** 4:1
63. 20 to 3　　**65.** 26 to 3　　**67.** 4 to 1　　**69.** 71 to 12
71. 5 to 12　　**73.** 1:48　　**75.** $\frac{4}{6}, \frac{6}{9}, \frac{8}{12}$
77. $\frac{2}{20}, \frac{3}{30}, \frac{4}{40}$　　**79.** 18:1　　**81.** $\frac{1}{3}$
83. 15 to 40 = $\frac{15 ÷ 5}{40 ÷ 5}$ = 3 to 8

Section 6.2 (p. 355)

1. 10 engines per 3 cars　　**3.** $40 per 3 hr
5. 50 mi per 3 gal　　**7.** 50 mi per 1 hr, or 50 mph
9. 16 customers per 7 toys　　**11.** 30 books per 7 readers
13. 20 mi per 1 gal, or 20 mpg　　**15.** 235 books per 1 hr,
or 235 books per hr　　**17.** 9 to 5　　**19.** 1 mi per 2 min

21. 60¢ per 1 lb, or 60¢ per lb   **23.** 20 min per 1 box, or 20 min per box   **25.** 125¢ per 3 oz
27. a. $3 per 5 dozen   **b.** 60¢ per 1 dozen, or 60¢ per dozen   **c.** 5¢ per 1 egg, or 5¢ per egg   **29. a.** $5 per 1 lb, or $5 per lb   **b.** 500¢ per 1 lb, or 500¢ per lb
c. 125¢ per 4 oz   **31. a.** $360 per 1 acre, or $360 per acre   **b.** $1 per 121 square feet   **c.** 100¢ to 121 square feet   **33.** 15   **35.** 1 mi per 12 min

Section 6.3 (p. 361)

1. 2 to 1   **3.** 2 to 3   **5.** 5 to 66   **7.** 25 to 36
9. 20:21   **11.** 6 to 1   **13.** 5 to 1   **15.** 5 to 1
17. 10 to 9   **19.** 2 to 1   **21.** 9 to 2   **23.** 1 to 1
25. $1\frac{1}{4}$ to $\frac{5}{8} = \frac{5}{4} \div \frac{5}{8} = \frac{2}{1}$ or 2 to 1
27. 23 mi per 8 hr   **29.** 24 to 1   **31.** 20 to 3
33. 1 to 3,000   **35.** 500 to 9   **37.** 3 to 1
39. 10 to 9   **41.** 7 to 1   **43.** 100 to 1   **45.** 1 to 10
47. 1 to 300   **49.** 18 to 25   **51.** 50 to 3   **53.** 1 to 21
55. 29 gallons per $35   **57.** $x = 6$ **59.** $t = 10$
61. $\frac{7}{2} \div \frac{7}{1} = \frac{\overset{1}{\cancel{7}}}{2} \times \frac{1}{\cancel{7}} = \frac{1}{2}$ or, 1 to 2

Section 6.4 (p. 367)

1. 3.75 to 1   **3.** 15 to 1   **5.** 2.1 children per family
7. $6\frac{2}{3}$ dozen eggs per chicken   **9.** $376\frac{2}{3}$ people to 1 millionaire   **11.** 250 to 1   **13.** $3/lb   **15.** $1.40/gal
17. $5/model   **19.** $1.27 \div \frac{1}{4}$ lb = $5.08/lb
21. $28,500/acre   **23.** 11.0¢/1 oz   **25.** 7.4¢/egg
27. 12.0¢/oz   **29.** 7   **31.** 2.6   **33.** 5   **35.** 5
37. 3.4   **39.** 1.5   **41.** $6\frac{1}{4}$ **43.** $1\frac{1}{6}$
45. $0.58; 58¢ in losses were paid out for every $1 collected in premiums
47. 12 ft to 15 ft = $\frac{12 \text{ ft} \div 15}{15 \text{ ft} \div 15} = 0.8$ to 1

Section 6.5 (p. 373)

1. $3,500 to Jim, $1,500 to Sue   **3.** 520 gal to W, 40 gal to Z   **5.** $4,000 to Meg, $3,200 to Millie   **7.** 350 marbles to Cal, 250 marbles to Don   **9.** $4.95 to Lynn, $2.55 to Lou   **11.** $1,200 to Alice, $800 to Betty, $400 to Cath   **13.** 1,612 employed by A, 1,209 employed by B, 806 employed by C   **15.** 26 pints of gas, 6 pints of oil
17. a. $51,750 to $20,250 reduces to 23 to 9
b. 23 + 9 = 32, 160 ÷ 32 = 5 hours, 23 × 5 = 115 hours on the bridge, 9 × 5 = 45 hours on the building   **19. a.** Al gets $3,467.80, Bob gets $2,477, Cy gets $1,981.60
b. Al pays $1,492.75, Bob pays $1,066.25, Cy pays $853
21. first gets $105,000, second gets $60,000, third gets $15,000   **23.** Meg plants 22 per hour. Susan plants 24 per hour. Susan plants 2 more per hour.   **25.** Last year she scored 19 points per game. This year she scored 21 points per game. She scored 2 more points per game this year.
27. a. $1.15 per gallon for Jayne, $1.25 per gallon for her friend. Jayne paid the better price.   **b.** Jayne paid 10¢ less per gallon.   **29. a.** 6.6¢/ounce for Mike, 6.0¢/ounce for Debbie. Debbie made the better buy.   **b.** Debbie paid 0.6¢ less per ounce.   **31. a.** Six-pack sells for 2.5¢/ounce, 2-liter sells for 1.5¢/ounce. 2-liter is the better buy.
b. You pay 1.0¢ less per ounce with the 2-liter bottle
33. a. 34 people per sq mi in the former Soviet Union, 304 people per sq mi in China.   **b.** China has about 9 times more people per sq mi.   **35.** $\frac{13}{30}$ **37.** $3\frac{4}{9}$
39. a. 18.2 mpg before the tune-up, 23.7 mpg after the tune-up   **b.** 5.5 more miles per gallon after the tune-up
c. save about $110

Section 6.6 (p. 379)

1. true   **3.** true   **5.** false   **7.** true   **9.** false
11. true   **13.** true   **15.** false   **17.** true
19. false   **21.** true   **23.** false   **25.** false
27. false   **29.** true   **31.** true   **33.** 1.632
35. 1.33   **37.** No, $\frac{\frac{1}{2} \text{ in}}{6 \text{ ft}} = \frac{3\frac{5}{8} \text{ in}}{38 \text{ ft}}$ is a false proportion
39. $\frac{\overset{1}{\cancel{5}}}{8} \times \frac{3}{\underset{2}{\cancel{10}}} = \frac{3}{16}$ Do not cross multiply when you multiply fractions.

Section 6.7 (p. 385)

1. $x = 16$ **3.** $t = 15$ **5.** $h = 6$ **7.** $y = 81$
9. $s = 32$ **11.** $x = 63$ **13.** $x = 18$ **15.** $x = 8$
17. $x = 20$ **19.** $y = 21$ **21.** $x = 8\frac{4}{7}$ **23.** $v = 300$
25. $x = 90$ **27.** $x = 18$ **29.** $x = 44$ **31.** $h = 30$
33. $x = 22\frac{2}{9}$ **35.** $x = \frac{3}{4}$ **37.** $x = 4$ **39.** $y = 6$
41. $x = 4$ **43.** $t = 2.7$ **45.** $x = 0.2$
47. $s = 0.125$ **49.** $k \doteq 0.642$ **51.** $\frac{1}{4} < \frac{5}{16} < \frac{3}{8}$
53. $\frac{2}{3} < \frac{7}{9} < \frac{5}{6} < \frac{11}{12}$ **55.** $\frac{x}{16} = \frac{12}{5}$
$$5 \cdot x = 12 \cdot 16$$
$$x = \frac{192}{5}$$
$$x = 38.4$$

Section 6.8 (p. 393)

1. 336 mi   **3.** 21 favors   **5.** $12\frac{1}{2}$ ft
7. 15,750 voters   **9.** 9 lb   **11.** 12 hr
13. 118.125 mi   **15.** $\frac{264 \text{ mi}}{14.1 \text{ gal}} = \frac{9,300 \text{ mi}}{x \text{ gal}}$
$$264 \cdot x = 9,300 \cdot 14.1$$
$$x \doteq 497 \text{ gal}$$
17. 96 games   **19.** about 17,400,000 lb
21. 1 lb = 16 oz,   **23.** 200 mg
$$\frac{20 \text{ oz}}{\$1.69} = \frac{16 \text{ oz}}{x}$$
$$x \doteq \$1.352 \text{ per pound}$$
25. a. $1\frac{1}{4}$ tsp salt, $3\frac{3}{4}$ cups of water   **b.** 5 servings

27. 57 votes per 100 **29.** 0.6 **31.** 0.625

33. about 300 ft **35. a.** $x = 3\frac{3}{4}$ ft high

b. about 560 ft **c.** about 3,300 ft
37. a. 156 beats per min **b.** 114 beats per min
c. He recovered $156 - 114 = 42$ beats in the first minute

Review (p. 397)

1. a. 163 to 17 **b.** 17 to 4 **c.** 21 to 163

3. 61 to 27, 61:27, $\frac{61}{27}$ **5.** 10 to 7 **7.** 4/5 **9.** 3/5

11. 4 to 1 **13. a.** 10 to 13 **b.** 24 to 5 **c.** 3 to 1
d. 15 to 8 **15.** 18 to 1 **17.** $3 per toy **19.** $29
per 3 hr **21.** 7 min per mi **23.** 75¢ per gal
25. 25 min per box **27.** 4 to 3 **29.** 16 to 1
31. 80 to 1 **33.** 16 ft per min **35.** 20 to 1
37. 39¢ per lb **39.** $1.386 per lb **41.** Fran made the
better buy. Fran paid 7.2¢/oz, Bob paid 7.9¢/oz.
43. 16.8 mpg with A, 18.1 mpg with B. Gas B gave the
better mileage. **45.** $345 for Melissa, $115 for Jason
47. false **49.** true **51.** $x = 4$ **53.** $x = 7.5$

55. $x \doteq 5.7$ **57.** $x = 2\frac{21}{32}$ **59.** 10 oz **61.** $1,872

63. 21 ft **65.** about 6.6 gal

Chapter Test (p. 401)

1. a. 8 to 1 **b.** 7 to 2 **c.** 9 to 56

2. 3 to 4, 3:4, $\frac{3}{4}$ **3.** 5 to 21 **4.** 625 packages per

hour **5.** 2 lb per 5 min **6.** 116 to 33 **7.** 34 to 5
8. 14.2¢ per oz **9.** $4,900 per acre **10.** 11.7¢/oz for
A, 11.4¢/oz for B. B is the better buy. **11.** $45 to Millie,

$25 to Nick, $10 to David **12.** true **13.** $x = 3\frac{5}{9}$

14. $y = 1\frac{1}{2}$ **15.** $n = 20$ **16.** $h = 31.25$

17. $3\frac{5}{6}$ in **18.** 210 pounds

Cumulative Test (p. 403)

1. 9 **2.** 16 **3.** $w = 12$ **4.** 10 **5.** 84 **6.** 6

7. $6\frac{1}{24}$ **8.** $\frac{17}{24}$ **9.** $7\frac{1}{5}$ **10.** $1\frac{5}{8}$

11. a. 1 share lost $\frac{1}{3}$ of $12\frac{3}{8} = \frac{1}{\overset{}{\underset{1}{3}}} \times \frac{\overset{33}{99}}{8} = \$4\frac{1}{8}$

b. 1 share is now worth $12\frac{3}{8} - \$4\frac{1}{8} = \$8\frac{1}{4}$

12. 109.733 **13.** 0.503 **14.** 51.084 **15.** 0.596
16. 6 to 1 **17.** $w = 15$ **18.** 80 to 3
19. $480, $360 **20.** $3.44

CHAPTER 7

Pretest (p. 406)

1. 42, 100 **2.** $33\frac{1}{3}$, 100 **3.** 76% **4.** 48%

5. 0.092 **6.** $87\frac{1}{2}$ **7.** $\frac{24}{25}$ **8.** R% = 47%,

B = $300, A = $141 **9.** R% = 27%, B = ?, A = 13.5
10. $27 **11.** 20% **12.** 40 **13.** 94.8%
14. 1.2 million **15.** 75 gallons of milk **16.** $18,000
17. a. $135 **b.** $315 **18.** 40% gain in profit
19. $33.60 **20.** $4,400

Section 7.1 (p. 411)

1. 40, 100

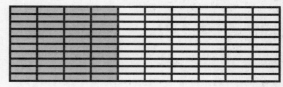

3. 85

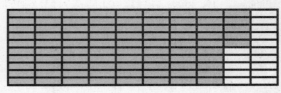

5. 12.5

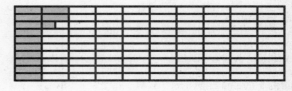

7. $\frac{1}{4}$, 100

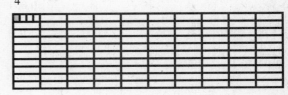

9. 30% **11.** 39% **13.** 20% **15.** $16\frac{2}{3}$%

17. $99\frac{1}{2}$% **19.** 99.92% **21.** 90 **23.** 3.9

25. 18 **27.**

29.

31.

33. $\frac{16}{25}$ **35.** $\frac{3}{50}$ **37.** $\frac{56}{100}$% or 0.56%

39. $100\% - 0.6\% = 99.4\%$

Section 7.2 (p. 415)

1. 62% **3.** 5% **5.** 21.2% **7.** 1.8%
9. 6.08% **11.** 10.04% **13.** 30% **15.** 50.001%
17. 400% **19.** 250% **21.** 1,200%
23. 75,000% **25.** 300.07% **27.** 1,255%

29. $62\frac{1}{2}\%$, or 62.5% **31.** 450% **33.** 0.48

35. 0.05 **37.** 0.722 **39.** 2 **41.** 2.347 **43.** 40
45. -0.0004 **47.** 0.000006 **49.** 0.8091 **51.** 0.023
53. 2.9 **55.** 22.637 **57.** $0.06\frac{1}{2}$, or 0.065

59. $0.35\frac{1}{4}$, or 0.3525 **61.** 0.12375 **63.** 0.8

65. 0.7 **67.** $0.004\% = .00\ 004\% = 0.00004$

Section 7.3 (p. 423)

1. 25% **3.** 80% **5.** $62\frac{1}{2}\%$ **7.** $36\frac{4}{11}\%$

9. $41\frac{2}{3}\%$ **11.** 90% **13.** 140% **15.** $187\frac{1}{2}\%$

17. 225% **19.** 150% **21.** 402.5% **23.** $633\frac{1}{3}\%$

25. $21\frac{2}{3}\%$ **27.** $4\frac{2}{3}\%$ **29.** $31\frac{1}{4}\%$ **31.** 5%

33. $7\frac{11}{27}\%$ **35.** $\frac{11}{20}$ **37.** $\frac{9}{10}$ **39.** $\frac{69}{100}$

41. $\frac{2}{25}$ **43.** $1\frac{4}{5}$ **45.** 7 **47.** $\frac{2}{3}$ **49.** $\frac{5}{6}$

51. $2\frac{3}{8}$ **53.** $\frac{1}{500}$ **55.** $\frac{1}{700}$

57. $5\frac{1}{3}\% = \frac{4}{75}$. $4 out of every $75 tax dollars pays for

sanitation.   **59.** $82\frac{1}{2}\% = \frac{33}{40}$. 33 out of every 40 students

studies a foreign language.   **61. a.** 2% **b.** $\frac{1}{50}$

63. $\frac{11}{500}$ **65.** $\frac{13}{80}$ **67.** $\frac{17}{40}$ **69.** $\frac{5}{8}$ **71.** $\frac{81}{50,000}$

73. $\frac{1}{5,000,000}$ **75.** $1\frac{81}{200}$ **77.** $\frac{157}{3,000}$ **79.** $\frac{513}{1,000}$

81. $\frac{9}{1,000}$ **83.** $\frac{13}{160}$ **85.** $\frac{9}{25}$, 36% **87.** 1.25,

125% **89.** $1\frac{1}{2}$, 150% **91.** 0.6, 60% **93.** $\frac{1}{125}$,

0.8% **95.** $1\frac{2}{3}$, 1.6667% **97.** $\frac{249}{400}$, 62.25%

99. $b = 3.5$ **101.** $y \doteq 555.56$ **103.** 63%; 88%

105. $\frac{2}{25} \to 25\overline{)2.00}\ \ \begin{array}{r}.08\\ \underline{2\ 00}\\ 0\end{array} = 8\%$

Section 7.4 (p. 429)

1. $R\% = 65\%$, $B = 300$ cars, $A = 195$ cars
3. $R\% = 8\%$, $B = \$35$, $A = \$2.80$ **5.** $R\% = 0.07\%$,
$B = 6,500$ kg, $A = 4.55$ kg   **7.** $R\% = 120\%$, $B = 0.540$
oz, $A = 0.648$ oz   **9.** $R\% = 600\%$, $B = \$50$, $A = \$300$
11. $R\% = 44\%$, $B = ?$, $A = 35$ **13.** $R\% = 20\%$,
$B = ?$, $A = 12$ **15.** $R\% = 38\%$, $B = 65$ pounds, $A = ?$
17. $R\% = 90\%$, $B = \$550$, $A = ?$ **19.** $R\% = ?$,
$B = 63.2$, $A = 21.8$ **21.** $R\% = ?$, $B = 88$, $A = 55$
23. $R\% = ?$, $B = 40$, $A = 128$ **25.** $R\% = ?$, $B = 180$,
$A = 450$ **27.** $R\% = 34\%$, $B = \$60.50$, $A = ?$

29. $1\frac{7}{15}$ **31.** $5\frac{11}{18}$

33. $R\% = ?$
$B = \$36$ (follows "percent of")
$A = \$50$

Section 7.5 (p. 439)

1. $A = 252$ **3.** $A = 330.6$ **5.** $A = \$4.68$
7. $A = 80$ men   **9.** $A = 13.2$ lb   **11.** $A = \$1,320$
13. $B = 80$ **15.** $B = 50$ **17.** $B = 90$
19. $B = 20,000$ people   **21.** $B = 120$ **23.** $B = 10$
liters   **25.** $R\% = 30\%$ **27.** $R\% = 30\%$
29. $R\% = 23\%$ **31.** $R\% = 60\%$ **33.** $R\% = 160\%$

35. $R\% = 33\frac{1}{3}\%$ **37.** $B = 350$ **39.** $A = 6$

41. $R\% = 75\%$ **43.** $B = 1\frac{2}{3}$ **45.** $R\% = 80\%$

47. $A = \$64$ **49.** 38.67 **51.** 2.028

Section 7.6 (p. 445)

1. $A = 2,090$ people   **3.** $B = 230$ women
5. $R\% = 71\%$ **7.** $B = \$880$
9. a. $R\% = ?$, $B = 1,500$, $A = 645$,

$R\% = \frac{A}{B} = \frac{645}{1,500} = 43\%$

b. $100\% - 43\% = 57\%$ opposed

11. $R\% = \frac{2}{3}\%$, $B = 600$ lb, $A = ?$,

$A = R\% \times B = \dfrac{\overset{2}{\cancel{\frac{2}{3}}}}{\underset{1}{\cancel{100}}} \cdot \dfrac{\overset{6}{\cancel{600}}}{1} = 4$ lb of gold

13. $A = \$23,200$ **15.** $R\% = 40\%$ **17.** $A = \$551$

19. $R\% = 92\%$, $B = ?$, $A = 138$ lbs, $B = \frac{138}{0.92} = 150$ lb

21. $A \doteq 372$ mg   **23.** What percent of $180 is $65;

$R = 36\frac{1}{9}\%$ **25.** 70% of $1,500 is what number;

$A = 1,050$ men   **27.** 20% of what number is $35,000;
$B = \$175,000$ **29.** 5% of $80 is what number; $A =$
4 plants died   **31.** 15% of $37.60 is what number;
$A = \$5.64$ **33.** 40 **35.** 0.47 **37.** $B = \$2,250$
39. $\$1.05/lb$

Section 7.7 (p. 457)

1. $A = \$4,324$ **3.** $B = \$8,000$ **5.** $A = \$5,920$
7. $B = \$54,500$ **9.** $R\% = 5\%$
11. $R\% = 12\%$, $B = \$5,620$, $A = ?$,
$A = R\% \times B = 0.12 \times \$5,620 = \$674.40$, July
salary $= \$200 + \$674.40 = \$874.40$
13. $\$920 - \$200 = \$720$ in commissions.
$R\% = 4\%$, $B = ?$, $A = \$720$,

$B = \frac{A}{R\%} = \frac{720}{0.04} = \$18,000$ in car sales

15. a. $A = \$315$ **b.** $\$585$ **17. a.** $\$19.20$
b. $\$6.90$ **c.** $\$149.99$ **19.** $\$280.32$
21. $R\% = ?$, $B = \$340$,

$A = \$340 - \$238 = \$102$, $R\% = \frac{\$102}{\$340} = 30\%$

23. a. $B = \$865$ **b.** $\$692$

25. R% = ?, B = $55.60, A = $58.38 − $55.60 = $2.78, R% = $\frac{\$2.78}{\$55.60}$ = 5%

27. R% = 10%   **29. a.** R% = 25% increase
b. R% = 17.5% actual increase   **31. a.** $7,020
b. $16,380   **33. a.** $720   **b.** $2,880

35. R% = $144\frac{4}{9}$%   **37. a.** R% = $16\frac{2}{3}$% drop in weight

b. R% = 20% increase in weight   **39.** $21,627.50
41. M = 42   **43.** I = 12   **45. a.** A = $12,250
b. Each received $6,125   **47. a.** $32,000 and $19,200
b. She lost $3,200 on the transaction   **49.** about a 99.6% decrease

Section 7.8 (p. 469)

1. I = $80   **3.** I = $1,440   **5.** I = $186.91
7. I = $640   **9.** I = $63
11. I = $5,000 × 0.15 × 2 = $1,500   **13.** I = $81
15. I = $243.75   **17.** $380 per month   **19.** $550 per month   **21.** Borrow $14,600 − $2,600 = $12,000.
I = $12,000 × 0.10 × 2 yr = $2,400. Monthly payment = ($12,000 + $2,400) ÷ 24 = $600
23. a. $4,896   **b.** $419.33   **25. a.** $20,000
b. $1,000   **c.** $66,000   **27.** P = $2,000

29. T = 4 years   **31.** R = $\frac{\$75}{\$1,250 \cdot 1.5}$ = 4%

33. I = $173.25   **35.** R = 6.5%   **37.** P = $6,400

39. I = $5.30   **41.** T = 2 years   **43.** R = $12\frac{1}{2}$%

45. T = 4 years   **47.** P = $10,000   **49.** T = 16 years
51. R = 9%   **53.** S = $898.88, I = $98.88
55. S ≐ $2,701.22, I = $301.22   **57.** S ≐ $3,941.29,
I = $191.29   **59.** S ≐ $49,847.28, I = $9,847.28
61. a. S ≐ $5,630.81   **b.** I = $630.81
63. a. S ≐ $3,253.82   **b.** I = $1,253.82

65. 4   **67.** $6\frac{2}{5}$ **69.** I ≐ $373.48

71. a. S = $1,120; I = $120   **b.** S = $1,123.60;
I = $123.60   **c.** S ≐ $1,125.51; I = $125.51
d. S ≐ $1,126.83; I = $126.83   **e.** You earn more
interest the more often you compound.
73. a. S = $1,030; I = $30   **b.** S = $1,060.90;
I = $60.90   **c.** S = $1,092.73; I = $92.73
d. S = $1,125.51; I = $125.51   **e.** No. Amount of
interest earned in each additional quarter is more than
previous quarter, but not by a fixed amount.

Review (p. 475)

1. 30, 100   **3.** 99%   **5.** 72%   **7.** $77\frac{7}{9}$%

9. 37%   **11.** 140%   **13.** 400%   **15.** 0.91

17. 2.15   **19.** 0.072   **21.** 60%   **23.** $233\frac{1}{3}$%

25. 90%   **27.** $\frac{9}{20}$ **29.** $\frac{8}{9}$ **31.** $\frac{11}{8}$ or $1\frac{3}{8}$ **33.** $\frac{33}{400}$

35. $\frac{1}{300}$ **37.** R% = 35%, B = 400, A = 140

39. R% = 10%, B = 380, A = 38   **41.** R% = ?,
B = 60, A = 50   **43.** R% = 40%, B = ?, A = 20
45. R% = 106%, B = 45, A = ?   **47.** B = $350

49. R% = $71\frac{3}{7}$%   **51.** A = 54   **53.** R% = $37\frac{1}{2}$%

55. A = 285 yachts   **57.** B = $1,625   **59.** A = 924
plants   **61.** $660   **63.** $845   **65.** $126
67. 4.5% increase   **69. a.** $1,451.52   **b.** $36,011.52
71. I = $1,440   **73.** I = $108.75   **75.** $116 per month
77. P = $2,000   **79.** T = 4 years   **81.** R% = 4.5%
83. $64.93   **85.** $431.01

Chapter Test (p. 479)

1. 7.2, 100   **2.** 71   **3.** 7%   **4.** 150%

5. 0.058   **6.** 0.0006   **7.** $41\frac{2}{3}$%   **8.** 287.5%

9. $\frac{21}{25}$ **10.** $\frac{9}{250}$ **11.** B = $130

12. A = 3,000 bushels   **13.** R% = 80%
14. B = $12,000   **15.** R% = 42.5%   **16.** A = 570
boys   **17.** 8% decrease   **18.** $75
19. B = $60,000   **20.** A = $3,240   **21.** I = $176
22. P = $3,000   **23.** R% = 10%   **24.** T = 1.5 yr, or
18 mo   **25.** S = $7,168   **26.** $112 per month
27. $594.50, to the nearest cent   **28.** S = $7,986

Cumulative Test (p. 481)

1. two and six hundred three thousandths   **2.** 5,000.0005
3. a. 510   **b.** 0.05   **c.** 7.0   **4.** 42.007
5. 37.246   **6.** 90.24   **7.** 14.3   **8.** 44.01
9. $84.62   **10.** 20 bags   **11.** 1.875   **12.** 0.636

13. $\frac{23}{50}$ **14.** $2\frac{3}{40}$ **15.** 4.59

16. 7 < 7.0077 < 7.07 < 7.087 < 7.107   **17.** 11 to 3
18. 8 miles per gallon   **19.** 8 to 3   **20.** 1 to 80
21. 15.2¢/oz   **22.** 7.8¢/egg   **23.** Tom gets $32,232,
Dick gets $12,087, Harry gets $28,203   **24.** x = 15

25. k = 168   **26.** $823.50   **27.** $256   **28.** $\frac{7}{10}$

29. 20%   **30.** 0.156   **31.** $26\frac{2}{3}$%   **32.** 2.65

33. A = $360   **34.** B = $800   **35.** R% = 5%
36. A = $7,068   **37.** R% = 62%   **38.** R% ≐ 9.5%
39. $160   **40.** 160%   **41.** I = $1,104
42. P = $880

CHAPTER 8

Pretest (p. 484)

1. a. 15   **b.** 8   **c.** none   **d.** 48   **2. a.** 24
b. 22.5   **c.** none   **d.** 30   **3.** 58.8 gal

4. 8.4 gal   **5.** $\frac{22\%}{3\%}$ = $\frac{0.22}{0.03}$ ≐ 7.3 times

6. 11.2 gal   **7.** $6 million   **8.** $1 million less profit
9. about 2.7 times more profit   **10.** $5.5 million
11. José, Nick, Steve, Ron

Section 8.1 (p. 491)

1. 4.5, 4, none, 6   **3.** 38.4, 40, none, 25
5. 2, 1.5, 0, 6   **7.** 4, 5, 5, 3   **9.** 41, 20.5, 10, 124
11. 2.5, 2.5, bimodal 2 and 3, 1   **13.** 25, 25, none, 30
15. 6, 6, 6, 0   **17.** 0.686, 0.2, 0.1, 2.6   **19. a.** 100,
100, none, 20   **b.** 100, 100, none, 40   **c.** 100, 100,
none, 100   **d.** The only measure that changed was the
range. The mean, median, and mode were the same.

21. $76   **23.** 78

25. a.

Number of children	Frequency
0								
1										
2													
3									
4			
5		
6		

b. 1.9 children per family   **c.** 2, 2, 6   **27. a.** 76%
b. 24%   **c.** about 60,800 people scored lower
29. $x = 8$ **31.** $x = 10.5$ **33. a.** 8, 8   **b.** 35, 8
c. 305, 8   **35.** $29,620, $28,000, $28,000, $30,000
37. 5, 5, 5, 9, 9, 9   **39.** mean $= \dfrac{20}{5} = 4$
43 a. yes   **b.** yes   **c.** no   **d.** no

Section 8.2 (p. 499)

1. a. 9,000,000 square miles   **b.** 48 people per square
mile   **3.** South America   **5.** Australia and Europe
differ by 1,000,000 square miles. Europe and Antarctica
differ by 1,000,000 square miles.   **7.** 100,000 cars per
month   **9.** about 5 times more   **11.** 100,000 more cars
per month   **13.** 265 million people   **15.** Europe
17. Australia and the Pacific Islands, increased by about
11,000,000 people   **19.** $29,403   **21.** $10,123 more
per year   **23.** 10% or less   **25.** $336   **27.** $168
more   **29.** $56 more per month   **31.** 7 to 1
33. 12.5 hrs   **35.** 1.7   **37.** 0.054

Section 8.3 (p. 509)

1. about 5%   **3.** 1990   **5.** 1945, about 2%
7. a. about 3.2% and 23.5%   **b.** about 7 times greater
9. 2,900   **11. a.** 19 to 22   **b.** 2,900   **c.** increase
from 11 to 19, the same from 19 to 22, decrease from 22 to
76+   **13. a.** 19 to 50 (19–22 and 23–50)   **b.** 800
calories per day   **15.** 3,650   **17.** 51%   **19.** 1940
to 1950; dropped 17%   **21. a.** 41%   **b.** 360,000
23. 81% to 19% is about a 4 to 1 ratio   **25.** $20,000
27. a. August   **b.** $60,000   **29.** $30,000 more
31. a. Jan to June, July to Aug   **b.** June to July, Aug to
Nov   **c.** Nov to Dec   **33.** 30% of $65,000 =
$0.3 \times \$65,000 = \$19,500$ profit   **35.** 58%
37. 207.5%   **39.** 400%

Section 8.4 (p. 515)

1. Pat, Mary, Cathy, Laura   **3.** Tom, Bill, Hank, Frank
5. Pat is the secretary. Chris is the singer. Nancy is the
teacher.   **7.** Bob is the foreman. Henry is the chemist.
George is the stockclerk. Lou is the salesperson.
9. $\dfrac{3}{10}$ **11.** $\dfrac{1}{4}$

Review (p. 517)

1. 10, 9, none, 11   **3.** 8, 8.5, 10, 17
5. 5, 2.5, 2, 21   **7.** 183 lb   **9. a.** 10, 10, none, 20
b. $36\dfrac{2}{3}$, 10, none, 100   **c.** $336\dfrac{2}{3}$, 10, none, 1,000
11. $6 million   **13.** 4 times more   **15.** miscellaneous

17. 247,995   **19.** about 4 times more   **21.** 4,000
23. 290 more   **25.** $\dfrac{1}{2}$ **27.** tuberculosis
29. heart disease   **31.** $39   **33. a.** May to June
b. $9 per share   **35.** about $1,440   **37.** Sally owns
the plane. Pam owns the boat. Terry owns the hot air
balloon.

Chapter Test (p. 521)

1. a. 83.125   **b.** 84   **c.** 84   **d.** 40   **2.** 7.525 in,
8 in, 8 in, 6 in   **3. a.** 12.5   **b.** 16   **c.** 0   **d.** 20
4. 94   **5.** 1950 and 1960   **6.** 3.1 years   **7.** 20.3
years   **8.** 1900 and 1990   **9. a.** yes   **b.** yes
10. 1900   **11.** 8   **12.** 5 fewer   **13.** 100% more
14. 6, 5.5, none, 9   **15.** 14 more   **16.** increasing

Cumulative Test (p. 523)

1. 1,791   **2.** 64   **3. a.** 100,000   **b.** 113,760
4. 9   **5.** $1\dfrac{7}{60}$ **6.** $96   **7.** $1\dfrac{3}{20}$ **8.** 0.571
9. 3.408   **10.** 60.61   **11.** $s = 103.7$ **12.** $w = 15$
13. 16.5   **14.** $0.0079 < 0.077 < 0.7 < 0.707$
15. 46 to 15   **16.** $4,102, $1,172   **17.** $y = 6$
18. 7%   **19.** 1.82   **20.** A = $210.80   **21.** B = 5
22. R% = 60%   **23. a.** 2   **b.** 2   **c.** 2   **d.** 4

CHAPTER 9

Pretest (p. 526)

1. $2\dfrac{1}{4}$ gal   **2.** 270 min   **3.** 192 oz   **4.** $9\dfrac{1}{6}$ ft
5. 3,000 ml   **6.** 8,600 g   **7.** 54 ℓ   **8.** 25 cm
9. 2.409 km   **10.** 0.69 g   **11.** 1.062 kg   **12.** 15.1 ℓ
13. 11.0 inches   **14.** 54 kg   **15.** 120 ft

Section 9.1 (p. 531)

1. 7 in < 4 ft   **3.** 3 oz < $\dfrac{1}{2}$ lb   **5.** 1 qt = 2 pt
7. 6 nickels = 3 dimes   **9.** 5 in ≯ 1 ft
11. 8 qt ≠ 8 gal   **13.** 8 ft   **15.** cannot do
17. 20 cents   **19.** 5 men   **21.** cannot do   **23.** 13
25. 14 miles   **27.** 3 ft   **29.** 14 ft   **31.** $90
33. 60   **35.** 60%   **37.** 5 qt − 2 qt = 3 qt
39. 14p **41.** 14x **43.** cannot do   **45.** 18y
47. 2d **49.** 80r

Section 9.2 (p. 533)

1. 3 ft   **3.** 36 in   **5.** 21 ft   **7.** 7 yd   **9.** $6\dfrac{1}{2}$ ft
11. 138 yd   **13.** $4\dfrac{1}{2}$ mi   **15.** $2\dfrac{1}{2}$ min   **17.** 210 min
19. $2\dfrac{1}{2}$ lb   **21.** $11\dfrac{1}{2}$ qt   **23.** $\dfrac{3}{4}$ ft   **25.** 12 oz
27. $1\dfrac{3}{4}$ c   **29.** $4\dfrac{1}{2}$ lb   **31.** $13\dfrac{1}{3}$ oz   **33.** $6\dfrac{1}{4}$ gal
35. 2,700 sec   **37.** 216 in   **39.** 40 oz
41. $663,962.88   **43.** $2^2 \cdot 5$ **45.** prime
47. 21.0 mph

Section 9.3 (p. 545)

1. 400 cm   **3.** 0.58 m   **5.** 6 km   **7.** 15.2 m
9. 4.2 km   **11.** 0.78 km   **13.** 2.5 dam   **15.** 82 mm
17. 145,000 m   **19.** 300,000,000 m per sec, or 300 million m per sec   **21.** 2.25 m tall   **23.** 3,000 ml
25. 0.58 ℓ **27.** 80 hl   **29.** 25.9 ℓ **31.** 7,500 ml
33. 0.724 kl   **35.** 700,000 ml   **37.** 36 ℓ **39.** 92 ℓ
41. 2,000 ml   **43 a.** 240,000 ml   **b.** 240 ℓ
c. $60,000   **45.** 4,000 mg   **47.** 0.51 g   **49.** 30 hg
51. 38 g   **53.** 17,500 mg   **55.** 0.354 kg
57. 71.8 dag   **59.** 0.096 g   **61.** 890,000 g
63. 2,500 g   **65. a.** 7,500 mg   **b.** 7.5 g
67. $\frac{3}{5} < \frac{2}{3} < \frac{7}{10}$ **69.** $\frac{7}{12} < \frac{2}{3} < \frac{3}{4} < \frac{5}{6}$

Section 9.4 (p. 551)

1. 35.56 cm   **3.** 45.45 kg   **5.** 90.72 g
7. 12.88 km   **9.** 1.27 m   **11.** 27.43 m   **13.** 100 ℓ
15. 1,637.50 g   **17.** 37.74 ℓ **19.** 181.82 kg
21. 15.9 qt   **23.** 262.47 ft   **25.** 10.97 m
27. 179.50 mi   **29.** 30.68 ft   **31.** 42.18 km
33. 874.89 yd   **35.** 13.76 gal   **37.** 36.2¢ /liter
39. 6 to 1   **41.** 12 to 1   **43.** 61°F   **45.** 11 km/sec
47. 19.7 ft high, 19.7 ft wide, 39.4 ft long
49. about 9 mi

Review (p. 554)

1. 10 gal   **3.** 20 mi   **5.** 3 pounds   **7.** 100 pt
9. $3\frac{1}{2}$ ft   **11.** $3\frac{3}{4}$ lb   **13.** 7 mi   **15.** 2,400 sec
17. 4 hr   **19.** 276 mi/hr   **21.** 1.357 m   **23.** 2,850 g
25. 5.8 kg   **27.** 8,000 ml   **29.** 700 mg
31. 53 mm   **33.** 11.48 ft   **35.** 8.48 qt
37. 15.84 lb   **39.** 11.38 gal   **41.** 13.41 m

Chapter Test (p. 556)

1. 14 qt   **2.** 585 sec   **3.** $6\frac{1}{4}$ ft   **4.** $16\frac{1}{8}$ lb
5. 1.6 cm   **6.** 5,200 g   **7.** 1.4 ℓ **8.** 40 ml
9. 1,850 m   **10.** 0.45 g   **11.** 0.64 kg   **12.** 40.6 cm
13. 12.7 pt   **14.** 15.8 lb   **15.** 30.8 ft

Cumulative Test (p. 556)

1. 3,506,000   **2.** 11   **3.** 400 + 80 + 3   **4.** 5,616
5. $12\frac{2}{15}$ **6.** $4\frac{32}{45}$ **7.** 39.31   **8.** $\frac{81}{125}$ **9.** 18.342
10. 0.05488   **11. a.** 99 gallons   **b.** $118   **12.** $\frac{16}{25}$
13. 8.5%   **14.** 20   **15.** $84   **16. a.** 39   **b.** 40
c. 40   **d.** 12   **17.** $9,600,000   **18.** 3 to 1
19. $\frac{3}{10}$ **20.** $12,800,000 more   **21.** $2,800,000 more
22. $7\frac{1}{2}$ ft   **23.** $9\frac{1}{4}$ lb   **24.** 81.3 cm   **25.** 393.7 ft
26. a. 26 pieces   **b.** 2 cm

CHAPTER 10

Pretest (p. 560)

1. acute   **2.** straight   **3.** right   **4.** obtuse
5. 28 m   **6.** 42 ft   **7.** 27 in^2 **8.** 128 ft^2
9. $688   **10.** 96 yd^2 **11.** A = 78.5 cm^2, C = 31.4 cm
12. 240 ft^3 **13.** 226.08 ft^3

Section 10.1 (p. 563)

1. perpendicular   **3.** parallel   **5.** parallel
7. $L_1 \parallel L_2$, L_1 and L_2 are both horizontal
9. $L_1 \perp L_2$, L_3 is vertical   **11.** acute   **13.** obtuse
15. obtuse   **17.** acute   **19.** acute   **21.** acute
23. obtuse   **25.** straight   **27.** 125   **29.** 1

Section 10.2 (p. 571)

1. 20 ft   **3.** 36 cm   **5.** 40 m   **7.** P = 12 m
9. 7 ft   **11.** 120 ft   **13.** 62 ft 8 in
15. cost = $20.28   **17.** $40   **19.** $m = 150$
21. $s = 90$ **23. a.** 48 ft   **b.** 96 ft   **c.** 100%
25. $51.17   **27.** $P = 2 \cdot 8 \text{ ft} + 2 \cdot 3 \text{ ft}$
$= 16 \text{ ft} + 6 \text{ ft}$
$= 22 \text{ ft}$

Section 10.3 (p. 581)

1. $A = 150 \text{ ft}^2$ **3.** $A = 81 \text{ cm}^2$ **5.** $A = 9 \text{ ft}^2$
  $P = 50 \text{ ft}$ $P = 36 \text{ cm}$ $P = 13 \text{ ft}$
7. $A = 48 \text{ cm}^2$ **9.** $A = 552 \text{ ft}^2$ **11.** $A = 125 \text{ in}^2$
  $P = 30 \text{ cm}$ $P = 108 \text{ ft}$ $P = 58 \text{ in}$
13. about 1,600 m^2 **15.** about 60 ft^2 **17. a.** 4.5 ft × 2.75 ft = 12.375 ft^2 **b.** 54 in × 33 in = 1,782 in^2
19. $42.12   **21.** $A = 24 \text{ ft} \times 18 \text{ ft} = 432 \text{ ft}^2$; 432 ft^2 ÷ 9 ft^2/yd^2 = 48 yd^2; cost = 48 yd^2 × $23/yd^2 = $1,104
23. $258   **25.** $1,123.50   **27.** 0.18 acres
29. Area = 135 ft × 68 ft = 9,180 ft^2; acres = 9,180 ft^2 ÷ 43,560 ft^2/ac = 0.21 acres   **31.** 1.32 acres
33. I = $80   **35.** T = 3.5 yr   **37.** 264 in^2
39. 72 in^2 **41.** 12'9" = 153 in, 14'3" = 171 in; Area = 153 in × 171 in = 26,163 in^2; tiles needed = 26,163 in^2 ÷ 81 in^2/tile = 323 tiles; cost = 323 tiles × $1.26/tile = $406.98
43. a. 4,700 ft^2 **b.** about 0.11 ac   **c.** 9 courts
45. $A = L \times W = 5.5 \text{ ft} \times 8.75 \text{ ft} = 48.125 \text{ ft}^2$

Section 10.4 (p. 588)

1. 40°   **3.** 35°   **5.** 60°   **7.** 60 ft^2 **9.** 13 yd^2
11. 162 m^2 **13.** 1.5 mi^2 **15.** 121.5 in^2
17. 35 m^2 **19.** $A = \dfrac{4 \text{ m} \times 10 \text{ m}}{2} = 20 \text{ m}^2$
  $P = 10 \text{ m} + 5 \text{ m} + 8.1 \text{ m} = 23.1 \text{ m}$
21. 14.0625 ft^2 or 2,025 in^2 **23.** $T = 2$
25. $A = 37.68$ **31. a.** 60°   **b.** 70° each   **c.** Yes
d. Yes   **e.** No   **f.** 30° and 120°   **g.** 0°   **h.** Yes

Section 10.5 (p. 597)

1. 78.5 ft   **3.** 56.52 cm   **5.** 12.56 ft   **7.** 200.96 ft
9. a. 4 ft 6 in = 54 in, $C = \pi D = 3.14 \times 54 \text{ in} = 169.56 \text{ in}$
b. 4 ft 6 in = 4.5 ft, $C = \pi D = 3.14 \times 4.5 \text{ ft} = 14.13 \text{ ft}$

11. 125.6 ft   **13. a.** 549.5 ft   **b.** \$13,627.60
15. a. 12.56 in   **b.** 25.12 in   **c.** 50.24 in
d. 100.48 in   **e.** The circumference doubles.
17. a. $C = 3,000$ ft $+ 3,000$ ft $+ 2 \cdot 3.14 \cdot 500$ ft
$\quad = 3,000$ ft $+ 3,000$ ft $+ 3,140$ ft
$\quad = 9,140$ ft
b. 9,140 ft $\div$ 5,280 ft/mi $\doteq$ 1.7 mi   **19.** 254.34 ft^2
21. 19.625 ft^2 **23.** 12.56 in^2 **25.** 706.5 yd^2
27. a. 19.625 ft^2 **b.** 2,826 in^2 **29.** 11.0325 ft^2
31. a. $A = 3.14 \cdot (120$ ft$)^2 - 3.14 \cdot (95$ ft$)^2$
$\quad = 45,216$ ft$^2 - 28,338.5$ ft$^2 = 16,877.5$ ft^2
b. $\dfrac{\$70}{100 \text{ ft}^2} = \dfrac{x}{16,877.5 \text{ ft}^2}, x = \dfrac{\$70 \cdot 16,877.5 \text{ ft}^2}{100 \text{ ft}^2} =$
\$11,814.25   **33.** 1,130,400 mi^2
35. a. 3,106.5 yd^2 **b.** \$232.99, rounded to the nearest
cent   **37.** 5.61   **39.** 0.07   **41. a.** 24,900 mi
b. polar circumference = 24,819 miles; equatorial
circumference is 81 miles longer
43. 9″, 64 in^2, 8.8¢; 12″, 52%, 113 in^2, 77%, 7.5¢; 15″,
34%, 177 in^2, 57%, 6.4¢   **a.** percent increase in area
b. 15″ pizza   **45.** 13.76 ft^2 **49. 1.** perimeter
2. Circumference   **3.** three   **4.** obtuse   **5.** 90°

Section 10.6 (p. 608)

1. 135 ft^3 **3.** 24 m^3 **5.** 160 ft^3
7. $V = 25$ ft $\times 18$ ft $\times 8$ ft $= 3,600$ ft^3
9. $V = (8$ ft$)^3 = 512$ ft^3 **11.** 9,600 ft^3
13. $V = 18$ ft $\cdot 9$ ft $\cdot \frac{1}{3}$ ft $= 54$ ft^3; 54 ft$^3 = \dfrac{54 \text{ ft}^3}{1} \cdot \dfrac{1 \text{ yd}^3}{27 \text{ ft}^3} =$
2 yd^3; cost $= 2$ yd$^3 \cdot \$90/yd^3 = \$180$
15. a. 1,536 ft^3 **b.** 96 crates
17. a. $V = 81$ ft $\cdot 40$ ft $\cdot \frac{4}{12}$ ft $= 1,080$ ft^3;
1,080 ft$^3 \div 27$ ft^3/yd$^3 = 40$ yd^3 of topsoil
b. cost $= \$26 \cdot 40 = \$1,040$ **19.** $\frac{3}{5}$ **21.** 36
23. a. 128 ft^3 **b.** 144 pieces   **c.** 32 ft^3 **d.** $\frac{1}{4}$
e. \$25   **25. a.** $3\frac{9}{32}$ ft^3 **b.** about 205 lb
27. a. 16 in long, 12 in wide, 4 in deep   **b.** $V = 16$ in $\cdot$
or $V = 12$ in $\cdot 4$ in $= 768$ in^3

Section 10.7 (p. 615)

1. $V = 401.92$ ft^3 **3.** $V = 1,130.4$ m^3
5. $V = 552.64$ in^3 **7. a.** 62.8 ft^3 **b.** $V = 469.7$ gal
9. $V = 3.14 \cdot (10$ ft$)^2 \cdot (\frac{1}{3}$ ft$) \doteq 104.7$ ft^3;
104.7 ft$^3 \div 27$ ft^3/yd$^3 \doteq 3.9$ yd^3
11. a. 1,413 ft^3 **b.** about 10,569 gal
13. a. 201 in^3 **b.** 75 in^3 **15.** $V \doteq 267.95$ ft^3
17. $V \doteq 523.33$ yd^3 **19.** $V \doteq 4.187$ ft^3
21. $V \doteq 33,493$ ft^3 **23. a.** $V \doteq 47.7$ ft^3
b. $V \doteq 82,400$ in^3 **25. a.** $V \doteq 0.0082$ in^3 **b.** 122
27. 17.8   **29.** 7.69   **31. a.** $V \doteq 4.54$ ft^3
b. 34 gal   **c.** 17 min   **33. a.** $V \doteq 113.04$ ft^3
b. $V = 904.32$ ft^3 **c.** 8 times

Review (p. 620)

1. $L_1 \perp L_2$, L_1 is horizontal, L_2 is vertical.   **3.** $L_2 \parallel L_3$,
$L_1 \perp L_2$, $L_1 \perp L_3$, L_1 is horizontal, L_2 and L_3 are vertical.
5. obtuse   **7.** right   **9.** obtuse   **11.** acute
13. 21 ft   **15.** 72 m   **17.** $P = 2 \cdot 12$ ft $+ 2 \cdot 13$ ft $=$
50 ft; cost $= \$1.29 \times 50 = \$64.50$ **19.** $A = 56$ ft^2; $P =$
30 ft   **21.** $A = 48$ cm^2; $P = 34$ cm   **23.** 225 in^2
25. $A = 12$ ft $\cdot 15$ ft $= 180$ ft^2
$\quad$ cost $= \$0.79 \cdot 180$ ft$^2 = \$142.20$
27. 1.15 ac   **29.** 55°   **31.** 31.5 ft^2 **33.** 123 ft^2
35. $A = \dfrac{40 \text{ ft} \cdot 9 \text{ ft}}{2} = 180$ ft^2 **37.** 43.96 ft
39. 706.5 in^2 **41.** $C = 2 \cdot 3.14 \cdot 50$ ft $= 314$ ft
43. 78.5 in^2 **45.** 432 ft^3 **47.** 527 in^3
49. $V \doteq 75.4$ m^3 **51.** $V \doteq 4.2$ yd^3 **53.** 3.14 m^3
55. $V \doteq 0.268$ cm^3

Chapter Test (p. 625)

1. obtuse   **2.** acute   **3.** straight   **4.** right
5. $P = 146$ ft   **6.** $P = 24$ in   **7.** $P = 42$ cm
8. $A = 189$ in^2 **9.** $A = 49$ m^2 **10.** $A = 82$ ft^2
11. $C = 40.82$ in   **12.** $A \doteq 136.8$ m^2
13. $A = 72.25$ ft^2 **14.** \$93.08
15. $A = 21$ ft $\cdot 15$ ft $= 315$ ft^2; 315 ft$^2 = \dfrac{315 \text{ ft}^2}{1} \cdot \dfrac{1 \text{ yd}^2}{9 \text{ ft}^2} =$
35 yd^2; cost $= \$18.75/yd^2 \cdot 35$ yd$^2 = \$656.25$
16. \$3,712.80   **17.** $V = 12.25$ ft $\cdot 4.75$ ft $\cdot 5.5$ ft $\doteq$
320 ft^3 **18.** $V \doteq 38.8$ m^3 **19.** 2,198 ft^3
20. \$13,400

Cumulative Test (p. 627)

1. 58 R 52   **2.** \$3,893   **3.** $2\frac{13}{24}$ **4.** $1\frac{1}{2}$
5. a. 20 doses   **b.** $\frac{5}{6}$ of a dose   **6.** 0.217
7. 4.065   **8.** $\frac{22}{25}$ **9.** 2.6   **10.** 50 to 1
11. \$3,315, \$1,275   **12.** $w \doteq 18.67$
13. A = \$165.20   **14.** B = 48 oz   **15.** R% = 8%
16. I = \$162
17. a.

number of children	frequency
0	III
1	III
2	IHI I
3	II
4	I
5	I

b. 2   **c.** Yes   **d.** 1.875   **18.** 6.3 g
19. 4,500 cl   **20.** 0.5 kg   **21.** 6.8 cm   **22.** 250 m
23. 360 ft^2 **24.** $C = 12.56$ ft   **25.** $A = 98$ cm^2

CHAPTER 11

Pretest (p. 630)

1. 26   **2.** +3, or 3   **3.** −6   **4.** −11   **5.** −4
6. 32   **7.** 4   **8.** 5   **9.** −20   **10.** −60
11. 35   **12.** −2   **13.** 4   **14.** 24

15. mean = $-4°$, range = $34°$ **16.** 25   **17.** -64
18. 4   **19.** -12 **20. a.** 26   **b.** 92
21. a. 531,000   **b.** 0.00207   **22. a.** 6.04×10^{-4}
b. 2.87×10^9

Section 11.1 (p. 635)

1. $+40°$ **3.** -35 ft   **5.** $+\$110$ **7.** -23
9. $-\$4,000$ **11.** 13   **13.** 15   **15.** 8   **17.** 16
19. 10   **21.** $>$ **23.** $<$ **25.** $<$ **27.** $>$
29. $=$ **31.** $>$ **33.** $>$ **35.** $<$ **37.** -8
39. 45   **41.** -6 **43.** 6   **45.** -7 **47.** 75%
49. $83\frac{1}{3}\%$ **51.** $|-9| + |-6| = 9 + 6 = 15$

Section 11.2 (p. 643)

1. 8 plus 5; 13   **3.** negative 6 plus 4; -2 **5.** 3 added
to negative 4; -1 **7.** negative 5 added to 16; 11
9. 6 plus negative 4; 2   **11.** 8 plus negative 8; 0
13. negative 7 added to negative 13; -20 **15.** negative 9
added to positive 5; -4 **17.** 0   **19.** -6 **21.** -12
23. -5 **25.** 0   **27.** 7   **29.** -1 **31.** 11
33. -22 **35.** 6   **37.** 3   **39.** -5 **41.** -18
43. 16   **45.** -37 **47.** -10 **49.** -13 **51.** 6
53. 0   **55.** -13 **57.** -15 **59.** -1 **61.** 0
63. -1 **65.** -58 **67.** $12°$ **69.** 9   **71.** $-\$15$
73. 120 mm   **75.** $50,000 \ell$ **77.** $-5 + 10 = +5$

Section 11.3 (p. 649)

1. 3   **3.** -4 **5.** -10 **7.** -10 **9.** 11
11. -5 **13.** -20 **15.** 0   **17.** -5 **19.** 1
21. -8 **23.** 18   **25.** 0   **27.** -97 **29.** 100
31. 30   **33.** 8   **35.** 6   **37.** 50   **39.** 1
41. 0   **43.** 15   **45.** 0   **47.** 3   **49.** 10
51. -6 **53.** 8   **55.** $\$197$ **57.** $\$1$
59. rise = $17° - (-21°) = 17° + (+21°) = 38°$ **61.** 65 ft
63. 64   **65.** 1   **67.** 10

Section 11.4 (p. 655)

1. 35   **3.** 6   **5.** -30 **7.** -16 **9.** -20
11. -9 **13.** 0   **15.** 56   **17.** 15   **19.** 36
21. -24 **23.** -27 **25.** 0   **27.** 27   **29.** -72
31. 24   **33.** -16 **35.** 48   **37.** -30 **39.** 0
41. -420 **43.** -48 **45.** 35   **47.** -6 **49.** -8
51. 5   **53.** -8 **55.** -12 **57.** Not Possible
59. -8 **61.** -5 **63.** 9   **65.** 0   **67.** 35
69. -8 **71.** -6 **73.** Not Possible   **75.** 8
77. 0   **79.** -7 **81.** 7   **83.** -20 **85.** 15
87. $\frac{2}{3}$ **89.** $\frac{8}{3}$ **91.** $(-2)(-3)(-7) = (+6)(-7) = -42$

Section 11.5 (p. 661)

1. 25.   **3.** 36   **5.** 100   **7.** -36 **9.** -27
11. 16   **13.** -27 **15.** -16 **17.** -36 **19.** 3
21. 54   **23.** -25 **25.** -8 **27.** 0   **29.** -21
31. 45   **33.** 6   **35.** 8   **37.** -4 **39.** 19
41. 32   **43.** -1 **45.** 20   **47.** 1   **49.** 1
51. 4   **53.** 10   **55.** 13   **57.** 5   **59.** 96
61. 49   **63.** 7   **65.** 24   **67.** -10 **69.** -14
71. -2 **73.** 1   **75.** -12 **77.** 7,7,7,7
79. 27.5, 27.5, bimodal 25 and 30, 5   **81.** 10

83. $(-4)(-4)(-4) = (+16)(-4) = -64$
85. $(-100) \div (10) \cdot (-10) = (-10) \cdot (-10) = 100$

Section 11.6 (p. 667)

1. 460   **3.** 592   **5.** 80,000   **7.** 993,200
9. 60,020,000   **11.** 7,300,000,000   **13.** 25,000
15. 5.6×10^1 **17.** 9.02×10^2 **19.** 5.8×10^3
21. 3.86×10^1 **23.** 6×10^4 **25.** 7.8×10^0
27. 4.65×10^5 **29.** 2.55×10^8 **31.** 1.86×10^5
33. 4.5×10^9 **35.** 9.26×10^7 **37.** 4.3×10^{-1}
39. 6.3×10^{-2} **41.** 9.53×10^{-3} **43.** 5.2×10^{-5}
45. 1×10^{-1} **47.** 7×10^{-6} **49.** 6.045×10^{-3}
51. 5.6×10^{-4} **53.** 6.1×10^{-6} **55.** 9.25×10^{-2}
57. 1.57×10^{-5} **59.** 7.5×10^{-2} **61.** 6.624×10^{-27}
63. 0.036   **65.** 0.00703   **67.** 0.1026
69. 0.000 090 4   **71.** 0.048   **73.** 0.000 850 7
75. 0.000 000 000 0054   **77.** 0.000 001 077
79. 0.000 000 000 000 000 000 000 026 372   **81.** 19
83. 9

Review (p. 671)

1. $-37°$ **3.** 3   **5.** 16   **7.** 6   **9.** 22   **11.** 2
13. -10 **15.** -14 **17.** -45 **19.** -22
21. -11 **23.** 22   **25.** $-16°$ **27.** 1   **29.** 0
31. 30   **33.** -17 **35.** -10 **37.** -51 **39.** -7
41. $43°$ **43.** 25   **45.** 21   **47.** -80 **49.** 36
51. -60 **53.** -1 **55.** -20 **57.** 24
59. -450 **61.** 8   **63.** 5   **65.** -2 **67.** 8
69. -8 **71.** 9   **73.** 12   **75.** 81   **77.** $-1,000$
79. -19 **81.** 3   **83.** -48 **85.** 70,500
87. 0.000 018   **89.** 60.25   **91.** 9,400
93. 1.5×10^4 **95.** 9×10^{-1} **97.** 5.07×10^5
99. 4.8×10^{-6} **101.** 6.4×10^3 **103.** 1.1×10^{-4}

Chapter Test (p. 675)

1. 150   **2.** 26   **3.** 70   **4.** 13   **5. a.** $<$
b. $>$ **c.** $=$ **d.** $<$ **6.** 5   **7.** -17 **8.** -27
9. 27   **10.** -50 **11.** -4 **12.** -100 **13.** -12
14. 120   **15.** 5   **16.** -8 **17.** -5 **18.** -60
19. 9   **20.** $69°$ below average   **21.** $\$829.43$
22. -216 **23.** -16 **24.** 7   **25.** 86   **26.** -16
27. 46   **28. a.** -8 **b.** 60   **c.** -12 **d.** 3
29. mean = -1, range = 25   **30.** 4,600
31. 0.000 793   **32.** 100,600   **33.** 0.8
34. 0.000 003 55   **35.** 50,000   **36.** 5.3×10^{-2}
37. 9×10^4 **38.** 4.23×10^2 **39.** 7×10^{-1}
40. 2.35×10^7 **41.** 5.04×10^{-4}

Cumulative Test (p. 677)

1. 161,734   **2.** 34   **3.** $w = 9$ **4.** 45   **5.** 43
6. $1\frac{5}{48}$ **7.** $2\frac{5}{7}$ **8.** $2\frac{5}{18}$ **9.** 42   **10.** 45.237
11. 4.2   **12. a.** 9   **b.** $\$0.53$ **13.** 35.56
14. $\$82,600$ **15.** $\$310.50$ **16.** $\$4,000$
17. $\$1,000$ **18.** 1988 to 1989; $\$1,000$ **19.** 1986 and
1987   **20.** 2,500 per week $\times$ 52 weeks = $\$130,000$
21. 320 cm   **22.** 0.32 km   **23.** 60 ft   **24.** 360 ft^2
25. -1 **26.** -7 **27.** -4 **28.** 35   **29.** -10
30. 36   **31.** 0.04   **32.** 1,800   **33.** 4.05×10^5
34. 7.2×10^{-2}

CHAPTER 12

Pretest (p. 680)

1. Terms are $5x$, $-3y$; 5 is the coefficient of x; -3 is the coefficient of y. **2.** Terms are $8x$, $-y$, -6; 8 is the coefficient of x; -1 is the coefficient of y; $+6$ is the constant term. **3. a.** No **b.** Yes **c.** No **4. a.** No **b.** No **c.** Yes **5.** $x = -3$ **6.** $x = 17$ **7.** $x = -4$ **8.** $x = -2$ **9.** $8x + 36$ **10.** $-15 + 21x$ **11.** $24 - 6x + 42y$ **12.** $5x + 12y$ **13.** $9y$ **14.** $6p + 4$ **15.** $x = 3$ **16.** $x = 1$ **17.** $x = 3$ **18.** $7p$ **19.** $v - 8$ **20.** $7t + y$ **21.** $x = 2$ **22.** $a = 12$ cm

Section 12.1 (p. 685)

1. terms are $6x$, 2
6 is coefficient of x
2 is constant term
3. terms are $5w$, -4
5 is coefficient of w
-4 is constant term
5. terms are $-8x$, $4s$
-8 is coefficient of x
4 is coefficient of s
7. term is $3v$
3 is coefficient of v
9. terms are x, 1
1 is coefficient of x
1 is constant term
11. terms are $-2x$, $4c$, -2
-2 is coefficient of x
4 is coefficient of c
-2 is constant term
13. terms are 8, $-g$, y
8 is constant term
-1 is coefficient of g
1 is coefficient of y
15. terms are x, $-2y$, $-3z$, 1
1 is coefficient of x
-2 is coefficient of y
-3 is coefficient of z
1 is constant term
17. a. No **b.** Yes **c.** No **19. a.** Yes **b.** Yes **c.** No **21. a.** Yes **b.** Yes **c.** Yes **23. a.** No **b.** No **c.** Yes **25. a.** No **b.** Yes **c.** No **27. a.** No **b.** No **c.** No **29. a.** No **b.** No **c.** Yes **31. a.** No **b.** Yes **c.** No
33. $x = 12$ **35.** $y = 11\frac{2}{3}$

Section 12.2 (p. 689)

1. $x = 7$ **3.** $x = -11$ **5.** $x = 1$ **7.** $x = -13$ **9.** $x = 13$ **11.** $y = -3$ **13.** $w = 7$ **15.** $z = 8$ **17.** $v = -10$ **19.** $t = 0$ **21.** $a = -12$ **23.** $b = -12$ **25.** $k = -1$ **27.** $x = -24$ **29.** $x = 11$ **31.** $x = \frac{1}{2}$ **33.** $t = -1$ **35.** $d = -1\frac{3}{7}$ **37.** $x = -1.05$ **39.** $x = 28.35$ **41.** $d = 0.12$ **43.** -15 **45.** -3 **47.** 3 **49.** 7
51.
$$x - 5 = 8$$
$$x + (-5) = 8$$
$$x + (-5) + (+5) = 8 + (+5)$$
$$x + 0 = 13$$
$$x = 13$$

Section 12.3 (p. 697)

1. $x = 5$ **3.** $x = -7$ **5.** $x = -4$ **7.** $x = 8$ **9.** $x = -8$ **11.** $x = -4$ **13.** $x = 30$ **15.** $x = -49$ **17.** $x = 3$ **19.** $x = -3.1$ **21.** $x = 5$ **23.** $x = 4$ **25.** $x = -1$ **27.** $x = 2$ **29.** $x = -3$ **31.** $x = -3$
33. $x = -1$ **35.** $x = -4$ **37.** $3\frac{3}{4}$ **39.** $3\frac{1}{3}$
41. $5x = -20$
$$\frac{5x}{5} = \frac{-20}{5}$$
$$x = -4$$
43. $\frac{x}{5} = -40$
$$\frac{5}{1} \cdot \frac{x}{5} = \frac{-40}{1} \cdot \frac{5}{1}$$
$$x = -200$$

Section 12.4 (p. 707)

1. $6x + 14$ **3.** $12x - 18$ **5.** $24 - 12x$ **7.** $-18x - 3$ **9.** $49x - 21$ **11.** $-6 + 5x$ **13.** $-8x - 24$ **15.** $6x + 4y + 8$ **17.** $63w - 28t + 14z$ **19.** $x = 10$ **21.** $x = 40$ **23.** $x = -3\frac{13}{24}$ **25.** $x = 4\frac{2}{5}$ **27.** $x = 3$ **29.** $x = -2$ **31.** $x = -1.3$ **33.** $15x$ **35.** $4w$ **37.** 0 **39.** $1x$, or x **41.** $-5x$ **43.** $-18x$ **45.** $0x$, or 0 **47.** $-7y$ **49.** $10w$ **51.** 0 **53.** $8x + 5$ **55.** $-3x + 7$ **57.** $-3x + 15y$ **59.** $-3x + 2$ **61.** $x = 4$ **63.** $x = -1$ **65.** $x = -1$ **67.** $x = 1$ **69.** $y = 1$ **71.** $x = -17$ **73.** $x = 7\frac{1}{2}$ **75.** 41 **77.** $5 + 3x$, cannot combine terms. It stays $5 + 3x$.
79.
$$4(2x + 5) = -3$$
$$8x + 20 = -3$$
$$8x + 20 + (-20) = -3 + (-20)$$
$$8x = -23$$
$$x = -\frac{23}{8}$$
81. $-3(y - x) = -3(y + (-x))$
$$= -3 \cdot y + (-3)(-x)$$
$$= -3y + 3x$$

Section 12.5 (p. 713)

1. $p + q$ **3.** $45 + x$ **5.** AB **7.** $b - \frac{1}{2}$ **9.** $x + 3 + z$ **11.** $t - k$ **13.** $\frac{k}{8}$ **15.** $24 \div g$ **17.** $w - 9$ **19.** $5t$ **21.** $2x + 4 + 3m$ **23.** $(x + y) + 5$ **25.** $n - (2 + x)$ **27.** 8 **29.** -1 **31.** 12 **33.** width 54 ft, length 66 ft **35.** 26 years old **37.** Greg is 43, Dave is 44 **39.** width = 22 ft, length = 44 ft **41.** 28.5 in and 41.5 in **43. a.** \$14,000 **b.** \$840 **45.** 25 **47.** 65
49. $35° + x + 35° + x = 360°$
$$2x + 70° = 360°$$
$$x = 145°$$

Section 12.6 (p. 721)

1. $x = +1$ and $x = -1$ **3.** $x = +8$ and $x = -8$ **5.** $x = 30$ and $x = -30$ **7.** $x \doteq 5.745$ and $x \doteq -5.745$ **9.** $c = 15$ in **11.** $b = 10$ cm **13.** $a = 12$ m **15.** $c \doteq 12.21$ in **17.** $b \doteq 7.55$ ft **19.** $a \doteq 1.99$ in **21.** $b \doteq 6.63$ m **23.** $c \doteq 316.23$ ft **25.** $a \doteq 6.24$ m **27.** 13.5 mi **29.** 34.6 ft **31.** 223.6 mi **33. a.** 17.0 ft **b.** \$221.40 **35.** 0.6 **37.** 0.167 **39. b.** 12 in

Review (p. 725)

1. terms are 4, $7t$
7 is the coefficient of t
4 is the constant term
3. terms are x, $-y$, 1
1 is the coefficient of x
-1 is the coefficient of y
1 is the constant term

5. a. No **b.** No **c.** No **7. a.** No **b.** No
c. Yes **9. a.** Yes **b.** No **c.** No **11.** $x = -2$

13. $x = 0$ **15.** $x = 10.5$ **17.** $x = -4\frac{1}{2}$

19. $w = -15$ **21.** $x = 6.2$ **23.** $x = -1$
25. $x = 3$ **27.** $12x + 20$ **29.** $-5x + 15y - 30$
31. $11x$ **33.** $4x$ **35.** 0 **37.** $2x + 8$
39. $15x + 5z$ **41.** $x = 5$

43. $x = -\frac{3}{2}$ or -1.5 **45.** $x = \frac{1}{2}$

47. $4v$ **49.** $3(t + 2)$ **51.** $7z + y$ **53.** 4
55. -6 **57.** Marie is 29, Nancy is 32 **59.** 9 yrs old
61. $x = 7$ and $x = -7$ **63.** $x \doteq 8.66$ and $x \doteq -8.66$
65. $a \doteq 5.29$ m **67.** $b = 15$ yd **69.** $b = 12$ ft
71. 75 ft **73.** 50 ft

Chapter Test (p. 729)

1. $x = -3$ **2.** $x = -7$ **3.** $x = 5$ **4.** $x = -0.5$
5. $x = -7$ **6.** $x = 6$ **7.** $x = 4$ **8.** $x = -3$
9. $x = 6$ **10.** $x = 7$ **11.** $24x + 20$
12. $-21 + 28x - 7y$ **13.** $-5 + 10y$ **14.** 1
15. $x = 5$ **16.** $x = -1$ **17.** $x = 1$ **18.** $x = -2$
19. 8 **20. a.** length = 29 in; width = 21 in
b. 609 in^2 **21.** 45 years old **22.** Dave is 40; Ralph
is 38 **23.** $c = 17.5$ ft **24.** $a \doteq 9.75$ ft
25. 127.3 ft **26.** 120 ft

Cumulative Test (p. 731)

1. 128 **2.** 16 **3. a.** 456 **b.** Not possible
c. 461,000 **d.** 0 **4.** $11\frac{11}{12}$ **5.** $\frac{5}{6}$ **6.** $26\frac{3}{10}$

7. $1\frac{11}{42}$ **8.** 61.7321 **9.** 0.96 **10.** 57.6
11. 4.09 **12. a.** 98.5 gal **b.** \$116 **13.** \$417.60
14. 80% **15.** 4,750 registered voters **16.** \$32.20
17. a. 6 **b.** 7 **c.** 10 **d.** 10 **18.** 25 mm
19. 50 in^2 **20.** 3,846.5 ft^2 **21. a.** -2 **b.** 3
c. -4 **d.** -45 **e.** 5 **f.** -20
22. 5.06×10^{-4} **23.** 3.5×10^5 **24.** $w = -10$
25. $t = -1$

calculating principal, rate, or time,
463–465
Solution for an equation, 683
Sphere, 613
  hemisphere, 614
  volume of, 613
Square
  area, 575
  perimeter, 567
Square units of measure, 573, 574, 580
Square root, 76
  using a calculator, 78
  related power sentence, 77
Squares, 77
Standard notation, 9
Statistics, 485
Stock market, 205, 245
Substituting for a variable, 67
Subtraction
  and borrowing, 28, 235
  checking, 26
  of decimals, 280, 281
  defined, 25
  difference, 25
    estimating, 38, 281
  of fractions, 209, 233–236
  of integers, 645–648
    key words indicating, 43
    minuend, 25
  of mixed numbers, 209, 234–236
    related addition, 26
    repeated, 101
    subtrahend, 25
  of whole numbers, 25
  zeros in, 29
  zero property of, 26
Subtrahend, 25
Sum, 15
  estimating, 37, 280

T

Table of data, reading a, 496
Table of fraction, decimal, percent
    equivalents (see Appendix)
Tables
  addition facts, 15
  multiplication facts, 52

Tally, 487
Tax, sales, 307
Temperature, 549, 550
Teminating decimal, 316
Terms
  defined, 681
  like, 702
Tick, 503
Time, units of, 533
Ton, 533
Translating word sentences into
  algebraic expressions, 709, 710
  arithmetic expressions, 127
  equations, 710–712
Trapezoid, 583
Trends, 504, 506
Trial divisor, 104
Triangle
  area, 586, 587
  defined, 566
  equilateral, 590
  isosceles, 590, 720
  perimeter, 566
  Pythagorean Theorem and, 715–720
  right, 585
  scalene, 590
  sum of angles, 585
Truncating, 276
2-3-5-7 rule, 163

U

Unit fractions, 533
Unit price, 364
Unlike fractions, 227
Unlike terms, 702

V

Variables, 6, 129, 681
  and computer programs, 684
  using letters to represent, 5
  substituting for, 67
Vertex point, 562
Vertical line, 561
Volume, 603
  of a cone, 614
  of a cube, 606

  of a cylinder, 611
  of a rectangular solid, or box, 603–606
  of a sphere, 613
  units of, 603

W

Week, 533
Weight
  English units, 533
  English/Metric conversions, 547
  Metric units, 543
Whole numbers, 3
  addition of, 15
  average of, 112
  carrying and, 18
  division of, 101–105
  estimation, 37, 38, 64, 104, 106
  expanded notation, 11
  finding divisors of, 161
  multiplication of, 61–63
  order of operations with, 66, 119–122
  place value names of, 9
  reading, 9, 10
  rounding, 35
  standard notation, 9
  subtraction, 25
  word names for, 10, 11
  zero as a place holder, 268
Word names
  for decimals, 266, 267
  for whole numbers, 10, 11

Y

Yard, 533

Z

Zero
  addition property of, 16
  as a base, 76
  in division, 96, 151
  as an exponent, 75
  in a fraction, 151
  identity property of, 16
  in multiplication, 53
  as a place holder, 268
  reciprocal of, 179
  in subtraction, 26, 29